Wastewater Treatment and Disposal

POLLUTION ENGINEERING AND TECHNOLOGY

A Series of Reference Books and Textbooks

EDITOR

PAUL N. CHEREMISINOFF

Associate Professor
of Environmental Engineering
New Jersey Institute of Technology
Newark, New Jersey

1. Energy from Solid Wastes, *Paul N. Cheremisinoff and Angelo C. Morresi*
2. Air Pollution Control and Design Handbook (in two parts), *edited by Paul N. Cheremisinoff and Richard A. Young*
3. Wastewater Renovation and Reuse, *edited by Frank M. D'Itri*
4. Water and Wastewater Treatment: Calculations for Chemical and Physical Processes, *Michael J. Humenick, Jr.*
5. Biofouling Control Procedures, *edited by Loren D. Jensen*
6. Managing the Heavy Metals on the Land, *G. W. Leeper*
7. Combustion and Incineration Processes: Applications in Environmental Engineering, *Walter R. Niessen*
8. Electrostatic Precipitation, *Sabert Oglesby, Jr. and Grady B. Nichols*
9. Benzene: Basic and Hazardous Properties, *Paul N. Cheremisinoff and Angelo C. Morresi*
10. Air Pollution Control Engineering: Basic Calculations for Particulate Collection, *William Licht*
11. Solid Waste Conversion to Energy: Current European and U.S. Practice, *Harvey Alter and J. J. Dunn, Jr.*
12. Biological Wastewater Treatment: Theory and Applications, *C. P. Leslie Grady, Jr. and Henry C. Lim*
13. Chemicals in the Environment: Distribution · Transport · Fate · Analysis, *W. Brock Neely*
14. Sludge Treatment, *edited by W. Wesley Eckenfelder, Jr. and Chakra J. Santhanam*
15. Wastewater Treatment and Disposal: Engineering and Ecology in Pollution Control, *S. J. Arceivala*

Additional Volumes in Preparation

Wastewater Treatment and Disposal

Engineering and Ecology in Pollution Control

S. J. Arceivala

Regional Office for Southeast Asia
World Health Organization
New Delhi, India

MARCEL DEKKER, INC. · New York and Basel

Library of Congress Cataloging in Publication Data

Arceivala, S. J.
    Wastewater treatment and disposal

  (Pollution engineering and technology ; 15)
  Includes index.
  1. Sewage disposal. 2. Sewage--Purification.
I. Title. II. Series
TD741.A8   628.3   81-1521
ISBN 0-8247-6973-2   AACR2

MARCEL DEKKER, INC.

270 Madison Avenue, New York, New York 10016

Current printing (last digit):
10 9 8 7 6 5 4 3 2 1

PRINTED IN THE UNITED STATES OF AMERICA

PREFACE

This textbook is organized in two parts, waste treatment and waste disposal. An attempt has been made to look at the whole field including disposal to rivers, lakes, sea, and land, all in one book. This has made it possible to adopt a unified approach. Several common threads run through the whole subject matter. For example, ecological and health considerations pervade all pollution control activities, and topics such as mass transport and dispersion appear in the design of practically all systems, as do reaction kinetics, recycling, accumulation, and a host of other topics. It should, therefore, help the reader to see the whole field in a common perspective.

Subjects such as land disposal and discharge to the sea, which are gaining in importance in many countries, have as yet hardly appeared in any exhaustive manner in textbooks. Such subjects have been dealt with mainly in special reports and multiauthored books of an advanced nature. Therefore, full chapters on both these subjects have been included.

This book should be of equal interest to students and practicing engineers in developed and developing countries since, when it comes to waste disposal, all possible environmental consequences have to be technically examined in much the same manner no matter where one is. Again, when it comes to waste treatment, everyone is first interested in the relatively simpler, less sophisticated methods such as aerated lagoons and oxidation ponds and, then, in activated sludge, extended aeration, and such other methods which have been covered at length in this book. In the design of treatment units, extensive use has been made of the dispersed flow model and other modern reactor design principles. A separate chapter covers the rational design of sludge dewatering systems.

Wherever possible some useful data from developing countries have been included together with corresponding data from developed ones, pointing out essential differences. The book is full of practical examples (nearly a hundred solved examples have been included), many of which have been culled from the author's own experiences in both warm and cold climates.

iii

A work of this nature spread over some years would not have been possible without the periodic "shots" of encouragement received from many colleagues and friends. My grateful thanks are first due to Professor J. Carrel Morris of Harvard University whose suggestion to revise my earlier book on "Simple Waste Treatment Methods" led to this greatly enlarged version.

My thanks are also due to my many professional colleagues and students in parts of Southeast Asia, the Middle East, Europe, and South America, whose questions and doubts taught me a great deal. Special mention must be made of Professor Dr. Mutlu Sumer who offered useful comments on dispersion topics in Chapter 7.

My grateful thanks must also be expressed to the World Health Organization which kindly permitted publishing of my earlier book and this sequel to it.* Last, but not the least, my thanks are due to my family and particularly to my wife who cheerfully kept things going while "the professor" was poring over his books on innumerable weekends and holidays!

S. J. ARCEIVALA

*The opinions expressed in this book are solely those of the author and do not reflect in any manner the views and policies of the World Health Organization.

CONTENTS

Part II

WASTEWATER TREATMENT

Wastewater Treatment and Disposal

Part I

ECOLOGICAL AND ENGINEERING ASPECTS OF WASTE DISPOSAL

Chapter 1

THE MICROBIAL CELL

1.1 INTRODUCTION

The environment has a <u>biotic</u> component consisting of living matter,
and an <u>abiotic</u> component consisting of nonliving organic and inorganic
matter such as water, gases, minerals, etc. The biotic and abiotic components function together as an ecological system or an ecosystem, displaying a complex interaction between themselves and existing at an
equilibrium under given conditions. Any change brought about by man's
activities or other circumstances places a "stress" upon the system,
disturbing the equilibrium, evoking a suitable "response" from it, and
in turn, leading to a new equilibrium condition which may or may not be
beneficial to the original components of the system, including man.
Ecology is the study of these interrelationships between the biotic and
abiotic components and among the different elements of which they are
constituted.

Although this book deals mainly with water pollution and its control,
there are, in fact, no boundaries between the soil, water, and gaseous
phases of the environment. Pollution occurring in one phase can be transferred to another, and a broad perspective must be maintained in approaching any pollution control problem.

Today's need for an integrated outlook on environmental management
is better met by the "community" approach of ecology than by the "organism"
approach of traditional biology. First we shall briefly examine an individual microbial cell and, second, see the biotic component as a whole. A
few aspects of ecological significance have been discussed at length to
assist the reader in appreciating some later topics concerning wastewater
treatment, particularly in lagoons and ponds, and the disposal of wastes
to fresh water, to marine water, and on land.

1.2 THE MICROBIAL CELL

Microorganisms are grouped into three kingdoms: animal, plant, and
Protista.* The latter includes most of the microorganisms of interest in
wastewater treatment. A few examples from each kingdom are given as
follows:

<u>Animal kingdom</u>

Vertebrates: humans, animals, fish, amphibians

Invertebrates: molluscs (mussels, snails)

arthropods (flies, mosquitos)

worms (segmented: annelids
            thread: nematodes
            flat: platyhelminths)

Rotifers (animalicules)

Porifers (sponges)

others

<u>Plant kingdom</u>

Spermophyta   (seed-bearing)

Pteridophyta   (ferns)

Bryophyta   (mosses)

<u>Protista</u>

Higher (eucaryotic): algae, protozoa, fungi, slime molds

Lower (procaryotic): blue-green algae, bacteria

Briefly, animals and plants are multicellular and show tissue differentia-
tion; Protista are unicellular or multicellular but without tissue differ-
entiation. The cell is the basic unit of life in each kingdom regardless
of the complexity of the microorganism.

A typical diagram of a microbial cell is given in Figure 1.1. The
cytoplasm is a colloidal suspension of proteins, carbohydrates, and other
complex compounds with a well-defined nucleus in higher Protista and a
relatively poorly defined nucleus in the lower Protista. The cytoplasm is
enclosed by a cell mambrane. Some microbes show a distinct capsule around
them, and display flagella which enable their movement. RNA is contained
in the cytoplasm and mainly assists in synthesis of proteins, while DNA is
contained mainly in the nucleus and is the key to cell reproduction. The
average specific gravity of protoplasm in general is 1.045 to 1.050. It

----

*First or primitive.

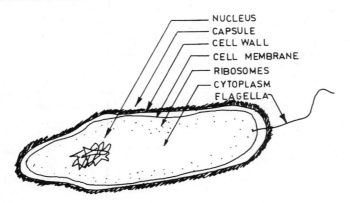

NUCLEUS
CAPSULE
CELL WALL
CELL MEMBRANE
RIBOSOMES
CYTOPLASM
FLAGELLA

Figure 1.1 A typical bacterial cell.

is somewhat increased by silicified or cellulose walls and decreased by
the presence of vacuoles.

Multiplication of microbial cells is by binary fission, a given cell
yielding two identical cells. The rate of multiplication depends on several factors such as available nutrition, energy, and temperature. Since
a cell has to receive its nutrition through its outer surface, there is a
limit to the amount that can penetrate and diffuse in the cell. As the
cell grows in bulk, its nutritional requirements cannot be met through the
cell surface and the cell multiplies into two cells, thus providing a more
favorable ratio of surface area to volume. Binary fission of bacterial
cells may occur every few minutes under favorable conditions.

Cellular multiplication in the case of a unicellular organism leads
to an increase in the number of individuals. Cellular multiplication in
the case of multicellular organisms leads to an increase in size of the
individual.

Microorganisms can be classified as photoautotrophs, chemoautotrophs,
or chemoheterotrophs depending upon their energy and carbon source as shown
in Table 1.1. Besides carbon, all organisms need nitrogen, phosphorus, and
several other nutrients which are essential to growth. Microorganisms are
also divided into groups according to their requirements for oxygen. Aerobic microorganisms are those which thrive in the presence of free oxygen,
whereas anaerobic microbes are capable of taking their oxygen from the
breakdown of chemical substances and, thus, do not need free oxygen. Facultative microbes are those that can function under either condition.

Table 1.1 Energy and Carbon Sources for
Different Types of Microorganisms

Classification	Energy source	Carbon source	Typical examples
Photoautotrophs	Sunlight	CO_2	Algae, plants
Chemoautotrophs	Inorganic, through chemical reactions	CO_2	Sulfur bacteria, nitrifying bacteria
Chemoheterotrophs	Organic, through biochemical reactions	Organic carbon	Common biological waste treatment organisms

1.3 NUTRIENT CONTENTS

Dissolved salts vital to life are called biogenic salts. Salts of nitrogen and phosphorus are of major importance as are salts of potassium, calcium, magnesium, and sulfur. The elements and their compounds needed in relatively large amounts are referred to as macronutrients. Other elements and their compounds which are essential for living systems but are required only in extremely minute quantities are called micronutrients.

Some elements known to be essential to plants in traces are: Fe, Mn, Cu, Zn, B, Na, Mo, Va, and Co. Most of these are essential for invertebrates, and a few others, such as iodine, for vertebrates. Some of the micronutrients resemble vitamins in that they act as catalysts.

A bacterial cell is composed of about 80% water, 18% organic matter, and 2% inorganic matter. The dry protoplasm of bacterial cells contains the following approximate concentrations of the major elements:

Element	Content (%)
C	49.0
H	6.0
O	27.0
N	11.0
P	2.5
S	0.7
Na	0.7
K	0.5
Ca	0.7
Mg	0.5
Fe	0.1
others	balance

The empirical formulae for the organic matter constituting the cell vary with the type of cell as shown below, and can be determined as illustrated in Example 1.1:

Organism	Constituent empirical formula
Bacteria	$C_5H_7O_2N$
Algae	$C_5H_8O_2N$
Protozoa	$C_7H_{14}O_3N$
Fungi	$C_{10}H_{17}O_6N$

Among the macronutrients, carbon, nitrogen, and phosphorus are most important. For the optimal operation of any biological process involving microbial growth, sufficient quantities of nitrogen and phosphorus must be available in accordance with the general composition of the cell. New cells synthesized from the degradation of organic matter have been shown to contain, on dry weight basis, 12.3% nitrogen and 2.6% phosphorus at maximum levels. The nonoxidizable residue after extended aeration or aerobic stabilization processes contains only 7% nitrogen and 1% phosphorus; hence, cell age is an important factor in estimating nutrient requirements. The mean stoichiometric relation between C:N:P for algal material is 41:7:1.

Microbial solids in activated sludge have shown an average empirical formula of $C_{118}H_{170}O_{51}N_{17}P$ from which their nutritional requirements of C, N, and P can be computed stoichiometrically. However, it is generally more convenient in the case of wastewater treatment to relate the nutrient requirement to the biochemical oxygen demand (BOD) of the waste (Sec. 4.6.1).

Example 1.1 If the percent dry weight of each element in a sample of algae is as given in column 3 of the following data, compute its empirical formula (the calculations are tabulated).

Element	Atomic weight	Percent dry wt.	Percent dry wt./at. wt.	Ratio of atoms[a]	Even number of atoms in empirical molecule
C	12	52.5	4.375	106.0	106
H	1	7.4	7.40	179.6	180
O	16	29.6	1.85	44.7	45
N	14	9.2	0.657	15.9	16
P	31	1.3	0.041	1.0	1
Total		100.0			

[a]Obtained by multiplying all numbers in column 4 by a factor which gives 1.0 for the smallest number of column 4.

From the tabulated data, the empirical formula for the algae in question
is $C_{106}H_{180}O_{45}N_{16}P$. Conversely, the percent dry weight content of each
element can be computed if the empirical formula is known. For example,
if the empirical formula for algal protoplasm is taken differently, say
$C_{106}H_{263}O_{110}N_{16}P$, the percent dry weights of C, H, O, N, and P are found
to be, respectively, 35.8, 7.4, 49.5, 6.3, and 0.87.

1.4 SUBSTRATE UTILIZATION AND ENZYMES

The solubilized medium on which a cell grows is called the substrate.
The substrate penetrates the cell wall, after which the microbial cell acts
on the substrate through its intracellular enzyme system, and draws its
nutritional requirements from it. Thus,

$$\text{Enzyme + substrate} \rightarrow \text{end products + enzyme + energy} \qquad (1.1)$$

The end products formed through the utilization of the substrate are diffused out of the cell wall. Substrate utilization kinetics will be discussed later.

Enzymes are organic catalysts manufactured by the cell. They are
composed of protein or protein with an organic or inorganic molecule; the
protein constitutes the apoenzyme whereas the nonprotein part constitutes
the coenzyme. Enzymes may be extracellular or intracellular; they are
highly substrate-specific and are affected by temperature, pH, and other
factors. They act very rapidly, even in minute quantities, and as shown in
the above equation are not destroyed in the reaction but are available for
repeated breakdowns of the substrate. There are various enzymes and en-
zymatic actions for which details can be found in texts dealing with microbiology. It is well to remember that both enzyme synthesis and activity
are extremely important in the regulation of the overall metabolic pathways.

1.5 MICROBIAL METABOLISM

The metabolism of any organism consists of two major pathways: catabolism, or the energy-generating pathway in which substrate breakdown or
degradation occurs, and anabolism, or the energy-consuming pathway in which
biosynthesis or growth occurs. Net growth is the result of catabolism and
anabolism in a cell.

1.5.1 Generation of Energy in Microbial Cells

A. Respiration

Respiration signifies any exothermic reaction, aerobic or anaerobic, having an energy value for living organisms; in heterotrophs it is also called biological oxidation. Although oxygen per se may not be involved, oxidation implies the loss of one or more electrons from the substance oxidized. Reduced inorganic compounds may also serve as electron donors. A mutual transfer of electrons must occur from the substance oxidized to the oxidizing agent. Transfer of H_2, for example, from a substrate to a hydrogen acceptor is similar to the transfer of electrons to an oxidizing agent. This occurs in dehydrogenation with enzyme dehydrogenase. Other methods of obtaining energy also exist.

In dehydrogenation, coenzymes play an important part, often becoming the intermediate H_2 acceptor until eventually, after several reactions, a final H_2 acceptor is found:

$$\text{Substrate} + \text{coenzyme} \xrightarrow{\text{dehydrogenase}} \text{oxidized substrate} + \text{coenzyme} + \text{energy} \qquad (1.2)$$

This is shown in Figure 1.2a.

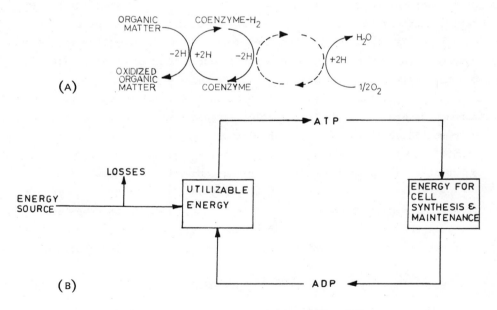

Figure 1.2a Steps in the biological oxidation of organic matter during respiration (see text). Figure 1.2b Schematic representation of the role of ATP as a coupling agent between catabolism and biosynthesis.

The final acceptor may be one of the following substances (shown in order of decreasing energy yield): (1) dissolved oxygen (under aerobic conditions), or (2) one of the following under anaerobic conditions: nitrates, sulfates, and carbonates. It is useful to qualify (1) above as aerobic respiration and (2) above as anaerobic respiration to differentiate between their final mode of H_2 acceptance.

When several acceptors are available, the system chooses that acceptor which yields the highest available energy. For this reason, dissolved oxygen is utilized first and, when it is exhausted, the system turns anoxic and shifts to the use of nitrates for hydrogen acceptance. When nitrates are exhausted, the system turns anaerobic and breaks down and utilizes the sulfates which, in accepting the hydrogen, get reduced to the odorous sulfides. As long as any of the higher energy-yielding substances are available, the lower one will not be utilized. Thus, nitrates are artificially added, for example, to overloaded holding lagoons to prevent the formation of sulfide odors.

The essential difference between aerobic and anaerobic microorganisms lies in the manner in which the coenzyme regenerates itself. In both cases, biological "oxidation" occurs, but different H_2 acceptors are involved. Strict aerobes need free oxygen for their metabolic respiration; the facultative microbes may use either free oxygen or nitrates as the final H_2 acceptor. Sulfates or carbonates are used as final H_2 acceptors by the strict anaerobes only; they cannot use aerobic respiration as an alternate means of energy-yielding metabolism.

Generally, more energy is released through aerobic reactions than through anaerobic ones. For example, aerobic breakdown of one mole of glucose yields about 688 kcal, whereas during anaerobic fermentation, only about 52 kcal are yielded. For this reason, aerobic breakdown of organic substrates generally proceeds faster than anaerobic breakdown.

Metabolic activity occurs continually in all cells since energy is required for cell maintenance. Respiration for chemoautotrophs [Eq. (1.3)] involves the oxidation of inorganic compounds. Respiration for chemoheterotrophs [Eq. (1.4)] involves the breakdown of an organic substrate:

$$2NH_3 + 3O_2 \xrightarrow{\text{chemoautotrophs}} 2HNO_2 + 2H_2O + \text{energy} \qquad (1.3)$$

$$C_5H_7O_2N + O_2 \xrightarrow{\text{chemoheterotrophs}} CO_2 + NH_3 + H_2O + \text{energy} \qquad (1.4)$$

Endogenous respiration or autooxidation involves the breakdown of organic matter contained in the cell itself in order to give energy for cell maintenance, not new cell growth. Thus, there is a fractional decrease in cell mass with time. While biosynthesis gives a "gross production" of organic matter, endogenous respiration destroys a part of this production, and the result is called net production. Under equilibrium conditions, new synthesis just replaces the cell mass destroyed. In all systems shown in Figure 1.3, the energy released by endogenous respiration is not available for new cell growth.

B. Photosynthesis

Another mode of energy-yielding metabolism is photosynthesis, in which light is used as a source of energy. The absorption of light by the photosynthetic pigment system of the organism is followed by a process of energy conversion through which a part of the absorbed light energy is converted into chemical bond energy. The generation of ATP (see Sec. 1.5.2) is now light-dependent, and is called photophosphorylation.

Molecular oxygen does not play a role in the energy-yielding reactions of photosynthesis; consequently, photosynthesis can occur under strictly anaerobic conditions. However, in plant photosynthesis (e.g., algae) the oxidation of water is an accompanying feature. Molecular oxygen is produced as a by-product and, because the photosynthetic organisms are growing in oxygen, they are aerobes. However, certain photosynthetic bacteria do not oxidize water, thereby producing no oxygen, and can, therefore, grow in strictly anaerobic conditions (e.g., purple-colored sulfur bacteria).

C. Fermentation

Another mode of energy-yielding metabolism is fermentation, in which organic compounds such as carbohydrates serve as both electron donors and electron acceptors. All fermentations proceed under strictly anaerobic conditions (except with lactic acid bacteria which can undergo fermentation even in the presence of oxygen).

1.5.2 Energy Transfer Mechanism

Briefly speaking, energy is released in a cell either by photosynthesis or by chemical oxidation of organic or inorganic substrates through an exothermic, energy-yielding reaction that is enzyme-catalyzed. The energy so released would be lost as heat except that adenosine diphosphate (ADP) contained in the cell is converted to adenosine triphosphate (ATP) through the addition of inorganic phosphate, thus storing energy in chemical form as energy-rich bonds (see Figure 1.2b).

The ATP now formed can be gradually reconverted by the cell to the ADP, releasing energy for cell maintenance and growth. In this manner, the energy level of the substrate decreases through cell action while the energy level of the cell material increases. However, the energy decrease and increase are not equal as there are losses. Thus, there is always a progressive decrease in the energy level of the substrate-cell system. This can be stated in terms of the second law of thermodynamics that, in any energy-yielding reaction, the total energy liberated is not fully available for the performance of work:

$$\Delta H = \Delta F + T(\Delta s)$$

where

ΔH = change in heat energy (where heat energy is the heat of combustion when a compound is fully oxidized)

ΔF = change in free energy

Δs = entropy change (energy not available for work)

T = absolute temperature

As the substrate is acted upon by the cells, there is a reduction in its heat energy. A part of the energy given off is lost as entropy change and is not available for respiration purposes. The major role of the utilizable energy is to convert ADP to ATP and, thus, remain stored for later use. This conversion is known as substrate-level phosphorylation.

Energy is required for the entry of nutrients into the cell, for their conversion in the cell to low molecular weight intermediaries (amino acids, nucleotides, etc.), and finally for their polymerization to produce proteins, polysaccharides, lipids, etc., which are the major constituents of cell material. In fact, only a small part of the substrate energy is thus converted to net cell growth.

The energy required to convert the available nutrients into a suitable form to facilitate their participation in synthesis reactions is variable. The major nutrients in question are carbon, nitrogen, and sulfur, of which carbon contributes more to cell material on a weight basis than does nitrogen or sulfur. The energy requirement varies, depending on the difference

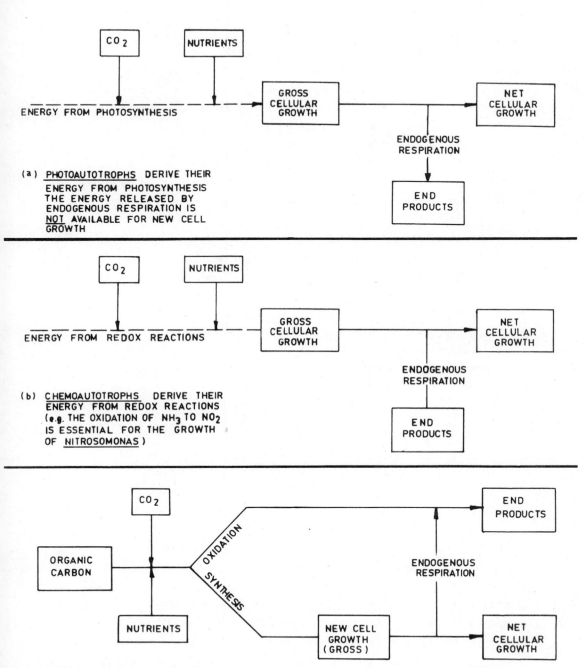

Figure 1.3 Energy and nutrient requirements of microorganisms.

in oxidation state between the nutrients and the final products of biosynthesis.

With respect to carbon, the overall state of oxidation of the cell is approximately that of carbohydrate (symbolized by CH_2O). Hence, chemoheterotrophic organisms that use organic carbon sources on this oxidation level for the synthesis of cell material require little or no net input of energy for the conversion of the carbon sources to an assimilable level. However, an organism such as a chemoautotroph or photoautotroph that uses CO_2 as its source of carbon must first use energy to convert CO_2 to the oxidation level of CH_2O. For this reason, a given weight of substrate produces less weight of cells of the chemoautotrophic type (e.g., <u>Nitrosomonas</u> or <u>Nitrobacter</u>) than of the heterotrophic type.

Similarly, energy is required for the assimilation of nitrogen and sulfur. Most of the nitrogen in cell material is at the oxidation level of NH_3. Hence, no net input of energy is required when nitrogen is available to the cell in the form of NH_3, amino acids, or peptones. However, energy is required if nitrogen is available in the form of nitrates and must first be reduced to the NH_3 form. For sulfur, the cell material is in the same oxidation level as sulfide.

In aerobic respiratory metabolic processes, the availability of oxygen as a terminal electron acceptor makes possible, in principle, complete oxidation of organic substrates to CO_2. However, a certain fraction of the substrate carbon is assimilated and used along with other nutrients for the synthesis of new cell material, the rest usually being converted entirely to carbon dioxide (see Figure 1.3c). Thus, the organic carbon supplies both the energy and the carbon needed for building protoplasm.

1.6 MICROBIAL GROWTH

As stated in Sec. 1.1, binary fission of microbes can occur every few minutes under favorable conditions. The time required for each fission is called the generation time. The growth of microbial populations is, however, limited by one or more factors, such as the availability of nutrients or the changes produced in the microenvironment by the microbes themselves. It is often possible to characterize four principal phases as shown in Figure 1.4.

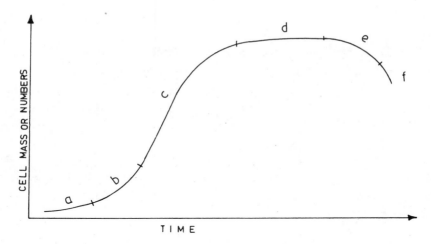

Figure 1.4 Typical microbial growth pattern.

The first phase shown in Figure 1.4 is the lag phase designated "a"
and occurs when microorganisms in the medium have not yet acclimated to
the local environment. The next phase is the autocatalytic* growth phase,
made up of logarithmic and first-order growth (b + c) in which the growth
rate reaches the constant maximal value as determined by the generation
time of the particular organism. The third phase is the stationary population phase d, which occurs when limiting conditions are approached and
cell production rate equals cell death rate. The last is the death phase
which is again logarithmic (e + f); the death rate is now faster than the
production rate, and the numbers decline with time.

In terms of the mass of cells, the second phase in Figure 1.4 can be
referred to as the log-growth phase. The next phase may be called the
declining growth phase and the last one, the endogenous phase. In the
endogenous phase, the available nutrition is at a minimum, resulting in
the destruction of a part of the cell protoplasm in order to yield energy
for cell maintenance.

_____

*The term <u>autocatalytic</u> refers to a reaction whose velocity spontaneously
increases with time. A <u>retardant</u> reaction is one in which the velocity
decreases with time.

1.6.1 Cell Doubling Rate

For any initial population, say N_0 of unicellular microbes, the population size in successive generations can be computed as follows:

after 1 generation, $N_1 = 2^1 N_0$

after 2 generations, $N_2 = 2^2 N_0$

after n generations, $N_n = 2^n N_0$

Similarly, the number of generations that have occurred within a given time t can be computed from the initial size of the population N_0 and the size after time t, N_t:

$$N_t = N_0 2^{k't}$$ (1.5)

where

k' = number of doublings of the population, or the growth rate

t = time

The reciprocal of k' gives the mean doubling time.

Temperature has a considerable effect on cell metabolism, growth rate, and multiplication time; this will be discussed later.

Example 1.2 A microbial population increases from 10^2 to 10^9 cells in 10 hr. Find k' and the mean doubling time.

$$k' = \frac{\log_{10}N_t - \log_{10}N_0}{\log_{10}2 \times t}$$

$$= \frac{9 - 2}{0.301 \times 10}$$

$$= 2.33 \text{ generations per hour}$$

$$\text{mean doubling time} = \frac{1}{k'} = \frac{1}{2.33} = 26 \text{ min}$$

1.6.2 Specific Growth Rate

Since microorganisms multiply as a result of binary fission, their growth is a function of their numbers (or mass) at any given instant. Thus, in studies of growth kinetics of continuous cultures, the instantaneous growth rate per unit organism, namely, the specific growth rate μ is often used, where

$$\mu = \frac{dX/dt}{X} \tag{1.6}$$

where

μ = specific growth rate, time^{-1}

X = microorganism mass or concentration

dX/dt = rate of change of microorganism mass or concentration with time

The manner in which μ changes in the constant and declining growth rate regions was given by Monod [1] in the form of an empirical equation developed in 1942 from pure culture studies in which μ varies in accordance with a hyperbolic function (see Figure 1.5). This accounts for a lower specific growth rate when there is nutrient limitation since growth is restricted if any one of the nutrients required for cell formation is missing or insufficient. Figure 1.5 shows that as the limiting nutrient (or substrate) concentration S increases, the specific growth rate μ also increases until, after a certain concentration, it levels off at the maximum possible value of μ, namely μ_{max}, at the given conditions. In other words, when there is no nutrient deficiency there can be a certain maximum specific growth rate at the given conditions (such as at a given temperature), no matter how abundant the substrate is.

The value of μ_{max} is likely to be different for different types of organisms, substrates, and conditions. For a given set of conditions, Monod's empirical function gives:

$$\mu = \mu_{max} \left[\frac{S}{K_s + S} \right] \tag{1.7}$$

where

μ_{max} = maximum specific growth rate, time^{-1}

S = concentration of rate-limiting nutrient or substrate. This could be, for example, organic carbon, nitrogen, or phosphorus.

K_s = "saturation concentration," namely, the concentration of the rate-limiting nutrient or substrate at which $\mu = \frac{1}{2}(\mu_{max})$ (Figure 1.5).

Thus, for a given set of conditions, one can determine experimentally the values of μ at different values of S and then use Eq. (1.7) to evaluate μ_{max} and K_s, as explained in Chapter 12.

Where two limiting nutrients are involved, Eq. (1.7) can be rewritten as follows to take into account the effect of both on the specific growth rate:

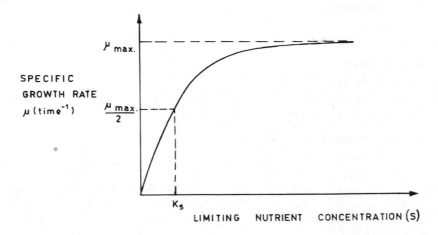

Figure 1.5 Specific growth rate and limiting nutrient concentration [1].

$$\mu = \mu_{max}\left[\frac{S_1}{K_{s1} + S_1}\right]\left[\frac{S_2}{K_{s2} + S_2}\right] \qquad (1.8)$$

Often, in natural systems as well as in biological treatment processes, the actual specific growth rate μ may be lower than the maximum value, depending on the relative values of S and K_s in the given situation. This is illustrated in the following example, and in Example 3.1; applications to field studies are discussed in relevant chapters.

Example 1.3 For a mixed-culture activated sludge growing on glucose, the value of K_s = 0.2 mg/liter, while for growth on domestic sewage, K_s = 45 mg/liter. If the substrate concentration is 10 mg/liter in both cases, compute μ in terms of μ_{max} for each.

For glucose,

$$\mu = \mu_{max}\left[\frac{10 \text{ mg/liter}}{0.2 \text{ mg/liter} + 10 \text{ mg/liter}}\right]$$

Hence, $\mu \simeq \mu_{max}$

For sewage,

$$\mu = \mu_{max}\left[\frac{10 \text{ mg/liter}}{45 \text{ mg/liter} + 10 \text{ mg/liter}}\right]$$

Hence, $\mu = 0.18 \; \mu_{max}$

1.7 GROWTH INHIBITION

Apart from lack of nutrients, microbial growth may also be inhibited or reduced by other factors (e.g., presence of certain chemicals) which interfere with or inhibit in some manner the enzyme system or the general metabolism of the microbial cell. Such substances are regarded as toxic substances insofar as the particular cell is concerned. Their toxic effect can occur even though the substrate is rich in nutrients and other biogenic factors. Toxic substances may also be of a physical nature, such as heat (thermal pollution) or ionizing radiation.

The extent of inhibition depends on the nature of the toxic substance and the microorganism in question, and growth may be totally inhibited in some cases. Toxic effects can be studied quantitatively, for example, by observing the specific growth rates, the oxygen consumption rate (in the case of aerobic organisms), the gas production rate (in the case of anaerobic organisms), or through bioassay tests on higher forms of life, like fish, as discussed in succeeding chapters.

REFERENCES

1. Monod, J., Recherche sur la Croissance des Cultures Bacteriennes, Herman et Cie., Paris, 1942.

Chapter 2

THE MICROBIAL COMMUNITY IN AN ECOSYSTEM

2.1 THE MICROBIAL COMMUNITY

Chapter 1 was concerned with the microbial cell and its characteristics; but individuals have to live in a community, affecting it and in turn being affected by it. The factors that affect the occurrence and growth of individual organisms must now be briefly examined in the context of ecological communities in order to understand how an ecosystem as a whole can respond to various external stimuli. Survival depends on the ability to tolerate the large complex of conditions which can be broadly grouped as follows:

1. Availability of energy, nutrients, and trace elements necessary for growth and effect of biogenic and toxic factors as seen earlier with regard to cell metabolism (Chapter 1)
2. Effects of interactions among biota in an ecosystem
3. Effects of environmental factors of a physical nature such as temperature, pressure, humidity, light, currents, etc., on the microbial community and its spatial distribution in an ecosystem

2.2 EFFECTS OF INTERACTION BETWEEN ORGANISMS

A biotic community is an assemblage of populations living in a prescribed area or "physical habitat." It is a loosely organized unit insofar as it has characteristics in addition to its individual and population components. Various new factors come into play when an organism has to live in a community since communities display a definite functional unity with characteristic trophic (nutritional) structures and patterns of energy flow.

Interactions among species and groups of populations can be described by listing and defining the eight important interactions which can take place between populations of two species. These are generally based on their nutritional requirements, the effects of their end products, or other factors.

1. Neutralism, in which neither population group is affected by association
    with the other; their growth rates are neither increased nor decreased.
    This is often the case with many chemoheterotrophs that grow on organic
    substrates such as activated sludge.

2. Competition, in which each population group adversely affects the other
    (namely the growth rate of each is reduced) owing to struggle for food,
    nutrients, space, or other common needs. An example of this interac-
    tion is the degradation of glucose and galactose in an activated sludge
    system.

3. Mutualism, in which both populations benefit (growth rates are in-
    creased); in fact, in this kind of interaction, the relation is oblig-
    atory (neither can survive without the other).

4. Protocooperation, in which both populations are benefited as in mutual-
    ism but relations are not obligatory.

5. Commensalism, in which one is benefited but the other is not affected.
    This perhaps represents the first step toward the development of bene-
    ficial relations. The growth of photosynthetic algae and chemohetero-
    trophic microbes in waste stabilization ponds is a commensal interaction.
    Protocooperation and mutualism can be regarded as succeeding steps in
    an evolutionary series.

6. Amensalism, in which one is inhibited (growth rate is reduced) but the
    other is not affected.

7. Predation, in which one adversely affects the other by direct attack
    and is, in fact, dependent on the other for its existence. Figure 2.2
    shows the successive growth of predators when an organic substrate is
    aerated.

8. Parasitism, which is similar to predation in that it is obligatory for
    one and inhibiting for the other. Examples of this kind of interaction
    are well known.

2.3 EFFECTS OF ENVIRONMENTAL FACTORS OF A PHYSICAL NATURE

    The abiotic component of the ecosystem consists of water, gases, etc.,
on which various physical factors can have a direct bearing and, thus,
affect the biotic component of that system.

    In regard to physical factors such as temperature, light, etc., it is
necessary to evaluate the condition in the immediate vicinity of a microbial community which may not always be the same as the general conditions
in a locality. The term microenvironment signifies the immediate environmental area occupied by an organism or a community, and this concept has
been developed in recent years in response to the observation that the
environment of a particular organism at a particular time is not necessarily
the same within a few micrometers as it is a few meters away. There can be
important local differences. Organisms often adjust their rate functions

to the microenvironment; these locally adapted variants are called ecotypes. Because of the differences between the environment and microenvironment, the data obtained at one site should be extrapolated with great caution in order to apply them to other sites where the conditions may be different. Data obtained in a laboratory under controlled conditions must be used with great care in the design of full-scale systems meant to function under widely varying natural conditions.

Besides nutrients, trace elements, and other biochemical factors already seen, there are various environmental factors of a physical nature which can affect the development and spatial distribution of a microbial community. These include temperature, light, pressure, humidity, and currents.

2.3.1 Temperature

All organisms are sensitive to temperature, as it affects their metabolism and growth. Water temperature and its variation in time and space are of critical importance to many physical and physiological phenomena, and have significant ecological consequences. Certain varieties of organisms grow best within certain temperature ranges as shown in the following table:

Temperature Range (°C)	Group
10-20	Psychrophilic ("cold-loving")
20-40	Mesophilic ("moderation-loving")
40-65	Thermophilic ("heat-loving")

Temperature is often responsible for the zonation and stratification of plants and microorganisms in both water and land environments. Organisms possess the ability to compensate or acclimate to changes provided they are within certain limits of tolerance; but the luxuriant growth of an organism is likely to be found only where the temperature is favorable for that organism. Any displacement of the organism from its favored habitat may disturb its metabolism or even kill it. Temperature effects on waste treatment processes are discussed further in Chapter 12.

Temperature also has a substantial effect on various physical and chemical processes in a waterbody. Solubility, saturation concentration,

diffusion, the exchange of gases from the atmosphere, the diffusion of
nutrients released from the breakdown of organic matter and dead biota from
the bottom of a waterbody to the top, etc., are all affected by temperature.
Temperature plays a central role in physical limnology since temperature is
the main factor affecting density, which in turn influences the motion or
stratification of water within a waterbody, and thus its quality.

Density and viscosity affect settling, coagulation, flocculation,
currents, and other phenomena involving motion. Freshwater is densest at
$4°C$ (density of 1.000).

The density of seawater depends not only on the temperature but also
on the salinity. The temperature at which saline water has maximum density
and the temperature at which it freezes are different from those for freshwater, and will be discussed later in Chapter 8.

2.3.2 Light

Ecologically, the nature of light (wavelength or color), the intensity
(measured as energy), and the duration (length of day) are all known to be
important in their effects on biotic communities especially with regard to
those populations which depend on photosynthesis. Low light intensities
can be limiting to growth. Photosynthesis rates generally increase as
light intensity increases until an optimum is reached, after which higher
intensities may again prove limiting. Whole communities of algae may show
characteristic patterns of growth and distribution depending on the availability of light.

Communities tend to stratify according to their environmental and
other requirements in order to maximize their utilization of energy. Some
communities are known to migrate to the water surface at night and move
downwards at day.

The amount of light received at a water surface is dependent upon the
latitude, elevation, cloud cover, and other factors. It varies from day
to day, and is minimal in the winter and maximal in the summer. The incident light is sharply reduced as it penetrates a body of water. The depth
to which light will penetrate in a lake depends on the transparency of the
water. Even in distilled water light does not penetrate to great depths;
the intensity is reduced to nearly half the original value at a depth of
17.75 m and to 1% of the original value at about 118 m. The reduction in

intensity for natural lake waters is even more rapid with depth and can be described by the following equation, which is based on Lambert's Law:

$$I_d = I_0 e^{-K_e d} \tag{2.1}$$

where

I_d = light intensity at depth d

I_0 = light intensity at surface

 d = depth of penetration (m)

K_e = light extinction coefficient, per meter

Light intensities may be measured as illuminance in lux units* and the reader is referred to Chapter 8 for field determinations of the coefficient K_e. A graphical plot of log (I_d/I_0) versus depth d gives a straight line whose slope equals K_e. Visible light ranges between wavelengths of 3900 to 7600 Å, and each wavelength has different penetration characteristics. The rate of absorption of solar radiation in water is different for different wavelengths; some values of K_e are shown in Table 2.1.

Blue has been shown to have the smallest coefficient of extinction; other colors are extinguished at more shallow depths. As sunlight penetrates more deeply, it is progressively stripped of its longer and shorter wavelengths until the only component remaining to be scattered upward to

---

*10.83 lux units = 1 footcandle = 41.86 ergs/cm²-sec
12.57 lumens/m² = 1 candela/m².

Table 2.1 Extinction Coefficient for Light of Different Colors

Color	Wavelength (Å)	Coefficient of extinction in distilled water $K_e (m^{-1})$
Violet	4078	1.03×10^{-3}
Blue	4730	4.80×10^{-4}
Green	5040	1.03×10^{-3}
Yellow	5650	3.96×10^{-3}
Orange	6125	2.35×10^{-2}
Red	7200	9.75×10^{-2}

the eye is blue. The cleaner the water, the deeper is the penetration and
the stronger the blue color.

Natural waters differ very greatly in transparency and an approximate
comparison of the illumination conditions can be obtained from the values
of their extinction coefficients. Table 2.2 gives some values of the coefficients for a few waterbodies with respect to the central part of the
visible spectrum (5000 to 6000 Å).

A very simple device often used for getting an indication of the light
penetration characteristics of a body of water is the Secchi disc which is
essentially a flat plate (20 cm diameter) painted with alternating black
and white quadrants (see Figure 2.1). The plate is suspended in water and
the depth at which the disc is no longer visible is noted; the clearer the
water, the deeper is the disc lowered. It is evidently not possible to be
as accurate as with illuminometers, but the disc is extremely simple to
use and valuable in comparing different waterbodies, or the conditions at
different times in the same waterbody.

The transparency of inland and coastal waters is affected from season
to season by differences in the amount of stirring, silt, and turbidity
brought in by inflows, and the growth of plankton. Even within a given
time, the transparency may vary with depth owing to stratification. Transparency variations combined with seasonal variations in intensity of light
received at the surface and in the length of the day result in quite substantial fluctuations in the light received at different levels in a waterbody. High incident light combined with high transparency may give a
light intensity that is higher by as much as three or four orders of magnitude compared with the periods when both light intensity and transparency

Table 2.2 Depth of Light Penetration in Different Waterbodies

Water	$K_e (m^{-1})$	Depth for 1% of surface light (m)	Secchi disc depth (m)
Distilled water	0.039	118	44
Caribbean Sea	0.041	110	41
Crater lakes		60-80	
Coastal waters (typical)		5-35	2-10
Most inland lakes		3-30	2-10

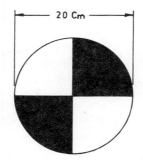

Figure   2.1   A Secchi disc

are low at a given subsurface level. Changes in the magnitude of illumination in the aquatic environment are, therefore, often much greater than
in the terrestrial environment.

2.3.3 Other Physical Factors

Pressure and humidity have a role to play in regulating the activity
of organisms, and affect their distribution in the environment. Effects
are most noticeable in the macroenvironment (e.g., tropics, high altitudes,
etc.).

Other factors such as velocity and currents can affect the nature and
location, as well as disruption, of an aquatic community; this explains the
difference between a stream and a small pond community. Runoff may carry
with it pollutants and eroded soil, the latter affecting the life in a
river or other waterbody by the suspended solids that cover and destroy a
benthic community. Agitation and turbidity in the water may also exclude
or reduce light penetration, thereby affecting productivity and, thus, the
whole ecological community. The biotic community in a mechanically aerated
lagoon is different from that in a waste stabilization pond.

2.4 LIMITS OF TOLERANCE

As early as 1840, Liebig stated that "growth of a plant is dependent
on the amount of foodstuff which is presented to it in the minimum quantity."
However, Liebig's Law of the Minimum is not adequate since not only may too
little of something be a limiting factor but also too much (e.g., heat and
light). Thus, it would be more correct to say that all organisms have

limits of tolerance rather than just limits to the minimum. This was stated
by Shelford in 1913, and thereafter expanded upon by others to make it more
comprehensive as follows [1]:

> The presence and successful growth of an organism or a group
> of organisms depends upon a complex of physical, chemical and
> biological conditions, and any condition which approaches or
> exceeds the limits of tolerance can be called a limiting condi-
> tion or a limiting factor.

It is true that organisms will have a different range of tolerances for
different factors, wider for some and narrower for others. There is evidence to show that, when conditions for an organism or a group of organisms
are not optimum with regard to one factor, the limits of tolerance with
respect to some other factors may also be reduced (synergistic effects).
The limits of tolerance may vary both geographically and seasonally.

It is easy to see that organisms with a wide range of tolerance for
all factors are likely to be the most widely distributed [1]. For example,
in waste treatment lagoons and ponds, facultative microbes would be more
widely distributed than strict anaerobes which may be confined only to the
bottom layers. Other examples can be given from the waste treatment field
to illustrate the question of tolerance to various conditions. Chemoautotrophic nitrifying bacteria are slower growing and more sensitive to fluctuating conditions. Consequently, carbonaceous matter is first stabilized
by the faster-growing and more tolerant chemoheterotrophs and the nitrogenous matter is stabilized afterwards, and only if pH, temperature, and other
factors are right. Nitrifying organisms as a group have narrower ranges
of tolerances than other organisms involved in waste treatment. On the
other hand, while denitrification can be brought about by facultative microbes which may use either free oxygen or nitrate as the final H_2 acceptor
in respiration, the reduction of sulfates to sulfides involves strict
anaerobes and, thus, requires more consistent anaerobic conditions.

Several examples from natural aquatic environments can also be given
to illustrate the effect of tolerance on the relative occurrence of different species. For example, salinity differences in marine and estuarine
waters cause marked variations in the species and the number of individuals
per species in the flora and fauna in such waters. Another example is the
asphixiation level of oxygen for fish which is known to be synergistically
dependent on temperature and on the presence of certain chemicals in the
aquatic medium.

Tolerance considerations also come into play in the choice of crops
to be grown on a wastewater irrigation farm. Wastewaters high in salinity
or in certain specific elements such as boron may tend to build up their
concentrations in poorly drained soils and, thus, be suitable for use only
with plants which have a high tolerance for them.

2.5 SPECIES DOMINANCE

The tolerance of an organism affects both its distribution and its
relative dominance within an ecosystem. The organism which finds the given
conditions best suited to it is likely to predominate within a community.
The relative predominance of different species can best be illustrated with
reference to an example from wastewater treatment using a batch aerated
system.

Figure 2.2 illustrates how growth depends upon individual metabolic
rates, the nature and availability of food, competition for certain components of the food, the physical environment, and the predatory nature of
the organism. In the batch aerated system that starts off with a high
concentration of biodegradable organic substrate, the relative predominance
of organisms changes with time. Just as in the cases given later in Sec.
3.14, a series of temporary communities prepare the way until the system
eventually develops a stable community in harmony with the given conditions.
In this regard, the example also illustrates ecological succession. Sar-
codina and phytoflagellates predominate in such systems just starting and
having high substrate concentration. Zooflagellates and bacteria increase
as phytoflagellates decrease. Free-swimming Ciliata are found when there
are a large number of bacteria.

Stalked ciliates are found when the substrate concentration has been
reduced to a low value and the number of bacteria is below the level necessary to support free-swimming Ciliata. Rotifers finally indicate very
low levels of substrate concentration.

In each case there is a time lag between the peak growth of one organism and the peak growth of its predator. The ciliates and rotifers are
major predators. As the substrate concentration decreases, the energy
level also decreases. Free-swimming ciliates operate at a relatively high
energy level, stalked ciliates at a lower level, and rotifers at the lowest
energy level. This sequence of organism predominance relates to a waste

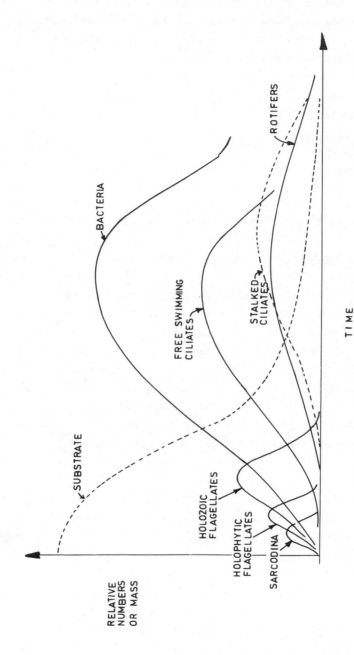

Figure 2.2 Relative dominance of organisms with time in batch aeration of an organic substrate.

treatment system in batch operation which starts off with a high substrate
concentration and low numbers of microorganisms. As the numbers or mass
of microorganisms increases, the mass of substrate within the given volume
decreases.

A stable activated sludge plant effluent will show only a few stalked
ciliates and no other protozoan forms. In fact, the presence of ciliates
would reflect the efficiency of operation of the plant with considerable
accuracy. Similarly, an extended aeration plant would display the presence
of rotifers. If, in either case, the plant operation is upset for some
reason and the system has to make a fresh start, it will pass through the
previous stages before it stabilizes once again.

REFERENCES

1. Odum, E., _Fundamentals of Ecology_, Third Edition, W. B. Saunders, 1971.

Chapter 3

THE BIOTIC COMPONENT OF THE ECOSYSTEM

3.1 THE BIOTIC COMPONENT

To be able to study the environment and the interaction between its
biotic and abiotic components, it is helpful to see how energy and nutrient
flow is achieved in the ecosystem and how man's activities in general, and
the discharge of wastes in particular, affect the system [1]. What happens
in the biosphere also happens within the confines of a small ecological
subsystem such as a lagoon, pond, wastewater treatment unit, or an irrigation field; the basic principles are the same.

The biotic component of any ecosystem or subsystem consists of the
following:

1. Primary producers which are photosynthetic autotrophic organisms,
   mainly green plants and algae, which can grow on simple inorganic
   substances (CO_2, NH_3, etc.) in the presence of sunlight (see Table
   1.1).

2. Consumers which are heterotrophic organisms which ingest other organisms or particulate organic matter. They include, for example, zooplankton, worms, fish, animals, and man.

3. Decomposers which are again heterotrophic organisms, mainly bacteria
   and fungi which break down the complex compounds in dead organisms
   and other nonliving organic matter.

3.2 PRIMARY PRODUCERS

The chief source of energy for the ecosystem is solar energy in the
form of both direct and diffuse sunlight. A part of this light is reflected back, a part is absorbed and emitted again as heat, and a part is
used in photosynthesis by all green chlorophyll-containing organisms.

Photosynthesis, as seen earlier, is a mode of energy-yielding metabolism which results in the manufacture of new organic matter in the form of
cell material. Hence, the chlorophyll-containing plants are called primary
producers and consist of

1. Terrestrial vegetation: plants.

2. Phytoplankton: unicellular species of algae floating in the upper
    layers, or euphotic* zone, of waterbodies including the seas.

3. Phytobenthos: plants associated with the bottom of shallow waters
    and littoral regions; the aquatic vegetation. In shallow ponds the
    littoral zone constitutes a relatively larger portion of the total
    euphotic zone than it does in deeper bodies of water.

The new organic matter (primary production) helps to sustain all the
consumers and decomposers who depend on it. Their energy requirements are
met from the energy stored in the new organic matter and, consequently,
their growth is limited by the availability of the latter. A closer look
at algal production in waterbodies will be useful at this juncture.

3.2.1 Algae

Unicellular species of phytoplankton such as algae are of extreme importance both in natural bodies of water and in certain waste treatment
systems. Most species of interest to water engineers are microscopic in
size and contain green chlorophyll; they may also contain additional pig-
ments, giving them a blue-green, yellow-green, or other characteristic
coloration. Many algae that perform normal photosynthesis in light, using
CO_2 as the carbon source, can grow well in the dark. A variety of organic
compounds are used to shift from photosynthetic to respiratory metabolism
in order to obtain their energy; the shift is determined by the presence
or absence of light. Thus, in such cases photosynthesis and respiration
predominate alternately [2].

Blue-green algae are photoautotrophs and, in most cases, are obligate
ones unable to grow in the dark. However, out of all photosynthetic organisms, only certain blue-green algae are able to obtain their nitrogen
requirements directly from atmospheric nitrogen by a process called nitrogen fixation. Thus, they are able to flourish in all parts of the world
and in soil, water, and oceans.

Table 3.1 gives the typical characteristics of some algae. Their
temperature preferences, their nutritional needs, light, and other requirements give them a seasonal variation in occurrence within a waterbody,

---

*Well lighted. The euphotic zone is the zone in which sufficient light
penetrates to let algae grow.

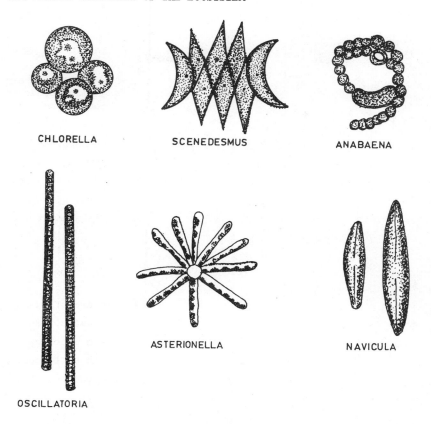

CHLORELLA   SCENEDESMUS   ANABAENA

OSCILLATORIA   ASTERIONELLA   NAVICULA

EUGLENA

Figure 3.1 A few algal species.

although some forms are practically perennial. Various factors affect
their seasonal succession in a system. An insight into the requirements
for algal growth can be had from its empirical formula which is expressed
as $C_{106}H_{180}O_{45}N_{16}P_1$. Figure 3.1 shows a few algal species.

Table 3.1 Typical Characteristics of Some Algae

Protista	Division	Algae color	Favorable temperature (°C)	Characteristic odor	Other characteristics	Some examples
Higher[a]	Chrysophyta	Yellow-green	8-13	Vegetable to oily	Cold weather organisms; abundant in early spring; worst group of offenders among algae.	Ochromonas
Higher[a]	Chrysophyta (class Bacillariophyceae; the diatoms)	Yellow-green to brownish	13-18	Aromatic to fishy	Prominent in spring season; produce odor troubles; usually prefer clean waters; present in oxidation ponds also.	Navicula, Synedra, Cyclotella, Asterionella
Higher[a]	Chlorophyta	Green	18-35	Grassy to fishy	Very active photosynthesizers; common in freshwater and stabilization ponds; multicellular forms can give floating mats.	Chloroella, Scenedesmus, Chlamydomonas, Euglena[c]
Lower[b]	Cyanophyta	Blue-green	27-40	Grassy to moldy	Primarily warm weather organisms; generally pollution and temperature tolerant; some toxic; some give odors and nuisances.	Anabaena, Oscillatoria, Aphanizomenon

[a]Having a well-defined nucleus (eucaryotic).

[b]Having a poorly defined nucleus (procaryotic).

[c]May be shown separately as division Euglenophyta.

36

Therefore, the production of algae by photosynthesis requires nitrogen, phosphorus, and carbon besides water in accordance with the following equation:

$$106CO_2 + 90H_2O + 16NO_3^- + PO_4 + light \rightarrow \underset{(2427)}{C_{106}H_{180}O_{45}N_{16}P_1} + \underset{(4944)}{154.5O_2}$$

$$(3.1)$$

If nitrogen is taken up by algae from ammonia (NH_3) instead of NO_3^-, Eq. (3.1) can be rewritten as follows:

$$106CO_2 + 66H_2O + 16NH_3 + PO_4 + light \rightarrow \underset{(2427)}{C_{106}H_{180}O_{45}N_{16}P_1} + \underset{(3800)}{118.5O_2}$$

$$(3.2)$$

There is some uncertainty regarding the form of nitrogen preferred for new algal growth. All algae can use either ammonium, nitrites, or nitrates as a nitrogen source at suitable conditions. Although there is some evidence to show that the ammonium ion is often used preferentially when both are available, there are certain examples where nitrates are preferred. Generally, both ammonium and nitrates are utilized to about the same extent in laboratory cultures [6].

On a stoichiometric basis, the weight of oxygen produced per gram of organic algal matter produced is 2.04 g (i.e., 4944/2427) where nitrate is utilized by algae, and 1.58 g (i.e., 3800/2427) when ammonia is utilized. Experimental work under practical environmental conditions has shown this value to vary from 1.25 to 1.75 g O_2 per gram of algal dry matter produced. These values vary somewhat with the type of algae and their chemical composition, and can be determined from the composition of organic matter when newly formed.* [11].

Observations on pure cultures of phytoplankton have generally indicated that respiration consumes less than 10% of the oxygen produced in photosynthesis. Values ranging from less than 1% to over 14% have been observed [9]. In the cases where percentage is low, net phytoplankton

---

*Algal growth can also be represented as follows:

$$106CO_2 + 16NO_3^- + HPO_4^{2-} + 122H_2O + 18H^+ + light \rightarrow$$
$$C_{106}H_{263}O_{110}N_{16}P_1 + 138O_2 \qquad (3.3)$$

Accordingly, 1 g of P and 7.2 g of N are required for the production of 140 g of O_2 through photosynthesis, and the oxygen-to-algae ratio is 1:25.

production is nearly equal to gross production; this is likely when photo-synthesis is rapid. But there may be other times when, for various other reasons, photosynthesis may be slow. On an average for a whole year, res-piration may destroy almost 25%, and in some regions even up to 50%, of the gross production in a lake. Low values would apply to waste stabiliza-tion ponds where lighting conditions are more uniform during the seasons.

Field determinations of phytoplankton respiration rates generally give higher values than those observed for pure cultures since the former also include respiration loss owing to the presence of zooplankton along with phytoplankton in the sample. Toms et al. [10], in studying sewage treat-ment plant effluent polishing lagoons, observed the initial respiratory demand of algal solids to be about 0.007 mg O_2/hr per mg of algal solids. In terms of chlorophyll a, they found that respiration and photosynthesis affected O_2 as follows:

$$R = 0.8B + 0.02 \tag{3.4}$$

$$P_s = 11.6B - 0.06 \tag{3.5}$$

where

R = respiration rate, mg O_2/liter-hr
P_s = photosynthetic production, mg O_2/liter-hr
B = mg of chlorophyll a in algal biomass/liter-hr

Thus, respiration was found to consume less than about 7% of the photosyn-thetic production of oxygen at their test lighting and temperature condi-tions. Oswald and Gotaas [11] noted that sewage-grown algal cells have a respiratory oxygen demand of about one-tenth of the weight produced per day at 25° C.

Phytoplankton respiration rates are decreased by a decrease in oxygen tension; thus, they are lower in standing water than in well-aerated flow-ing waters and supersaturated oxygen conditions tend to increase respiration rates; decreases in temperature also decrease phytoplankton respiration rates, but the decrease is much greater than the decrease in photosynthesis, so that at low temperatures the quotient assimilation/respiration is higher. The pH also affects respiration rates which are sometimes optimum at a cer-tain pH, and fall off as pH rises or falls [8]. It is the release of oxygen simultaneously with photosynthesis that has led to the engineering use of photosynthesis for wastewater treatment in stabilization ponds (Chapter 16). Its effect on the oxygen balance in lakes is discussed in Chapter 6.

Net phytoplankton production is obtained by subtracting the algal biomass destroyed in respiration from the gross biomass produced in the system. Corresponding net oxygen production rates may vary from about 1.0 to 1.7 g O_2 per gram of net algal solids (dry weight) produced. The net production of new algal cell material or new terrestrial or aquatic vegetation through photosynthesis helps to store solar energy in organic form which becomes the potential food for all consumers who must depend on it. The energy value of algae varies on a rough basis from 5 to 6 kcal/g dry weight of algae on ash-free basis. A gram of dry algal matter roughly contains 0.5 g of carbon.

That fraction of new organic production which is not consumed by consumers is subject to attack by decomposers (saprophytes) who break down the organic matter to its earlier, simple inorganic constituents such as nitrates, phosphates, CO_2, etc. It is interesting to note that the oxygen requirement of the aerobic decomposers is exactly the same as the oxygen produced in photosynthesis [namely, Eqs. (3.1) to (3.3) can be written in reverse]. Thus, if all organic matter produced in the world were to undergo decomposition, the net oxygen production would be zero. However, all organic matter produced is not destroyed, some is stored without decomposition (e.g., oil). This slight excess of production over decomposition is, in fact, responsible for the buildup of oxygen in the atmosphere.

Phytoplankton are the principal producers, the first link in the food chain, and the amount of consumers at each subsequent trophic level (see below) will be greatly dependent on the amount of potential food energy available to them. Any factor which affects the condition of the water may affect its productivity and have consequences of far-reaching importance to the balance in the ecosystem under consideration. A few such factors are discussed below.

Light influences biological growth, reproduction, locomotion, pigmentation, and other life processes. Among green plants, light is required for the production of chlorophyll in the chloroplasts. Pigmentation is adversely affected by too little as well as too much illumination. As we have seen, illumination diminishes rapidly with depth even in relatively clear water. Benthic plants are confined to the shoreline or the littoral zone shown in Figure 3.2. Beyond the littoral zone lies the limnetic zone which extends down to a certain depth, as will be explained later. Both the littoral and limnetic zones together constitute the euphotic zone.

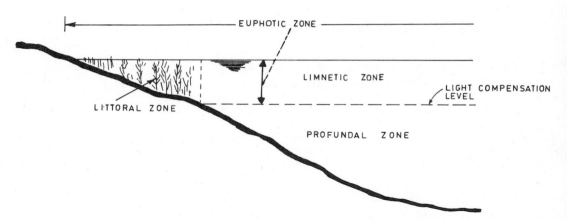

Figure 3.2 Different zones in a waterbody (adapted from Ref. 3).

Phytoplankton can be present in both the littoral and the limnetic zones, extending down to a level called the compensation level, the level at which the gain in energy by photosynthesis just equals the loss of energy by respiration in the phytoplankton. The depth of this layer in a lake can be determined as explained in Sec. 3.6.1.

In the surface layer, with adequate light available, the algae may produce 10 to 20 times more oxygen than is consumed in respiration during the same time. As the depth increases, photosynthesis decreases while respiration remains approximately the same. If phytoplankton are to grow, photosynthesis during the day must build up enough organic matter to offset the material lost by respiration during the day and night. If photosynthesis is insufficient, as it would be below the light compensation level, the phytoplankton would be gradually destroyed by respiration. Hence, phytoplankton can be expected to grow only down to this level; below this level lies the profundal zone.

Table 2.2 gives the depth in meters for light to be reduced to 1% of the surface value; this often signifies the light compensation level for lakes in temperate regions. The level varies from season to season and is affected by any factor which would affect light intensity or the transparency of the water, thereby affecting the growth of the phytoplankton. For many algae, a light intensity greater than 20,000 ergs/cm^2-sec* may adversely affect growth. Thus, optimum algal growth may occur not at the water surface but several centimeters below it where the optimum light intensity exists.

---

*41.86 ergs/cm^2-sec = 1 footcandle.

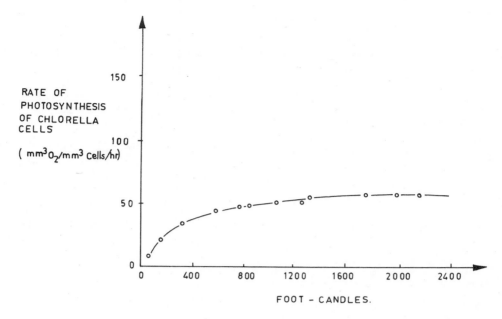

Figure 3.3 Effect of light intensity on the rate of photosynthesis (adapted from Ref. 8).

Different algae exhibit different light saturation levels (Figure 3.3). In high light intensities, an increase in temperature raises the point at which light saturation occurs. High-temperature strains of <u>Chlorella</u> have been isolated which will tolerate light intensities over 120,000 ergs/cm^2-sec. Also, algal cultures in logarithmic growth phase tend to have a higher level of light saturation than older cells. For many algae the light compensation level is reached when the intensity is around 1000 ergs /cm^2-sec. Natural daylight intensities vary considerably with the latitude, seasons, cloud cover, etc., and are discussed further in Chapter 16.

The effect of nutrient concentration on the algal growth rate can be described using a Monod-type equation (see Chapter 1) to take into account the actual growth rate as a function of the specific growth rate, and the algal biomass present at any instant. Thus, if the algal biomass is expressed in terms of "particulate phosphorus" (or chlorophyll, if desired), one can write:

$$\frac{dP}{dt} = \mu_{max}\left[\frac{S_{(P)}}{K_{S_{(P)}} + S_{(P)}}\right]\left[\frac{S_{(N)}}{K_{S_{(N)}} + S_{(N)}}\right]P \qquad (3.6)$$

where

$$P = \text{algal biomass (expressed as particulate phosphorus)}$$

$$t = \text{time}$$

$$\mu_{max} = \text{maximum specific growth rate, time}^{-1}$$

$$S_{(P)} \text{ and } S_{(N)} = \text{substrate concentration of dissolved phosphorus and nitrogen, respectively}$$

$$K_{S_{(P)}} \text{ and } K_{S_{(N)}} = \text{saturation constants for phosphorus and nitrogen, respectively}$$

The value of μ_{max} is also determined by limiting factors such as temperature and light and varies with the time of year. It also varies from layer to layer of water, but often, for simplicity in computation, one value may be assumed for the epilimnion as a whole and adjusted for the time of year. Zero growth is assumed below the light compensation level.

The constants $K_{S_{(P)}}$ and $K_{S_{(N)}}$ are sometimes assumed as 10 and 50 mg per liter, respectively, for large waterbodies. Proposals have been made [7] for a continuous bioassay test for the determination of K_S values for specific test organisms such as algae and given biostimulants.

Example 3.1 Estimate the instantaneous growth rate of the algal biomass in a lake epilimnion in which the dissolved phosphorus and nitrogen is 50 and 250 µg/liter, respectively, and the algal biomass expressed as particulate phosphorus is 4 µg/liter. Assume μ_{max} = 0.5 per day at site conditions.

From Eq. (3.6),

$$\frac{dP}{dt} = 0.5 \left[\frac{50}{10 + 50} \right] \left[\frac{250}{50 + 250} \right] (4.0 \text{ µg/liter})$$

$$= (0.347 \text{ per day})(4.0 \text{ µg/liter}) = 1.39 \text{ g/liter-day}$$

It will thus be observed that the actual specific growth rate at given conditions is 0.347 per day against the maximum specific growth rate of 0.5 per day.

Canale and Vogel [4] have collected the large amount of data available in the literature and reduced it to "specific growth rates" by assuming the general stoichiometric equation

$$CO_2 + H_2O \xrightarrow{\text{photosynthesis}} CH_2O + O_2 \qquad (3.7)$$

and that 0.7 mg of chlorophyll a is equivalent to 45 mg of carbon and 100 mg of dry algal weight. The calculated specific growth rates (limited to cases where optimum light and nutrient conditions were presumed to be present) were plotted versus temperature to discover major temperature-

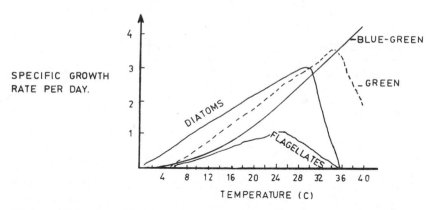

Figure 3.4 Specific growth rates of algae as a function of temperature (adapted from Ref. 4).

growth trends, if any, for a large number of algal varieties grouped into
greens, blue-greens, diatoms, and flagellates* as shown in Figure 3.4.

The results are interesting and bear out the observations of other
workers regarding the optimum temperatures beyond which one group of algae
are replaced by another. For example, the flagellates group shows an optimum growth rate at about $25°$ C, the diatoms at about $30°$ C, the greens at
about $35°$ C, and the blue-greens at some point around $40°$ C.

Introduction of toxic chemicals or nutrients and other biogenic substances in wastewater can also affect their growth. Thermal pollution
would have similar consequences. The presence of toxic substances may
reduce the chlorophyll content of algae [5] which may correspondingly reduce oxygen production. At high concentrations algal growth may be completely inhibited.

3.3 CONSUMERS AND TROPHIC LEVELS

The new plant or organic matter generated by photosynthesis constitutes the potential food and, hence, energy source for all other forms of
life. The concept of trophic levels is connected with nutrition and is

---

*Greens represent all of the Chlorophyta. Blue-greens represent the class
Myxophyceae of the division Cyanophyta. Diatoms represent the class
Bacillariophyceae of the division Chrysophyta. The flagellates are an
artificial group composed of all classes of divisions Euglenophyta, Pyrrho-
phyta, and Chloromonadophyta and two classes of Chrysophyta--the Chryso-
phyceae and Xenthophyceae.

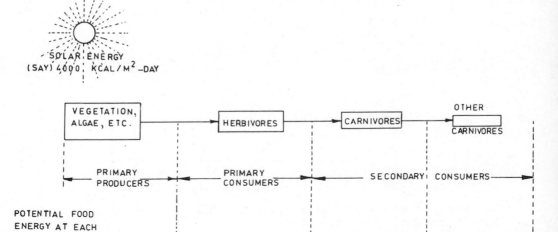

Figure 3.5 Trophic levels.

based on the generalization that herbivores feed on the primary producers while carnivores feed on the herbivores and smaller carnivores, as shown in Figure 3.5.

Primary producers constitute the first trophic level. Herbivorous consumers who feed directly on them are called <u>primary consumers</u>; they consist of zooplankton (water fleas, rotifers, etc.), zoobenthos such as bloodworms (chironomids) and other bottom feeders, nekton (swimming organisms and some fish), and larger animals such as ruminants. The carnivorous consumers are called <u>secondary consumers</u> as they feed on other consumers and include crustaceans, insects, and larger fish. These are followed by higher levels of consumers such as man and animals who may, in fact, occupy more than one trophic level.

3.3.1 Energy Transfer Efficiency

Only about 1% of the total energy received from the sun is converted by photosynthesis into chemical energy in the form of living organic matter. In fact, for oceans as a whole this efficiency is more nearly 0.18%, while for rich tropical forests it may be as much as 3%. These are averages over

large areas and actual efficiencies in some localized favorable cases (e.g.,
oxidation ponds) may be two to three times more.

The efficiency of energy transfer from one trophic level to another is
also very low, being less than 15% at each step. Thermodynamically, the
energy transfer situation apparently is not efficient and a considerable
portion of the energy is converted into heat at each step in the transfer
of energy. It is similar to entropy loss in thermodynamics. An animal,
for example, takes in chemical energy of food and converts a large part
into heat while reconverting only a small part into chemical energy in the
form of new protoplasm in its body. Respiration accounts for most of the
energy losses. This may be on the order of 25 to 75%. Further loss may
occur because some of the food which the animal ate was not assimilable and
some was converted into animal biomass which was not utilizable as food by
man.

It is clear, therefore, that only a very small percentage of the original energy fixed in the primary producers flows on to the succeeding trophic levels. The total metabolic assimilation (i.e., production of biomass
and respiration) can be stated in terms of energy as kilocalories* and the
flow of energy can be shown to be sharply reduced or decreasing from one
trophic level to another.

If the light energy received is on the order of 3000 to 4000 kcal/m^2-
day*, the energy contained in the primary producers may be on the order of
30 to 40 kcal/m^2-day as shown in Figure 3.5, while it would be only about
3 and 0.3 kcal/m^2-day, respectively, at the next two trophic levels. Thus,
it is seen that the energy flow tapers off substantially as we pass from
one trophic level to another. The biomass also consequently tapers down.†

---

*Kilogram-calorie (kcal) = 1000 gcal or cal = amount of heat necessary to
raise 1 kg (or liter) of water by 1°C at 15°C. 1.0 Btu = 0.252 kcal.
Joule (J) = 0.24 gcal = 10^7 ergs.

†For biomass, in general, a very rough estimate is 2 kcal/g live (i.e.,
wet) weight since most living organisms contain 2/3 or more water and minerals. On dry weight basis their heat energy content is approximately

Biomass	kcal/g dry wt.
Algae	4.9-6.0
Invertebrates (excluding insects)	3.0
Insects	5.4
Vertebrates	5.6

For phytoplankton, 1 g carbon ≃ 2 g dry matter ≃ 10 kcal.

(One result of this in our daily lives is that meat generally costs more per kilogram than vegetables.) Generally, not more than four or five steps are to be seen in the food chain, otherwise too low an overall efficiency would be obtained to be biologically competitive.

The tapering biomass can be visualized as a pyramid with the primary producers at the base giving an estimated net production of 100×10^9 tons of dry organic matter annually (1/3 in oceans and 2/3 on land) or roughly 10^{17} to 10^{18} kcal of net energy fixation per year with each superimposed layer being about one order of magnitude less.

The energy involved in man's metabolism amounts to an estimated 10^{14} kcal/yr which is only a thousandth part of the net primary productivity and, therefore, man does not influence the ecosphere substantially in an overall sense. However, man dissipates substantially more energy (10 to 100 times) than required for his metabolism [1] and, furthermore, this is not evenly spread in the ecosphere; "pockets" exist where the ecological subsystem is adversely affected by these discharges as can be seen in the case of heavily polluted rivers, lakes, and other waterbodies.

3.4 DECOMPOSERS

Any individual from any trophic level whose death is not the result of being caught alive and eaten will be eventually consumed by decomposers; decomposers are, in fact, microconsumers. They are heterotrophic organisms, mainly bacteria and fungi, referred to as saprobes or saprophytes ("decay organisms"). Decomposers are microscopic, relatively immobile (usually imbedded in the medium being decomposed), and have high rates of metabolism and turnover.

Decomposers break down the complex organic components of dead organisms and obtain their energy by dissimilation of some of the decomposition products. In the process they form simple inorganic substances usable by producers. One of the important characteristics of decomposers as a group is their ability to function under a variety of environmental conditions. In the presence of free oxygen, for example, aerobic microbes thrive; in the absence of free oxygen, anaerobic microbes take over; facultative microbes function under either condition.

Organic matter essentially contains carbonaceous (starch), nitrogenous (protein), and sulfurous (oils and fats) material together with phosphorus,

iron, and other elements. Microbial decomposition gives various end products depending on whether the decomposition is aerobic or anaerobic; this
is shown diagrammatically in Figure 3.6.

In aerobic decomposition, carbohydrates which may be expressed as CH_2O
are broken down by aerobes to yield simple end products such as CO_2 along
with energy for their growth:

$$6(CH_2O)_x + 5O_2 \rightarrow (CH_2O)_x + 5CO_2 + 5H_2O + \text{energy} \tag{3.8}$$

In anaerobic decomposition of carbonaceous matter, volatile acids are
first produced, mainly acetic, propionic, and butyric, which are then acted
upon by <u>methane bacteria</u> to produce methane and CO_2. The first step in
which putrefaction, or organic acid formation, occurs can be demonstrated,
for example, by the following reaction in which acetic acid is produced:

$$5(CH_2O)_x \rightarrow (CH_2O)_x + 2CH_3COOH + \text{energy} \tag{3.9}$$

In the second step in which methane fermentation occurs we may get:

$$2.5CH_3COOH \xrightarrow{\text{methane bacteria}} (CH_2O)_x + 2CH_4 + 2CO_2 + \textbf{energy} \tag{3.10}$$
(acetic acid)

It has been found from tracer studies that around 70% of the methane
finally produced from the anaerobic decomposition of carbohydrates, proteins, and fatty acids comes from acetic acid as shown in Eq. (3.10).
Another major source of methane is propionic acid (CH_3CH_2COOH) which is
actually first converted to acetic acid, CO_2, and CH_4 by a special group
of methane bacteria; the acetic acid is then converted by another group of
methane bacteria to give more CO_2 and CH_4. Butyric acid (C_3H_7COOH) also
yields CO_2 and CH_4 with the help of yet another group of methane bacteria.
Thus, several species or groups of methane bacteria are involved.

Nitrogenous material, such as the protein contained in organic matter
undergoing aerobic decomposition, is generally first converted to ammonia,
then to nitrites, and finally to nitrates which are the stable form as long
as conditions remain aerobic. The conversion of ammonia involves the action
of chemoautotrophic organisms such as <u>Nitrosomonas</u> and <u>Nitrobacter</u> which
obtain their energy from oxidation at each step, and their carbon from CO_2:

$$2NH_4^+ + 3O_2 \xrightarrow{\text{Nitrosomonas}} 2NO_2^- + 2H_2O + 4H^+ \tag{3.11}$$

$$2NO_2^- + O_2 \xrightarrow{\text{Nitrobacter}} 2NO_3^- \tag{3.12}$$

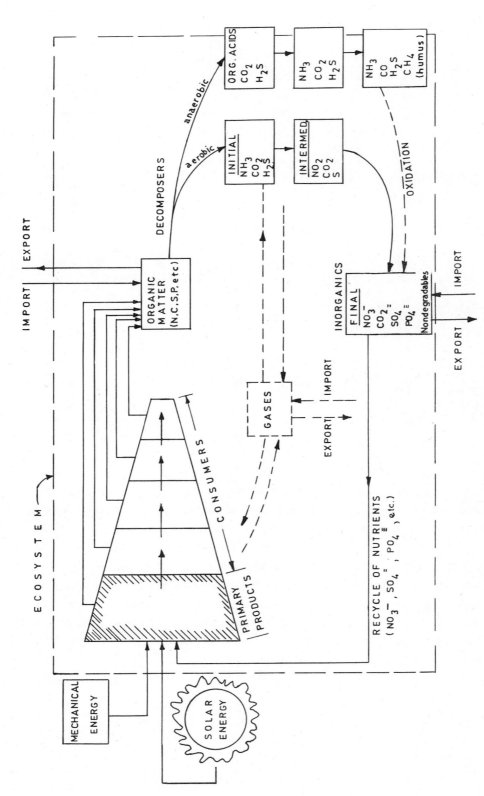

Figure 3.6 Energy flow and recycling of nutrients in an ecosystem.

In anaerobic decomposition, nitrogenous material ends up in the ammonia or nitrogen form. Reduction from nitrate to nitrogen involves the action of facultative heterotrophic organisms and is referred to as denitrification. Owing to the heterotrophic nature of the organisms, energy has to be obtained from an organic carbon source. Thus, gradual denitrification can occur using an endogenous food source, sewage, or methanol, etc.

The uptake of nitrates and ammonia by plants enables the recycling of nitrogen in the ecosystem. Nitrogen is a constituent of cell protoplasm; it passes through the food chain and eventually, when the animal or plant cell dies, the nitrogen is converted to ammonia and then to nitrates.

The sulfurous material in organic matter is gradually oxidized in aerobic decomposition, converting sulfides to sulfur and finally to stable sulfates. Specialized microorganisms are responsible for converting sulfides to sulfates and reducing sulfates back to sulfides (for example, colorless, green, and purple sulfur bacteria).

Heterotrophic microorganisms help to convert organic sulfur to sulfates under aerobic conditions and to H_2S under anaerobic conditions. H_2S is also formed under anaerobic conditions through the action of autotrophs such as Desulfovibrio bacteria. Thus, under anaerobic conditions sulfurous material remains in the sulfide form. Deep in soils and sediments, a sulfide pool exists, so to speak, consisting mainly of iron sulfides and some H_2S. Their oxidation yields sulfates which enter waters as solutions and are also taken up by plants. The reduction of sulfates to sulfides restores the latter in the pool.

Other substances contained in organic matter are also acted upon similarly by the decomposers, producing different end products depending on aerobic or anaerobic activity. The end products again enter the system through the synthesis of new plant and animal protoplasm. The example of phosphorus must also be included here. Phosphorus is contained in all protoplasm and bones. Dead protoplasm is acted upon by phosphatizing bacteria to yield dissolved phosphates, part of which enter bottom sediments as precipitates and part are converted to protoplasm. Phosphate rocks form a pool of phosphorus; rock erosion also yields phosphates.

Generally speaking, products of anaerobic decomposition such as ammonia and sulfides are not in the oxidized state, but are capable of oxidation to give stable end products such as nitrates and sulfates; however, conditions must change from anaerobic to aerobic to permit oxidation to occur. The more resistant products of bacterial decomposition of organic matter end up

as humic substances. They are not easy to characterize chemically owing
to their diverse origins. In general, humic substances are formed by the
combination of aromatic compounds (having benzene ring and side chain bond-
ing) with the decomposition products of proteins and polysaccharides.
Thus, they are relatively refractory in nature. The various end products
formed at each step as a result of either aerobic or anaerobic activity
are shown in Figure 3.6 together with other activities occurring within a
given ecological subsystem.

The numbers of decomposers are generally large although their total
mass is small compared with the macroconsumers discussed earlier. However,
they compensate for their small size by their high rate of metabolism.
Metabolism is accompanied by heat dissipation to the environment; for example,
high temperatures are soon reached in a heap of compost [3].

A population of decomposers can be severely affected by the presence
of toxic substances and other unfavorable conditions in the environment.
It is essential to note that the presence of decomposers is essential to
continuity of life within an ecosystem since, in their absence, all the
nutrients would be tied up in dead organic matter and no new life would be
produced (see Sec. 3.5).

3.5 ENERGY FLOW AND THE RECYCLING OF NUTRIENTS
     IN AN ECOSYSTEM

Figure 3.6 shows how the flow of energy and nutrients takes place in
a typical ecosystem receiving light energy and perhaps some mechanical
energy as well. The primary producers and consumers must die eventually
and join the mass of dead organics. These organics contain carbon, hydrogen, oxygen, sulfur, and phosphorus and are attacked by decomposing bacteria and fungi either aerobically or anaerobically.

In the aerobic system, microbial decomposition of organic matter as
described in the previous section gives the initial, intermediate, and final
products as shown in Figure 3.6. The final products are all inorganics
such as nitrates, sulfates, CO_2, and phosphates; each one of these is useful
for new growth and, hence, is shown as "nutrients" cycled back to form more
primary producers with the help of additional energy inputs. Thus, the
figure illustrates how nutrients are conserved and recycled, but energy is
not; energy is lost at each trophic level, and in the respiration of decomposers.

As shown in Figure 3.6, microbial decomposition yields somewhat different products in the anaerobic system. Mineralization or decomposition is then relatively incomplete; its products include methane, amines, ammonia, and hydrogen sulfide, besides CO_2 and humus. Some of these are capable of further oxidation as mentioned earlier.

Figure 3.6 also shows the gases exchanged in photosynthesis and respiration. Gaseous exchange is an important indicator of material exchange in the ecosystem and, in fact, can be used as a measure of such activity. Energy flow can be measured indirectly by measuring O_2 or CO_2 or the biomass itself. Furthermore, organics imported into the ecosystem (in the form of pollution, etc.) as well as the organics exported or lost from the system (in the form of downstream flow in a river or in the form of harvested crops or algae) can be included in the figure in the box marked "organics." Similarly, in the box marked "inorganics" one could show either imports into the system (as from pollution, etc., entering the system) or exports (in the form of leaching, for example). The entry of toxins into the system may reduce or completely eliminate one or more of the producers, consumers, or decomposers, thus damaging the ecosystem. All inputs and outputs can be stated in terms of energy flow in the same units of measurement, such as kilocalories.

A lake, river, or waste stabilization pond could all be considered as good examples of ecological subsystems to which the above concepts can be applied. Nutrients may be imported into the system in the form of surface runoff, waste discharges from the surrounding community, and in other ways. Assuming that conditions are favorable for algal growth, one could expect algae to flourish in the upper or euphotic zone where sunlight is available. When the algae so produced die and/or otherwise sink lower down in the waterbody, they constitute organic matter which is attacked by decomposing bacteria, and the nutrients contained in the algae are released into the water. Nutrients used up in bacterial growth are also released when the bacteria die. Hence, the nutrients remain available for repeated use for producing algae and bacteria in the waterbody unless removed or exported from the system through discharge from the waterbody or through other ways such as crop and algae harvesting.

All the essential elements of living matter tend to circulate in the biosphere in characteristic paths from environment to organisms and back to the environment. These are known as inorganic-organic cycles or biogeochemical cycles. It is evident that, if the biosphere is to remain in operation, the biologically important materials must undergo cyclical changes so that after utilization they are put back, at the expense of some solar energy, into a reusable form.

According to Odum [3] the pattern of nutrient cycling in the tropics is in some ways different from that in the temperate zone. In cold regions a large portion of the organic matter and available nutrients is in the soil or sediment at all times; in the tropics a much larger percentage is in the biomass. This makes it necessary to adapt agricultural and other practices to suit the natural nutrient cycling pattern in each zone. It is also evident that, in the tropics, land disposal of sewage and sludge would be a desirable way of supplying nutrients to a soil relatively poor in them.

The artificial injection of some elements into water, air, or land is occurring much faster these days owing to industrial pollution. Many metal ions are released into the environment at rates exceeding the natural rate of cycling, resulting in adverse effects on the ecosystem. New cycles have come into being that may distribute very widely, and in toxic quantities, elements such as lead and mercury as well as pesticides, defoliants, etc., and adversely affect the green mantle of the earth [12].

3.6 PRODUCTIVITY IN A SYSTEM

Productivity, in general, is defined as a gain in energy through the production of organic matter by photoautotrophs and chemoautotrophs. Consider the growth of a single autotroph such as an algal cell. Out of the the total energy fixed by the cell, or its gross production (GP), some part is used in respiration for cell maintenance while the rest is "stored" within the cell. The net amount thus stored is called the net production (NP); see Figure 3.7. Net production is measurable as the growth of a plant or organism and can be expressed as dry weight or in other ways. If the part used for respiration by the autotrophs is referred to as R_A, we get

$$NP = GP - R_A \tag{3.13}$$

In temperate forest regions, R_A has been consistently found to be 0.55 GP [12]. In the tropics the ratio of R_A/GP is higher while at higher altitudes, it is lower. Even in a given location it changes with the seasons.

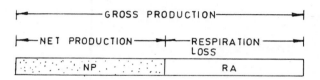

Figure 3.7 A diagrammatic representation showing that part of the total energy that is fixed is lost in respiration.

This relationship also holds for an entire plant and animal community or an ecosystem. If we consider not only the plants but also the consumers of plants, we must take into account the respiration losses of each consumer that is sustained by the original quantum of producers. That is what happens as an ecosystem matures; consumer populations increase substantially, adding to the total respiration losses of the system. If the energy lost in the respiration of all the consumers or heterotrophs is denoted by R_H, the total respiration loss is $R_A + R_H$. For an ecosystem, the difference between gross production and total respiration losses constitutes the energy stored within the system, or in other words, the net ecosystem production (NEP). Thus, from Figure 3.8 we see

$$NEP = GP - (R_A + R_H) \tag{3.14}$$

Based on this concept, a distinction can be drawn between a developmental or successional ecosystem and a mature or climax one. In the successional system the total respiration is less than the gross production, leaving surplus energy that is built into the structure, so to speak, in the form of NEP, and adding to the resources of the site. On the other hand, in a climax system all the energy fixed is used in the combined respiration of the plants and the consumers, thus giving NEP = 0. There is no energy left over and no net annual storage. Climax ecosystems probably represent an efficient use of the resources of a site.

In research spanning the better part of a decade, studies were made in an oak-pine forest at the Brookhaven National Laboratory [12]. A major problem in this study was measuring the total respiration of the forest. Taking advantage of the frequent inversions of temperature occurring at night, the rate of accumulation of CO_2 during these inversions was used as a direct measure of the total respiration, since at night the effect of photosynthesis was eliminated. The air column remained stable for several hours, thus affording an opportunity to see how the carbon dioxide buildup

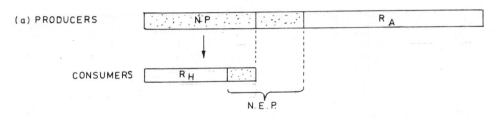

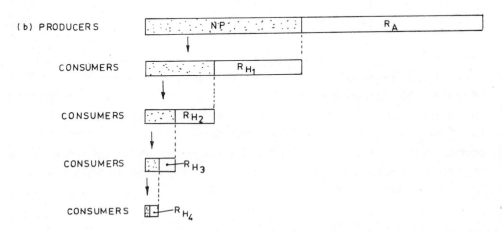

Figure 3.8 (a) A successional ecosystem in which some of the net production is stored as growth (NEP = net ecosystem production). (b) A climax or mature ecosystem in which all the energy originally fixed is used in respiration (gross production = R_A + R_{H1} + R_{H2} + ... + R_{Hn}).

took place. Other methods of estimation were also used and the results showed that per m^2 per year the GP was 2650 g of dry organic matter. The respiration requirements of the autotrophs were 1450 g and those of the heterotrophs were 650 g. Hence, net primary production was 1200 g (2650 - 1450 = 1200) and the NEP was equal to 550 g as shown in Figure 3.9 [NEP = GP - (R_A + R_H) = 2650 - (1450 + 650) = 550 g]. In this forest, the value of R/GP is 55% which compares well with the values noted earlier for temperate forests. Since net storage is taking place, the forest can still be considered as "immature." The ratio of (R_A + R_H)/GP = (1450 + 650)/2650 = 80%. Hence, the forest is at about 80% of climax or "late successional" [12]. The net production of 1200 g/m^2-year is in the low middle range for forests and is typical of the productivity of small-statured forests. Incidentally, the efficiency of solar energy use for the above forest was found to be only 0.9%. As an illustration of the range of productivity to

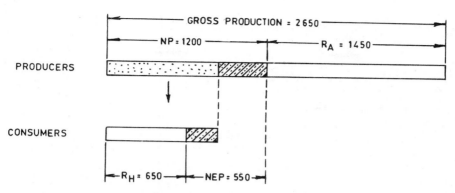

Figure 3.9 Net ecosystem production in an oak-pine forest (adapted from Ref. 12).

be expected in different situations, some examples are given in Tables 3.2 and 3.3.

From Tables 3.2 and 3.3 it is evident that the productivity of the seas is much lower than that of land; in fact, deep oceans are comparable to deserts in their productivity. Large areas usually show considerably less productivity than small favorable areas. In large waterbodies, pro-

Table 3.2 Primary Productivity of Various Ecosystems

Type of ecosystem	Gross primary productivity[a] (10^3 kcal/m^2-year)
Deserts, semiarid grasslands, deep oligotrophic lakes	< 0.5
Open oceans	< 1.0
Continental shelf waters, deep lakes, grasslands, mountain forests, some agriculture	< 1.0-3.0
Shallow eutrophic lakes, moist forests and grasslands, most agriculture[b]	3.0-10.0
Some estuaries, coral reefs, evergreen forests, terrestrial communities on alluvial plains, and intensive (energy-subsidized) agriculture	10-25
Upper limit for gross photosynthesis	50

[a] One gram of carbon production is roughly equivalent to 10 kcal. Net primary productivity averages about half of gross productivity.

[b] Estimated maximum net primary productivity of wheat, corn, and rice ranges between 4.4 and 5.5 x 10^3 kcal per m^2 per year.

Source: adapted from Ref. 3.

Table 3.3 Secondary Productivity as Measured in Fish Production

Type of ecosystem	Net secondary productivity[a] (kcal/m²-year)
Unfertilized natural waters	
World marine fishery average	0.3
U.S. fish ponds with stocked carnivores	9.0-34.0
German fish ponds with stocked herbivores (carp)	22.0-80.0
Fertilized waters	
U.S. fish ponds with stocked carnivores	45-112
German fish ponds with stocked herbivores (carp)	200-336
Fertilized waters with outside food added	450-900

[a]Determined by use of harvest methods.

Source: adapted from Ref. 3.

ductivity may be further limited by poor light penetration. Coastal waters
and estuaries generally show higher productivity than that of the open ocean
owing to the greater amount of nutrients washed down into them from adjacent
lands. Between deserts and open oceans on the one hand and rich tropical
forests on the other, the difference in productivity can range over three
orders of magnitude. Sewage ponds, polluted waters, and lands used for
intensive agriculture show characteristically high levels of productivity
owing to a greater availability of nutrients. The difference in productivity
between oligotrophic and eutrophic lakes can also be considerable as shown
in Table 3.2. Differences can also be seen in the same ecosystem between
summer and winter. Various factors affecting productivity are further discussed in Chapter 6.

Secondary productivity when measured in terms of fish yield is naturally
much lower than primary productivity (Table 3.3). Productivity for herbivorous fish is greater than for the carnivorous fish which are one trophic
level higher.

The world marine fish potential is estimated to be on the order of
100 million tons per year, and, at the present rate of consumption and its
expected growth rate, the full potential may be reached by 1977. Although
organics and other pollution drained into rivers and coastal waters tend to
provide the nutrients by which productivity may actually be increased, the

benefits are not necessarily derived by man since pollution makes the environment relatively unfavorable for the more delicate varieties of fish that man relishes. Furthermore, the presence of toxic substances can sharply reduce the overall productivity in water (and on land), and the accumulation of toxic substances in the food chain can make the food useless to man. If the world is to support an ever-increasing population, every effort will have to be made to find new ways of increasing productivity and curbing the activities which reduce it.

3.6.1 Measurement of Primary Productivity

Productivity measurements are based on some indirect quantity such as the amount of substance produced [protoplasm, total organic carbon (TOC), chlorophyll, etc.], the amount of raw material used (e.g., nitrogen or phosphorus), or the amount of by-product released (e.g., oxygen or carbon dioxide); radioactive techniques are also used. It is necessary to know whether the method used measures the gross or the net primary productivity. Table 3.4 lists some of the methods used; each has some advantages and some disadvantages [3].

Primary productivity can be determined in an aquatic environment from the production and consumption of oxygen at any level in the euphotic zone in the following manner. Samples are taken at the desired level in three bottles. The initial dissolved oxygen concentration (DO) is determined for one bottle by the Winkler method while the other two bottles are closed and lowered to their original level after one bottle has been opaqued or

Table 3.4 Methods for Measurement of Primary Productivity

Method	Productivity measured
Harvest measurement	Net
Oxygen measurement	Gross/net
Carbon dioxide measurement	Gross/net
Measurement of TOC	Net
Radioactive tracers	Net
Measurement of chlorophyll	Net

blackened to prevent photosynthesis while the other bottle remains clear.
The two bottles are kept suspended for some hours after which they are
withdrawn and their dissolved oxygen measured. When working at sea, it is
sometimes convenient to expose the bottles to the light and temperature
conditions of the sea on the deck of the ship instead of suspending them
in the water. The sum of the net oxygen produced in the clear bottle and
oxygen used in the dark bottle is the total or gross oxygen production
since, from Eq. (3.13)

$$GP = NP + R$$

or   GP = (final DO in clear - initial DO) + (initial DO
           - final DO in black)

or   GP = (final DO in clear) - (final DO in black)   (3.15)

A nonphotosynthetic microbial population (i.e., bacteria) is also present
along with algae in each bottle and some oxygen utilization occurs as a
result of bacterial respiration. This is assumed to be equal in both
bottles. Furthermore, the algal respiration rate is assumed to be the same
in light and dark conditions. If the test period is unduly long, all the
algae in the black bottle may die.

        If algal weight is measured before and after the test, one can estimate
the oxygen production rate at the given level and temperature in the lake
in terms of mg of O_2 per hour per mg of algal solids. If, for instance,
the dry weight of chlorophyll a is known for the algae in question, the O_2
production rate can be stated in terms of chlorophyll a. The TOC is also
often determined. Productivity could also be stated in terms of weight per
unit area and time. In a body of water, productivity varies with depth and
with the time of day and season. Values of productivity per unit volume
and time determined at different depths can be summed to obtain the average
productivity per m^2 of surface of the lake.

        The lake depth at which oxygen production just equals oxygen consump-
tion in the clear bottle, namely, the depth at which there is no net change
in the DO content of the clear bottle, is the compensation level.

        A more sensitive method of productivity measurement involves the use
of ^{14}C, a radioactive isotope. The experiment is set up in the manner already described except that a small amount of sodium bicarbonate containing
the isotope ($NaH^{14}CO_3$) is added to each bottle at the start of the experiment. At the end, the algal contents are filtered off and the amount of
^{14}C that they have built into their tissues is measured by means of a

Geiger counter. The greater sensitivity of this method makes it more desirable to use with waters of low productivity.

It should be borne in mind that both of these methods are measuring the plankton productivity and not the total productivity of the lake, which also includes littoral rooted vegetation. In case of large lakes, planktonic production is far greater than the production of littoral vegetation.

3.6.2 Relative Magnitudes of Production and Respiration

Depending upon the relative amounts of production and respiration within a system, waterbodies are classified as oligotrophic, mesotrophic, or eutrophic. The term trophic denotes nutrition, and oligo-, meso-, and eu- signify the nutritional status as little, moderate, or high (well), respectively. These terms are further discussed in Chapter 6.

If the production of O_2 through photosynthesis is denoted by P, while the consumption of O_2 through respiration is denoted by R, then

P > R denotes that primary producers (like algae) are in abundance, a state of autotrophy exists, more food and oxygen are produced than consumed, and there is net ecosystem growth.

P < R denotes that consumption is greater than production, a state of heterotrophy exists, consumers are in abundance, and oxygen resources tend to be depleted.

Figure 3.10 shows that if P and R are represented graphically, a self-sufficient community, which on an average consumes what it makes (P = R), would be located along the diagonal line. Seasonal variations or imbalances between P and R would be as shown by the dotted line. As long as the

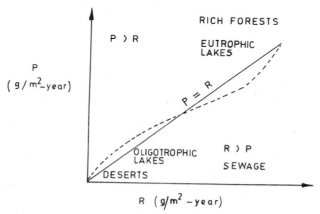

Figure 3.10 Relative magnitudes of production (P) and respiration (R) in different ecosystems.

overall chemical composition of the water is more or less constant, a sort
of steady state is maintained and P and R are in general balance. The entry
of pollution may change this chemical composition and disturb the balance
between P and R.* Oligotrophic lakes, barren streams, and deserts would plot
lower down on the diagonal line since both P and R would tend to be small and
more or less equal to one another. Eutrophic lakes and rich tropical forests
would generally plot higher up and to the left of the diagonal line since
both P and R would be high, but P would generally tend to be greater than R.
The problems caused under such circumstances are discussed in Chapter 6.

For waters just below a sewage outfall, respiration is much higher
than production. The discharge of organic matter encourages the growth of
decomposing organisms which shoots up the value of R, while algal growth
may be actually precluded in the initial stages owing to turbidity and a
reduction in sunlight penetration (see Figure 3.11). This state of affairs
may continue until mineralization occurs. Gradually, photosynthesis is
stimulated and more and more plant biomass is produced; hence, P is now
greater than R. This kind of alternating situation continues until equilibrium is reached and P tends to equal R. Meanwhile, some nutrient losses
also occur; some to the atmosphere and some are irretrievably lost in sediments.

Varying values of P and R were recorded in a study [13] covering a
distance of 320 km from a river outfall as shown in Figure 3.12. Here it
is also seen that, ultimately, P tends to equal R and stabilization of the
ecosystem can be said to be concurrent with the stabilization of the waste
itself.

---

*Stumm et al. [7] point out that an imbalance can also occur when P and R
organisms become physically separated as in a stratified lake.

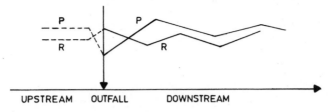

Figure 3.11 The effect of outfall discharge on relative production
(P) and respiration (R) values.

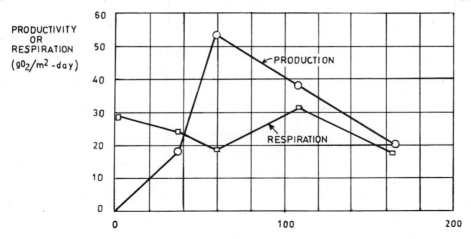

Figure 3.12 Pollution stabilization in White River, Indiana, as indicated by stabilization of the ecosystem (adapted from Ref. 13).

3.7 FOOD CHAINS AND WEBS

From the previous discussion it is seen that some of the potential food energy fixed by the primary producers is consumed by herbivores and a succession of carnivores at different trophic levels. The lines of organic transfer constitute food chains. Basically, there are three types of food chains: (1) predator or grazing chains, (2) parasitic chains, and (3) saprophytic chains. When several chains crisscross one another they form "webs."

3.7.1 The Predatory or Grazing Chains and Webs

This chain is the result of predation or grazing where one organism is eaten by another. The chain generally starts with the primary producers and progresses from smaller to larger consumers. For example, Figure 3.13 shows such a chain in which algae (phytoplankton) are eaten by zooplankton such as fleas, that are eaten by invertebrates, which are in turn eaten by fish, which are finally eaten by man.

Actually, a predator may not live on one type of food only; it may consume a variety of foods, thus producing food webs similar to those typically observed in waterbodies such as lakes or bays (Figure 3.14).

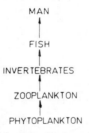

Figure 3.13 A predatory or grazing food chain.

The web is not necessarily confined to the aquatic medium; it may extend into the soil below and the atmosphere above.

Figure 3.14 shows how a material at a given trophic level may serve as food to various consumers at the next higher level. Man himself is the best example in this regard as he may feed on a variety of plants, herbivores, and carnivores; so do other herbivores and carnivores. Thus, the lines of organic transfer are not simple chains but often quite complex webs.

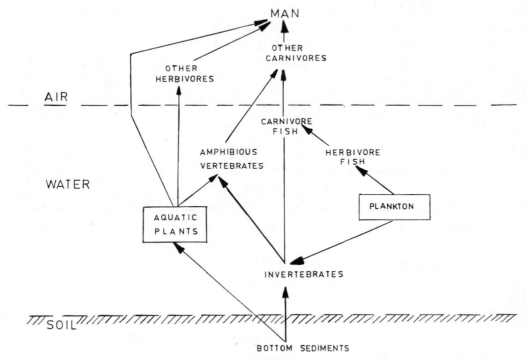

Figure 3.14 A food web interconnecting the biota in air, water, and soil.

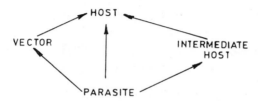

Figure 3.15 Parasitic food chains.

3.7.2 The Parasitic Chain

The parasitic chain is formed as a result of the transmission of organic food material from the host to the parasite with or without the help of an intermediary organism (Figure 3.15). For example, the transmission of malaria occurs through a food chain involving the malarial parasite, a vector such as the <u>Anopheles</u> mosquito, and the host which may be man. Various other parasitic infections involve similar food chains, the food in these cases generally passing from the larger to the smaller consumers. Of course, many parasites draw their nutritional requirements directly from the host's system without the need of an intermediary.

3.7.3 The Saprophytic or Decay Chain

The action of decomposers, either aerobic, anaerobic, or facultative, on organic matter or dead organisms may give rise to a saprophytic ("sapro" meaning decay) chain. In this chain, organic matter is transferred from the substrate to new organisms or cells produced as a result of decomposition.

All food chains and webs could, in fact, be interconnected. A carnivore may be a predator, a parasite may be thriving on this predator, and when it dies saprophytes may take over. These food chains, besides being interconnected, also form effective feed-back mechanisms to help give appropriate responses to outside interferences and ensure stability of the population in every trophic level (homeostasis*).

---

*Homeostatis comes from "homo" meaning same, and "stasis" meaning standing. It signifies the mechanisms by which an organism tends to resist change and continue in a state of equilibrium.

3.8 STABILITY OF A TROPHIC LEVEL

Among consumers, the stability of any trophic level depends on several factors which can be summarized as follows:

1. The rate at which food is produced in the previous trophic level: an adequate food supply ensures the succession of consumers in a system.

2. The number of food chains between the trophic levels: the stability increases with an increasing number of channels through which energy can flow (similar to a water distribution network).

3. The number of species present; since each species is part of a food chain, the more species present, the greater the number of food chains and, hence, the greater the stability. Therefore, biotic diversity is an indicator of stability.

4. The number of individuals in each species; if a particular species becomes too abundant, the stability of the whole community is reduced. However, predation provides a sort of control in this regard; a predator may eat a variety of animals but, since the most abundant kind will be caught most often, it is evident that there is a sort of automatic limit to abundance [1].

5. The extent of "homeostasis" operating in the system for response to outside interferences.

6. The extent of chemical buffer intensity to withstand the effects of external inputs [1].

3.9 ACCUMULATION IN FOOD CHAINS

An interesting observation that can be made at this stage is the possibility of transmitting a conservative (nondegradable) substance present at one trophic level to other trophic levels through the predator chain. As a large mass of "food" at one trophic level is needed in order to support life at the next higher trophic level, it is evident that the conservative substance originally contained in the large mass will now be concentrated into a smaller mass. Thus, the concentration will increase at each step as the conservative material passes along the food chain. If the efficiency of energy transfer is, for example, 10% at each trophic level, the concentration will increase tenfold at each level. There is, no doubt, some reduction in concentration owing to natural elimination processes of living organisms and animals (e.g., through their waste products) but the net accumulations may still be quite substantial. Biological concentrations of substances like DDT, mercury, cadmium, etc., have caused serious health problems. These and the mechanisms of concentration are discussed in Chapter 4.

3.10 SPECIES DIVERSITY AND STABILITY

As seen above, survival of an ecological community depends on its tolerance to a vast complex of physical, chemical, and biological conditions. The biota existing in an aquatic environment is the result of the total environmental complex to which the biota is subjected over a period of time.

Any "harmful" effect on plants and animals tends to cause a shift in the population from sensitive species to only the more resistant ones. In unfavorable environments, therefore, the number of species is reduced. The biocenosis* becomes poorer in species. However, this is often accompanied by an increase in the number of individuals per species (see Figure 3.16).

Investigations should be directed mainly to those biocenoses which have an important indicator value, and which can be quantified as far as possible using statistical methods. For example, benthic macrofauna are more suitable to study than phytoplankton or zooplankton populations. Results of simple qualitative surveys can be expressed in terms of the relative abundance of each species at a sampling site, or species dominance. Such surveys generally precede a quantitative investigation.

Quantitative surveys require that the organisms collected be taxonomically and quantitatively determined in the laboratory. The aim of taxonomic work is to determine the species as far as possible, but because this is often difficult, it may be practical to stop at the level of genus or family. Smaller organisms are counted with the aid of a microscope or a magnifying glass. The total mass of organisms is determined by weighing

---

*Biocenosis is the balanced relation between the plant and animal species in an ecosystem under given external conditions.

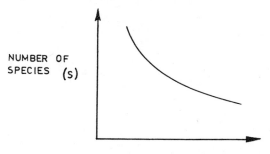

Figure 3.16 The relationship between the number of species and the number of individuals per species.

and is expressed as fresh weight (for living organisms), as wet weight (for preserved organisms), or as dry weight. The dominance and number of species are often expressed in terms of different diversity indices.

The entry of wastes into a waterbody, for example, may result in due course in an increase in the primary productivity of the waterbody by increasing nutrients and other biogenic compounds, which are brought in along with the wastes. With more potential food energy now available, the population and variety of consumers would increase but for the fact that the unfavorable condition precludes an increase in varieties. In fact, the number of species may reduce. Hence, within the remaining species the number of individuals must increase greatly to consume the additional food energy available, or else the number of decomposers would increase. Pollution may increase the total biomass (of fish, for example) available in a given site, but will most certainly reduce the number of species, killing the more sensitive ones first.

The concept of species diversity, originally suggested in 1922 by Gleason, is now being used more frequently as an indicator of pollution. The species diversity of a community in a given habitat can be found by noting the number of species met while counting a large enough number of individuals, say, between 500 and 3000. It is often found that if the cumulative number of species observed is plotted against the logarithm of the number of individuals counted, a straight line results. Sometimes cumulative species versus the square root of individuals gives a linear plot.

Sometimes, instead of a species-numbers relationship, a species per unit area relationship may be used. In both cases, a common observation is that new species are found in proportion to the logarithm or some exponential function of the number of individuals or of the area sampled. The larger the number or area, the greater the variety of species likely to be spotted; thus,

$$\frac{dI}{dS} = K_D I \qquad\qquad (3.16)$$

where

   S = number of species

   I = total number of individuals counted

   K_D = species diversity index

Species diversity indices are often estimated using one of the following methods:

1. K_D = number of species observed in counting **first** 1000 individuals $\qquad$ (3.17)

2. $K_D = \dfrac{S - 1}{\ln(I)}$ $\qquad$ (3.18)

3. $K_D = \dfrac{S}{\sqrt{I}}$ $\qquad$ (3.19)

Species diversity tends to be higher in older communities than in newer ones. Species diversity tends to be low for those systems where strong physical or chemical factors tend to control as, for example, thermal conditions, salinity, etc. On the other hand, species diversity tends to be relatively high in environments favoring biological activity (Figure 3.17). Table 3.5 gives a few typical values of K_D.

Table 3.5 Species Diversity in Some Systems

System	Number of species per 1000 individuals
Tropical sea/ocean bottom	75-90
Rain forest water	60-75
Stable streams	30-40
Bays	5-30
Polluted waters	< 5

Source: Adapted from Refs. 3 and 7.

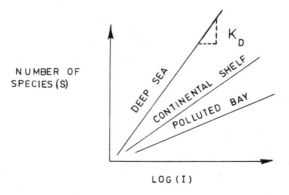

Figure 3.17 The relationship between the number of species and the number of individuals.

Shannon-Weaver proposed a different index which is reported to better reflect changes in community composition [3]:

$$D = -\sum_{s=1}^{s} \left[\left(\frac{i_s}{I}\right) \log_2 \left(\frac{i_s}{I}\right) \right]$$ (3.20)

where

> i_s = number of individuals in each species
> D = average diversity index (diversity per individual)
> I = total number of individuals

and logs are to the base 2.*

The Shannon index takes into account both diversity and evenness in the apportionment of individuals among the species. This index is somewhat less affected by sample size and generally shows normal distribution, allowing the application of tests for statistically significant differences between means.

The maximum value D_{max} occurs when there are an equal number of individuals i_s in each of the S number of species. For example, if $S = 4$ and $I = 64$, the maximum diversity occurs when the number of individuals per species is equal, namely, $i_s = 64/4 = 16$, and its value is given by

$$D_{max} = -\sum_{1}^{4} \left(\frac{16}{64}\right) \log_2 \left(\frac{16}{64}\right) = -\log_2 \left(\frac{1}{4}\right) = 2.0$$

In general terms,

$$D_{max} = -\log_2 \left(\frac{1}{S}\right) = \log_2(S)$$ (3.21)

The minimum value of the diversity index D_{min} occurs when there is only one individual in each species, except one in which all the remaining individuals are found. The value depends on the total number of individuals I and the number of species S as below:

$$D_{min} = -\sum_{1}^{s-1} \left(\frac{1}{I}\right) - \left[\frac{I - (S - 1)}{I} \log_2 \frac{I - (S - 1)}{I} \right]$$ (3.22)

Thus, for the example just given,

---

*$\log_2(X) = \dfrac{\log_{10}(X)}{\log_{10}(2)} = 3.32(\log_{10}(X)) = 1.44(\ln X)$

$$D_{min} = -\overset{3}{\underset{1}{\Sigma}} \left(\frac{1}{64} \right) \log_2 \left(\frac{1}{64} \right) - \left[\left(\frac{61}{64} \right) \log_2 \left(\frac{61}{64} \right) \right] = 0.93$$

It is desirable to report such indices in a "scaled" version to express their deviation relative to their maxima and minima, and thus enable a comparison between different communities. This is stated as

$$R = \left[\frac{D_{max} - D}{D_{max} - D_{min}} \right] \tag{3.23}$$

where R = 0 implies that the observed diversity D is as large as possible $(D = D_{max})$, while R = 1 implies minimum diversity $(D = D_{min})$ at given values of I and S.

Example 3.2   The results shown in columns 1 and 2 of Table 3.6 were obtained in a field study. Compute the Shannon-Weaver Diversity Index.

From Eq. (3.20) $D = -(-2.142) = 2.142$

From Eq. (3.21) $D_{max} = \log_2(S) = 2.584$

From Eq. (3.22) $D_{min} = 0.137$

From Eq. (3.23) $R = \left(\dfrac{2.584 - 2.142}{2.584 - 0.137} \right) = 0.180$

Table 3.6

Species	Individuals in each species (i_s)	$\dfrac{i_s}{I}$	$\log_2\left(\dfrac{i_s}{I}\right)$	$\left(\dfrac{i_s}{I}\right)\log_2\left(\dfrac{i_s}{I}\right)$
1	20	20/364 = 0.05	-4.32	-0.216
2	100	0.27	-1.89	-0.510
3	78	0.21	-2.25	-0.473
4	130	0.36	-1.47	-0.529
5	19	0.047	-4.41	-0.207
6	17	0.047	-4.41	-0.207
S = 6	I = 364			$\Sigma = 2.142$

Under conditions of light pollution, the number of species in the tolerant groups may remain the same (or even increase) whereas the number of relatively intolerant ones would decrease. Under heavy pollution all groups would decrease.

Thus, it is seen that species diversity is always lowered in a community under "stress." This reduction signifies an adverse effect on the

stability of the given ecosystem, as the stability at each trophic level
is decreased. Since each species is part of a food chain, the lesser the
species the lesser and shorter the food chains or, in other words, the
simpler the food webs. It is, therefore, apparent that any radical change
in the number of individuals in a species has a severe effect on the subsequent trophic levels because there is very little or no alternative choice
of food. There are more individuals, perhaps, but of only a few kinds. As
is well known from human experience, there is inherent instability in too
many of one kind. For this reason, a waste stabilization pond is also relatively unstable. It has few species with many individuals per species
with the consequence that there are simple and short food chains which can
be adversely affected by shock loads, variations in light energy, or temperature. The whole community is vulnerable to small disturbances.

All communities are not equally suitable for diversity studies. Some
communities are subject to periodic variations as a result of natural environmental fluctuations. In this regard, the pelagic ecosystem is generally unsuitable, while benthic communities on homogeneous, soft (muddy)
bottoms that are not subject to intertidal influences are most suitable
[14].

Thus, indices are useful in comparing one community with another in a
similar habitat, using comparable sample sizes, and ascertaining beforehand
whether data fit on a log, square root, or other basis. The indices are
also useful in assessing the environmental impact of waste disposal systems.
Measurement of indices at a given site before and after the construction of
new outfalls can give useful objective information (Fig. 3.18).

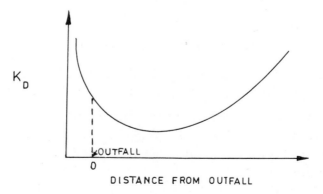

Figure 3.18 The reduction in the species diversity index (K_D) downstream of an outfall.

3.11 THERMAL POLLUTION

In previous discussions, references have been made from time to time
to thermal effects and it will be advantageous to focus our attention once
again on this subject. A change in thermal equilibrium also places a
"stress" on an ecosystem just as any pollutant does and, hence, is often
referred to as thermal "pollution."

Any change in the thermal status of an aquatic environment has several
consequences, the severity of which depends upon the magnitude and sudden-
ness of the change. The principal effect is related to fish life. The
dissolved oxygen concentration in water drops with increased temperature,
while at the same time the oxygen required for respiration by fish and
other biota increases. The increased rate of biochemical activity which
follows increased temperature also leads to a reduction of dissolved oxygen
in the water. What may have earlier been a case of low dissolved oxygen
may now become clearly a case of anaerobic conditions.

The finer varieties of game fish need a minimum of about 4 mg/liter
DO, whereas the coarser varieties may get by with a DO concentration as low
as 2 mg/liter. Sudden changes in temperature and high temperatures are
also bad for fish. For trout, the temperature of the stream has to be
lower than 15° C; for pink salmon, a temperature in excess of 23° C can be
lethal. There are various other fish of commercial value which survive
well even at temperatures near 30° C. Fish feeding, growth, and reproduction
are affected by temperature, and undue harm to fisheries can usually be
prevented by ensuring that stream temperatures do not rise by more than 3
to 5° C above their normal temperatures, except in limited zones of mixing
downstream of an outfall.

In the event of increased temperatures and reduced DO concentrations,
toxic chemicals are often likely to prove dangerous at lower concentrations
than when temperatures are low and DO is high. Thus, in this sense, temperature, DO, and chemical concentrations (as well as pH) are synergistic
in their effects. Furthermore, a species may be destroyed at less than its
lethal temperature if other factors such as DO favor the growth of compet-
itors, predators, and parasites which may adversely affect the species in
question.

Thermal effects are also seen to result in a reduction of species diversity, simplification of food webs, and general development of biotic
instability within a system. When an ecosystem is disturbed, not only may

the species diversity be reduced, but a reduction in the population of consumers or grazers may also occur, enabling the decomposers to take over.
The potential food energy in the system may get diverted from the useful
food chain to the decay chain. In such cases, diversion of energy takes
place from more intricate food webs to simpler systems involving smaller
organisms, plants, and shorter food chains, thus inducing instability.

The average temperature of the Rhine in Europe is expected to increase
by 3.8° C over its present value owing to the establishment of power plants
on that river. It is feared that beyond 28° C, the ecological balance of
the Rhine may be upset. Thermal pollution in any stream tends to reduce
the species diversity, promote biotic instability, and affect plant zonation and stratification. If a cold water stream is thermally polluted to
the point where it resembles a tropical stream in temperature, the system
will respond, not by developing a luxuriant, tropical biota, but with a
reduction in the number of its original species and the flourishing of only
the coarser and hardier ones. Available energy in the system would now be
more likely channeled through gulls and crabs rather than exotic tropical
fish. The standards for the control of thermal and other forms of water
pollution are discussed later.

3.12 ECOLOGICAL SUCCESSION

Just as there is evolution in biology, there is succession in ecology.
Ecological succession is a systematic sequence of replacements of one community by another in a given area until a relatively stable community is
evolved which is in equilibrium with the local conditions. As an example,
consider a newly constructed lake. Typically, in an ecosystem, community
development begins with pioneer stages which comprise mainly primary producers (like algae) and decomposers (heterotrophic microorganisms). The
macroconsumers or animals are, at first, fewer in proportion to the producers. The consumer animals have more complicated life histories and
habitat* requirements than green plants; hence, animal populations require
a longer period of time for development. Thus, the total energy fixed by
the primary producers is initially more than the requirement of the community of consumer animals, and the portion of energy not consumed by them

---

*Habitat is the place where an organism or an entire community lives.

is diverted to the decay chain. Most aquatic media, however, provide a nutritionally complex habitat for a large variety of metabolic types of organisms and, therefore, in due course the consumer animals invade and claim their share of the total potential food energy. After a series of temporary communities have prepared the way, the lake eventually develops a stable or climax community [3].

Succession results from the action of a community on the habitat, which tends to make the area less favorable for itself and more favorable for other sets of organisms until the equilibrium or climax state is reached. As stated earlier, in a pioneer community there may be a surplus of potential food energy. This may change the habitat physically and be a source of growth of consumer animals, thus influencing the composition of the community. A newly exposed rock or sand surface, a plowed field, and a batch of wastewater undergoing aeration (Sec. 2.5) all constitute communities undergoing succession.

The numbers and types of organisms undergo a change with time; this kind of regulatory adjustment is probably the result of competition. A dead animal or a log of wood in the forest supports a community which changes with its state of decay; the order of attack being fungi, bacteria, nematodes, and finally other soil invertebrates. In general, ecological succession refers to changes which are brought about over a period of time by the action of the community on the substrate and its microclimate while the external factors remain the same.

Succession is directional and, therefore, predictable. It proceeds in the direction in which the available energy is consumed most efficiently, namely, in the direction in which entropy losses are reduced. This results in a higher and higher degree of biological diversity in the species and types constituting the community until entropy losses are least; i.e., the total biomass within a given system increases with succession. The respective stages of succession are characterized by a decrease in the flow of energy per unit mass and time. They tend finally to reach optimum metabolic efficiency.

Every organism in a community has a functional status or position; this is called the ecological niche. By analogy, a niche is like the "profession" of an organism, while the habitat is like its "address." The niche occupied by an organism tells us something of its activities, its nutrition and energy source, its rate of metabolism and growth, its effects

on other organisms, and the extent to which it is capable of modifying the
ecosystem. Each species can be said to be a profession. Ecological succession proceeds in the direction of filling vacant niches, so to speak,
eventually producing a highly diverse community with increased stability.
This development of increased stability gives the system greater capability
for survival and is perhaps the most important reason for the occurrence of
ecological succession [3].

As the organization of the ecosystem becomes more and more complex,
an ever greater proportion of the energy is used to maintain the biocoe-
nosis. At the climax stage, all the energy is utilized in respiration, on
the average, and the net ecosystem production is zero. When undisturbed,
succession is unidirectional and produces a stable community; this may, of
course, take a considerable amount of time.

The foregoing discussion is summarized in Figure 3.19 which graphically
depicts the relative change in species diversity and net productivity with
time as an ecosystem progresses toward maturity. As is well known, evolution in general biology leads toward increased complexity of order. Similarly, succession leads toward high diversity in a community. During this
development, energy usage improves as entropy losses diminish. In a highly
diverse community, all the members play their part in cycling matter and
energy; they are interlocked by various feedback loops and, because of the
complicated food web, each species has multiple relationships with others.
This forms a network of checks and balances which affords a high degree of
stability in the ecosystem [1].

Succession may be stopped or reversed as a result of unfavorable
changes in the environment such as temperature changes and the introduction
of toxins, pollutants, radiation, etc. They may lead to a destruction of
some of the sensitive species along with a disruption of the fine checks
and balances which prevailed earlier in the system. A different community
of a simplified and less stable structure may be set up with a reduced
number of energy pathways. Any unfavorable stimulus would cause a shift
from the right to the left in Figure 3.19. A waste stabilization pond or
any body of water which receives a slug of pollution shows a reversal in
its succession process; its species diversity decreases while net production
and energy flow per unit mass and time increase.

In many instances, the fish, birds, and trees that man finds most
valuable are members of the developmental community rather than of the
climax community [3]. For example, the "youngest" ponds contain relatively

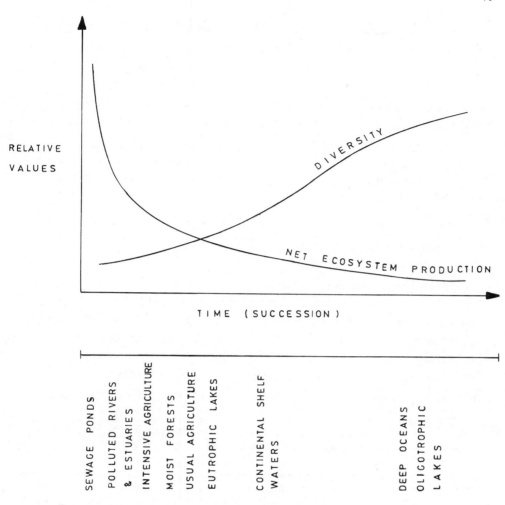

Figure 3.19 The relative change in species diversity and net production as an ecosystem progresses toward maturity.

little rooted vegetation and are likely to have bass and bluegills. "Older"
ponds are more and more choked with vegetation and become smaller in volume
as the actions of plants and animals fill up their basins with sediments.
The bass and bluegills may now be replaced by catfish, carp, and other
coarser varieties. Eventually, shrubs and trees move in, the pond becomes
a swamp, and finally dry land. Succession is a basic characteristic of all
aquatic and terrestrial communities. Sometimes man disturbs succession by
introducing excessive amounts of nutrients (e.g., eutrophication).

In a sense, agriculture also disturbs the natural productivity of land
as it was when left to itself. To increase productivity, man must induce
a shift from the right to the left in Figure 3.19, eliminate diversity, and
work toward a monoculture. This is done through energy inputs in the form
of plowing, fertilizers, etc. A sewage pond where algae harvesting or
fish culture is the objective must be operated to discourage rooted plants
and bottom sediments (to avoid side food chains) in order to channel as
much of the basic energy as possible into the floating algae. Again, a
kind of monoculture is practiced (with its inherent instability) in the
interest of maximizing output [1].

3.13 ECOLOGICAL INDICATORS OF POLLUTION

Since the ecological community is determined by the environment, we
can use this information in the opposite sense to judge the environment
from the community present; this would, in fact, be a method of bioassay.
Communities and individual organisms can be used as indicators in assessing
the quality of an environment.

From the earlier discussion about the limits of tolerance (Sec. 2.4),
one could generalize as follows with regard to ecological indicators:

1. Those species which have narrow limits of tolerance (stenotypes) would
    make better indicators of specific situations rather than those with
    wide limits (eurytypes) which would naturally have a wider geographic
    distribution.

2. Larger species would make better indicators than smaller species; the
    former are in a relatively more stable stage of succession. It is in-
    teresting to note here the observation of Rawson as quoted by Odum [3]
    that algae, for example, do not make good indicators of lake types.

3. A community is a more reliable indicator than a single species (e.g.,
    an activated-sludge mixed-liquor community rather than a single spe-
    cies).

Pollution of a waterbody or the entry of toxic material in a waste treatment unit may be continuous, periodic, occasional, or a one-time accident.

Depending upon the extent of "stress" applied to an ecosystem by an
unfavorable stimulus, the biota may undergo a process of adaptation by
either of two methods: genetic change or adaptive plasticity [1]. The
method of adaptation would, among other things, depend on the length of
time for which an environmental stress is imposed on a species in relation
to its life span. Generally, macrosystems cope with their environmental

stresses by adaptive plasticity, or developing physiological capability to withstand the stress. Microbial communities, on the other hand, tend to respond by genetic change owing to their capacity for fast multiplication. This is what happens in waste treatment provided the microorganisms have an opportunity to acclimatize. It would be interesting to observe here that the ability to mutate sometimes enables microorganisms to grow even on highly toxic substances in wastewater, such as cyanide and phenol, and on antibiotics such as penicillin. Gradual acclimatization is, of course, essential as is the need to have dilute solutions of more or less uniform concentration. Sudden inflows of high concentrations (slugs) can readily wipe out the microbial community.

The biota existing in a waterbody is, therefore, the result of the total environment and is, in a way, the "average" result of the variety of environmental stresses and their duration. The biological community in a "clean" aquatic environment will be different from that in a polluted one, and more so in a highly polluted one. The presence or absence of certain species serves as a biological indicator of pollution. This method is often informative and is certainly better averaged in time.

Chemical analysis, however, has the advantage of giving quantitative results, whereas biological examination in mainly qualitative. On the other hand, chemical analysis reflects the situation at the time of sampling only and the effects of "slugs" or transient or periodic pollutants may be missed altogether. It is, therefore, advantageous to combine observation of biological indicators along with the chemical analysis to assess water quality and wastewater treatment systems. Bioassay tests are discussed in Chapter 4.

REFERENCES

1. W. Stumm, The Land-Water Exchange — The Ecological Viewpoint, World Health Organization, WHO/W. POLL/72.9, Geneva.

2. P. Stanier, N. Doudoroff, and E. Adelberg, The Microbial World, Prentice Hall, Third Edition, 1970.

3. E. Odum, Fundamentals of Ecology, Third Edition, W. B. Saunders, 1971.

4. P. Canale and A. Vogel, Effects of temperature on plankton growth, JASCE, Env. Eng. Div., Tech. Notes, 1974.

5. D. Thirumurti and E. Gloyna, Relative toxicity of organics to Chlorella pyridinoisa Univ. Texas, Austin, Texas, Report 11-6503, 1965.

6. C. Forsberg, Nitrogen as a growth factor in fresh water, First AWPR. Conference on Nitrogen as a Water Pollutant, Copenhagen, 1975.

7. R. Mitchell, Water Pollution Microbiology, Wiley-Interscience, 1972.

8. F. E. Round, The Biology of the Algae, Edward Arnold, London, 1965.

9. Sverdrun, Johnson, and Fleming, Oceans, Prentice Hall, 1963.

10. I. Toms, M. Owens, J. Hall, and M. Mindenhall, Observations on the performance of polishing lagoons, Jour. Inst. Water Poll. Conf., U.K., 74, 1975.

11. W. Oswald, and H. Gotaas, Photosynthesis in sewage treatment, Trans. ASCE. 122, 1957.

12. Scientific American, 1971.

13. N. Armstrong, E. Gloyna, and B. Copeland, Ecological Aspects of Stream Pollution, Advances in Water Quality Improvement, Vol. 1, University of Texas, Austin, Texas, 1968.

14. Stirn et al. Selected biological methods for assessment of marine pollution, In Marine Pollution and Marine Waste Disposal, E. Pearson and de Franjipani, Eds., Pergamon Press, 1975.

Chapter 4

NATURE OF POLLUTANTS AND THEIR EFFECTS

4.1 NATURE OF POLLUTANTS

The various pollutants which can enter any ecological subsystem may belong to one of the following three groups:

1. Degradable
2. Nondegradable (conservative)
3. Biologically accumulative

Degradable pollutants include a large number of complex, organic substances which undergo biological degradation through the activity of various types of decomposers, yielding simpler, inorganic end products such as CO_2, nitrates, etc., as discussed in Sec. 3.4. Microbial concentrations of such organisms as coliforms, pathogens, etc., also undergo reduction with time owing to natural die-off; when they die, they become degradable organic matter. Fish and higher animals similarly undergo biological degradation when dead.

There is another group of degradable substances, radioactive isotopes, which undergo physical degradation or decay. Some of the radioactive isotopes decay rapidly while others decay at extremely slow rates. For example, the isotope uranium 238 takes 4.5 billion years for its radiation intensity to be reduced to half its original value and can be considered as nondegradable for most practical purposes; iodine 131 takes only 8 days for the same reduction. For both types of degradable substances, degradation is a time-rate process affected by various factors.

Nondegradable substances are inert to biological action with time and are also called conservative substances. Chlorides and various salts, for example, fall in this group. So do certain organics which are referred to as refractory substances.

Various metals and trace elements in nature are also nondegradable. Some of the nondegradable substances, however, undergo accumulation in

living organisms as will be discussed later. Their capability for biological accumulation **often** makes them dangerous to those who consume them.

Those pollutants which can be accumulated in the food chain are often
referred to as <u>persistent</u> substances. The ability of a substance to be
biologically accumulated is independent of its degradation characteristics,
although over a period of time the net accumulation in an organism will be
less where decay is taking place simultaneously with accumulation. Some
examples of persistent substances are mercury, cadmium, lead, arsenic, manganese, and some pesticides; radioactive isotopes may also show biological
accumulation. The biota of interest to environmental engineers are algae,
mollusks, fish, aquatic and terrestrial vegetation, animals, and man as the
ultimate consumer of all food. Wastewater treatment in ponds and lagoons,
followed by fish ponds and land irrigation, involves all the above biota.

4.2 EFFECTS OF POLLUTANTS

Figure 4.1 shows the fate of pollutants discharged by man as a result
of his industrial, agricultural, and urban activities. As was stated
earlier, there are no boundaries between the soil, water, and gaseous phases
of the environment; pollution occurring in one phase can get transferred to
another and vice versa, as shown in the figure. Pollutants entering the
environment may degrade or decay with time or may remain inert and, thus,
undiminished with time. Hence, depending on the nature of the substance,
a resultant concentration will be observed in the environment. This condition will have various economic and health effects as shown in the figure.
The health effects may be due to the total body burden received, partly
through direct intake with air and water and partly through the ingestion
of persistent substances that have accumulated in the food chain. Figure
4.1 shows the role of irrigated crops, fish, and the various consumers in
this regard.

The remaining part of this chapter is a discussion of the major items
of interest to us, namely,

1. Accumulation of pollutants in waterbodies
2. Biological accumulation of persistent substances and the development of
    health criteria
3. Disease transmission from pathogens, and natural die-off of microorgan-
    isms

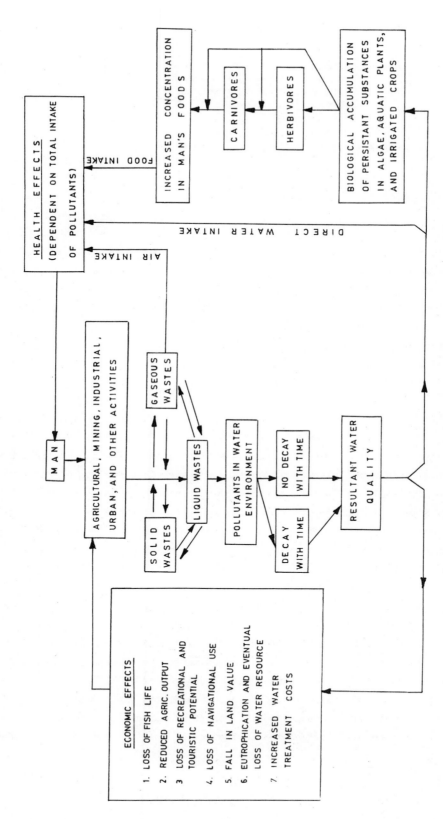

Figure 4.1 The relationship between man and the water environment.

4. Oxygen consumption due to aerobic bacterial activity on
    a. Carbonaceous matter (BOD, COD, and TOC)
    b. Nitrogenous matter (nitrogenous BOD)

5. Eutrophication

6. Water quality tests

7. Environmental impact assessment

4.3 ACCUMULATION OF POLLUTANTS IN WATERBODIES

The fate of conservative, nonconservative, and persistent substances
in a flowing body of water such as a river can be illustrated as in Figure
4.2 which shows such indicators of pollution as coliforms, nutrients, dissolved oxygan (DO), etc. Figure 4.2, part a shows three towns situated on
a river; they draw their water supply from the river and discharge their
wastewaters into it. Figure 4.2, part b shows how conservative and persistent pollutants tend to build up as the river flows past each town.
Agriculture runoff from the lands between the towns would also contribute
to this buildup (not shown in the figure).

The effect of degradable organic matter on the DO concentration in the
river water is shown in Figure 4.2, part c. The concentration of oxygen is
the result of two different activities progressing simultaneously in the
river: oxygen is used by bacterial activity while oxygen is replenished in
the flowing water by re-aeration from the turbulent surface of the river.
The net result is that the oxygen concentration goes up and down as the
river passes from town to town. The mechanics of these two simultaneous
processes are described fully in Chapter 7. As far as the fate of the organics themselves is concerned, their degradation leads to a gradual reduction in their concentration between the towns as shown by the curves in
Figure 4.2 part d.

Figure 4.2d can also be used to show how bacteria, which are again
nonconservative, die after each town is passed. All these are time processes and sufficient time between two towns must elapse for bacterial
die-off to be appreciable and dissolved oxygen to be restored to a desired
value. In a river such as the Rhine reuse occurs 52 times [1].

The cost of water treatment increases as the quality of water downstream in a river becomes worse. Operations research or optimization techniques can be used to find the best solution that is least costly to the

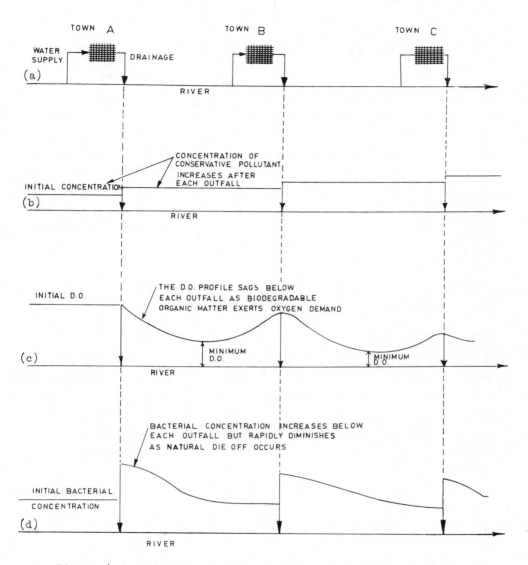

Figure 4.2 Pollution of a river showing the buildup of conservative
substances, the self-purification of organic matter, and the bacterial die-
off occurring downstream of successive outfalls.

nation while fulfilling the desired environmental and health objectives.
This is discussed in Chapter 7. Table 4.1 gives the range of concentrations
of certain substances as observed in some polluted rivers. These values may

Table 4.1 Comparison of WHO European Drinking Water Standards (1970) with Analysis Results of Some Rivers Polluted with Urban and Industrial Wastes

Characteristics	WHO European drinking water standards (1970)	Natural dune waters (Netherlands)	Rhine at Dutch-German border	Ruhr (Essen)	Delaware River (USA)
Temperature, °C			12.1-13.9		23.2-25.3
BOD_5, mg/liter			5.2-9.0		2.6-4.6
Oxygen, %	a		50-68	86-94	26-31
pH			7.3-7.7		7.0-7.1
NH_4^+, mg/liter	0.05		1.1-3.7	1.0-1.6	1.0-1.2
NO_3, mg/liter	<50[b]		7.5-12.9	12.3-16.7	2.1-3.1
Chlorides, mg/liter	300[c]		117-188		11-49
Detergents, mg/liter	0.2[d]		0.26-0.52		0.2-0.32[f]
E. coli per 100 ml.	e		1,100-4,600		12,900-44,700
Hg, µg/liter	f	0.5	1.4	0.03-0.17	
Cd, µg/liter	10				
As, µg/liter	50	2.0	9.0		
Pb, µg/liter	100	4.0	19.0	10-270	
Cr, µg/liter	50	<1.0	9.0		
Cn, µg/liter	50				
Cu, µg/liter	50[g]	5.0	14.0	10-100	
Zn, µg/liter	5,000[h]	14.0	155.0	10-280	
Se, µg/liter	10	4.0	6.0		
Ni, µg/liter	f			10-160	

Fe, µg/liter	$100^{g,h}$		20-700
Mn, µg/liter	50^h		10-200
Phenolics, µg/liter	1.0^h	19-23	
CCE, µg/liter	200		
Radioactivity, pCi/liter			
Gross alpha, 3			
Gross beta, 30			

[a]Preferably at least 5 mg/liter.

[b]Recommended less than 50 mg/liter but acceptable up to 100 mg/liter.

[c]This limit may be relaxed up to 600 mg/liter in certain cases.

[d]Preferable limit for anionics.

[e]Ideally, coliform organisms must be absent from any water entering a distribution system. However, in 5% samples taken over a year, one or more coliforms may be allowed, provided that positive results are not obtained in two or more consecutive samples and at least 100 samples are examined over a year.

[f]Limits not prescribed.

[g]At pumping station.

[h]Higher concentrations may cause trouble.

Source: Refs. 2, 3, and 11.

be compared with those observed in natural dune waters and with the World
Health Organization European Drinking Water Standards.

Estuaries, bays, lagoons, and lakes all accumulate pollutants; they
are, in a sense, pollution "traps." Movements of waterbodies and pollutants are governed by such factors as currents, eddies, diffusion, dispersion, tides, and stratification caused by temperature, density, and salinity.
At times, the movements may be restrictive to the free exchange of water
masses, and pollutants washed in will be trapped and increase in concentration with time. Hence, all waterbodies, whether flowing or stagnant, or
inland or coastal, tend to show increases in the concentrations of conservative substances with time.

The accumulation of nutrients and other substances in waterbodies
such as lakes that are subject to evaporation and other losses can be
estimated over a period of time by using suitable mathematical models as
described in Chapter 6.

Seawater is the result of evaporation, precipitation, and runoff over
a very long period of time. Hence, accumulations of persistent pollutants
have to be measured against this natural background accumulation. Table
4.2 gives the average values of various constituents in seawater free from
pollution [4-6]. Coastal waters receive freshwater runoff from the adjoining land masses and, therefore, have lower concentrations of overall
salinity than the open sea, but perhaps higher concentrations of some
specific substances washed into them. The Black Sea is an interesting
example. It receives pollution from large areas drained into it by the
Danube, the Volga, the Sakarya, and other rivers passing through several
countries. The average salinity of the Black Sea as it flows out through
the Bosporus Straits is 17,180 mg/liter compared with the salinity of an
"average" sea of 34,320 mg/liter (Table 4.2).

For proper comparisons to be made between concentrations of trace
elements in "average" seawater and in the Black Sea, for example, it is
necessary to calculate the ratio of the concentration of an element to
the concentration of salinity in each sea, and then compare the two ratios
in order to obtain the accumulation factor. This method accounts for the
differences in salinity. For example, copper is present in the surface
flow of the Bosporus (and, therefore, in the Black Sea) at 18 μg/liter [7],
whereas in the "average" sea it may be approximately 3 μg/liter (Table 4.2)
thus giving the ratio in each sea as

$$\text{Ratio} = \frac{\text{concentration of element, μg/liter} \times 10^{-6}}{\text{concentration of salinity in the sea, g/liter}}$$

Table 4.2 Some Chemical Characteristics of Polluted
and Unpolluted Seawater

Characteristic	Average concentration in seawater [a]	Trace elements in Black Sea measured as surface flow in the Bosporus	
		Concentration [b]	Accumulation factor
Total salinity, mg/liter	34,320	17,180	
Major inorganic elements, 100 mg/liter and more			
Sodium	10,770		
Chloride	19,353		
Sulfur (as sulfate)	2,712		
Magnesium	1,294		
Calcium	413		
Potassium	387		
Minor elements, 1-100 mg/liter			
Carbon	0.2-3.0		
Boron	4.6	2.6	1.2
Silicon	3.0		
Bromine	67.0		
Fluorine	1.0		
Strontium	8.0	2.48	0.62
Common trace elements, <1000 µg/liter			
Nitrogen	500		
Phosphorus	70		
Iron	1.7-150	14	2.8
Barium	10-63		
Iodine	60		
Rubidium	120		
Lithium	170		
Some other trace elements, <10 µg/liter			
Molybdenum	4-12	72	14.5
Selenium	4-6		
Arsenic	3		
Nickel	2	<5	-
Zinc	1.5-10	32	6.5
Aluminum	1-10	88	17.6
Lead	0.6-1.5	<10	-
Copper	0.5-3.5	18	11.5
Mercury	0.15-0.27		
Chromium	0.13-0.25		
Cadmium	0.11	5-9	168
Manganese	0.1-8.0		
Uranium	3.0		
Radium	1 x 10		

[a] Concentrations of many elements can vary considerably with location, time, season, and biological activity. Data from Refs. 4-6.

[b] Results of limited sampling done in 1967, taken from "Feasibility Studies and Master Plan for Water Supply and Sewage for Istanbul Region" by DAMOC Consortium for WHO/UNDP.

Source: Refs. 4-7.

The ratio is approximately equal to 1.0×10^{-6} for the Black Sea and 0.087×10^{-6} for the "average" sea. The "accumulation factor" is, therefore, equal to 11.5 (1.0/0.087). Similarly calculated accumulation factors for other elements are given in Table 4.2.

Compared with the concentrations in "average" seawater, the Black Sea shows particularly high accumulation; cadmium accumulation is 162 times higher than in "average" seawater, and a moderate accumulation of 6 to 17 times is observed for copper, aluminum, molybdenum, and zinc in the Black Sea. These data were obtained in 1967 [7] and give an indication of the extent of pollution in the Black Sea.

Biological concentration of a substance in the food chain is an additional factor to be considered and is discussed in the following section.

4.4   BIOLOGICAL ACCUMULATION

The overall accumulation of a contaminant within an ecosystem is the result of accumulation in the abiotic component of the system (e.g., water, soil, and air) and its progressive accumulation in the biotic component at each step in the food chain. Overall accumulation is reduced to the extent that biological and/or physical degradation of the pollutant, and its excretion from the biota, occurs.

As stated in the previous section, movements of waterbodies and the pollutants within them are governed by various factors, and the concentration of an element in a given locality may be several times greater than in another locality.

Accumulation of pollutants and salts within soils irrigated by wastewaters depends upon the nature of the soil, its drainage characteristics, and various other factors discussed more fully in Chapter 9. Accumulations in soil may give concentrations anywhere from 2 to 10 times or more than the concentrations in the water used for irrigation. Various physical and chemical interactions such as hydrolysis, colloid formation, adsorption, precipitation, etc., may transfer the pollutants from the water or soil to the surface of plants, algae, silt, or organic matter, may form organic complexes and enter the biosystem through food chains, or may accumulate in bottom sediments and then enter the food chain.

Within the organism, accumulation proceeds simultaneously with elimination from the system through natural biological functions (e.g., elimination occurs from human beings through sweat, urine, feces, and falling hair). Furthermore, during the time a substance resides in an organism or organ, physical decay or chemical breakdown may occur. Radioactive isotopes entering a system undergo physical decay during the time they reside in the system. Chemical compounds may undergo breakdown or transformation in the system, which may reduce or increase the toxicity of the original compound or produce other end products. However, certain persistent substances (trace elements and heavy metals) may remain unaffected between entry and elimination; their concentration in the system is reduced only through biological elimination. Depending upon the nature of the substance, different mathematical models can be used to describe the net accumulation of a substance in an organ.

Elimination is correlated with the total accumulation and a definite fraction of the latter is eliminated per unit of time. If A_0 is the total accumulation at the time when further exposure to a contaminant is stopped, then the total accumulation is reduced gradually owing to biological elimination and the value after time t becomes A_t, where

$$A_t = A_0 e^{-\beta t} \tag{4.1}$$

In this typical first-order equation, β = the elimination rate constant (the fraction of total accumulation eliminated per unit time). It differs from organism to organism and from substance to substance in the same organism.

The elimination rate β is constant for a given case at all values of the total accumulation. Thus, when the total accumulation is more, the total elimination is also more. As accumulation is reduced, the quantity eliminated is also reduced. It is possible that in some cases β itself may vary. For example, the elimination capacity of an organ may become impaired as accumulation increases.

In general, for the total accumulation to become half its original value, the time required is called the "half-life" and is obtained from the above equation by substituting $A_t = 0.5A_0$, thus giving the biological half-life T_B as

$$T_B = \frac{0.693}{\beta} \tag{4.2}$$

For methyl mercury, the half-life is about 70 days in humans and β is equal to 0.01 per day. A few other half-lives are as follows [3, 10].

Substance	Biological half-life (total body)
Hg	70 days
^{54}Mn	37 ± 7 days
Pb	~4 years
Cd	> 30 years
PCB	50-200 days (in birds and animals)

When repeated exposure to a substance has begun and biological elimination is the principal mode of removal from the organism (namely, physical decay or chemical breakdown are not involved), the course of accumulation can be expressed by an exponential function of the following type:

$$A_t = \frac{a}{\beta} (1 - e^{-\beta t}) \tag{4.3}$$

where a = dose or exposure rate per unit time.

With exposure, the accumulation in the organism increases, but as accumulation increases, the eliminated quantity also increases until steady-state condition is reached (see Figure 4.3).

The total accumulation on reaching steady state is determined by the intensity of exposure, but the time required to reach steady state is dependent upon the elimination rate. The greater the elimination rate, the more rapidly a steady state is reached. Some substances may take a few days or a few months, whereas others may take a year or more to reach steady-state conditions.

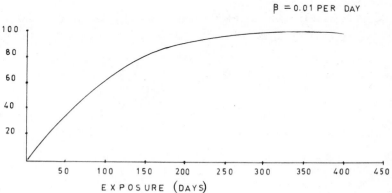

Figure 4.3 A theoretical course of accumulation of the total body burden of man, at steady state, after an exposure to MeHg (from Ref. 8).

Where there is physical decay, as in the case of radioactive contaminants, there is additional loss within the residence time in the organism owing to decay. Spontaneous disintegration in the case of radioactive isotopes, and physical decay in general, follows first-order kinetics and can be denoted by the following equations:

$$N_t = N_O e^{-\lambda t} \qquad\qquad (4.4)$$

where

N_t and N_O = final and initial disintegrations per unit time, respectively

λ = radioactive decay constant $(time^{-1})$

Again, the half-life resulting from physical decay can be denoted by T_p, where

$$T_p = \frac{0.693}{\lambda} \qquad\qquad (4.5)$$

Table 4.3 lists the half-life resulting from radiological decay in a few isotopes commonly encountered in waste disposal.

When a substance is subject to both physical decay and biological elimination, the effective half-life is less than the half-life obtained from either one of these modes. The effective half-life T_e is given as follows:

Table 4.3 Half-lives of a Few Isotopes

Isotope	Type of emission	Radiological half-life (T_p)	Biological half-life (T_B)	Effective half-life (T_e)
^{131}I	β, γ	8 days	180 days	7.7 days
^{32}P	β	14.3 days	1,200 days	14.0 days
^{60}Co	β, γ	5.3 years	8.4 days	8.4 days
^{90}Sr	β	24 years	3,900 days	2,700 days
^{226}Ra	α, γ	1,600 years	16,000 days	16,000 days
^{14}C	β	5,000 years	180 days	180 days
^{239}Pu	α	2.4×10^4 years	360 days	360 days
^{238}U	α	4.5×10^9 years	30 days	30 days

$$T_e = \frac{T_p T_B}{T_p + T_B}$$ (4.6)

It is interesting to note that if either one of the above two half-lives
is much larger than the other, it tends to eliminate itself in computing
T_e. For example, if T_B = 1 year and T_p = 10,000 years, the value of T_e
also equals approximately 1 year. Table 4.3 lists the effective half-lives
of some isotopes. From the effective half-life of any substance residing
in any organism, the value of the effective elimination rate constant β_{eff}
can be computed using an equation similar to Eq. (4.2). Thereafter, accumulation at any time t can be estimated from the dose using Eq. (4.3).

As noted in the earlier discussion in Sec. 3.12, as an organism is
consumed at each successive trophic level, further accumulation occurs in
each successive organism. For example, a large number of algae contaminated
with a persistent substance may be consumed by one herbivorous fish; thus,
this one fish consumes, and perhaps accumulates, the total mass of contaminant contained in all the algae. Similarly, one large carnivorous fish
may consume several herbivores and thus have a still greater mass of the
contaminant accumulated in it than in the previous fishes. Hence, a gradual
buildup occurs at each successive trophic level.

Field measurements of accumulations can give the integrated effect of
the various factors which affect biological accumulation and elimination
and which are not easy to separate individually in a theoretical study.
Drawing upon the example of a conservative substance such as DDT in river
water, Odum [9] has quoted an example where the concentration of DDT, although extremely small initially, increased considerably at each step along
the food chain as summarized in Table 4.4.

Table 4.4 DDT Accumulation in a Food Chain

Item	DDT (mg/liter)
Water	0.00005
Plankton	0.04
Minnows (fish)	0.2-0.9
Predatory fish	1.3-2.0
Scavenger bird	6.0
Fish-eating duck	22.8

Source: Ref. 9.

Similar biological concentrations of mercury, cadmium, etc., have been known to occur. The use of such data in setting water quality standards will be further described later.

The <u>concentration factor</u> (CF) for any contaminant contained in a living organism is expressed as a ratio of its concentrations in the organism and in the medium (e.g., water or air) in which the organism lives. For example, for the aquatic medium, CF is defined as follows:

$$CF = \frac{\text{concentration in organism or plant (wet weight)}}{\text{concentration in surrounding water}}$$

CF is also referred to as the "enrichment factor."

The mechanism of biological concentration or enrichment is dependent on such various factors as the following:

1. The nature of the chemical substance (its physical decay, chemical breakdown, and biological elimination).

2. The requirements of the biological species for that chemical substance.

3. Temperature.

4. The concentration of the chemical substance in the aquatic environment.

5. The diluent effect, in the case of radioactive isotopes, due to the concentration of the stable isotope in the water (for example, ^{90}Sr goes to the bones and stable Ca^{2+} also goes to the bones. The human system does not differentiate between the isotope and the stable form, and hence Ca^{2+} has a "diluent" effect in this situation).

Observed CF values in the edible parts of aquatic foods have been found to range widely from 1 to 10,000, as listed in Table 4.5 for some typical cases, and indicate the severity of the problem for humans consuming such fish and other aquatic foods. Instances of mercury and cadmium poisoning serve as very good examples.

A study of the concentration or enrichment factors of some aquatic plants in the Ruhr River [11], for which the quality of the surface water is as shown in Table 4.1, revealed that the most intense enrichment takes place in the mosses (Bryophyta). In the case of more highly developed plants (Spermatophyta) the enrichment is dependent primarily on the species of plant and its specific selectivity. Some examples are given in Table 4.6.

Similarly, enrichment takes place in food grains and other agricultural products. For example, rice is rather efficient at concentrating substances from the soil and water used to grow it. Samples of Japanese rice have been found to contain as much as 200-1000 µg/kg of mercury. According to some authors, the levels of lead (Pb) in milk have risen slightly over the past 20 to 30 years. Human intake of cadmium is chiefly through the food chain.

Table 4.5 Observed Concentration Factors in Some Biota

Contaminant	Edible red algae	Mollusks	Crustaceans	Fish
Radioactive				
^{90}Sr	0.1-1.0	0.1-1.0	0.1-1.0	0.1-1.0
^{137}Cs	1-10	0.01-10	10-100	10-100
^{106}Ru	1,000	1-1,000	1-1,000	1-10
^{95}Zr	100-1,000	10-100	100	1-10
^{65}Zn	100	1,000-100,000	10-10,000	1,000-10.000
^{60}Co	100	10-1,000	10-1,000	10-100
^{131}I	1,000 in seaweeds			
^{54}Mn	1,000	1,000-10,000	100-10,000	100-1,000
^{55}Fe	1,000-10,000	100-10,000	100-10,000	100-10,000
^{32}P				10,000
Nonradioactive				
Mercury				>3,000
Cadmium				>4,500 in some marine animals
Zinc				100-500
Manganese				200-2,000
Phosphorus				10,000
DDT		1,200-9,000		
Endrine and Dieldrine		500-1,500		
Pesticides		up to 70,000		

Source: Refs. 6 and 11.

Table 4.6 Concentration Factors for Some Plants in the Ruhr River

Plant	Average enrichment factors for stated metals						
	Hg	Fe	Mn	Ni	Zn	Pb	Cu
Spermatophyta							
Ranunculus fluitans	120	210	1,420	330	2,000	575	350
Nuphar luteum	>430	100	1,400	550	165	100	78
Sagittaria sagittifolia	>230	580	-	120	-	-	220
Myriophyllum spicatum	200	460	1,000	320	1,400	182	870
Bryophyta							
Fontinalis antipyretica	740	3,000	20,800	1,500	9,400	3,200	1,050
Hygroamblystegium	>930	3,200	15,300	770	2,500	4,600	2,800

Note: Ruhr River water quality is given in Table 4.1.
Source: Ref. 11.

Information of this nature is essential for the development of water quality
and health criteria, and is also useful in evaluating the possible effects
of the use of domestic and industrial wastewaters for irrigation purposes
(Chapter 9).

4.4.1 Development of Health Criteria

It will be useful at this stage to illustrate how information of the
type just discussed concerning bioaccumulation is utilized in developing
water and food quality criteria for persistent substances. The example of
methyl mercury, for which much work has been done in recent times, will be
examined, but a similar approach is feasible for other substances also.

Figure 4.4 gives the essential considerations in the formulation of
health criteria. Toxicological studies include identification of the routes
of entry of the contaminant into the human system (e.g., respiratory tract
and alimentary canal) and its distribution pattern in the body. Tracer
experiments or autopsies may need to be performed. Certain organs may show
a preference for certain contaminants; accumulation in an organ can be expressed as a percentage of the total accumulation in the body ("total body
burden") for that contaminant. The critical organ must also be identified
and the concentration at which malfunction or disease occurs must be determined. The determination of this requires a study of epidemiological

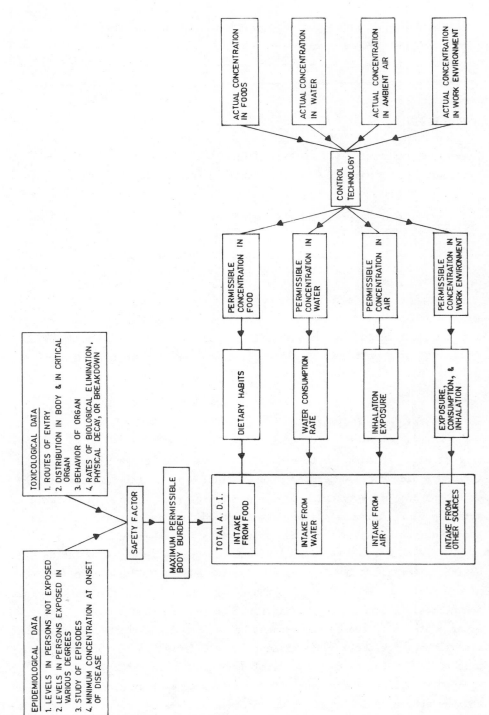

Figure 4.4 Essential steps in the development of health criteria and standards.

evidence at the time of the actual episodes to arrive at the total body burden observed at the onset of the disease. Going backwards from this information, the exposure rate (dose of contaminant) at which the total body burden is likely to reach the critical concentration can be determined from Eq. (4.3) if the value of the effective elimination rate constant β_{eff} is known.

Hopefully, the "acceptable daily intake" (ADI) of the contaminant from all sources together (air, water, food, and others) can then be specified after applying a suitable safety factor. It is then necessary to make a realistic estimate of the likely intake of the contaminant from each of these sources. This requires a study of the general conditions and dietary habits of the most vulnerable sections of the community (e.g., coastal communities are likely to have more fish in their diet) in order to arrive at some limits for the contaminant in the item(s) of food most concerned with that contaminant. In the case of water-borne contaminants, for example, water consumption is often assumed to be 2.2 liters per person per day.

The following data for mercury have been extracted from various sources [6, 8, 10]:

Routes of entry: skin, respiratory tract, and alimentary canal.

Critical organ: central nervous system (leading to numbness, paralysis, loss of sight and smell, and even death).

Distribution in the body: 10% of the total body burden is carried mostly in the brain. Accumulation in blood cells is twice that in the whole blood.

Excretion routes: feces, urine, and hair.

Biological elimination rate: the half-life is 70 to 90 days calculated for the whole body. This is somewhat shorter in the blood and longer in the brain and follows a single-phase exponential course. The effective value of β is 0.01 per day.

Epidemiological evidence:

1. Lowest reported levels of methyl mercury in the brains of persons who died in Japan was about 5 μg per gram of brain weight at the onset of the disease.

2. Other data suggest that the clinically manifested poisoning of adults sensitive to methyl mercury may occur at 0.2 μg Hg/g blood (0.4 in blood cells), a level that, from Japanese data, seems to be reached on exposure to about 0.3 mg Hg, as methyl mercury intake, per day through fish (or about 4 μg Hg/kg body weight per day).

Some typical mercury contents:

Surface waters meeting WHO 1971 International Drinking Water Standards contain < 1.0 μg/liter.

> The air of large industrial cities (24 hr average) contains <1.0 μg/m^3.
>
> The expected mercury intake from average U.S. diets (EPA, 1971) is 10 μg/day.
>
> Marine fish from uncontaminated waters (Sweden) contain 0.01-0.10 μg/g fresh fish.
>
> Marine fish from various Mediterranean waters contain 0.2-0.84 μg/g fresh fish.
>
> Acute toxicity in the fish itself is manifested at ~20 μg/g fish.

From the epidemiological evidence given, the total weight of methyl mercury in an average-sized brain of 1.6 kg is estimated to be 5 μg/g × 1600 g = 8 mg. If this constitutes 10% of the total body burden, the latter would amount to 80 mg. The daily elimination rate is 1% of the total body burden. Thus, mass eliminated daily equals 80 × 0.01 = 0.8 mg/day. Hence, for steady-state conditions to be maintained, the total methyl mercury intake should not exceed 0.8 mg/day. This theoretical maximum figure would, however, need to be checked with other epidemiological evidence and a safety factor must also be applied.

From the epidemiological evidence given, a daily intake of only 0.3 mg/day as methyl mercury in fish resulted in clinically manifest poisoning in susceptible adults. The intake of mercury from water and air is very small, indeed, as it is from other items of food in the general diet. Thus, the maximum daily intake of methyl mercury should preferably be limited to 0.3 rather than 0.8 mg/day. Applying a safety factor of 10 to this figure, the ADI through fish may be fixed at 0.03 mg/day as methyl mercury.

The joint FAO/WHO Expert Committee on Food Additives (1972) has suggested a provisional tolerable weekly intake of 0.3 mg total Hg, of which no more than 0.2 mg may be in the form of methyl mercury. The latter corresponds to the value of 0.03 mg/day as computed earlier.

From a knowledge of the dietary habits of the people, a "derived limit" can be placed on the content of methyl mercury in fish flesh. For example, based on the dietary habits in Sweden,* and assuming that methyl mercury intake is mainly from fish consumption, it has been estimated that a concentration of 0.2 mg/kg fish flesh would be relatively safe [8]. Another "derived limit" would be that for methyl mercury in the fresh or marine water in which fish are cultured. If the concentration factor is, for

---

*The per capita fish consumption averages 210 g/week.

example, 1000 for the fish in question, the permissible limit in water is
computed as 0.2 μg/liter.

The purpose of giving this illustrative computation was not to justify
the limits suggested for mercury, but rather to show the complex and widely
different types of toxicological and epidemiological studies involved in
developing health criteria. Such studies could not have been undertaken
some years earlier owing to lack of microanalysis techniques.

The development of "<u>criteria</u>" is different from the development of
"<u>standards</u>." The latter is a political decision in which a certain element
of risk is balanced against costs. For example, in the case of mercury just
described, it is possible that, if the derived limit of 0.2 mg/kg fish flesh
is specified in a country, it may cause unemployment for a sizable section
of the fishing community in some of its regions. On the other hand, if the
limit is relaxed to 0.5 mg/kg fish flesh, a small percentage of the population (the heavy fish eaters) may just cross the acceptable daily intake but
not the toxic level. It would then clearly be a political decision whether
to encroach on the safety factor or not. Thus, standards may vary from
nation to nation within a relatively narrow margin since health cannot be
subjected to a purely market economy. In the coming years, one can expect
to see an all-around greater refinement in standards and also some upgrading of standards by those who can afford the extra costs.

4.5 DISEASE TRANSMISSION THROUGH WATERBORNE PATHOGENS

The contamination of fresh and marine waters by pathogenic microorganisms from insanitary modes of disposal of human excreta and waterborne
wastes have been well known for a long time. Some of the diseases so transmitted are shown in Table 4.7. Infected shellfish and sewage-irrigated
crops can spread some of the diseases listed.

The discharge of organisms to the aquatic environment is generally
followed by a fairly rapid rate of natural die-off. Where insufficient
time exists for natural die-off to occur before possible use of water downstream, artificial disinfection by chlorine or other means may be necessary.

A few studies have shown a correlation between the total numbers of
bacteria and the 5-day BOD value of running and stagnant waters of different
degrees of pollution as indicated by the BOD test. Log-log plots of data
appear to give straight-line correlations. Using direct counts with membrane

Table 4.7 Some Common Waterborne Diseases

Disease	Pathogen
Amoebic dysentery	Entamoeba histolytica
Bacillary dysentery	Shigella spp.
Bacterial enteritis	Salmonella spp.
Cholera	Vibrio cholerae
Infectious hepatitis	Hepatitis virus
Typhoid and paratyphoid	Salmonella spp.

filters, the following correlation was obtained for both stagnant and running waters (reservoirs, rivers, treated sewage, etc.) with a correlation
coefficient of 0.917 for 195 samples [12a].

$$\log N = 1.165 \log (BOD_5) + 5.262 \tag{4.7}$$

where N = numbers of total bacteria per ml. From this correlation, one
can estimate, for example, that the total number of bacteria range between
10^5 and 10^6 per ml of water when the BOD_5 ranges approximately from 0.5 to
5.0 mg/liter. However, such estimations only give an idea of the order of
magnitude involved.

4.5.1 Natural Die-Off of Organisms

Coliforms and other organisms of intestinal origin tend to exhibit
natural die-off or decrease in numbers when exposed to environmental conditions which may differ from those within the human system (e.g., temperature, pH, physical forces), and when exposed to predators. Die-off may be
artificially accelerated by the presence of toxic substances and other
grossly bactericidal agents (e.g., chlorination), which is discussed under
appropriate waste treatment topics.

Die-off rates can be studied in batch-type laboratory tests by suspending the desired organisms in clean water (or water obtained from the
river or marine site of interest) subjected to controlled conditions of
temperature and light (to simulate the site conditions) and observing their
decrease in numbers with time. Die-off rates can also be studied at the
site itself, in which case more representative results would be obtained
as they would take into account the effects of predation and physical

forces such as mixing, sedimentation, etc. However, field conditions are not uniform and introduce additional factors which give wide variations in results and make comparisons difficult (see Sec. 7.1).

Coliform or E. coli die-off rates are often used in water pollution studies instead of the die-off rates of individual enteric pathogens. The former, although not pathogenic, are always present in much greater numbers (thus having greater statistical probability of being sampled) and, generally speaking, reflect a die-off pattern comparable with that of the enteric pathogens, with the exception of certain viruses which may outlive the coliforms.

Bacterial die-off is often approximated by Chick's Law (1908) which states, in effect, that die-off follows first-order kinetics.

$$\frac{dN}{dt} = K_b N \qquad (4.8)$$

where

N = number of organisms

K_b = bacterial die-off rate constant, per unit time

Integrating, we get the following relationship between N_0, the initial number of organisms, and N_t, the numbers remaining at any time t:

$$N_t = N_0 e^{-K_b t} \qquad (4.9)$$

This can be rewritten as

$$\ln\left(\frac{N_t}{N_0}\right) = -K_b t \quad \text{or} \quad \log_{10}\left(\frac{N_t}{N_0}\right) = -k_b t \qquad (4.10)$$

where $K_b = 2.3 k_b$. A graphical plot of $\log (N_0/N_t)$ versus time t on semilog paper should give a straight line if the data fit first-order characteristics. The slope of the line so obtained gives k_b (see Figure 4.5b). From Eq. (4.10) the time required for a given percentage of removal is obtained as

$$t_{90} = \frac{1}{k_b} = \frac{2.3}{K_b}$$

$$t_{99} = \frac{2}{k_b} = \frac{4.6}{K_b}$$

Data can also be plotted on probability paper to obtain the likely statistical frequency of a given bacterial concentration occurring in a given case.

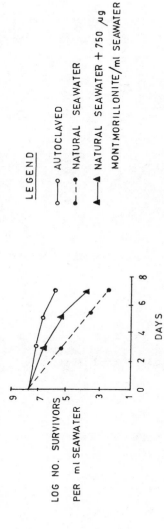

LOG NO. SURVIVORS

PER ml SEAWATER

DAYS

LEGEND

○—○ AUTOCLAVED

●--● NATURAL SEAWATER

▲—▲ NATURAL SEAWATER + 750 μg
MONTMORILLONITE/ml SEAWATER

Figure 4.5a The effect of the addition of montmorillonite on the survival
of E. coli in natural seawater; o ——— o autoclaved; ●- - - ● natural seawater;
▲ ——— ▲ natural seawater + 750 μg montmorillonite/mliter seawater (from Ref. 12).

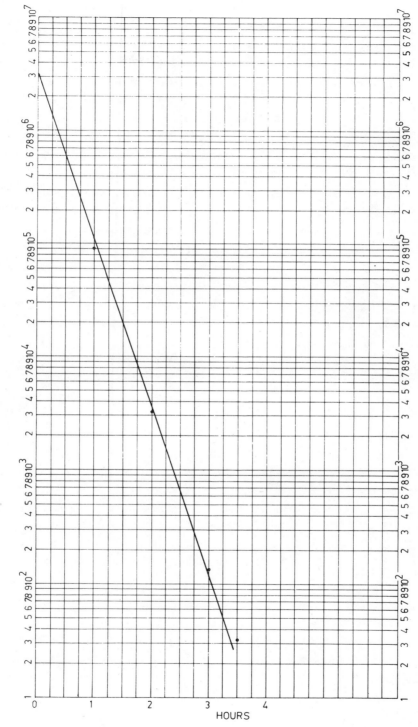

NUMBER OF COLIFORMS

HOURS

Figure 4.5b The rate of coliform die-off observed during a field test.

103

Microorganism die-off data generally show different values of K_b (and, therefore, of t_{90}) depending upon the nature of the organism and the prevailing conditions in the aquatic environment. For example, die-off in natural waters is faster at tropical latitudes than at temperate ones. Turbulent streams show faster die-off rates than do slow and sluggish ones, although the final numbers depend upon the initial density and the distance to be traversed. First-order kinetics imply that the greater the microorganism density, the higher the removal; hence, die-off may be expected to be more rapid in polluted streams than in cleaner ones.

Similarly, organisms tend to show faster die-off in marine waters than in freshwaters; this is apparently not due to the salts contained in the seawater but to certain organic, heat-labile substances, predation, and other factors [12,13]. Bitton and Mitchell [12] performed laboratory experiments to show the protective effect of a colloidal material such as montmorillonite clay on the decline of Escherichia coli in natural seawater. Apparently montmorillonite provides a colloidal envelope around E. coli cells which possibly protects them from the lytic action of marine predators (see Figure 4.5a).

A few typical values of natural die-off rates in fresh and marine waters are shown in Table 4.8 together with those for oxidation ponds for the purpose of comparison. As stated earlier, values obtained through field tests show larger variations owing to the effect of several physical factors such as radiation, mixing conditions, temperature, etc.

Viral die-off rates have generally been observed to be slower than those of E. coli under identical conditions. Besides temperature, the presence of a microbial population in water affects virus survival. Thus, viruses have been observed to survive longer in distilled water or autoclaved river or seawater compared with natural waters. Coxsackie virus in some river waters could not be isolated after 16 days at $8°$ C and 6 days at $20°$ C, but in sewage they survived beyond 40 days at $20°$ C [31]. Virus inactivation can also occur as a result of adsorption on clay particles. Continued improvements in virus isolation techniques have often shown that earlier studies were quantitatively unreliable.

The effect of temperature on the growth rate of organisms can be formulated in terms of the Van't Hoff-Arrhenius equation (see Chapter 12) as follows:

$$k_{b_{(T° C)}} = k_{b_{(20° C)}} \theta^{T-20} \qquad (4.11)$$

Table 4.8 Typical Values of the Die-off Rate Constant K_b and the Time Required for 90% Removal of E. coli

Water	Nature of test	Temp. (°C)	t_{90} (days)	K_b (day^{-1})	Reference
Tap water + 0.1% sewage	lab	20	2.5	0.92	31
River water	field	winter	2-5	0.46-1.15	14, 15, 31
Ground water	field	10	4.58	0.5	
Oxidation ponds	field	20	1.91-2.3	1.0-1.2	See Chapter 16
Seawater natural	lab	room temp.	1.2	1.9	Estimated from
autoclaved	lab	room temp.	5.0	0.46	Reference 12
Seawater	field	15-28	0.046-0.40 (1.1-9.6 hr)	5.5-48	See Chapter 8

The value of θ = 1.072 as observed for Salmonella typhosa gives a temperature quotient of $θ^{10}$ = 2.0 or, in other words, the rate would double for every 10°C rise in temperature. For Entameba histolytica, θ = 1.12 and, thus, $θ^{10}$ = 3.0, or the rate trebles for every 10°C rise in temperature. For E. coli in oxidation ponds, Marais (Chapter 14) gives a value of θ = 1.19 from field studies. This signifies that the rate increases or decreases 5.7 times for every ±10°C change in temperature. For nitrosomonas, Downing gives θ = 1.128, namely, the rate is affected 3.33 times [16,29].

The coefficient θ for microbial activity, expressed as respiration rate, has been reported as 1.074 and 1.085 by two different investigators [17]. Identical values of θ are implied in the general observation that growth rate is often doubled for every 10°C rise in temperature. This is, of course, valid only within a certain temperature range. Generally, above 35°C the growth rates of mesophilic organisms tend to show inhibition. At temperatures less than 12°C, E. coli have been observed to show a reduction in metabolic activity [15,16]. The natural die-off rates of coliforms and other microorganisms on irrigated crops are given in Chapter 9.

In studying organism die-off rates under disinfection by chlorination, additional factors such as disinfectant concentration and time of contact have been shown to be of importance [14]. This has led to the suggestion by Hom [32] of a modified form of Chick's differential equation.

Example 4.1 Raw sewage diluted with seawater in the ratio of 1:100 was placed in several polythene bags and exposed to natural conditions in inshore waters for different time periods to study the natural die-off rates of coliforms. From the observed summertime data of $28.5°C$ seawater temperature, estimate t_{90} and k_b.

Time, hours:	0	1	2	3	3.5
Number of coliforms:	3.8×10^6	9×10^4	3.16×10^3	1.4×10^2	32

From the semilog plot shown in Figure 4.5b, k_b at $28.5°C = 1.2$ per day.

4.6 OXYGEN CONSUMPTION

Among the environmental effects of pollutants, one of the most crucial is oxygen consumption due to aerobic bacterial activity on organic matter. Several tests have been devised to determine the oxygen "demand" of organic matter. The BOD, COD, and TOC tests are discussed briefly below.

4.6.1 Biochemical Oxygen Demand (BOD)

Standard methods for conducting BOD tests are available [19] and will not be repeated here. It would be more meaningful to discuss the nature of the test and interpretation of its results. The BOD test is an indirect method for determining the carbonaceous concentration of degradable organic substrates. All organics are not degradable, but those that are would support microbial growth which in turn would consume oxygen. Thus, the BOD test is essentially a standardized bioassay test for observing the rate and extent of oxygen consumed during degradation. If all organic matter was equally biodegradable, only a chemical method of analysis would suffice.

The general microbial growth curve was discussed earlier in Sec. 1.6. A typical growth curve was seen in Figure 1.4. The BOD test also gives similar curves (see Figure 4.6) since oxygen uptake is the result of microbial activity which shows an initial rapid "synthesis" phase followed by a slower one. By diluting the wastewater sample with aerated, nutrient-containing dilution water, it is ensured that carbon will become the limiting factor while the supply of oxygen and nutrients will be in surplus throughout the test.

When the cumulative oxygen consumed is plotted against time as shown in Figure 4.6, it is seen that the rate is high at the start for about 1 to 2 days and, thereafter, levels off gradually, until after several days there may again be a sharp but small rise in oxygen consumption. The total time until this latter rise occurs may be on the order of 10 to 20 days, depending mainly on the nature of the wastes and the temperature. This is the first-stage BOD, or the oxygen demand of carbonaceous matter in the waste. As the carbonaceous demand is satisfied, the slower-growing nitrogenous organisms demand further oxygen for nitrification ($NH_3 \rightarrow NO_3^-$). This is the nitrification oxygen demand (NOD) which is discussed in Sec. 4.7.1. The standard BOD test is performed at $20°C$ for a period of only 5 days, from which the ultimate first-stage demand is estimated. Typical BOD values for different wastewaters are given in Chapter 5.

Utilization of complex substrates often follows first-order kinetics. Hence, oxygen consumed also shows similar first-order kinetics in which the rate of removal of substrate ds/dt, at time t, is proportional to the concentration of substrate at that time. Thus

$$\frac{dS}{dt} = KS \qquad\qquad (4.12)$$

where K = reaction rate constant, per unit time. In this case, the substrate

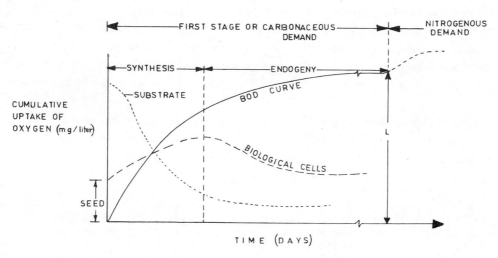

Figure 4.6 A typical BOD curve.

in question is organic matter. However, it is not the organic matter at a given time that is measured by the BOD test, but the oxygen already consumed up to a given time. Thus, it is evident that, since the concentration of organic matter at any time is proportional to the oxygen demand yet to be exerted, namely (L - y), Eq. (4.12) can be rewritten in the well-known form

$$\frac{dy}{dt} = K(L - y) \tag{4.13}$$

where

y = total oxygen consumed up to time t

L = total oxygen demand in the first stage (also called ultimate first-stage demand)

t = time from start

Integrating, we get

$$y = L(1 - e^{-kt}) \tag{4.14}$$

The reaction rate K is an overall rate constant ($time^{-1}$), but it is evident from the shape of the curve in Figure 4.5 that, within the first 1 or 2 days when rapid synthesis of new cell material occurs, the value of K is high, whereas for the remaining endogenous period it is relatively low. The former has been shown to be often 10 to 20 times higher than the latter. However, what the above equation gives is an overall or average value of K.

There are several methods available for determination of the reaction rate constant K provided that a series of BOD values, $\bar{y}_1$, y_2, y_3 ... y_n, are available for the corresponding time values t_1, t_2, t_3 ... t_n. Although BOD is taken as a first-order reaction, the method given in Chapter 12 for finding the K value for first-order reactions cannot be used since in BOD reactions both K and L are unknown.* Hence other methods are used [14,21, 22].

Raw waste with a greater quantity of organics can support a greater amount of new cell synthesis than treated or relatively clean waters. Hence, the overall K rate for the former would be likely to be higher than for the latter. The value of K depends on

---

*Gaudy [31] points out that use of the first-order equation in fitting oxygen uptake data from a BOD bottle is erroneous since in the latter there is generally an initial log growth phase up to the point of inflection, followed by a first-order retardant growth phase. The former is not covered by the first-order equation.

1. Nature and quantity of organics (raw or treated, extent of degradability, etc.)
2. Temperature
3. Presence of inhibiting substances

For this reason, the K value determined from BOD bottle tests cannot be directly used in making an oxygen-sag analysis in rivers and other methods must be used (see Chapter 7). Also, since the K value itself depends on the concentration of organics, it is a moot point whether the usual BOD test can be called a truly standard test since the use of different dilutions in making the test can give different K values for the same waste [31]. A few typical values of K are given in Table 4.9 for some wastewaters. Variation of K with temperature is given in accordance with the Van't Hoff-Arrhenius equation which has been discussed more fully in Chapter 11.

As stated earlier, the BOD test is performed for a duration of 5 days only. This is a seemingly arbitrary choice which started with the British Royal Commission in the early part of this century, but its continuance has its roots in the general observation that the statistical precision of the 5-day test is better than that for the same test at a shorter duration.

Several attempts have been made, and are continuing, to find shorter-duration BOD tests with reasonable reliability and precision. One such attempt [20] has been to measure the oxygen consumed up to its "plateau" (1.5 to 2 days) during which synthesis of new cells occurs, and then to compute theoretically the subsequent oxygen demand as equal to 1.42 times the weight of cells. Thus, the total first-stage oxygen demand is the sum of the two. This test involves the determination of the plateau and an accurate weighing of cells, both of which may be difficult in some cases.

Table 4.9 Typical K Values for Some Wastewaters

Source	K at $20°C$	k^a at $20°C$
Untreated domestic wastewater	0.23-0.50	0.10-0.28
Food industry wastes	0.23-0.50	0.10-0.28
Well-treated effluents	0.14-0.23	0.06-0.10
Rivers with "clean" waters	0.10-0.20	0.04-0.08

[a]Denotes values when logs to the base 10 are used ($k = 0.4343K$).

4.6.2 Chemical Oxygen Demand (COD)

Chemical oxidation of carbonaceous matter using an oxidant like dichromate ($K_2Cr_2O_7$) in the presence of a catalyst (Ag_2SO_4) can give a relatively quick estimation of the carbonaceous contents of a sample (within a matter of a few hours) compared to the BOD test which normally takes 5 days. However, the basic difference between the two tests is that the latter measures the oxygen demand resulting from bacterial activity while the former measures the oxygen demand owing to chemical oxidation. All chemically oxidizable compounds are not necessarily biodegradable and, hence, the two tests are not identical. Dichromate is not reduced by any ammonia in a waste, nor by any ammonia liberated from the proteinaceous matter that may be present in the waste.

The COD values of many organic compounds are practically 95 to 100% of the theoretical oxygen demand. Using the above catalyst, short-chain alcohols and acids are oxidized 85 to 95% or more. Benzene, toluene, and pyridine are not oxidized at all [19]. When wastes such as domestic sewage contain only readily biodegradable organics, and no toxic matter, the COD value can be used to approximate the BOD_u. For industrial wastes, the results may be highly variable. For example, in pulp and paper mill wastes, lignin will show COD but very low BOD owing to its relatively nondegradable character.

As most wastes have a varying proportion of degradable and nondegradable substance in them, the only practical way in which the COD test can be used for an estimation of BOD is to develop a specific correlation between COD and BOD_5 values for each case. Wastes undergoing treatment will exhibit a higher ratio of COD/BOD_5 for treated wastes than for raw wastes; treatment removes the more readily biodegradable substances. In recent years, the use of gamma radiation from cobalt 60 has been shown to alter the character of substances and affect their biodegradability, thus changing the COD/BOD_5 ratios before and after irradiation [26]. Table 4.10 gives some indication of the likely ratios for different wastes.

It is interesting to note that by its very nature the COD test has a higher precision than the BOD test. The standard deviation of the COD test has been found to be about 8% as compared to nearly 20% for the BOD test. The COD test can be very useful in the day-to-day control of treatment plant performance. In drawing composite samples at a sewage treatment plant, the BOD test requires the samples to be refrigerated until tested.

Table 4.10 Ratio of COD/BOD_5 Observed for Certain Wastes

Waste	Ratio of COD/BOD_5
Raw settled sewage[a]	1.68-2.43
Treated sewage	3.0-20
Milk wastes (raw)	1.62
Textiles (no seeding)	2.21-2.52
Refinery (with phenol)	1.83

[a]Average of 156 samples from 11 sewage treatment plants in Holland. The mean value of the ratio was 1.90, and the median was 1.77 (Source: Ref. 25).

Sampling for the COD test can be organized without refrigeration if acid is added to each sample to keep the pH less than 2.0. A composite sample can thus be accumulated without refrigeration over a period of a day or even a week if desired.

For certain toxic industrial wastes, where the BOD test would not work, the COD test is very useful in determining the carbonaceous content of the waste. For compounds of known composition, the theoretical oxygen demand can be computed as shown in Example 4.2.

Example 4.2: Calculate the theoretical oxygen demand of a 100 mg/liter solution of glucose.

$$C_6H_{12}O_6 + 6O_2 \rightarrow 6CO_2 + 6H_2O$$

Taking molecular weights,

$$\frac{\text{oxygen}}{\text{glucose}} = \frac{192}{180}$$

Hence, COD of solution $= \frac{192}{180} \times 100$ mg/liter $= 106.7$ mg/liter.

4.6.3 Total Organic Carbon (TOC)

The total organic carbon test involves complete combustion of all organic carbon to CO_2 and H_2O, and measuring CO_2 with the help of a specially sensitized infrared analyzer. The results can be obtained very rapidly and recorded on a chart if desired. The TOC test is becoming increasingly popular as it lends itself to autoanalysis.

Theoretically, it could be said that

$$\frac{COD}{TOC} = \frac{mol.~wt.~of~O_2}{mol.~wt.~of~C} = \frac{32}{12} = 2.66$$

The actual ratio has been shown to vary from zero, for substances resistant to dichromate oxidation, to 5.33 for methane, and slightly higher for inorganic reducing agents. Values ranging from 4.15 for raw sewage to 2.20 for treated sewage effluents have been observed [17]. Similarly, from Sec. 4.6.2, COD = 1.68 to 2.43 x BOD_5 for raw settled sewage. Hence,

$$\frac{BOD_5}{TOC} = \frac{2.66}{1.68~to~2.43} = 1.1~to~1.58$$

Eckenfelder reports that Mohlman and Edwards found values of BOD_5/TOC ranging from 1.35 to 2.62 for raw domestic waste, and that Wuhrman found values of 1.87 for raw sewage and 1.0 to 1.2 for treated sewage [17].

4.7 NITROGEN AS A POLLUTANT

Like carbon, nitrogen is also an important constituent of all living matter. Nitrogen gas constitutes the major fraction in air, yet nitrogen can be a pollutant owing to some of the undesirable effects it can create under certain conditions. For example, some nitrogen compounds can place considerable demands on the oxygen resources of waterbodies (nitrogenous BOD), nitrogen under certain conditions can lead to eutrophication in natural waterbodies, and in the ammonia form it may be directly toxic to fish. When nitrogen is applied to land, some of it can seep through the soil as nitrates in the leachate and pollute groundwaters. Excessive concentrations of nitrates in drinking water have been associated with public health problems like methymoglobaenemia.

Nitrogen can exist in several forms in the aquatic environment such as the following:

1. Dinitrogen (molecular nitrogen, N_2)
2. NH_3, NH_4^+, NO_2^-, and NO_3^-
3. Organic nitrogen
    a. Dissolved (amino acids, urea, etc.)
    b. Particulate (bacteria, phytoplankton, and zooplankton)

Thus, nitrogen is found in various redox states ranging from -3 in the most reduced form (ammonia) to +5 in the most oxidized form (nitrate).

Transformations can occur from one state to another and the major reactions are (1) assimilation of NH_3 and NO_3^- to give organic nitrogen, (2) occurrence of ammonification when organic nitrogen is converted back to NH_3, (3) nitrification from NH_4^+ to NO_3^-, (4) denitrification from NO_3^- to N_2, and (5) nitrogen fixation in which dinitrogen is reduced to ammonia and then to organic nitrogen. All the above transformations are biochemical in nature and, therefore, their occurrence and the speed with which the transformations occur are affected by a vast array of factors, thus making the nitrogen cycle a complex one. Figure 4.7 shows how the various transformations are related to each other and to the growth of bacteria, phytoplankton, and zooplankton. The figure does not show separately the effect of aerobic and anaerobic conditions.

Widely varying rates of nitrogen fixation under both aerobic and anaerobic conditions have been reported in literature, the rates are generally much lower under anaerobic conditions. The nitrogen fixers are mainly some blue-green algae and some bacteria.

Most bacteria, fungi, algae, and higher plants have the ability to use organic nitrogen, NH_4^+, and NO_3^- as a source of nitrogen; their preference is generally in the order stated although there are several differences. NO_2^- can also be used as a source of nitrogen in very low concentrations [28].

Ammonia exists in solution either as NH_3 or as NH_4^+, the relative proportion depends greatly on pH (see Figure 4.8). At pH less than 8.0, practically all is present as NH_4^+ in solution, at a little over pH 9.0 the dissociation is about half and half, while at about pH 11.0 practically all of it is in the NH_3 form. NH_3 is highly soluble in water and toxic to fish even in small concentrations.

The oxidation of NH_4^+ first to NO_2^- and then to NO_3^- is referred to as nitrification. This is discussed later in detail owing to its significant effect on the oxygen balance of rivers, lakes, and marine waters. Eutrophication will be discussed more fully in Chapter 6 along with nitrogen transformations in lake waters. The nitrogen problem in the case of land disposal of wastewaters and sludges will be discussed in Chapter 9.

Laboratory procedures for determining the different forms of nitrogen are available [19]. Determination of NH_3 involves raising the pH of the sample and distilling off the ammonia with steam. The NH_3 in the condensate is finally determined colorimetrically. Also, nitrites and nitrates are generally determined colorimetrically; nitrites are often present in very small concentrations.

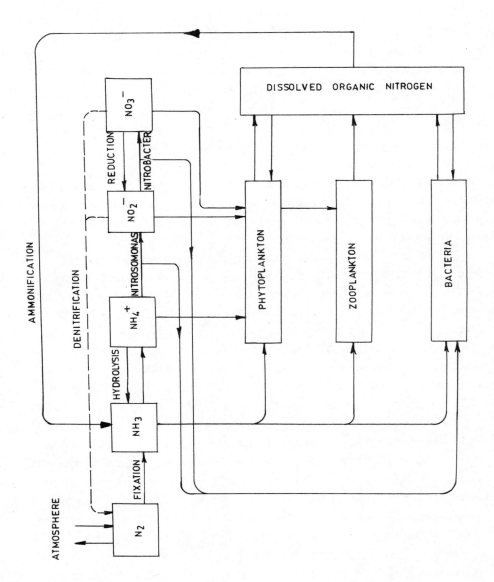

Figure 4.7 Nitrogen transformations

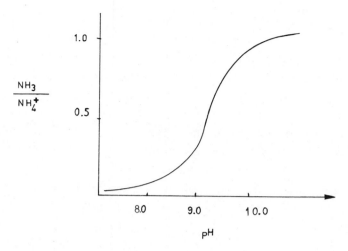

Figure 4.8 The proportion of NH_3 to NH_4^+ at different pH values.

Organic nitrogen is determined by the Kjeldahl method [19] in which
the sample is first boiled to drive off free ammonia and then digested to
convert organic nitrogen to ammonia. When free ammonia is not driven off
first, the result gives what is called "total Kjeldahl nitrogen" (TKN).
Total nitrogen (TN) is the sum of all forms of nitrogen. Thus,

TKN = NH_3-N + organic N

TN = TKN + NO_2-N + NO_3-N

The nitrogen content of raw domestic wastes has been indicated in
Table 5.1, from which it can be seen that organic nitrogen is about 40% of
the total nitrogen. In treated wastes, the organic nitrogen may be less
than 10% of the total nitrogen.

4.7.1 Nitrogenous BOD

Ammonia contained in wastewaters may be progressively oxidized to
nitrites and nitrates through the activity of aerobic microorganisms referred to as nitrifying bacteria. These organisms are mainly chemoautotrophs using CO_2 as a source of carbon, and deriving their energy from
redox reactions. Oxidation is essential for their growth. Some heterotrophs may also be involved.

It will be recalled from Sec. 1.4.2 that, where carbon is obtained by a cell from CO_2, first some energy has to be spent in converting it to the same oxidation level as the cell (such as CH_2O) before it can be made assimilable. This reduces the yield of new cells per unit weight of substrate oxidized. Hence, the growth rate of nitrifying organisms and the corresponding oxidation of nitrogenous material is slower than that of carbonaceous material. Furthermore, the nitrifying organisms are sensitive to pH (optimum is 7.5 to 8.5), to the concentration of dissolved oxygen, and to the presence of various other substances.

Generally, nitrification begins after the major portion of the carbonaceous BOD is satisfied (see Figure 4.5). The oxidation of ammonia takes place in two steps, first to the nitrite (NO_2^-) state through the agency of "Nitrosomonas,"* and then to the final nitrate (NO_3^-) state brought about by "Nitrobacter." The reactions which occur are as follows and oxygen requirements can be computed from them without the need for conducting any laboratory tests as required for carbonaceous BOD.

1. $2NH_4^+ + 3O_2 \xrightarrow{\text{Nitrosomonas}} 2NO_2^- + 2H_2O + 4H^+$ $\hspace{2cm}$ (4.15)

2. $2NO_2^- + O_2 \xrightarrow{\text{Nitrobacter}} 2NO_3^-$ $\hspace{3cm}$ (4.16)

Stoichiometrically, 4.57 mg/liter of dissolved oxygen are required to oxidize 1 mg/liter of NH_4^+ to the nitrate form. However, the actual oxygen demand is about 4.33 mg/liter because all nitrogen is not oxidized; a part is used for cell synthesis [27].

Observations show that NO_2^- concentrations are generally low in wastewaters undergoing nitrification. This is because conversion of NO_2^- to NO_3^- is rapid compared to the conversion rate from NH_4^+ to NO_2^-. In other words, the latter is rate limiting. Nitrification kinetics have been studied by various workers in order to determine the nitrification rate K_n in biological waste treatment processes, in rivers, and in soils using appropriate mathematical models. Temperature and pH also have a considerable effect on nitrification rates as discussed in the appropriate chapters.

4.8 PHOSPHORUS

Phosphorus can be present in water in a variety of forms, both organic and inorganic. Generally its three principal forms are

---

*Nitrocystis oceanus may be more important in seawater.

1. <u>Orthophosphate ion (PO_4^{3-})</u>: Inorganic phosphates enter water from the soil strata, certain industrial wastes, detergents, and other sources. Orthophosphates are also formed as a result of the degradation of organics.* Phosphates are utilized for biological growth.

2. <u>Polyphosphates</u>: These are similar to polymers of phosphoric acid from which water has been removed. Their hydrolysis results in the formation of orthophosphates.

3. <u>Organic phosphorus</u>: This is present in a variety of compounds and in living cells. As stated above, the decomposition of organic matter yields orthophosphates (Figure 4.9).

In waste treatment and natural self-purification systems, both poly-phosphates and organic phosphorus are converted to orthophosphates. Thus, in stabilized wastewaters and many natural waters, the phosphorus present is principally in the orthophosphate form; however, our interest lies in knowing the total phosphorus. Thus, samples to be tested for total phosphorus are first acid-digested in the presence of a strong oxidizing agent to convert any polyphosphates and organic phosphorus present in the sample to the orthophosphate form which is then generally measured colorimetrically [19].

The contribution of phosphorus likely from various sources such as domestic sewage and agricultural and other runoff is detailed in Chapter 5. Removal in conventional biological treatment generally does not exceed 50% of the incoming concentration unless additional treatment (often in the form of chemical precipitation) is incorporated in the plant. The role of phosphorus in eutrophication is considered at length in Chapter 6 and its removal and retention in irrigated soils in Chapter 9.

*The actual form of orthophosphate depends on the pH of the water. In typical municipal wastewater, the predominant form may be HPO_4^{2-}.

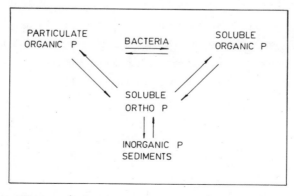

Figure 4.9 Phosphorus transformations.

4.9 EUTROPHICATION

The term "eutrophication" comes from "eutrophic" meaning well-nour-
ished.* Eutrophication refers to the natural or artificial addition of
nutrients such as nitrogen and phosphorus in a form which increases the
productivity of water and brings about consequent changes in plant and
animal life in the body of water, perhaps reducing its utility and beauty
and threatening its very existence in the course of time. In this sense,
the term eutrophication is applicable to lakes or bodies of still water
which can exhibit "age" or ecological succession and even "die" of old age
so to speak. Rivers do not age in the same sense as lakes, although increases in nutrients can give corresponding increases in productivity and,
therefore, the term eutrophication cannot properly be applied to running
water. This is overcome by making a distinction between (1) eutrophication in the strict sense (increase in nutrient supply) and (2) the effects
of eutrophication.

The effects of eutrophication may sometimes be desirable and a deliberate increase in nutrients may be brought about as in the case of fish
ponds which are purposely fertilized with sewage to increase their fish
yield. In this regard, the reader is referred to the discussion in Sec.
3.12 where it was mentioned that the quality of fish becomes coarser as
eutrophication increases; quantity is obtained at the expense of quality.
Judicious enrichment of marine waters to increase their productivity together with improvements in harvesting technology remain a desirable goal
of current research efforts.

When the effects of eutrophication are undesirable, as when the value
of a lake as a source of water supply and recreation is diminished, eutrophication is considered as a form of pollution.

Eutrophication of lakes is, in fact, a natural process which can affect
all lakes over a long period of time. Some lakes may be potentially more
eutrophic than others as a result of various natural conditions. The process of eutrophication can, however, be accelerated by man, and relatively
rapid changes can be brought about in water quality which would detract

---

*The term "eutrophic" was first introduced by the German limnologist Weber
in 1907.

from its optimum use. Engineers are interested in learning the causes of
this accelerated form of eutrophication, the factors which affect its rate,
the effects on water quality, and the possible control measures. The
freshwater resources of any country are limited and one cannot afford to
lose them.

4.9.1 Undesirable Effects of Eutrophication

Much interest lies in curbing the undesirable effects of eutrophication and it may be well to list them at this stage. Some of the undesirable
effects concern the day-to-day treatment of water for drinking, industrial,
and other purposes and others concern the waterbody as a water resource in
the broadest sense of the term.

1. Increased costs for water treatment stemming from the excessive presence
   of algae in the raw water. This may be due to the necessity for

   a. Taste and odor removal

   b. Color removal

   c. Increased consumption of treatment chemicals

   d. More frequent backwashing of filters

2. Rejection of water for drinking and animal watering purposes owing to
   the presence of toxic secretions from certain algae, which may be in
   sufficient concentrations to harm health.

3. Increased costs for an industrial water supply owing to similar reasons
   as given above, and owing to the likelihood of algal slimes in cooling
   waters.

4. Diminished use of water in recreation, bathing beaches, navigation,
   and general reduction in touristic beauty owing to

   a. Frequent algal blooms and unsightly floating mats

   b. Prolific weed growth

   c. Mosquito and insect nuisance

   d. Fish kills

5. Changes in the quality and quantity of fish of commercial value.

6. Gradual loss of the lake itself as, with advanced eutrophication,
   autochthonous accumulation increases, littoral vegetation proliferates,
   and the lake becomes shallower and shallower until eventually it ends
   up as a swamp or marsh.

For a detailed discussion of the factors affecting eutrophication and
some possible control measures, the reader is referred to Chapter 6.

4.10 WATER QUALITY TESTING

The extent of pollution of a waterbody can be measured and assessed by carrying out various tests of which some of the more important ones can be grouped as follows:

<u>Physical</u>

> Temperature
> Color
> Odor
> Turbidity
> Solids (total, dissolved, and suspended)
> Conductivity
> Radioactivity

<u>Chemical</u>

> Dissolved oxygen (DO)
> Oxygen demand (BOD, COD, and TOC)
> pH, alkalinity, and acidity
> Nitrogen
> Phosphorus
> Chlorides
> Sulfates
> Salinity
> Heavy metals* (Hg, Cd, Pb, etc.)
> Other toxic substances (As*, pesticides, halogenated organic
>   compounds*, PCB*, etc.)
> Oils
> Surface active agents
> Specific gases (CO_2, H_2S, etc.)
> Carbon-chloroform and alcohol extracts

<u>Biological</u>

> Coliforms and <u>E. coli</u> (fecal coliforms)
> Specific pathogens (e.g., <u>Salmonella</u> and worms)
> Viruses
> Algae
> Species diversity index
> Bioassay

The methods of analysis and the precautions to be taken in sampling are beyond the scope of this publication. Details are available elsewhere [19, 30].

4.11 BIOASSAY TESTS

Bioassay tests are performed to study the effects of small quantities of toxic substances, singly or in <u>synergistic</u> (combined) action. The tests

---

*UNESCO/WHO places these items on a so-called black list.

essentially consist of subjecting selected specimens of fish to a known concentration of a toxic substance or substances in water and estimating the median tolerance level (TL_m). The median tolerance level is that concentration of the waste in the experimental water at which 50% of the test organisms (fish) have survived for the given period of the test. The period of the test may be 24 to 48 or even 96 hours, during which time the fish may be exposed to the given concentration in a continuous flow or batch-type experiment. The former is preferred for several reasons and is shown in Figure 4.10.

It is important that the test conditions suit the local conditions with regard to temperature, water, DO, pH, and, of course, the fish used. Generally, for simulating freshwater conditions, the fish used are bass, trout, minnow, carp, etc., while for estuarine and salt waters, sticklebacks, killifish, and the members of the genus <u>Fundulus</u> are used. Usually, the

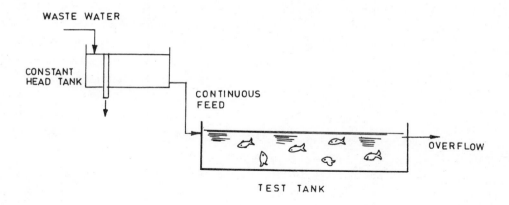

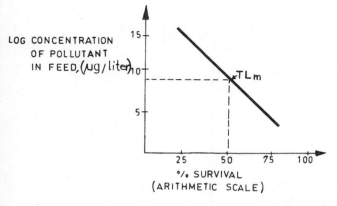

Figure 4.10 Continuous flow experiment for bioassay test.

water tank is loaded with about 10 fish after acclimatization (which is important), or approximately 2 g weight of fish for every liter of water in the tank. The fish should be of more or less uniform size and growth. Generally, the fish should not be more than 7.5 cm in length and the longest fish should not be more than 1.5 times longer than the smallest one.

At the end of the test period, the number of fish succumbing, as well as the number of fish in "distress," are recorded. The results can be plotted graphically on semilog paper as shown in Figure 4.10, from which

$$\log (TL_m) = \log C_2 + \frac{p_2 - 50}{p_2 - p_1} (\log C_1 - \log C_2)$$

where C_1 and C_2 are different concentrations of waste (chosen close to the median value on the graph) and p_1 and p_2 are the corresponding percent survivals.

This is a very useful test to perform in the case of certain industrial wastes containing toxic chemicals in small concentrations either singly or together. In the latter case, a simple chemical analysis would not indicate the true situation if the presence of two or more chemicals leads to a magnification of the toxic effect owing to synergism. When synergistic conditions exist, doubling the concentration of two pollutants would more than double the resulting effect; this would become evident only in a bioassay test.

On the other hand, the bioassay test is a simplification of the field conditions. There are several factors which affect the tolerance levels of fish. For example, the presence of calcium reduces toxicity of lead and zinc, and an increase in temperature reduces the asphyxiation level of fish. Even small changes in pH may have marked effects on the median tolerance levels. pH changes may affect the dissociation of some chemicals and thereby affect tolerance levels. For example, the ammonium ion (NH_4^+) predominates at low pH and is less harmful than ammonia (NH_3) which predominates at higher pH values.

The 48-hour TI_m values for certain chemicals, including pesticides, are given in Table 4.11. These values are only indicative since the test results depend greatly on the fish used, the duration of exposure, and various environmental factors as described above.

The term "toxicity units" is sometimes used.

$$\text{Toxicity units} = \frac{100}{\% \ C}$$

Table 4.11 Median Tolerance Levels of Fish

Compound	48-Hour TL_m values (mg/liter)
Chlorinated hydrocarbons	
Aldrin	0.02-0.04
Endrin	0.20
DDT	0.10-0.60
BHC	2.0
Texaphene	3.0
Organophosphorous pesticides	
Fention	0.03
Parathion	0.2-1.0
Ronnel	5.0
Chemicals	
Alkyl benzene sulfonates (ABS)	3.0-12.0
Ammonia (NH_3)	2.0-3.0
Chlorine	0.05-0.2
Cyanide (CN^-)	0.04-0.1

Source: Refs. 6, 9, and 10.

where %C = percent concentration of wastewater in the receiving water at which there is 50% survival of the test species after a 96-hour exposure in a bioassay procedure. When there is greater than 50% survival in 100% wastes, toxicity units are calculated as

$$\text{Toxicity units} = \frac{\log\,(100 - s)}{1.7}$$

where s = % survival in 100% wastewater.

The actual values of toxic chemicals to be permitted in rivers would have to be smaller than the ones obtained by such tests in order to allow for the various factors discussed above. Depending upon the conditions, the laboratory values may have to be reduced by a factor of 10 or even more. In prescribing the limits of concentration in industrial effluents, discharged to the river, the low flow values likely to recur once in 5 or 10 years or more may be taken into consideration.

It should be noted that the bioassay test merely indicates the lethal dose for a given species of fish but does not indicate anything concerning the accumulation of the substance in edible plants and fish.

4.12 ENVIRONMENTAL IMPACT ASSESSMENT

Figure 4.1 illustrated the fate of pollutants discharged to the water environment. A similar fate awaits those discharged to air or soil, with interaction between the different phases, depending on the nature of the pollutant and the dispersive and the absorptive and accumulative capacity of the biotic and abiotic components of the ecosystem. Thus, in a system where pollutants can spread far and wide and have various "boomerang" effects on health, the ecosystem, and the economy, as shown in Figure 4.1, some assessment of their likely environmental consequences must be undertaken <u>before</u> their discharge at a particular site.

Environmental impact assessment (EIA) is an orderly method, a structured approach, for studying the probable changes in and the consequences to the various physical, chemical, biological, and even socioeconomic characteristics of the environment brought about by a proposed "action" (e.g., establishment of an industry or construction of a large multipurpose dam). Their interrelationships are defined as "impacts." They are not confined to engineering elements, but include sociological and economic ones also.

Identification of impacts is somewhat complicated by the fact that primary impacts may be followed by secondary and tertiary ones. For example, the primary impact of a waste discharge may be reduction of DO in an estuary, while the secondary effect may be reduced spawning of fish, followed by further effects on the whole fishing and tourist industry in the area. In some instances, the secondary and tertiary effects of a development project may be due to induced population and other growth with induced energy and land-use patterns.

Ideally, impacts should be possible to quantify and evaluate in such a manner as to be readily understandable by the decision makers. It generally requires considerable time and funds to be able to obtain sufficient data, and is yet unable at times to provide satisfactory answers to all questions. Nevertheless, such an exercise may be desirable, even essential, for a large project with serious consequences.

Normally, when an engineer performs, for example, an oxygen sag analysis for a river stretch, models DO depletion and nutrient buildup in a bay, or models air pollution dispersion or salt buildup in soil, he takes an essential step in environmental impact analysis. To that extent many environmental engineers are already performing a part of the EIA exercise. However, to be comprehensive, the EIA exercise should include a wider range of characteristics of the environment — not only physical and biochemical but also socio-economic — and include community participation in the decision making process.

EIA is not limited to developed countries only. Even in developing countries, conducting an EIA (even on a relatively limited scale) can be of great value since in its simplest form EIA is an orderly process of thinking which can help incorporate environmental considerations into development planning. From the point of view of the developer, EIA can be of great help in proper site selection to minimize environmental damage and/or costs for remedial action (e.g., waste treatment). It can also help sometimes in the selection of the very process for manufacturing, that is, selecting the one with minimum residuals polluting the environment and benefitting from newer technologies in this regard. EIA also helps governmental authorities in promoting orderly development without impeding economic growth, and helps them in planning realistically their monitoring and data collection programs and exercising follow-up vigilance.

4.12.1 EIA Procedures

Various approaches have been followed in different countries. In some countries (e.g., the United States) EIA is legally required for all major federal actions. In some others (e.g., the United Kingdom) an EIA may be required only for certain developments and the authorities can demand it under an existing system for development control such as land use planning. In yet others, EIA consists mainly of ecological modeling to assist in a centralized planning system. The nine European Common Market countries are developing a harmonized approach between themselves.

Essentially the steps followed in any system for EIA must include at least the following:

1. Preliminary activities to narrow down the scope of EIA studies (to eliminate costly acquisition of useless information)

2. Description of proposed project, and of any reasonable alternatives
    (location, production, residuals, pollutants, employment potential and
    infrastructure requirements, etc.)

3. Description of present environmental features likely to be signifi-
    cantly affected (e.g., air, water, soil, climate, flora and fauna,
    built-up environment, and landscape)

4. Assessment of the likely effects on the environmental, economic, and
    social components indicating the nature of the effects (direct, in-
    direct, cumulative, short-, medium-, or long-term, permanent or tempo-
    rary, positive or negative) and identification of their sources
    (physical presence of project, use of resources, emission of pollu-
    tants, accidents, etc.)

5. Description of measures envisaged to eliminate, reduce, or compensate
    adverse impacts on the environment.

The environmental impact assessment which is normally a technical
exercise is generally required to be followed by an environmental impact
statement (EIS) in which the results of the assessment are discussed in
terms of beneficial and adverse impacts for the benefit of decision makers
and the general public.

4.12.2 Methods of Impact Assessment

Preparing a description of present environmental features (item 3
above) requires much local data. Land use planning departments may possess
much useful information in the form of maps showing centers of population
and population distribution, geomorphological information, soils, drainage,
watersheds, highways, airport approaches, transmission lines, underground
pipes, flora and fauna, agricultural areas, sanctuaries and areas of special
importance, etc. These can be very useful in preliminary selection of
project sites or alignment of linear developments so as to minimize environmental damage.

Air and water quality data may be obtainable if some agencies are
involved in environmental monitoring. If not, qualitative estimations and
short-term surveys may be necessary at the preliminary stage until need
for more costly, long-term surveys is established. Predesign surveys for
coastal waters are further discussed in Chapter 8.

Thereafter, assessment of likely effects on the environment (item 4
above) may be done qualitatively using checklists or matrices to give a
preliminary screening of project alternatives. In some instances, public
authorities may provide specific checklists of items which must be given

consideration for specific types of projects. For example, there may be a
specific checklist for water resources development projects, another for
oil installations, another for fertilizer industries, and so on.

Items included in such checklists are often of a broad nature. For
example, a checklist for screening water resources projects may include
items such as the following on which likelihood of impact is to be determined: tourism, recreation, fisheries, agriculture, ecology, health, siltation, erosion, eutrophication, earthquakes, fog formation, malaria, displacement of people, economic activity, downstream effects, landscape, etc.
Impacts on the above kind of items can be stated by expert judgment as
beneficial or adverse, reversible or irreversible, short-term or long-term,
local or widespread.

In a preliminary screening of different alternatives for a project,
it is useful to include "No Plan" also as an alternative, since sometimes
not going ahead with a project may prove to be worse than going ahead with
one. Checklists should also cover the construction phase of a project besides its operational one since at times the former may prove more damaging
to the environment than the latter.

Instead of checklists, one could use <u>matrices</u> which could be tailor-
made to suit the needs of any type of project to be evaluated. Matrices
are prepared by listing all proposed activities and subactivities of a
project (e.g., land development, dredging, filling, labor housing, provision of services, transport, processing, product spillage, wastes discharge, gaseous emissions, etc.) to give a vertical column for each activity.
The various components of the environment (e.g., air, water, soil, fisheries,
bird life, etc.) as well as of the socioeconomy (e.g., employment, health,
landscape, public attitudes, etc.) are listed vertically to give rows.

This gives a matrix of intersecting cells. Those activities which are
likely to have an impact on any component of the environment can then be
identified by placing a cross-mark in the corresponding cells. Again, this
is based on expert judgment.*

Environmental components are generally interconnected and form webs
or networks, and an ecological approach is often demanded in identifying

---

*If desired, judgment can be extended to denote the "magnitude" and "im-
portance" of the impact in each cell using numerals ranging from 1 to 10
for each characteristic. For example, a score of 2/9 would imply an impact
of small magnitude but much importance. The scoring system, however, is
subjective and could lead to controversies.

the secondary and tertiary impacts. Thus, <u>networks</u> first need to be identified. With increasing detail, networks may become unwieldy, but a certain
amount of network identification is implied in the preparation of appropriate matrices.

In comparing alternative schemes, it is useful if the extent of the
impact is indicated qualitatively using such parameters as number of people
likely to be affected adversely or favorably, the extent of the area likely
to be developed or lost (e.g., by flooding), the relative magnitude of concentration, or numbers, or volume involved (e.g., scheme A may give twice
as much of the same pollutant as scheme B) and other such parameters that
enable comparisons to be drawn. Checklists and matrices are often supplemented by explanations and reasoned arguments.

Both checklists and matrices are widely used and are useful as aids
for identifying activities which have an impact on the environment and for
presenting the information to decision makers and the public. They are
most valuable in identifying potential impacts that need to be studied in
a detailed manner.

Finally, where the environmental impacts identified by use of checklists and matrices need to be investigated in detail, and quantified if
possible, <u>field surveys</u> and the use of <u>mathematical models</u> may become necessary. Although ecological systems per se are difficult to model, increasing attention is being focused on the use of simulation models at the planning stage to assess impact on air and water quality and ultimately on
health, fisheries, land use, and so forth. Development of models requires
obtaining sufficient baseline data, preparing inventories of existing and
future pollutant discharges, modeling the dispersive, absorptive, and accumulative capacities of the ecosystem in question, calibrating the simulation model and finally using it to predict future situations.

Simulation models can be used to advantage by planners to evaluate
different possible strategies for controlling and minimizing adverse impacts
at minimum costs. Their use with lakes, rivers, coastal waters, and land
is discussed further in the relevant chapters.

4.12.3 Development Permits

Based on an impact assessment and an impact statement (which have been
made available for public scrutiny and/or subjected to a public hearing)
the planning authority may issue a permit or consent specifying any necessary

conditions. Consent conditions specifying production and waste emission levels may require the developer to take additional protective measures as might be considered necessary. In some cases, consent conditions may include follow-up monitoring of the quality of specific components of the environment (e.g., air or water) to ensure that prescribed standards and "derived limits" (see p. 96) are met, and to help generate the necessary feedback data on which a periodic review of the consent conditions can be based. An element of monitoring and feedback of information is useful to include in all important developments. Follow-up monitoring requirements for sea outfalls are further discussed in Chapter 8.

REFERENCES

1. The Rape of the Rhine, Time Magazine, June 1973; a report on work by H. Sontheimer of Karlsruhe Univ., Germany.

2. Report of the Special Commission on the Pollution of Surface Waters, International Water Supply Assoc., Vienna Conf., Vienna, Austria, 1969.

3. WHO European Drinking Water Standards, Revised, 1970.

4. Chemical Oceanography, Vol. I, Academic, New York, 1965.

5. Encyclopedia of Oceanography, Reinhold, New York, 1966.

6. Control of Pollution in Coastal Waters, WHO, Geneva, EP/71.4, 1970.

7. Master Plan for Water Supply and Sewerage for Istanbul Region, prepared by DAMOC consortium for WHO/UNDP, 1970.

8. Methyl Mercury in Fish, Report from an Expert Group, Nordisk Hygienisk Tidskrift, Stockholm, 1971.

9. E. Odum, Fundamentals of Ecology, Third Edition, W. B. Saunders, 1971.

10. Health Hazards of the Human Environment, WHO, Geneva, 1972.

11. F. Dietz, The Enrichment of Heavy Metals in Submerged Plants, Advances in Water Pollution Research, 6th International Conference, Jerusalem, Israel, 1972.

12. G. Bitton and R. Mitchell, Protection of E. coli by montmorillonite in seawater, Jour. ASCE, 100, 1974.

12a. Straskraba and Straskrabova, Eastern European Lakes, in Eutrophication, Nat. Acad. Sci., Washington, 1969.

13. B. Ketchum, C. Carey, and M. Briggs, Preliminary studies on the viability and dispersal of coliform bacteria in the sea, in Limnological Aspects of Water Supply and Waste Disposal, Amer. Assoc. Adv. Sci., Washington, 1949.

14. G. Fair, J. Geyer, and D. Okun, Water and Wastewater Engineering, John Wiley and Sons, New York, 1968.

15. L. Reid and D. Carlson, Chlorine Disinfection of Low Temperature Waters, Jour. ASCE, 100, 1974.

16. A. L. Downing, Population Dynamics in Biological Systems, Proceedings Third International Conference on Water Pollution Research, Munich; WPCF, USA, 1966.

17. W. W. Eckenfelder, Water Quality Engineering for Practicing Engineers, Barnes and Noble, New York, 1970.

18. J. Ingraham, The Bacteria, Vol. IV, Academic, New York, 1961.

19. Standard Methods for Examination of Water Supplies and Wastewaters, APHA, 1971.

20. Busch, 15th Ind. Wastes Conf., Purdue Univ., 1961.

21. E. W. Moore, H. A. Thomas, and W. B. Snow, Simplified method for analysis of BOD data, Sew. Ind. Wastes, 22, 1950.

22. H. A. Thomas, Graphical determination of BOD curve constants, Water Sew. Works, 97, 1950.

23. Y. Fujimoto, Graphical use of BOD equation, Jour. WPCF, 36, 1964.

24. D. Bagchi and N. Chaudhuri, Graphical method of evaluating the constants of BOD curve, Ind. Eng., 1965.

25. A. Pasveer, Personal communication.

26. Investigations on BOD and color removal by ^{60}Co irradiation, at Auburn Univ., USA, Mech Eng., 92, 1970.

27. Wezernak and Gannon, Appl. Microbiol., 15, 1962.

28. Nitrogen as a Water Pollutant, WPCF Conf., Copenhagen, 1975.

29. A. L. Downing, Factors to be Considered in Design of Activated Sludge Plants, Advances in Water Quality Improvements, Vol. I, Univ. Texas, Austin, Texas, 1968.

30. Standard Methods for Water Quality Examination for Member Countries of the Council for Mutual Economic Assistance (CMEA).

31. R. Mitchell, Ed., Water Pollution Microbiology, Wiley-Interscience, 1972.

32. L. Hom, Kinetics of chlorine disinfection, Jour. ASCE, Sanit. Eng. Div., 1972.

33. B. D. Clark, K. Chapman, R. Bisset, and P. Wathern, Methods of Environmental Impact Analysis, Built Environment, 4, No. 2, 1979.

34. J. Golden, Environmental Impact Data Book, Ann Arbor Science Publishers, Ann Arbor, Michigan, 1979.

35. Urban Jain and Stacy Jain, Environmental Impact Analysis, Van Nostrand Reinhold Company, New York, 1977.

Chapter 5

CHARACTERISTICS OF WASTES FROM DIFFERENT SOURCES

5.1 CHARACTERISTICS OF DOMESTIC WASTEWATER

Domestic wastewater contains organic and inorganic matter as suspended, colloidal, and dissolved solids. Their concentration in the wastewater depends on the original concentration in the water supply, and the uses to which the water has been put. The climate and the wealth and habits of the people have a marked effect on the wastewater characteristics. The presence of industrial wastes in the public sewers can substantially alter the nature of the wastewater. Concentrations are also affected by the amount of water used per person, since in many communities the amount of solids added per person vary within relatively narrow limits. Thus, wastewater characteristics vary not only from city to city but also from season to season and even hour to hour within a given city.

Raw domestic wastewater characteristics are shown in Table 5.1. The range of values given are also typical for municipal wastewaters which are predominantly domestic in character. The wastes are mainly organic in nature, containing carbon, nitrogen, and phosphorus among others, with relatively high concentrations of microorganisms. They are readily putrescible, and the biological degradation of organic matter proceeds even as the wastes flow through sewers. This alters some characteristics as time passes. The values given in Table 5.1 refer to relatively fresh wastes [1]. Furthermore, all values given in Table 5.1 are stated in terms of g/capita to facilitate their use for design purposes, and for comparison between different communities. Values given in terms of mg/liter are obscured by the fact that water usage varies markedly between communities, especially between those in developed and developing countries.

Values of biochemical oxygen demand (BOD) generally average around 54 g per person per day where the sewage collection system is reasonably efficient. In some developing countries, the BOD values may be only 30 to 40 g per person per day as all the sewage produced may not enter the

Table 5.1 Domestic Wastewater Characteristics

Item	Range of values contributed in wastes (g/capita/day)
BOD_5	45-54
Chemical oxidation demand (dichromate)	1.6 to 1.9 x BOD_5
Total organic carbon	0.6 to 1.0 x BOD_5
Total solids	170-220
Suspended solids	70-145
Grit (inorganic, 0.2 mm and above)	5-15
Grease	10-30
Alkalinity, as $CaCO_3$	20-30
Chlorides	4-8
Nitrogen, total, as N	6-12
Organic nitrogen	~0.4 x total N
Free ammonia[a]	~0.6 x total N
Nitrite	—
Nitrate	~0.0 to 0.05 x total N
Phosphorus, total, as P	0.6-4.5
Organic phosphorus	~0.3 x total P
Inorganic (ortho- and polyphosphates)	~0.7 x total P
Potassium, as K_2O	2.0-6.0
Microorganisms present in wastewater	(Per 100 ml wastewater)
Total bacteria	10^9-10^{10}
Coliforms	10^6-10^9
Fecal Streptococci	10^5-10^6
Salmonella typhosa	10^1-10^4
Protozoan cysts	up to 10^3
Helminthic eggs	up to 10^3
Virus (plaque forming units)	10^2-10^4

Note: Values given in the table represent approximate daily per capita contributions. To these must be added the contents in the original water supply. Per capita daily water usage ranges from 80 to 300 liters in most sewered communities, although consumption in some U.S. and other communities is even higher than 500 liters. The pH of wastewater generally ranges from 6.8 to 8.0, again depending on raw water quality.

[a]The major nitrogen compound in domestic waste is urea which is readily hydrolyzed to NH_3 and CO_2 by the enzyme urease present in sewage. Hence, NH_3 constitutes the major fraction of total nitrogen in domestic sewage.

Source: Ref. 1.

sewer system. Where combined sewers are used, the BOD values may be about 40% higher, namely 77 g per person per day. In the case of offices, factories, schools, etc., where there is part-time occupancy, the BOD values are generally taken as half of 54 g per person per day, or less. For restaurants and cafeterias each meal served may be taken to contribute one-fourth of 54 g of BOD. Theaters and cinemas may be taken to contribute one-sixth of 54 g of BOD per seat. Hotels and hospitals, on the other hand, may contribute as much as 1.5 to 2.5 times more than the usual 54 g of BOD per person per day. Domestic sewage is the primary source of nutrients such as nitrogen and phosphorus; most industrial wastes (except the food and fertilizer industries) are relatively minor sources of these nutrients in municipal wastewaters. In many developing countries the amount of nitrogen and phosphorus in wastewaters may lie at the lower end of the ranges shown in the table.

Since interest in many countries often lies in irrigational and industrial reuse of wastewater, it is interesting to know the mineral pick-up from domestic usage of water. Table 5.2 gives some typical values reported as national averages in the U.S. [2], while Table 5.3 gives some values reported from Israel [3].

Table 5.2 Mineral Pick-up from Domestic Usage of Water in the U.S.

Item	Pick-up from one cycle of domestic water usage (mg/liter)
Chlorides as Cl	20-50
Sulfates as SO_4	15-30
Nitrates as NO_3	20-40
Phosphates as PO_4	20-40
Sodium as Na	40-70
Potassium as K	7-15
Calcium as $CaCO_3$	15-40
Magnesium as $CaCO_3$	15-40
Boron as B	0.1-0.4
Total dissolved solids	100-300
Total alkalinity as $CaCO_3$	100-150

Source: Ref. 2.

Table 5.3 Mineral Pick-up in Sewage from Towns
and Agricultural Settlements in Israel

Item[a]	Municipal sewage[b]	Agricultural settlements[c]
Nitrogen as N	5.18	7.0
Potassium as K	2.12	3.22
Phosphorus as P	0.68	1.23
Chlorides as Cl	0.54	14.65
Boron as B	0.04	0.06
Sodium as Na	0.60	14.75
Total hardness as CaCO$_3$	2.50	6.25
Total dissolved solids	40	78
Electrical conductivity, μmhos/cm	600	470
Sodium absorption ratio, meq/liter	2	1.5

[a]Given in g/capita-day except where noted.

[b]Average of 62 municipalities surveyed (2.1 million total population).
Sewage flow at 100 liters/capita-day.

[c]Includes sewage and livestock water; total flow at 250 liters/capita-day. Population surveyed was 96,880 in 267 settlements.
Source: Ref. 3.

5.2 INDUSTRIAL WASTEWATER CHARACTERISTICS

Characteristics of industrial wastewaters vary widely from industry
to industry, and even within the same industry depending on raw materials
used, processes employed, and various other factors. It is, therefore,
difficult to generalize, although a few typical values are shown in Table
5.4. A detailed discussion of the nature of wastes from industries is
beyond the scope of this book.

One characteristic of most industrial wastes that must be mentioned
at this stage is the likelihood of wide fluctuations in both flow and constituents of a waste. For this reason, it is best to depict such cases in
terms of statistical probabilities of reaching certain values. For example,
wastewater flow and the concentration of any desired constituent (e.g.,
phenol, BOD, etc.) can be readily plotted on arithmetic probability paper
from which the percent of time a value is likely to be equalled or exceeded

Table 5.4 Water Requirement and Wastewater Characteristics of Some Selected Industries

Industry	Water requirement	Wastewater characteristics			Nature of pollutants and characteristic effects
		Quantity	BOD	Other Items	
Food					
Beet sugar	27 m³/ton without reuse 3 m³/ton with reuse	89.5% of intake	1 kg BOD$_5$ per ton of beets		Food industry wastes, in general, contain much organic matter whose degradation leads to depletion of dissolved oxygen in streams and estuaries and likely effects on aquatic life such as fish. Odorous and anaerobic conditions may be created.
Cane sugar		500 liters/ton	0.6 kg BOD$_5$/ton		Some food industries operate seasonally only; they often produce solid wastes also.
Meat slaughter houses	5 m³/1000 kg live wt. of slaughtered animals				
Meat slaughtering and packing houses	30 m³/1000 kg live wt.	96.8% of intake	From packing only, 15–20 kg/1000 kg of meat packed	3–9 kg N/1000 kg live wt.	
Fruit and vegetable canning	8–80 m³/ton	67% of intake			
Beverages					
Beer		10–15 liters per liter beer	8 g/liter of beer	0.1–0.5 g N per liter of beer	Similar in effects to food industry wastes and domestic sewage, but may have very high BOD.
Milk		2–10 liters per liter of milk	0.1–0.2 kg/100 kg of milk		Wastes from food and beverage industries are often disposed on land, where possible, by crop irrigation.

135

Table 4.5 Continued

Industry	Water requirement	Wastewater characteristics			Nature of pollutants and characteristic effects
		Quantity	BOD	Other items	
Whisky		20 liters per liter of whisky			
Soft drinks		2-5 liters per liter of product	600-2000 mg/liter		
Pulp and paper					
Pulp manufacture		40-200 m³/ton			Large in volume; problems often caused by solids, fibers, and color. May need nutrient supplementation in treatment.
Pulp bleaching		80-200 m³/ton			
Paper making		40-120 m³/ton			
Integrated pulp and paper mill		190-230 m³/ton	60-165 kg/ton		
Textile					
Cotton spin, weave, and processing	120-750 liters per kg goods	93% of intake	150 kg/1000 kg of goods	7-15 kg N per 1000 kg of goods	Color problem from dyes; high pH's due to use of NaOH in processing; fluctuating character.
Wool		500-600 liters per kg goods	300 kg/1000 kg of goods		Dyes, grease, high BOD.
Rayon		25-58 liter per kg of goods	30 kg/1000 kg of goods		High pH, sulfides, and Zn.
Nylon		100-150 liters per kg goods			
Polyester		67-133 liters per kg goods	200 kg/1000 kg of goods		
Chemical					
Petroleum refining	200-400 liters per 117-liter barrel		45 g/barrel	Phenol: 4 g/barrel	Problems caused by oil emulsions, phenol, and sulfides.

136

Source	Water volume	Waste characteristics	Effects
Soap manufacture	200 m³/ton		Taste and odor, toxicity, thermal pollution.
Detergents	13 m³/ton		Grease, solids, and pH. May contain phosphates and nondegradable organics.
Fertilizers	Per ton of NH_3: Carbon slurry, 2500 liters; Scrubber, 100–800 liters; NH_3 plant, 2000–7000 liters; Urea plant, 3000–5000 liters	10–25 kg NH_3 per ton NH_3 manufactured 6–22.5 kg urea/ton NH_3	High nitrogen content toxic to fish; also promotes eutrophication. Phosphatic fertilizers give P in wastes.
Metal plating	1–25 liters per liter of plating solution	1–15 mg CN per liter 3–100 mg Cr per liter 0–25 mg Ni per liter	May have toxicity problem and sludge, CN^-, Cd, Cr, Zn, etc. Affect stream life. Some substances persist in food chain.
Miscellaneous			
Tanneries	2.0–8.0 m³/100 kg of hides	9 kg/100 kg of hides Susp. solids 22–30 kg per 100 kg Total dissolved solids: 35–40 kg/100 kg of hides	Hair, solids, sludge, high pH, high BOD, and odor.

Note: Values included in this table have been taken from various sources, and indicate order of magnitude values. Actual values may vary substantially from plant to plant.

137

can be read off. For most process design and sludge handling purposes, 50%
(i.e., the average value) is used, whereas for hydraulic design purposes,
99% or a higher value may be used. To obtain meaningful values, the data
must cover a long enough period of survey.

Flow measurement, as well as sampling, often has to be done in industries in such a manner that individual streams from different departments
or even different machines can be evaluated in order to assess possibilities
for reuse of water, waste reduction, or segregation of wastes to facilitate
treatment and/or disposal. Flow measurement is often done using a V-notch
weir installed temporarily for this purpose in an existing channel or sewer.
For a V with a $90°$ notch, the following approximation may be used [4]:

$$Q \text{ m}^3/\text{sec} \cong 1.4(\text{head in meters})^{2.5}$$

Various other methods of flow measurement are also available.

The duration of sampling for industrial purposes is very important if
a composite and representative sample is to be obtained. Generally, samples
can be taken every 15 min or 1/2 hr, depending on variability, and compos-
ited over a total period long enough to cover one shift of operation in the
industry, or one clear cycle of operation of a unit. Temperature and pH
are measured before compositing. It is preferable that the sampling point
be kept at the middepth of a sewer or channel, and all sources contributing
to it should be known on a sewer diagram of the industry. Data so collected
can be used to prepare a materials balance and flow diagram.

In comparing industrial wastes and in referring to their pollutional
load on water resources, the term "equivalent population" is often used.
A BOD_5 value of 54 g per day is generally taken to be equivalent to the
sewage from one person. Thus, 54 g BOD_5/capita-day is referred to as the
"population equivalent" and the actual BOD_5 of the industrial waste is
divided by this figure to obtain the equivalent population. Other parameters besides BOD_5 may also be used to estimate the equivalent population,
but must be clearly stated.

It would not be out of place to mention here that the presence of
industrial wastes along with domestic sewage in municipal sewers is not
harmful per se. In fact, domestic sewage often affords welcome dilution
to the industrial wastes, provides nutrients for biological growth, and
seeds the wastewater with microorganisms. Thus, industrial wastes which
may sometimes be more difficult to treat by themselves may, in fact, become more readily treatable when mixed with domestic sewage. The latter

also has a good amount of alkalinity present in it and, thus, helps to
buffer the wastewater against pH changes resulting from industrial wastes.
For example, domestic sewage with an alkalinity of 150 mg/liter as $CaCO_3$
is capable of accepting 13.5 kg of H_2SO_4 per 1000 m^3 sewage flow without a
significant change in pH. In many cities in industrialized countries, more
than 30% of their municipal wastewater may consist of industrial wastes
without any undue effect on their treatability.

The presence of high amounts of toxic substances may render the mixed
wastes less treatable and cause other problems as well. The desirability
of admitting industrial wastes into public sewers with or without pretreatment must, therefore, be considered carefully. Some industrial wastes may
also damage sewers and make their maintenance more difficult. These difficulties can be overcome by the enforcement of proper standards for the discharge of industrial wastes to public sewers.

Example 5.1 Estimate the flow, BOD, and total nitrogen received at
a waste treatment plant serving the following:

1. 180,000 persons with water supply at 120 liters/capita-day
2. Meat slaughter and packing, 5000 kg live weight/day
3. Brewery, 80,000 liters of beer/day
4. Cotton integrated mill, 50,000 kg/day

The estimated values are shown tabulated below along with the assumed
basis of calculation in each case.

| Item | Assumed basis | | | Estimated values | | |
	Flow	BOD_5	Total N	Flow (m^3/day)	BOD_5 (kg/day)	Total N (kg/day)
180,000 pop-ulation	100 liters per cap-ita-day	54 g/cap-ita-day	8 g/cap-ita-day	18,000	9,700	1,440
Meat	29 m^3 per 1,000 kg	50 kg per 1,000 kg (slaugh-ter and pack)	6 kg per 1,000 kg	145	250	30
Beer	10 liters per li-ter beer	8.0 g/li-ter beer	0.3 g per li-ter beer	800	587	24
Cotton inte-grated mill	500 liters per kg	150 kg per 1,000 kg	10 kg per 1,000 kg	25,000	7,500	500
			Total:	43,945	18,037	1,994

Hence, $BOD_5 = \dfrac{18,037 \text{ kg/day} \times 10^6}{43,945 \times 10^3} = 410 \text{ mg/liter}$

and, total $N = \dfrac{1,994 \times 10^6}{43,945 \times 10^3} = 45.4 \text{ mg/liter}$

Example 5.2 A wastewater flow of 10,000 m^3/day is received at a
sewage treatment plant. The population served is 40,000 and there are a
few industries in addition. If the BOD_5 of the wastewater is 400 mg/liter,
estimate the BOD due to (1) domestic sewage and (2) industries.

$$\text{Total } BOD_5 = \dfrac{400 \text{ mg/liter} \times 10,000 \text{ } m^3/\text{day} \times 10^3}{10^6} = 4,000 \text{ kg/day}$$

$$\begin{array}{l}\text{Estimated } BOD_5 \text{ due} \\ \text{to domestic usage}\end{array} = \dfrac{40,000 \times 54 \text{ g/capita-day}}{10^3} = \begin{array}{l}2,160 \text{ kg/day} \\ \text{(i.e., 55\% of total)}\end{array}$$

$$\begin{array}{l}BOD_5 \text{ due to} \\ \text{industries}\end{array} = 4,000 - 2,160 = 1,840 \text{ kg/day (i.e., 45\% of total)}$$

5.3 CHARACTERISTICS OF RAINFALL AND RUNOFF

5.3.1 Direct Rainfall

Even direct rainfall can contribute some pollutants to an area, their
concentration depends on the nature of the atmosphere from which they are
washed out. Nitrogen, for example, contained in direct rainfall over urban
areas may exceed 1 mg/liter as total nitrogen, whereas over rural areas the
concentrations may be much lower. This is due to a greater nitrogen content in urban atmospheres from the use of fossil fuels. Nitrogen fixation
through lightning is a relatively minor source. Areas with alkaline soils
(not capable of absorbing NH_3 from the atmosphere) have also been associated with relatively high levels of nitrogen in rain water. Phosphorus
concentrations in rainfall range about one-tenth of the nitrogen value, or
less. Suspended solids and COD also depend on the nature of the atmosphere.
Table 5.5 gives some values for a few items in direct rainfall. Not much
data are available concerning atmospheric precipitation of heavy metals.
As an example, one may cite the mass balance prepared for heavy metals in
two Canadian lakes, for which the Cr, Hg, and Zn contents in atmospheric
precipitation were estimated at 7.0, 0.01, and 41.5 mg/m^3, respectively
[5].

Table 5.5 Typical Values of Some Nutrients in Rainfall and Runoff from Urban, Forest, and Agricultural Lands

Source	Units	Total nitrogen as N	Total phosphorus as P	Carbon		Suspended solids	References
				COD	BOD		
Direct rainfall	mg/liter	0.5-1.5	0.004-0.03	10-20		10-20	6-11
Urban storm water runoff (separate sewers)							
a. Cincinnati[a]	kg/km²/year	875	105	31,150	4,725	64,050	6
b. Europe	kg/km²/year	952	90				7
Forest runoff							
a. Several U.S. studies	kg/km²/year	143-357	2.6-12.8				7, 8
b. Swiss pre-Alps	kg/km²/year	840	4				7
c. Diverse forests, Finland	kg/km²/year		17-27				8
Pastures in Swiss pre-Alps							
Natural	kg/km²/year	1,650	74				7
Fertilized	kg/km²/year	1,940	102				7
Agricultural runoff			7-105[b]				6
a. 3 U.S. areas (Sawyer)	kg/km²/year	784	45				6
b. Swiss plateau (mixed agricultural use, fertilized)	kg/km²/year	1,400-3,000	21-50				7
c. 7.5-acre U.S. field (Weibel)	kg/km²/day of rainfall	20	1.7				6
d. Irrigation return flows							
(1) Surface	kg/km²/year	274-2,690	103-434				7, 9
(2) Subsurface	kg/km²/year	4,250-18,600	280-906				9, 10

[a]Average numbers of bacteria per 100 ml were: total coliforms 58,000; fecal coliforms 10,900; fecal Streptococci 20,500.

[b]0.05-1.0 mg/liter as total P, of which 15-50% may be as soluble orthophosphate [11].

5.3.2 Storm Water Runoff from Urban Areas

The quality of storm runoff is the result of water contact with land, the amount of physical erosion of the soil, the extent of chemical solutionizing, and the transporting power of the flowing water. The nature of the land surface is, therefore, important and runoff quality varies significantly depending on land usage (urban, agricultural, forest, etc.).

Urban storm water runoff is also dependent on the nature and efficiency of the sewage system. If sewage is treated, the total quantity of pollutants reaching a watercourse may be less for an urban area served by separate sewers than one served by combined sewers. Interceptor sewers in a combined sewage system generally carry only from 3 to 6 times the normal dry weather flow, the balance is allowed to overflow directly to watercourses without treatment. In the U.S. the annual loss of sewage through such overflow is estimated at about 3% [6].

Where separate urban storm sewers are installed, an indication of the quality of runoff can be obtained from a field study made sometime ago in a 27 acre, separately sewered residential, light commercial section of Cincinnati [6]. The values obtained in this study are shown in Table 5.5, along with similar values given by Wuhrmann [7]. It must be emphasized that the areas studied represented relatively well-planned, neat urban land use. In the Cincinnati study, the storm water runoff coefficient was only 0.37 owing to several residential lawns and gardens included in the area. The values shown in Table 5.5 would, therefore, not be directly applicable to other areas, particularly the crowded urban areas of some of the developing countries where all usual city services (such as the collection of refuse and animal droppings, etc.) are not at full strength.

The quality of storm water runoff from combined sewers can be estimated by adding to the above values for separate sewers those resulting from sewage overflow. The fraction of raw sewage lost in overflow has to be estimated depending on the frequency and duration of storms generating overflows in a city.

5.3.3 Storm Water Runoff from Agricultural and Forest Areas

As stated earlier, runoff quality depends on land usage and the type of soil. Precipitation supplies the basic inputs to an area and these are subsequently modified by the selective retention of some substances in the soil and the preferential release of others in accordance with the principles of hydrogeochemistry.

A few typical values of total nitrogen and total phosphorus observed in the case of forest runoffs are given in Table 5.5. Most of these values are obtained by measuring the flow and concentration regularly in streams draining known areas. Phosphorus values are generally low, reflecting the usual phosphorus retention properties of soils. However, areas under the influence of high rates of soil erosion show correspondingly high values of phosphorus in their runoff.

Pastures and grasslands can give substantial contribution of nitrogen and phosphorus in their runoff depending upon the extent of the fertilizers used. Similar is the case with agricultural values for nitrogen and phosphorus. Agricultural and forest runoffs are of a diffuse nature and, thus, their nutrient contributions are generally expressed in terms of weight per unit area and time (e.g., kg/km^2-year). The extent of nutrient runoff is also a function of the frequency and duration of rainfall. Therefore, Weibel [6] has attempted to give nitrogen and phosphorus values in terms of kg/km^2 per day of rainfall. Evidently, there are also other factors, but the values given in Tables 5.1, 5.4, and 5.5 should suffice for preparing inventories of BOD, nitrogen and phosphorus inputs to lakes, rivers, estuaries, and other water bodies.

REFERENCES

1. S. Arceivala, Simple Waste Treatment Methods, Middle East Tech. Univ., Ankara, Turkey, 1973.

2. Metcalf and Eddy, Inc., Waste Water Engineering, McGraw Hill, 1972.

3. A. Feinmesser, Survey of sewage utilisation for agricultural purposes, Advances in Water Pollution, Int. Ass. Water Poll. Res., 1970 Conf., Vol. I.

4. M. Suess, Water and Wastes Engineering, pp. E-10, 1970.

5. J. Patterson and P. Kodukula, Heavy metals in the Great Lakes, WHO Water Qual. Bull., 3 (4), 1978.

6. S. Weibel, Urban drainage as a factor in eutrophication, Eutrophication: Causas, Consequences, Correctives, Nat. Acad. Sci., Washington, D.C., 1969.

7. K. Wuhrmann, Eutrophication, Seminar at the Univ. Texas, Austin, Texas, 1966.

8. C. Cooper, Nutrient output from managed forests, Eutrophication: Causes, Consequences, Correctives, Nat. Acad. Sci., Washington, D.C., 1969.

9. J. Biggar and R. Corey, Agricultural drainage and eutrophication, Eutrophication: Causes, Consequences, Correctives, Nat. Acad. Sci., Washington, D.C., 1969.

10. R. Engelbrecht and T. Morgan, Land drainage as a source of phosphorus in Illinois surface waters, in Algae & Metropolitan Wastes, USPHS, SEC TR W61-3.

11. R. Mitchell, Ed., Water Pollution Microbiology, Wiley-Interscience, 1972.

Chapter 6

POLLUTION CONTROL OF LAKES

6.1 QUALITY OF WATER IN LAKES

Lakes may be natural or artificial and the quality of the water in a
lake at any given time is the sum total of the effects of various physical,
chemical, and biological conditions to which it has been subjected up to
that time. As seen earlier, ecological succession is a basic characteristic
of all aquatic and terrestrial communities. Man's activities may speed up
or reverse succession and alter the water quality, even making it unfit for
supporting its various legitimate functions. The purpose of this chapter
is to see how the quality of water in large waterbodies such as lakes can
be managed more beneficially by understanding some of the factors which
affect it.

Pollution may be natural or artificial. The natural drainage from
adjoining lands (catchment area) flowing into a river or lake may bring
with it various substances in suspended and dissolved form. The transport
of soil, dead vegetation, and other debris from purely natural sources may
wash into a waterbody many tons of organic and inorganic substances each
year. To this may be added the substances leached from the soil with which
the body of water is in contact. Man's industrial and other activities in
the vicinity of a waterbody may introduce yet more organics and inorganics
which would be included under artificial sources of pollution. The relative
proportion of pollutants from natural and artificial sources would vary from
one situation to another and, even for a given case, the proportion would
vary with the passage of time as habitation increases and development takes
place within its catchment area. All organic material, whether natural or
man-made, which has its source outside the lake and finds its way into it
is referred to as <u>allochthonous</u> material.

Part of the allochthonous material entering a lake is carried off in
the overflow from the lake and part in the seepage or percolation underground. The amount of organic matter remaining in a lake is increased by
new organic matter produced in the lake, the latter being referred to as

autochthonous material. All substances contained in a lake are subject to
physical, chemical, and biological processes such as adsorption, chemical
precipitation, oxidation and reduction, biochemical degradation, etc.,
besides being subject to the important effects of exposure to sunlight,
wind, temperature, and evaporation.

Two terms are often used to describe the zones in a lake from the
point of view of productivity. The euphotic zone, where photosynthetic
activities take place and primary "food" is produced, is referred to as
the trophogenic zone, while the bottom zone, where the organic matter is
degraded, is called the tropholytic zone.

6.2 LIMNOLOGICAL ASPECTS

The following discussion gives a brief review of certain limnological
aspects of special interest to engineers for understanding the factors
which affect the productivity of lakes, leading to their possible eutrophication, and the methods available for control and management.

6.2.1 Thermal Stratification

Temperature variation in a lake may be expected over the seasons as
indicated in Figure 6.1 which gives the typical temperature cycle and the
relative disposition of the three zones: hypolimnion, epilimnion, and the
thermocline.

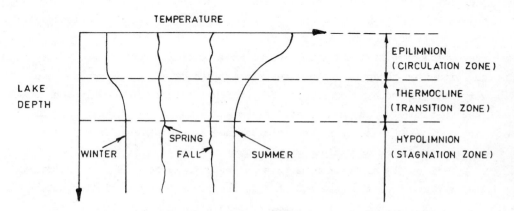

Figure 6.1 Lake stratification zones and seasonal variations in
water temperature.

Generally, the lower zone (hypolimnion) of any deep lake or waterbody undergoes smaller variations in temperature between summer and winter as compared to the upper zone (epilimnion) where there is greater heat transfer to and from the atmosphere, and wind mixing. The latter may extend over a depth of about 5 to 15 m and constitutes the zone of circulation. Circulation accounts for the more or less uniform temperature (and uniform density) in the epilimnion at any given time. Temperature can, however, vary substantially from season to season in this zone as seen in Figure 6.1.

The transition zone between the epilimnion and the hypolimnion is called the thermocline. It is the zone in which the change of temperature per unit depth is relatively rapid, about 1° C or more per meter depth. Thus, there are three thermal zones in a deep lake or waterbody. The epilimnion and hypolimnion have lesser rates of temperature change with depth, while in the thermocline the change of temperature with depth is rapid.

In the winter the lake surface water cools, its density increases, and it sinks lower. If this continues for a sufficient period of time, the lower water reaches a temperature of 4° C, the temperature of maximum density, while the surface water may continue to become even colder and consequently "float" on the top because it is less dense. The temperature profile for winter conditions is also shown in Figure 6.1. During the spring the surface waters tend to get warmer than the lower waters and again "float" on the top. During the summer, this condition would be observed even more as shown in the figure. At all times, a slight diurnal variation in temperature would also be noticed in the epilimnion.

Owing to density differences between the upper and lower layers of water in a lake, pronounced stratification can take place in the winter and summer. This stratification precludes intermixing between the layers, resulting in stagnation. In fact, also within the hypolimnion there is not likely to be any substantial vertical circulation of water because of more or less uniform density within the zone. For this reason, the hypolimnion is often referred to as the stagnation zone. The epilimnion normally has wind-induced circulation except during periods of ice cover when vertical and horizontal movements are suppressed.

In the spring the temperature profile changes from the winter condition to the summer condition, the temperature profile eventually becomes more or less vertical all the way from the top to the bottom, and there is practically no stratification or stagnation. This condition also occurs in autumn (see Figure 6.1). This is the time when lake overturning is possible provided sufficient wind energy is available as will be discussed later.

During the summer, the relatively rapid drop in termperature within the thermocline produces a stable stratified condition which is not easily disturbed by the wind. Only some "distortion" in the stratification pattern is caused by high winds.

6.2.2 Lake Overturning by Wind

Water driven by strong winds to the windward shore can build up a head which generates a return current. As shown in Figure 6.2, a shearing plane can be visualized which divides the wind-induced surface currents from the return currents. When the temperature decreases with depth as in winter and summer, this shear plane will lie near the surface because the deeper (and, therefore, colder and denser) waters offer much resistance to displacement by the warmer and lighter surface waters. Thus, the column of water that can be overturned is small. During spring and autumn when the temperature gradient is practically absent and no stratification exists, the column of water that can be overturned with the same energy is larger.

On the assumption that the temperature gradient is more or less uniform, the density is a function of depth and the work to be done by the wind in mixing the water is proportional to the difference in specific weights of the upper and lower strata of the water column [1]. The difference is greater in the spring than in the autumn with the result that lake waters are more difficult to put into circulation by spring winds than by autumn winds. Spring overturn is relatively more difficult and extends

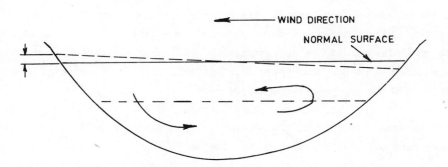

Figure 6.2 A conceptual diagram of a lake overturning.

over a shorter period of time compared to the autumn overturn which takes
place more readily and continues longer under favorable wind conditions.
In tropical climates, stratification patterns may persist throughout the
year, thus discouraging lake overturn.

In terms of circulation patterns, lakes are sometimes referred to as
holomictic or meromictic. Holomictic lakes are those which display total
circulation extending to the entire depth of the lake, so that all the
water mechanisms are in balance from top to bottom. Meromictic lakes, on
the other hand, are those which have no total circulation and the bottom
waters may remain isolated from the upper ones. Thomas [7] distinguishes
between meromictic lakes which have no total circulation and others which
have circulation at some time of the year. He refers to the latter as
"facultative meromictic," having maximum or main circulation generally in
the spring.

Again referring to Figure 6.2, it can be seen that above the shear
plane there is a wind-driven current while just below it is a return current running counter to the wind. This is a condition of shear flow in a
stratified fluid with the shear plane at the thermocline, so to speak.

If the shear is expressed as a velocity gradient (dv/dz) and the stratification is expressed as a density gradient (dp/dz), both in terms of the
vertical axis z, then the balance between the stabilizing force of the density gradient and the disturbing force of the shear flow can be expressed
in terms of the nondimensional Richardson Number R_i as follows:

$$R_i = \frac{g(d\rho/dz)}{\rho(dv/dz)^2} \qquad\qquad (6.1)$$

where

g = gravity constant

ρ = density

As discussed by Mortimer [2], if R_i is less than 1/4, the supply of
energy from the shear flow is greater than the loss of energy experienced
by turbulent motions in doing work against gravity on the density gradient.
Stirring and mixing is then likely to take place. As long as the Richardson Number is greater than 1/4 there will be very little mixing across the
thermocline at that wind speed.

If a lake basin is long enough and wind speed great enough, a distortion of the thermocline can occur with some aeration of the hypolimnion
at the upwind end and some nutrient enrichment of the epilimnion through

the wind drift. When the wind disturbance has subsided, the average shape
of the density profile may be restored to its shape prior to disturbance,
but pushed a little deeper [2].

6.2.3 Light Penetration

Light penetration is important for algal growth in a lake. The depth
to which light penetrates and its intensity at any depth depend on the
transparency of the water and can be measured by use of an illuminometer
or a simple Secchi disc (see Sec. 2.3.2). Light penetration is reduced by
the turbidity in the lake caused by surface runoff or by resuspension of
settled material under turbulent conditions. Algal growth depends on light
penetration, but the growth itself reduces light penetration and thus is
self-limiting. The use of long-term data on Secchi disc depths in assessing eutrophication trends is discussed in Sec. 6.9.1.

In studying 55 Florida lakes, some clear and some colored, Shannon et
al. [17] found the following correlation between Secchi disc depths and the
color and turbidity values of the lake waters:

$$\frac{1}{SD} = 0.003(COL) + 0.152(TUR) \tag{6.2}$$

where

    SD = Secchi disc depth, m
    COL = color, mg/liter
    TUR = turbidity units

The above relationship was found to be significant at the 99% confidence level. The depth at which the net productivity of phytoplankton is
zero is usually taken as the level of 1% light penetration. In some approximations, 1.5 times the Secchi disc level is assumed to give the light compensation level. From Eq. (6.2) it is evident that turbidity affects light
penetration much more than does color.

6.3 NATURE AND QUANTITY OF INFLOWING SUBSTANCES

The nature and quantity of various substances draining into a lake
depend on the climate, the ecology of the area, the natural vegetation,
and the agricultural, urban, and industrial activities within the catchment

area. Hence it is necessary to know the use to which land is put at present
and likely to be put to in the future. It is an axiom that water quality
management goes hand in hand with land-use planning.

The substances draining into lakes can be many and varied and, generally, it is not possible to identify all of them; some are degradable,
whereas some may be conservative, persistent, or toxic. Typical values of
some of the constituents of domestic sewage, agricultural and natural runoff, and certain industrial wastes have been given in Chapter 5. Appropriate values could be selected to prepare an approximate inventory of the
total quantity of each major constituent flowing annually into the lake
from each source. Agricultural and other similar sources are of a diffuse
nature, whereas others such as large industries are point sources. Such
inventories give order of magnitude values and, for the major sources and
substances, data could be refined by actual field measurements, if desired.
Example 6.5 shows how an inventory of the quantities of wastes from different sources can be prepared.

Special mention sould be made of the increasingly frequent reports
concerning heavy metal pollution in lakes. By far the largest contribution
come from air-borne pollutants settling on the lake and its catchment area.
A relatively smaller, almost negligible contribution is received from direct
industrial and municipal discharges [30].

6.4 NET ACCUMULATION OF SUBSTANCES IN A LAKE

A certain amount of physical accumulation of substances occurs in a
lake owing to differences between total inflows and outflows of substances
to and from a lake. Inflows may be due to direct rainfall, to surface runoff, or in some cases to the ground water charge entering the lake.

Similarly, the outflows from the lake will be due to evaporation,
seepage, and storm overflows. Substances are not lost as the water evaporates since dissolved substances are left behind. Seepage, however, carries
out all the dissolved substances, leaving behind only the suspended materials.
Storm overflow, on the other hand, flushes out material in suspension and
solution, generally affecting more the substances contained in the epilimnion rather than in the hypolimnion. All of these factors have an effect
on the net accumulations within a waterbody and the net accumulation from
time to time is either positive or negative. The relative proportion of

overflow, seepage, and evaporation is different for each lake, and also
varies with the seasons.

Furthermore, the accumulated material may be conservative or degradable.
If it is degradable, a reduction in the accumulated quantity occurs depending on the mean residence time of the material in the lake and its degradation or decay rate constant K per unit time. A material may also be "withdrawn" from a system when it undergoes biochemical transformations (e.g.,
carbon to CO_2) or physical trapping in some part of the lake (e.g., bottom
sediments).

Generally, lakes provide long retention time to incoming materials,
and for many organics their degradation rate constants are such as to pre-
clude any extensive accumulation in the water under favorable conditions.
Undoubtedly, radioactive wastes have long half-lives, but they are normally
not discharged to lakes. Interest often lies in estimating the buildup of
nutrients and certain other substances. Nitrogen buildup is difficult to
estimate owing to possible loss of nitrogen through denitrification and
gain through nitrogen fixation. Phosphorus buildup and distribution within
a lake is also affected by several factors, for example, sediments often
act as traps for precipitated phosphorus.

In order to estimate the buildup of a conservative substance, a materials balance needs to be prepared. This can be done more readily on
the assumption of complete-mixing conditions within a lake, although this
is not strictly true even at the time of lake overturning. When a steady-
state condition is reached,

$$\begin{bmatrix} \text{weight of material incom-} \\ \text{ing from all sources} \end{bmatrix} = \begin{bmatrix} \text{weight of material out-} \\ \text{going by all routes} \end{bmatrix}$$

The various flows to and from a lake, and the corresponding concentrations, can be shown as in Figure 6.3a where

Q_i = quantity inflowing (precipitation, catchment drainage, etc.)

Q_e = quantity lost by evaporation

Q_s = quantity lost by seepage

Q_o = quantity removed from lake by overflows (spillway losses
         and supplies through headworks)

C_i = influent concentration

C_L = concentration in the lake and, therefore, also in the
         overflows and seepage as complete mixing is assumed

V = lake volume

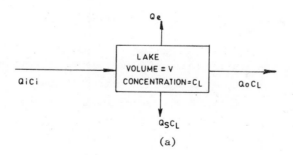

(a)

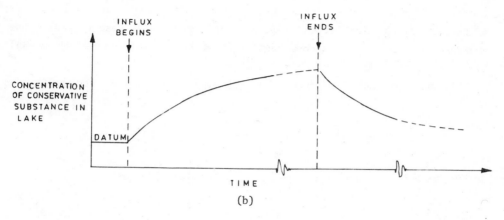

(b)

Figure 6.3 (a) Mass balance around a lake. (b) The concentration of conservative substances during and after influx in a lake.

Thus,

$$Q_i = Q_o + Q_s + Q_e \qquad\qquad (6.3)$$

and, at steady-state conditions,

$$Q_i C_i = Q_o C_L + Q_s C_L + Q_e \times 0 \qquad\qquad (6.4)$$

Hence, we get the <u>concentration ratio</u> C_L/C_i as follows:

$$\frac{C_L}{C_i} = \frac{Q_o + Q_s + Q_e}{Q_o + Q_s}$$

or,

$$\frac{C_L}{C_i} = 1 + \frac{Q_e}{Q_o + Q_s} \tag{6.5}$$

Thus, the concentration ratio of conservative substances depends upon the ratio of the quantity evaporated and the total quantity leaving the lake through overflow and seepage. Lakes having high evaporation rates, as in warm, arid regions, and little overflow would tend to show a relatively high net accumulation of incoming materials.

A lake may also exhibit annual variations in the concentration of a conservative substance due to rainfall fluctuations. In a particularly wet year ($Q_o \gg Q_e$), lake concentrations may drop owing to a flushing out of the material. In a particularly dry year or a succession of a few dry years, $Q_e \gg Q_o$, and lake concentrations would show a buildup. The vagaries of climate would be reflected in lake water quality from year to year. The average concentration of a conservative substance would also be determined by the quality of the incoming waters.

If seepage losses are negligible ($Q_s = 0$), Eq. (6.5) would simplify to

$$\frac{C_L}{C_i} = 1 + \frac{Q_e}{Q_o} \tag{6.6}$$

If evaporation is also neglected ($Q_e = 0$), Eq. (6.5) gives

$$C_L = C_i \tag{6.7}$$

Example 6.1 What will be the steady-state concentration of a conservative substance entering a lake at 100 mg/liter in a warm, semiarid area where the quantity of water lost through evaporation is equal to the quantity withdrawn from the lake annually? Neglect seepage.

$$\frac{C_L}{C_i} = 1 + \frac{Q_e}{Q_o} = 1 + 1 = 2.0$$

Hence,

$$C_L = 2.0 \times 100 = 200 \text{ mg/liter}$$

Another method of estimation can be used where the decay constant K or the half-life t_{50} of a degradable substance is known.* As before, it

_____

*Half-life, $t_{50} = 0.693/K$(base e). For example, for an organic waste, if K = 0.23 per day, $t_{50} = 0.693/0.23 = 3.0$ days.

is necessary to assume complete-mixing conditions in a lake, but it is now necessary to make one more assumption, namely, that degradation follows first-order kinetics. Thus, preparing a mass balance for the substance in a lake, we get

$$V \frac{dC_L}{dt} = Q_i C_i - Q_o C_L - VC_L K \tag{6.8}$$

where K = first-order decay rate constant per unit time. Assuming $Q_i = Q_o$, integration gives

$$C_L = \frac{QC_i}{Q + VK} \left\{ 1 - e^{-t[(1/t_o) + K]} \right\} + C_o e^{-t[(1/t_o) + K]} \tag{6.9}$$

where

t_o = mean hydraulic residence time

C_o = concentration in lake at t = 0

At steady state (t → ∞), the lake concentration C_{ss} is obtained as

$$C_{ss} = \frac{QC_i}{Q + VK} = \frac{C_i}{1 + Kt_o} \tag{6.10}$$

Equation (6.10) is the same as the well-known equation for complete-mixing conditions and first-order kinetics. The lake is like a complete-mixing reactor.

If, instead of steady-state conditions, our interest lies in knowing the time required to reach, for example, 95 or 99% of the steady-state value, we can obtain the following from Eq. (6.9):

$$t_{95} = \frac{3}{(1/t_o) + K} \tag{6.11a}$$

$$t_{99} = \frac{4.6}{(1/t_o) + K} \tag{6.11b}$$

Equations (6.11a) and (6.11b) are valid both when a new influx has begun and when an existing influx is stopped. Furthermore, if a substance is nondegradable (i.e., K = 0), we get its buildup as shown in Figure 6.3b. Equation (6.11a) then simplifies to give

$$t_{95} = 3t_o \tag{6.12}$$

and, similarly, we get

$$C_{ss} = C_i \tag{6.13}$$

However, equations involving the use of the decay constant K can be used only in cases where K is known under lake conditions. Decay may be physical as in the case of radioactive wastes, or biological as in the case of degradable organics. In either case "decay" in a lake may also be due to various other factors such as chemical precipitation.

6.4.1 Phosphorus Residence Time

As stated above, the term t_o denotes the hydraulic residence time in the lake. For conservative dissolved substances, the residence time would be the same as that of water. But, for substances which are nonconservative or behave nonconservatively owing to various biological, chemical, or physical processes leading to their removal or trapping in the system (e.g., phosphorus), the mean residence time can be estimated in a manner similar to the one used to determine the mean cell residence time or solids retention time in the biological treatment of wastes [20]. For example, the phosphorus residence time for a lake can be estimated as follows:

$$\text{Phosphorus residence time (years)} = \frac{\text{mean annual steady-state phosphorus content in the lake}}{\text{annual phosphorus load received by the lake}}$$

The phosphorus residence time* is difficult to determine and has been done for only a few lakes so far [21]. It is based on the assumption that the lake is at steady state at the time of sampling and the load received by the lake is constant over the sampling period. Some estimated values of both hydraulic and phosphorus residence times for a few lakes are given in Table 6.1.

It is interesting to note from Table 6.1 that the phosphorus residence times for the four cases shown vary within a relatively narrow range. Generally, one may expect t_p for lakes to be less than 1.0. For estuaries and fjords the value ratio may be greater than 1.0 owing to efficient nutrient trapping and recirculation [12].

Instead of reaching steady-state conditions which would theoretically take very long, it may suffice to know the time required to reach, for instance, 95% of the steady-state value. Thus, from Eq. (6.12) we get

---

*Phosphorus residence time is the hypothetical time necessary to bring the phosphorus concentration of a lake to its present value, starting from zero concentration.

Table 6.1 Estimated Mean Hydraulic and Phosphorus Residence Times
for Some Lakes

Lake	Hydraulic residence time t_o (years)	Phosphorus residence time t_p (years)	Reference
Washington	3.2	0.8	20
Minnetonka	25.0	0.9	20
Sebasticook	3.5	1.4	20
Mendota	4.5	1.0	21

Source: Refs. 20 and 21.

t/t_o = 3.0. Similarly, for flush-out conditions we get the same value of
t/t_o = 3.0. In other words, the time that must elapse, after either the
start or end of influx, to reach 95% of the corresponding steady-state value
in each case is three times the value of t_o. Of course, for nonconservative
substances t_o should be replaced by t_p.

As an illustration, in Lake Minnetonka with a hydraulic residence
time t_o = 25 years, a conservative substance will need 75 years (i.e.,
3 x 25) after the start of influx to reach 95% of the steady-state value,
whereas phosphorus with its lower residence time will take only 2.7 years
(3 x 0.9) to reach 95% of its steady-state value. The residence time is
lower because phosphorus is removed by other mechanisms besides being
flushed out with water displaced from the lake.

Thus, the relative magnitudes of the mean residence times for water
and an added substance determine the time required to reach steady-state
conditions. Lakes or reservoirs where the hydraulic residence time is
small respond quickly to increased or decreased inputs as they displace
the old contents with some rapidity. Thus, the nature of the material
(conservative or otherwise) does not matter much. Lakes with long detention times and low flush-out rates respond more slowly with regard to conservative substances, but with substances which are degradable or are affected by other mechanisms of removal the residence time of the substance
governs. For four hypothetical lakes, the time required in each case to
reach 95% of the steady-state value is shown in Table 6.2. The results are
equally applicable to cases where buildup occurs after the start of influx
or where flushing takes place after stoppage.

Table 6.2 Effect of Mean Residence Time on the Time Required
to Reach 95% Steady State

Lake	Mean residence time (years)		Time required to reach 95% steady state (years)	
	Water and conservative materials	Assumed nonconservative material	Conservative materials	Assumed nonconservative material
A	30.0	1.0	90.0	3.0
B	10.0	1.0	30.0	3.0
C	1.0	1.0	3.0	3.0
D	0.5	1.0	1.5	1.5

If the hydraulic residence time of a lake is relatively small, or of an order of magnitude similar to that for phosphorus or other nonconservative material, the dominant mechanism of removal is flushing from the lake and the lake approaches steady state at about the same time for both conservative and nonconservative materials. Thus, not much error is apparent in case a nonconservative material has been assumed as a conservative one. But the error would be significant in the case of large lakes with long hydraulic residence times.

Computed values using the above methods could be verified by field sampling to check whether the lake has improved as expected after the inflow of substances such as phosphorus has been reduced or stopped through treatment or diversion. Studies on Madison lakes by Sonzogni et al. have borne out the validity of such estimations [21].

Example 6.2 A degradable material with a half-life of 60 days enters a lake with a volume of 3000×10^6 m^3. The inflowing quantity is 900×10^6 m^3/year with a concentration of 25 mg/liter. Estimate the concentration in the lake at the end of 1/2 year and its effective residence time at steady state. Assume that the earlier concentration in the lake was $C_o = 0.5$ at t = 0.

$$K = \frac{0.693}{t_{1/2}} = \frac{0.693}{60} = 0.0115 \text{ per day} = 4.21 \text{ per year}$$

$$t_o = \frac{3000 \times 10^6}{900 \times 10^6} = 3.33 \text{ years}$$

1. **At t = 0.5 year**

    Substituting in Eq. (6.9) we get

    $$C_L = \frac{900(10^6)(25 \text{ mg/liter})}{900(10^6) + 3000(10^6)(4.22)} \left\{ 1 - e^{-0.5[(1/3.33)+4.22]} \right\}$$

    $$+ (0.5)e^{-0.5[(1/3.33)+4.22]}$$

    $$= 1.54 \text{ mg/liter}$$

2. **At steady state**

    $$C_L = \frac{QC_i}{Q + VK} = 1.66 \text{ mg/liter from above}$$

3. **Effective residence time at steady state**

    $$= \frac{\text{material in lake}}{\text{incoming material/year}}$$

    $$= \frac{(1.66 \text{ mg/liter})(3000 \times 10^6 \times 10^3)}{(25 \text{ mg/liter})(900 \times 10^6 \times 10^3)} = 0.22 \text{ years}$$

    Example 6.3 A substance is discharged at a concentration of 100 mg per liter in a stream entering a lake in which the hydraulic residence time is 105 days including other runoff from the catchment area. Neglecting evaporation and seepage, determine the likely concentration in the lake at steady state if (1) degradation rate constant K = 0.023 per day (half-life = 30 days) and (2) the substance is nondegradable (K = 0).

1. From Eq. (6.10)

    $$C_L = \frac{C_i}{1 + Kt}$$

    $$= \frac{100}{1 + (0.023)(105)}$$

    $$= 29.2 \text{ mg/liter}$$

2. From Eq. (6.13)

    $$C_L = C_i = 100 \text{ mg/liter}$$

6.5 BIOLOGICAL ACTIVITY IN A LAKE

A lake is an excellent example of an ecological subsystem which is shown diagrammatically in Figure 3.6. It has producers (e.g., algae

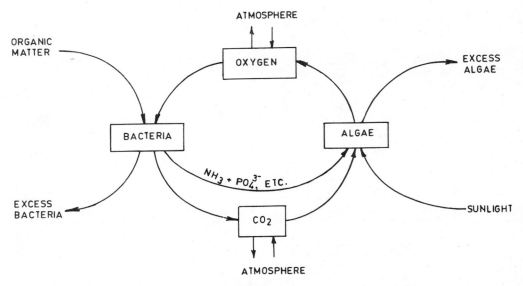

Figure 6.4 Algal and bacterial activity in a lake.

and littoral vegetation), a host of consumers of higher and higher trophic levels (ending, generally, with fish) that live on this primary production, and a vast population of decomposers. Substances affecting biological activity are brought in with lake inflow, leached from the soil, and transferred from the atmosphere. Important among these are the nutrients carbon, nitrogen, phosphorus, and others.

Figure 6.4 shows how bacteria and algae help each other along. The presence of organic matter and the availability of nutrients and sunlight are necessary to sustain primary production. The average phytoplankton production, measured as carbon production, may vary from less than 25 g of carbon per m^2 per year in oligotrophic lakes to over 700 in eutrophic ones (see Table 6.4). Daily production rates can be high under favorable conditions, even leading to what are called "algal blooms."

The entire population of consumers depends on phytoplankton productivity. A part of the phytoplankton is ingested by the herbivores (zooplankton and fish) while much of it sinks below the euphotic zone and dies. Dead phytoplankton, zooplankton, and other organic matter settle slowly as debris and are referred to as "seston." The supply of seston is reduced through decay and zooplankton grazing, and augmented through the death of zooplankton. The seston sinking rate is on the order of 5 to 15 cm/day, but varies considerably and may require a few days to reach the bottom. Meanwhile, biological degradation proceeds in the sinking seston.*

Primary productivity combined with temperature stratification determines the nature and quantity of food available at the bottom. During spring and autumn the phytoplankton (mainly green algae and diatoms) are rapidly distributed to the bottom by lake circulation. During summer, the blue-green algae which may predominate in the epilimnion may sink slowly owing to stratification; while slowly sinking, the algae will undergo degradation and reach the bottom invertebrates in that condition [3]. Thus, the value of phytoplankton as food for the bottom fauna varies with the seasons, especially in the case of deep lakes.

The action of decomposers breaks down organic matter into simple, inorganic end products such as CO_2, nitrates, phosphates, sulfates, etc., which tend to diffuse back to the epilimnion. The nutrients are thus recycled, and with a fresh input of solar energy they could reconstitute into new phytoplankton in the upper layer. The presence of toxic, humic, and other harmful substances (e.g., thermal pollution) may adversely affect organic growth in any form.

6.6 OXYGEN BALANCE IN LAKES AND RELATED EFFECTS

Another important factor affecting biological activity in a lake is the presence or absence of dissolved oxygen (DO) in its different parts. Oxygen is received by a lake from gas transfer at the water-air interface, from inflows of freshwater into the lake, and from algal photosynthesis. The distribution of oxygen in a lake is affected by stratification, circulation, diffusion, various types of biological activity, and other phenomena which result in the creation of different microenvironments within the lake besides showing seasonal and diurnal variations.

Gas transfer at the surface is a function of the deficit in oxygen concentration from the saturation value; the greater the deficit, the faster the rate of gas transfer as in a first-order equation. The change in oxygen concentration with time is related to the oxygen deficit as follows:

---

*The degradation rate was estimated in one case at k = 0.02 per day [24].

$$\frac{dC}{dt} = K_2(C_s - C) \tag{6.14}$$

$$= K_2 D \tag{6.15}$$

where

C_s = concentration of oxygen at saturation conditions at a
given temperature

C = actual concentration at time t

K_2 = re-aeration coefficient per unit time

D = deficit from saturation at any time t

Integration gives

$$D = D_a e^{-K_2 t} \tag{6.16}$$

where D_2 = initial deficit.

The re-aeration coefficient K_2 is a function of the mass transfer coefficient of the liquid film at interface K_L and the interfacial area available for mass transfer per unit volume of liquid. Thus,

$$K_2 = K_L \left(\frac{A}{V} \right) \tag{6.17}$$

where

A = area of the water surface through which gas transfer occurs
(effective area is increased by wave action)

V = volume of the body of water

It is evident that, other conditions being equal, the larger the surface area per unit volume, the faster the re-aeration rate. However,

$$V = Ah$$

where h = height of the waterbody. Hence, for a quiescent body of water,

$$K_2 = \frac{K_L}{h} \tag{6.18}$$

Values of K_L reported for some estuaries and shallow lakes range between 10^{-4} and 10^{-6} m/sec [11a].

The re-aeration rate for a body of water is also proportional to the turbulence, which in turn is proportional to the wind energy dissipated over the waterbody, resulting in generation of surface currents. McKee [11b] has expressed this in terms of G, the mean temporal velocity gradient as used in flocculation studies. G is proportional to the surface current velocity and water viscosity. For lakes and estuaries, G often lies between 2.0 and 5.0 per second and is related to K_2 empirically as follows:

$$K_2 = \frac{2.0 + 0.87G}{3.28h} \qquad (6.19)$$

where

  h = depth of water (meters) to which wind currents are effective

  K_2 = re-aeration coefficient, per day

As an example, if G = 3.6 per second and h = 6 m, we get $K_2 = 0.257$ per day.

Banks [11a] studied the effect of wind on the mass transfer coefficient K_L for oxygen, applying the results to Lake Chapala, Mexico. For "medium" wind velocities (perhaps 5 to 7 m/sec), the relationship between wind speed and K_L was found to be linear.

$$K_L = (1.80 \times 10^{-6})U \qquad (6.20)$$

in which the units of K_L and U are in m/sec. For larger values of U, the relationship was

$$K_L = (0.32 \times 10^{-6})U^2 \qquad (6.21)$$

Applying this to Lake Chapala, which is about 80 km long and 6 to 25 km wide and has an average depth of 8 m, Banks estimated the value of K_L as 6.8×10^{-6} m/sec, corresponding to an average wind speed U of 2.6 m/sec. Thus, the value of K_2 was estimated from Eq. (6.18) as

$$K_2 = \frac{K_L}{h} = \frac{6.8 \times 10^{-6}}{8} = 0.85 \times 10^{-6} \text{ per sec}$$

$$= 0.073 \text{ per day}$$

All above equations for K_2 show that shallow waterbodies tend to have higher re-aeration rate coefficients. Equations (6.20) and (6.21) are not strictly applicable to polluted bodies of water where the presence of oil, detergents, and other surface active agents may reduce the value of the mass transfer coefficient of a liquid.

Values of K_2 applicable to lakes and ponds have been traditionally taken to range from 0.12 to 0.36 per day (see Table 7.2) depending on depth, extent of surface area, wind, etc. The effect of temperature on K_2 is given by

$$K_{2(T°C)} = K_{2(20°C)}\theta^{T-20} \tag{6.22}$$

where θ, according to various investigators, ranges from 1.016 to 1.037.

Oxygen transfer from the atmosphere accounts for only a relatively small amount of the total oxygen consumed in a lake. Algal production of oxygen can, in fact, be several times more. In Example 6.4 it is seen that surface re-aeration rate is on the order of 1.8 mg/liter per day. Generally, the oxygen deficit in the epilimnion is not as much as has been assumed in the problem; thus, the re-aeration rate is even less. Often, during favorable times, the epilimnion is supersaturated with oxygen produced through algal activity, and oxygen is then given off to the atmosphere rather than the other way around.

Example 6.4 After a lake turnover, the DO in the epilimnion is only 2.0 mg/liter. Estimate the natural surface re-aeration rate per day at 10°C, assuming K_2 = 0.20 per day at 10°C (C_s = 11.33 mg/liter at 10°C).

If no biological or chemical demand of oxygen is exerted, estimate the time required to reach 90% of saturation concentration in the surface waters.

From Eq. (6.14)

$$\frac{dc}{dt} = K_2(C_s - C)$$
$$= 0.20(11.33 - 2.0)$$
$$= 1.866 \text{ mg/liter-day}$$

This is the maximum rate at the start; thereafter, as C increases, the rate decreases. Hence, using Eq. (6.16) we find the time required to reach 90% saturation

$$D = D_a e^{-K_2 t}$$

where the initial deficit D_a = 11.33 - 2.0 = 9.33 mg/liter. At 90% saturation, the deficit would be 10% of C_s, or D = 1.133 mg/liter. Hence,

$$1.133 = 9.33e^{-(0.20)t} \quad \text{and} \quad t = 10.5 \text{ days}$$

The photosynthetic production of oxygen was discussed in Sec. 3.2.1 where it was shown that, stoichiometrically, 1.25 to 1.75 g of O_2 are produced per gram of dry algal matter. Thus, knowing the algal production rate, one can estimate the oxygen production rate. Photosynthesis, however, exhibits diurnal as well as seasonal variation depending on light, temperature, etc. A corresponding variation in the oxygen production rate must be expected.

On the other hand, oxygen is consumed at all times by the respiration of the organisms living in the water and on the lake bottom. The biochemical degradation or decomposition of organic matter supports a large population of decomposers which add to the oxygen requirement for their respiration. The DO profile in a lake or other waterbody is the net result of oxygen availability and oxygen consumption at different depths (see Figure 6.7).

While the epilimnion is generally well oxygenated, the hypolimnion is more critically placed in this regard. It is shut off from the atmosphere and receives its oxygen supply by downward diffusion through the upper layer of water, by photosynthesis if light penetration permits, or through a mass exchange of water. Water circulation can be quite considerable during floods and lake overturns, but negligible during periods of stagnation in winter and summer. At all times, however, the hypolimnion is the recipient of all the settled organic matter undergoing degradation. The downward diffusion of DO through the thermocline is a function of the vertical diffusion coefficient and the vertical gradient of DO concentration. A method of estimating the downward diffusing flux of DO has been given in Chapter 8 with regard to marine conditions.

Bella [23] and others [24] have shown that the concentration of DO in a lake can be adequately described by a one-dimensional diffusion equation, and Bella has concluded, on the basis of his study, that oxygen concentrations in the hypolimnion were influenced by (1) diffusion rates, (2) respiration rates, (3) the photosynthetic production of oxygen, and (4) atmospheric re-aeration. Out of these, he found that (1) and (2) had a greater effect than (3) and (4) on the resulting concentrations. Newbold et al. [24] developed a mathematical model to study oxygen depletion in Lake Cayuga, New York, and concluded that the method had its limitations owing to the use of a one-dimensional model and the lack of data for modeling biological processes (see Figure 6.5). In Lake Cayuga they found that benthic demand predominated in the lowest parts of the lake, while in the upper parts respiration, decay, and benthic demand all played important parts and interacted with the diffusion (also see Sec. 8.10 for pollution modeling for DO and nutrient accumulation in stratified waterbodies).

Based on stoichiometric considerations, one can diagrammatically show the effect that 1 µg of available phosphorus in the epilimnion has on the

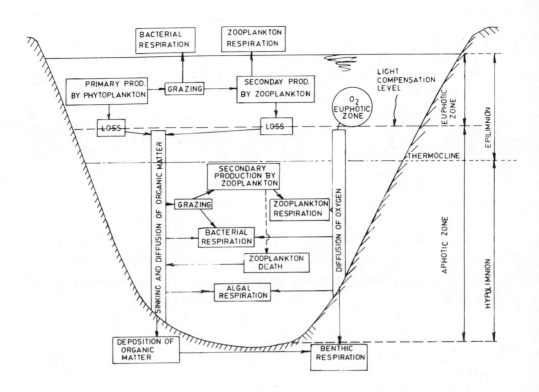

Figure 6.5 An oxygen model for a lake. Tentative coefficients are:
(1) grazing rate = 0.71 g/m³-day, (2) digestive efficiency = 85%, (3) seston
sinking rate = 7.5 m/day, (4) decay coefficient = 0.02/day, and (5) respiration rate/day = 0.228 e(0.61°C + 0.965) (from Ref. 24).

concurrent production of algae and oxygen, and on the oxygen required thereafter for aerobic degradation of the above algae when they sink into the
hypolimnion and die (see Figure 6.6).

In accordance with the empirical formula, algae contain 1.3% phosphorus
as P. Hence, 1.0 µg of P would be contained in 77 µg of algae. The photosynthetic growth of 77 µg of algae (or 40 µg carbon on the assumption that
carbon is 52% of dry algal weight) would give a concurrent production of, e.g.,
140 µg of O_2 (Sec. 3.2.1) which would be mainly confined to the epilimnion,
supersaturating it and even giving off some to the atmosphere. Some O_2
would slowly diffuse downward through the thermocline to the hypolimnion.
Algal decomposition by aerobic organisms would break down the algal matter
to give back 1 µg of P, besides NO_3, CO_2, water, and inert sludge (nondegradable part). Decomposition, as seen earlier, requires about 1.8 µg

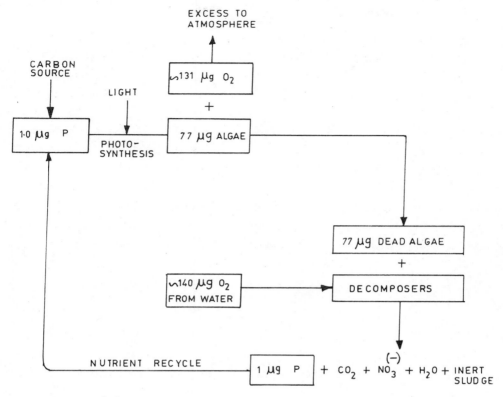

Figure 6.6 The oxygen production and utilization that accompanies nutrient recycling in a lake.

of O_2 for 1.0 µg of algae. Thus, the decomposition of 77 µg of algae would need about 140 µg O_2 from the water in the hypolimnion (i.e., about 3.5 g of O_2/g of carbon fixed).

The above denotes what happens in one cycle of nutrient utilization and release. The released nutrient is again available for utilization provided all other conditions are favorable. Nutrient recycling rates can vary from a few months to a few days or even less (see Sec. 6.7.3). In fact, even if the nutrient recycling rate were known, any estimation similar to the above would be quite complicated by the fact that algal production is not uniform but has its peaks in time and space. DO is also consumed by the respiration needs of other microorganisms besides decomposers in the hypolimnion, and DO consumption and replenishment <u>rates</u> are **often** more critical than the total quantities.

Temperature also affects the rate of biochemical decomposition of organic matter. Warm temperature combined with relatively shallow depth can give a hypolimnion with much putrefactive activity, leading to a considerable decrease in oxygen and even to a complete absence of oxygen, during stagnant periods. The amount of oxygen in the hypolimnion depends on the amount of decomposable material entering it, the lake depth, and the length of time that the bottom water is cut off from the atmospheric oxygen supply. The dissolved oxygen profile, therefore, has a marked seasonal character (Figure 6.7).

Under anaerobic or anoxic conditions, the organisms break down nitrates to obtain further oxygen (Sec. 1.5). In eutrophic lakes, some nitrogen is eliminated from the system in this manner (denitrification). If still more organic matter is present, the organisms now turn to sulfates and finally to CO_2 to obtain their oxygen. Breakdown of sulfates leads to the production of sulfides such as H_2S which make the environment odorous and disagreeable.

Lakes are normally not loaded to such an extent that sulfide is produced, but sewage ponds may reach this stage if overloaded, especially in warm climates. If a eutrophic lake produces sulfides and also has soluble iron (Fe^{2+}) in the lower part of the hypolimnion, black deposits of ferrous sulfides may be formed.

Dissolved oxygen availability and water temperature are important factors which determine the rate of biological activity. Undegraded organic matter may lie at or near the lake floor during the winter as if it were in a refrigerator until spring circulation moves it upward. Rapid biological activity at that time may lead to a depletion of oxygen with consequent effects on the lake ecosystem, unless the algal production of oxygen is enough in that layer.

Oxygen uptake by benthic material has been observed to vary from 0.1 g O_2/m^2-day to 1.0 g of O_2/m^2-day or even higher; the rate probably doubles for every 10°C rise in temperature [24]. Over the long term, total oxygen uptake can be no more than what can be accounted for by the quantity of oxygen-demanding materials settling into the benthic mud.

Even when the whole hypolimnion contains oxygen, it penetrates only a few millimeters below the sediments on the floor of the lake. Farther down in the sediments, reducing conditions predominate. The change from oxidizing to reducing conditions can be detected by redox potential measurements [12]. In fact, bottom sediments play an important role in the oxygen

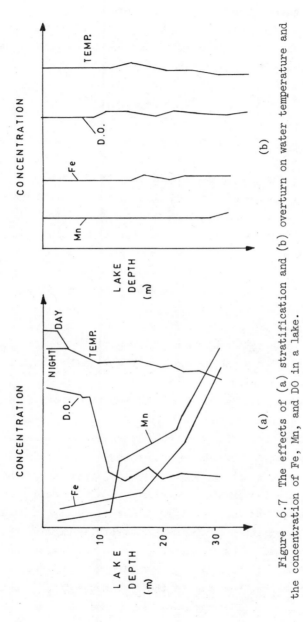

Figure 6.7 The effects of (a) stratification and (b) overturn on water temperature and the concentration of Fe, Mn, and DO in a lake.

budget of eutrophic lakes. When the oxygen available in the hypolimnion is used up, the "interface" between oxidizing and reducing conditions ascends to the thermocline level.

The oxidation-reduction conditions explain the variation in concentration of ferrous and ferric, and of manganous and manganic ions with depth. The upper waters having oxygen tend to oxidize the ferrous and manganous ions (Fe^{2+} and Mn^{2+}) brought up by lake circulation to give ferric and manganic ions (Fe^{3+} and Mn^{3+}) which are relatively insoluble at the usual pH values in lake waters. Therefore the ferric and manganic ions are precipitated down as insoluble oxidized complexes, and surface waters consequently do not exhibit appreciable concentrations of iron and manganese.

However, as the precipitated complexes travel down and reach reducing conditions, they are reduced to the more soluble ferrous and manganous states and remain in dissolved form in the bottom water. Hence, iron and manganese concentrations tend to be greater in the lower parts of a lake during stratification conditions as shown in Figure 6.7. As the redox potential continues to fall during summer stratification, the simultaneous decrease of ferrous iron and sulfate concentrations in the water suggests the formation of ferrous sulfide precipitates at the low potential.

During spring and autumn overturn, oxygen is carried to all parts of the lake and, owing to circulation, the concentration of all substances tends to become uniform with depth (see Figure 6.7). This is important to bear in mind when sampling in a lake.

An interesting use has been made of the fact that, with the onset of reducing conditions in a lake, the iron precipitates earlier and the manganese precipitates later as the redox potential falls farther. Consequently, if the ratio of Fe:Mn is determined in the bottom sediments, it will be high at the start of oxygen depleting (reducing) conditions and decrease when fully anaerobic conditions are established. Thereafter, as the season changes and the conditions reverse, the ratio will again tend to increase. Examination of sedimentary profiles from a lake floor may reveal such seasonal peaks in the Fe:Mn ratio both before and after the period of maximum oxidation [2]. These kinds of observations lend credence to the hypothesis that the loss of these elements from a lake is sometimes brought about more by alternating oxidizing and reducing conditions than by variations in the supply of these elements from the drainage system of the lake.

Phosphate concentration also shows a variation with redox conditions in a lake. As seen earlier, while oxygen is available at the sediment surface, iron and manganese are precipitated as insoluble oxidized complexes. These complexes apparently "scavenge" the water environment for phosphates and other materials depending on pH and other factors. The materials are mobilized in reducing soils and coprecipitated during stagnation. The concentration of phosphate rises in the bottom water along with reduced forms of iron and manganese while, at the fall overturn, reintroduction of oxygen in the bottom waters leads to the disappearance of reduced forms of iron and manganese and to a reduction in the phosphate concentration.

Phosphorus also forms precipitates with aluminum, calcium, and iron (in the presence of oxygen). Thus, the ultimate fate of a portion of the phosphorus in lakes is burial in the sediments. Some phosphorus is released back to the water but, over an annual cycle, the net movement of phosphorus is from water to the sediment. For this reason, all lakes act as "traps" and inputs exceed output. In oligotrophic lakes, phosphorus retention can exceed 99% [7] but the retention may be reduced to 50% or even less in shallow eutrophic lakes and in others receiving little influx of phosphate-binding materials.

The concentration values of phosphates (and nitrates) in the surface water of a lake tend to show a characteristic pattern; they represent the changes in the balance between "input" and "output" at any given time. The concentrations of the water are generally maximum at the time of spring turnover and, thereafter, tend to diminish rapidly during the growing season of algae as the nutrients get locked up in the new algal growth. Hence, minimum values are generally observed in the summer when algal growth rate is at its highest.

Stumm et al. [28] point out that information on the nutrient concentration in the interstitial water may be more important than on the total nutrient content of sediments. Sediment samples from various eutrophic lakes in the western U.S. show 0.08 to 10.5 mg soluble phosphorus/liter in interstitial water. Immediately following spring and fall overturns, the hypolimnion concentration may be less and phosphorus may be released from the sediments to the overlying water. Also, during periods of lake overturn when conditions become uniform and homeothermic, the concentrations tend to become more or less uniform at all depths. Thus, a record of the

sampling time, depth of sampling, and condition of the lake at the time of
sampling are all necessary to facilitate interpretation of observed results.

The alkalinity, pH, and temperature of the water determine the free
CO_2 concentration in it. The relation between these factors is given as

$$\text{Free } CO_2 = \frac{aH^2}{K_1(H + 2K_2)} \qquad (6.23)$$

where

  a = meq bicarbonate-carbonate alkalinity

  H = hydrogen ion activity, 10^{-pH}

  K_1 = first dissociation constant of carbonic acid
       at concerned temperature

  K_2 = second dissociation constant of carbonic acid
       at concerned temperature

In nature the pH rises as CO_2 is removed by algae. From empirical data,
King [26] suggests that green algae become limited if the free CO_2 concentration falls much below 7.5 μmoles/liter. Furthermore, King suggests
that the relatively large reserve of inorganic carbon remaining as HCO_3^-
and CO_3^{2-} at that point is available to the blue-green algae which appear
to be more efficient at the lower free CO_2 values. In fact King states
that the inorganic carbon contained in the $HCO_3^--CO_3^{2-}$ buffer system could
support significant blooms of blue-green algae without the need for the
CO_2 supplied by bacterial respiration. The latter, if available, would
only accelerate bloom development.

Seasonal variations in pH and alkalinity are also observed in a lake.
During summer when photosynthetic activity is likely to be maximum in the
euphotic zone, the pH shows a diurnal variation that is higher at midday
and lower at night. Photosynthesis uses up carbon dioxide, thus raising
the pH; alkalinity declines because of the precipitation of calcium carbonate. Diurnal variations are well known to engineers concerned with
waste stabilization ponds.

In the hypolimnion, conditions are the reverse, namely, pH is low due
to the production of carbon dioxide through bacterial decomposition of
organic matter, but alkalinity is relatively high. The greater the organic
load on a lake and the greater the availability of nutrients, the sharper
will be the difference between the surface and bottom values of pH and
alkalinity.

In newly constructed reservoirs, the quality of water tends to show a
gradual change over the years. Generally, new reservoirs are more pro-

ductive in the early years since natural terrestrial vegetation which is
killed when the reservoir is first filled, undergoes decomposition, releasing nutrients and favoring algal and bacterial growths. Color may also
be somewhat stronger at the start. Gradually, over a period of a few years,
conditions tend to stabilize [1].

6.7 FACTORS AFFECTING PRODUCTIVITY IN LAKES

It would be naive to think that nutrient content is the only factor in
determining the level of primary production. There is, in fact, no direct
relationship between nutrient content and lake water quality. Various
factors are involved and it is desirable to list those factors below as the
control measures to be adopted and taken into account. The factors affecting productivity can be grouped under climatic, geomorphological, and
biochemical factors as follows:

Climatic: among the factors which affect the microclimate of a lake
can be listed

Radiation

Temperature

Wind

Evaporation

Precipitation

Geomorphological:

Nature of soil

Soil seepage

Ratio of epilimnion to hypolimnion

Extent of littoral area and its morphometry

Ratio of runoff volume to lake volume

Biochemical:

Nutrient inflows

Nutrient recycle rate

It should be stressed at this juncture that various factors affect the
nature of water in a lake, but only a few have been observed sufficiently
to warrant making any conclusion about their modifying influence, particularly on primary productivity and eutrophication. The various factors
listed above have been shown diagrammatically in Figure 6.8 to illustrate
how they affect the productivity of a lake.

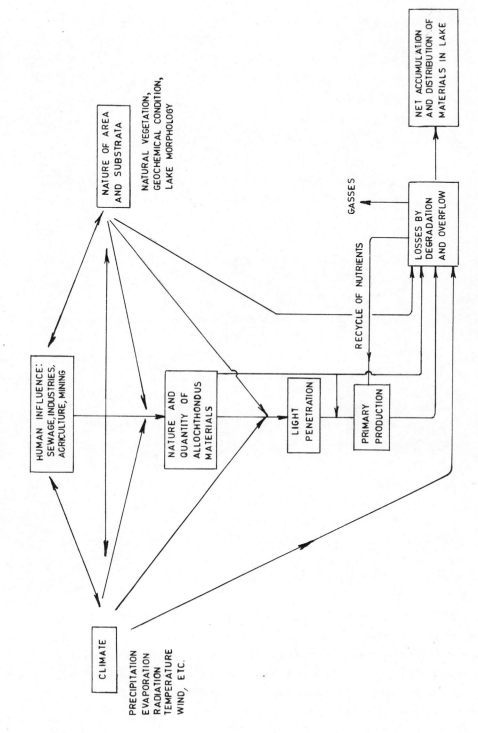

Figure 6.8 Factors affecting the trophic nature of a lake.

6.7.1 Climatic Factors

Radiation, temperature, wind, and evaporation phenomena are relatively
well known and, generally, data are available for specific localities.
Their effects on lake conditions and water quality were discussed earlier
in Chapter 2. The buildup of salts as a result of evaporation was seen in
Sec. 6.4.

6.7.2 Geomorphological Factors

The geological character of the soil in the drainage area and in the
lake basin should have considerable effect on the lake water quality, but
the complexity of the problem and lack of data make it difficult to correlate the two except in a qualitative manner. Generally, the minerals and
substances leached from the soil would be present in the water; however,
some substances are more amenable to leaching and washing away from the
top soil than others. For example, nitrogenous compounds are more readily
washed out than phosphates from soil. Similarly, nitrates are present in
larger amounts than are phosphates in direct rainfall over a lake. However,
sediments physically washed into a lake can bring in the phosphorus bound
on them. Thus, the relative proportion of nitrogen and phosphorus in a
lake may be affected by the extent of rainfall, runoff, and soil erosion.
The nitrogen and phosphorus contents expected in surface runoff have been
given in Chapter 5.

The geological character of the drainage area and its morphological
and vegetational characteristics determine what fraction of the total rainfall in the area will seep into the soil and what fraction will run off
into the lake. A lake may also receive groundwater charge, and the quality
of the groundwater fraction may be quite different from the quality of the
surface water owing to several sorbitive and solutionizing activities during
ground flow.

The physical characteristics of the sediments flushed down with the
storm runoff will affect their sedimentation rates; clays, silts, and
mineral particles will all settle at different rates and some may not
settle at all. Finely divided matter will increase the turbidity of the
water and reduce light penetration. Soil erosion will similarly reduce
light penetration and affect the productivity of the water even though
nutrients may be ample.

Borman [29] explains the results of a full-scale experiment conducted by the U.S. Forest Service which cut down the forest of a watershed of 15.6 ha in order to study the effect of deforestation on stream flow and its quality. The output of stream water increased to about 1.4 times the earlier average, the output of particulate matter increased 4.3 times, and the net output of dissolved inorganic substances increased 14.6 times. Deforestation reduced the flow of ammonia nitrogen to higher plants, but increased its flow to nitrifying organisms, thus considerably increasing the concentration of nitrates in the stream. Nitrification-released hydrogen ions, which replaced metallic cations on the soil exchange surfaces, and Ca, Mg, Na, and K concentrations in the stream water increased almost simultaneously with the increase in nitrates.

The geological characteristics of the lake basin and especially its littoral region will determine whether vegetation can take root in it or not. Steep, rocky edges will discourage littoral growths whereas softer and flatter edges will afford good ground for benthic vegetation to flourish far out into the lake, provided light and nutrients are adequately available.

Shallow lakes will exhibit a greater proportion of epilimnion to hypolimnion and be more prone to eutrophication, provided other physical and biological conditions are also favorable for it. Similarly, the ratio of lake volume to the total volume of storm runoff will determine the lake renewal time and determine how close the lake concentrations will be to the inflow concentrations. The longer the time that the inflow remains in the lake or waterbody, the wider the difference between the quality of inflow water and lake water. An extreme example is the sea or ocean. Renewal times in lakes and other large waterbodies vary from a month or two to several years (Table 6.1).

6.7.3 Biochemical Factors

A. Nutrient Inflows

The most important among the biochemical factors is the nature and quantity of substances, whether of natural or man-made origin, that drain into the lake from its catchment. Generally, the major interest lies in the inflows of carbon, nitrogen, and phosphorus. Where applicable, the typical values of nitrogen and phosphorus given in Chapter 5 could be used to prepare a rough inventory of their annual inflows into a given lake if the extent and nature of the catchment area and other details are known.

Such inventories are essential in evaluating the possible alternative strategies that could be followed for control of eutrophication to cut down maximum inflows of nutrients at minimum cost. Data for preparing inventories could be refined by actual field measurements of the major items. It should be stated here that nitrogen budgets for lakes are complicated by nitrogen gain through fixation and loss through denitrification in a lake. Toxic substances can affect bacterial and algal growth.

B. Nutrient Recycle Rate

Nutrients are recycled from plankton to water and back to plankton unless removed by overflow, seepage, or the harvesting of algae, vegetation, and fish from the lake. An increase in bacterial activity shortens the regeneration time of nutrients within the lake ecosystem and produces more growth-promoting metabolites. Circulation and diffusion help to transport these substances from the tropholytic zone back to the trophogenic zone.

In view of the rapidity with which nutrients are used by a growing plant population, the supply of nitrate or phosphate in a lake water would soon be exhausted unless regeneration kept pace with utilization. In this sense, an adequate rate of regeneration is just as necessary as an adequate concentration of the nutrients for high productivity.

There are two aspects to regeneration of nutrients. First, biochemical decomposition and breakdown of organic matter must occur to release the nutrients; herein lies the work of the decomposers. Biochemical decomposition rate is greatly affected by the temperature and availability of organic matter; the warmer the waters, the faster the breakdown. Summertime is the most favorable season for the release of nutrients. The second factor is the transport of the released nutrients from the tropholytic zone to the trophogenic zone or the epilimnion where they can again participate in the photosynthetic production of plant matter. This vertical transport is brought about by mass circulation of the water during autumn and spring overturn and by slow eddy conduction and diffusion at other times. The presence of a thermocline during summer and winter stagnation retards the upward movement.

Regeneration of phosphorus in Chespeake Bay, for example, was observed by sampling for nitrate nitrogen and phosphorus. Between April and September 1965, the nitrate concentration in the upper bay was reduced from a rather uniform concentration of 45 µg at. per liter to less than 1 µg at. per liter. The total phosphate phosphorus values in that period ranged from 1 to 2 µg at. per liter with 1.5 µg at. per liter as a frequent value.

With an average composition of algal cells as given earlier, the nitrogen-phosphorus atom ratio would be 15:1. Hence, utilization of nitrogen of about 45 µg at. per liter would require about 3 µg at. per liter of phosphorus, or more than the observed values at any time. Consequently, the authors concluded that phosphorus was cycled at least twice during the period from May through August 1965 [4].

In another study on the same bay, radioactive carbon productivity measurements were used to find that the phosphorus turnover rate in the upper waters was between 1 and 4 days. This was possible because of rapid grazing of the algae by the zooplankton, which in turn were being decomposed quickly to recycle the nutrients [4].

6.8 CLASSIFICATION OF LAKES

In the early days of ecology, lakes were classified according to their geographical sites such as tropical, alpine, etc. In the early part of the century (1907-1921) European limnologists such as Weber, Naumann, and Thiememann laid the foundations of ecological classification of lakes by using phytoplankton as a yardstick to compare the productivities of different lakes, observing heterotrophic activities, and emphasizing the physical, chemical, and biological aspects of the problem.

In Chapter 3, trophic levels were discussed when explaining the flow of energy and biomass, both of which taper like a pyramid when passing from one trophic level to another. The word "trophy" is also used to describe the supply of organic matter in a lake and its consequent ecological status. Thus, five types of lakes are broadly distinguished and a general description rather than a definition of each type is given below:

1. Oligotrophic (poorly nourished)
2. Mesotrophic (moderately nourished)
3. Eutrophic (well nourished)
4. Dystrophic (badly nourished)
5. Mixotrophic (dystrophic but productive)

A. Oligotrophic Lakes

Oligotrophic lakes are the geologically younger lakes. They are normally deep, have clear water, and exhibit low productivity. Phytoplankton production is of a low order, algal density is low, and algal blooms are

Table 6.3 Approximate Range of Primary Productivity
in Lakes in Temperate Climates

Item	Lake trophic status		
	Oligotrophic	Mesotrophic	Eutrophic
Total phosphorus at spring turnover, mg/m^3	< 10-20	20-50	> 50
Algal cells per ml	0-2,000	2,000-20,000	> 20,000
Algal biomass dry weight, mg/liter	0.1	1-10	> 10
Chlorophyll a — peak values in photic zone, mg/m^3	0-3	3-20	> 20
Phytoplankton production:			
Annual rates, g carbon per m^2-year	10-50	50-200	200-> 750
Mean daily rates in growing season, g carbon/m^2-year	0.03-0.2	0.2-1	> 1
Secchi disc depth, m (relatively)	several	few	1-2 or less
DO in hypolimnion of deeper lakes during stratification period	present	present	absent
Species diversity	relatively low	high	lower than in oligotrophic state

rare. There may not be much diversity in species, thus giving a moderate
species diversity index (Table 6.3). Oligotrophic lakes are relatively
poor in phosphorus, nitrogen, and calcium. Electrolytes are low and organic materials are few in suspension or are on the lake bottom.

In oligotrophic lakes, oxygen is present in the hypolimnion during
periods of summer stagnation, and complete mineralization of organic substances, both allochthonous and autochthonous, is ensured. Under aerobic
conditions and at the pH values of most such lakes, the phosphate ions
thus liberated form insoluble complexes with the iron and manganese. This
gives a continuous withdrawal of the phosphorus from the lake water to the
sediments.

The oxygen concentration in the hypolimnion has a decisive influence
on the phosphorus turnover in the lake. The critical stage of the "aging"
process occurs when the amount of oxygen in the hypolimnion is no longer
sufficient for the degradation of organic matter. When a lake reaches this

stage (and anaerobic conditions appear in the hypolimnion) the eutrophication process sets in, and the lake does not normally recover of its own
accord. As the process of eutrophication advances, at first these symptoms
merely become more pronounced; subsequently, however, severe changes occur,
adversely affecting the lake's flora and fauna.

Eutrophic lakes, shallow or deep, are typically rich in plant nutrients
and in organic materials. Electrolytes are available in abundance. They
exhibit relatively higher rates of productivity and may also have prolific
weed growth (littoral vegetation). Eutrophic lakes may have much organic
matter (autochthonous and allochthonous) in suspension and on the lake
bottom. The hypolimnion is relatively small and shallow and oxygen is
depleted seasonally (clinograde distribution) or may be entirely absent.
Not only is algal growth abundant as stated above, but algal blooms also
tend to be common. The species diversity index is relatively low (Table
6.3).

The large mass of algal material and littoral vegetation produced in
these lakes makes them highly autochthonous or autotrophic. Excessive
plankton turbidity prevents penetration of light to the lake bottom and
growth of rooted plants, thus depriving many invertebrates of food and
shelter. This further increases plankton growth and makes the lake less
suitable for fish. Littoral vegetation also invades larger and larger
areas of the lake and, in the course of time, shallow lakes tend to end up
as swamps or marshes. They may then be called "senescent" (bordering on
extinction). The lakes may eventually become extinct and a water resource
lost to the country. Several lakes in different parts of the world are
reported to be in a highly eutrophic state and perhaps well on the way to
extinction.

A eutrophic lake such as Lake Erie is reported to be producing over
4.54×10^9 kg of algae per year. When algae die and sink, 1 kg of dead
algae in the form of degradable organic matter consumes 1.4 to 1.8 kg of
oxygen in its stabilization. This places a severe demand on the oxygen
available in the hypolimnion which is consequently, more often than not,
devoid of oxygen. The areal rate of oxygen consumption in the hypolimnion
is several times more than in that of oligotrophic lakes.

The essential feature to bear in mind about highly eutrophic lakes
is that more organic matter of plant and animal origin is received as well
as produced in these lakes than is possible for bacteria to degrade in the
same period of time. Thus, undegraded and partially degraded organic matter

rapidly accumulates besides the relatively small quantities of inert residues of degradation. This slowness in bacterial degradation is especially pronounced in the cold climates. Degradation is also retarded by anaerobic conditions in the hypolimnion. Problems arise because there is an imbalance between organic matter available and organic matter that can be broken down by bacterial activity in a lake. Similar happenings in sewage-irrigated soils is referred to as "soil sickness."

B. Mesotrophic Lakes

Mesotrophic lakes lie between oligotrophic and eutrophic.

C. Dystrophic Lakes

Dystrophic lakes occur principally in bog surroundings and old mountains. They may have phosphorus, nitrogen, calcium, and organic materials, but the growth of most organisms is limited by the occurrence of high concentrations of humic substances. In dystrophic lakes, the plankton biomass as well as the species diversity index is relatively low. Algal blooms are also less frequent and oxygen may be absent in deeper waters. Such lakes are usually shallow and may also ultimately end up as swamp or marsh, but resulting more from allochthonous materials or allotrophic conditions than from autotrophic conditions.

D. Mixotrophic Lakes

Some lakes studied in Finland have been found to be rich in humic matter and yet quite productive [9]. They have been called mixotrophic lakes. If we distinguish between the two different sources of organic matter supply, allochthonous and autochthonous, and express their rates of action by the corresponding degree of "allotrophy" and "autotrophy," the four major trophic lake types can be indicated as in Figure 6.9 [9].

It should be pointed out that most of the lakes studied in some detail have been located in temperate and cold climates. Very little data are available from tropical areas [13a] and it is a moot point whether the conclusions drawn from studies in colder climates can be applied directly to lakes in warmer areas where natural processes may be operative at high rates throughout the year, resulting in a different set of factors affecting eutrophication.

Lake succession is generally visualized as passing from an oligotrophic to a eutrophic state. In regard to pollution, lake eutrophication is unidirectional; as time passes, an oligotrophic lake becomes eutrophic. How-

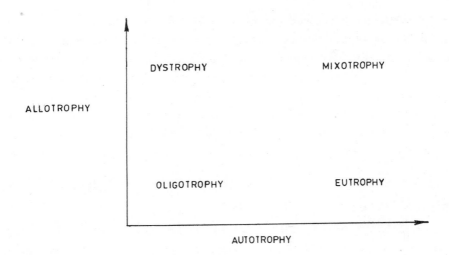

Figure 6.9 A classification of lakes based on the degree of auto-trophy and allotrophy (from Ref. 9).

ever, in the natural, historical setting, paleolimnological studies have shown that lakes have at times alternated between oligotrophic and eutrophic states in response to differing geoclimatological conditions. Hence, the mechanism is reversible and the lake can be viewed as "an unstable microcosm" capable of displaying fluctuations in lake trophy [6].

Hutchinson [5] discussed the possibility of reversal from a eutrophic to an oligotrophic state. When for any reason the influents to a drainage lake (in eutrophic condition) disappear and the lake becomes a seepage lake with the influent water now largely derived directly from rain or from groundwater that has passed through peat or other sorptive material, the lake may gradually turn oligotrophic. Shoji Hori [6] also makes several observations of alternating oligotrophy and eutrophy in the case of some Japanese lakes studied.

6.9 INDICATORS OF THE TROPHIC STATUS OF LAKES

With progressive eutrophication, various changes occur in a lake. These changes can be observed by assessing the changes with time in certain physical, chemical, and biological parameters or indices.

At the outset, it must be mentioned that not many data are available on this subject and very few lakes have been observed over a long enough period of time and in such detail to provide all the answers. Furthermore, these lakes have been situated mostly in colder climates. However, sufficient information has been accumulated to give valuable insights into this aspect and to serve as a base from which limnologists can go further in both cold and warm climates.

To assess the eutrophication process, it would be best if the present chemical and biological condition of a lake could be compared with its primitive state wherever such comparison is available. In the absence of biochemical data, paleolimnological investigation of lake sediments can be usefully undertaken, especially when lakes are undergoing natural eutrophication.

It is desirable that, in studying lake waters, the lake be studied with its complete drainage basin and the lake sediments (the whole lake ecosystem) to understand the various activities within it. Only then would it be possible to assess the potential of the lake ecosystem for eutrophication. A nutrient such as phosphorus may be in the water phase or it may be locked up in the algal growths or in the bottom mud. Isolated studies would not reveal the total available supply in all forms, nor the rate of turnover.

The various indicators of the trophic status of lakes can be grouped as follows under three broad categories:

1. Physical
   a. Transparency of the water
2. Chemical
   a. Concentration of nutrients in the lake
   b. Nutrient loading
   c. Dissolved oxygen in the hypolimnion
   d. Other chemical indices
3. Biological
   a. Primary productivity
   b. Other biological indices such as
      Species diversity index
      Frequency of algal blooms
      Extent of littoral vegetation
      Presence of characteristic phytoplankton and zooplankton groups
      Changes in bottom fauna
      Effects on fish life

Biological indices are more meaningful if supporting physical and chemical
data are simultaneously obtained in field surveys. Some of the indices are
discussed below in greater detail along with a few case studies.

6.9.1 Light Penetration

     With increasing eutrophication, algal density increases, and the
simplest mode of measurement is with the Secchi disc. Reductions in Secchi
disc readings over a period of years become an indirect measure of the increase in net primary productivity of a waterbody. Figure 6.10a gives an
example from Japan [6] where the transparency of the water was determined
monthly from as early as 1927; this enabled comparison with values obtained
in recent years. The 1961 values show a reduction in transparency over all
months of the year compared with 1930 values, thus indicating a possible
increase in primary productivity, color, turbidity, or all three over the
years. Evidence from other parameters discussed below would help to decide
the main causes.
     Dillon and Rigler [32] have shown with data for a large number of
lakes in southern Ontario studied over the stratification period (June to
September), that Secchi disc depths are inversely correlated to the algal

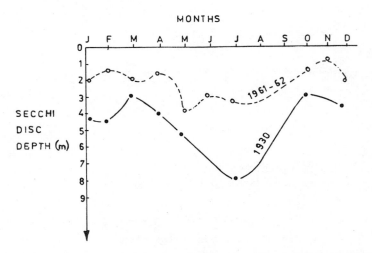

Figure 6.10a The change in transparency in Lake Yogo-ko, Japan,
from 1930 to 1962 (adapted from Ref. 6).

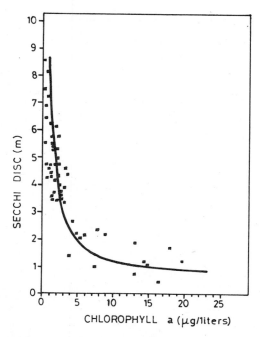

Figure 6.10b Relationship between Secchi disc depth and chlorophyll a concentration for a number of lakes in Canada (adapted from Ref. 32).

density measured as chlorophyll a concentration in lakes (see Figure 6.10b) since generally

$$\text{Algal mass} = 67(\text{chlorophyll a mass}) \qquad (6.24)$$

A similar correlation between Secchi disc depth and chlorophyll a can be obtained for lakes in any other region. If Secchi disc depths are not allowed to reduce beyond a given minimum depth, one can estimate the corresponding chlorophyll a concentration which must not be exceeded during the summer months. The chlorophyll a concentration is similarly dependent on the phosphorus concentration in the lake (see next section), and phosphorus inputs to the lake may need to be controlled.

6.9.2 Concentration of Nutrients in a Lake

Sawyer [13] reported that in eutrophic lakes the phosphorus and nitrogen concentrations during spring overturn were observed to be greater than

100 µg/liter and 300 µg/liter respectively. Stumm et al. [28] stated that unpolluted lakes may show total phosphorus values ranging between 10 and 40 µg/liter, whereas eutrophic lakes may show widely ranging values from 30 to 1500 µg/liter.

Phosphorus concentrations in lakes at holothermic conditions during spring turnover have shown a good correlation with algal concentrations (measured as chlorophyll a) during the following summer season. Dillon and Rigler [32] as well as others, pooling data from a large number of Japanese, European, and North American lakes, have shown a correlation between these two parameters, giving a straight line on a log-log plot of the data:

$$\text{Log (chlorophyll a, mg/m}^3) = 1.45 \log (\text{P, mg/m}^3) - 1.14 \qquad (6.25)$$

The very high degree of correlation shown by this equation (Figure 6.10c) would indicate that phosphorus is indeed the limiting nutrient for algal growth. Controlling the latter in such lakes would require reducing

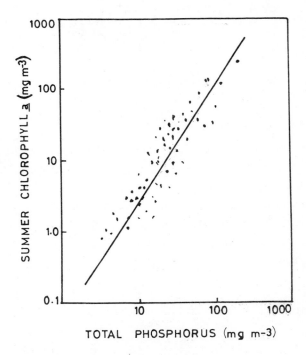

Figure 6.10c Typical relationship between total phosphorus concentration (at spring turnover time) and average chlorophyll a concentration in summar.

the phosphorus concentration which is a function of phosphorus inputs and
lake morphology (see next section).

While precise limiting values of nitrogen and phosphorus are not easy
to establish in view of the various factors involved, the current view in
the U.S. seems to be in favor of keeping phosphorus* values at not more than
50 µg/liter in waters directly entering lakes and not more than 100 µg/liter
in waters entering rivers [14].

The desire to limit the phosphorus concentration to ensure aerobic
conditions at all times in the hypolimnion may not be quite feasible in
warmer tropical climates owing to faster nutrient recycle times and other
factors; also, lake turnovers may not occur as frequently. Moreover, in
warm climates more organic material of allochthonous origin is likely to
be brought into a lake coupled with increased metabolic activity.

Some observed nitrogen and phosphorus values together with DO in the
hypolimnion of a few lakes are given in Table 6.4. The utility of measuring nutrient concentrations is enhanced when carried out over a number of
years to establish trends (see Figure 8.6). Several authors [15,17,28]
caution that no one nutrient or parameter can be considered in isolation.

The algal requirement of nitrogen and phosphorus is approximately in
the ratio of 16:1 as stated earlier. In oligotrophic lakes the phosphorus
concentration is low and tends to exhaust first [15]. Nitrogen is often
in excess at the time phosphorus becomes depleted. In eutrophic lakes, on
the other hand, especially those receiving sewage pollution with values of
nitrogen:phosphorus (N:P) lower than 16:1, the opposite may occur; nitrogen
may be exhausted while some phosphorus still remains during the algal growing season. The N:P ratio in the lake surface waters during a rapid algal
growth season may be another evidence of the trophic status of a lake. The
presence of free nitrates in the epilimnion during the summer stagnation
period could well be a sign of oligotrophication.

Sometimes in a given lake the ratio N:P may change after a heavy rainfall. Agricultural runoff may bring into a lake much more nitrogen compared
with phosphorus. Little phosphate may be washed from the soil in the catchment area and the lake may displace more phosphorus than was brought in by
runoff. Hence, the N:P may change, especially after a very wet year. The
time of sampling in a lake must, therefore, be watched and generally confined to a short but rapid algal growth period to determine the ratio at
limiting conditions. Thus, the seasonal variation in a lake resulting from

_____

*Of which up to 30% may be present as soluble orthophosphate.

Table 6.4 Observed Data for a Few Lakes

Lake	Surface area (km²)	Max. depth (m)	Nitrates as NO3⁻ (µg/liter) Holothermic	Nitrates as NO3⁻ (µg/liter) Summer minimum	Phosphates as PO4³⁻ (µg/liter) Holothermic	Phosphates as PO4³⁻ (µg/liter) Summer minimum	Oxygen content (mg/liter) At 50 m depth	Oxygen content (mg/liter) At bottom	Remarks	Reference
Switzerland										
Walensee	24	145	1,800	500	<10	0	9.5	6.8	Oligotrophic	7
Zurichsee	68	138	3,200	0-400	290	0	4.7	0	Mesotrophic-eutrophic (algal masses first appeared on the beaches in 1949 when the PO4 concentration was 80-90 µg/liter).	7
Greifensee	8.6	32	8,000	0-1,500	1,110	0	Zero at 20 m	0	Eutrophic (1966 value of PO4³⁻ is 5.5 times the 1950 value).	7
Germany										
Tegernsee	9.1	71	1,800		70		1.7	0		7
Italy										
Lago di Como	1.6	410	2,200	1,700	60	—	8.6	5.8		7
Turkey										
Iznik	300	67			25	<1.0	5.5		Values for 1953-1954	19
Japan										
Yogo-ko				20		3.6			Increase in productivity observed over last 40 years (see Figure 6.10a for changes in Secchi disc values).	6
USA										
Lake Erie	25,718	17.7 (mean)	1,200 inorganic (winter maxima)		50 total P, 11 dissolved P (winter maxima)		<3.0 in hypolimnion		Eutrophic. Primary productivity 0.1-7.3 g C/m²-day. Also algal blooms and many oligochaetes and midges in benthos.	6
Lake Sebasticook									Algal nuisances, cell mass = 40 mg/liter	6

the above factors is quite substantial and is revealed by weekly or bi-monthly samples. Long-term comparisons are possible when data are obtained over a number of years. The trends become meaningful when nitrogen and phosphorus inputs from different sources are also known from year to year.

It is necessary to restate at this juncture that some lake ecosystems are by their very nature more susceptible to eutrophication than others. The source of nutrients is not always man and his industrial and agricultural activities. The sources may even be purely natural resulting from the geology and vegetation of the area. The waters of eastern Europe have a generally high degree of natural eutrophy while the waters of northern and central Europe in their natural state tend to be oligotrophic [8].

6.9.3 Nutrient Loading

A quantifiable relationship has been shown to exist between the amount of nutrients reaching a lake and its trophic status [31-33]. This is particularly true for lakes in which productivity is controlled by phosphorus (as limiting nutrient), and implies that the trophic nature of a lake may change (from oligotrophic to eutrophic) if the phosphorus discharged to it exceeds a certain critical level. Several models have been proposed for determining the critical levels [31-33]. Of these, Vollenweider's model [31] is relatively simple to use as it requires data only on phosphorus inputs and the lake's area, mean depth, and detention time. His empirical equation is:

$$L = P_{sp} \times q_s (1 + \sqrt{\bar{Z}/q_s}) \tag{6.26}$$

where

L = phosphorus load on lake, mg P/m^2-year

P_{sp} = concentration of total phosphorus in lake water at time of spring turnover, mg total P/m^3

$\bar{Z}$ = mean depth of lake, m

q_s = overflow rate = $\bar{Z}/t_o$ = Q/A, m/year where

> t_o = theoretical detention time
>
> Q = annual inflow rate
>
> A = lake area

From limnological experience, Vollenweider suggests that if the value of P_{sp} is not to exceed 10 mg total P/m^3 at spring turnover so as to ensure

oligotrophic conditions, Eq. (6.26) can be rewritten to give the <u>critical</u>
phosphorus load L_c on the lake as:

$$L_c (\text{mg P/m}^2\text{-year}) = 10q_s(1 + \sqrt{\overline{Z}/q_s}) \qquad\qquad (6.27)$$

Eutrophic conditions may be expected in the lake when the actual phosphorus load L equals two to three times L_c. For any given lake, therefore,
it is necessary to determine the mean depth and area at the time of spring
turnover and the annual water inflow rate to be able to compute from Eq.
(6.27) the critical phosphorus load and compare it with the actual load to
predict its trophic state. Vollenweider has shown that data from over 60
lakes fitted well when values of L were plotted against q_s (see Figure
6.11a), except for a few lakes that seemed misplaced. He has also given a
nomogram based on the above equation, as shown in Figure 6.11b.

As an illustration, if the mean depth $\overline{Z}$ of a lake = 84 m, and its
overflow rate q_s = 10.6 m/year, the critical loading of phosphorus L_c is
seen from the nomogram to be about 400 mg/m^2-year. Eutrophic conditions
would be likely if the actual load were about two to three times this
value.

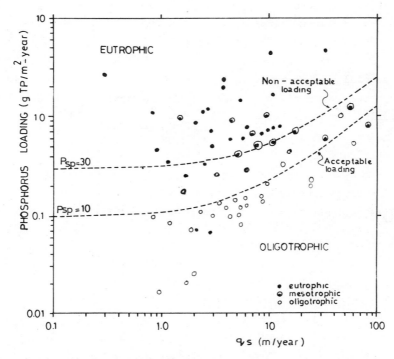

Figure 6.11a Actual trophic condition of several lakes in relation
to their phosphorus load and overflow rate. (Adapted from Ref. 31.)

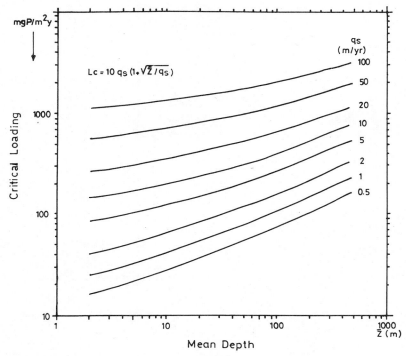

Figure 6.11b Nomogram for estimation of critical loading L_c for phosphorus. [See Eq. (6.27).]

6.9.4 Dissolved Oxygen in the Hypolimnion

The presence or absence of oxygen in the hypolimnion, especially during summer stratification, is a good indicator of the extent of degradation of organic matter in a lake. The greater the amount of organic matter present (from allochthonous and autochthonous sources), the greater the bacterial activity and consumption of oxygen. Figure 6.12 shows the possible seasonal change in the oxygen profiles of an oligotrophic and a eutrophic lake. Table 6.4 also shows evidence from a few cases of each type.

The oxygen demand for biochemical degradation of organic matter settling into the hypolimnion of a lake was dealt with in Sec. 6.6. Hutchinson [5] found by measurements that the hypolimnic loss of oxygen per unit area and time was, as would be expected, greater in eutrophic lakes than in oligotrophic ones. The rate at which oxygen consumption progressed in

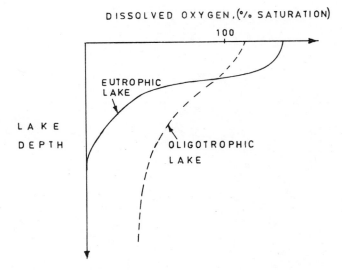

Figure   6.12   Lacustrine dissolved oxygen profiles.

the hypolimnion of oligotrophic lakes was found to be within the range of
0.04 to 0.3 g O_2/m^2-day.   In eutrophic lakes, the rates ranged between 0.5
and 1.4 g O_2/m^2-day.

6.9.5 Other Chemical Indices

Various other chemical indices are being sought in identifying the
trophic state of lakes both in the present and past.   Some of these have
already been mentioned earlier, such as the relative deposition of Fe and
Mn in bottom sediments [2].   The percent of nitrogen in bottom sediments
has also been used to assess the trophic status of lakes [6].   Precipitation of lime during oxygen depletion conditions in summer stagnation can
sometimes provide corroborative evidence when bottom sediments are examined [7].

Pearsall [18] has suggested the cation ratio $(Na^+ + K^+)/(Mg^{2+} + Ca^{2+})$
as a parameter showing inverse correlation with the trophic level.   Other
observers have also noted high productivity and occurrence of algal blooms
in lakes with hard waters, namely, those high in Ca^{2+} and Mg^{2+}.   Shannon
and Brezonik [17] have found a similar correlation in the case of several
Florida lakes.   These and other chemical indices are valuable to hydrogeo-

chemists. More data are required before we can learn to interpret them correctly. A detailed discussion of these indices is beyond the scope of this book.

6.9.6 Biological Indicators (Primary Productivity)

Direct measurement of the primary productivity of water would be a positive method of determining the increase in eutrophication from year to year in a lake and of classifying it (see Sec. 3.6.1 for methods of measurement).

While describing the different types of lakes, reference was made to their productivity. Some typical values for different waterbodies are given in Table 6.3 for the sake of illustration. Mesotrophic lakes lie between oligotrophic and eutrophic ones.

Expressing productivity rates in terms of the weight of carbon fixed per unit area is not entirely satisfactory. Seasonal and even daily variations in the area and depth of the euphotic zone make it difficult to permit easy comparison between different lakes. Algal productivity per se may not always show a correlation even with nutrient levels in a lake. Shannon and Brezonik [17] have quoted an instance in Florida where the lake nutrient contents were high but plankton productivity low; this was due to profuse growth of water hyacinths.

In eutrophic lakes, "algal blooms" are frequently observed. During bloom periods large numbers of algal cells may grow within the short period of a day or so. This would naturally require that sufficient amounts of nutrients be present in the water and that light, temperature, and other conditions be favorable (see Figure 6.13). Although nitrogen and phosphorus are recycled in a lake ecosystem, carbon is not. Carbon is the source of energy and is consumed. Thus, carbon must enter the ecosystem through atmospheric gas exchange, in the drainage, or through both pathways.

For an algal bloom to occur such that, for instance, 200,000 cells/ml are produced, which is approximately the equivalent of 56 mg/liter dry weight of algae, the carbon required is about 29.4 mg/liter (= 52% × 56 mg/liter) as organic carbon, which, if expressed as CO_2 amounts to 110 mg per liter [= 29.4(44/12)]. If all the carbon requirement is to be met through atmospheric gas transfer, it would imply a CO_2 transfer rate of

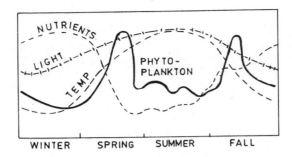

Figure 6.13 Yearly phytoplankton blooms in temperate zone ponds and lakes (adapted from Ref. 2).

of 110 mg/liter-day or an average of about 5 mg/liter-hr. Generally, this is not feasible as atmospheric transfer can account for less than 1 mg/liter-hr under favorable conditions (pH <9.0). Thus, some workers have suggested that organic inflows to a lake are essential so that bacterial activity may release the necessary amounts of CO_2 to sustain a "bloom" [16]. Others, however, maintain that nitrogen and phosphorus along with inorganic carbon from $HCO_3^- - CO_3^{2-}$ in water are sufficient to sustain significant algal blooms (see Sec. 6.6).

6.9.7 Other Biological Indicators

The abundance of planktonic, bacterial, benthic, and fish populations as well as their species diversity change as eutrophication progresses. Eutrophic lakes develop dense algal growths, but, although the individuals are more numerous, the species are few. For example, only a few species of blue-green algae may develop in a eutrophic lake.

Presence of humic substances, toxic materials, and an increase in salinity have similar consequences on a waterbody. Hence the species diversity index by itself is not an index of eutrophication, and corroborative evidence from other indices would be necessary to establish it.

When there are few species and many individuals per species, the food chains become fewer and shorter. Variations in environmental conditions now have an exaggerated effect on the aquatic population, their vulnerability being demonstrated in the disappearance of a species for short or long periods of time and the development of blooms (population explosions) of some species of plankton.

Thus, the frequency of algal blooms becomes a kind of index to the trophic condition of a lake. The more eutrophic the condition, the more frequent the blooms.

Where lake morphology permits, some littoral vegetation is likely to show up at the shoreline. With an increasing supply of nutrients, the littoral vegetation may flourish. The depth to which such vegetation (benthic plants and macrophytes) will take root will be governed by light penetration. As eutrophication increases the productivity of a lake, the transparency of the water diminishes, and, therefore, it stands to reason that the depth to which littoral vegetation can flourish now decreases. Hence, in this sense, the planktonic productivity and extent of submerged vegetation are somewhat related. As eutrophication proceeds, submerged vegetation would be found only at very shallow depths [3].

The presence of characteristic phytoplankton and zooplankton groups has long been used by biologists as an indicator of aquatic conditions, and especially of pollution. Progressive eutrophication can also be sensed from observations of changes in biota and their distribution in a waterbody. It is beyond the scope of this book to go into details of different phytoplankton and zooplankton groups which would help to characterize the lake conditions.

The effect of eutrophication on bottom fauna has been discussed by Jonasson [3]. From the viewpoint of temperature, the bottom fauna lives at a cold temperature for some months, with the substratum devoid of oxygen, low in pH, and high in alkalinity; they also live in complete darkness. The communities drop in diversity as eutrophication progresses owing to (1) their dependence on substratum (burrowing-type animals are encouraged), (2) a need for respiratory adaptation (this causes some groups to disappear and new ones to appear), and (3) utilization of food as a function of lake rhythm (this affects their life cycle among other things). The bottom animals receive more food during the spring and autumn circulation, less food (and that partly decayed) in the summer, and very little food during winter stratification. In some stages of eutrophication, benthic invertebrates may even be eliminated.

In general, oligotrophic lakes would be inhabited by stenoxybiont benthic fauna, namely those organisms which have a relatively narrow tolerance range for oxygen variations. Eutrophic lakes, on the other hand, would be inhabited by euryoxybiont organisms or those with a wider range of tolerance (see Sec. 3.15).

Bacteria are essential for the breakdown of organic matter and the return of organic nutrients to the aquatic medium. Bacteria themselves constitute food for protozoa, which in turn are food for higher metazoa, leading ultimately to fish in the lake. Eutrophic lakes would support relatively higher populations of bacteria and, thus, of protozoa than oligotrophic lakes, although the effect of eutrophication on the relative numbers and kinds of bacteria is not known. The effects on nitrification and nitrogen fixation in lakes are also insufficiently known.

Fish are not particularly good indicators of gradually creeping eutrophication because they are at the top of the trophic pyramid and are likely to be least affected by relatively small changes. Fish, however, do respond to lake fertilization by exhibiting higher growth rates, and this may provide a fairly useful index to the measurement of an increase in lake eutrophication. A closer examination of the physiology of such fish may reveal some differences in their fat and water contents. Fish exposed to certain pollutants may pick up a characteristic taste in their flesh or show higher concentrations of fat-soluble substances (such as DDT).

Changes in dissolved oxygen, temperature, or turbidity of the water may affect not only the spatial distribution of fish but even their very survival in a lake. Spawning grounds in littoral regions of lakes may be affected by turbidity, for example, with a consequent reduction in the reproduction rates of affected fish.

Generally, as the productivity of a lake increases, one may expect an increase in fish yield for the obvious reason that more fish food (algae) is now available. This knowledge is put to commercial use when purposely fertilizing fish ponds with sewage or artificial nutrients. Some values of fish production in fertilized ponds have been given in Table 3.3.

Unfortunately, however, the species of fish and their relative numbers undergo a change as fertilization in the commercial sense, or eutrophication in the pollution sense, progresses. Gradually, fine fish are replaced by coarser varieties. This change appears to take place over a relatively narrow range of mesotrophy. In several European lakes which were once oligotrophic, a large percentage of the fish catch consisted of fine fish varieties such as trout, pike-perch, eel, etc., while today in the same lakes, after eutrophication, the predominant catch consists of coarser varieties of fish such as carp, bass, and so on. In this sense, fish can also be considered as indicators of eutrophication.

6.9.8 Development of a Trophic State Index (TSI)

The large number of variables involved in eutrophication make it difficult to develop a deterministic model at the present time. However, an attempt has been made in the U.S. to develop an empirical model based on data from 55 north and central Florida lakes situated in a subtropical climate [17]. Sampling was done over a 1 year period (1969-1970). Lake measurements included Secchi disc transparency, temperature, turbidity, color, organic nitrogen, total PO_4^{3-}, specific conductivity, Na^+, K^+, Ca^{2+}, Mg^{2+}, primary productivity, and chlorophyll a. Primary productivity was measured using a laboratory light box procedure rather than at the site in order to standardize light and temperature conditions and have a more uniform basis of comparison among the 55 lakes, some of which had clear and some colored waters. Lake catchments and land use patterns were studied in each case using aerial photographs and topographic maps. Population data were grouped according to the nature of sanitary services and the mode of disposal in order to correlate land use patterns and the trophic state of the lake in each case.

Using multivariate statistical techniques, the authors developed a "trophic state index" (TSI) based on several selected parameters for which correlation matrices were developed and the method of principal component analysis was used (see Table 6.5).

The development of a TSI was facilitated by the use of computers. A hypothetical index scale was developed with the minimum value as zero, indicating zero productivity and zero value for other indicators. The 55

Table 6.5 Parameters Used in the Development of a TSI

Parameter	Mean annual value	Standard deviation
Secchi disc depth, m	1.19	1.3
Conductivity, μmhos/cm	93.1	101.30
Total organic nitrogen, mg/liter as N	1.02	0.82
Total phosphates, mg/liter as P	0.125	0.177
Primary productivity (in terms of lake volume), mg of C/m^3-hr	44.8	82.30
Chlorophyll a, mg/m^3	16.9	19.80
Cation ratio, $(Na^+ + K^+)/(Ca^{2+} + Mg^{2+})$	0.68	0.61

lakes studied by the authors were separated according to their TSI values
using classic trophic state terminology as follows:

Trophic state	Value of TSI
Hypereutrophic	> 10
Eutrophic	7-10
Mesotrophic	3-7
Oligotrophic	2-3
Ultraoligotrophic	< 2

A word of caution must be interjected here since engineers often prefer to use indices, namely that eutrophication is a complex phenomenon and,
as the authors themselves are the first to point out, universal indices
could not be developed to cover very diverse conditions ranging from the
arctic to the tropics. However, such indices have the value that they
provide baseline data against which future trends can be assessed in terms
of one single parameter. In this respect, they are similar to the air
pollution indices often used by cities. Under valid conditions, trophic
state indices can also be used to compare different lakes and study the
effects of land use patterns and populations in the catchment area. A
high degree of correlation was obtained between the latter factors and
lake TSI values for the Florida lakes [17].

6.10 CONTROL OF EUTROPHICATION

The various undesirable effects likely to result from eutrophication
were discussed in Sec. 4.9. Briefly, these could be such that they affect
the beneficial uses of water such as public water supplies, recreational
use, and, in extreme cases, the loss of a water resource itself after a
long enough period of time. Out of all the factors which affect eutrophication, there is only one which is within the scope of human intervention:
this is nutrient supply. The rest of the factors are essentially natural
in character and very little can be done about them.

The first step in any control program is, of course, good and regular
monitoring with regard to the various indices discussed earlier from which
the trophic status of a lake can be assessed and long-term trends evaluated.

When planning a sampling program, it would be well to remember that a lake
and its catchment are inseparable and must be studied together. The next
step would be to prepare an inventory of both qualitative and quantitative
inflows and especially to know the sourcewise contribution of nutrients.
Other hydrogeological data are also needed as discussed earlier. These
data would provide valuable inputs for developing a short-term and long-
term program of zoning and land management to attain the desired objectives
of water quality.

The reduction of nutrient supply to a waterbody can be brought about
by various methods involving either physical diversion of the nutrient-
bearing waters away from the waterbody to be protected, or the treatment
of wastewaters to reduce their nutrient content. It is not possible to
develop specific limits of nutrient concentrations in treated wastewaters
since several factors affect the trophic status of a lake. The general
approach, therefore, would be to follow the "best available practice" in
waste treatment where necessary and proceed step by step (e.g., first conventional, then tertiary treatment) in accordance with long-term goals.
In some cases, nutrients already present in a lake may need to be removed
or washed out by suitable means together with a program for reduction of
new inflows. Some of the possible control methods are discussed below.

6.10.1 Diversion of Nutrients from a Lake

Where domestic sewage and industrial wastes constitute major sources
of nutrients, their diversion from the lake can be engineered by constructing conduits or canals, provided topography and terrain permit, to carry
the nutrient-rich wastewaters from the source directly to a point downstream of the lake.

The wastewaters may be discharged downstream

1. To a river served by the same or other catchment area, provided the
    dilution flow and other requirements are met. Generally, a river can
    handle more BOD and nutrient content owing to better aeration and
    mixing. It can be argued that we are merely transferring the pollution
    from one site to another, but there is no harm in doing so provided
    the river is capable of handling it. This is part of any water manage-
    ment program.

2. To an estuary, which may have an even greater potential than a river
    for handling wastewater owing to tidal flushing.

3. To coastal waters where there may also be a good potential for dilution and disposal.

4. For irrigational use on land where climate, soil, the nature of waste-water, and other conditions permit. Agricultural soils have a high capacity for accumulating and retaining mineral ions, especially phosphorus, from the wastewater.

Lake Washington near Seattle [10] provides a good example of such a study because it was investigated before it became enriched and because it is still under study (Table 6.6). Sewage inflows which brought in nutrients have now been diverted and an improvement in lake water quality has been observed. The earlier study on Lake Washington was made in 1933 and the nitrate N, phosphate P, and chlorophyll values are shown month by month in Figure 6.14. The figure shows how from year to year the nutrients began to accumulate and the chlorophyll in surface waters (productivity) correspondingly increased. The figure also shows how the concentration of nutrients drops when the algal growth season starts. Sewage effluents were diverted between 1963 and 1968, and Figure 6.14 shows how the nutrient levels and algal peaks have already begun to diminish in this period. It is expected that, since the Cedar River has low nutrient concentrations and since the volume of water entering the lake is large (31% or more of the lake volume per year), a dilution of the existing storage of nutrients in the lake should take place rapidly [10].

6.10.2 Removal of Nutrients from Wastewater

Wastewater treatment is generally confined to the removal of carbon, nitrogen, and phosphorus by using methods which are now fairly well known. Treatment methods are discussed later in this book. In control of eutrophication, interest often lies in the removal of phosphates since their

Table 6.6 Nitrogen and Phosphorus Entering Lake Washington in 1957

Source of flow	Total nitrogen (inorganic nitrogen and total Kjeldahl nitrogen)	Total phosphorus	Ratio of N:P
Streams	1,395,300	39,600	35.2
Sewage effluents	182,700	49,900	3.67

Source: Adapted from Edmonson [10].

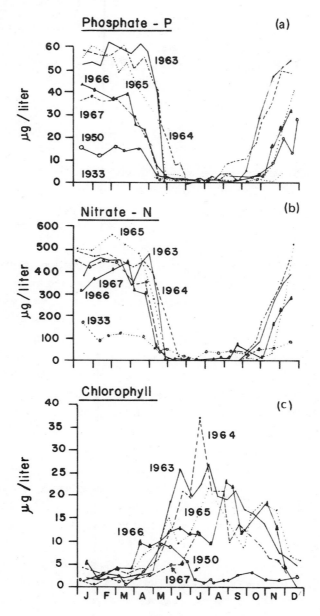

Figure 6.14 The seasonal changes in (a) phosphate, (b) nitrate, and (c) chlorophyll in the surface water of Lake Washington during various years. Source: Ref. 10.

natural inflows from land drainage and rainfall are generally less than those of nitrogen except when subject to pollution by the influence of man. Even naturally oligotrophic lakes show the presence of free nitrate ions

throughout the year and, therefore, elimination of sewage nitrogen may not
be meaningful. Where phosphorus removal is necessary, it would make good
sense to discourage the use of phosphatic detergents by the public. In
fact, their use should be generally discouraged.

6.10.3 Manipulation Within or Removal of Nutrients from Lakes

Removal of nutrients from a lake is possible, but prevention would be
better than a cure. Prevention of the entry of nutrients would be a more
fruitful and durable method of control than periodic removal of nutrients
from a lake since, as soon as they are removed, their buildup starts again
unless their influx into the lake is stopped.

Nutrients are locked up in fish, in the vegetation (macrophytes), and,
of course, in the algae besides being in the water and sediments. Removing
fish and the macrophytes, especially when the lake is at low level, and exposing as much as possible of the littoral area would help to remove some
nutrients. Dredging of sediments would also help, although it would prove
very expensive as a nutrient removal measure. Dredging might be more
feasible where simultaneous deepening of the lake is also desired, although
reduction of nutrients in this manner is no guarantee of success. Even
small quantities of phosphorus with rapid recycle rates can sustain much
productivity. Removal of nutrients by harvesting fish and vegetation may
be more meaningful when followed by flushing the lake and preventing further inflows. The use of chemical toxicants is not recommended as the
nutrients would not be removed thereby, nor is enough known about their
aftereffects on the ecosystem and the food chain. Artificial mixing of
reservoirs to destroy stratification has been tried but, again, the above
remarks apply.

Flushing out of an existing lake with freshwater relatively low in
nutrients may prove useful provided large volumes of water from adjoining
rivers, impoundments, or catchments can be diverted to pass through the
lake and flush out the original nutrient-rich water.

Increasing the flow through a lake would increase the overflow of
nutrients and the quantity of flow required for this purpose can be estimated by using equations given earlier in Secs. 6.3 and 6.9.3. The feasibility of this method lies in the easy availability of water for this purpose. The water quantity requirements are reduced if flushing is preceded

by the removal of nutrients as described above and by prevention of further
influx.

6.10.4 Zoning and Land Management

Zoning and land management is an extremely valuable tool in water management. In fact, the two are inseparable as long-term measures for controlling the water resources of a country. The use to which land is put
in the vicinity of a lake will determine the nature of the drainage water
[17,29]. Setback of properties on the lake edge and protection of shorelines from deforestation, erosion, etc., would help. Selection of suitable
sites for industries, urban developments, agricultural fields, forests,
etc., have a bearing on water quality in lakes as the reader must find
evident from the previous discussions.

6.11 LAKE WATER QUALITY OBJECTIVES

In summarizing the discussions presented thus far, one could state
that, essentially, there are two objectives to be aimed at in protecting
lake water quality: first, one must protect the natural state of the lake
and not accelerate its aging process through eutrophication. Second, during its useful life, one must protect the water quality for whatever beneficial uses the lake can serve (e.g., drinking water supply, recreation,
irrigation, etc.). Lakes and the streams that feed them form integrated
systems, and the reader is referred to Sec. 7.8 for a more extensive discussion of water quality objectives and standards.

Example 6.5 Prepare an inventory of nitrogen and phosphorus from
different sources entering a lake. The average phosphorus concentration
presently in the lake = 0.15 mg/liter as P at holothermic conditions.
Comment on the situation, recommending any measures you consider desirable
to reduce productivity in the lake.

Data

Lake waterspread = 10,000 ha (average)

Lake depth = 30 m (average)

Lake volume = 3000 x 10^6 m^3 (average)

Total catchment area = 1 x 10^6 ha, made up of lake waterspread,
forests, mixed agricultural areas, and urban-suburban areas as
shown in Table 6.7.

Table 6.7 Sourcewise Inventory of Nitrogen and Phosphorus Inflows to Lake for Data Given in Example 6.5

Source	Area (ha.)	Nitrogen			Phosphorus		
		Basis	kg/year	% of total	Basis	kg/year	% of total
Direct rainfall	10,000 (average water-spread)	1 mg/liter in rain	30,000	0.4	0.1 mg/liter in rain	3,000	0.5
Forest runoff	790,000	300 kg/km^2-year	2.37×10^6	33.9	6 kg/km^2-year	0.474×10^5	8.0
Mixed agricultural areas	100,000	1,000 kg/km^2-year	1×10^6	14.3	50 kg/km^2-year	0.5×10^5	8.5
Urban storm runoff	100,000	900 kg/km^2-year	0.9×10^6	12.8	90 kg/km^2-year	0.9×10^5	15.3
Sewered population[a]		12 g/capita-day in raw sewage and 50% removal in treatment	1×10^6	14.4	3 g/capita-day	2.5×10^5	42.5
Unsewered population[b]		30% of raw	0.39×10^6	5.6	30% of raw	0.98×10^5	16.7
Major industries	—	Effluent analysis	1.3×10^6	18.6	Effluent analysis	0.5×10^5	8.5
		Total nitrogen:	6.99×10^6	100	Total phosphorus:	5.88×10^5	100

[a] 500,000 persons.
[b] 300,000 persons.

Sewered population = 500,000 people

Sewage flow = 150 liters/capita-day (discharged after conventional biological treatment)

Unsewered population = 300,000 people, with only 30% of untreated sewage flow draining into lake

Major industrial sources (effluent analysis):

Nitrogen = 1.3×10^6 kg/year

Phosphorus = 0.5×10^5 kg/year

Average rainfall = 750 mm (year)

Runoff coefficient = 0.2 (average)

An inventory of nitrogen and phosphorus inflows has been prepared (see Table 6.7) using norms given earlier in Chapter 5 for different types of wastewaters.

The total volume of annual inflows to the lake, neglecting ground flow, can be estimated as follows:

1. Storm runoff = $\left(\dfrac{750 \text{ mm/year}}{10^3} \right) (10^6 \text{ ha} \times 10^4)(0.2)$

$= 1500 \times 10^6$ m^3/year

2. Sewage flows = $(500{,}000 \times \dfrac{150}{10^3} \times 365) + (300{,}000 \times 0.3 \times \dfrac{150}{10^3} \times 365)$

$= 32.3 \times 10^6$ m^3/year

Hence, the average incoming concentration C_i of phosphorus is

$$C_i = \frac{5.88 \times 10^5 \text{ kg/year} \times 10^3}{(1500 + 32.3)10^6 \text{ m}^3/\text{year}} = 0.38 \text{ mg/liter}$$

On a long-term basis, assuming that all inflows are in balance with outflows, the steady-state concentration of a conservative material would theoretically be the same as that in the inflow in terms of averages. If phosphorus was conservative, its concentration in the lake at steady state would be 0.38 mg/liter. However, it is not strictly conservative and its actual concentration in the lake as recorded averages only 0.25 mg/liter. Thus,

$$\text{Phosphorus residence time} = \frac{\text{phosphorus in lake}}{\text{incoming P per unit time}}$$

$$= \frac{(0.15 \times 10^{-6})(3000 \times 10^6 \times 10^3)}{5.88 \times 10^5/\text{year}}$$

$$= 0.77 \text{ years}$$

$$\text{Hydraulic residence time} = \frac{V}{Q} = \frac{3000 \times 10^6}{1532.3 \times 10^6/\text{year}} = 1.96 \text{ years}$$

The average holothermic concentration of 0.15 mg/liter in the lake seems

on the high side and in all probability should result in high algal productivity with its attendant difficulties. Actual productivity and transparency (Secchi disc) data give corroborating evidence. In fact, the conditions outlined in this case are excessively severe. If the annual inflows
of N and P are calculated in terms of g/m^2-year, we find

$$\text{Annual load of N} = \frac{(6.99 \times 10^6)10^3}{10,000 \times 10^4} = 69.9 \text{ g/m}^2\text{-year}$$

$$\text{Annual load of P} = \frac{(5.88 \times 10^5)10^3}{10,000 \times 10^4} = 5.88 \text{ g/m}^2\text{-year}$$

These loads are several times greater than the values indicated in Figures
6.11a and 6.11b and, unless drastic reductions in nutrient inputs are
feasible, the lake would be condemned to severe eutrophication.

Upon a review of the N and P inventory given in Table 6.7, it is seen
that phosphorus reduction could be substantial if domestic sewage and industrial wastes are tackled for treatment to remove phosphorus before discharge to the watercourses. Another avenue to explore would be the feasibility of diverting the urban storm water away from the lake if topography
and other features permit.

If 50% reduction in phosphorus influx is achieved by means of diver-
sions/treatment, the incoming concentration would be

$$0.38(0.5) = 0.19 \text{ mg/liter as P}$$

Now, if it is assumed that the mean residence time of phosphorus will continue to remain the same, namely, 0.77 year even after the above reductions, we can conclude that the average phosphorus concentration in the
lake will also reduce by a corresponding 50% over the earlier value of
0.15 mg/liter. Thus, the lake may show an average concentration of 0.075
mg/liter as P (0.25 x 0.5). Similar results can be obtained by using
Vollenweider's method [Eq. (6.25)].

The time required to reach 99% of the new steady-state value after
the influx of P is reduced will be obtained from Eq. (6.10).

$$\text{At} \quad \frac{C}{C_m} = 0.99 \quad \text{we have} \quad \frac{t}{t_o} = 4.5$$

Hence, the time required = 0.77 x 4.5 = 3.47 years.

REFERENCES

1. G. M. Fair, J. C. Gayer, and D. Okun, Water and Wastewater Engineering, John Wiley & Sons, New York, 1968.

1a. E. Odum, Fundamentals of Ecology, Third Edition, Saunders, 1971.

2. C. Mortimer, Physical factors with bearing on eutrophication in lakes, in Eutrophication: Causes, Consequences, Correctives, Nat. Acad. Sci., Washington, D.C., 1969.

3. P. Jonasson, Bottom fauna and eutrophication, in Eutrophication: Causes, Consequences, Correctives, Nat. Acad. Sci., Washington, D.C., 1969.

4. J. Carpenter, D. Pritchard, and R. Whaley, Observations on nutrient cycles in some coastal plain estuaries, in Eutrophication: Causes, Consequences, Correctives, Nat. Acad. Sci., Washington, D.C., 1969.

5. G. Hutchison, Eutrophication, past and present, in Eutrophication: Causes, Consequences, Correctives, Nat. Acad. Sci., Washington, D.C., 1969.

6. S. Horie, Asian lakes, in Eutrophication: Causes, Consequences, Correctives, Nat. Acad. Sci., Washington, D.C., 1969.

7. E. A. Thomas, Eutrophication in central European lakes, in Eutrophication: Causes, Consequences, Correctives, Nat. Acad. Sci., Washington, D.C., 1969.

8. M. Straskraba and V. Straskrabova, Eastern European lakes, in Eutrophication: Causes, Consequences, Correctives, Nat. Acad. Sci., Washington, D.C., 1969.

9. W. Rodhe, Crystallisation of eutrophication concepts in N. Europe, in Eutrophication: Causes, Consequences, Correctives, Nat. Acad. Sci., Washington, D.C., 1969.

10. W. Edmonson, Eutrophication in North America, in Eutrophication: Causes, Consequences, Correctives, Nat. Acad. Sci., Washington, D.C., 1969.

11. A. Beeton, Changes in the Great Lakes, in Eutrophication: Causes, Consequences, Correctives, Nat. Acad. Sci., Washington, D.C., 1969.

11a. R. Banks, Some features of wind action on shallow lakes, Jour. ASCE, 101, 1975.

11b. J. McKee, in Camp, Harris, and McKee, Feasibility and Master Plan for Izmir City, Turkey, 1970.

12. W. Stumm and J. Morgan, Aquatic Chemistry, Wiley-Interscience, New York, 1970.

13. C. Sawyer, Basic concepts of eutrophication, Jour. Water Poll. Con. Fed., 38 (5), 737, 1966.

13a. C. Serruya, Water quality management of a warm ecosystem, WHO Water Qual. Bull., 1977.

14. K. Mackenthun, The phosphorus problem, Jour. AWWA, 1968.

15. C. Forsberg, Nitrogen as a growth factor in fresh water, IAWPR Conference on Nitrogen as a Water Pollutant, Copenhagen, 1975.

16. L. Keuntzel, Jour. Water Poll. Cont. Fed., 1969.

17. E. Shannon and P. Brezonik, Eutrophication analysis: A multivariate approach, Jour. ASCE, Env. Eng. Div., 98, 1972.

18. W. Pearsall, Jour. Ecol., 9, 1922.

19. Personal communication.

20. R. Megard, Eutrophication and the phosphorus balance of lakes, Amer. Soc. Agric. Eng., Winter Meeting, Chicago, Illinois, 1971.

21. W. Sonzogni and G. Fred Lee, Diversion of wastewaters from Madison lakes, Jour. ASCE, Env. Eng. Div., 100, 1974.

22. J. Edzwald, Discussion on Reference 21, Jour. ASCE, Env. Eng. Div., 101, 1975.

23. D. Bella, Dissolved oxygen variations in stratified lakes, Jour. ASCE, Env. Eng. Div., 96, 1970.

24. J. Newbold and J. Liggett, Oxygen depletion model for Cayuga Lake, Jour. ASCE, Env. Eng. Div., 100, 1974.

25. E. Fruh, Measurement of eutrophication and trends, Jour. WPCF, 38, 1966.

26. D. King, Carbon limitation in sewage lagoons, quoted in discussion on Carbon and nitrogen as regulators of algal growth, Jour. ASCE, Env. Eng. Div., 100, 1974.

27. G. Fuhs, Nutrients and aquatic vegetation effects, Jour. ASCE, Env. Eng. Div., 100, 1974.

28. R. Mitchell, Ed., Water Pollution Microbiology, Wiley-Interscience, 1972.

29. F. H. Borman, An ecosystem approach to problem solving, NATO Conference, Istanbul, 1973.

30. J. Patterson and P. Kodukula, Heavy metals in the Great Lakes, WHO Water Qual. Bull., 3 (4), 1978.

30a. Wagenot, Grenney, and Jurinak, Environmental transport model of heavy metals, Jour. ASCE, Env. Eng. Div., 104, 1978.

31. R. Vollenweider, Advances in Defining Critical Loading Levels for Phosphorus in Lake Eutrophication, OECD Co-operative Programme for Eutrophication, Canada Centre for Inland Waters, Box 5050, Burlington, Ontario, Canada, 1976.

32. P. Dillon and E. Rigler, A Simple Method for Predicting the Capacity of a Lake for Development Based on Lake Trophic Status, Journal Fisheries Res. Board, Canada, 32, No. 9, September 1975.

33. D. Imboden, Phosphorus Model of Lake Eutrophication, Limnol. Oceanogr., 19, 297, 1974.

Chapter 7

DISPOSAL OF WASTES TO RIVERS AND ESTUARIES

7.1 DISPOSAL OF WASTES TO RIVERS

A river is in many ways like a lake, and yet not quite the same. As a flowing body of water, a river often has a much shorter life-span relative to a lake. It also has some essential differences imposed by the dynamics of its flow conditions. Hence, a river is not just a long lake. From its source to the sea, a river may take from a few days to a few weeks to flow. A lake on the other hand may provide such a long residence time to the water that it can exhibit more clearly the effects of "age" discussed in Chapter 6, which a river cannot in the same sense.

The velocity of flow of a river also has several important consequences from an ecological standpoint. Velocity determines what particles will settle and what will not. Velocity also affects turbulence, aeration, turbidity, erosion, and transport of materials in a river. Thus, velocity has a significant influence on the nature and disposition of the benthic and aquatic communities within a river. The generally long length of a river exposes it to a large catchment area often widely varying in nature, and in its polluting potential, from one part to another. Thus, the whole river ecosystem consisting of the river water, its sediments, and its catchment area must be considered together and estimation of the waste assimilating capacity of a river must be done with great care in order to protect and promote the various beneficial uses to which a river can be put. Among these uses may be included drinking water supplies, the culture of fish and other aquatic products, and agricultural, industrial, navigational, and recreational uses of the river water. The objectives of river pollution control must invariably focus on the protection and promotion of these uses.

7.1.1 Organic Pollution

Historically, rivers were the first to which waste disposal had to be
controlled by the promulgation of suitable standards. As far back as 1912,
the United Kingdom found it necessary to adopt what came to be called the
"Royal Commission Standards" in order to protect their relatively small
streams passing through populated and rapidly industrializing areas. These
standards were mainly for the purpose of controlling organic pollution and
required that wherever the streams afforded a dilution to the wastewaters
of more than 8:1 but less than 150:1, the effluent needed to be treated
such that its biological oxygen demand exerted over 5 days (BOD_5) was not
greater than 20 mg/liter and its suspended solids not greater than 30 mg
per liter. Where the dilution was more than 150, preliminary treatment
such as screening was considered sufficient.

It is interesting to note that this so-called "20/30 standard" was
based on some common sense observations by the Royal Commission. It found
that most British streams would be able to ensure a dilution of 8:1 to the
wastewaters from their towns and communities (not true today). It also
noted that, where the BOD_5 of a stream was 2 mg/liter or less, the stream
could be regarded as "clean," and that the conditions downstream of an outfall were fairly satisfactory as long as the BOD_5 of the stream after admixture with the sewage just below the outfall did not exceed 4 mg/liter
[1]. Based on these observations, one could estimate the allowable BOD_5
of the effluent assuming stream flow as 8 liters for every liter of sewage
flow. Thus, if effluent BOD_5 is x mg/liter, one can make a simple materials
balance as follows for milligrams of BOD_5 upstream and downstream of the
outfall:

$$(2 \text{ mg/liter} \times 8 \text{ liters}) + (x \text{ mg/liter} \times 1 \text{ liter})$$

or $$\leq (4 \text{ mg/liter} \times 9 \text{ liters})$$

$$x \leq 20 \text{ mg/liter}$$

If the raw sewage had a BOD_5 of 200 mg/liter, it had to be treated sufficiently, generally involving biological treatment, to give a final value
of 20 mg/liter, or a BOD removal efficiency of $(200 - 20)/200 = 90\%$.

This was the simple basis of the Royal Commission Standard to protect
British streams. Sometimes, in the absence of an alternative standard, it
was also indiscriminately followed by other countries under an entirely
different set of conditions.

In the context of modern industrial and urban development, river pollution control today has become more complex and has to include various parameters besides BOD and suspended solids to preserve water quality. However, it is true that even today one of the most important parameters to consider is the dissolved oxygen (DO) content of the river and the effect of wastewater discharge on it. Oxygen is essential for fish and certain other aquatic life. Oxygen ensures "freshness" of the water and imparts to it a quality which is difficult to describe in words. Thus, oxygen constitutes an important asset of the river ecosystem and directly affects its various uses. A considerable portion of the discussion in this chapter is, therefore, devoted to oxygen balance in rivers.

In present times, concern with organic pollution is confined not only to oxygen balance in streams. Organic wastes also introduce nutrients like nitrogen and phosphorus and thus affect the productivity of the water. Heavy algal growths in downstream sections of a stream may increase difficulties in water treatment for downstream users and lead to other conditions typical of eutrophication in lakes, although eutrophication in the real sense of ecological succession may not occur in a stream owing to the limited "age" of the waters by the time they enter the sea. Streams entering lakes must, no doubt, be watched more carefully with regard to their nutrient levels as the stream and lake together with their catchments constitute one system.

Nitrification, when it occurs, places an additional demand on the oxygen contained in the stream water. This is discussed in Sec. 7.2.3.

7.1.2 Chemical Pollution

Another aspect of river pollution which increasingly concerns more and more countries is the effect of chemical and other discharges on water quality with regard to its hygienic and organoleptic properties, accumulation in the food chain, and other such aspects. This is discussed later with particular reference to the formulation of river pollution control standards.

The reader is referred to Figure 4.2 for a brief overview of the fate of conservative and nonconservative materials discharged to a river serving three towns in a row. Conservative materials brought in from each town merely accumulate in the river water. Nonconservative materials discharged into the river from one town undergo degradation and reduction in concentration with time until a fresh charge is received from the next downstream town.

It is interesting to note here that large and long rivers serving a number of towns and major industries located on their banks may, in fact, result in a natural, although unintentional, reuse of water over and over again, with each community downstream having to do with a progressively worsening quality of raw water. Some indication of this problem was given in Sec. 4.3. The accumulation of detergents has indeed attracted considerable attention.

Various biochemical reactions occur during the residence time of the chemical pollutants in the riverine ecosystem. Biochemical reactions imply entrance into the food chain at some trophic level. Thereafter, as Wuhrmann [1a] describes, two mechanisms are responsible for a concentration decrease from the water medium.

1. Oxidation in the energy metabolism leading to a permanent loss of the substance from the system. A typical example of this type of loss is biodegradation of organic matter.

2. Concentration of the substance, with or without change in its chemical composition, in progressively fewer but larger organisms up the food chain.

Some persistent substances such as the metals Hg, Cd, Mn, etc., as well as radioactive substances and nondegradable organics (e.g., certain pesticides) may even enter the food chain and show high concentrations in algae, fish, and some aquatic plants. The reader is referred to discussions in Sec. 4.4 in which several examples have been included.

Pollution invariably results in a reduction of the species diversity. The effect on the total biomass within the system, however, depends on the nature of the pollutant. Toxic pollutants may reduce while degradable organics may increase the total biomass production. Field evaluation of species diversity is difficult owing to the fact, even in the unpolluted condition, that diversity varies continually along the length of a stream.

A parameter of water quality which is finding increasing application, especially in industrialized countries, is the "carbon-chloroform extract" (CCE). CCE is a measure of the various organics and other adsorbable materials present in raw water in too small a concentration to be enumerated individually but which can be collectively significant in their effect on hygienic and organoleptic properties of the water. Another test used is

the polycyclic aromatic hydrocarbon (PAH) test for substances such as benz-
pyrenes.

7.1.3 Microbial Pollution

Microbial die off of enteric pathogens and nonpathogens occurs in
streams under natural conditions which are not favorable, nutritionally
and environmentally, for their growth. Microbial concentrations in streams
are generally measured in terms of indicator organisms such as coliforms
rather than specific pathogens. High concentrations are invariably observed
for some distance downstream of any outfall. Thereafter, the concentrations
may be low and mainly due to runoff from agricultural and other areas since
man is not the only source of coliforms. Bacteriological data are often
well supported by a sanitary analysis of the water, which includes ammonia,
nitrites, and nitrates among other items that throw light upon the past
history of a waterbody and the adjacent land usage.

Coliform die off rates have been discussed in Chapter 4. In several
U.S. streams they have been reported to show die off rate constants K_b
ranging from 1.0 per day in large streams to 1.8 per day in medium-sized
ones, as measured in field surveys [5]. At these rates, about 1 to 2 days
may be required in a stream to have 99% die off of coliforms (also see
Table 4.8). Temperature also has a considerable influence on die off rates
as discussed in Chapter 4.

Relatively little information is as yet available concerning the die
off rates of viruses in streams. Viruses can survive for prolonged periods,
and the coliform test is not a good indicator of virus hazard. Engineering
evaluations of the hazard have been made [2a].

Natural die off is essential to return the stream waters to their
original low microbial levels before they are reused downstream with or
without treatment. In cold climates and where density of development does
not permit enough time for die off before reuse, artificial disinfection
of effluents may be essential.

Where public water supplies are drawn from a stream, the degree of
treatment to be given depends on the physical, chemical, and biological
quality of the raw water. Where treatment is limited to simple chlorination of raw water, U.S. experience indicates that it is desirable to limit
the coliform concentration in raw water to about 50 coliforms per 100 ml;

but where treatment includes chemical coagulation, flocculation. filtration, and disinfection, a coliform concentration of even 5000 per 100 ml
can be routinely handled and brought down to drinking water standards.
Higher concentrations would demand more extensive treatment.

    Example 7.1 Estimate the distance downstream at which a large river
will be likely to have an average coliform concentration of 5000 per 100 ml
when the concentration just downstream of a sewage outfall is 150,000 per
100 ml. Assume K_b at 20° C = 1.0/day and river velocity is 12 m/min.

    Assuming plug-flow conditions, $N_t/N_o = e^{-K_b t}$

    $$\frac{5,000/100 \text{ ml}}{150,000/100 \text{ ml}} = e^{-1.0t}$$

    $$t = 3.4 \text{ days}$$

and

    $$\text{river distance} = \frac{12 \times 60 \times 24 \times 3.4}{10^3} = 59 \text{ km}$$

7.1.4 Self Purification

    The term "self purification" is often used to describe the various
physical, chemical, and biological interactions that go on within a stream
water and its sediments whereby the condition of a polluted stream gradually
returns to its prepollution state. Adapting from Wuhrmann [1a], one can
state that self purification is the sum of all those processes which lead
to a decrease of mass transport of organic and inorganic substances within
a given flow distance, and may be accompanied by transfer from an immature
to a mature ecosystem.

    A stream is like a plug-flow reactor in which there is no feedback of
information from the downstream to the upstream section. Each section is
like an individual batch in which the water quality is constant, and the
biocenosis is in a state of equilibrium with the ecological conditions in
that section [1a]. Thus, each section has its own ratio of photosynthetic
oxygen production (P) and community respiration (R) and a natural biological gradient, so to speak, exists along the length of the stream. As a
secondary effect, an oxygen gradient also exists. The characteristic of
any stream is a continually moving P:R ratio.

Working with artificial channels in which a preset biocenosis had been developed, Wuhrmann [1a] has demonstrated that so-called "pure water" biocenoses consisting exclusively of phototrophic microphytes are largely inactive in removing organic pollutants. Thus, clean water does not have an a priori self-purification capacity. As pollutants enter and hetero-trophs such as Sphaerotilus natans develop, the self-purifying capacity rapidly increases. The stream acquires a capacity to "assimilate" pollu-tion. This capacity, in fact, increases up to a point with the amount of pollutants discharged and, therefore, can be evaluated only when some boun-dary conditions (e.g., DO levels) are preset.

The well-known Monod relationship between growth rate and concentration of limiting substrates also applies to streams. A relationship exists be-tween the type and concentration of biodegradable substances at any section in a stream and the quantity of heterotrophs utilizing the compound. Con-sequently, the heterotrophic population is high immediately downstream of an outfall. The phototrophic population, on the other hand, depends on average light energy received at a section and on its penetration and uti-lization efficiency under given conditions.

If organic pollution is also stated in terms of energy imported into the system, the ratio of the two forms of energy E_s/E_L (i.e., substrate energy to light energy) can be shown to correlate with the ratio of photo-trophs to heterotrophs, P/H. From limited data, Wuhrmann [1a] has observed that when $E_s/E_L > 50$ to 100, heterotrophs completely overgrow phototrophs. When $E_s/E_L < 10$ to 20, the biomass is mainly phototrophic. The ratios of E_s/E_L and P/H vary with the seasons, the temperature, and the pollution load besides various other factors.

7.2 OXYGEN BALANCE IN STREAMS

It will be useful to first itemize the various sources and sinks of oxygen with reference to a watercourse since oxygen production and con-sumption continually change in time and space.

Oxygen Sources

1. Natural atmospheric re-aeration
2. Photosynthesis
3. Oxygen contained in tributary flow or incoming wastewater, if any

Oxygen Sinks

1. Respiration of all biota in water and bottom sediments

2. Immediate chemical oxygen demand (COD) of wastes discharged to a river system

3. Biological oxidation/decomposition of organic matter in water and bottom deposits

These sources and sinks just mentioned are shown in Figure 7.1 with reference to a section of a typical watercourse. Microbial respiration progresses practically at a constant rate throughout the day and night, whereas photosynthesis is limited to only daylight hours. Hence, as stated in Sec. 3.2.1, the net oxygen production (i.e., P - R) varies with time (Figure 7.2a). This is often neglected in conducting stream re-aeration studies and only the atmospheric re-aeration rate is taken into account. It is generally possible to assume that the latter occurs at a constant rate over a 24-hour cycle (Figure 7.2b).

Thus, even in a stream under natural conditions, DO shows a diurnal variation with lower values of DO early in the morning, and perhaps full or even supersaturation in the afternoon (see Figure 7.2c). Waters with high productivity display a larger amplitude between maximum and minimum values.

For a stream receiving pollution, biodegradation of the organic matter also consumes oxygen; this tends to give lower DO values on the diurnal profile. Engineering interest in this regard is generally not confined to

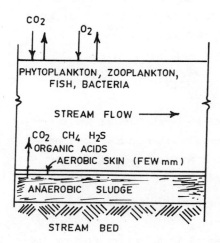

Figure 7.1 Organic sources and sinks in a typical watercourse.

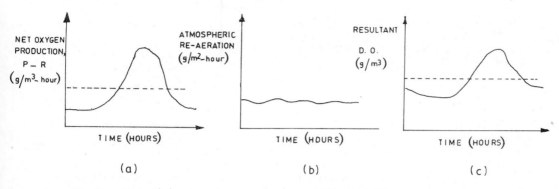

Figure 7.2 (a) Diurnal variations in DO as a result of net oxygen production, (b) DO present as a result of atmospheric re-aeration, and (c) DO concentration resulting from the action of net oxygen production and atmospheric re-aeration.

diurnal variations in a small section of the river, but rather to an over-all DO profile for a long stretch of the river downstream of an outfall. This is shown in Figure 7.3 where the horizontal axis denotes the time of flow or distance traveled downstream from the wastewater outfall; the vertical axis denotes the DO concentration, the maximum value of which is given by the saturation concentration. The actual concentration of the stream as it arrives at the outfall point may be equal to, or less than, the DO saturation value depending on its antecedent conditions. Once the wastewater enters the stream, biological activity increases and biodegradable substances are broken down to simpler end products, releasing the nutrients contained in the wastes.

Thus, in terms of our earlier presentation in Sec. 3.6.2 and Figure 3.11, we find a rapid increase in the respiration rate R of the biotic community. This accounts for a reduction in the overall DO content just downstream of the outfall.

Bacterial degradation of organic matter helps to release nutrients such as nitrogen and phosphorus into the stream water. Figure 7.3 shows the progressive conversion of ammonia from organic matter to nitrates in the river. These nutrients are then available for new organic growth through photosynthesis in downstream sections. Thus, an increased rate of productivity P may be observed in a stream as it flows towards the sea (Figure 3.11). Under most natural conditions P is in balance with R; but organic pollution tends to alter the P/R ratio.

The relative predominance of different micro- and macroorganisms in a stream as it flows and undergoes "self purification" is also shown in

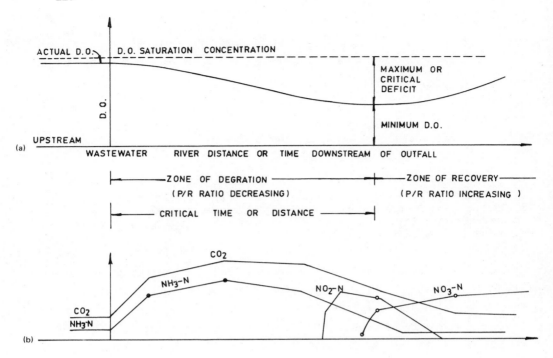

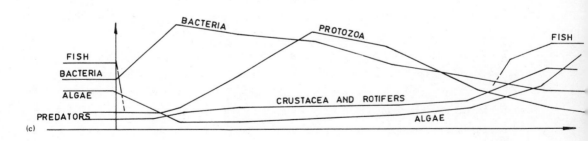

Figure 7.3 The DO profile downstream from an outfall resulting from
(a) Self purification, (b) the presence of nitrogen and carbon dioxide,
and (c) the dominant micro- and macroorganisms.

Figure 7.3. As explained in Sec. 3.11, predator organisms pick up in numbers gradually. Finally, fish life tends to flourish provided other conditions are favorable for growth.

The DO profile is, in fact, the result of two different phenomena:
some DO is consumed (BOD) by biotic respiration, while some DO is replen-

ished in the system from natural atmospheric re-aeration. Since BOD rates
are faster at the start owing to first-order kinetic patterns, the overall
DO profile declines for some distance until degradation and respiration
slow down giving re-aeration an opportunity to catch up, with a consequent
increase in DO to a new equilibrium level. This gives the well-known oxygen
sag curve, a spoon-shaped curve which describes the so-called "self-purification" process in a stream. Our main interest lies in the low point
reached in the curve.

The low point in the oxygen-sag curve occurs at some distance downstream of the outfall. This point is critical since the minimum DO in the
stream is reached here, and this minimum must be sufficient for the kind
of fish life desired to be protected in the stream. Field tests for DO
along a stream can confirm the position of the low point. Its position is
determined by various factors such as stream flow and velocity, temperature,
relative magnitudes of pollutional loads, their degradation rates, and the
stream re-aeration rate. The low point in a given case does not always
occur at the same geographical position, but may occur earlier or later
depending on the factors just enumerated. It is this occurrence of the
low point, often several kilometers downstream of the outfall, that accounts
at least partly for the apathy of the polluter who suffers not from his own
actions.

The DO profile may also be traced in terms of "deficits" rather than
actual DO. The deficit is the difference between the saturation value and
the actual DO. The point of lowest DO now becomes the point of maximum or
critical deficit. Thus,

C_s = oxygen saturation concentration, mg/liter
C = actual oxygen concentration, mg/liter at any time t
C_o = initial oxygen concentration, mg/liter
D_t = deficit from saturation, mg/liter at any time t
D_a = initial DO deficit, mg/liter = $(C_s - C_o)$
D_c = critical or maximum deficit, mg/liter

In determining the DO profile along a stream, Figure 7.4 can be utilized to express graphically three cases, two representing boundary conditions and one the intermediate condition of general interest. Case 1 represents the first boundary condition in which there is only re-aeration

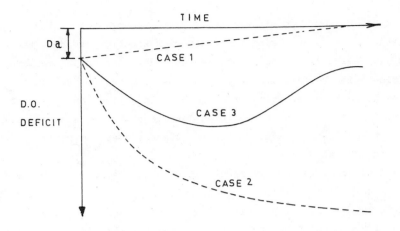

Figure 7.4 The DO deficit in a stream, with time, under three assumed conditions (see text for explanation).

and no BOD. Case 2 represents the other boundary condition in which BOD is exerted, but there is no re-aeration. Finally, Case 3 covers the condition as is generally experienced, namely, BOD is exerted and re-aeration also occurs.

Case 1

Case 1, involving only re-aeration, quickly enables the waterbody with an initial oxygen deficit D_a to reach saturation concentration and, thus, have zero deficit with time. The reader is referred to Sec. 6.6 concerning re-aeration in lakes from which the following equation was derived:

$$D = D_a e^{-K_2 t} \tag{7.1}$$

in which K_2 = re-aeration coefficient per unit time. Equation (7.1) is a typical first-order reaction in which the rate of reduction in deficit diminishes with time. Depending upon the value of K_2, the time required to bring about a desired change in the DO deficit can be estimated (Example 6.4).

Case 2

Case 2 concerns the other extreme, when BOD is being exerted but re-aeration is totally absent in the stream. Hence, the deficit increases rapidly as shown in Figure 7.4. This is, in fact, similar to making the very safe assumption of no re-aeration while estimating the waste assimilating capacity of a stream.

Since no re-aeration occurs, the DO contained in the upstream water is the only DO available for meeting the BOD of the waste. A typical computation can be illustrated as follows:

Find the stream flow required per 1000 persons discharging raw sewage with BOD_u at 75 g/capita-day if the upstream DO is at saturation at $20°$ C and the minimum DO downstream is to be not less than 3 mg/liter.

BOD_u of waste = 75 x 1000 = 75,000 g/day

DO available from upstream water = 9.17 - 3.0 = 6.17 mg/liter
$$= 6170 \text{ mg/m}^3 \text{ flow}$$

Hence,

$$\text{Stream flow required per 1000 persons} = \frac{75,000 \times 1000}{6170}$$
$$= 12,155 \text{ m}^3/\text{day} \ (\cong 12 \text{ m}^3/\text{capita-day})$$

Sewage flow (at an assumed rate of, for example, 120 liters/capita-day) amounts to 120 m³/day. This signifies a dilution ratio of about 100:1 as a rule-of-thumb approximation with a safe assumption of no aeration.

Incidentally, a dilution flow of about 12,000 m³/day per 1000 people is equivalent to the background figure of 4 ft³/sec per 1000 persons (range: 2.5 to 10 ft³/sec) often used in the U.S. in the past to make a rough-and-ready estimate of the dilution requirements for raw sewage discharged to streams [2]. It is evident that this method would give higher estimates of dilution requirements in most cases.

Case 3

Case 3 represents the usual situation in which a waste with BOD is introduced into a stream with re-aeration. Mixing occurs within a certain distance of the outfall depending on velocity. If samples of the stream water are taken at three points, A, B, and C, spaced sufficiently downstream to allow for mixing and biodegradation, and tested for BOD_u, the results may indicate a situation similar to the one shown in Figure 7.5. Each of the three samples would show a progressively lower value of L, the ultimate BOD.

If the values of L at each station are plotted against time as shown in Figure 7.5, it is seen that several different curves are possible depending upon the flow conditions in the stream. The condition most commonly assumed in dealing with stream flows is the normal plug-flow condition which lends itself readily to mathematical analysis. This curve is shown in the figure. The other curves shown in Figure 7.5 refer to different conditions such as

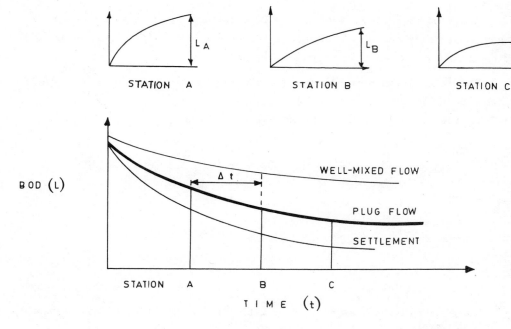

Figure 7.5 The decrease in ultimate BOD in a stream, with time, under different mixing conditions.

1. Excessive longitudinal mixing (but not quite complete mixing)
2. Channel scour
3. Sludge or solids settlement
4. Large amount of biological growths

It is thus evident that the BOD bottle cannot represent all these types of situations and simplifying assumptions need to be made. At just stated, the most common assumption made is that of plug flow. Several attempts are now being made to estimate the longitudinal mixing coefficient and to use appropriate equations which take into account the fact that streams do not have an ideal plug-flow regime. Data can be better fitted using the dispersed flow model or the dead zone model (Sec. 7.2.2), when significant longitudinal intermixing occurs.

On the assumption of plug flow and first-order kinetics, it can be shown from Chapter 12 that

$$L = L_a e^{-K_r \Delta t} \tag{7.2}$$

from which

$$K_r = \left(\frac{1}{\Delta t}\right) \ln\left(\frac{L_a}{L}\right) \tag{7.3}$$

where

K_r = BOD removal rate constant for stream per unit time

Δt = time elapsed between two adjacent stations

L_a and L = initial and final BOD_u, respectively, after Δt

At this juncture, it may be helpful to interject that K_r is, in fact, larger than the constant K for BOD removal in a bottle. It reflects the aquatic and benthic removals specific to the contact opportunity and mixing conditions in a stream, which operate in such a manner as to help the overall reaction go faster than in a BOD bottle.

The value of K_r may often be high for some distance downstream of an outfall, and low thereafter (for example, K_r'), although this is not always observed as it depends on the strength of the waste, the solids contained in it, and the relative proportion of waste and stream discharges (see Figure 7.11). The rate constant K_r' refers more to soluble BOD removal in a stream, whereas K_r includes the effects of adsorption, sedimentation, and immediate demands, if any. The appropriate value applicable to a stream section must be inserted in Eq. (7.2). The value may be different after each outfall into a stream. Section 7.3.5 gives empirical derivations of K_r' depending on stream characteristics.

Organics deposited on the stream bottom undergo anaerobic breakdown, releasing products which consume oxygen from the upper layers. The rate of oxygen consumption is expressed by the constant K_3 per unit time. This constant is generally not determined since it is included in the value of K_r determined from field tests.

The presence of benthic deposits becomes important when material that can settle is contained in the wastes discharged to a stream, or when the velocity of the stream flow is less than 6 to 12 m/min, promoting settlement of organic matter. If anaerobic flotation or channel scour occurs at high floods, the deposits may be dislodged and carried downstream.

Typical values of K_r and K_r' found in stream surveys are listed in Table 7.1; values for K_2 are listed in Table 7.2.

Combining the DO uptake rate with the re-aeration rate in a river, the rate of change in oxygen deficit with time can be stated in terms of the classic Streeter-Phelps equation as

$$\frac{dD}{dt} = -K_2 D + K_r L \tag{7.4}$$

Table 7.1 Typical Values of Rate Constant K_r for Streams

Stream section	Typical values per day (base e), 20° C (stream flows less than 20 m^3/sec)
Immediately downstream of outfall (K_r)	0.30-2.0
Subsequently downstream (K_r')	0.30-0.80

Source: Refs. 2, 3, and 5b.

Substituting from Eqs. (7.1) and (7.2), we get

$$\frac{dD}{dt} = -K_2 D_a e^{-K_2 t} + K_r L_a e^{-K_r t} \tag{7.5}$$

Integration gives

$$D_t = \frac{K_r L_a}{K_2 - K_r}(e^{-K_r t} - e^{-K_2 t}) + D_a e^{-K_2 t} \tag{7.6}$$

in which all terms have been defined previously.* Equation (7.6) is the well-known form of the oxygen-sag equation and enables the typical spoon-shaped curve to be developed from it if the values of all the terms on the right-hand side of the equation are known. Values of K_r and K_2 are affected by temperature (see Sec. 7.2.4).

---

*In this chapter, the use of capital letter K denotes values to the base e, while the use of small letter k denotes values to the base 10.

Table 7.2 Typical Values of Re-aeration Rate Constant K_2 for Some Waterbodies

Waterbody[a]	Re-aeration constant K_2 per day (base e), 20° C	
	Deep	Shallow
Small ponds	0.12	0.24
Sluggish streams, large lakes	0.24	0.36
Large streams of low velocity	0.36	0.46
Large streams of normal velocity	0.46	0.69
Swift streams	0.69	1.15
Rapids and waterfalls	1.15 and higher	

[a]For comparison, diffused-air activated-sludge plants exhibit values of K_2 over 8.0.
Source: Ref. 2.

7.2.1 Critical Deficit

The time at which the critical or maximum deficit occurs, t_c, can be obtained from Eq. (7.6) as

$$t_c = \left(\frac{1}{K_2 - K_r}\right) \ln\left[\frac{K_2}{K_r}\left(1 - \frac{D_a}{L_a K_r}^{K_2 - K_r}\right)\right] \tag{7.7}$$

If the BOD of the stream at the critical point is L_c, we get from Eq. (7.2)

$$L_c = L_a e^{-K_r t_c} \tag{7.8}$$

and D_c is given by

$$D_c K_2 = L_c K_r \tag{7.9}$$

or

$$D_c = \frac{K_r}{K_2} L_a e^{-K_r t_c} \tag{7.10}$$

Use of Eqs. (7.7) and (7.10) can enable one to estimate the time at which the critical deficit occurs and the magnitude of the critical deficit itself. If the stream velocity downstream of outfall is known, the distance at which the critical deficit occurs below the outfall can also be estimated (see Example 7.15). In the development of Eqs. (7.6) through (7.10), the effects of nitrification on the oxygen resources of the waterbody have been neglected.

If the oxygen-sag equation [Eq. (7.6)] is differentiated and equated to zero, as was done by Thomas [2], one can directly obtain the value D_c at the point of inflection. This yields a rather fearful looking but useful equation:

$$\frac{L_a}{D_c} = \left(\frac{K_2}{K_r}\right)^{1+\{[1-(D_a/D_c)^{0.418}]/[(K_2/K_r)-1]\}} \tag{7.11}$$

This equation is particularly helpful in evaluating the two boundary conditions with regard to the stream initial deficit conditions. The "best" case is when the stream arrives at the outfall point fully saturated with DO, namely, D_a is equal to zero. The other boundary case is when the initial deficit itself is equal to the critical deficit, namely, the stream has considerable antecedent BOD in it. Now, if the term K_2/K_r is replaced by f in Eq. (7.11), the two boundary cases yield the following results:

Case 1: When $D_a = 0$,

$$\frac{L_a}{D_c} = f^{(f/f-1)} \qquad\qquad (7.12)$$

Case 2: When $D_a = D_c$,

$$\frac{L_a}{D_c} = f \qquad\qquad (7.13)$$

Example 7.2 The conditions of flow in a stream are as given below. Estimate the permissible value of L_a (i.e., BOD_u of stream-sewage mixture just downstream of outfall). Saturation DO = 9.2 mg/liter at a stream temperature of $20°$ C. D_a = 2 mg/liter. Minimum DO downstream $\nleq$ 4 mg/liter. k_r = 0.1, k_2 = 0.24 (base 10).

Critical deficit D_c = 9.2 - 4.0 = 5.2 mg/liter

$$\frac{L_a}{D_c} = \left(\frac{0.24}{0.10}\right)^{1+[1-(2.0/5.2)^{0.418}/(0.24/0.1)-1]} = 4.0$$

or

$$L_a = 4.0 D_c = 4.0 \times 5.2 = 20.8 \text{ mg/liter}$$

Note: This gives the permissible ultimate BOD (not BOD_5, which will be less).

Example 7.3 For the data of Example 7.2, estimate L_a at the two boundary conditions (1) $D_a = 0$, and (2) $D_a = D_c$.
From the data,

$$f = \frac{0.24}{0.10} = 2.4 \qquad D_c = 5.2 \text{ mg/liter}$$

1. At $D_a = 0$,

$$\frac{L_a}{D_c} = f^{(f/f-1)} = 2.4^{(2.4/2.4-1)} = 4.45$$

$$L_a = 4.45 \times 5.2 = 23.2 \text{ mg/liter}$$

2. At $D_a = D_c$,

$$\frac{L_a}{D_c} = f$$

$$= 2.4$$

$$L_a = 2.4 \times 5.2 = 12.5 \text{ mg/liter}$$

Fair [2] has simplified the use of Eq. (7.11) by the development of charts (see Figure 7.6) from which the values of permissible L_a can be read corresponding to different values of the ratios $f = K_2/K_r$ and D_a/D_c.

Having found the permissible value of the ultimate BOD, L_a, in the stream-sewage mixture just downstream of the outfall, one needs to know

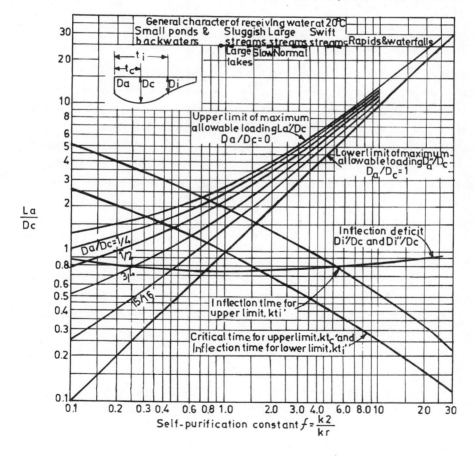

Figure 7.6 The allowable loading of receiving waters (adapted from Ref. 2).

the stream flow and the BOD_u already present upstream, as well as the wastewater flow, in order to compute the permissible BOD_u of the proposed wastewater discharge. This computation is illustrated in Example 7.3.

Example 7.4 The permissible value of L_a calculated for a stream is 10 mg/liter. If the stream flow is 1 m^3/sec, with an antecedent BOD_u of 4 mg/liter and the wastewater flow is 10,000 m^3/day, compute the permissible BOD_u of the wastewater.

$$(84,400 \ m^3/day \times 4 \ mg/liter) + (10,000 \ m^3/day \times x) = (94,400 \ m^3/day \times 10 \ mg/liter)$$

Hence,

$$x = \frac{(94,400 \times 10) - (84,400 \times 4)}{10,000} = 60 \ mg/liter$$

7.2.2 Dispersed Flow

Our previous discussion was based on the simplifying assumption of ideal plug-flow conditions in streams. However, in reality the flow is not ideal but dispersed, some parts flowing faster and some slower than the mean-flow velocity. Hence, in recent years much attention has been paid to the development of models which take this nonideal condition into account to enable a better estimation of the spread and concentration of pollutants discharged to streams, estuaries, and other bodies of water.

When a slug of tracer is injected into a static mass of water, diffusion or spreading occurs as a result of random molecular motion. In a static fluid, this is described by Fick's Law, according to which the rate of mass transfer of tracer per unit area (the mass flux) resulting from molecular diffusion is proportional to the concentration gradient of the tracer. In a flow with laminar velocity, spreading of the tracer occurs from the combined action of diffusion and velocity variations; this combined action results in dispersion. If the flow is turbulent, molecular diffusion is neglected since dispersion results mainly from the mixing property of turbulence.

Taylor described turbulence as "an irregular motion which in general makes its appearance in fluids when they flow past solid surfaces or even when past neighboring streams of the same fluid flow or over one another." Qualitatively, turbulence has also been described as a hierarchy of eddies along with the mean motion. Energy is transferred from larger eddies to

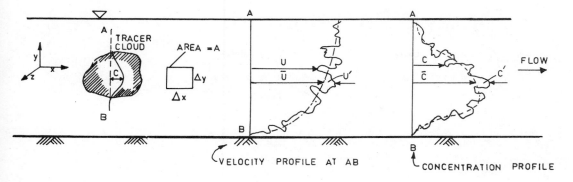

Figure 7.7 Mass transport during two-dimensional flow (adapted from Ref. 42).

smaller ones and finally dissipated via viscous action. An essential feature of turbulence is its random motion. The concept of turbulence as a random phenomenon has also led to the use of statistical representations as we shall see later.

In order to be able to predict the concentration of a substance at any distance from its source, its transport must be studied using a three-dimensional approach since diffusion and dispersion occur in that way (x, y, and z directions, Figure 7.7). However, often for the sake of simplicity when dealing with open-channel flow as in rivers, variations in parameters are assumed to occur only in the x and y directions; no velocity and concentration gradients are assumed to exist in the z direction. Thus, when a tracer is injected in an open channel, the "cloud" is assumed to spread out first in only the x and y directions and, after the cloud has completely filled the cross section in the y direction, the primary variation in concentration is taken to be in just one direction (x) so that, thereafter, a one-dimensional model is assumed to apply.

This approach is found to be reasonably satisfactory in practice where the real conditions do not deviate substantially from the assumed ones. But, for example, in meandering rivers and estuaries with islands and sandbars present, lateral dispersion (i.e., in the z direction) may have a considerable effect on the overall situation and use of the one-dimensional model may not be satisfactory [42].

A.   Molecular Diffusion

In accordance with Fick's Law, diffusion gives the mass flux of a substance relative to the mass average velocity in a given direction as follows:

$$Jx = -D_m \left(\frac{\partial C}{\partial X} \right) \qquad\qquad (7.14a)$$

where

J_x = mass flux relative to mass average velocity in the x direction

D_m = coefficient of molecular diffusion ($L^2 T^{-1}$)

C = concentration or mass density of the substance (ML^{-3})

X = distance (L)

In order to find the molecular diffusion in laminar flow, it is desirable to transform the above statement of Fick's Law from relative coordinates to stationary coordinates in which case it can be shown that the following relationship results:

$$n_x = CU_x + J_x$$

or

$$n_x = CU_x - D_m \left(\frac{\partial C}{\partial X} \right) \qquad\qquad (7.14b)$$

where

n_x = mass flux in the x direction with reference to stationary coordinates

U_x = mass average velocity in the x direction with reference to stationary coordinates

Let us now examine two-dimensional open-channel flow by taking the specific case of a tracer introduced as a slug into this flow (Figure 7.7). By assumption, the cloud spreads with time in the x and y directions only, and at any position there will be a distribution of concentration in the cloud as shown at section AB in the figure. As the cloud enlarges, the maximum concentration in the cloud decreases.

A mass balance or conservation equation can be applied to the tracer entering and leaving through the sides of an element of area A which is fixed relative to a stationary coordinate system and has sides Δx and Δy. Applying in the x direction we can state:

Rate of change of   input - output + generation
amount in element = or decay within element

If the tracer is a conservative substance, the last term can be dropped and the change in concentration with time $\partial C/\partial t$ can be equated to change in mass flux n_x with distance as the thickness of the element tends to zero. Thus,

$$\frac{\partial C}{\partial t} = -\frac{\partial n_x}{\partial x} \tag{7.14c}$$

Substituting from Eq. (7.14b) into Eq. (7.14c) gives

$$\frac{\partial C}{\partial t} = -\frac{\partial}{\partial x}\left(CU - D_m \frac{\partial C}{\partial x} \right) \tag{7.14d}$$

From the equation of continuity, $\partial U/\partial x = 0$ (since velocity in the vertical direction is zero for the configuration shown in the figure). Equation (7.14d) can be rewritten as

$$\frac{\partial C}{\partial t} = -U\left(\frac{\partial C}{\partial x} \right) + D_m \left(\frac{\partial^2 C}{\partial x^2} \right) \tag{7.15a}$$

Similarly, accounting for molecular motion in the y direction, Eq. (7.15a) can be enlarged to include a term as shown:

$$\frac{\partial C}{\partial t} = -U\left(\frac{\partial C}{\partial x} \right) + D_m \left(\frac{\partial^2 C}{\partial x^2} \right) + D_m \left(\frac{\partial^2 C}{\partial y^2} \right) \tag{7.15b}$$

Thus, the rate of change of concentration with time at any point is partly due to convection [i.e., $U(\partial c/\partial x)$] and partly due to molecular diffusion (terms with D_m).

B. Turbulent Diffusion

Referring back to Figure 7.7, we see the concentration and velocity distributions along AB as would be obtained for turbulent flow if actual measurements were made.* The values u and c denote the actual values at any point at any instant, while $\bar{U}$ and $\bar{C}$ indicate the time-averaged values at that point (i.e., up to the dashed lines).

Fluctuations from the average are denoted by U' and C'. It will be recalled that turbulent flow always shows random fluctuations. It is now necessary to write

$$u = \bar{U} + U'$$

$$c = \bar{C} + C'$$

and V' for velocity fluctuations which occur in the y direction even though the main flow is in the x direction. Thus, following a similar approach as used in deriving Eq. (7.15b), we get, in terms of time-averaged quantities,

---

*Velocities can be determined at any depth using a pitot tube or current meter.

$$\frac{\partial \overline{C}}{\partial t} + \overline{U}\left(\frac{\partial \overline{C}}{\partial x}\right) = D_m\left(\frac{\partial^2 \overline{C}}{\partial x^2}\right) + D_m\left(\frac{\partial^2 \overline{C}}{\partial y^2}\right) + \frac{\partial}{\partial x}\left(-\overline{U'C'}\right) + \frac{\partial}{\partial y}\left(-\overline{V'C'}\right) \quad (7.15c)$$

The last two terms of Eq. (7.15c) denote the net convection resulting from U', V', and C'. However, the transport resulting from this convection has often been shown to follow a diffusive type law analogous to Fick's law. Thus, by analogy to molecular diffusion, turbulent (eddy) diffusion coefficients e_x and e_y are introduced with reference to the x and y directions, respectively, to give

$$\overline{U'C'} = -e_x\left(\frac{\partial \overline{C}}{\partial x}\right) \qquad (7.15d)$$

$$\overline{V'C'} = -e_y\left(\frac{\partial \overline{C}}{\partial y}\right) \qquad (7.15e)$$

and, hence, Eq. (7.15c) is now written as

$$\frac{\partial \overline{C}}{\partial t} + \overline{U}\left(\frac{\partial \overline{C}}{\partial x}\right) = \frac{\partial}{\partial x}\left(D_m + e_x\right)\frac{\partial \overline{C}}{\partial x} + \frac{\partial}{\partial y}\left(D_m + e_y\right)\frac{\partial \overline{C}}{\partial y} \qquad (7.15f)$$

Equation (7.15f) is the mass balance equation for turbulent flow in terms of time-averaged quantities.

C. Longitudinal Dispersion

It was stated earlier that, as the tracer cloud moves downstream, it spreads out in both the x and y directions until the cloud completely fills the channel cross section from top to bottom. There is then only a slight variation of $\overline{C}$ in the y direction, and the main variation is now in the x direction. This marks the end of what is called the "convective" period and the start of the "diffusive" period. During the diffusive period, Eq. (7.15f) is still valid but can be replaced by a simpler one-dimensional model to solve for the concentration distribution in the x direction.

Again, Taylor was the first to show that the convection associated with velocity variations from the average at a cross section is proportional to the longitudinal gradient of the concentration [43]. Consequently, the one-dimensional equation used for estimating the transport of the tracer in a channel of uniform cross section takes the following well-known form:

$$\frac{\partial C}{\partial t} + U\left(\frac{\partial C}{\partial x}\right) = D\left(\frac{\partial^2 C}{\partial x^2}\right) \qquad (7.16)$$

where

C = cross-sectional average concentration
U = cross-sectional average velocity (mean-flow velocity)
D = dispersion coefficient, area/time

By definition, both molecular and turbulent diffusion are included in the dispersion coefficient D which indicates the overall effect of the mixing process. In turbulent flow, molecular diffusion is negligible and even turbulent diffusion in pipes has been shown by Taylor to be about 1% or less of the total dispersion. Mixing resulting from velocity and concentration variations from the average in the longitudinal direction is primarily responsible for dispersion. Thus, D is also often referred to as the longitudinal dispersion coefficient. However, as discussed further in Sec. 7.10 concerning estuaries, lateral dispersion (i.e., in the z direction) which is neglected in the above derivation may in some cases have a considerable effect on the overall dispersion and give much larger values of D than anticipated. The equation gives the rate of change of the concentration with time at any cross section of flow as a result of dispersion <u>and</u> convection. If in a given case the dispersion rate is small enough for the dispersion term to be dropped from Eq. (7.16), the resulting equation will be similar to a typical plug-flow equation.

The terms diffusion and dispersion have sometimes been used interchangeably. Generally, the transport associated with molecular and turbulent action has been referred to as diffusion and the transport associated with velocity variation across the flow section as dispersion. Dispersion is the major convective transport mechanism of interest in stream studies.

Typical values of D in streams and estuaries are of the following order of magnitude:

Waterbody	D (m^2/hr)
Small streams	10-10^3
Large streams and rivers	10^3-10^5
Estuaries	10^5-10^6 and over

Fisher [24a] states that D in streams and estuaries should be proportional to the square of the width of the stream and the inverse of the depth. Thus, relatively shallow and wide streams should give higher values of D, whereas deeper and narrower ones should give lower values. Section 7.3.7 gives various field and analytical methods of estimating D and discusses further the effect of lateral velocity variations on the longitudinal dispersion coefficient.

Hays [26] has developed a different model which takes into account the process of trapping and slowly releasing a tracer in a stream. He divides a stream into two sections, a main stream and a dead zone, with mass transfer taking place across the interface. This model enables one to obtain a better fit of observed stream data than is possible by using the dispersed flow model. However, Hays's model can be used provided the ratio of dead spaces to total stream volume is known or can be estimated, and one has prior knowledge of some other factors applicable to the stretch of stream in consideration. Furthermore, the solution procedure is complex and requires the use of a computer [22].

When heavier-than-water pollutants are present (e.g., sand or silt with radioactivity or organics adsorbed on their surfaces) in a discrete form in a stream, they will remain in suspension [46,47] provided that

$$\frac{V_s}{U_*} < A$$

where

$\quad V_s$ = settling velocity of particles in quiescent fluid
$\quad U_*$ = shear velocity = $\sqrt{gRS}$ where
$\qquad\qquad$ R = hydraulic radius
$\qquad\qquad$ S = slope of water surface
$\quad A$ = constant, generally between 0.5 and 1.5 [47]

Such particles in suspension tend to move in the form of a "cloud" with a mean velocity which is less than the mean flow velocity, and their dispersion coefficient is greater than that for neutrally buoyant particles, depending on the parameter V_s/U_*. For example, if $V_s/U_* = 0.3$, the dispersion coefficient of the heavier-than-water pollutants is five times as great as that of the water itself. The general trend is that the heavier the particles, the greater the dispersion coefficient [45-46]. This is not so important if the dispersion is due mainly to transverse variations of velocity rather than to vertical variations.

C. DO Profile under Dispersed Flow Conditions

The effect of having a relatively highly dispersed flow instead of plug flow is to alter the magnitude, time, and location of the critical DO deficit downstream of an outfall (see Figure 7.8). The difference may be substantial for high values of D. Another result of dispersion is to spread out the incoming waste so that the effects of the waste on DO and

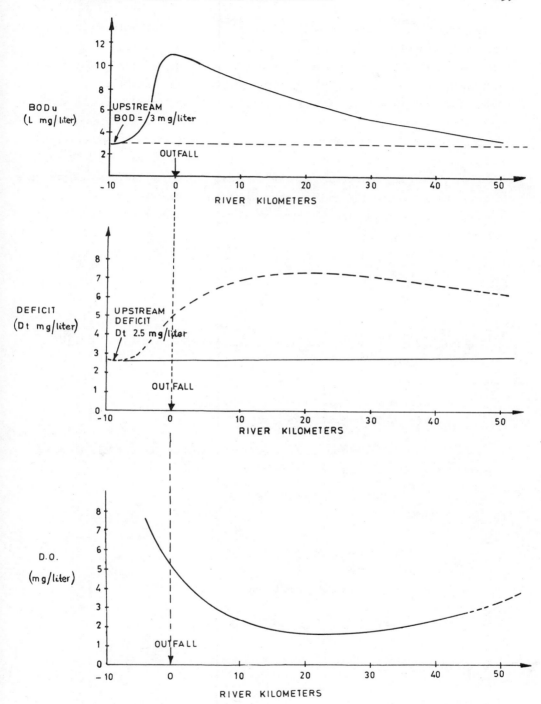

Figure 7.8 Profiles for DO, BOD_u, and DO deficit along a river with significant mixing conditions (data from Example 7.5).

other items are also felt for some distance <u>upstream</u> of the discharge point besides the usual downstream effects (see Figure 7.8).

In order to enable an estimation of the DO profile under dispersed flow conditions, O'Connor [23] has developed a solution based on the Streeter-Phelps equation, but has introduced the term D to take into account longitudinal mixing. For the steady addition of, for instance, W units of BOD/day to a stream at a point x_1, the DO deficit D_t and L can be estimated at any distance x upstream or downstream by using the following equations. (It may be noted that, as the dispersion coefficient D approaches 0, the equations give the same answers as the Streeter-Phelps equations for plug-flow conditions.) The BOD_u at any distance upstream or downstream of outfall is given by

$$ L = \left(L_o + \frac{W}{Qa_1} \right) e^{b_1} \tag{7.17} $$

and the deficit D_t is similarly given by

$$ D_t = \frac{K_r}{K_2 - K_r} \left(L_o + \frac{W}{Q} \right) \left[\left(\frac{1}{a_1} \right) e^{b_1} - \left(\frac{1}{a_2} \right) e^{b_2} \right] + D_o e^{b_2} \tag{7.18} $$

where

L_o = upstream BOD_u, mg/liter

D_o = upstream DO deficit, mg/liter

L_o and D_o should be measured sufficiently upstream of the outfall so as not to be affected by longitudinal mixing.

W = BOD_u discharged to stream, kg/day

Q = stream flow, m^3/day

U = stream velocity, m^3/day

K_r = BOD removal rate, per day, in stream

K_2 = stream re-aeration rate, per day

D = longitudinal mixing coefficient, m^2/day

$$ a_1 = \sqrt{1 + \frac{4K_r D}{U^2}} $$

$$ a_2 = \sqrt{1 + \frac{4K_2 D}{U^2}} $$

$$ b_1 = \left[(1 \pm a_1) \frac{U}{2D} (x - x_1) \right] $$

$$b_2 = \left[(1 \pm a_2) \frac{U}{2D} (x - x_1) \right]$$

In computing b_1 and b_2, the <u>negative</u> sign is used for downstream positions, and the <u>positive</u> sign for upstream positions. The term $x - x_1$ gives the distance from the outfall and is, therefore, negative for upstream positions and positive for downstream ones. Example 7.5 illustrates the use of this method.

Example 7.5   A town located on the estuarine part of a river discharges 100,000 kg BOD_u/day. The freshwater flow is 100 m^3/sec at an average summer temperature of 20° C. The average velocity $U = 4.6$ m/min (6624 m/day). The estimated values of K_r and K_2 at 25° C are 0.2 and 0.15 per day, respectively, while D has been estimated at 10^6 m^2/hr (24×10^6 m^2/day). The far upstream value of L_o = 3 mg/liter and D_o = 2.5 mg/liter. Compute the values of L and D_t at 3 km upstream, and 3 km and 20 km downstream of the outfall (saturation DO at 20° C = 9.2 mg/liter).

From the given data, we first compute

$$a_1 = \sqrt{1 + \frac{4K_r D}{U^2}} = \sqrt{1 + \frac{(4)(0.2)(24 \times 10^6)}{(6624)^2}} = 1.2$$

$$a_2 = \sqrt{1 + \frac{4K_2 D}{U^2}} = \sqrt{1 + \frac{(4)(0.15)(24 \times 10^6)}{(6624)^2}} = 1.15$$

$$\frac{W}{Q} = \frac{100,000 \times 10^6}{100 \times 86,400 \times 10^3} = 11.57 \text{ mg/liter}$$

At 3000 m upstream of the outfall:

$$b_1 = \left[(1 + a_1) \frac{U}{2D} (x - x_1) \right] = \left[(1 + 1.2) \frac{6624}{2(24 \times 10^6)} (-3000) \right]$$
$$= -0.91$$

$$b_2 = \left[(1 + a_2) \frac{U}{2D} (x - x_1) \right] = \left[(1 + 1.15) \frac{6624}{2(24 \times 10^6)} (-3000) \right]$$
$$= -0.89$$

$$L = \left(L_o + \frac{W}{Qa_1} \right) e^{b_1} = \left(3 + \frac{11.57}{1.2} \right) e^{-0.91} = 5.1 \text{ mg/liter}$$

Farther upstream the value of L will approach L_o (i.e., 3 mg/liter). In order to estimate deficits,

$$D_t = \frac{K_r}{K_2 - K_r}\left(L_o + \frac{W}{Q}\right)\left[\left(\frac{1}{a_1}\right)e^{b_1} - \left(\frac{1}{a_2}\right)e^{b_2}\right] + D_o e^{b_2}$$

$$D_t = \frac{0.2}{0.15 - 0.2}(3 + 11.57)\left[\left(\frac{1}{1.2}\right)e^{-0.91} - \left(\frac{1}{1.15}\right)e^{-0.89}\right] + 2.5e^{-0.89}$$

$$D_t = 2.52 \text{ mg/liter}$$

At 3000 m downstream of the outfall:

$$b_1 = \left[(1 - 1.2)\frac{6624}{2(24 \times 10^6)}(3000)\right] = -0.0828$$

$$b_2 = \left[(1 - 1.15)\frac{6624}{2(24 \times 10^6)}(3000)\right] = -0.0621$$

$$D_t = \left(\frac{0.2}{0.15 - 0.2}\right)(3 + 11.57)\left[\left(\frac{1}{1.2}\right)e^{-0.0828} - \left(\frac{1}{1.15}\right)e^{-0.0621}\right]$$
$$+ 2.5e^{-0.0621}$$

$$D_t = 5.3 \text{ mg/liter}$$

Similar computations can be made for other distances downstream of the outfall. The results are given below and also shown graphically in Figure 7.8.

Distance (km)	BOD_u L_1 (mg/liter)	Deficit D_1 (mg/liter)	DO (mg/liter) (20°C)
-3	5.1	2.5	6.7
3	11.5	5.3	3.9
20	7.3	7.2	2.0
50	3.2	6.6	2.6

From Figure 7.8 it can be observed how, in the case of a river flow with a fair degree of longitudinal mixing, the effect of the waste discharge is felt even some distance upstream of the outfall; downstream the position and magnitude of the critical deficit can be seen. Use of the conventional Streeter-Phelps equations assuming plug flow would have given a different DO profile in this case.

7.2.3 Nitrification

In our discussions thus far we have dealt with the oxygen demand resulting from stabilization of the carbonaceous substances in organic matter. Nitrogenous substances contained in organic wastes entering a stream may also place an additional demand on its oxygen resources. It will be recalled from Chapter 4 that nitrogenous substances yield ammonia (NH_3) which is oxidized to nitrites (NO_2^-) with the help of chemoautotrophs of the genus Nitrosomonas, and eventually to the fully oxidized nitrates (NO_3^-) with the assistance of Nitrobacter. The actual requirement of oxygen is 4.33 mg/liter for every mg of NH_4-N/liter, which is slightly less than the stoichiometric requirement [12].

Although oxygen is consumed in nitrification, it is not lost from the nitrate form and remains available for use when the free dissolved oxygen is depleted and denitrification occurs. But nitrification demand can push the DO sag curve lower and retard the recovery of a stream to a desired DO level downstream. Thus, nitrification demand should not be neglected where it is likely to occur.

The question then posed is, when is nitrification likely to occur in a stream and at what rate? Several factors affect the occurrence of nitrification since it is brought about through the action of rather sensitive organisms which demand that environmental conditions be held within relatively narrow limits to favor their growth.

Initially, the carbonaceous demand in a stream becomes low before nitrification starts. Thus, a heavily polluted stream does not undergo nitrification until the carbonaceous demand is first satisfied. In that sense, nitrification is a problem associated with relatively cleaner streams. Nitrifying organisms are also sensitive to temperature and pH. Stream temperatures around 25 to 28° C, as in warmer climates or where thermal pollution is occurring, are optimum for their growth, and pH limits must range between 7.6 and 8.0. Dissolved oxygen is generally required to be over 2 mg/liter in the water. The generation times of nitrifying organisms have been observed to range between 10 and 30 hr depending on temperature, oxygen content of the water, and the initial concentration of nitrifying bacteria [12,13]. Courchaine [13] ascribes nitrification in streams to a combination of three factors: established populations of nitrifiers from well-treated effluents of upstream waste treatment plants, high nitrogenous demand relative to carbonaceous BOD, and optimal stream temperatures.

The per capita contribution of nitrogen in sewage ranges from 6 to 12 g/day (Table 5.1). Normally, biologically treated effluents contain about 10 to 25 mg/liter oxidizable nitrogen, thus having a potential oxygen demand of 43 to 108 mg/liter. This value is in the same range as, if not more than, that of the carbonaceous BOD_u of effluents from such treatment plants. Clearly, therefore, nitrification demand can have a considerable effect on the oxygen-sag curve.

The kinetics of nitrification have been described by some workers as following first-order kinetics:

$$\frac{dN}{dt} = -K_N N \tag{7.19}$$

where

N = concentration of ammonia nitrogen, mg/liter

K_N = nitrification rate constant, per day

When integrated, this has given the equation

$$N = N_0 e^{-K_N t} \tag{7.20}$$

where N_0 = initial concentration of ammonia nitrogen, mg/liter. Others have preferred to assume that K_N is not constant but dependent on the momentary substrate concentration. Thus

$$\frac{dc}{dt} = -K_N N_0 (N_0 - N) \tag{7.21}$$

In this case, after integration the expression can be written as

$$\frac{N}{N_0 - N} = e^{-K_N N_0 (\alpha - t)} \tag{7.22}$$

where α = time for half completion of reaction.

Liebich [12] also includes a term in the above formulations to take into account the initial bacterial concentration which he feels should not be neglected in the case of nitrification in streams. Autocatalytic equations have been shown to fit better the data observed from streams.

The expression for DO deficit in a stream given earlier in the differential form [Eq. (7.4)] needs to be modified to take into account the oxygen demand due to nitrification. Unfortunately, this is not possible since the

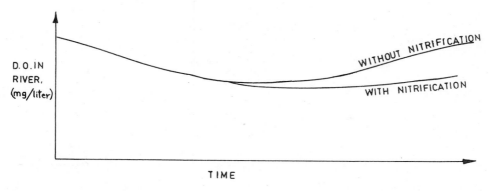

Figure 7.9 The DO profile along a stream with and without nitrification.

modified differential equation cannot be solved. Hence, the method generally followed is to calculate separately for carbonaceous and nitrogenous demands and superimpose the two for small, successive intervals of time. The nitrification rate is converted to the oxygen consumption rate by multiplying by 4.33.

The increases or decreases in oxygen or oxygen deficits are additive. They can be plotted individually and the net oxygen concentration curve can be computed [3]. A typical oxygen-sag curve with and without taking nitrification into account is shown in Figure 7.9. The estimation of K_N is dealt with in Sec. 7.3.5.

7.2.4 Temperature Effects on Rate Constants

Rate constants for BOD removal, bacterial die off, nitrification, and re-aeration in streams are all affected by temperature. The general equation correlating the rate constants at different temperatures has been derived in Chapter 12. It is of the type

$$K_{T^\circ C} = K_{20^\circ C}\theta^{T-20} \tag{7.23}$$

in which the values of θ corresponding to different rate constants are given in Table 12.4. The values of θ corresponding to the rate constants involved in stream self-purification studies are given below for ready reference.

Rate constant	θ
K_r	1.01-1.075
K_r'	1.06-1.075
K_2	1.024-1.028
K_N	1.106
K_b	1.075

The critical DO, and its point of occurrence, both shift with temperature. In any field survey one must invariably note the temperature of the stream water at the time of survey.

7.3 EVALUATION OF PARAMETERS AFFECTING OXYGEN BALANCE

From the equations presented thus far, it is evident that the allowable pollutional load on a stream is determined by the relative magnitudes of the following parameters:

1. Initial DO content (or deficit from saturation) of the stream
2. Minimum DO downstream (or permissible critical deficit)
3. BOD of upstream water
4. Stream flow
5. Rate constants K_r, K_2, and K_N
6. Temperature
7. Longitudinal dispersion in the stream

A brief discussion of each of these parameters will be useful in order to describe their effect on the oxygen balance of a stream and the manner in which field data are to be obtained and analyzed.

7.3.1 Initial DO in Stream

The initial DO content in a stream is the result of various conditions affecting deoxygenation and re-aeration upstream of the outfall. At best, the stream may be fully saturated with DO as it arrives at the outfall. The DO concentration at saturation conditions C_s is given by

$$C_s = \frac{0.678(P - V_P)}{t^\circ C + 35} \tag{7.24}$$

where

C_s = oxygen saturation concentration, mg/liter

P = pressure in mm Hg

V_P = vapor pressure in mm Hg

t = temperature (between 0 and $35°$ C)

C_s values calculated at 760 mm pressure and different temperatures are presented in Table 13.1. At high elevations these values need to be multiplied by $P/760$ to account for reduced pressure.

Thus, it is evident that streams in warmer regions and at high elevations are inherently in an unfavorable condition compared to streams in temperate climates or on flat plains. The temperatures in which pollution control engineers are interested are the summer temperatures. In the temperate regions the summer temperatures of streams are often around $20°$ C (C_s = 9.17 mg/liter), but in the warmer countries they may range around 25 to $30°$ C and even up to $35°$ C in the smaller streams in very hot climates. Consequently, streams in warmer countries may have 10 to 20% less DO to begin with, in addition to the rapid rate of deoxygenation at higher temperatures. Similarly, mountain streams have less DO at saturation owing to lower pressures.

Usually, streams do not have DO at saturation levels, but somewhat less. In the absence of field data, the initial DO in even a clean stream as it arrives at an outfall point is assumed to be at 70 to 90% of the saturation value.

7.3.2 Minimum DO in Stream

The minimum DO content that should be present in a stream is dictated by the desirability of protecting its fish culture potential. For example, the recommended minimum DO values for fish culture are

Fish	Minimum DO (mg/liter)
Trout	5
Salmon and other game fish	4
Some coarse fish	2

The decision to ensure a minimum DO content in a stream above 2 mg/liter is mainly a commercial one. Where fisheries of commercial interest are to be protected, the minimum DO may be specified around 4 mg/liter, and in the special case of trout, around 5 mg/liter. With reduced DO, fish become more susceptible to metallic poisons and other hazards. Fish avoid low DO areas and their migration patterns from fresh to marine waters and vice versa may be disrupted (Sec. 8.6.5).

Temperature also has an effect on fish tolerance to low DO. Generally, the colder the temperature, the lower the rate of asphyxiation (Sec. 3.13).

7.3.3 BOD of Upstream Water

The BOD of upstream water is the result of natural drainage into the stream and discharges from urban and industrial activities in the concerned catchment area. It will not be out of place here to refer to the "condition" of a stream in terms of its BOD_5 as is often done in the U.K. [1]:

Condition	BOD_5, 20° C, of stream
Very clear	1
Clean	2
Fairly clean	3
Doubtful	5
Bad	10

Such a mode of expressing the condition of a stream is neither necessarily universally applicable nor essential. What is necessary is to know the BOD_5 or BOD_u of the upstream water during the critical summer season of low flow, or to be able to estimate it with reasonable accuracy. There are several polluted streams and rivers in many countries where the BOD_5 values are known to range between 10 and 100 mg/liter or even more. A formidable job of cleaning up such rivers faces the concerned communities.

7.3.4 Stream Flows

As is well known, dilution plays an important role in waste disposal. Besides the direct effect of dilution at high flows, there are several other effects such as the effect of flows on the re-aeration rate constant K_2, on benthic deposits, and on the stream ecosystem as a whole, which must be taken into account.

Interest lies in the minimum stream flow or the minimum dilution afforded to wastes. Minimum flow fluctuates from year to year and may sometimes even be equal to zero. Thus "minimum" flow must be given a statistical definition to be meaningful. This is an important aspect of water quality management, especially in perennial streams, because the application of standards for limiting concentrations (mg/liter) of different substances becomes impossible unless the "minimum" flow is precisely stated. A network of flow-gauging stations may need to be established at suitable locations in a given region as discussed in Sec. 7.7.

Both flood and drought flow curves generally tend to be skewed to the right. Low-flow frequency analysis can be readily performed using usual statistical analysis procedures as suggested by Gumbel [4], Velz [5], and others.

It is preferable, as with all statistical analysis, that the data cover as long a period of time as possible. Flood records spanning a period of 25 years or more would be more reliable, but often one has to make do with less. A short period of record of, for example, 10-15 years may be adequate for a stream with less variability of flow from year to year; a record of 30 years may be necessary to ensure equal reliability where drought flows are highly variable. Statistical analysis methods may need to be adapted to deal with short period records [4]. There are other approaches by which such situations can be handled, such as correlating the data with those of other adjacent streams of similar hydrological characteristics for which long-term data are available, or synthetically generating data using other methods [6].

Stream flow data are often recorded as 1-day averages, but in picking out the minimum flow value occurring in a given year, the 1-, 7-, 15-, or even 30-day average value is used. As the averaging period is lengthened, the value reduces since very low flows do not last for long in perennial streams. For water pollution control purposes, the average low flow over a number of consecutive days equal to the time of passage through the critical reaches of the stream is considered as the desirable average period to adopt. This varies from stream to stream and in U.S. experience it ranges from 3 to 15 days or more [5]; New York State, for example, requires that the minimum average 7-day consecutive flow be considered. In some countries only the daily flows are subjected to a statistical analysis.

The data so picked out for each year are now arranged in order of magnitude. The frequency of occurrence of each value is computed as shown in Example 7.6, plotted on logarithmic-probability paper, and the line of best fit is drawn. If n number of values are arranged in ascending order, and if a particular value is, for instance, mth in this series, then

$$\text{Frequency of occurrence} = \frac{m}{n+1} \tag{7.25}$$

and

$$\text{Its recurrence interval} = \frac{n+1}{m} \tag{7.26}$$

Thus, a low-flow value which has, for example, a frequency of occurrence interval of 20% (i.e., a 20-percentile value) will have a recurrence interval of once in 5 years.

The design flow for waste assimilation capacity studies is sometimes taken to be the one with a recurrence interval of 5 to 10 years or more. The longer the recurrence interval, the lower the stream flow implied, and thus the stricter the treatment required. Some agencies specify a recurrence interval of only 2 years (i.e., average low flow),* some vary it depending on the nature of the substance involved. Factors involved in deciding upon this period are discussed in Sec. 7.8 which is concerned with setting pollution control standards. Example 7.7 illustrates the effect that the choice of recurrence interval has on the degree of treatment required to meet given control standards. The method of analysis just described is, of course, inapplicable to cases where stream flows are artificially controlled through upstream reservoirs.

The flow in a stream is sometimes stated in terms of "yield" per unit drainage area. Thus, the yield may be stated, for example, as m^3/sec per km^2 of drainage area tributary to a gauging station. The minimum yield is then stated with reference to its averaging period (1, 7, or 15 days, or whatever is desired) and the return period (once in 2, 5, or more years). This enables a comparison to be made between different streams and at different stretches in a given stream with regard to their yield.

A further parameter enabling evaluation of the drought characteristics of streams is the "variability ratio"; some streams have a more variable

---

*In a study of two rivers in Turkey [5a] it has been shown that the five-percentile value obtained from a log-normal plot of all flows is very similar to the 50-percentile values obtained from a plot of the minimum daily flows, and minimum consecutive 7-day averages observed over several years.

flow than others. This can be expressed through the variability ratio,
which is generally the ratio of the once-in-10-years low flow to the average low flow of a river. The variability ratio may, in fact, be more
accurately termed the uniformity ratio since a high value for the ratio
denotes greater uniformity of flow from year to year. The values may range
from 0.1 or less to 0.85 or more for perennial streams.

In tropical countries with a distinct rainy season extending over only
2 to 4 months, many streams are of a seasonal character and carry no appreciable flow for some time in the dry months. Often the flow carried by
them may be the result of only wastewater discharges by upstream users.
The dilution afforded by freshwater is practically zero and a higher degree
of waste treatment or alternative methods of disposal may need to be sought.

Example 7.6 Data are available for a stream from 1944 onward. Inspection shows that the low flows generally occur in August or September
each year. The flows in August 1944 were recorded as follows in m^3/sec for
each day:

Day	Daily average flow	Day	Daily average flow	Day	Daily average flow
1	4.60	11	4.05	21	4.16
2	4.38	12	3.94	22	4.05
3	4.38	13	3.94	23	4.05
4	4.16	14	3.72	24	3.94
5	4.16	15	3.72	25	3.94
6	3.94	16	3.94	26	3.83
7	3.94	17	3.94	27	3.83
8	4.05	18	4.16	28	3.94
9	4.16	19	4.16	29	3.94
10	4.16	20	4.16	30	3.94
				31	3.94

For the above month, the minimum daily average flow was 3.72 m^3/sec.
However, the minimum average flow for 7 consecutive days is found by inspection of the data recorded between August 11 and 17, 1944; the average
value was 3.89 m^3/sec. Similarly, the value of minimum average flow for
7 consecutive days for other years have been found from 1944 to 1970 and
are tabulated in ascending order to enable the plotting of data as shown
in Figure 7.10.

Order number (or m)	Month/year	Flow (m^3/sec)	Frequency percent [$1/(n + 1)$]
1	Aug. 1945	1.05	5.6
2	Aug. 1949	2.01	11.1
3	Aug. 1947	2.07	16.7
4	Aug. 1951	2.74	22.2
5	Aug. 1952	3.13	27.7
6	Aug. 1967	3.67	33.3
7	Aug. 1944	3.89	38.8
8	Sep. 1946	4.13	44.5
9	Sep. 1948	4.18	50.0
10	Oct. 1950	4.45	55.5
11	Nov. 1963	4.92	61.1
12	May 1962	5.38	66.7
13	Dec. 1968	6.03	72.2
14	Feb. 1964	8.55	77.8
15	Oct. 1965	8.90	83.3
16	Sep. 1969	9.11	88.9
17	Sep. 1970	9.68	94.5

From the line of best fit drawn on log-normal distribution paper, it is seen that, for the river in question, low flows recur as follows:

Low flows (m^3/sec)	Percent of time the stated value or less occurs	Recurrence interval (years)
1.93	10	10
2.21	14.3	7
2.55	20	5
4.30	50	2

Note: If a recurrence interval of only 2 years is specified, the designer can count upon a (statistical) "minimum" flow of 4.30 m^3/sec for dilution of wastes, whereas if a recurrence interval of 10 years is specified, the "minimum" flow for design purposes is less than half this value.

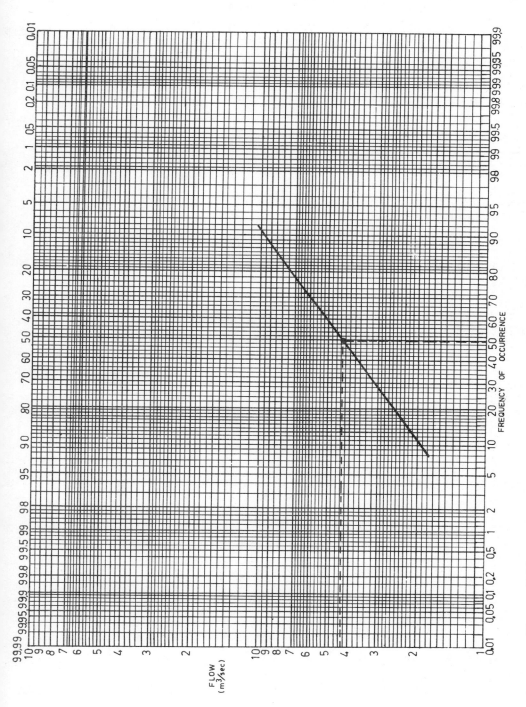

Figure 7.10 The frequency of low flows at a river gauging station (see Example 7.5).

Example 7.7 A town proposes to discharge sewage at approximately 17,000 m^3/day to the stream whose flow characteristics are as evaluated in Example 7.6. Estimate the permissible BOD_5 in treated effluent if the minimum DO downstream is to be kept at not less than 4 mg/liter at low flow in the stream with a recurrence interval of (1) once in 2 years and (2) once in 10 years. Assume K_2/K_r = f = 2.0; temperature of stream and sewage = $20°$ C; DO saturation concentration C_s = 9.2 mg/liter; upstream water is at 50% saturation and has a BOD_5 = 2 mg/liter; raw sewage BOD_5 = 300 mg/liter.

Case (1) From Example 7.6, for a recurrence interval of once in 2 years, low flow Q = 4.3 m^3/sec.

$$D_a = 9.2 - 0.5(9.2) = 4.6 \text{ mg/liter}$$

Assume D_a is not substantially affected by sewage discharge.

$$D_c = 9.2 - 4.0 = 5.2 \text{ mg/liter}$$

To enter Figure 7.6,

$$\frac{D_a}{D_c} = \frac{4.6}{5.2} = 0.885 \quad \text{and} \quad f = 2.0$$

Hence,

$$\frac{L_a}{D_c} \simeq 2.2 \quad \text{and} \quad L_a = 2.2 \times 5.2 = 11.4 \text{ mg/liter}$$

Taking BOD_5 = 0.7 BOD_u, we get a permissible BOD_5 of river-sewage mixture = 0.7 × 11.4 ≃ 8 mg/liter. Hence, by a materials balance, for sewage at approximately 0.2 m^3/sec,

$$(\text{Sewage } BOD_5 \times 0.2) + (2 \text{ mg/liter} \times 4.3) = [8 \text{ mg/liter} \times (4.3 + 0.2)]$$
$$\text{Permissible sewage } BOD_5 = 137 \text{ mg/liter}$$
$$\text{Treatment efficiency required} = \frac{300 - 137}{300} = 54\%.$$

Case (2) For a recurrence interval of once in 10 years, Q = 1.93 m^3/sec. Assuming that other conditions remain unchanged, the permissible BOD_5 of river-sewage mixture = 8 mg/liter as before. But, making a fresh materials balance, permissible sewage BOD_5 = 66 mg/liter. Hence,

$$\text{Treatment efficiency required} = \frac{300 - 66}{300} = 78\%$$

The example illustrates how the choice of a recurrence interval affects the treatment requirements.

7.3.5 Determination of Rate Constants

Determination of rate constants is clearly the most difficult of all the tasks involved in formulating the oxygen-sag equation for any stream. All the constants K, K_r, K_2, and K_N are affected by temperature. The values of the constants are also affected by the nature of the waste, the stream flow and turbulence, and such other factors. Their values are, therefore, variable from section to section of any stream, and even in a given section they are affected by seasonal and other factors. The constant K_N is similarly affected by various factors but is applicable only when conditions are favorable for nitrification.

Furthermore, the shape of the oxygen-sag curve is quite readily affected by even slight errors in the determination of the rate constants. Therefore, a method often recommended in this regard is to obtain the actual DO-sag curve by sampling along a stream at the same time that field tests for rate constants are done. In this manner, the theoretically calculated DO-sag curve and the actual one can be matched and, if necessary, the calculated values of the constants can then be slightly adjusted until the two curves match reasonably well (see Example 7.15).

Thus, the determination of constants involves field work in the form of sampling and DO analysis along with obtaining required hydrological data for the stream in question. The work can be minimized by limiting it to the critical (summer) season only. Re-aeration rate constants can be determined solely from hydrological data, if available, and verified with actual field values.

A. Rate Constant K

The rate constant K pertains to the BOD test performed in the laboratory for a given wastewater. Its method of estimation is given in Sec. 4.6.1, along with typical values for sewage and a few other wastewaters.

B. Rate Constant K_r

The rate constant K_r signifies the BOD removal rate under stream conditions. Its value depends on the nature and strength of the waste, the stream temperature, biomass, contact surfaces, mixing conditions, and other factors [3]. The deoxygenation rate in a stream can be determined by taking samples along a stream, estimating therefrom as illustrated in Example 7.8. Plots on semilog paper tend to give straight lines (see Figure 7.11).

In the stretch of stream just below an outfall, a faster rate of BOD removal may be observed (especially if organics tend to settle out) than farther downstream where the BOD removal rate K_r may decrease and level off at some lower value. If BOD values are measured at two points A and B along a river and the time of travel t noted, then from Eq. (7.3) one can obtain the value of K_r at the stream temperature as

$$K_r = \frac{1}{t} \ln (L_A/L_B) \tag{7.27a}$$

in which L_A and L_B are the BOD values at the upstream and downstream sampling stations, respectively. Several stations would have to be sufficiently close together if the change in deoxygenation rate along the stream below an outfall is to be traced.

In order to facilitate work when field surveys are difficult to undertake, attempts have been made to correlate empirically in-stream deoxygenation rates to channel characteristics (such as depth, velocity, Froude and Reynolds numbers, flow, and wetted perimeter). Wright and McDonnell [5b] have shown (utilizing data from 23 river systems) that K_r 20° C, base e, correlates well with stream flow Q (correlation coefficient = 0.926) as follows:

$$K_r = 10.3(Q)^{-0.49} \tag{7.27b}$$

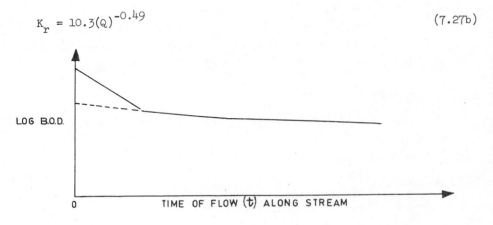

Figure 7.11 The rate of BOD removal in a stream.

where Q = stream flow in cubic feet per second. Furthermore, they observed that for flows of less than about 22 m^3/sec (800 ft^3/sec), K_r values were generally quite high (from 0.3 per day to 4.25 per day), but for larger flows the deoxygenation rates ranged only from 0.3 per day down to 0.08 per day, and were consistent with typical BOD bottle rates. Empirical equations (if shown to be in reasonable agreement with field observations) are valuable for use in predicting stream DO profiles at different flows.

Example 7.8 A river survey shows a reduction of 80% BOD in a 1-day time of flow downstream from an outfall. Estimate K_r.
From Eq. (7.27a),

$$K_r = \frac{\ln (L/0.2L)}{1.0} = 1.61 \text{ per day}$$

Example 7.9 Estimate the percent of change in K_r value for doubling of the stream flow. Use Eq. (7.27b).

$$\frac{K_{r_1}}{K_{r_2}} = \frac{10.3/(Q_1)^{0.49}}{10.3/(2Q_1)^{0.49}} = 1.4$$

Hence K_{r_2} value reduces to 71% of the K_{r_1} value as the stream flow doubles.

B. Re-aeration Constant K_2

Re-aeration in a stream is the result of gas transfer from the atmosphere at the air-water interface followed by its diffusion through the waterbody. The re-aeration rate constant K_2 per unit time is, therefore, affected by several factors such as those discussed in Chapter 13 and include the following:

1. Area of air-water interface in relation to the volume of water in a given stretch of a stream. If stream width is W, the length L, and the average depth H, its area per unit volume A/V is proportional to 1/H since

$$\frac{A}{V} = \frac{L \times W}{L \times W \times H} = \frac{1}{H} \tag{7.28}$$

2. Velocity and turbulence of the stream. The effective surface area of the stream is affected not only by (1) above but also by the turbulence and the resulting surface disturbances which constantly renew the water surface.

3. Partial pressure of O_2 above the water surface.

4. Coefficient of mass transfer and diffusion. This coefficient is in turn affected by temperature, increasing in value as the temperature increases. The presence of surface active agents (e.g., detergents) decreases the value of the transfer coefficient.

5. Deficit from saturation DO in the stream. The mass transfer is proportional to the DO deficit (C_s - C) at any instant and, thus, acts like
   a first-order reaction; the transfer is high when the deficit is high.

   The value of the re-aeration constant K_2 can be determined from field
tests over a given stretch of a stream. It varies from stretch to stretch
in a stream depending on velocity, depth of flow, etc., in that stretch.
Furthermore, K_2 varies seasonally, daily, diurnally, and even hourly as
temperature, flow, and other factors change. The presence of photosynthetic activity during the daytime results in an apparently higher value
of K_2 during the daytime compared with the nighttime. This can be estimated [7] although, generally, for all practical purposes, algal and other
photosynthetic production of oxygen in a stream is neglected. The use of
a gas tracer technique developed by Tsivoglou et al. [7a] is reported to
give better estimates of K_2.

   In summer, higher temperatures aid re-aeration, but low flows and low
turbulence tend to offset that advantage to some extent. Often, in fact,
K_2 increases as flow decreases (Figure 7.12). Thus, it is essential to
record all the stream flow conditions, temperature, etc., at the time a
field test is done as described below.

   Attempts have been made in recent years to derive empirical correlations between K_2 and hydraulic parameters such as velocity and depth of
flow to make it possible to readily estimate K_2 for different stretches of
a stream. What is more important is that this would enable predictions of
K_2 values at different flows to make computer simulations of stream water
quality possible at different assumed conditions.

   Neglecting the re-aeration that is due to photosynthesis, we get the
generalized expression for natural aeration as

$$\frac{dc}{dt} = K_L \frac{A}{V} (C_s - C) \tag{7.29}$$

in which K_L = mass transfer coefficient (liter)(t^{-1}) for the liquid phase
which governs the process; the other notations have been defined earlier.
For a stream, from Eq. (7.28), A/V = 1/H and C_s - C = deficit D, mass/vol-
ume. Thus,

$$\frac{dc}{dt} = \left(\frac{K_L}{H} \right) D = K_2 D \tag{7.30}$$

   In general terms, K_2 can be correlated inversely with the stream
depth H and velocity V since velocity affects turbulence and re-aeration.

Table 7.3 Values of Constants[a] in Eq. (7.31)

C	n	m	θ	Reference
2.25	1.0	4/3	1.028	8
2.30	1.0	5/3	1.024	9

[a]In S.I. units corresponding to U, m/sec, and H, m.
In Ref 9, Churchill et al. take the hydraulic radius of
the stream (area/wetted perimeter) instead of average
depth H.

Suitable constants can be introduced into this general correlation to account for different stream conditions. Temperature effects can also be
included as shown below.

$$K_2 = \frac{CU^n}{H^m}\left(\theta^{T-20}\right)$$
(7.31)

where

K_2 = re-aeration rate constant, base e, (t^{-1}) at $20°C$

H   = average depth of stream, L

U   = velocity of stream flow, $L(t^{-1})$

C   = constant depending on stream water quality characteristics

m and n = constants depending on stream physical conditions

θ   = temperature effect coefficient

T   = temperature, °C

The values of the constants in Eq. (7.31) as estimated by Langbein and
Durum [8] by plotting available data from various sources are given in
Table 7.3, together with those obtained by Churchill et al. [9], after
converting to S.I. units.

Mention must be made of the stream re-aeration prediction model developed from theoretical considerations by O'Connor and Dobbins [10]. Brown
[16] performed a statistical evaluation of some re-aeration prediction
equations and found that the O'Connor-Dobbins model performed better than
those obtained from empirical correlations of observed coefficients with
the average values of hydraulic variables (e.g., equation of Churchill et
al.). The O'Connor-Dobbins model identifies two different flow regimes,
isotropic and nonisotropic, for which the equations for predicting K_2 are
as follows:

1. For shallow, turbulent streams, with Chezy's C > 20,

$$K_2 = \frac{2.303 \times 480 D_L^{1/2} S^{1/4}}{H^{5/4}}$$ (7.32)

2. For deep, less turbulent streams, with Chezy's C < 20,

$$K_2 = \frac{(D_L U)^{1/2}}{H^{3/2}}$$ (7.33)

where

D_L = mixing coefficient resulting from diffusion (0.81×10^{-4} ft^2 per hr at 20° C)

S = slope of water, ft/ft

H = average depth, ft

U = average stream velocity, ft/hr

K_2 = re-aeration rate constant, per hr (base e)

Other models have also been developed which take into account the longitudinal mixing coefficient D_L in predicting K_2 [17]. Such models are likely
to be preferred as they take into proper account the mixing conditions in
a river. Efforts are also being made to develop models taking into account
the net productivity of the stream water [18].

In order to be able to use Eqs. (7.31), (7.32), or (7.33) for an estimation of K_2 at different stream flows, it is necessary to know how the
stream depth and velocity fluctuate with flow. Typical correlations take
the following forms derived from log/log plots:

Depth H   = aQ^P (7.34)

Velocity U = bQ^q (7.35)

Width W   = cQ^r (7.36)

in which Q = stream flow, m^3/sec, or yield, m^3/sec-km^2, and a, b, c, p, q,
and r are constants for a given stream. Since Q = W x H x U, we get

$$(cQ^r)(aQ^P)(bQ^q) = Q^{1.0}$$ (7.37)

Thus, the sum of exponents p + q + r = 1.0 and the product of coefficients
abc = 1.0.

For several rivers in the U.S., correlations have shown typical values
of coefficients p, q, and r as below [3]:

H varies as $Q^{0.3-0.5}$ (7.38)

U varies as $Q^{0.5-0.7}$ (7.39)

W varies as $Q^{0.1-0.3}$ (7.40)

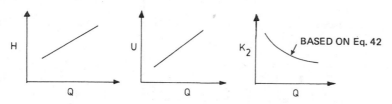

Figure 7.12 Typical relationships between Q and H, U, and K_2 in a stream (see text for explanation).

Such correlations can also be shown graphically as illustrated in Figure 7.12, in which the concerned parameter is plotted against the stream flow or yield.*

Example 7.10 Assuming that Langbein and Durum's generalized equation applies to the conditions in a river basin, estimate K_2 per day at 25° C when the velocity of stream flow is 0.75 m/sec, and average stream depth is 4 m. Also derive a correlation between Q in m^3/sec and K_2 if, for the river in question, H in meters = $0.8Q^{0.4}$ and U in m/sec = $0.10Q^{0.55}$, where Q is in m^3/sec.

1. From Eq. (7.31) and using the values of the constants given in Table 7.3, we get

$$K_2 = 2.25 \cdot \frac{U}{H^{4/3}} (1.028)^{T-20} \qquad (7.41)$$

$$= 2.25(0.75) \frac{1}{(4)^{4/3}} (1.028)^{25-20}$$

$$= 0.30 \text{ per day}$$

2. Substituting given relations of H, U, and Q in Eq. (7.41), we get the following relationship:

$$K_2 = \frac{(2.25)(0.10)(Q)^{0.55}}{[(0.80)(Q)^{0.4}]^{4/3}} (1.028)^{T-20}$$

or, $K_2 = 0.30Q^{0.017}(1.028)^{T-20}$

In some instances, K_2 has been found to vary as the negative

---

*As an illustration, Pyatt and Grantham [11] found the following correlations in the Farmington River basin in Massachusetts and Connecticut, where Q is given in ft^3/sec, and H, U, and W in ft, ft/sec, and ft, respectively:

H = 0.241 $Q^{0.33}$
U = 0.113 $Q^{0.56}$
W = 36.68 $Q^{0.11}$

power of Q, signifying that K_2 reduces as flow increases. Thus, for example, a value observed in one river was

$$K_2 \propto Q^{-0.13} \tag{7.42}$$

The constant K_2 can also be determined along with K_r from field studies. If a stream is sampled at, for instance, two stations, A and B, at a known distance apart, to measure the DO, BOD_u, temperature, and stream flow at each station, one can estimate the oxygen deficits from saturation. The value of K_r is then obtained as seen earlier [Eq. (7.27a)].

Furthermore, if the DO deficits at stations A and B are D_A and D_B, respectively, the average change in deficit per unit time is the result of the difference in the oxygen consumption rate and the re-aeration rate. Thus, from Eq. (7.8), over a short distance,

$$\frac{D_B - D_A}{t} = K_r \bar{L} - K_2 \bar{D} \tag{7.43}$$

where $\bar{L}$ and $\bar{D}$ refer to average values between stations A and B. Rearranging Eq. (7.43), we get

$$K_2 = K_r \left(\frac{\bar{L}}{\bar{D}} \right) - \frac{D_B - D_A}{\bar{D}t} \tag{7.44}$$

The average change in DO deficit per unit time will be more rapid for a shallow, fast-flowing stream than for a deeper, sluggish stream. The values of K_r and K_2 obtained by using Eqs. (7.27) and (7.44) are specific for the stream stretch and the flow and temperature conditions at which the survey was performed. The value of K_2 so obtained can be compared with those obtained by using Eq. (7.31), (7.32), or (7.33) solely with hydraulic data.

Example 7.11   Two stations, A and B, are 7.5 km apart on a stream whose average velocity of flow in the stretch is 0.2 m/sec. The other data obtained in the survey are presented in Table 7.4. Estimate K_r and K_2 per day for the stream. Saturation DO at 23° C = 8.67 mg/liter.

$$\text{Time of flow} = \frac{7500 \text{ m}}{0.2 \text{ m/sec}} = 0.438 \text{ days}$$

From Eq. (7.27), $K_r = \dfrac{1}{0.438} \ln \left(\dfrac{11,000}{6,300} \right) = 1.27$ per day

Table 7.4 Observed Data in Stream Survey

Station	Flow (m³/day)	Temperature (°C)	BOD_u (mg/liter)	BOD_u (kg/day)	DO Deficit (mg/liter)	DO Deficit (kg/day)
A	1830×10^3	23	6.0	11,000	6.27	11,300
B	1900×10^3	23	3.31	6,300	3.81	7,000
			Average:	8,650		9,150

From Eq. (7.44),

$$K_2 = K_r \left(\frac{\bar{L}}{\bar{D}}\right) - \frac{D_B - D_A}{\bar{D}t}$$

$$= 1.27 \frac{(8650)}{(9150)} - \frac{(7000 - 11300)}{(9150)(0.438)}$$

$$= 2.272 \text{ per day}$$

C. Constant K_N

Field determination of the constant K_N for nitrification rate, per unit time, can also be undertaken in a manner similar to the one for estimating K_r. In fact, sampling for both can be included in the same survey program.

Nitrification is generally measured either as gain in nitrates (NO_3^-) along a river stretch, or as reduction in ammonia (NH_3) or ammonium ion (NH_4^+), and plotting the data to find the slope of the line of best fit.

Example 7.12 Based on nitrate production along a watercourse, 90% oxidation was observed to be achieved within 0.62 day travel time. Estimate K_N under these conditions, assuming first-order kinetics.

$$K_N = \frac{\ln\left(\frac{N_o}{N}\right)}{t} = \frac{\ln\left(\frac{N_o}{0.1N_o}\right)}{0.62} = 3.71 \text{ per day}$$

7.3.6 Stream Temperature

In view of the importance of temperature on all the rate constants seen earlier, prediction of dissolved oxygen concentrations in streams requires prior knowledge of temperature. The critical temperature in this

regard is the summer temperature. Stream temperature can be readily determined at several stations established within a river region to cover the
main flow and its major tributary flows. Often stream temperature data are
not readily available and other methods must be used.

Kothandaraman [14] and Song et al. [15] have attempted to develop
water temperature models applicable to streams. The use of such models is
particularly useful where water temperature data are not available over a
sufficiently long period of time and have to be deduced from air temperature
data which are more often available. Song et al. gave consideration to the
effects of air temperature fluctuations and precipitation on the water temperature change, but soon found that precipitation had very little significance and could be dropped from consideration. Air temperature showed high
serial correlation with water temperature; both exhibited fluctuations that
were serially correlated up to 7 days. Dimensional considerations lead to
a suggestion that the sensitivity of the stream water temperature to the
air temperature departure was inversely proportional to the square of the
average depth of the stream. Thus, deeper streams were less sensitive to
fluctuations in air temperature than were shallow ones.

However, for most natural streams with no significant heat sources or
sinks, the average stream temperature is equal to the average air temperature over a long period of time. Thus, where insufficient water temperature
records are available, it may be possible to use the average air temperature
in place of the average water temperature. But where fluctuations in the
temperature of a stream are needed to be known, as in predictive studies
or water quality simulation studies on a computer, a stochastic model can
be used to generate synthetic temperature data [15].

7.3.7 Determination of the Dispersion Coefficient

Presently, two approaches are available for the determination of the
dispersion coefficient D in streams and estuaries: (1) tracer tests in the
field and (2) analytical methods based on certain stream parameters.

Field Methods

Tracer tests using dyes or radioactive isotopes take into account the
realities of the field situation but are expensive and sometimes difficult
to execute. The method involves the addition of a tracer as a slug (pulse

input) followed by measurement of the concentration at several time periods
at a point sufficiently downstream to obtain a time—concentration curve.
If this curve were truly Gaussian in nature, the relationship between time
t and concentration C at any cross section X would be given by

$$C = \frac{M}{A\sqrt{4\pi Dt}} \, e^{-[(x-Ut)^2/4Dt]} \qquad (7.45)$$

where

M = total mass of tracer added

A = cross-sectional area of flow normal to X

U = mean velocity of flow

Thus, from the observed time—concentration curve, the value of the
dispersion coefficient D can be obtained. However, the curves observed in
the field are not quite Gaussian, but are skewed especially at short distances downstream from the injection point. In keeping with Eq. (7.16),
the one-dimensional model is applicable after about six dimensionless time
units downstream, but a much longer time is required in order to let the
curve become nearly Gaussian. Most field observations are made at less
than 40 dimensionless time values for practical reasons since, otherwise,
a large initial dose of tracer and a long enough reach of uniform channel
would be required. The time—concentration curves, therefore, often exhibit
some degree of skewness, and use of methods based on Gaussian properties
must be limited to only those cases where the "long-tail" effect is minimum.

Methods based on Gaussian properties take into account the basic relationship that the dispersion coefficient is a measure of the rate of change
of variance of the tracer cloud. First published by Newton in 1905 in a
theory of Brownian motion, the relationship can be stated as

$$D = \frac{1}{2} \frac{d}{dt} (\sigma x^2) \qquad (7.46)$$

in which σx^2 = the variance of the concentration distribution with respect
to distance along the stream. This equation is derived from the diffusion
equation without regard to the initial concentration distribution and can,
therefore, be used within the diffusive period (see Sec. 8.8.4 concerning
the estimation of D in coastal waters). Furthermore, the equation is
valid even if time—concentration curves are used instead of distance—concentration curves [24a].

Thus, if time—concentration curves are obtained for two downstream
stations, A and B, both sufficiently downstream of the injection point,

the dispersion coefficient D can be estimated from the rate of change of
variance, namely

$$D = \frac{1}{2}U^2 \left(\frac{\sigma_B^2 - \sigma_A^2}{\bar{t}_B - \bar{t}_A} \right) \tag{7.47}$$

where

σ_B^2 and σ_A^2 = variances of the time—concentration curves at A and B

$\bar{t}_B$ and $\bar{t}_A$ = mean times of passage of the tracer cloud past each station

U = mean velocity of flow

The method of Levenspiel, based on estimations of the variance of a
single time—concentration curve, may give erroneous results owing to long-
tail effect since the tail has a longer moment arm than the region near the
peak and the estimation of variance is based on taking second moments.

The approach used by Harris [25] is based on statistical considerations but involves mean travel time rather than the variance of the curve
and gives

$$\bar{U} = x \cdot \frac{1}{n} \Sigma \frac{1}{t} \tag{7.48}$$

$$\bar{U} = x \left[\frac{\Sigma \left(\frac{c}{t} \right)}{\Sigma c} \right]$$

$$\bar{t} = \frac{\Sigma ct}{\Sigma c} \tag{7.49}$$

where x = distance and c and t are concentration and time values picked out
from the time—concentration curve plotted using equal intervals of time.
It may be noted that the average velocity $\bar{U}$ can also be estimated if two
stations are established downstream of the tracer injection point. The
time between the centroid of two curves so obtained gives a better average
of the flow time of the tracer between the two stations.

Example 7.13 In a field test conducted on a small stream using a
dye (Rhodamine WT), the results obtained are given below for a station located 200 m downstream from the injection point. Estimate D in m^2/hr.

t, min	C, µg/liter	ct	c/t
0	0	0	—
5	12	60	2.40

t, min	C, µg/liter	ct	c/t
10	26	260	2.60
15	41	615	2.73
20	40	800	2.00
25	32	800	1.28
30	27	810	0.70
35	23	805	0.66
40	19	760	0.47
45	16	720	0.35
50	13	650	0.26
55	11	605	0.20
60	8	480	0.13
Σ	$\overline{268}$	$\overline{7365}$	$\overline{13.98}$

From Eq. (7.49) we get

$$\overline{t} = \frac{\Sigma ct}{\Sigma c} = \frac{7365}{268} = 27.48 \text{ min}$$

$$\frac{1}{t} = \frac{\Sigma \frac{c}{t}}{\Sigma c} = \frac{1398}{268} = 0.052 \text{ min}^{-1}$$

$$\overline{U} = x \cdot \frac{1}{t} = 150 \text{ m} \times 0.052 = 7.8 \text{ m/min}$$

$$D = \frac{\overline{U}[(\overline{U}\,\overline{t}) - x]}{2} = \frac{7.8[(7.8 \times 27.48) - 200]}{2} = 56 \text{ m}^2/\text{min} = 3360 \text{ m}^2/\text{hr}$$

Many methods have been reported in the literature for estimating D from time—concentration curves. Thackston et al. [19] have summarized well the pros and cons of some of the methods commonly used and have presented a method of least squares estimation which involves the use of a computer to arrive at a reliable estimate of D.

Whether or not any method has given a correct coefficient may be checked by an integration procedure given by Fisher [24a]. In this procedure, time—concentration curves are obtained at two stations. The upstream observed curve is used as the initial tracer distribution and from it a curve for the downstream station is predicted using the one-dimensional equation [Eq. (7.11)]. The predicted and actual curves for the downstream station are compared. If the comparison is not good, a new dispersion coefficient is selected and the calculation is repeated until the best possible comparison is obtained. This procedure is termed the "routing procedure" because of its similarity to flood routing.

The computations are facilitated by the use of a computer program
which eventually gives a D value that fits within approximately ±2.5% of
the observed data. The coefficient obtained by the routing procedure is,
by definition, the one producing the best possible match between prediction
and observation. It is therefore, the best possible coefficient and one
against which coefficients obtained by other methods can be compared.

Analytical Methods

Field determinations of dispersion coefficients, though more realistic,
suffer from several handicaps. Besides being time consuming and expensive,
they reflect only the dispersion conditions at the site and time of measurement. At other sites along a river and at other times when flows are different, the dispersion coefficient can be quite different from the value
determined on a given occasion. Furthermore, for predicting the dispersion
of a pollutant under assumed flow and other conditions, computer programs
need to be run, and one or two observed values of the dispersion coefficient
at a site are not enough. Thus, the use of analytical methods without the
need for field studies is naturally attractive.

Analytical methods are often empirical and based only on a knowledge
of certain stream parameters such as flows, depth, width, etc. They are
similar in some ways to the analytical methods used for estimating the
stream re-aeration constant K_2. Several analytical methods for determining
the dispersion coefficient have become available but only a few are mentioned below. Not all methods are able to predict the coefficient equally
well since the actual conditions in a stream may vary widely from those
assumed or implied in a given analytical equation. Consequently, it is
useful if a few field measurements are made to determine which analytical
method agrees reasonably well with the true conditions in a given stream.
Fisher [24] observes that, in practice, it is usually sufficient for D to
be correct within a factor of about four, as concentration distributions
are generally not too sensitive to the D value used in computations. Hence,
simple analytical methods are of great potential value in such studies.

One of the earliest analytical methods was the one of Taylor for dispersion in pipes [43]. Elder [44] similarly gave an equation applicable
to flow in an infinitely wide two-dimensional channel

$$D = \alpha H U_*$$
(7.50)

where

H = depth of flow

U_* = shear velocity = $\sqrt{gRS}$

α = 5.93; (Sayre [45] and others have shown that α varies
between 3 and 24)

Holley et al. [37] showed that Elder's equation tends to grossly
underestimate the true D value in the case of streams and estuaries where
curves in channels and the presence of sandbars or islands could lead to
the existence of transverse velocity variations otherwise neglected in the
usual one-dimensional model. Thus, observed D values may be several times
greater than those predicted by Eq. (7.50). Fisher has given a more universally applicable method which requires the field measurement of point
velocities along any cross section to know the deviations of point velocities from mean velocity. The river cross section is divided into several
slices and a summation procedure is used [24a]. He also gives an approximate equation derived from this method by making certain simplifying assumptions typical of streams:

$$D = \frac{0.011U^2(W^2)}{HU_*} \qquad\qquad (7.51)$$

in which W = half channel width and all other terms have been defined earlier.

McQuivey and Keefer [21] gave another simplified relationship between
D and stream parameters which is different in concept but reported to give
a better fit to observed data (standard error of 30%) than that obtained
by use of Eq. (7.51). The equation is based on 40 time-of-travel studies
in 18 streams under widely ranging flow conditions:

$$D \cong 0.058 \frac{Q_o}{S_o W_o} \qquad\qquad (7.52a)$$

where

Q_o = discharge at steady base flow

W_o = stream channel width at steady base flow

S_o = slope of stream energy gradient at steady base flow

Liu [49] has recently developed another equation which is particularly
useful for natural streams with a lateral (transverse) variation of velocity
and in which the longitudinal dispersion D is mainly the result of a lateral
rather than a vertical gradient of velocity:

$$D = \beta \frac{Q^2}{U_* R^3} \tag{7.52b}$$

where

    Q = stream discharge

    R = hydraulic radius

    U_* = frictional or shear velocity

    β = dimensionless coefficient

The value of β can be determined from

$$\beta = 0.18 \left(\frac{U_*}{U}\right)^{1.5} = 0.18 \left(\frac{\sqrt{gRS}}{U}\right)^{1.5} \tag{7.52c}$$

where S = slope of energy gradient.

Using the same field data, Liu checked the accuracy of results obtained by using the equations of McQuivey and Keefer [Eq. (7.52a)] with those of Fisher [Eq. (7.51)] and his own [Eq. (7.52b)] for prediction. Liu showed that, while the largest error produced by using Eq. (7.52a) was 715 times, the corresponding largest error from using Eq. (7.51) was 18 times the measured value, and the error from his own Eq. (7.52a) was only 6 times the measured values. Thus, his method is claimed to be an improvement on the prediction accuracy.

Example 7.14 Over a fairly uniform length of stream, the cross section is 63 ft wide with average depth 2.43 ft and an average velocity of 1.82 ft/sec. Furthermore, the hydraulic radius is 4 ft, while slope of the energy gradient is 0.00085. Estimate the dispersion coefficient D using Eqs. (7.51), (7.52a), and (7.52b).

From Eq. (7.51):

$$D = \frac{0.11(1.82)^2 \left(\frac{63}{2}\right)^2}{(2.43)\sqrt{32 \times 4 \times 0.00085}} = 451 \ \text{ft}^2/\text{sec}$$

From Eq. (7.52a):

$$D = \frac{0.058(63 \times 2.43 \times 1.82)}{(0.00085)63} = 302 \ \text{ft}^2/\text{sec}$$

From Eq. (7.52c):

$$\beta = 0.18 \left(\frac{\sqrt{32 \times 4 \times 0.00085}}{1.82}\right)^{1.5} = 0.01389$$

From Eq. (7.52b):

$$D = \beta \, \frac{Q^2}{U_* R^3} = \frac{(63 \times 2.43 \times 1.82)^2 0.01389}{\sqrt{32} \times 4 \times 0.00085 (2.43)^3} = 228 \; \text{ft}^2/\text{sec}$$

7.4 FIELD STUDIES FOR DETERMINING OXYGEN PROFILE

A field study can be performed in such a manner that the various rate constants and the oxygen profile for a given stretch of a stream can be determined. Sampling stations should be set up only after making a preliminary inspection to take into account

1. Points of entry of wastes along a stream

2. Points of entry of freshwater tributaries

3. Points where the hydraulic character of a stream undergoes a significant change such as in width, velocity, slope, or depth

The parameters usually included for measurement in stream surveys are

1. Flow of the stream at each station and its velocity, average depth, and width

2. Temperature of stream water

3. Dissolved oxygen and BOD_u at each station (nitrogen and phosphorus estimations may also be included if desired)

4. Distance between stations

Sampling has to be synchronized with flow as far as possible so that the same batch of pollution is sampled as it passes each station. The flow can be observed by the use of a tracer (e.g., dye or a float). Dilution gauging techniques can also be used. Data so collected may be summarized where possible in the form of a few statistical parameters. In some cases, graphical correlations may be prepared to facilitate later use of the data in making computations at different flows.

Example 7.15 illustrates how, after a field survey is done, the actual and theoretically calculated DO profiles can be matched.

Example 7.15 An industry discharges its wastewater at 43,000 m³/day with a BOD_u of 350 mg/liter into a stream having a BOD_u = 2 mg/liter and in which the flow at the time of the field survey is 10 m³/sec with an average velocity of 12 m/min and a temperature of 20° C. The stream, together with a tributary, serves as a source of water supply to a town located 45 km downstream (see Figure 7.13). The survey results are shown in

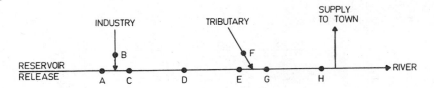

Figure 7.13 The position of sample points during a stream survey (see Example 7.15).

Table 7.5. Calculate the theoretical DO profile and see how it matches with actual data. K_2 has been estimated independently to be 0.5 per day over the entire stretch (C_s 20° C = 9.2 mg/liter).

1. BOD_u is reduced from 18.5 mg/liter at station C to 8.7 mg/liter at E in 1.74 days' travel time. Hence the fraction remaining = 8.7/18.5 = 0.47.

    $$K_r = \frac{\ln (0.47)}{1.74} = 0.433 \text{ per day}$$

2. The initial DO deficit from saturation (20° C) at station C = 9.2 - 6.9 = 2.3 mg/liter. The DO deficits D_t at stations D, E, G, and H are computed in Table 7.6 using Eq. (7.6). Each component of Eq. (7.6) is shown separately in the table in order to systematize the computations.

    It is seen from Table 7.6 that computations are started from station C and carried through station E. Then, from station G, a fresh start is made since the conditions of flow and BOD are changed. Whenever the values of K_r or K_2 change in a stream section, a similar fresh start has to be made to take into account the revised conditions. Column 13 gives the final computed DO values at each station. These can be compared with actually

Table 7.5 Observed Data in Stream Survey (Example 7.15)

Station	A	B	C	D	E	F	G	H
Distance, km	-0.1		0	15	30		30.1	45
Flow, m³/sec	10	0.5	10.5	10.5	10.5	1.5	12.0	12.0
Time of flow, days			0	0.87	1.74		1.74	2.25
Observed DO, mg/liter	7.3	0	6.9	3.2	2.2	7.2	2.8	3.1
Observed BOD$_u$, mg/liter	2	350	18.5		8.7	5.0	8.23	7.0

Table 7.6 DO Computations for Data Given in Example 7.15

(1) Station	(2) t (days)	(3) $e^{-K_r t}$	(4) $e^{-K_2 t}$	(5)[a]	(6) L_a (kg/day)	(7)[b]	(8) D_a (kg/day)	(9)[c]	(10)[d] D_t (kg/day)	(11) Q (m³/sec)	(12)[e] D_t (mg/liter)	(13)[e] DO (mg/liter)
C	0	1	1	0	16,783	0	2,086	2,086	2,086	10.5	2.3	6.9
D	0.87 (from C)	0.686	0.647	0.039	16,783	4,228	2,086	1,350	5,578	10.5	6.15	3.0
E	1.74 (from C)	0.470	0.419	0.051	16,783	5,529	2,086	874	6,403	10.5	7.05	2.14
G	0	1	1	0	8,532	0	6,635	6,635	6,635	12.0	6.4	2.80
H	0.51 (from G)	0.802	0.775	0.027	8,532	1,488	6,635	5,142	6,630	12.0	6.38	2.82

[a] Column (3) - column (4).
[b] $[K_r/(K_2 - K_r)]$ x column (6) x column (5).
[c] Column (8) x column (4).
[d] Column (7) + column (9).
[e] At 20° C.

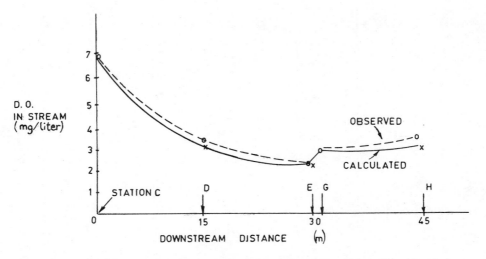

Figure 7.14 The observed and calculated DO profiles for the data given in Example 7.15.

observed values to see how well they match. In case of small discrepancies, the values of the coefficients K_r and K_2 may be slightly adjusted until the calculated and observed profiles match. Figure 7.14 shows the two profiles illustrating that the fit is good except for the portion between stations G and H. This may be due to the fact that K_2 has been assumed to have a constant value throughout the stretch. It is possible that K_2 (and perhaps K_r also) is affected downstream of the confluence point of the tributary and the main stream.

From the results seen in this case, it is evident that the DO reaches a rather low value of 2.2 mg/liter in the stream just ahead of the tributary. The relatively cleaner and fresher water brought in by the tributary helps to give an upward kink in the DO profile shown in Figure 7.14. The observed BOD_u of the stream as it arrives at station H is 7.0 mg/liter. This is also somewhat high as the water at this point is being taken to serve a town. A more detailed analysis of the water would be desirable in order to assess its suitability for supply after treatment.

However, from the viewpoint of stream conditions it may be desirable to take some steps so that DO depletion is reduced, especially if fish life and commercial and recreational uses are involved.

7.5 ALTERNATIVE STRATEGIES FOR STREAM QUALITY CONTROL

The usual approach to stream quality control has been to lay down standards governing the stream water quality and effluents discharged into it. The bases on which this is done are discussed in Sec. 7.8. To meet whatever standards are laid down, there are two approaches from which to choose.

The conventional approach to stream quality control has been to require waste treatment at each individual outfall. Thus, every town and industry which has an outfall into a stream must meet the standards. This may require many of them to install and operate some form of waste treatment facility.

The approach often recommended in present times is to undertake a regional approach for the river basin as a whole to meet the desired water quality objectives rather than to follow a piecemeal approach. A greater variety of alternative strategies become available for water quality control when a regional approach is adopted. Often it leads to overall economy and greater assurance of achieving quality objectives. A suitable organizational structure is, no doubt, a prerequisite for undertaking any water quality control program, the more so when it involves a regional approach.

Among the alternative strategies available are the following:

1. Allocation of stream uses. This is, in fact, one of the most important strategies followed to optimize waste treatment, as far as possible. Streams are generally classified according to the major use to which they are likely to be put: drinking, fish culture, irrigation, etc. Depending on the purpose, water quality standards are applied. Thus, for example, more lax standards may be applicable to the lower reaches of a stream, especially where it is tidal.

    Sometimes, where topography permits, the lower portion of a stream may be allowed for waste discharge purposes while an adjoining tributary may be reserved for water supply and recreational purposes. Allocation of stream uses must be evidently made to optimize a region's water resources for various uses.

2. Stream flow regulation. This generally consists of providing impoundment for low flow augmentation. Upstream flows may be held in specially provided reservoirs, or additional storage capacity my be provided in reservoirs constructed for other purposes. Multipurpose projects for irrigation, hydroelectric, and other purposes can be expanded to include storage for the release of additional quantities downstream during the critical months to ensure more dilution and a higher DO content. Stream flow regulation can also be brought about in certain favorable situations by interbasin diversion of waters if topography permits.

Another form of stream flow regulation is the provision of weirs in a stream to influence velocity, flow time, re-aeration, etc., to improve downstream characteristics. Ecological consequences of any form of stream flow regulation are complex and difficult to predict. Construction of the Aswan Dam in Egypt is a case in point.

Example 7.16 illustrates how some of the above alternatives can be evaluated for their effect on downstream oxygen content.

3. <u>In-stream treatment</u>. Another possibility which exists, and can be taken into consideration if a suitable organizational and legal framework exists, is to provide for in-stream treatment which generally consists of aeration at a suitable downstream point to prevent the DO content from falling below the prescribed minimum at the critical time of the year. The advantage comes from the fact that the stream's assimilative capacity is fully utilized at high flows, and costly aeration is limited to only a small period of time. This amounts to a kind of collective treatment and involves cost allocation between the various beneficiaries. Principles of aeration are discussed in Chapter 13. Their application to in-stream aeration is discussed in the literature [27, 35].

4. <u>Waste treatment</u>. Individual or collective waste treatment <u>before</u> discharge to a stream may be necessary in some cases and, in fact, may be the method of choice after considering all the above alternatives as well as some more, such as land disposal or ocean disposal of wastes.

<u>Example 7.16</u> From an analysis of the data given in Example 7.15, it is seen that the stream suffers considerable depletion of its DO content downstream of the industrial discharge. In order to ensure that the minimum DO will not fall below 4 mg/liter, the following alternative methods are proposed to be evaluated for which the following must be estimated:

1. Degree of effluent treatment to be done by the industry to ensure that the minimum DO = 4 mg/liter at station E. Assume K_r, K_2, and flow are unaffected and have the same values as given in Example 7.15.

2. Effect of 25% increase in stream flow (by augmentation from an upstream reservoir) on the DO content at station E. Correlation of flow with velocity and re-aeration constant K_2 is as follows. Assume K_r remains unchanged.

   Velocity V in m/sec = $0.06(Q \ m^3/sec)^{0.50}$

   Re-aeration, K_2/day = $0.67(Q \ m^3/sec)^{-0.13}$

3. Power required for in-stream aeration assuming aerators can deliver 1.3 kg O_2/kWh at stream conditions.

Case 1

Since flow is unchanged, the time of travel from C to E is also unchanged, namely, t = 1.74 days. Initial deficit D_a = 2.3 mg/liter. Permissible deficit D_t = 9.2 - 4.0 = 5.2.

$$\text{Deficit } D_t = \frac{K_r L_a}{K_2 - K_r}\left[e^{-K_r t} - e^{-K_2 t}\right] + D_a e^{-K_2 t}$$

or

$$5.2 = \frac{0.433 L_a}{0.5 - 0.433}\left[e^{-0.433 \times 1.74} - e^{-0.5 \times 1.74}\right] + 2.3 e^{-0.5 \times 1.74}$$

Hence,

$$L_a = 12.85 \text{ mg/liter}$$

$$\text{Upstream } BOD_u = 2 \text{ mg/liter} \quad \text{and} \quad \text{flow} = 10 \text{ m}^3/\text{sec}$$

By materials balance, permissible BOD_u in industrial waste = 230 mg per liter.

$$\text{Percent BOD removal necessary} = \frac{350 - 230}{350} = 34\%$$

Case 2

Stream flow arriving at station C is now increased to 13.0 m³/sec.

$$V = 0.06 Q^{0.50} \quad \text{hence,} \quad V = 0.06(13)^{0.50} = 0.216 \text{ m/sec (13 m/min).}$$

$$K_2 = 0.67 Q^{-0.13} \quad \text{hence,} \quad K_2 = 0.67(13)^{-0.13} = 0.48 \text{ per day.}$$

$$\text{Now,} \quad \text{travel time } t = \frac{30 \text{ km} \times 10^3}{13 \text{ m/min}(60 \times 24)} = 1.60 \text{ days.}$$

Initial upstream DO = 7.3 mg/liter (as before) but, owing to increased flow, DO after mixing with waste will now be 7.03 mg/liter. Thus, deficit D_a = 2.17 mg/liter.

Similarly, with the discharge of untreated waste but with greater dilution by stream flow,

$$L_a = 14.88 \text{ mg/liter}$$

$$D_t = \frac{0.433(14.88)}{0.5 - 0.433}\left[e^{-0.433 \times 1.6} - e^{-0.48 \times 1.6}\right] + 2.17 e^{-0.48 \times 1.6}$$

$$= 4.46$$

Hence, DO at station E = 9.2 - 4.46 = 4.74 mg/liter. (A flow slightly less than 13 m³/sec might suffice. This flow would be required in the critical season only, and must be available as storage in an upstream reservoir.)

Case 3

In-stream aeration would be needed at some point sufficiently upstream of station E so that the stream DO is never less than 4 mg/liter. The oxygen transfer required can be roughly estimated at

$$\frac{10.5 \text{ m}^3/\text{sec} \times 86,400 \times 10^3 \times (4 - 2.14)}{10^6} = 1687 \text{ kg/day}$$

Assuming that floating aerators are capable of delivering 1.3 kg per kWh at stream conditions,

$$\text{Power required} = \frac{1687}{24} \times \frac{1}{1.3} = 54 \text{ kWh}$$

(Aeration will be limited to critical periods only.)

Note: The capital and operating costs of all the three above alternatives can be estimated in order to choose the most economic alternative. Other alternatives besides the three seen above may also need to be considered.

7.6 APPLICATION OF SYSTEMS ANALYSIS

Owing to the fact that many variables are involved in water quality management, decision makers have increasingly taken to applying mathematical modeling techniques programmed for computer analysis to assess the technical and economic consequences of their decisions. Thus, alternative strategies for handling wastewaters generated in a river basin are evaluated for their ability to meet predetermined technical criteria and are compared costwise to arrive at an optimum solution. Forecasts of water quality in streams are also made.

The successful application of systems analysis methods to water quality management problems depends on (1) the ability to develop a mathematical model taking into account the interrelationships between the components of a system (at times a system may be too complex to be modeled and simplifying assumptions may reduce the accuracy of the results) and (2) the availability of reliable input data. The latter constraint is often the more crippling one, even more so in developing countries. Extrapolations from experience in developed countries and different climates may not be valid. Even where data are available there are several shortcomings due to unknown heterogeneities of a system, unknown time dependence of parameters, and approximations introduced. Nonetheless, systems analysis techniques can be of value if applied with a clear understanding of their limitations.

Often after a mathematical model is developed to represent a system, it needs to be "calibrated" by actual field testing to enable its subsequent application with some confidence. In water quality management, forecasts of water quality are generally needed over a period of time if any meaningful planning of water resources is to be done. Here is where computer simulation techniques are most useful. During simulation, models are subjected to various inputs or environmental situations to explore the nature of results obtained as they reflect the results likely to be obtained in the real system if the model is a valid one. Thus the general simulation procedure involves the following:

Identification of system components
and their interrelationships

↓

Formulation of computer program

↓

Calibration

↓

Program operation

↓

Results

↓

Alternative inputs

It is not within the scope of this book to go into the details of programming; excellent references exist in the literature illustrating applications to rivers and estuaries [28-33]. However, it will be useful if some of the practical requirements of simulation and optimization are discussed here, such as the nature of data required to facilitate such studies. Often, with little additional cost, the data collection system can be so reorganized as to yield data of value to systems analysts, and this opportunity should not be missed by developing countries newly embarking on control programs.

In preparing a simulation model for a river basin, a topographic map of the area is first required. A 1/25,000 scale contour map is often satisfactory to outline the watersheds of the main river and its tributaries. Then, the river is divided into segments or reaches, each of which is assumed to have constant properties in itself. A reach terminates and another starts where there is a significant change in the system such as when a tributary joins or a major waste discharge point occurs. Of course, waste discharges (e.g., agricultural runoff) also occur as distributed loads all along the length of a river, and can be taken into account in a simulation. The different reaches are numbered starting from the downstream end, and their areas are measured. Storm runoffs are correlated to the areas drained.

Locations of existing stream gauge stations are also determined. Using Eq. (7.37), relationships between (1) rate of flow and velocity, and (2) rate of flow and depth of flow must be developed for each reach in order to compute time of flow in the reach. The deoxygenation and re-aeration rate constants can be found using equations given earlier. A relationship between K_r, K_2, and the rate of flow Q can thus be developed for each reach. Where stream gauge data are not readily available for a long enough period, the historical data can be used to generate synthetic data preserving their statistical properties [6], or other methods suggested in Sec. 7.3.4 can be used. River temperature data are also essential.

Simulation is often limited to dissolved oxygen conditions in the river and its tributaries. The classic Streeter-Phelps equation is modified to use with a computer program by including an "error term" to account for factors which affect assumed values of K_r and K_2 [31]. This term is later evaluated in field calibration tests. Other terms can also be included to account for photosynthesis, respiration, etc., if desired. Nitrogenous oxygen demand can be taken into account if conditions warrant. The cumulative effect on oxygen concentrations can be determined from reach to reach by use of the program. A subprogram for simulation of the behavior of conservative materials may be added to the model, if desired, although this is often not done in river quality simulations.

The organic (and other) load progressively received by each reach of the river must be determined. As the load will vary with each passing year, any simulation over a number of years requires that the load data be obtained with care. Here it becomes apparent that, if a land use plan is available for a river region, some attempt can be made to estimate the load. In the absence of a plan, not much purpose is served by making such estimates. However, assumed values of loads can yield a forecast of the river water quality from which land use planning agencies can benefit by knowing the load limits which must not be exceeded if a desired water quality objective (like minimum DO) is to be met.

The simulation model can be coupled with an optimization procedure for a cost-benefit analysis of different water quality levels. This, no doubt, requires that a considerable amount of realistic cost data be available, or else erroneous conclusions may be reached. Constraints such as nonavailability of sophisticated equipment or skilled staff must be taken into account when selecting alternative methods of treatment.

It is well known that "benefits" are difficult to assess in monetary terms, especially where aspects such as health and well-being are concerned. In practice, therefore, the assessment has to be confined only to cost optimization to attain desired water quality levels or comply with predetermined standards.

Thus, the technological and economic consequences of different alternatives such as those discussed in Sec. 7.5 can be assessed. Simulation can become an invaluable basic tool in forecasting and planning for the management of water quality systems.

In the absence of computers, manual computations are necessary and the work is often limited to an estimation of stream water quality in a given reach after a desired period of time. Other, more rough-and-ready methods of estimation are also used depending on the extent of data available for forecasting and planning.

7.7 WATER QUALITY MONITORING

Before deciding whether monitoring should be short term or long term, manual or automatic, consideration should be given to the objectives of a monitoring program for a river basin. These objectives may include one or more of the following:

1. To assess the immediate water quality situation in a river basin.
2. To establish water quality baselines and trends.
3. To enforce water quality standards, detect offenders, and ascertain improvements owing to abatement measures.
4. To predict future water quality resulting from development in the region and the adoption of various control measures.
5. To provide an early warning system for downstream users about adverse water quality conditions approaching their water intakes.
6. To determine timing for flow augmentation or in-stream treatment, etc., where such methods of water quality control are provided for.

A manual sampling program is generally the first to be established. It is best adapted to short-term surveys and preliminary assessments of the existing situation in a river, or in a specific reach of a river, around a major pollution source such as a town or industry. Manually executed programs are also often used to make periodic checks in a river to ascertain water quality trends, to enforce standards, detect offenders, etc.

Manual monitoring programs involve a minimum of instrumentation which may be a distinct advantage in developing countries. Reasonably effective programs have been so established using mobile laboratories and simpler facilities. Apart from their higher manpower requirements, they suffer from a few disadvantages such as the natural limitations imposed on frequency of sampling, time lag between collection and analysis, and the general inability to detect sudden changes in water quality caused by industrial spills, storms, etc.

Automatic water quality monitoring programs have advantages and disadvantages just in reverse to those mentioned above. Equipment availability and dependability are factors of prime importance. Two kinds of systems are in use: robot and telemetric. Robot stations are those which automatically monitor a parameter, but the results are recorded on a chart at site. Telemetric stations are connected to a central station to which the data are telemetered for further processing either by a computer or special staff.

A telemetrically linked system has the advantage of giving data on real-time basis, if desired, and of being able to know quickly if a station fails to transmit information owing to equipment fault or damage resulting from flood or vandalism. Repairs can be undertaken rapidly. A robot station would remain "dead" until so discovered during a periodic site visit, with the result that much data may be lost.

For both types of systems, periodic site visits are needed. Even telemetrically linked stations need periodic inspection and recalibration of equipment as many of them are stable only for periods of 1 to 4 weeks or so; they also need temperature compensation. Robot stations need more frequent site visits as recording charts have to be replaced. For both systems the usual components involved at each station are (1) pumps and flow tubing to lift a sample from the river, (2) sensors, and (3) an electronic system. Clogging of pumps and tubes, biological growths on sensor surfaces, loss of calibration, and power failure are some of the common causes for loss of data. Humidity and heat are additional causes of special

importance in tropical countries. Experience in the U.S. and U.K. seems
to indicate that robot monitors operating in isolation may give only about
50% return of required data, whereas telemetry can raise the return to 90
or 98% [34].

Depending on availability of staff, equipment, financial resources,
and the nature of data to be obtained, it is necessary to (1) select the
parameters for monitoring, (2) determine the desirable frequency of sampling
for each parameter, and (3) locate the monitoring stations in optimum positions to best identify changing conditions and their likely causes. It may
be noted that often, especially in developing countries, the locations may
also be quite dependent on accessibility, power availability, and equipment
safety.

Selection of parameters for manually conducted surveys is not as crucial as for automatic systems since, in the former, one can adjust with
experience. But with automatic systems, considerable financial investment
is involved. Assuming dependable equipment is available for measuring a
parameter, its inclusion in the program must be strongly justifiable as
all too often one is likely to generate an excessive amount of data which
may be interesting but not essential for meeting the objectives of the
program.

The most commonly used parameters at continuously monitoring stations,
and for which reasonably dependable equipment are available, are [34]

1. Temperature
2. Meteorological data (rainfall, runoff, etc.)
3. Conductivity
4. Turbidity
5. pH
6. Dissolved oxygen
7. Oxidation-reduction potential
8. Chlorides

These parameters are capable of automatic in situ measurement by means of
sensors or electrochemical transducers. Although one can generally expect
that from 10% to 37% of the data will be lost owing to some form of mal-
functioning, a number of new developments in sensor technology are constantly taking place which improve their sensitivity and stability. Several specific ion electrodes have also been developed. Trace metals analysis
is now possible with anodic stripping voltametry techniques and the equip-

ment is automated, using continuous-flow quartz cells. Thus, various other parameters capable of inclusion in monitoring programs include: Cu, Pb, Zn, Bi, Co, Ni, CN⁻, F⁻, S⁻, and NH_3. Radioactivity can also be measured continuously.

Parameters such as COD and TOC require wet analysis methods which are now automated. Such equipment can be installed in mobile laboratories and used where turbidity and other factors do not interfere with analysis. Daily supervision is, however, required. Coliform tests may be needed occasionally and are performed by manually operated methods.

Monitoring technology is indeed rapidly advancing and, undoubtedly, it will be possible to monitor several new parameters in the future. The question of which parameters really need to be included in a program will then remain as pertinent as now.

Different parameters may be measured at different stations depending upon their relevance at a site. Frequency of sampling may also be determined accordingly (e.g., some parameters may be important to measure only at low flows). In manual monitoring programs, samples must be preserved carefully during transport.

Before embarking on a new monitoring program for a river basin on a routine basis, it is highly recommended that experience first be gained in the specific region through short-term, manually conducted surveys. These short-term surveys (repeatedly performed, if necessary) serve as pilot surveys and greatly help in making a judicious selection of parameters to be included in the routine program, their frequency of analysis, and the location of the monitoring stations.

The provision of a mobile laboratory is an asset for any control agency. In most countries, control functions are performed through manual rather than automatic monitoring, and mobile labs are extremely useful. It is advantageous to provide such a lab in a trailer pulled by a jeep. The jeep can then be used for other purposes (obtaining samples, for instance) once the trailer is parked at a suitable site along a river bank.

Automatic monitoring systems are expensive. The annual cost of collecting data on a continuous basis by instrumental methods is on the order of $5000 (1970 costs) per parameter per year, and a reduction in sampling frequency would have little bearing on this figure [34]. The considerable amount of data generated by an automatic system calls for the installation of a correspondingly sophisticated data storage and retrieval system for the effective performance of any desired water quality forecasting and management functions.

7.8 STREAM STANDARDS

Based on current knowledge of the health, ecological, and economic effects of water quality, it is possible to spell out certain desirable water quality objectives at which to aim. Standards are formulated to enable control agencies to meet these objectives. The general philosophy behind the formulation of health criteria and the establishment of standards was discussed in Sec. 4.4.1 using the example of mercury accumulation in fish.

Water quality objectives for freshwater streams take into account the several major uses to which freshwater is put, and the fact that all streams, or all portions of a stream, are not necessarily required to meet all potential uses. This has led to the concept of classification of streams, which indicates that their quality has to meet the requirements of one or more of the following potential major uses:

1. Public water supplies used for drinking
2. Fish culture
3. Agriculture (including animal watering)
4. Industrial uses
5. Recreational and navigational uses

For each typical use, water quality objectives are established that take into account the special constraints on quality imposed by that use. Thus, public water supplies are generally required to conform to the severest quality standards from the health viewpoint. Others are required to take other aspects into account; these are summarized in Table 7.7.

To attain the desired water quality objectives, two types of standards may be called for. One type, called "effluent standards," is applicable to the quality of the municipal or industrial wastewater discharges to a watercourse. The other type, called "stream standards," is applicable to the quality of water in a stream. It may be necessary to apply both types of standards in a given situation. If stream standards are clearly defined, it is possible to develop effluent standards from them in such a manner as to utilize to the maximum possible extent the natural diluting and assimilating capacity of a stream. This, of course, implies that the effluent standards applicable to a given town or industry need revision from time to time as a greater quantity of pollutants is generated for disposal to the same stream.

Table 7.7 Typical Stream Classifications and Water Quality Objectives

Stream classification based on its use	Stream water quality objectives
1. Drinking water supply	Water supplies from the stream should be capable of meeting drinking water standards of the country with regard to sanitary quality, toxic substances, etc., after a degree of treatment likely under the technological and economic resources of the region, noting that some substances are hardly removable by usual treatment.
2. Fish culture and aquatic life	Dissolved oxygen, pH, and temperature should be maintained within a suitable range depending on the species of fish, etc., to be protected. Other factors include ammonia, other toxic organic and inorganic pollutants, floating oil, high turbidity, etc., which affect fish mortality, fish taste, or biological accumulations in the food chain in objectionable quantities. In many cases, stream waters meeting classification 1 above are satisfactory for fish culture, provided DO and NH_3 are within limits for fish life.
3. Agriculture (including animal watering)	Important items to look for are: (1) salinity (limits depend on the nature of the soil, climate, and crop tolerance), (2) presence of phytotoxic elements (e.g., boron and aluminum) and persistent substances (trace metals and herbicides) which enter food chains. See Chapter 9.
4. Industrial uses	Widely varying requirements exist; least critical for cooling waters, most critical for high-pressure boilers, food, beverages, and pharmaceuticals. Generally, waters meeting any of the above three classifications can be treated further by industries themselves as necessary.
5. Recreational uses	For contact sports: coliforms, pH, turbidity, color, freedom from infective parasites, and aesthetically objectionable substances (oil, foam, floating debris, odor, excessive algal and weed growths).

This approach is an advisable one to follow, especially in an area of low density of development where each of the few towns and industries located along a stream can be readily asked to conform to a calculated limiting value for any constituent in the effluent so that the overall stream quality is maintained. Thus, the stream quality standards are always met while the derived effluent standards are reviewed, and possibly made stricter, as more and more development occurs in the region. The effluent standards eventually stabilize, so to speak, when development stabilizes.

For certain items (oil content, for example), only effluent standards have to be prescribed since they are independent of the volume of flow in a stream. For others, where dilution helps and where degradation with time or distance along the stream occurs, the critical flow at which the prescribed standard is to be met must be specified.

For stream quality parameters such as DO, temperature, etc., the critical flow is generally the minimum daily flow or minimum consecutive 7-day average with a recurrence interval of once in 2 years (i.e., average minimum flow) or once in 5 or more years if the fisheries are of important commercial value. Alternatively, the 5-percentile value obtained from a log-probability plot of all daily flows may be used. Where the stream is likely to have very low (or even zero) flows for extended periods of time, the stream standards for nondegradable and toxic substances have to be practically equivalent to drinking water standards to protect downstream users.

For persistent substances that enter the food chain, biological accumulation to objectionable levels may occur even at the low concentrations in waters acceptable for drinking purposes. Thus, their discharge to watercourses should be avoided or reduced at the source as far as is practicable.

An attempt is made below to outline the rationale behind a typical standard which would be applicable to many developing countries where it is desired to preserve the quality of inland surface waters for public water supplies, fishing and aquatic life, and other such purposes which demand a relatively high quality of raw water. Such waters are commonly referred to in the literature as "class I waters." (It is presumed that stream classification has already been accomplished.)

Before proposing any standards, it must be emphasized that this attempt is merely for the purpose of illustrating the subject and, evidently, cannot take into account the large number of variables and specific situations that may be encountered in a real case.

Furthermore, it is assumed that public opinion in the developing country would not wish to support an unduly high level of standards. Unduly strict standards may be difficult to comply with or to enforce, even with the best of intentions. It may also be feared that they will slow down the industrialization program of the country. Thus, it is desired to concentrate on what is attainable rather than what is ideal.

The proposed standards can be grouped under two categories, effluent and stream, as shown in Table 7.8. Effluent standards (items 5 to 9 in Table 7.8) must be met irrespective of the flow in a stream; all the other standards pertain to stream water quality. Knowing the critical flow appropriate for each item, the controlling authority can calculate the maximum concentration allowable in the effluents of the concerned polluters. This amounts to a sort of apportioning of the stream's assimilative and dilution capacity between its polluters. The whole process must be updated as conditions change. In determining the stream water quality standards shown in Table 7.8, it has been assumed that WHO International Drinking Water Standards will have to be met by downstream water users with or without treatment.

This is, indeed, a simplification of the normally complex task of setting standards, but may be valid as a starting point for discussions and refinement at the level of a developing country which is just embarking on a control program. It is recognized that other approaches may be more valid in some cases. Many developing countries are undergoing rapid industrialization, often importing quite sophisticated technology and producing a wide variety of wastes. Thus, for comprehensive control measures to be adopted in such countries, standards must be developed not only for dealing with sanitary wastes (e.g., BOD, DO, coliforms), but also with some of the more common toxic and other substances. Standards also have to be considered from the viewpoint of background values resulting from natural causes. Finally, economic considerations enter where health is not the main reason for a standard. Some balancing of costs and risks is invariably implied in setting standards.

Several countries have established quite comprehensive standards for their specific situations. A few examples are included in Table 7.9.

Table 7.8 Typical Standards for Control of Quality of Inland Surface Waters Used for Public Water Supplies,
Fish, Aquatic Products (Class I Waters)

Item	Typical standards applicable to Effluent	Typical standards applicable to Stream	Critical stream flow recurrence interval	Remarks
1. Dissolved oxygen	-	Minimum 5 mg/liter for trout and not less than 4 mg/liter for other game fish during at least 16 out of 24 hr, but not less than 3 mg/liter at any time. Minimum 2 mg/liter any time if game fish are not important.	Once in 2 or 5 years [a]	Limits may be specified in terms of percent of saturation DO concentration (e.g., 60%) if desired. Natural respiration, especially in summer, may bring DO values down during the night. Hence it is desirable to know the background levels and the type of fish to be protected.
2. BOD	-	-	-	No standard need be specified since the maximum permissible value of effluent BOD will have to be determined such that the DO standard given in (1) above is not exceeded at the critical stream flow (see text).
3. pH	-	6.5-8.5	-	Stream water must meet drinking water standards and be suitable for fish life.
4. Temperature	-	Stream temperature shall not be raised by more than 3°C (compared with upstream temperature) after mixing.	Once in 2 or 5 years	Additional requirements may need to be specified in the case of sensitive species (e.g., trout) where applicable.
5. Oils and grease (extractable matter)	Not more than 10 mg/liter from industries and 30 mg/liter from domestic sources.	-	-	Additional requirements may be specified in the case of certain specific types of oils. Oil removal may be required using "best available technology."

285

Table 7.8 (Continued)

Item	Typical standards applicable to Effluent	Stream	Critical stream flow recurrence interval	Remarks
6. Floating solids	Not larger than 2 cm	-	-	Floating solids may be unsightly and, if large, obstructive.
7. Settleable inorganic solids	Not larger than 2 mm		-	To avoid deposition in streams and formation of sludge banks.
8. Ammonia-N	Not more than 0.5 mg/liter		-	Important where fish life is to be preserved.
9. Free chlorine	Not more than 0.5 mg/liter		-	Affects fish in vicinity of outfall.
10. Coliform bacteria	Disinfection needed if downstream users are located near outfall.		-	Coliforms undergo fairly rapid die off in streams, especially in warm temperatures. Coliform concentrations at intakes of downstream public water supplies shall be less than 5000 per 100 ml as a monthly average (20% of samples not to exceed 5000 and 5% of samples not to exceed 20,000 per 100 ml). Where chlorination is done, residual shall not exceed value given in item 9.
11. Toxic substances:	(see remarks)		Once in 2 to 5 years if stream standards are used[a]	Limits for all the items except chromium and cadmium listed herein are based on WHO International Drinking Water Standards (1971) on the assumption that downstream untreated waters may be used or treatment may not be adequate to remove these substances. Chromium standard is based on WHO European Drinking Water Standards (1970). Additionally, standards may be prescribed for copper
Arsenic	0.05 mg/liter			
Cadmium	0.005 mg/liter			
Chromium	0.05 mg/liter			
Cyanide	0.05 mg/liter			
Lead	0.10 mg/liter			
Mercury	0.001 mg/liter			
Selenium	0.01 mg/liter			

286

(1.5 mg/liter) and zinc (15 mg/liter). Although not toxic, they affect water acceptability.

An alternative approach which has some merit owing to the likely synergistic effect that these substances may have on fish and on biological accumulation on the food chain is to lump all these items together and prescribe a limit of, for example, 1 mg/liter not to be exceeded either individually or collectively in the <u>effluent</u>. This is also often easier from an enforcement point of view.

12. Other substances affecting health or acceptability:

Item	Value	Remarks	Frequency
Nitrates	45 mg/liter	WHO European Drinking Water Standards (1970) consider water as "acceptable" up to 100 mg/liter as NO_3.	Once in 2 or 5 years[a]
Radioactivity	(gross α) 3.0 P Ci/liter (gross β) 30.0 P Ci/liter	Should preferably be absent in effluents.	Once in 2 or 5 years[a]
Phenols	0.5 mg/liter 0.002 mg/liter	Whichever gives a lower value in the stream. (A stricter limit of 0.001 mg/liter in stream water may be adopted if desired.)	Once in 2 or 5 years if stream standards are used[a]
Detergents (anionic)	2.0 mg/liter 1.0 mg/liter	Whichever gives a lower value in the stream.	Once in 2 years

---

[a] The critical stream flow requirement may be made less strict in the early years of enforcement, especially for those items which are not likely to have serious health consequences. Later, with better experience, finance, and public opinion, the requirement can be made stricter (also see Sec. 7.3.4).

287

Table 7.9 Typical Receiving Water Standards

Determinants	U.S., EPA	European Common Market	CMEA (Eastern Europe)
Cadmium	0.004	0.005	0.005
Cyanides	0.005	0.05	0.02
Organic phosphorus compounds	0.005-0.05 g/liter	0.001	-
Lead	0.01 of LC_{50}-96	0.05	0.1
Chromium (VI)	-	-	0.1
Arsenic	-	0.05	0.05
Mercury	0.0001	0.001	0.001
PCB	0.001, g/liter	0.0001	-
pH	6.5-9.0 (sea, 6.5-8.5)	6.5-8.5	6.5-8.5
BOD_5	-	5	5
COD	-	-	20
Suspended solids	not reduce photosynthesis	80	50
Extractable matter	Oils, 0.01 of LC_{50}-96	0.2	0.3
Phenols	0.001	0.005	0.002
Copper	0.1 of LC_{50}-96	0.05	1
Zinc	0.01 of LC_{50}-96	0.2	0.2
Iron	1.0	1.0	1
Manganese	0.1	0.1	0.3
Chromium (total)	0.1	0.05	1
Fluoride	-	1.2	-
Coliforms, MPN/ml	fecal 43/100 ml	50/ml	10/ml
Dissolved oxygen	5	50%	5
Ammonia-N	0.02 (free NH_3)	2.0	0.5
Beryllium	1.1	0.001	-
Free chlorine	0.01	-	-
Nickel	0.01 of LC_{50}-96	0.05	2
Chlorinated hydrocarbons, µg/liter	0.001-0.004, µg/liter	0.001, µg/liter	-
Sulfide (H_2S)	0.02	-	0
Temperature, °C		According to natural conditions	According to natural conditions
Total phosphorus	-	0.3	-
ANA detergents	-	0.2	0.5

Note: Measurements are given in mg/liter except where noted.

7.9 DISPOSAL TO ESTUARIES

In many countries, dense concentrations of populations and industries
have developed around their numerous estuaries. Consequently, many of them
have come to be grossly polluted. Since the estuaries also serve these
communities by way of their (fish) food supply and recreational, navigational, and other needs, estuarine pollution control becomes a matter of
prime concern for these communities.

Essentially, an estuary is a body of water over which there is a
noticeable effect of the sea resulting in a rise and fall in water level
owing to river flow and the tide, with salinity variations over the average
tidal cycle. Generally, dispersion is significant in estuaries, while advection may or may not depend on the rate of freshwater flow and the cross-
sectional area.

Hence, estuarine dynamics and problems are in some respects different
from those of rivers and seas. A few of their special aspects are, therefore, being considered here.

7.9.1 Estuary Mixing Types

Estuaries can be grouped in accordance with their mixing conditions
which govern their vertical circulation patterns. The amount of freshwater
which an estuary receives from its catchment undergoes a certain degree of
mixing with the seawater entering it to maintain density equilibrium.

The freshwater, being less dense than seawater, tends to flow out to
the sea in the surface layer of the estuary. The heavier seawater tends
to flow into the estuary along the lower layer. Vertical mixing between
the two occurs in various degrees, giving salinity gradients. The intrud-
ing seawater tends to penetrate the estuary in the form of a wedge or
tongue (see Figure 7.15) and is often referred to as the saline wedge.
The extent to which the wedge penetrates into the estuary is a function of
the channel depth, the freshwater discharge, and the density differential
between the sea and the freshwater.

Many factors affect vertical mixing, and a rough approximation of the
mixing type can be had by dividing the volume of freshwater which enters
the system during a half tidal cycle $Q_f(T/2)$ by the volume of the tidal

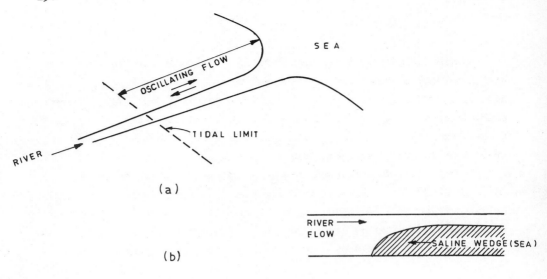

(a)

(b)

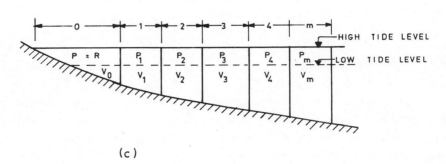

(c)

Figure 7.15 (a) The tidal reach, (b) salt wedge, and (c) sectioning for a modified prism for a typical estuary.

prism.* Thus, three types of estuaries are identified with regard to their mixing conditions and consequent salinity variations:

Type	$\left(\dfrac{Q_f(T/2)}{\text{Prism volume}}\right)$	Remarks
1. Well mixed	0.1	Typically small river and large tidal difference. Gentle salinity gradient, or homogeneous.

---

*The tidal prism is defined as the volume of water contained in the estuary between mean high water and mean low water. It is a function of the tidal range and estuarine geometry.

Type	$\left(\dfrac{Q_f(T/2)}{\text{Prism volume}}\right)$	Remarks
2. Partially mixed	0.2-0.5	Increasing salinity gradient
3. Stratified	0.7	Small tidal effect; river relatively large. Salinity gradient shows considerable stratification

In well-mixed estuaries, salinities decrease gradually from the mouth to the upper reaches. Bottom salinities normally exceed surface salinities in a section by about 15 to 25%, thus giving a relatively gentle gradient. In partially mixed estuaries, the presence of an interface between fresh-water and seawater is increasingly indicated by salinity or velocity pro-files as the estuary becomes more and more stratified. Then, vertical mixing occurs within each layer and also across the interface as long as one or both layers are turbulent.

An estuary may change from one type to another by large variations in freshwater discharge resulting from either natural or man-made causes (e.g., flow diversion) or velocity and other changes (e.g., deepening by dredging may increase stratification).

Vertical mixing patterns can be observed by measuring current veloci-ties over complete tidal cycles at various locations and depths [36]. Es-tuarine mixing conditions also have a considerable effect on their current patterns.

7.10 DISPERSION IN ESTUARIES

Dispersed flow models derived earlier for rivers (Sec. 7.2.2) can be suitably modified to apply to estuaries. In general, the various factors affecting diffusion and dispersion in natural estuaries can be listed as follows:

1. Molecular diffusion
2. Stratification
   a. horizontal
   b. vertical
3. Eddy turbulence

a. Due to free turbulence resulting from (1) tidal flows and (2) advective flows

b. Due to departure from one-dimensional flow caused by (1) horizontal (longitudinal and lateral or transverse) velocity gradients and (2) vertical velocities

At the concentration gradient generally present in rivers and estuaries, molecular diffusion is considered insignificant as a transport mechanism. In well-mixed estuaries, both (1) and (2) above are considered insignificant.

The major difference between flow in rivers and estuaries is that the flow oscillates in estuaries. Dispersion in oscillating flow depends on the ratio of the period of oscillation and the time-scale for mixing in a given direction [37]. Generally, for _wide_ estuaries this ratio is much higher in the vertical direction than in the transverse one and the dispersion coefficient D for such estuaries can be approximated from Elder's equation seen earlier.

$$D = 5.93HU_* \qquad\qquad\qquad (7.50)$$

where H = depth of flow (ft) and U_* = shear velocity (ft/sec) as given by

$$U_* = 3.9 \left(\frac{n}{R^{1/6}} \right) \left(\frac{2}{\pi} U_T \right) \qquad\qquad (7.53)$$

where

n   = Manning's coefficient (often assumed as 0.03 to 0.035)

R   = hydraulic radius of the channel (for wide estuaries, $R \simeq h$) in feet

U_T = amplitude tidal velocity

For _narrow_ estuaries and for those where large-scale circulatory mixing is induced as a result of islands, sandbars, or uneven tides, the transverse velocity gradients may be more important to the dispersion process and may cause the observed value of D to be on the order of 10 times greater than that estimated from Eq. (7.50). Holley et al. [37] present criteria for identifying such estuaries and describe a method for estimating D. However, they state that calculated concentration distributions do not seem to be very sensitive to this possible increase in D provided the tracer (or pollutant) release is on a continuous basis. For practical purposes, it is often quite justifiable to use empirical estimates of overall dispersion coefficients, as long as it is noted that they are applicable to (1) well-mixed estuaries and (2) conditions of constant density as would be likely in the upper freshwater portions and in the lower seaward end of estuaries.

Example 7.17 Estimate the dispersion coefficient D for an estuary 2000 ft wide and 16 ft deep in which records show the value of U_T to be 0.6 ft/sec. The estuary is reasonably uniform in cross section, is homogeneous and well mixed, and vertical rather than transverse velocity gradients prevail. Assume n = 0.035.

From Eq. (7.53),

$$U_* = 3.9 \left(\frac{0.035}{(16)^{1/6}} \right) \frac{2}{\pi} (0.6) = 0.033 \text{ ft/sec}$$

Hence, from Eq. (7.50)

$$D = (5.93)(16)(0.033) = 3.15 \text{ ft}^2/\text{sec} = 1150 \text{ m}^2/\text{hr}$$

For well-mixed, uniform estuaries, the salinity of chlorides concentration in the system itself can be used as if it were a continuously added conservative tracer. This is sometimes called the salinity intrusion method. The concentrations C measured at various locations x are related as:

$$C = C_0 e^{U(-x)/D} \tag{7.54}$$

where

C_0 = concentration at x = 0

U = velocity (determined from freshwater flow and channel area A, or measured at site)

The results are plotted as logarithms of concentration versus distance measured from the sea end of the estuary, and D is calculated from the slope of the line obtained since

$$\ln \left(\frac{C}{C_0} \right) = \left(\frac{U}{D} \right) \left(-x \right) \tag{7.55}$$

Example 7.18 In a well-mixed, uniform estuary the velocity is estimated to be 0.6 km/day and the salinity at its sea end to be 20 g/kg. The salinity at a distance of 3 km upstream is 10 g/kg. Estimate the value of the dispersion coefficient D.

From Eq. (7.55),

$$D = \frac{U(-x)}{\ln \left(\frac{C}{C_0} \right)} = \frac{(600 \text{ m/day})(-3000 \text{ m})}{\ln \left(\frac{10}{20} \right) (24)} = 1.1 \times 10^5 \text{ m}^2/\text{hr}$$

Several measurements at different distances from the sea should in fact be made to get a more accurate value of D.

Generally, as stated in Sec. 7.2.2, the value of the dispersion coefficient for estuaries ranges around 10^4 to 10^7 m^2/hr and over, although in some cases much higher or lower values may be observed. The differences between observed and empirically computed values of D depend upon the extent of departure from the one-dimensional model assumed. Values of D considerably larger than those predicted may be observed when transverse velocity gradients exist, as they most certainly do in many estuaries and narrow straits. In some cases the dispersion coefficients are a function of the advective flows; as the flows increase, so do the coefficients.* In others, where tidal mixing greatly overshadows the advective mixing, no such relationships may be discerned [38.40].

7.11 DISTRIBUTION OF A POLLUTANT IN AN ESTUARY

The fate of pollutants discharged to an estuary, and their distribution within it, depend on

1. The nature of the pollutant (degradable, conservative, etc.)
2. The type of estuary (well mixed, stratified, etc.)
3. Location of discharge point (surface or bottom)
4. The relative volumes of fresh, saline, and wastewaters involved
5. The effective mixing characteristics of the estuary as reflected by its mixing or dispersion coefficient

It is evident that the pollutant distribution within an estuary will differ with varying environmental and hydrological conditions.

Degradable substances undergo a loss with time; therefore, their concentrations are considerably dependent upon the mean residence time in the estuary. A certain amount of nutrient buildup can also occur in an estuary. A modification of the method described in Sec. 8.10 for bays can be used to estimate nutrient buildup in estuaries.

Generally speaking, pollutant distribution along the length tends to be more uniform in a well-mixed estuary than in a stratified one. In studying the distribution of a pollutant in a stratified estuary, Ketchum [39] observes that at flood tide the entire water mass may move inland; at

---

*In northern San Francisco Bay, the diffusion coefficients were found to vary approximately as the 3/4 power of the advective velocity [38].

ebb tide in such estuaries the net seaward flow may be limited to the surface layer only.

Hence, if a pollutant is discharged into a stratified estuary, it is
carried upstream like the salt in seawater and is carried downstream like
freshwater. Thus, at some point in the estuary, a maximum value is observed
with a falling off both upstream and downstream. Figure 7.16 shows the
distribution of phosphorus in the Hudson River estuary, New York [41].

Various mathematical models have been developed for estimating the
concentration of degradable and conservative substances in time and space.

Tidal Prism Method

One of the earliest methods developed was the tidal prism method of
Ketchum [39] first given in 1950. The "tidal prism" is defined as the
volume of water held within the banks of an estuary between the low and
high tide levels. The method assumes that (1) unpolluted water enters the
estuary at high tide and mixes homogeneously with the "polluted" water
present in the estuary and (2) the amount of polluted water leaving the
estuary at low tide depends on the ratio of volume of tidal prism to total
water volume. Thus,

$$P = R + T \tag{7.56a}$$

where

P = volume of tidal prism

T = volume of water entering estuary during high tide

R = volume of water entering estuary from land sources (streams,
    sewers, etc.) during high tide

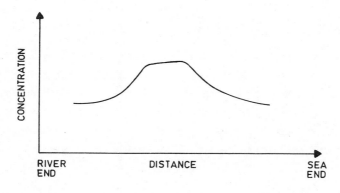

Figure 7.16 The distribution of a pollutant in a stratified estuary
(adapted from Ref. 41).

If V = volume of water in the estuary during low tide, we can write

$$\frac{V}{V + P} = \frac{m_1}{m_0} = r \qquad\qquad (7.56b)$$

where

m_1 = mass of pollutant left in estuary after a tidal period

m_0 = mass of pollutant in estuary before flush out

r = fraction of pollutant left in estuary after one tidal period

By assuming that in each tidal period the same amount of water enters the estuary, the fraction of pollutant left in the estuary can be estimated after n tidal periods as

$$K = r + r^2 + r^3 + \ldots + r^n$$

or

$$K = \frac{r(1 - r^n)}{(1 - r)} \qquad\qquad (7.56c)$$

where K = fraction of pollutant left in estuary after n tidal periods.

As $n \to \infty$, we have

$$K = \frac{r}{1 - r} \qquad\qquad (7.56d)$$

In words, Eq. (7.56d) states that the mass of pollutant present in an estuary depends on the mass present at the start and on the fraction remaining after a flush out, which in turn depends on the ratio of $V/(V + P)$. This method of estimation is evidently too simple to give any accuracy, particularly since mixing of incoming water is not homogeneous. For this reason the modified tidal prism method was developed.

Modified Tidal Prism Method

Here the estuary is divided into several parts along its length and each part is considered separately. Each part is defined in such a way that the mixing process would be complete in a tidal period. Thus, the volume of water of the part at the seaward end of the estuary at high tide is made equal to the volume of water in the next part upstream during low tide. This definition can be expressed mathematically as

$$V_{i+1} = V_i + P_i \qquad\qquad (7.56e)$$

where

where

V_i = water volume of part i at low tide

P_i = tidal prism volume of part i

V_{i+1} = water volume of part (i + 1) at low tide

Hence, referring to Figure 7.15, for each part of the estuary from the river
end to the seaward end one can write

$$V_1 = V_0 + P_0 = V_0 + R$$

$$V_2 = V_1 + P_1 = V_0 + R + P_1$$

$$V_3 = V_2 + P_2 = V_0 + R + P_1 + P_2$$

$$V_m = V_{m-1} + P_{m-1} = V_0 + R + \sum_1^{m-1} P_i \qquad (7.56f)$$

During low tide the polluted water is carried to the sea and we may
write

$$q_i = 1 - r_i \qquad (7.56g)$$

where

q_i = amount of water carried from one part to another part toward the
      sea

r_i = the fraction of polluted water left in the part

The values of r_i can be written as in Eq. (7.56b) to give

$$r_i = \frac{V_i}{P_i + V_i} \qquad (7.56h)$$

Substituting from Eq. (7.56h) to (7.56g) we get

$$q_i = \frac{P_i}{P_i + V_i} \qquad (7.56i)$$

The mass of pollutant left in each part may then be calculated mathematically by

$$m_{t(i)} = \frac{m_0}{q_i} = \frac{m_0(V_i + P_i)}{P_i} \qquad (7.56j)$$

Harlemann [36] warns that Ketchum's segment approach is likely to
give good predictions only in the case of well-mixed estuaries.

Over a period of time, the volume of seawater entering the system equals the volume of seawater flowing out. There is no net augmentation and net seaward flow is determined by the river flow. However, the entrained seawater increases the diluting capacity of the mixed water. This increase in capacity is proportional to the relative amounts of freshwater and seawater in the brackish mixture. Thus, if an estuary is divided into segments and complete mixing is assumed within each segment at high tide, Ketchum [cited in Ref. 20] gives:

$$F = \left(1 - \frac{S_m}{S_s}\right) \qquad\qquad (7.57)$$

where

F = fraction of freshwater in the mixture in an estuary segment

S_s and S_m = salinity of seawater and mixture, respectively

and

$$\text{Dilution volume} = \left(\frac{\text{river flow}}{F}\right) \qquad\qquad (7.58)$$

For example, if S_s = 30 g/kg and S_m = 27 g/kg, $F = 1 - (27/30) = 0.1$. Thus, dilution volume in the segment is 10 times the river flow. In this manner, the volume available for dilution in a segment increases in the seaward direction.

Dispersion Method

The distribution of pollutants in an estuary can be estimated using the 1-dimensional model [Eq. (7.16)] discussed earlier, after adapting it to flow conditions in an estuary. It was shown by Stimer [48] that material distribution in 1-dimensional flow having any cross-sectional form with superimposed secondary flows can be expressed by the 1-dimensional diffusion equation. Consider an elemental segment of an estuary with length dx as shown in Figure 7.17 in which the following symbols are defined as

A = cross-sectional area of the segment

U = instantaneous water velocity

C = incoming pollutant concentration

D = effective mixing coefficient in the longitudinal direction

M = mass of pollutant added to the segment per unit length and time

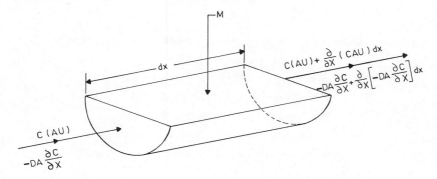

Figure 7.17 An elemental segment of an estuary (adapted from Ref. 41).

Writing a mass balance equation for a conservative pollutant we get

(Rate of change   =   (incoming rate) - (outflowing rate)
 in segment)   + (rate of addition)

Thus,

$$\frac{\partial}{\partial t} C(A\ dx)\ dt = C(AU)\ dt - \left[C(AU) + \frac{\partial}{\partial x} C(AU)\ dx \right] dt + M\ dx\ dt$$

$$+ \left(-DA\ \frac{\partial C}{\partial x} \right) dt - \left[-DA\ \frac{\partial C}{\partial x} + \frac{\partial}{\partial x} \left(-DA\ \frac{\partial C}{\partial x} \right) dx \right] dt$$

from which Karpuzcu [41] has shown

$$\frac{\partial}{\partial t} AC = - \frac{\partial}{\partial x} \left(CAU - DA\ \frac{\partial C}{\partial x} \right) + M \qquad (7.59a)$$

where

  U = average velocity of flow in segment

  C = average concentration of pollutant

If the pollutant is degradable, another term which characterizes the decay of the substance with time should be added. Thus,

$$\frac{\partial}{\partial t} AC = \frac{\partial}{\partial t} AC = - \frac{\partial}{\partial x} \left(CAU - DA\ \frac{\partial C}{\partial x} \right) + M - KCA \qquad (7.59b)$$

If C_i is the concentration in any segment, the concentration in the next segment C_{i+1} is given for a conservative tracer (K = 0) as

$$C_{i+1} = \frac{2(\Delta x_i + \Delta x_{i+1})}{A_i[2D_i - U_i(\Delta x_i + \Delta x_{i+1})]} \left[\frac{A_i D_i}{\Delta x_i + \Delta x_{i+1}} C_{i-1} \frac{A_i U_i}{2} C_{i-1} \right.$$

$$\left. - \Delta Q_i (C_{s(i)}) \right] \qquad (7.59c)$$

where

Δx_i and Δx_{i+1} = length of segments i and (i + 1), respectively

A_i = area of segment i

D_i = mixing coefficient for segment i

U_i = average velocity in segment i

Q_i = average flow through segment i

C_i = concentration of tracer in segment i

$C_{s(i)}$ = concentration in flow entering segment i

Equation (7.59c) can be used to estimate the effective mixing coefficient D_i if a conservative tracer is used and its measured concentration in a segment, and other variables listed above, are known. In this manner D_i may be calculated separately for each segment of an estuary. Then, by setting boundary conditions, the concentration of any pollutant can be calculated longitudinally for the whole estuary. This method can be used with various types of estuaries [41].

REFERENCES

1. L. Klein, Aspects of River Pollution,

1a. R. Mitchell, Ed., Water Pollution Microbiology, Wiley-Interscience, 1972.

2. G. M. Fair, J. Geyer, and D. Okun, in Water and Wastewater Engineering, R. Mitchell, Ed., John Wiley and Sons, N.Y., 1968.

2a. Engineering evaluation of virus hazard, Jour. ASCE, 96, 111, 1970.

3. W. W. Eckenfelder, Water Quality Engineering for Practicing Engineers, Barnes and Noble, N.Y., 1970.

4. E. J. Gumbel, Ann. Math. Stat., 12, 1941.

5. C. J. Velz, Applied Stream Sanitation, Wiley-Interscience, 1970.

5a. Ozturk Izzet, "Statistical Analysis of Low Flows in Rivers," M.S. thesis, Istanbul Technical University, 1979. (In Turkish.)

5b. R. Wright and A. McDonnell, In-stream deoxygenation rate prediction, Jour. ASCE, Env. Eng. Div., 105, 1979.

6. A. Maas, H. A. Thomas, and M. Fiering, Design of Water Resource Systems, Harvard Univ. Press, 1962.

7. G. Hornberger and M. Kelly, Atmospheric reaeration in a river using productivity analysis, Jour. ASCE, Env. Eng. Div., 101, 1975.

7a. E. Tsivoglou, Jour. WPCF, 40, 1968.

8. Langbein and Durum, Aeration Capacity of Streams, U.S. Geol. Surv. Circ. 542, Washington, D.C., 1967.

9. M. Churchill, H. Elmore, and R. Buckingham, Jour. ASCE, 3199, 1962.

10. Dobbins and O'Connor, Jour. ASCE, 123, 1958.

11. Pyatt and Granthum, Jour. ASCE, 97, 1971.

12. D. Liebich, Influence of Nitrification on the Oxygen Balance of Streams, Inst. für Siedlungswasserwirtschaft, Tech. Univ., Berlin, Federal Republic of Germany.

13. R. Courchaine, The significance of nitrification in stream analysis effects on the oxygen balance, Proc. 18th Purdue Univ. Indus. Wastes Conf., 1963.

14. V. Kothandaraman, Analysis of water temperature variations in large rivers, Jour. ASCE, San. Eng. Div., 97, 1971.

15. C. Song, A. Pabst, and C. Bowers, Stochastic analysis of air and water temperatures, Jour. ASCE, Env. Eng. Div., 99, 1973.

16. L. Brown, Statistical evaluation of reaeration prediction equations, Jour. ASCE, 100, 1974.

17. E. Thackston and P. Krenkel, Reaeration prediction in natural streams, Jour. ASCE, San. Eng. Div., 95, 1969.

18. G. Hornberger and M. Kelly, Atmospheric reaeration using productivity analysis, Jour. ASCE, Env. Eng. Div., 101, 1975.

19. E. Thackston, J. Hays, and P. Krenkel, Least squares estimation of mixing coefficients, Jour. ASCE, San. Eng. Div., 93, 1967.

20. E. Thackston and K. Schnelle, Predicting effects of dead zones on stream mixing, Jour. ASCE, San. Eng. Div., 96, 1970.

21. R. McQuivey and T. Keefer, Simple method for predicting dispersion in streams, Jour. ASCE, Env. Eng. Div., 100, 1974.

22. J. Hays and P. Krenkel, Mathematical modelling of mixing phenomenon in rivers, Advances in Water Quality Improvement, Vol. I, Univ. of Texas, Austin Texas,

23. D. O'Connor, D.O. analysis in a flowing stream, in Advances in Biological Waste Treatment, Proceedings Univ. of Texas, Austin, Texas (1966) and Jour. ASCE, 1965.

24. H. B. Fisher, Discussion on Ref. 23, Jour. ASCE, Env. Eng. Div., 101, 1975.

24a. H. B. Fisher, Dispersion predictions in natural streams, Jour. ASCE, San. Eng. Div., 94, 1968.

25. E. Harris, A new statistical approach to the one-dimensional diffusion model, Inter. Jour. Air Water Poll., 7, 1963.

26. J. Hays, Mass transport mechanisms in open channel flow, Doctorate thesis, Vanderbilt Univ., Nashville, Tenn., 1966.

27. L. Yu Shaw, Aerator performance in natural streams, Jour. ASCE, San. Eng. Div., 96, 1970.

28. R. V. Thomann, Mathematical model for D.O., Jour. ASCE., 89, 1963.

29. R. Thomann and M. Sobel, Estuarine water quality management and forecasting, Jour. ASCE, 90, 1966.

30. D. O'Connor, J. St. John, and D. DiToro, Water quality analysis, Delaware River estuary, Jour. ASCE, 94, 1968.

31. Grantham, Schaake, and Pyatt, Water quality simulation model, Jour.
     ASCE, 97, 1971.

32. S. Alfi, Y. Argaman, and G. Shelef, Mathematical model for prediction
     of D.O., in Advances in Water Pollution Research, 6th Inter. Conf.,
     Jerusalem, Israel, Pergamon, 1972.

33. T. Sueish and W. Yoshikoshi, Water quality management model for river
     underground flow system, 6th Inter. Water Poll. Res. Conf., Israel,
     Pergamon, 1972.

34. Automatic water quality monitoring, WHO Report on a conference held
     at Crackow, Poland, 1971.

35. K. Imhoff In-stream aeration, NATO-ASI, Biological Wastes, Istanbul,
     1976.

36. A. T. Ippen, Estuary and Coastline Hydrodynamics, McGraw-Hill, N.Y.,
     1966.

37. E. Holley, D. Harleman, and H. Fisher, Dispersion in homogeneous
     estuary flow, Jour. ASCE, Hyd. Div., 1970.

38. B. Glenne, Diffusive processes in estuaries, SERL, Report 66-6, Univ.
     of California, 1966.

39. B. Ketchum, Eutrophication of Estuaries, in Eutrophication: Causes,
     Consequences, Correctives, Nat. Acad. Sci., Washington, D.C., 1969.

40. Mathematical and hydraulic modelling of estuarine pollution, Water
     Poll. Res. Tech. Paper 13, HMSO, London, 1972.

41. M. Karpuzcu, Studies on the Golden Horn estuary, Ph.D. thesis, Istanbul
     Tech. Univ., Istanbul, Turkey, 1975.

42. E. R. Holley, Unified view of diffusion and dispersion, Jour. ASCE,
     Hyd. Div., 1969.

43. G. I. Taylor, The dispersion of matter in turbulent flow through a
     pipe, Proc. Royal Soc. London, 223A, 1954.

44. J. W. Elder, The dispersion of marked fluid in turbulent shear flow,
     Jour. Fluid Mech., 5, 1959.

45. W. W. Sayer, Dispersion of mass in open channel flow, Colorado State
     University Hydraulic Papers No. 3, 1968.

46. B. M. Sumer, Longitudinal dispersion of heavy particles in turbulent
     open channel flow, Jour. Fluid Mech., 65, 1974.

47. B. M. Sumer, Dispersion of suspended particles in open channel flow:
     A review, The Danish Centre for Applied Mathematics and Mechanics,
     Technical University, Denmark, Report No. 130, 1977.

48. B. M. Sumer, On the theory of turbulent dispersion of soluble matter
     in flows of irregular cross-sections, Proc. 14th Congress of Inter.
     Assoc. Hyd. Res., Paris, Vol. 1, Paper A-10, 1971.

49. H. Liu, Predicting dispersion co-efficient of streams, Jour. ASCE,
     Env. Eng. Div., 103, 1977.

Chapter 8

DISPOSAL TO COASTAL WATERS

8.1 DISPOSAL TO COASTAL WATERS

The special characteristics of the marine ecosystem resulting from its various physical, chemical, and biological features in complex interactions make any in-depth discussion beyond the scope of this book. Only a few of its essential features with special reference to waste disposal in coastal waters will be discussed here.

Physical differences from freshwater systems are caused inter alia by the prevalence of tides, waves, and currents. Chemical differences are due primarily to high salinity and the fact that all elements known today are present in marine waters. Biological differences result from the others just stated.

At the outset, a few commonly used terms may be mentioned. The continental shelf (called insular shelf in the case of islands) is the submerged border of a continent or island extending from the shore line to the depth at which the sea floor begins to descend steeply toward the bottom of the sea or ocean basin (see Figure 8.1). Conventionally, in the past the edge of the shelf has been taken to extend up to 100 fathoms (182.88 m)

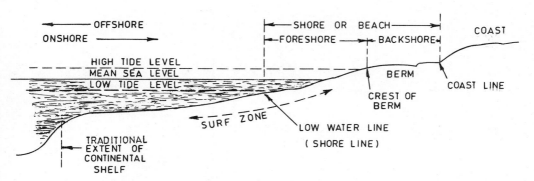

Figure 8.1 A definition sketch of a coastal area.

Less than about 5% of the ocean area is thus involved but is most important from the point of view of wastes disposal.

Seas are secondary in size to oceans. They are more or less landlocked and generally form part of, or connect with, an ocean or a larger sea. Bays and gulfs are a part of the sea or ocean extending into a recess in the land which forms a curve around it. A bay is smaller than a gulf but is of the same general character (for estuaries, see Sec. 7.8).

In the case of partially landlocked seas and bays, the hydrography of the connecting passage to the open sea or ocean is of considerable importance and affects the water quality, current disposition, and other characteristics of the enclosed waterbody. The Mediterranean Sea is an enclosed one with only 1% of its water flowing out per year (see Figure 8.2). Figures 8.1 and 8.2 help to define the extent of the shore or beach and a few other terms often used. The surf zone is the zone in which the effect of wave action is felt along the sea floor. This may extend up to a water depth of several m or more from the shore. It is reported to extend to about 10 m in the Mediterranean Sea and 15 m in the Atlantic.

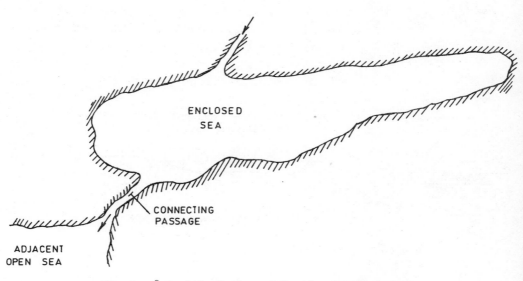

Figure 8.2 A typical example of a landlocked sea.

8.2 CHARACTERISTICS OF SEAWATER

8.2.1 Salinity, Temperature, and Density Profiles

Seawater contains inorganic and organic matter in suspension and solution along with various gases. A detailed analysis of "average" seawater has been given in Table 4.2 from which it is seen that the major cations and anions are the following:

Cations (g/kg)		Anions (g/kg)	
Na^+	10.770	Cl^-	19.353
Mg^{2+}	1.294	SO_4^{2-}	2.712
Ca^{2+}	0.413	HCO_3^-	0.142
K^+	0.387	Br^-	0.067
	12.864	I^-	0.060
			22.334
			12.864
		Total:	35.198 g/kg

Salinity levels in most oceans are high and generally range between 34 and 38 g/kg.* Seas and coastsl waters may show more widely varying values depending mainly on the extent of mixing with freshwater. Salinity variation is also observed over the depth of a marine waterbody. There is not much variation in the open oceans, but often quite considerable variation in coastal areas and bays.

Surface salinity is lowered by precipitation, ice melting, and incoming river discharges; salinity is increased by evaporation and ice formation. Thus, around the equator where precipitation is greater than evaporation, lower salinities are observed than in the tropics where the situation is reversed.† Again, at higher latitudes there is relatively less evaporation and, consequently, salinities are lower. Coastal waters are, of course, considerably affected by incoming river discharges.

---

*Also denoted by the symbol ‰ signifying parts per 1000 parts.
†In the Mediterranean Sea, for example, evaporation exceeds the entry of freshwater by about 4% [20].

Table 8.1 Salinity and Density of Some Waters

Water	Salinity (%)	Density (kg/m^3)	Density parameters (kg/m^3)
Drinking water standard	0.5 (max. 1.0)[a]	1000 (4° C)	0
Sewage	0.5-1.5[a]	1001-1003	1-3
Brackish water	1.5-3.0[a]		
Baltic Sea	~ 8.0	1006	6
Black Sea (surface)	17-18	~ 1013	~ 13
Mediterranean Sea	~ 34	~ 1028	~ 28
Indian Ocean	34.6-35.5		
Most oceans	34-38	1024-1030	24-30

[a]Total dissolved solids.

In the case of seas, bays, and other enclosed bodies of water, the
extent of the connection that exists between the enclosed body and the open
ocean affects salinity. Where this connection is narrow, mass transfer
characteristics are poorer and seasonal influence on fresh and sea water
balance is large. A few samples given in Table 8.1 illustrate the range
of values of salinities observed.

As mentioned, salinity also varies with depth at every site. This
depthwise variation is generally of a seasonal character. Figure 8.3 illustrates a typical salinity profile for coastal water with less salinity
in the upper layers, more in the lower layers, and a transition zone between called the halocline. The magnitude of salinity variation between
the upper and lower layers is less for the open oceans. The effect of
salinity on density and stratification is discussed later.

For purposes of analysis, salinity is defined as the total amount of
solid material contained in seawater when all the

1. Carbonate has been converted to oxide,
2. Bromine and iodine have been replaced by chlorine, and
3. Organic matter has been completely oxidized.

Since Cl^- is the major ion, salinity can be approximated by titration
with $AgNO_3$

$$S \ (‰) = 0.03 + 1.8050(Cl \ ‰) \tag{8.1}$$

where

    S = salinity, g/kg

    Cl = chlorinity, g/kg (equals Cl⁻ after Br and iodine have been
       converted to chlorine)

The electrical conductivity method of determining salinity is, however,
preferred. Salinometers for field use are available (see Table 8.7).

    The temperature of seawater also shows a seasonal variation with depth.
A typical temperature profile is also shown in Figure 8.3; it is similar to
the one for lakes discussed in Chapter 6. Uniformly warm temperature of
the upper layers indicates wind-induced mixing in that zone. This should
extend to several meters from the top. The <u>thermocline</u> generally shows a
change in temperature of more than $1°$ C per meter depth. The temperature
profile shows marked seasonal variations. It is interesting to know that
a sort of main thermocline exists between $50°$ N and $50°$ S separating the cold
waters from the warm waters; on this is imposed the local temperature profile.

    Both temperature and salinity affect the density of seawater. The
temperature t_m at which seawater has maximum density is related to the
salinity S ‰ by

$$t_m = 4 - 0.216(S ‰) \qquad\qquad\qquad (8.2)$$

    The temperature t_f at which seawater freezes is related to the salinity
by

$$t_f = -0.054(S ‰) \qquad\qquad\qquad (8.3)$$

A plot of t_m and t_f versus salinity shows that the two lines have different
slopes and intersect at S = 24.7‰. Thus, if salinity is greater than 24.7‰,

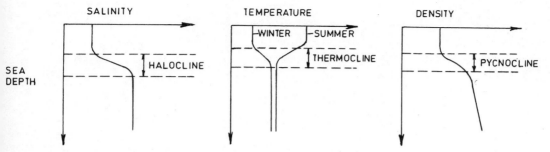

Figure 8.3 Typical variations in salinity, temperature, and density,
with depth, in a sea.

t_m is below t_f and the whole water column from surface to bottom has to be cooled to freezing point before ice is formed. This makes it difficult for a sea to freeze. But at S less than 24.7‰, particularly in the case of freshwater, the surface layers may readily freeze at low temperatures.

The density of freshwater at $4°$ C is 1 g/ml or 1000 kg/m^3. The density of seawater is always greater than 1000 kg/m^3 at all temperatures above $4°$ C. Thus, a convenient way of expressing it is by use of the term <u>density param-</u> <u>eter</u> to show the extent to which seawater exceeds this freshwater density.

Density parameter = (actual density, kg/m^3) - (1000, kg/m^3)

For example, if the actual density at a temperature is 1020 kg/m^3, the density parameter equals 20.

Knowing the salinity, and temperature, density parameter and, thus, the density can be readily found from Table 8.2. A few illustrative examples of density parameters have been included in Table 8.1.

Density variations with depth in the sea can also be plotted to give density profiles as shown in Figure 8.3. The surface waters generally show lower densities, while beyond the transition zone called **pycnocline** a higher density is observed. The higher density in the lower layers is due to both high salinity and low temperature.

Density affects the turbulence level and vertical exchange of water. Even slight numerical differences are sufficient to establish stratification and require considerable energy for vertical mixing.

Table 8.2 Values of Density Parameter (kg/m^3) Corresponding to
Different Values of Salinity and Temperature

Salinity (‰)	Temperature (° C)				
	0	5	10	15	20
0	-0.1	0.0	-0.3	-0.9	-1.8
5	4.0	4.0	3.7	3.0	2.1
10	8.0	8.0	7.6	6.9	5.9
15	12.0	11.9	11.5	10.7	9.7
20	16.1	15.9	15.3	14.5	13.4
25	20.1	19.8	19.2	18.3	17.2
30	24.1	23.8	23.2	22.2	21.0
35	28.1	27.7	27.0	26.0	24.8

Salinity, temperature, and density all vary in time and space, requiring field measurements over different seasons and at different distances from a shoreline in the vicinity of a proposed site for an outfall.

8.2.2 Other Characteristics of Seawater

Among the other characteristics of seawater which are of interest in waste disposal are the solubility of gases, particularly dissolved oxygen (DO). The saturation concentrations of DO in saline and fresh waters are related and can be approximated as follows:

$$C_{ss} = C_s - \frac{2.65 \, (Cl^-)}{33.5 - t^\circ C} \qquad (8.4)$$

where

C_{ss} = DO saturation concentration, mg/liter, in saline water
C_s = DO saturation concentration, mg/liter, in freshwater
Cl^- = chloride ion concentration, g/liter
t = temperature, $^\circ$C

Oxygen concentrations tend to decrease with increasing temperature and density. Oceanographers prefer to report concentrations of dissolved gases in terms of ml/liter at NTP, or mg-atoms/liter. Table 8.3 gives some values of DO in seawater at different conditions.

Because seawater is alkaline (pH = 8.0-8.2), it has a total CO_2 content that is much greater than would be computed from the partial pressure of CO_2 gas above a water surface. In oceans, total free CO_2 averages between 45

Table 8.3 Saturation Values of Oxygen in Seawater
from Normal Dry Atmosphere, ml/liter

Temperature (‰)	Salinity (‰) Chlorinity (‰)	27.11 15.0	30.72 17.0	34.33 19.0	36.11 20.0
0		8.55	8.32	8.08	7.97
10		6.77	6.60	6.44	6.35
20		5.63	5.50	5.38	5.31
30		4.74	4.63	4.52	4.46

Note: ml/liter x 0.08931 = mg-atoms O_2 per liter
       ml/liter x 1.429 = mg/liter
Source: Adapted from Ref. 2.

and 55 ml/liter, but in less saline seas such as the Baltic it is only about 22 ml/liter.

The amount of carbon present in oceanic water in inorganic compounds is between 2.1 and 2.5 mg-atoms per liter (25 to 30 mg/liter) depending upon salinity, temperature, and other factors. Total organic carbon, on the other hand, averages 0.2 mg-atoms per liter (2.05 mg/liter) in oceans, which is about one tenth of the amount present in inorganic form.

The maximum values of oxygen consumption for surface samples of un-filtered ocean water range between 1.5 and 2.0 ml/liter (2.14-2.86 mg/liter). Thus, BOD_5 values for seawater range between 1.0 and 3.0 mg/liter in most cases except in the vicinity of polluted shorelines where they may be higher. Bottom waters in some polluted areas have shown BOD values even higher than 6 mg/liter.

Nitrogen and phosphorus generally show seasonal variations depending on temperature, extent of biological activity, and other factors. The effect of upwelling is given in Sec. 8.2.5.

8.2.3 Tides

Tides are caused by forces resulting from (1) the relative motion of earth and moon, (2) the relative motion of earth and sun, and (3) the universal attractive force between earth, moon, and sun. Tides give rise to oscillating flows, currents, etc.

During a new moon and full moon, the tide-producing forces are additive and, hence, the tides produced at that time are of the greatest magnitude. These greatest tides, which can occur twice in a lunar month, are called spring tides. During the first and last quarters of the moon, the tide-producing forces of the sun and moon are in opposition and, thus, the resulting tides are of the minimum magnitude. These minimum tides are called neap tides.

Three types of tides and tidal currents are generally observed: semi-diurnal, diurnal, and mixed. The most important tide-producing forces are those of nearly semidiurnal (12.42 hr) and nearly diurnal (24.84 hr) period, for which reason the nature of the tide in any locality depends mainly upon the relative heights and phase angles of the partial tides corresponding to these forces [1,2]. At low latitudes where the semidiurnal components gen-

erally dominate, spring and neap tides come at intervals of 14.3 days and
the ratio between the ranges of the two tides may be fairly close to 2.77.
The diurnal feature predominates at higher latitudes.

In case of mixed tides, diurnal inequalities are often observed where
one of the two high waters of a day may be much higher than the other, and
one of the two low waters much lower. Also the time intervals may not be
equal. Tides of the Pacific coast are of this mixed type, while tides of
the Atlantic coast tend to be semidiurnal [2].

Tidal variations in open oceans often range from 0.5 to 1.5 m between
low and high tides. In bays and in seas connected to the ocean through a
relatively narrow opening, the tide may differ from that in the ocean because the shape of the bay or sea may amplify the oscillations because of
a decrease in water depth or width, or because of resonance. Thus, tidal
variations observed in coastal areas may range from just a few centimeters
to even several meters (maximum 15.4 m).

A special tide recording apparatus can be established at any desired
site to measure only long period tidal fluctuations, eliminating the effects
of wind generated waves. About 370 days of tidal record is necessary to
determine the characteristics of the various semidiurnal and diurnal components constituting the resultant tide. Summation could be performed on
a computer or by the more traditional method of using a tide predicting
machine.

Mean sea level is the plane about which the tide oscillates. It is
the average of the hourly heights of the tide measured over several years.
Mean high water and mean low water levels are the average heights of all
high waters and of all low waters, respectively, measured over several
years.

8.2.4 Currents

Ocean currents are generally influenced by one or more of the following factors: (1) Density differences, (2) tidal forces, (3) wind stress,
(4) wave action, (5) freshwater inflows, and (6) nearshore circulation
patterns. The actual magnitude and variation of currents at a site are the
result of complex interactions between these influencing factors and other

modifying factors such as coriolis forces*, friction, bottom topography, etc., which make it essential that the currents be measured over suitably long periods to enable a proper evaluation of the situation at a desired site.

Density Currents

Density differences resulting from temperature and salinity differences are responsible for the large scale currents observed in oceans (e.g., Gulf Stream current). On the average, relatively more solar energy is received, with consequently greater heat absorption, in equatorial areas than in temperate ones. This excess energy is equalized through currents and winds. Thus, more or less permanent currents circulate in open oceans from the equator to the polar regions. Where large scale mass movements occur, the relative magnitude of current speeds may be on the order of 0.07 to 0.3 m per sec. In enclosed seas such currents are prevented and excess energy is transferred more through the mechanism of winds and wind-induced currents.

Density differences between two adjacent areas lead to level differences since a higher column of low density water is needed to balance a shorter column of water of greater density. Hence, sea level tends to be higher where density is lower. The sloping sea surface affects the water with a force which, if bottom friction is neglected (in deep waters), must be balanced by the coriolis force. A current is so generated (called a geostrophic current) that the coriolis force acting on it is directed against the pressure gradient. The current follows the contour of the sea surface. In the northern hemisphere, the current flows in such a direction that the water of lower density is on the right-hand side of the current.

Tidal Currents

Tidal currents are horizontal movements of water resulting from the rise and fall of tides. They are distinguished from other currents by their periodic nature and are related to the character of the tide itself (diurnal, semidiurnal, or mixed; see Sec. 8.2.3). Tidal forces combined

---

*Coriolis forces caused by the earth's rotation affect all bodies in motion. The force acts perpendicular to current direction, but to the right in the northern hemisphere and to the left in the southern hemisphere. The force is proportional to the velocity of the current and to the latitude, and tends to deflect the currents somewhat. Consequently, in the northern hemisphere, circulatory currents are stronger along the western shores than along the eastern shores of the seas.

with topographical features may result in three general types of tidal currents: rotary, reversing, and hydraulic.

Rotating currents change direction and velocity continuously. They occur at short distances from the shore, mostly in open seas where coriolis forces are effective. In the northern hemisphere they generally rotate in a clockwise direction. With semidiurnal tide, the current will make two similar rotations in one lunar day. Over a lunar month periodic variations in the velocity of the rotating current will be observed and the velocity will be greater than average during spring tides. The magnitude of speed of the rotating currents may be on the order of 0.03 to 0.05 m/sec. A rotating current may be masked by the presence of a nontidal current, and a graphical plot of hourly currents may be useful in some cases to show the existence of the two components.

Reversing tidal currents are observed more in bays, estuaries, and enclosed seas. The current moves in one direction for about 6 hr followed by a movement in the opposite direction for a similar period, with a slack period in between. A circulation pattern may establish itself with net current in the shallower parts and a compensating current in the opposite direction in the deeper parts (Figure 8.4). The strength of reversing currents undergoes periodic changes corresponding to changes in tidal range as just described. Speeds may be on the order of 1 to 2 m/sec in restricted waterbodies, depending on the height of the tide, bottom topography, water depth, and other factors.

Hydraulic currents may occur in straits and sounds connecting two independent tidal bodies of water. Speeds may be on the order of 1 to 2 m/sec or more, again depending on the various factors enumerated above.

Wind-induced Currents

Wind blowing across a water surface exerts a stress on it, causing a drift or movement over a shallow depth. This movement alters the density distribution and leads to a current. The ratio between the velocity of the induced current and the wind velocity is called the "wind factor." Values of this factor in northern Europe have been found to range from 1/20 to 1/30 when surface current speeds were measured, but only 1/45 to 1/50 when average current speeds over a depth of 1.5 to 3 m were measured since wind drift decreases rapidly with depth. In 1902, Ekman gave an empirical relationship which also takes into account the latitude:

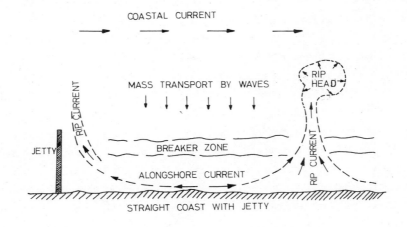

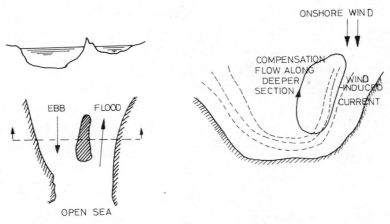

Figure 8.4 Nearshore circulation patterns.

$$\frac{U}{W} = \frac{0.0127}{\sqrt{\sin \varphi}} \qquad\qquad (8.5)$$

where

   U = current speed, m/sec, measured at about 2.4 m depth

   W = wind speed, m/sec

   φ = latitude

Thus, as an illustration, the ratio is equal to 0.03 at 10° N, and 0.016 at 40° N, signifying that the induced current speed is on the order of 1.6 to 3% of the corresponding wind speed between the given latitudes. Typically

observed values of current speeds range from 0.09 to 0.1 m/sec depending
on wind speed and location.

Ekman also deduced from theoretical considerations based on frictional
and coriolis forces that, in deep waters in the northern hemisphere, the
wind drift at the surface occurs at 45° to the right of the wind. Actual
measurements have shown this angle to be close to the theoretical value.
In shallow waters, however, the bottom friction may be more important than
the coriolis force and the net induced current may tend to be in the same
direction as the wind, and flowing through the relatively shallower regions
with a compensating flow in the opposite direction through the deeper parts
(Figure 8.4).

Wave-induced Currents

Wind blowing over a water surface also generates waves whose height
and period depend upon the wind speed and fetch length. When waves are
traveling in water the water particles move in a vertical plane in approximately circular or elliptical orbits depending on the depth of the water.
Since the velocity is generally greater at the top of the orbit than at the
bottom, the particles make a net forward progress with each cycle. Thus,
there is mass transport in the direction of the wave. Such wave-induced
currents are generally very small in the relative magnitude of their net
speeds, although appreciable volumes of water may be transported by this
phenomenon in shallow areas especially in the surf zone.

Freshwater Inflows

Currents may result in nearshore areas where water from rivers and
enclosed bays is discharged into the ocean in large quantities. Local effects may result from such conditions, such as refraction of waves by currents, changes in density distribution, and related effects.

Nearshore Circulation Patterns

The nearshore circulation patterns often tend to be markedly different
from those in deep waters beyond the surf line where a fairly uniform drift,
roughly parallel to the coastal contours in the middle and high latitudes,
may occur because of one or more of the factors discussed above. The nearshore circulation is considerably affected by bottom topography and other
factors such as the angle at which waves break along the coast.

The breaking of waves at an angle with the coast or a local rise in
water level resulting from mass transport by waves produces a hydraulic

head which must be balanced by a seaward return of water. Often, this
first generates an alongshore current moving parallel to the shore and
mostly within the surf line, followed by a high velocity flow seaward in a
relatively narrow zone which is known as a rip current. Figure 8.4 shows
an idealized nearshore circulation pattern. Because of their variable location and the short distance of flow, rip currents cannot be depended upon
for carrying away waste discharges at the shoreline, and on the contrary,
the discharges may be carried by alongshore currents for considerable distances, thus polluting the beaches.

     Waves and alongshore currents are also often responsible for the lit-
toral drift of beach sediments along a coast. Sand may be moved along the
bottom by stress resulting from waves and currents, or it may be suspended
by turbulence and moved by currents. The greatest movement of sand is
likely up to 10 m depth of water, although it could also occur in deeper
waters. Littoral drift of sand may considerably affect the safety and
useful life of an outfall and must be investigated in advance. Sand may
also be transported shoreward and seaward by currents, with cyclic changes
in sand depths corresponding to high and low waves.

8.2.5 Upwelling

     Water of the greatest density is usually formed in high latitudes with
cold temperatures. This water tends to sink and fill all ocean basins so
that deep bottom waters of all oceans (except a few isolated basins) are
cold. In general, it can be said that the deeper water in any vertical
column was once present at the surface somewhere in a higher latitude.
Sinking of surface water also occurs where converging currents are present,
but the amount of water that sinks in either case must be replaced by an
exactly equal amount of water that ascends.

     Ascending motion often occurs in regions of diverging currents, which
may be present anywhere in the sea, but which are especially seen along the
western coasts of continents in the northern hemisphere where prevailing
winds carry the surface waters away from the coasts. This ascending phenomenon which brings the heavier waters from lower levels to the surface
is known as upwelling. It is known to be particularly conspicuous along
the coasts of Morocco, southwest Africa, California, and Peru. It also
occurs on a large scale around the Antarctic continent [2].

Upwelling brings water of greater density and lower temperature up from depths extending up to 200 m or so, causing an overturning of the upper layer. Along the coast, the upwelled water replaces the light surface water transported away from the coast. This distribution of mass leads to an altered distribution of pressure which in turn gives rise to a current that flows in the direction of the wind.

The vertical motion resulting from upwelling can often help to explain the nature of observed temperature profiles, nutrient distribution, and some of their ecological consequences in a given environment. In coastal areas where upwelling is seasonal or intermittent, the nutrient content of the surface waters may show corresponding fluctuations superimposed on those owing to plant activity and often even masking them. For example, surface waters may show an increase in nutrient concentration instead of decrease in the season of maximum plant growth. Thus, it is evident that the effects of advection and diffusion must be taken into account when interpreting biological data.

Physicochemical conditions in the sea determine the organic production in different microenvironments, and these in turn influence the amount of benthic life and the nature of sediments (see Chapter 6 also), thus setting in motion a whole chain of ecological consequences.

8.3 ENVIRONMENTAL CONSIDERATIONS IN SEA DISPOSAL

Disposal of wastewater in coastal areas is achieved in two ways:

1. Inshore discharges, by extending the outfall sewer for a short distance, generally to discharge at or just beyond the low tide level.
2. Offshore discharges, generally at considerable depth, using properly engineered outfalls.

Environmental assessment in both cases must be based on a review of their health, aesthetic, and ecological implications. But, at the outset, it must be mentioned that system design includes waste treatment and outfall design. Both must be seen together, owing to their interaction, in arriving at optimum designs.

The water quality objectives to be met are essentially derived from a need to protect and promote the various beneficial uses to which the coastal waters in question can be put, and the likely impact that the wastewater disposal project may have on the microenvironment. Some of the important considerations include:

1. Health of the bathers who would be attracted to the beaches during favorable seasons.

2. Aesthetic effects such as those caused by odor, the presence of float-ables (e.g., oil and solid wastes), and discoloration of water, etc. They affect recreational and navigational uses, riparian land values, and some also have ecological consequences.

3. Ecological effects such as those caused by the following on the flora and fauna in the marine environment, some of which, again, have health significance:

    a. Oxygen depletion in localized areas affecting fish and other aquatic life

    b. Eutrophication and primary productivity

    c. Reduction in water transparency

    d. Reduction in species diversity

    e. Biological accumulation in the food chain affecting health and/or food taste

A critical examination of the above stated aspects is necessary to be able to assess the environmental impact of a waste disposal project (see Chapters 3 and 4).

As explained later in this chapter, the substances contained in waste-water discharged through a long outfall reduce in concentration because of the following:

1. Dilution owing to vertical rising from the outfall (see Figure 8.5) to the final level of the plume. The final level may be the water surface or some subsurface level.

2. Additional dilution owing to longitudinal dispersion in horizontal travel in the sea.

3. Reduction in concentration resulting from physical decay or biological degradation with time.

Conservative substances benefit only from dilution obtained by (1) and (2) above, whereas nonconservative degradable substances are reduced in concentration as a result of all the three factors stated above. Thus, dilution and travel time are both important in the case of degradable sub-stances such as organic matter, microorganisms, etc. Pretreatment reduces the extent to which dilution and dieoff time are required to meet water quality standards. Dissolved oxygen requirements have to be met especially in the vicinity of the discharge point where the plume is rising and flowing away. Coliform requirements have to be met at bathing beaches and shellfish grounds. Other requirements have to be met in the general disposal area.

HORIZON

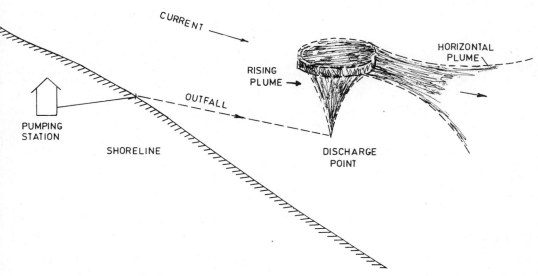

Figure 8.5 Plume formation after discharge through a long outfall. The vertical plume rises to its trapping level where the current fans it out horizontally and vertically.

Table 8.4 gives some typical recommended water quality objectives and standards based on experiences in the U.S. and Japan.

The <u>overall</u> reduction in the concentration of a substance is the product obtained by multiplying the individual reduction resulting from the above stated three factors. Thus, for example, if the initial dilution is 50 whereas the dilution owing to lateral dispersion is 10, <u>all</u> substances are reduced in their concentration by 50 x 10 = 500, thus far. If a degradable substance is further reduced by a factor of, for instance, 100 owing to decay with time, the overall reduction for that substance is 50 x 10 x 100 = 50,000. Coliforms present in high numbers in raw sewage often demand even greater reductions than the above to be able to meet beach standards.

Table 8.4 Coastal Water Quality and Effluent Discharge Standards/Guidelines of Some Countries

Parameter	For coastal water quality						For effluents discharged to sea			
	California		Israel		Japan	Denmark	California		Denmark[c]	
	50% of time	90% of time	Commercial fishing[a]	Recreation[b]			50% of time	90% of time	Enclosed bays and seas	Open bays and seas
Oil and grease (hexane extractable)	10 mg/m²	20 mg/m²					10 mg/liter	15 mg/liter	5 mg/liter	10 mg/liter
Floating solids	1.0 mg/m²	1.5 mg/m²							Not visible	
Dissolved oxygen, mg/liter			Depression ≤ 10% from natural background value	5% from background	7.5	Min. 5.0				
pH	Change ≤ 0.2		6.7-8.5	7.0-8.8	7.8-8.3		6.0-9.0	6.0-9.0	6.0-9.0	-
Temperature			Increase ≤ 3°C in cooler months and ≤ 1°C in warm months							
BOD5, 20°C, mg/liter						6.0	Monthly average ≤ 30; weekly average ≤ 45 (min. removal 85%)		100	400
NH3-N, mg/liter							40	60		
Total phosphorus, mg/liter									1.0	
Anionic detergents[d], mg/liter							0.5	1.0	5.0	10.0
Phenolic compounds, mg/liter							2.0	4.0	0.2	
Chlorinated hydrocarbons, µg/liter									Eliminate as much as possible	
Cyanide, mg/liter					None detectable		0.1	0.2	0.1	0.2
Organo-phosphorus					None detectable					

Parameter					
Toxicity, TU[e]	0.05	Bioassay test			
Ag, mg/liter	-	1.50	2.0	0.05	0.10
As, mg/liter	0.05	0.02	0.04	0.5	1.0
Cd, mg/liter	0.01	0.02	0.03	Eliminate as much as possible	
Cr (total), mg/liter	0.05	0.005	0.01	0.2	0.2
Cu, mg/liter	-	0.2	0.3	0.2	0.5
Hg alkyl, mg/liter	None detectable				
Hg total, mg/liter	None detectable	0.001	0.002	Eliminate as much as possible	
Ni, mg/liter	-	0.10	0.20	0.5	0.5
Pb, mg/liter	0.10	0.10	0.20	0.5	0.5
Zn, mg/liter	-	0.30	0.50	1.0	1.0

[a] Applies outside the immediate mixing and dilution zone.

[b] Minimum dilution required to be 1:700 if only preliminary treated sewage is discharged.

[c] Emission requirements based on minimum dilution of 10.

[d] 80% decomposable.

[e] TU = toxicity units = 100/96 hr TL_m. However, when there is more than 50% survival even in undiluted wastewater, the toxicity value is to be calculated as TU = $\log (100 - S)/1.7$ where S = percent survival in undiluted wastewater.

8.4 ALTERNATIVE STRATEGIES IN MARINE DISPOSAL

Some of the alternatives available for consideration in marine disposal projects are listed below. The choice depends on their ability to meet desired water quality objectives at minimum costs. In computing minimum costs, both capital and annual costs including loan repayment must be taken into account. A computer program can also be developed if desired for optimization purposes by including in the program the steps necessary for outfall design, waste treatment required, and costing of all systems to meet the desired objectives [19].

Some typical alternatives are as follows:

1. Effective source control (e.g., for pesticides and heavy metals) followed by preliminary treatment (screening, degritting, and grease removal if necessary) followed by direct discharge through an outfall of adequate length.

2. Source control plus preliminary treatment plus chlorination plus outfall of lesser length.

3. Treatment as in (2) above plus primary or secondary treatment plus outfall of still lesser length.

4. As in (3) above plus separate outfall for raw sludge disposal (smaller diameter but longer length than main outfall) or separate barging of sludge out to sea.

5. As in (3) above plus additional treatment for nutrient removal or relocation of outfall to avoid nutrient buildup in a given area.

6. As in (1) above but with land disposal of wastewater during bathing season (and also in other seasons favorable for irrigation).

8.4.1 Waste Treatment Prior to Sea Disposal

Two types of situations often occur: first, there are cases in which waste treatment and outfall length are interrelated so that the higher the degree of treatment that is given to a wastewater, the shorter is the outfall length required in order to meet specified water quality objectives. In such situations, an economic evaluation of the two alternatives can be carried out. However, the final choice may often be slanted in favor of a longer outfall, especially if it does away with the need for treatment which may cause problems in land, power, and equipment availability. Also, facilities for good operation and maintenance may not be assured. After all, the biological degradation that occurs in a treatment plant can also occur naturally in the marine environment.

With longer outfalls, BOD is generally not the main problem as considerable dilution is available in most cases. Shoreline discharges, however, cannot count on much dilution and their use must be limited to small flows, preferably after some treatment, or to areas where bathing is not feasible. Besides dilution of organic matter, there is a need to ensure compliance with the bacterial and viral quality of bathing waters, and outfall length has to be adequate to give both dilution and enough time for natural dieoff during travel to the shoreline. If outfall length is not adequate, some treatment including disinfection may be called for, again requiring a balancing of operating costs for treatment with capital cost for outfall construction.

The second type of situation met with is one in which some form of treatment may be desirable irrespective of a reasonable outfall length and the dilution obtainable. Such is the case, for example, with enclosed seas and bays in which water exchange with the adjacent open sea is not rapid, and considerable accumulation of pollutants can occur at steady-state conditions. Nutrients such as phosphorus can stay trapped in enclosed waterbodies and lead to eutrophication, in due course, with its attendant problems. Treatment may then be aimed at nutrient removal using physicochemical, biological, or other methods even though removal of organic matter per se and oxygen depletion may not be a serious problem.

Often the shoreline of a country may have enclosed sea areas and bays where water exchange is not likely to be rapid. If the beautiful blue waters which attract the tourists to such countries are to be preserved, their eutrophication must be prevented and the discharge of nutrients and other pollutants must be given special consideration.

Besides nutrients, treatment may also be essential for avoiding esthetic problems such as those caused by odors, discoloration, presence of oil, floatable solids, etc. These may be particularly important in disposal of industrial wastes and may need the employment of suitable mechanical methods (e.g., screening, skimming, and flotation).

Accumulation of a toxic substance or heavy metal in the food chain defeats the benefit obtained by increased dilution with a long outfall. Reduction in concentration of a substance because of dilution may be a hundredfold to a thousandfold in the vicinity of an outfall. However, if the substance is biologically accumulative, its buildup in algae, molluscs, shellfish, and other fish may also be on the same order of magnitude, thus nullifying the benefit of dilution.

Removal of these substances in waste treatment processes is generally difficult and expensive. Besides, their removal may only yield a sludge rich in these substances, whose disposal may still remain a problem. If the source of such substances is agriculture (e.g., DDT), control is even more difficult and the diffuse nature of the operation precludes any attempt at treatment. Thus, toxic substances and heavy metals are best controlled at the source itself. They should be substituted by nontoxic or nonaccumulative substances wherever possible. The gradual elimination of mercury usage in certain industries is a case in point. Sometimes, the manufacturing process itself may need some modification.

Finally, treatment may be essential for virus removal since recent studies by Katzenelson and Shuval [23] would indicate that t_{90} values of viral bodies are on the order of 48 hr compared with about 2 to 4 hr for coliforms. Hence, little reduction occurs by natural dieoff during travel from an outfall to the shoreline. Viral concentration of domestic sewage ranges from 100 to 10,000 PFU per 100 ml and, if the average concentration is assumed to be, for instance, 1000 PFU per 100 ml, the overall reduction resulting from dilution and dieoff should be on the order of 1:1000 since even 1 PFU can theoretically be infective. Such large dilutions may be difficult to ensure at all times and treatment may be necessary to protect bathers. Physicochemical methods of treatment may then have an important role to play. Viruses also concentrate in shellfish.

It will thus be seen that on many occasions some consideration has to be given to waste treatment for domestic and industrial wastes proposed to be discharged to the sea from small and medium size communities. The available treatment methods include the following from which a judicious choice would need to be made: (1) mechanical, (2) biological, and (3) physicochemical. Land disposal is, in fact, physical, chemical, and biological treatment all rolled into one and should be considered, even for seaside communities, wherever feasible.

Even for small installations, the minimum recommended treatment includes provision of screens. Grit removal is also recommended as it provides flexibility in adopting outfall design velocities and causes fewer maintenance problems. Screenings and grit so removed must be suitably disposed. Rotary screens appear to have greater utility with larger flows.

Both manpower and equipment requirement is highest for the activated sludge process, progressively lower for extended aeration and aerated lagoons, and practically negligible for ponds. Operational characteristics are also such that the highest skills are required for the activated sludge process and the least for ponds. Sludge treatment and disposal may often become a crucial factor in the choice of process, throwing the balance in favor of simpler methods such as lagoons, ponds, and ditches rather than the activated sludge process. Biological rotating discs are simple to operate but their sludge disposal problems may place them on a par with trickling filters except for very small installations where the sludge may be treated in Imhoff-type tanks or carted away for separate disposal. A combination of processes can be developed to give an optimum solution for a specific situation (Chapter 15).

Additional design aspects such as nutrient removal and microbial disinfection would need suitable modifications in the flow sheets. Nutrient removal can be achieved in two ways: either by process modification as just stated (for example, by alum or lime addition for phosphorus removal in activated sludge and extended aeration processes) or by following the lagoons and ponds with land irrigation.

Physicochemical methods appear to have a considerable potential for use with small and medium size communities, and in treatment of certain industrial wastes. Their land and power requirements are much lower than those of biological processes, they are not much affected by temperature, and they can be used on an intermittent basis, if desired. The chemicals generally preferred are lime or alum whose dosages may be on the order of 100 to 300 mg/liter or more. Phosphate removal upward of 90% can be achieved with concurrent reduction of BOD by 70 to 80% and heavy metals by 50 to 90%, depending on the particular metal (Harremöes [23]). The physicochemical treatment proposed in the Oslo area of Norway is expected to give 85% removal of particulate matter, 85% removal of phosphorus, and 70% removal of the organic matter, before discharge to the inner fjord (Baalsrud [23]).

Much of the phosphorus and organics remaining after chemical treatment can be removed by filtration through conventional rapid sand filters. Organohalogens and microconstituents of a refractory nature can be removed by adsorption in activated carbon beds which would follow the chemical or biological treatment steps. Use of activated carbon is often handicapped by lack of regeneration facilities.

The success of physicochemical methods depends largely on how the sludge is treated, dewatered, and disposed. More research is needed in the whole field of physicochemical methods to enable wider application.

8.5 MICROBIAL CONTROL OF BATHING BEACHES AND SHELLFISH AREAS

Internationally accepted criteria for the quality of coastal waters with regard to microbial contamination are not yet available. However, several countries have adopted standards for the control of coastal water quality, especially during the bathing season, by specifying limits for coliform counts in surface waters at the shoreline.

Generally, coliforms and E. coli have been taken as indicator organisms and, sometimes, Streptococcus faecalis has also been used for assessing the quality of bathing waters. The use of Clostridium perfringens has also been recommended. Coliforms and E. coli have the advantage of being in much larger numbers in seawater than any of the others, and well-standardized enumeration methods for them are available in most laboratories (also see Sec. 4.5.1).

The standards of some countries specify limits for E. coli but, according to Bonde [3], in hot climates the differential tests generally used for the identification of E. coli cannot be applied as other coliform types are heat tolerant and even indole-positive without belonging to the E. coli genus. In India, 60 to 70% of heat-tolerant aerogenes strains have been recorded as positive in the 44° C test for E. coli. Lack of fermentation of lactose at 44° C, which is typical for aerogenes strains isolated from waters in temperate climates, is not at all typical for the whole Klebsiella group, for example. Thus, in some countries coliform rather than E. coli standards have been used, although it is evident that many coliforms may have their origin elsewhere than in sewage. Higher coliform values than specified in the standards of some countries have been recorded along even relatively unpolluted beaches around Bombay.

A desirable approach to follow before a new outfall is installed is to obtain available data (supplemented by additional data, if necessary) to establish the present level of coliforms in the beach waters in question, expressing them through statistical parameters (80, 50, or 20 percentile values, for example). This would give background values against which future increases could be evaluated. Ideally, the natural levels should not be increased after installation of an outfall, but rather decreased if the earlier discharges were being improperly made. In the latter case, background values for beaches in the same region, but unpolluted by sewage discharges, should be sought.

Epidemiological studies on coastal pollution have given inconclusive results [4]. In a large scale survey of over 40 bathing beaches around the English coast that were known to be contaminated with sewage, only four cases of paratyphoid fever were recorded. These may have been due to bathing; they were reported from beaches with median coliform counts of more than 10,000 per 100 ml and with evidence of gross macroscopic fecal pollution.

It can be reasoned that a swimmer should not be exposed to any greater risk of pathogens from swimming than from drinking water. But, while the average consumption of drinking water is 2.2 liters per person per day, a swimmer may only swallow not more than approximately 10 to 20 ml or so of seawater (without feeling ill), and generally much less. Furthermore, statistically, the health risk would be in relation to the level of disease prevalent in the community and the length of the bathing season [5]. Thus, developing countries in warmer climates with higher incidence of enteric infection and with an almost year-round bathing season would in theory need to adopt stricter standards than those suitable for countries in temperate climates with better community health, especially if they wish to attract tourists from the latter.

Table 8.5 gives some typical examples of coliform/E. coli standards. Monitoring agencies must use proper statistical methods in analyzing and reporting their data.

Raw sewage coliform concentrations range from 10^6 to 10^9 per 100 ml (Table 5.1). If permissible values at bathing beaches are not to exceed an average of, for instance, 1000/100 ml, the removal required varies from 99.9 to 99.999%. This can be achieved only if the outfall is placed at sufficient length and depth to give both dilution of the waste and time for natural dieoff to occur before reaching the beach. The estimation of bacterial dieoff rates in seawater is discussed in Sec. 8.8.1.

Shellfish, including mussels, clams, oysters, scallops, etc., are often cultivated in saline waters of tidal estuaries, and danger to health arises from the fact that some of them may be eaten raw. They tend to "concentrate" bacteria and thus develop higher counts compared with the surrounding seawater. They have been implicated in several outbreaks of typhoid fever and infectious hepatitis. The U.S. Public Health Service recommends the following standards for waters used for shellfish cultivation:

Area	MPN Coliforms/100 ml not to exceed	
	Median	10% Samples
Approved	70	230
Restricted[a]	700	2300

[a]Shellfish from such areas may be marketed after "self-purification" in clean water or mildly chlorinated water.

Table 8.5 Coliform Standards for Bathing Beaches in Some Countries

Country or State	Organism	Values per 100 ml of seawater not to be exceeded the stated percent of time			
		50%	80%	90%	95%
European Economic Community (EEC)[a]	Total coliforms		500(G)		10,000(I)
	Fecal coliforms		100(G)		2,000(I)
	Fecal Streptococci			1,000(G)	
	Salmonellae (per liter)				0(I)
California (prior to 1977)	Coliforms	230	1,000[c]		
California (1977)	Fecal coliforms (MPN)	200 (in effluent)		400 (in effluent)	
Israel[b]	Coliforms		1,000		
Denmark	E. coli			100	1,000
Japan			1,000		

[a]EEC prescribes two levels: (G) denotes "guidelines" which member states should endeavor to observe, while (I) denotes mandatory requirements for all member states. Based on the observed statistical distribution in U.K. coastal waters, the (I) value of 10,000/100 ml (95 percentile) corresponds to a median value of about 800/100 ml, while the (G) value of 500/100 ml (80 percentile) corresponds to a median value of only about 150/100 ml for total coliforms. Samples are to be taken at places where average density of bathers is highest. Minimum sampling frequency is once every other week for total and fecal coliforms. Enumeration of fecal Streptococci and Salmonellae is required only when inspection shows they may be present or when the quality of water has deteriorated.

[b]A minimum of 10 samples per month is recommended for each sampling station.

[c]For many years the criteria for contact sports areas were that total coliforms should not exceed 1000/100 ml 80% of the time (roughly equivalent to a median value of 230) and the maximum value should be equal to or less than 10,000 MPN/100 ml. Samples are to be taken from surface waters up to 300 m from shoreline.

8.6 ECOLOGICAL EFFECTS ON THE MARINE ENVIRONMENT

8.6.1 Toxic Substances and Food Chains

From the point of view of pollution, some naturally occurring elements (Hg, Cd, Pb, Cu, Ni, Zn, Cr, and As), radioisotopes, pesticides, oils, and hydrocarbons are of particular interest. Upon discharge to the sea, they undergo various physical, chemical, and biological interactions such as the following: hydrolysis, adsorption on algae, silt and suspended organics, ion-exchange uptake in clay and other particles, formation of organic complexes, accumulation in silt and sediments, accumulation in the food chain, and gradual decay with time. The resultant concentrations in seawater are difficult to estimate owing to the extremely complex nature of these factors. Furthermore, their concentrations in seawater must be judged against the expected concentrations in "average" or unpolluted seawater to separate the effects of pollution from the effects of natural buildup of concentration that occurs in any seawater (see Sec. 4.3).

The subsequent biological accumulation of persistent substances depends on the nature of the organism and the substance, and other interactions in the food webs as described in Sec. 4.4. Objectionable concentrations can build up in fish consumed by man. The examples of mercury and cadmium toxicity from the consumption of fish are now well known.

In vitro tests carried out by Donnier [25] have shown the accumulation of heavy metals to occur in sea organisms as shown in Table 8.6. Accumulation of mercury has been dealt with at length in Chapter 4.

Considerable amounts of pesticides and chlorinated hydrocarbons drain into coastal waters from agricultural and industrial operations. Some of these are aldrin, dieldrin, parathion, malathion, lindane, and, of course,

Table 8.6 Concentration Factors for Some Metals
from Seawater to the Organism

Type of Food Chain	Item	Cu	Zn	Cr^6	Pb	Hg
Pelagic	Plankton	1000	1000	10-20	500	1000
	Fish	20-25	200	2	1	100
Neritic	Phytoplankton	1000	7000	1000	5000	3000
	Mollusks	120	1000	40	150	400

Source: Adapted from Ref. 25.

DDT and PCB (PCB is not a pesticide, the main sources of PCB are electrical, plastic, and other industries).

Oils and hydrocarbons find their way into the marine environment from land-based and marine transport operations. Oils dispersed on the sea surface form emulsions which may eventually disintegrate into characteristic tar lumps. In areas of major oil shipments, between 1000 and 2000 µg of tar lumps per m^2 may be observed [25]. The total hydrocarbon content in marine organisms does not always indicate pollution since some hydrocarbon is also synthesized naturally by some organisms. Careful analysis is needed to separate the natural from the man-made forms.

Hydrocarbons have been shown to accumulate in molluscs and fish in the sea [25,26]. In fish, the accumulation is mainly in the adipose tissues (liver and muscles). Some oysters grown in polluted waters have been found to contain as much as 1 mg hydrocarbons per kg wet weight of organism. Some hydrocarbons are carcinogenic.

Accumulations of PCB and DDT have also been observed. Recent studies have shown the ratio of PCB/DDT in oceanic biomass to be much higher than 1.0, even when the same ratio in discharges is lower [25]. This may be due to a slower degradation and a higher accumulation rate for PCB. Both DDT and PCB tend to concentrate more in surface waters.

Appreciable reductions in algal photosynthesis have been observed at about 10 µg/liter concentration of pp'DDT. Chlorinated pesticides and PCB also interfere with the growth of oyster shells [25]. Results of bioassay tests on fish need to be adjusted by a factor of 10^4 to 10^5 to account for likely deleterious effects on the larvae and eggs which are generally more sensitive than the adult fish.

According to one authority quoted by Bernhard and Zattera [25], a concentration of 10 µg pesticides per kg fresh weight of organism may be considered a low contamination of the environment, whereas 100 µg/kg fresh weight may indicate localized pollution.

Table 8.7 summarizes some data concerning typical observed concentrations of some elements, hydrocarbons, pesticides, etc., in seawater and fish. A few biological concentration factors from seawater to organisms such as plankton, mollusks, and fish are also given. Often it has been observed that the first trophic levels show the maximum enrichment. Thus, phytoplanktons, zooplanktons, and macrophytes show large concentration factors. Mollusks and crustaceans lie in about the same range. Further enrichment in the higher trophic levels may not occur to any appreciable

Table 8.7 Typical Concentrations of Selected Pollutants in Seawater and Some Organisms

Pollutant	Typical observed concentrations (μg/liter)			Typical concentrations in fish (μg/kg fresh wt.)		Some observed biological concentration factors from seawater to stated organism			Remarks
	Unpolluted seawater	Polluted seawater	Sediments (μg/kg dry wt.)	From distant ocean waters	From polluted coastal waters	Plankton	Mollusks	Fish	
Hg	0.01–0.2	Up to 3.6 (Minamata Bay)	Up to 3×10^3	40–110	200–4,000	100–2,000	400–2,000	100–5,000	Concentration up to 500 μg fresh weight of fish considered safe for regular human consumption.
Cd	0.01–0.1	Up to 10.5	Up to 10^4	<30	50–960	100–1,000			
Pb	0.02–1.5	Up to 5.0	Up to 2×10^5	100–500	500–1,800	500–5,000	150–	1–	
Cu	0.5–3.5	Up to 11	Up to 2×10^6	<500	200–9,000	1,000	120–	20–25	Little magnification occurs in going up the food chain.
Zn	1.5–10	Up to 100	Up to 2×10^6		3,100–40,000	1,000–7,000	1,000	200	
Cr[6]	0.04–0.25	Up to 2.5	Up to 1.2×10^6		960–11,000	10–1,000	40–	2–	
Ni	1–2	Up to 43	Up to 0.8×10^6		500–7,200				
As	2–3	Up to 35	Up to 10^6		300–11,500				Accumulation in the food chain generally occurs in the nontoxic As^{5+} form.
Mn	0.1–8.0				250–5,700			200–2,000	
Chlorinated hydrocarbons and pesticides									

Table 8.7 (Continued)

Pollutant	Typical observed concentrations (µg/liter)			Typical concentrations in fish (µg/kg fresh wt.)		Some observed biological concentration factors from seawater to stated organism			Remarks
	Unpolluted seawater	Polluted seawater	Sediments (µg/kg dry wt.)	From distant ocean waters	From polluted coastal waters	Plankton	Mollusks	Fish	
Aldrin				400-880		4600			
DDT		0.0023-0.19 in surface waters			17-1,280		1,200-9,000	8,500	Fish tissue values higher than 50 µg/kg fresh weight are considered hazardous.
Dieldrin				80-240					
PCB		0.2-19 (Mediterranean Sea)	800-9,000 (Mediterranean Sea)		2-2,250			12,000-72,000	Fish tissue values higher than 500 µg/kg fresh weight are considered hazardous.

Source: Adapted from Refs. 3 and 24-26.

extent and, in fact, fish-eating birds and marine mammals may show similar
or even lower concentration factors. Progressive increase along the food
chain is not always rapid.

In general, biological concentration factors may be expected to be
relatively lower in seawater than in freshwater where the natural concentration is itself lower. The reader is also referred to Tables 4.4 and 4.5
which give concentration factors for some radioisotopes and other substances.

Ideally, no measurable concentration of heavy metals, persistent organics, radioactivity, etc., should be present in any wastewater discharged
directly or indirectly to the sea. Substances such as cyanides and phenols
should also be excluded. Phenols have often been implicated in the tainting
of fish taste. One approach sometimes followed is to limit the concentration of the above substances in the seawater at the same level as permitted
in river waters. Wastes generally undergo much dilution in a sea, but biological accumulation of persistent substances in the food chain may cancel
out that benefit; thus, the desirability of controlling these substances as
far as possible at the source. In fact, source control of such pollutants
should be given the highest priority in any pollution control program.
Table 8.4 includes some typical receiving water standards for such substances. As further data become available it will be possible to improve
upon them.

Monitoring the seawater for very small concentrations of some of the
toxic substances may, indeed, be difficult owing to equipment and sampling
limitations and the fact that seawater is not so homogeneous all around.
This difficulty is sometimes overcome (as in the case of radioactivity
measurements [3]) by monitoring for a typical indicator organism up the
food chain, preferably a sedentary one, in the vicinity of the outfall.
Derived control limits are used for a substance in the specific organism,
thus benefiting from its averaging and concentrating mechanism to magnify,
so to speak, the low value of that substance in the water. Other methods
are also used.

8.6.2 Productivity and Eutrophication

Especially in the case of enclosed bays and seas, the discharge of
wastewaters into them may have an impact on the sewater quality in terms
of nutrients and productivity. No serious difficulties with regard to

dissolved oxygen may be caused if the waste is being sufficiently diluted, and the upper layers are well oxygenated, but nutrient accumulation and effects on primary productivity may be such as to possibly lead to eutrophication and other adverse ecological effects. In many ways the enclosed sea or bay may behave like a lake owing to its slow rate of exchange of water with the open ocean. The nature of the flora and fauna, of course, would be different in various respects from that in freshwaters.

Among the major nutrients are organic carbon, nitrogen, and phosphorus. The reader is referred to the more exhaustive discussion of this question presented in regard to lakes in Chapter 6. An inventory of nutrient inputs into the waterbody from all sources must first be prepared in a similar manner as recommended for lakes. Considerable difficulties may present themselves in preparing such an inventory and predicting future increases in incoming nutrients. Estimates may sometimes be only of an order of magnitude nature.

Simultaneously, it is necessary to know the average annual productivity of the receiving body of water at the present time (see Chapter 3). This information may be available or may have to be obtained from periodic surveys spread over different seasons of the year. Present-day productivity data also provide a base line against which future increases can be evaluated. It will be recalled from Table 3.2 that primary productivity in most coastal shelf waters averages 25 to 150 g of net carbon production per m^2 per year. Some polluted waters may exhibit much higher values.

If present productivity is stated in terms of grams of carbon per m^2-year and the average area of the waterbody is known, the total productivity in terms of tons per year can be estimated. Against this, the input of total organic carbon (TOC) can also be estimated, if the BOD of the wastewater is known or if TOC is directly measured, and estimates can be prepared for the design period. The input of organic carbon from wastewater discharge may be some fraction of the natural production of organic carbon in the waterbody. It is, of course, not possible to give any limiting values for this fraction as the impact of the input depends on various factors, the most important being the extent of mixing occurring with the open ocean and whether the discharge is to the upper or lower waters. Example 8.1 illustrates how an estimate of the type just described for organic carbon can be prepared.

Similar inventories for nitrogen and phosphorus can be prepared to estimate present-day and future likely inputs. Nitrogen balance for a waterbody is difficult to prepare owing to the occurrence of nitrogen fixation and denitrification. Precipitation of phosphates also makes it difficult to estimate buildup in a waterbody, although estimates of "phosphorus residence time" and likely buildup can be made using the methods described in Chapter 6. Some guidance concerning the permissible levels of phosphorus in enclosed seas, bays, and estuaries can also be obtained from the discussions presented in Chapter 6.

For oceans, approximately constant correlations have been observed between dissolved nitrate, phosphate, and carbonate carbon, the mole ratio being P:N:C as 1:16:106. This is reflected in the atomic compositions of algal protoplasm whose production in the photosynthetic zone requires the three nutrients in the above proportion and whose mineralization after death in the deeper layers releases nutrients in the same ratio [21]. In some oceans both N and P have been observed to be equally limiting, whereas in many estuaries and coastal waters nitrogen may be more limiting [21].

Phosphorus and nitrogen measurements are considerably affected by the time of year and depth of sampling owing to varying rates of nutrient uptake and recycling as discussed in Chapter 6. The increase in phosphate values from year to year in the Baltic Sea is shown in Figure 8.6.

Example 8.1 An outfall is expected to discharge daily 100^T of TOC, 20^T of nitrogen as N, and 4^T of phosphorus as P, to an enclosed sea of 1.2 x 10^4 km^2 area in which the natural organic carbon production rate is measured at about 80 g/m^2-year. Estimate the nutrient load placed on the sea.

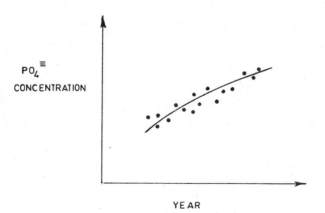

YEAR

Figure 8.6 The gradual increase in phosphates in the Baltic Sea (adapted from Ref. 21).

1. <u>Organic carbon load ratio</u>

$$\text{Natural productivity in sea} = (1.2 \times 10^4 \times 10^6)\left(\frac{80}{10^6}\right)$$

$$= 96 \times 10^4 \text{ tons/year}$$

$$\text{Organic carbon load from outfall} = (100)(365)$$

$$= 3.65 \times 10^4 \text{ tons/year}$$

$$\text{Ratio} = \frac{3.65 \times 10^4}{96 \times 10^4} = 3.8\% \text{ (assuming even distribution in the sea)}$$

The organic carbon load brought in by the outfall is low compared with the natural productivity of the seawater in question.

2. <u>Nitrogen and phosphorus load resulting from outfall</u>

$$\text{Annual nitrogen load} = \frac{(20 \times 10^6)(365)}{(1.2 \times 10^4 \times 10^6)} = 0.61 \text{ g/m}^2\text{-year}$$

$$\text{Annual phosphorus load} = \frac{(4 \times 10^6)(365)}{(1.2 \times 10^4 \times 10^6)} = 0.12 \text{ g/m}^2\text{-year}$$

As average values, assuming even distribution in the sea, these are not of a high order of magnitude (Sec. 6.8.2), but judgment must be reserved until contributions from other sources, and present levels of N and P in seawater, are evaluated.

8.6.3 Effects on Water Transparency

Light penetration is affected by color and turbidity of the seawater. Turbidity is, in turn, affected by phytoplankton concentration and incoming turbidity in land drainage; the incident light also varies. Thus, coastal waters display a continually varying transparency.

The use of the <u>Secchi disc</u> and illuminometers to measure light penetration, and the estimation of the rate of attenuation of light in water were discussed in Secs. 2.3.2 and 6.2.3. Light penetration data are extremely useful and seasonal records must be obtained. Long-term trends can be discerned and often correlated to productivity and other measurements (e.g., PO_4^{3-}). As productivity in a waterbody increases, light penetration decreases. For most coastal waters, Secchi disc readings average less than 10 m depth, with even less than 1 m depth in some highly productive areas in the warm seasons.

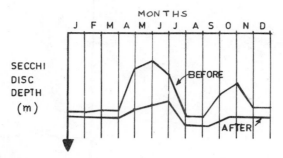

Figure 8.7 Secchi disc depths observed before and after the construction of an outfall.

The availability of light penetration data before and after construction of an outfall is often useful in assessing its environmental consequences and in interpreting other environmental data. Figure 8.7 shows how the Secchi disc readings varied at a site before and after construction of an outfall.

8.6.4 Effects of Species Diversity

The use of species diversity indices was discussed in Sec. 3.12 along with the factors affecting them. From some typical values given in Table 3.5, it is seen that a higher diversity can be expected in tropical seas (30 to 40 species per 1000 individuals) compared with bays where polluted waters may have even less than 5 species per 1000 individuals. Generally, the species diversity increases as one passes from the shoreline and estuarine locations to those in the open sea. A wide enough survey area must, therefore, be covered. Figure 8.8 shows the values of the index as observed at different locations in Texas Bay.

Again, species diversity becomes a useful parameter to assess the environmental impact of a waste disposal project if data are obtained before and after construction of an outfall. It also serves as a basis for determining water quality objectives. Ideally, the objective should be to ensure that the natural species diversity of a relatively unpolluted marine environment is not disturbed and reduced by waste disposal. Species diversity studies involve seasonal sampling and identification of the phytoplankton and zooplankton in the concerned microenvironment.

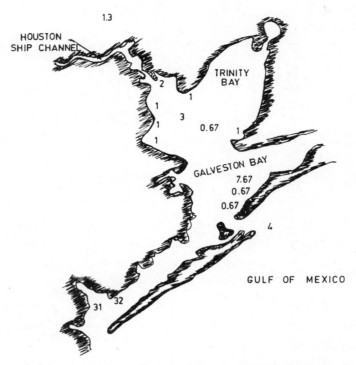

Figure 8.8 Species diversity indexes in selected Texas bays. The values of Kd are given and show a general increase from the polluted Houston ship channel to the Gulf of Mexico (adapted from Ref. 13).

8.6.5 Dissolved Oxygen Requirements of Fisheries

The DO requirements of fisheries depend on the species and their various life stages from hatching to adult growth. The requirements are complicated by the fact that some species migrate from fresh to saline waters, and vice versa, at certain periods of their lives. Thus, the entire system (river-estuary-sea) must be favorable for such fisheries to flourish.

Certain fish (e.g., salmon) migrate as adults from the sea up estuaries, with their young ones later returning to the sea. Some species undertake such migrations only irregularly. A few species (e.g., eel) migrate from the river to the sea, and their larvae return to the river. Some marine species (e.g., sole) spawn in estuaries, some (e.g., herring) use them as a nursery ground, some such as shrimp and shellfish spend their

Table 8.8 Summary of Field Observations on Minimum DO and
the Occurrence of Fisheries in Some Watercourses, U.K.

Watercourse	DO (mg/liter)	Species
River Trent	9 (50 percentile) } 5 (5 percentile) }	Game fish (trout)
	4 (50 percentile) } 2 (5 percentile) }	Coarse fish (mixed)
Rivers in U.S.	4	Mixed

Thames estuary	3-5	Flounder, lamprey,
		  smelt, herring,
		  and sprat

Source: Adapted from Ref. 6.

entire lives in the sea. Many species are, of course, limited to either a
marine or a freshwater habitat by salinity.

The microhabitat of different species of fish may also be different.
Some prefer the upper waters whereas others seek their food in the lower
levels. These preferences are due to differences in environmental conditions (temperature, DO, etc.) and food supply above and below the thermocline or halocline in the sea.

The DO present in coastal and estuarine waters is often highly variable, diurnally and seasonally, and sometimes even more so than in a river.
Thus, it is desirable to sample at different times and state the result in
terms of the statistical probability of a certain value occurring at a
given location. Correlation of such data with existing fisheries can give
some clue to the minimum DO percentile distribution over a year or perhaps
part of a year or, in some cases, even part of a tidal cycle [6]. From
data given in Table 8.8 it is seen that for most species of fish, minimum
DO values of 4 mg/liter are sufficient on the average. Minimum values of
even 2 mg/liter may be tolerable as long as their occurrence has a low
frequency. DO values less than 2 mg/liter are lethal for most fish, whereas
at less than 1.0 mg/liter noxious conditions may develop.

Other factors which affect fish life along with DO are temperature,
the presence of toxic substances, etc. These have been discussed in Chapter 3.

8.6.6 Dispersal Area

The wastewater emerging from the outfall mixes and flows with the sea-water forming what is called a "sewage field." In practice, the area and configuration of the sewage field are affected by (1) the rate of discharge, (2) outfall characteristics (e.g., its direction, depth, type of outlet), (3) density and stratification conditions, and (4) speed of water currents. The sewage field may spread over several hectares of sea surface and grad-ually blend into the seawater, losing its identity.

Discrete settleable solids contained in the wastewater tend to settle over a wide area depending on their size and on the current speeds. If the plume rises up to a height h meters from the sea bottom, the current velocity is u (cm/sec) and the solid particles are such that they have a settling velocity of v_s (cm/sec) in seawater at a given temperature. The resultant velocity and direction determine the distance x meters from the outfall at which the particles will touch the bottom. Thus,

$$\text{time,} \quad t = \frac{h}{v_s} = \frac{x}{u}$$

and

$$x = \frac{uh}{v_s} \tag{8.6}$$

The area over which the whole sewage field may extend, and not just where solids may settle, must be taken into account in determining the area over which preliminary surveys must be undertaken to obtain design data.

Example 8.2 Estimate the distance from an outfall over which discrete solids of 0.043 cm/sec settling velocity may settle in the direction of a current of 0.1 m/sec. Plume height = 25 m. From Eq. (8.6),

$$x = \frac{uh}{v_s} = \frac{10 \times 25}{0.043} = 5813 \text{ m}$$

8.7 SITE SURVEYS

Extensive site surveys need to be conducted in order to obtain data to enable (1) a study of the present environmental conditions, (2) assess-ment of the likely effects of wastewater discharges on them in the future, and (3) a design of an outfall taking into account various hydrographic and geologic factors. With these objectives in mind, one has to determine what physical, chemical, and biological parameters must be included in the survey, and the extent and duration of the sampling program.

The position of the site is more or less fixed by the sewerage plan of the city or the location of the industry desiring sea disposal. The extent of the survey area can be determined from a preliminary evaluation of the possible dispersal area and the local topographic and other features. As a rough indication, at least an area within about 1 km radius of the likely position of the diffuser, and 500 m on either side of the proposed outfall alignment should be included in the survey.

In preparing for such surveys, environmental engineers are often pleasantly surprised to find that much information, especially of a hydrographic nature, already exists with the navy and other organizations interested in the marine environment. Though not specifically obtained for the site in question, much of the available data and maps may throw useful light on planning further surveys, and in some cases (e.g., for contours, temperature, salinity, and current pattern), only a few correlating measurements may be needed.

Among the background information necessary to collect, one must also include the present sources of pollution of the sea and the available data on fisheries, utilization of beaches, etc. Thus, one may summarize as follows the kind of background data that would be useful to obtain:

1. Available topographic data, bathymetric data, and maps
2. Known hydrographic features (e.g., currents and tides)
3. Wind speeds and directions in vicinity of site
4. Salinity, temperature, and density profiles
5. Sources of pollution
6. Utilization of beaches and seawater
7. Fisheries, past and present
8. Available data giving characteristics of local fauna and flora

The data to be obtained in site surveys are conveniently grouped under:

1. Physical surveys
2. Chemical and biological surveys

In order to enable better interpretation of the data obtained, it is desirable that both the above types of surveys be carried out simultaneously, if possible. Alternatively, whenever chemical and biological data are obtained, the time of sampling and certain physical data such as temperature must be noted to enable interpretation of results in the light of the tides, currents, stratification, etc., prevailing at the time of sampling.

Ideally, sampling stations should form a three-dimensional network of a number of stations in the sea. However, sea surveys are expensive and only a few stations can generally be established from which samples for chemical and biological analysis can be periodically obtained. The designated stations can be identified at each visit by taking bearings from fixed landmarks at the shore.

Owing to the highly variable nature of many parameters, the period of observation must be sufficiently long, covering the seasons, and a statistical approach must be followed in expressing the results as far as possible. Thus, probabilities of occurrence of winds, currents, etc., should be stated at known confidence levels. In this manner, it should also be possible to state the eventual dilution afforded to the wastewaters, and the likely DO or coliform levels in terms of statistically probable values. Generally, in the colder regions, the critical season is the bathing season in the summer. DO for fish is also likely to be minimum in the lower waters in the summertime. In warmer climates where sea bathing may be done throughout almost all the year, the critical season would be one which would favor rapid travel of pollutants to the beach and/or affect fisheries through low DO.

Wind and tide data are generally available at meteorological stations over several years, but wind conditions at such stations, even though in the near vicinity, may be entirely different from those at the desired outfall site. A temporary station at the proposed site may be essential to see if any correlation does or does not exist between it and the permanent station. If a correlation does exist, the available data over a long period of time can be readily used.

The various parameters often included in site surveys are summarized in Table 8.9, which gives their frequency and method of sampling, their typical equipment requirements, precautions in the survey, etc., in order to give a general overview of this rather specialized field work in which many decisions depend on local conditions. Details are available in various references and standard oceanographic texts.

Often it may be desirable initially to carry out a _preliminary survey_ covering only a few selected parameters in order to be able to prepare

Table 8.9 Typical Parameters Included in Marine Disposal Surveys

Parameters	Location	Frequency or duration	Method	Remarks
Physical				
1. Tides	In general survey area	Continuous, or 5-7 days in a month	Level recorder with recording chart to note spring and neap tide levels in each month.	Tidal amplitudes in enclosed seas and bays may vary considerably from those in open seas.
2. Winds	In general survey area	To coordinate with currents and dispersion	Wind direction and speed recorders used from which "wind rose" diagrams are prepared similar to the one shown in Figure 8.9. Station location considerably affects results.	Long-term data available from a nearby permanent station of Meteorology Dept. may be used only if a correlation is first established with a station in the survey area.
3a. Currents	In survey area, at various depths from near surface to near bottom (see text).	For a month or more during the summer and winter seasons	Two types of current meters available: direct reading type and recording type. Former are generally of impeller type capable of swinging freely in a horizontal plane. Direction is determined by a built-in magnetic compass and current speed by number of impeller revolutions in a given time (generally 1-5 min). Boat is moored at site and meter lowered to desired depth. Remote reading possible from boat. Meter tends to move with boat. To reduce error, anchor boat at both fore and aft, and suspend meter from a davit equipped with an accumulator spring. Recording-type current meters are similar, but are moored in a fixed position and can give data recorded at 5-10 min intervals on magnetic tape for about one month at a time (see Figure 8.9d). Data can be treated further on a computer to obtain frequency distributions.	Speed range: 3-250 cm/sec ± 2%. Direction: 0-357° ± 5°. Preventive maintenance and periodic calibration is essential to avoid loss of data. Simultaneous measurement of temperature and conductivity is possible in the same instrument if desired.

343

Table 8.9 (Continued)

Parameters	Location	Frequency or duration	Method	Remarks
3b. Drogue movements	Near surface, mid-depth, or near bottom in general survey area.	On a few occasions generally prior to start of currents measurement program	"Drogues" are used to study general water-body movements in the layer in which the vane of the drogue is located (see Figure 8.9b). Winds are measured simultaneously. Useful in determining overall drift, different travel patterns of upper and lower layers of waterbody, presence of eddies, etc. Helps in selecting sites for installation of current meters. Dye spots may be used if desired to trace movement of surface layer.	Drogues tracked up to one tidal cycle after release at slack tide. Some unsuccessful owing to entanglement or hitting sea bed in shallow areas. Slight windage effects unavoidable.
4. Salinity, temperature, and density	At various depths 1-2 m apart to obtain profiles for different locations in general survey area	Six to eight times in summer, 2 to 3 times in winter	Temperature at any depth is measured by using reversing-type thermometers or electrical thermistors. Salinity can be measured by withdrawing a sample from selected depth (using a Ruttner or Universal sampler of the type shown in Figure 8.10a and titrating for chlorinity or from conductivity and temperature data. In situ salinometers enable both salinity and temperature to be measured simultaneously by probes. Equipment also available for continuous recording with output on magnetic tape. Equipment accuracy generally $\pm$ 0.1‰ for salinity and $\pm$ 0.1°C for temperature.	Density is calculated from simultaneous measurements of temperature and salinity. Hydrometers may also be used to find density.
5. Dispersion	In surface waters at proposed diffuser site	Over 4 tides in critical season	Drift cards, drogues, or dyes may be used to obtain measurements of dispersion and drift time to beach under known conditions of surface currents. See Secs. 8.8.3 and 8.8.4.	Information from items 1, 2, and 3 above is required at time of dispersion study.

344

No.	Parameter	Location	Frequency	Method	Remarks
6.	Light transmittance	In general survey area and in adjoining unpolluted areas	Seasonal	Secchi discs (see Figure 8.10e) or illuminometers are used. Latter also enable estimation of light extinction coefficients.	Also estimate depth of euphotic zone; results generally correlate with those of items 13 and 14.
7.	Submarine topography and geology	Along outfall alignment	—	Bathymetric maps (isobaths) prepared using echo sounding recorders. Trial borings taken to about 3 m depth every 100 m apart, along alignment. Seismic profiling done using side scanning sonar equipment and actual trial bore data for calibration. Soil tests performed for geological classification and design of foundations (shifting sands, erosion, siltation, littoral drift, etc., must be taken into account including need for trenching, sheet-piling, etc., during construction). Deeper exploration of soil is necessary if tunneling is contemplated.	Divers observations also recorded.

Chemical

No.	Parameter	Location	Frequency	Method	Remarks
8.	Dissolved oxygen	At various depths to obtain "profiles" at different locations in general survey area	Diurnal variations in different seasons	In situ DO measuring probes used, or samples obtained from different depths (using a Ruttner or Universal sampler) and DO found by Winkler method. The sampler is lowered in an open condition to the desired level when a weight (drop messenger) is released which triggers a spring mechanism by which the top and bottom caps are closed shut (Figure 8.10a).	For Winkler test, samples fixed on boat then titrated ashore. Note water temperature at time of sampling. A simple displacement-type sampler may be used up to 10 m depth only.
9.	Organic matter	Euphotic zone, different depths	Seasonal	TOC or BOD/COD tests (see Sec. 8.6.2 also) performed on samples obtained as in item 8.	Sample preserved during transport by acidifying.
	Organic matter	Bottom sediments	Occasional	Samples obtained using Ekman-type or other dredge (Figure 8.10d), dried at 105°C and loss of weight determined upon ignition at 600°C BOD can be determined in a Warburg respirometer.	Samples can also be analyzed for heavy metals and persistent substances (see item 12).

Table 8.9 (Continued)

	Parameters	Location	Frequency or duration	Method	Remarks
10.	Nitrogen and phosphorus	In surface waters generally	Seasonal	NO₃ and soluble phosphates generally measured. Latter more reliable since seawater constituents affect measurement of former. All results affected by extent of nutrients locked up in biota at time of sampling, and extent of upwelling if any. Ruttner sampler used (item 8).	Phosphates also measured in bottom water samples. Analysis performed within 3 hr of sampling (preserve by acidification).
11.	pH	Different depths	Occasional	pH or seawater is generally about 8.2 in deep seas.	
12.	Heavy metals, persistent substances	Bottom sediments, surface waters	Occasional	Appropriate methods involving use of atomic absorption, spectrophotometers, and other instruments capable of measuring at very low concentrations. Special water samplers used to avoid contact with any metal.	Performed only if considered necessary.
	Biological				
13.	Primary productivity	Euphotic zone	Seasonal	Samples obtained from different depths using Universal or Ruttner sampler. Light and dark bottle method or other methods then used for estimating productivity. Results at different depths averaged to give productivity as g of c/m²-yr and peak values (in summer) as g of c/m²-day.	
14.	Phytoplankton and zooplankton	In general survey area	Seasonal	Vertical and horizontal hauls made to determine species, and their general distribution pattern in survey area. Cell counts difficult to correlate to productivity data (item 13) and Secchi disc readings (item 6). Special conical plankton capturing nets used (50-60 μm mesh for phytoplankton and	Samples preserved in 5% formalin if necessary.

346

No.	Parameter	Location	Frequency	Method	Remarks
				300-500 μm for zooplankton) along with drop messenger operated mechanism to trap at different levels. Quantitative determination requires attachment of flowmeters to conical nets (see Figure 8.10c).	
15.	Fish	In vicinity of survey area	Occasional or seasonal	Echo sounding or beam travel methods. Latter used to study specieswise distribution of fish, although only a small percentage of the species present are sampled by it.	Sufficient data may be already available with fisheries departments.
16.	Benthic flora and fauna	Survey area	Occasional or seasonal	Dredges and trawls of various sizes used for collecting nonburrowing and free swimming benthos on or just above the bottom surface for qualitative studies. Quantitative benthic samples obtained using Van Veen or Ekman-type grabs, corers, or other more elaborate devices giving results per unit area of bed surface (Figure 8.10d).	Results obtained may correlate with DO and other data (e.g., certain species indicate polluted, anaerobic, or other conditions).
17.	Species diversity	Survey area (from shore to open sea)	Occasional or seasonal	Based on data obtained in items 14 and 16, a species diversity index may be calculated as given in Chapter 3 (also see Sec. 8.6.4).	Results obtained should correlate with salinity, pollution, or other data.
18.	Coliforms/ E. coli	Surface waters of beaches up to 300 m from shoreline	Bathing season	Membrane filter or other standard methods used. Value of t_{90} estimated as in Sec. 8.8.1.	Sample preservation during transport is essential. Multiple tube method preferred where samples are difficult to filter.

[a]For factors affecting the extent of the survey area to be included in a sampling program, and the choice of a "critical season" see text. Methods indicated for each item are only typical and depend upon the extent of accuracy desired and the nature of the equipment and facilities available.

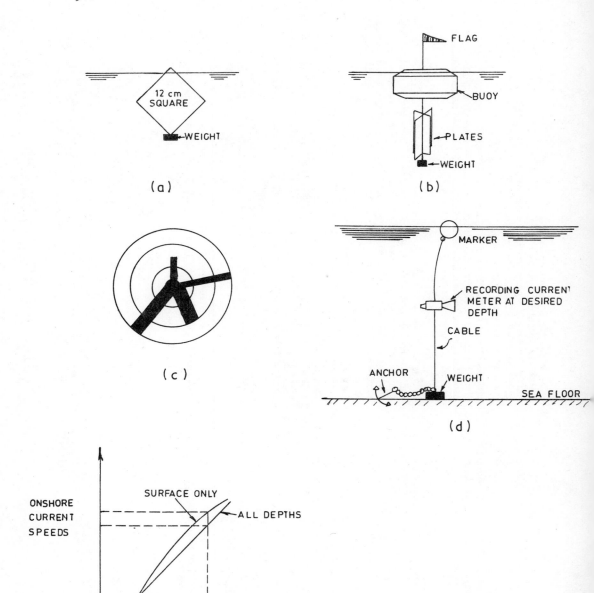

Figure 8.9 Apparatus for and data from current and drift time studies; (a) drift card, (b) drogue, (c) a current "rose," (d) installation of a recording current meter, and (e) probability plot of onshore current speeds.

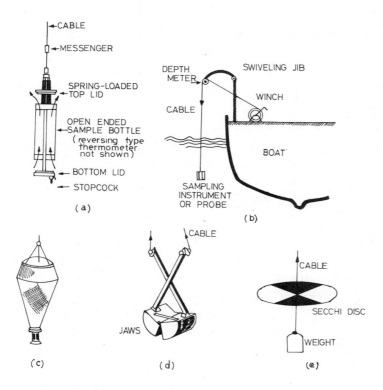

Figure 8.10 Examples of sampling equipment; (a) Ruttner water sampler in open condition during lowering, (b) typical arrangement for sampling
from a boat, (c) Hansen-type net for plankton sampling, (d) the Van Veen
grab for bottom sampling, and (e) Secchi disc.

tentative designs and costs. Such parameters would be, for example, salinity, temperature, currents, and a depth profile along the proposed outfall
alignment. This would perhaps constitute a bare minimum preliminary survey
program carried out on at least two or three occasions covering the critical
season just referred to as far as the hydrographic items are concerned (also
see Sec. 8.7).

Thereafter, a more exhaustive predesign survey program may be undertaken over at least a period of more or less 1 year. Samples are taken
during selected months to cover possible seasonal variations and include
selected physical, chemical, and biological parameters from Table 8.8 depending on whether the discharge is to an open sea or to an enclosed bay
with more critical conditions. It should also be borne in mind that oppor-

tunity comes but only once to obtain background data with which to compare the situation subsequent to the construction of the outfall.

In the case of enclosed waterbodies such as bays and estuaries with poor water exchange rates, an additional objective of the predesign survey may be to develop and calibrate some kind of mathematical model which can be used to simulate water quality in the waterbody under various pollutant loads and other assumed conditions. This may also need data to develop the hydrodynamic submodels based on water exchange rates and an understanding of the various sources and sinks of pollutants (Sec. 8.10). Much data concerning construction features are also required to be obtained in predesign surveys (e.g., geologic and seismic data, siltation, erosion, shifting sands, etc.).

Survey work at sea takes much more time and expense than corresponding work on land, even for relatively minor chores. Thus, the survey programs must include only the most essential items. Predesign surveys alone may cost 1 to 2% or more of the total cost of a reasonably sized project.

Periodic monitoring after an outfall is placed into operation may be performed by the control agency or imposed as a task on the polluter. The parameters included in postoperation surveys again depend on the nature of the receiving waterbody (open sea or enclosed bay, etc.), but generally involve biochemical items such as dissolved oxygen, nutrients, coliforms along bathing beaches and in shellfish areas, Secchi disc transparency, species diversity, etc., besides some special items such as organohalogens, hydrocarbons, and metals depending on the nature of discharges involved.

Agencies undertaking surveys are routinely equipped with their own boats and instruments with facilities for three or four scientists to live and work for several days at a time at sea. The boats must be capable of working in relatively shallow waters and also of anchoring whenever necessary in quite deep waters. They should have good radio communication and navigation equipment, a small laboratory, water supply, and sample storage facility at controlled temperature; all operating on the same voltage as on land. Experience also suggests that boats must be as vibration-free as possible where sensitive equipment is to be used on board, and all items including glassware be fixed on tables or kept in special pigeonholes [30]. Predesign surveys may need to be carried out using small boats rented from local fishermen where more elaborate vessels are unobtainable.

8.8 SOME FIELD DETERMINATIONS FOR DESIGN PURPOSES

A few site studies need to be explained in further detail. These include the estimation of E. coli or coliform dieoff rates, onshore current speeds, drift time of floatables, and the dispersion coefficient.

8.8.1 Estimation of E. coli/Coliform Dieoff Rate

Especially in the case of the surfacing type of plume, it is necessary to know how coliforms or E. coli will survive in their travel from the outfall to the beach. Dieoff time for coliforms was discussed in Sec. 4.5.1 with reference to various types of waters. From the data of Bitten and Mitchel, t_{90} values were estimated for E. coli (see Table 4.7) from batch studies in the laboratory at about 1.2 hr in natural seawater at room temperature. Several field studies undertaken in different seas have shown t_{90} to range from 1.1 to 10 hr, as given in Table 8.10.

Mitchell and Chamberlain [24] conclude from an examination of available laboratory and field data that solar radiation, sedimentation, and nutrient-related effects account for the bulk of observed coliform dieoff in the sea. In controlled experiments, other workers have also demonstrated similar results [34].

Table 8.10 Observed Values of t_{90} for E. coli in Seawater

Location	Nature of Waste	t_{90} (hours)	Reference
Los Angeles, Hyperion	Secondary effluent	9.6	32
	Primary effluent	4.1	
Los Angeles, Orange County	Primary effluent	1.8-2.1	32
Accra, Ghana	Raw sewage	1.3	32
Marmara and Aegean seas, Turkey	Raw sewage	0.4-1.3	27
Iskenderun Bay, Turkey[a]	Raw sewage	1.1	28
The Danish sound	Primary effluent	2.3	3
	Secondary effluent	10	

[a]Performed at site using polythene floating bags (see text).

From Table 8.10 it will be seen that the degree of pretreatment of sewage appears to have an effect on dieoff rates. Raw and primary settled sewages exhibit a faster dieoff rate of coliforms than do secondary treated effluents. This may be due to bacterial agglomerations in raw and primary sewages which sediment faster after discharge.

Gameson and Gould [24] found the following concerning coliform dieoff from raw sewage in the marine environment:

1. Shore based experiments in the dark showed a remarkable approximation to first-order kinetics and a pronounced effect of temperature.

2. Similar experiments in daylight showed greatly enhanced dieoff rates (t_{90} was about 20 min compared with about 1 day or more in the dark). More importantly, no effect of temperature could be found.

3. Lethal radiation extended up to 4 m depth in the sea studied (90% attenuation of radiation of 310 nm occurred at 4.2 m).

4. In situ experiments confirmed virtual cessation of coliform mortality in the dark. In other words, light played an extremely important role in the dieoff of coliforms.

Occhipinti [24] made over 50 in situ experiments near Sao Paulo, Brazil, also using raw sewage and found that no correlation between t_{90} and temperature could be established. He showed that t_{90} values obtained in his studies gave a log-normal distribution ranging from 35 to 220 min with a median value of 84 min. Thus, even at a given site, wide variations in t_{90} values were to be expected (e.g., between summer and winter).

Gameson, thereupon, plotted the data from his own studies and from those of Occhipinti and others to show that all the results were consistent with the hypothesis that solar radiation was perhaps the most important factor (Figure 8.11). These were three studies in which a series of experiments had been conducted at a given site to cover different seasons and other conditions. He found that, when t_{90} values were plotted latitude-wise on log-normal paper, both median values and the variability of t_{90} values increased with latitude and, therefore, explained the effect of radiation since the degree of radiation and its variability were both affected by latitude. The median values of t_{90} were 1.4 hr at 24° S (Brazil), 2.1 hr at 34° N (California), and 3.1 hr at 51° N (England).

An interesting conclusion that can be drawn from the above studies is that, for various reasons, the statistical variability of t_{90} values is quite high and, therefore, performing a few (often only one or two) field tests at a specific site can be of little value for design purposes. At the same time it may not always be feasible to arrange a large enough

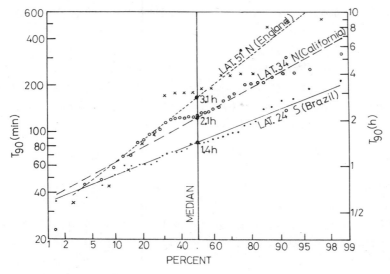

Figure 8.11 Logarithmic-probability plots of t_{90} values obtained in three series of in situ studies (adapted from Ref. 24).

testing program at every proposed site, especially for small and medium size communities. Thus, in many cases it should suffice to select a suitable t_{90} value by interpolation from Gameson's results shown in Figure 8.11, knowing the latitude of the proposed site.

Selecting the median t_{90} value would give the dieoff rate statistically likely to occur 50% of the time in a whole year in surface waters at the given latitude. A value, which is likely to occur 50% of the time when a whole year's data are considered, will occur a higher percentage of the time when only the summer bathing season is considered.

Thus, in selecting a 50% value from Figure 8.11, a lesser health risk is implied in the temperate countries than in the warmer ones where bathing may be possible for much of the year. Also, since the ratio of pathogens to coliforms may be high in some of the developing countries in the warmer regions, it may be more desirable to use the t_{90} value likely to occur 80% or 90% of the time rather than only 50%. Hence, a minimum t_{90} value of, for instance, 2 hr may be preferable to use, in warmer countries, even though a lesser value may be indicated from the figure.

Where performance of a field test is desired, two methods can be used depending on time and facilities available. In one method, samples of seawater taken from a given site are inoculated with varying concentrations of raw or treated wastewater and filled in several transparent polyethylene bags, which are then exposed to the natural conditions of light and temperature at the site (or at the shoreline) for different lengths of time. Coliform concentrations are determined at the start and at different times thereafter, and values of the dieoff rate k per unit time are determined from a semilog plot of coliform concentrations versus time (see Example 4.1). Such observations constitute batch (or ideal plug-flow) type studies, and

$$t_{90} \text{ (in hours)} = \frac{1}{k \text{ per hour}} \tag{8.7}$$

The other method of determining t_{90} involves field tests at existing outfall sites or releasing a batch of wastewater into the sea and sampling in the vicinity. This is more difficult to perform and requires that a dye also be released along with the wastewater so that the physical dilution occurring in mixing and travel in the sea is accounted for properly. Coliform reduction rate is due to both physical dilution and natural dieoff and needs to be adjusted for physical dilution in order to arrive at the true k value for natural dieoff rate from such studies. Where a batch of wastewater and dye is released at a test site, samples are taken every few minutes from a boat which follows the drifting dye batch. Samples are analyzed for dye and coliform concentrations and the k value is determined from a semilog plot of the ratio of the number of coliforms to dye concentration versus time [27]. Other approaches are also followed [31].

To estimate the coliform concentration C at any time t resulting from natural dieoff in the marine environment, the equation generally used is

$$C = C_o e^{-Kt} \tag{8.8}$$

where

C = coliform concentration after lapse of time t (hours)

C_o = coliform concentration initially, namely, after initial dilution or after vertical rise at outfall site

t = time elapsed after outfall discharge in hours (time of travel from outfall to beach or shellfish site)

K = $2.3k$ = $2.3/t_{90}$ = dieoff rate per hour

As explained in Sec. 8.3, the overall reduction of coliforms is the product of the reduction owing to initial dilution, subsequent dilution in horizontal travel, and natural dieoff (also see Example 8.11).

Viral dieoff rates are much slower than those of coliforms in seawater. Katzenelson and Shuval [25], studying viral inactivation in the laboratory using seawater from the Mediterranean coast of Israel and from the Red Sea, found t_{90} values for polio virus to be about 2 to 5 days at 15° C as compared with 1 hr for coliforms in the same waters. Mitchell [21] reported that enteric viruses have been isolated from shellfish at distances of more than 6.5 km below sewage outfalls. Consequently, the natural dieoff phenomenon can hardly be depended upon in the case of viruses and the safety of bathing beaches must be ensured through virus removal in waste treatment and disinfection followed by as large a physical dilution in the sea as is practicable.

As stated earlier, viral concentration in raw domestic sewage varies between 10^2 and 10^4 PFU (plaque-forming units) per 100 ml, with values commonly averaging between 10^2 and 10^3 PFU. In the absence of any treatment, this demands that physical dilution of the wastewater should be about 1000 or so to keep the health risk low since even a single PFU when swallowed may cause infection. It is evident that much attention will be focused on the subject of virus removal in the coming years.

8.8.2 Determination of Currents

Among the various field observations to be made in marine surveys, perhaps the single most crucial item is current speed and its direction. The occurrence of onshore currents are of particular interest as they tend to carry the wastes including all floatables toward the bathing beaches.

The current meters used are generally of two types: recording type and direct-reading type. In both types the instrument measures current speeds and directions; some details are given in Table 8.8 under "currents." Recording type instruments are more expensive and need to be installed at a site, using anchors, cables, and floats. Their advantage is that they can record variations in current speed and direction every few minutes on a magnetic tape for as long as a week or a month at a time for subsequent analysis with a computer. Figure 8.9d shows a typical installation. Their installation and recovery can be quite an expensive operation and, if not done well, the meter might be lost. Many meters have been lost owing to inexperience with sea conditions, poor moorings, damage by boats, and vandalism.

The direct-reading instruments are smaller, less expensive, and are generally lowered from a boat kept anchored at a given site. They give relatively instantaneous readings (1-5 min duration) of current and direction at any desired depth. They may be used along with recording-type meters if desired, although at some stations only direct-reading meters may be used that take measurements at 1/2 hr intervals over a tidal cycle.

Direct-reading current meters have the advantage of maneuverability, both horizontally and vertically in a given survey area. This is particularly useful in the preliminary survey phase. Their results are also known each day, unlike the recording meters where it is not uncommon for a survey team to be frustrated to find at the end of several days that, owing to some fault (e.g., a poor battery), no results were recorded at all and that a useful working season was lost. Preventive maintenance is, therefore, most essential. Periodic calibration of both direct and recording-type meters is also essential.

A couple of direct-reading current meters used in conjunction with a recording-type meter can yield much useful information. The latter may be erected at the proposed diffuser location to obtain a continuous record of current speeds and directions in the upper layer of the waterbody (about 1/3 to 1/4 of total depth below surface). If trapping of the wastewater is expected at or near the pycnocline, the recording meter may be placed at a lower level (about 2/3 to 3/4 of depth below surface). The period of measurement may range from a week to a month or more during the summer and winter seasons depending on the expected variations in currents.

The direct-reading meters are then used to cover the whole depth range in the vicinity of the fixed meter. In this manner, statistical correlations can be derived between the two and the data of the fixed, recording-type meter used (at least cost) to predict currents at other levels or locations. The occurrence of counter- and cross-currents relative to the current at the level of the fixed meter may also be observed.

Where wind-induced surface currents are important, a wind speed and direction recording station at the proposed site used in conjunction with a direct-reading current meter at 1 or 2 m depth can help to establish correlations between the two. If currents are density induced, it may not be possible to establish correlations with wind.

The results of current measurements can be statistically analyzed for speed and direction, and expressed in the form of "current roses" in which speeds are represented by vector lengths and frequencies by vector widths (see Figure 8.11c). Other conclusions that can be drawn from the data typically include (1) frequency and speed of current in a given sector (four or more sectors, as desired, could be formed from a circle with the

sampling station as its midpoint, and special attention given to the sectors
facing the beach), (2) frequency and duration of an uninterrupted current
in a given direction (e.g., onshore direction), and (3) correlation between
current and wind speeds and directions where such correlations exist. A
method for determining the dispersion coefficient from currents data has
also been developed [24].

Few onshore currents may be directly perpendicular to a beach in question. However, for design purposes all onshore currents within a sector
$30°$ on either side of the perpendicular drawn from the outfall point are
generally taken into consideration and the frequency of occurrence of currents of stated speeds are plotted on arithmetic probability paper, from
which it is possible to read off different percentile values for either
surface currents or those at all depths. For design purposes, the surface
current speed, which generally occurs 80% or less of the time, is taken
into account if a particular coliform standard has to be met 80% of the
time. Similarly, if the coliform standard has to be met 95% of the time,
the current speed used for design purposes is the one which occurs 95% of
the time (Figure 8.9).

Current studies can be usefully preceded by float studies to determine
the overall movement of the waterbody, the presence of eddies, etc.; based
on these, the proper location for current meters can be planned. In this
regard, the two methods are complementary. Current meters measure the flow
past a point, namely, they measure in the Eulerian sense, but they do not
indicate what happens to the fluid particles once they leave the point.
Floats, on the other hand, measure in the Lagrangian sense, and indicate
how drift occurs in response to currents, tides, wind, eddies, and other
forces, yielding information on how currents vary with time and place, and
how residual drift occurs.

Floats used for such studies are generally "drogues" (a sort of anchor
float) of which the vane type is shown in Figure 8.9b. By using different
cable lengths, the vane can be suspended at any desired depth in the sea.
It is thus possible to release several drogues, some for indicating the
flow in the upper layers and some for lower layers (which may be moving in
an entirely different direction from the former). Their paths are generally
followed over a tidal cycle, and observations are repeated on several occasions. All drogues suffer to some extent from the effects of wind on their
float and flag, however inconspicuous they may be, and it may be necessary
to take some correction into account. Tracer studies using dye spots may

also be undertaken, and a track of their movement kept from a boat or a
helicopter. Special drifters are used for studying the drift of the surface
layer and the seabed layer [22,30].

8.8.3 Estimation of Drift Time of Floatables

To understand how floatables (e.g., coliforms) will behave with surfacing plumes, it is necessary to conduct field tests and estimate the
drift time between the outfall and the beach. Such tests are performed
using drift cards, drogues, or other tracers such as Rhodamine B, bromine-
82, etc. They give a measure of the average and the shortest time of travel
under natural drifting conditions caused by surface currents together with
tidal and other forces.

Figure 8.9a and -b show a typical drift card and a drogue. In both
cases, projection above the water surface is minimal in order to avoid wind
effects. Drift cards can be made of polyethylene or other sheets that are
12 x 12 cm and weighted to make them float upright. The cards or drogues
are released from a boat at a point near the proposed outfall site, in
batches of, for instance, 50, every 1 or 2 hr over one entire tidal cycle
of about 25 hr. At the same time, hourly observations of ocean current
speeds and directions are made at the card release point using a current
meter as described in the previous section. These measurements are made
at about 1 m depth. Additional measurements may be made at other depths
if desired, although they are not essential for the drift time study. Wind
measurements may also be made at the same time. Thus, wind, currents, and
drift card data are obtained simultaneously over a tidal cycle.

The drift cards or drogues that succeed in reaching the adjoining
beach are recovered, carefully noting the arrival time and location of each.
Their drifting in the direction of the current, which may be at an angle to
the beach, spreads them out considerably, sometimes requiring that several
kilometers of beach are patrolled over a period of 2 to 3 days after their
release. Several of them are lost en route — even more than half the total
number at times — or end up beyond the reach of the collecting party.

The distribution of cards collected along the beach during all the
tests can be shown in the form of a histogram, and the center of gravity
of the distribution thus found. The results could also be expressed statistically on arithmetic or log probability paper and the standard deviation of
the spread calculated.

Since the time is noted at which each card is recovered, a statistical plot, generally on log probability paper, can be prepared for the percentage of cards recovered in a stated time or less. If desired, the speed at which each card traveled can be used instead of the time in plotting the data. From such a plot, the values of drift time or speed can be stated as 50 percentile, 80 percentile, or whatever other percentile is desired to be used for design purposes. Thus, the "minimum" design time and speed for floatables to travel from the proposed outfall point to the beach in question can be determined. Speed data obtained from drift card studies can be used to corroborate the speed data of surface currents determined by metering.

Example 8.3 illustrates how drift time computations can be made from field data. Dispersion coefficients can also be determined from such data, if desired (Sec. 8.8.4). It is desirable to repeat the whole experiment over at least 3 to 4 complete tidal cycles during the critical season.

Example 8.3 Data obtained from a drift card survey at a proposed outfall site are summarized in Table 8.11. Estimate the geometric mean

Table 8.11 Summary of Drift Card Survey Results (Example 8.3)

Station on beach	Number of cards recovered in stated drift time (hours)								Total at Station	Cumulative	
	0-6	6-12	12-18	18-24	24-30	30-36	36-42	42-54		Total	%
N_5	-	5	9	13	17	13	2	-	59	59	5.7
N_4	3	2	35	19	2	5	6	-	72	131	12.7
N_3	4	30	31	8	-	6	6	6	91	222	21.5
N_2	-	52	1	1	-	2	8	3	67	285	27.7
N_1	-	37	7	25	-	13	1	2	85	420	40.8
0	8	74	50	99	43	22	7	14	317	737	71.6
S_1	1	90	-	31	27	4	10	31	194	931	90.4
S_2	-	6	-	4	11	3	3	3	30	961	93.4
S_3	-	-	2	-	11	4	8	-	25	986	95.8
S_4	-	-	-	-	13	4	8	-	25	1011	98.3
S_5	-	-	-	-	3	9	2	4	18	1029	100

Total:	16	296	135	200	127	131	61	63			
Cumulative total:	16	312	447	647	774	905	966	1029			
Cumulative %:		1.55	30.3	43.4	62.8	75.2	87.9	93.9	100		

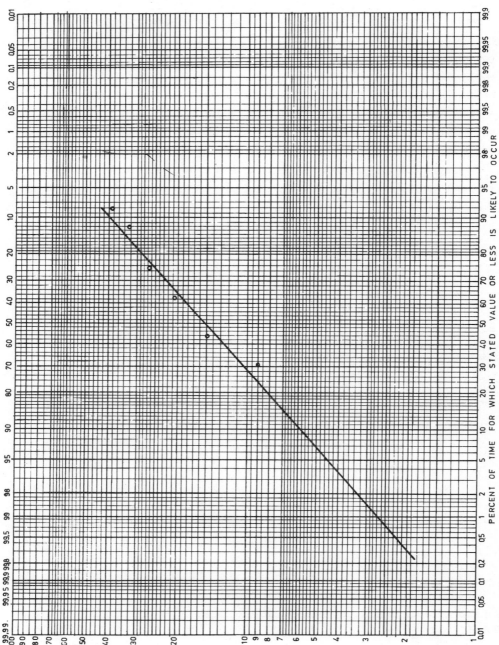

Figure 8.12 A probability diagram prepared from a drift time study (see Example 8.3).

drift time and the value likely to be equalled or exceeded 80% of the time. Each station is 1 km from the next along the beach.

A plot of the cumulative percentage of recovered cards versus average time period in which they were recovered is drawn on log probability paper as shown in Figure 8.12, from which the geometric mean drift time (50 percentile) is 15 hr. The 20 percentile value is 3.2 hr.

The total number of drift cards received at each station can be plotted in the form of a histogram if desired. The spread of the cards estimated statistically gives a standard deviation σ = 2.5 km on either side of the mean at station 0. Thus, in terms of Eq. (8.12), the dispersion coefficient ϵ = 5.78 x 10^5 cm^2/sec.

8.8.4 Estimation of Dispersion Coefficient

Where a rising plume is likely to surface, it is necessary to know how longitudinal dispersion will occur as the wastewater travels horizontally giving a horizontal plume at the sea surface. Dispersion gives some useful additional dilution with seawater which can be estimated by the method given in Sec. 8.16.

Dispersion in the sea is due primarily to eddy or turbulent diffusion, and in marine waste disposal literature it is often denoted by the symbol "ϵ" with the units of area/time. The reader is also referred to the discussion on this subject concerning dispersion in rivers and estuaries (Secs. 7.2.2 and 7.9.2) wherein dispersion has been designated by the symbol D. In case of seas, the Fickian diffusion model with a unidirectional transport velocity is generally applied.

The eddy dispersion or turbulent transport coefficient ϵ is actually not a constant. As a diffusing field becomes larger, an increasing scale of turbulence contributes to the mixing. Mixing takes place at the edges and, therefore, as the plume enlarges, the overall mixing rate increases. Thus, as shown by Sverdrup and others, ϵ varies with the scale of the phenomenon and with time.

In the early 1950s, Japanese oceanographers first showed that ϵ values obtained at different conditions and scales in different seas give a correlation as presented in Figure 8.13, to which the following equation can be fitted:

$$\epsilon = 0.01(b^{4/3}) \tag{8.9}$$

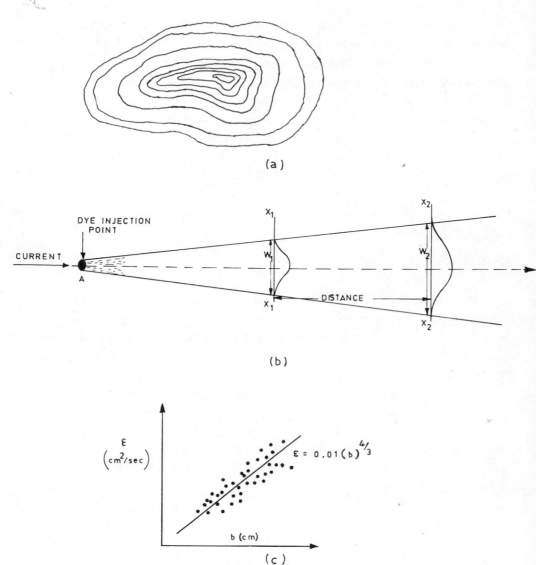

(a)

(b)

(c)

Figure 8.13 Eddy dispersion in coastal waters. (a) The dye patch formed in the absence of a pronounced current. The dye concentration "contours" shown can be developed from aerial photographs. (b) The plume formation after a tracer has been injected at point A. The tracer concentrations are measured at different points along the axes at X_1 and X_2 while current speed is measured simultaneously; the distance between the axes at X_1 and X_2 is also noted (see text for explanation). (c) The observed correlation between the eddy dispersion coefficient ε and the scale of phenomenon b.

where

        b = initial width of the horizontal plume, cm/sec

        ϵ = eddy diffusion coefficient, cm^2/sec

Other workers have also obtained essentially similar equations, but some with different values of the coefficients.

From Eq. (8.9) it is evident that the larger the initial width of the horizontal plume (i.e., the larger the length of the diffuser section in an outfall), the larger is the value of ϵ applicable to the phenomenon. If Eq. (8.9) is considered relevant to a given case, it can be used to readily estimate the value of ϵ. In fact, Eq. (8.9) has often been used to make an order of magnitude estimate of the dispersion coefficient in open seas, thus obviating the need for elaborate field tests which may be necessary where there is reason to believe that a substantially different value of ϵ may prevail.

In making an aerial photographic analysis of dispersion in the ocean at Oregon, Burgess [10] noted that there was no indication that the diffusion coefficient ϵ varied according to the 4/3 power of b. He observed that the shape of the plume depended on current velocity (and any residual buoyancy or momentum in the waste stream). At very low or zero velocity, the waste field merely ponded over the outfall area (see Figure 8.13); as the velocity increased, the plume elongated and its width was reduced, the minimum width being equal to the length of the diffuser section. Thus, he found that the one-dimensional Fickian diffusion model was reasonably accurate in predicting dilution at current velocities greater than about 0.12 m/sec. For lower velocities, a two-dimensional model was necessary to take into account the loss of waste (or tracer) to the lower layers by vertical mixing. This could be done by inserting a "decay" term in the equation and solving with the help of a computer. There are other modes of estimating vertical diffusion [3].

Assuming that the one-dimensional Fickian diffusion model is applicable to given conditions, ϵ can be estimated for a proposed outfall site by releasing a dye from a boat or dropping a dye marker from a plane and observing its dispersion with time. The width of the plume tends to increase gradually as shown in Figure 8.13. This can be observed in aerial photographs, or found by sampling the seawater concentrations along the two axes a distance D apart. Floats released at the time of study help the boat crew to trace the progress of the dye and its time of travel from the plane X_1-X_1 to X_2-X_2. Dye concentration "contours" can be developed for a drifting patch of dye. Within the plume, the dye concentration is assumed maximum along the centerline and decreasing on either side in a Gaussian fashion.

The coefficient of eddy diffusivity has been considered analogous to the diffusion coefficient of particles undergoing Brownian motion and was defined by Einstein in 1905 in terms of the variance σ^2 as $\epsilon = 1/2(d\sigma^2/dt)$ where σ = standard deviation of the Gaussian distribution. Thus, in its integrated form, applied to the two planes referred to above, we get

$$\epsilon = \frac{(\sigma_2)^2 - (\sigma_1)^2}{2(t_2 - t_1)} \qquad\qquad (8.10)$$

where $(t_2 - t_1)$ is the time elapsed during which the width of the plume increases from W_1 to W_2, and σ_1 and σ_2 are standard deviations of the measured concentrations normal to the direction of flow. Their squares give variances.

In conducting such an experiment in the sea, a dye is first released at a desired point A from a boat whose position is noted by triangulation from two fixed stations on the shore. The boat then moves and positions itself somewhere downstream along an axis X_1-X_1 while one or two boats may be anchored still further downstream along, for example, X_2-X_2. The time of arrival of the dye at each axis is judged from the travel of the floats. Current speed is also estimated at the same time by observing the positions of the floats from the two shoreline stations. Several water samples are taken along each axis.

A plot of the measured concentrations versus distances along each axis generally shows higher values in the midsection and a gradual tapering off on either side. Using statistical methods, the standard deviations of the curves are obtained at each axis and, thus, their variances can be found in terms of widths. Several possible sources of error exist in making such an experiment, apart from the difficulties of working at sea. Aerial photography can help to give pictures of the dye patch at different time intervals more readily. ϵ can also be estimated from observations of the patch at different times.

Assuming that standard deviations are related to the photographed plumed widths W_1 and W_2 as

$$\sigma_1 = \frac{W_1}{4} \qquad \text{and} \qquad \sigma_2 = \frac{W_2}{4}$$

we get

$$\varepsilon = \frac{(W_2)^2 - (W_1)^2}{32(t_2 - t_1)} \tag{8.11}$$

Example 8.4 In a single experiment with dye dispersion in a sea, aerial photographs showed that the width of the plume expanded from 46.4 to 69.7 m in 30 min. Compute ε in cm^2/sec, assuming Eq. (8.11) applies. From Eq. (8.11),

$$\varepsilon = \frac{(W_2)^2 - (W_1)^2}{32(t_2 - t_1)}$$

$$= \frac{[(69.7)^2 - (46.4)^2]10^4}{(32)(30 \times 60)} = 470 \ cm^2/sec$$

From Eq. (8.9),

$$\varepsilon = 0.01(b^{4/3})$$

Substituting, $b = 46.4 \times 10^2$ cm   we get $\varepsilon = 770 \ cm^2/sec$

The dispersion coefficient can also be estimated from drift time studies discussed in the previous section. In Eq. (8.10), substitution of $\sigma_1 = 0$ at $t_1 = 0$ at the point of release of the floats gives

$$\varepsilon = \frac{(\sigma)^2}{2t} \tag{8.12}$$

The value of σ is calculated as the standard deviation of the spread of drift cards in terms of distance along a beach, and the value of t is the mean drift time of the cards. The scale of the experiment is half the width in question. For example, from the data given in Example 8.3,

$$\varepsilon = \frac{(2.5 \ km \times 10^5 \ cm/km)^2}{2(15 \ hr \times 60 \times 60)} = 5.78 \times 10^5 \ cm^2/sec$$

Eddy diffusivity coefficients so determined are applicable to surface conditions and generally tend to decrease with depth; this is an important consideration in case of trapped plumes but is often neglected. Recently, Callaway has presented a method for predicting the subsurface horizontal dispersion of pollutants using current meter records [24].

Values of the dispersion coefficient computed from field tests are specific for the conditions at the time of the test. Thus, performing a single test is not sufficient. At least four tests at different times in the critical season should be performed. Observed values may also be compared with those obtained by reference to Figure 8.13 which, as stated earlier, is often taken to suffice for design purposes.

The estimation of dilution in horizontal travel is discussed further
in Sec. 8.16.

8.9 ESTIMATION OF DISSOLVED OXYGEN BALANCE

Seawaters are often stratified, giving a fairly well-oxygenated epilimnion and a relatively poorly oxygenated hypolimnion. For various reasons,
waste outfalls are often so engineered as to keep the rising plume submerged
at the pycnocline or some other level. If, in such cases, fish life in the
hypolimnionic water is to be protected, a proper estimation of its DO balance is called for. Often, the critical season in this regard is the summer
when respiration losses are maximum and oxygen-laden freshwater inflows are
minimum. Thus, some consideration must be given to the downward diffusion
of dissolved oxygen.

Downward diffusion of DO is somewhat slowed by the presence of the
pycnocline since diffusion has to take place against a vertical density
gradient. Koh and Fan [7] have given an empirical method for approximating
the downward diffusive flux which can be written as follows:

$$f = K\left(\frac{dc}{dy}\right) \times 10^{-7} \tag{8.13}$$

where

    f = downward diffusive flux of oxygen, kg/m^2-sec

    dc/dy = vertical gradient of DO concentration, mg/liter per meter
                depth

    K = vertical diffusion coefficient, cm^2/sec

It has been shown by them that K can be computed from the following observed
correlation:

$$K = \frac{10^{-4}}{e} \tag{8.14}$$

where e = relative density gradient per meter depth = $1/\rho_0(d\rho/dy)$ and
where $4 \times 10^{-7}/m \leq e \leq 10^{-2}/m$.

Thus, from a knowledge of the vertical density gradient and the DO
gradient in a given season, the downward flux of DO can be estimated. Example 8.5 illustrates how this can be done.

Generally, summer values of the downward diffusion of DO are higher as
the density gradient is less and the DO gradient more. In the summer the

upper waters tend to be well oxygenated while the lower waters tend to be
at their poorest in their DO content; thus, the DO gradient increases.
When organic waste is added to the lower layers through a submerged outfall,
further depletion of oxygen occurs as a result of increased bacterial respiration (and the immediate oxygen demand of the waste if present). This
increases the DO gradient still further with some corresponding increase
in downward diffusion of DO as the latter is directly proportional to the
former.

The maximum allowable DO gradient is, however, limited by the practical
constraint of requiring, for instance, 2 mg/liter minimum DO in the hypolimnion and the actual DO present in the upper waters just above the pycnocline. Estimating the difference between the downward flux of DO at its
existing gradient and at the maximum allowable gradient then becomes a mode
of computing the permissible BOD_u load in the hypolimnionic waters.

For example, if the DO content in the upper waters during summer is,
for example, 7.5 mg/liter and the lower waters at that time have 3.5 mg/
liter, the DO gradient is 4 mg/liter over the depth of the pycnocline. The
maximum permissible gradient over the same depth can be 7.5 - 2.0 = 5.5
mg/liter, giving a 37.5% increase over the existing gradient. Thus, the
downward DO flux will also increase by 37.5% over the present value if K
is assumed constant. This additional DO influx can be used to meet the
BOD_u of the waste provided the area over which it spreads can be estimated.
In the case of bays, some idea of the situation can be obtained by assuming
that the discharged waste spreads over the whole area of the enclosed waterbody.

As in the case of rivers, it is evident that the waste-assimilating
capacities of seas are also inherently lower in the warmer regions than in
the colder ones. The construction of an outfall into a large body of water
such as a sea does not significantly affect its inflows and outflows. Photosynthesis is also relatively unimportant as a source of oxygen at hypolimnion conditions. Thus, the principal mechanism of oxygen transfer is
the downward diffusion of DO as just discussed and its estimation must be
carried out, especially where the rising plume of wastewater from the outfall is likely to stay submerged at the pycnocline or other low level.

The DO content of natural seawater in its upper layers is generally
quite high, sometimes even supersaturated owing to photosynthesis. The
BOD_u of natural seawater generally varies from 1 to 3 mg/liter, sometimes
with higher values such as 6 mg/liter or even more in the bottom waters
(Sec. 8.2.2).

Example 8.5 At a proposed outfall site, the summer values of density
parameters versus depth show that at the pycnocline they change from 19.7
at 10 m depth to 21.5 at 30 m depth (see Figure 8.14). The DO profile similarly shows that DO drops from 6.5 mg/liter at 10 m depth to 3.5 mg/liter
at 30 m depth. Estimate the likely downward DO flux as kg/km²-day at the
stated conditions and when the minimum DO content at 30 m depth is allowed
to drop to 2 mg/liter.

$$\text{Vertical density gradient } e = \frac{1}{\rho_0}\frac{d\rho}{dy} = \frac{1}{1.0215}\frac{1.0215 - 1.0197}{30 - 10}$$

$$= 8.81 \times 10^{-5} \text{ per meter}$$

From Eq. (8.14) we have

$$K = \frac{10^{-4}}{e} = \frac{10^{-4}}{8.81 \times 10^{-5}} = 1.135 \text{ cm}^2/\text{sec}$$

$$\text{DO gradient } \frac{dc}{dy} = \frac{6.5 - 3.5}{10 - 30} = -0.15 \text{ mg/liter-m}$$

Diffusive flux of DO is obtained from Eq. (8.13) as

$$F = -K\left(\frac{dc}{dy}\right)10^{-7} = -1.35(-0.15)(10^{-7}) \text{ kg/m}^2\text{-sec}$$

$$= 0.17 \times 10^{-7} \text{ kg/m}^2\text{-sec} = 1470 \text{ kg/km}^2\text{-day}$$

When the DO at 30 m depth = 2.0 mg/liter, dc/dy = (6.5 - 2.0)/(30 - 10)
= 0.225 and F = 2206 kg/km²-day. Thus, BOD_u assimilation capacity = 2206

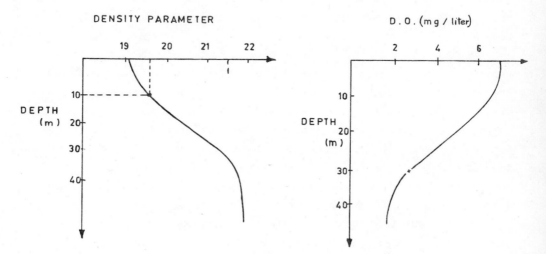

Figure 8.14 The changes, with depth, in the density parameter and
DO at an outfall site.

- 1471 = 735 kg/km^2-day. (Assuming uniform distribution of the waste over
an area of, for instance, 1 km^2, the BOD$_u$ assimilation capacity is 735 kg
per day.)

In the case of wastewater discharges, where the plume surfaces and the
waste field moves horizontally under the action of surface currents, the
Do balance is generally more favorable than in the previous case since the
epilimnion waters are better oxygenated.

8.10 NUTRIENT BUILDUP IN BAYS

Some form of mathematical modeling is often required for predicting
nutrient and oxygen concentrations, especially in enclosed seas and bays
to which wastewater is proposed to be discharged. This requires an understanding of some of the major factors and processes affecting the fate of
pollutants in the waterbody so that they can be simulated mathematically.
The purpose of the following discussion is to illustrate how a typical pollution balance model is formulated, the nature of the data required, and
the calibration and use of such models.

8.10.1 Pollution Balance Models

Considering the topography and stratification pattern of a waterbody,
a suitable pollution balance model can be developed representing it as a
set of two, three, or more interconnected "boxes" of known area and volume.
In its simplest form, a stratified waterbody can be depicted as a two-box
model with inflows and outflows as shown in Figure 8.15, together with an
exchange of water and substances between the upper and lower boxes. In
more elaborate models, the waterbody may be divided into sectors, again
depending on topography and other features, each sector being depicted as
above to give a string of interconnected boxes (it being understood that
in reality there are no clear-cut compartments, but only a continuum).

In a typical model, pollutants may enter and leave from any upper or
lower box in the system. During their residence time in a box, degradable
substances would undergo biological breakdown into simpler end products
which may diffuse downward along with settling debris or be transported to
any of the adjoining boxes. Bacterial respiration in each box would draw
upon the oxygen available in that box. The oxygen available in each box

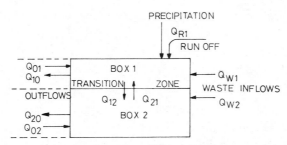

Figure 8.15 A typical two-box model for a stratified waterbody.
V = volume of box (m^3), Q = average flow (m^3/day), O = oxygen concentration
(g/m^3) in concerned flow, S = inflow of BOD (g/day), P = algal biomass,
expressed as particulate phosphorus (mg/m^3), μ = algal growth rate constant
(per day), μ_{max} = maximum growth rate constant (per day), K_P and K_N = half-
velocity constants (mg/m^3), a = constant (dimensionless), t = time (days),
R = respiration rate constant (per day), K_L = mass transfer coefficient
(m/day), U_z = vertical diffusion velocity (m/day), P_d = dissolved phosphorus
(mg/m^3), P_p = particulate phosphorus (mg/m^3), N_d = dissolved nitrogen ($mg/$
m^3), V_s = settling velocity (m/day), γ = phosphate exchange coefficient
(mg P/m^2-day).

would, in turn, depend on the various sources of oxygen (inflows of oxygen
laden water from adjoining boxes or from outside flows, photosynthesis if
any, gas transfer at the air-water interface, and downward diffusion) and
the so-called "sinks" of oxygen (outflows of oxygen laden water, all biotic
respiration losses, and chemical oxidation if any).

Algal productivity in each box would depend on the availability of nutrients and light. Consequently, algal productivity and simultaneous oxygen
production would be maximum in the upper boxes in favorable seasons and
perhaps negligible at all times in the lower boxes. Algal decay and respiration losses, on the other hand, may predominate in the lower boxes. Bacterial breakdown would regenerate nutrients (NH_3 and PO_4) which when transported to other boxes help further algal production in them. This recycling
of nutrients facilitates their accumulation in the system. Phosphorus is
generally the limiting nutrient and, therefore, the nutrient of greatest
concern. Its losses can occur only through outflows from the system and
precipitation in the bottom sediments under certain conditions [21]. For
each of the boxes, continuity equations for dissolved oxygen and nutrients
can be formulated as shown later, and concentrations of the concerned substances can be calculated by numerical integration.

8.10.2 Hydrodynamic Submodels

Besides knowing the area and volume of each "box," it is also necessary
to know the mean residence time in it and the water exchange occurring between the enclosed waterbody and the open sea. Thus, hydrodynamic submodel-
ing is first necessary to be able to estimate the net water received by the
system (precipitation, evaporation, natural runoff, and wastewater inflows),
the water transported from one box to the other, and, finally, the water
leaving the system through net outflows to the open sea. This submodel
must be used in conjunction with the pollution balance model [36].

The major factors affecting water circulation of large bodies are: (1)
amplitudes and periods of wind-induced seiches (oscillations) in the transition zone (pycnocline), (2) surface currents owing to wind, and (3) salinity variations. Other factors of relatively equal importance in the case of
large, enclosed waterbodies are water level variations resulting from tidal
effects, pressures, etc., occasional upwelling, and freshwater inflows.

Internal seiches can be identified by variations in the depth at which
a given salinity value occurs near the transition zone. Simultaneous wind
speed and direction measurements can help to establish statistical correlations between minimum wind speed and the duration at which an internal seiche
may be expected to occur. During prolonged windy periods, surface water is
pushed downward, thus causing a lowering of the transition zone. Salinity
measurements show the depth affected. The volume of water exchanged between
the upper and lower "box" at that time is computed from the surface area of
the box and the depth effected during the storm. The total quantity so
transported by all major storms is expressed as equivalent average flow over
the year. Mean residence time in a box is thus obtained by dividing the box
volume by the average flow rate.

8.10.3 Algal Oxygen Production

The nutrient content of algal matter usually varies with species and
age but is generally taken, on a conservative basis, to be 1% P and 7.2% N
by weight. Also production of 1 g of algae is generally accompanied by the
simultaneous production of 1.40 g of oxygen. Accordingly, 1 g of P and 7.2
g of N signifies an algal production of 100 g and an oxygen production of
140 g [21].

If P is the algal biomass (expressed as particulate phosphorus) and μ the growth rate constant, then the instantaneous growth rate can be stated as equal to μp in which μ can be expressed by a Monod-type equation:

$$\mu = \mu_{max}\left[\frac{C_p}{K_p + C_p}\right]\left[\frac{C_N}{K_N + C_N}\right] \tag{8.15}$$

in which all notations are as defined below Figure 8.15. The maximum growth rate constant μ_{max} is dependent on limiting factors such as water temperature and light and, thus, can be taken into consideration in a computer program as a function of time and depth. From literature values, the production rate in the upper boxes may be assigned any value μ_{max}^o up to a maximum of 1.0 per day, and zero in the lower boxes owing to a continuous absence of light, and made time dependent by an equation of the type

$$\mu_{max} = \mu_{max}^o\left(1 - a \cos\frac{2\pi t}{365}\right) \tag{8.16}$$

in which a is a constant with a suitably assigned value. The half-velocity constants K_N and K_p may also be assigned suitable values. Typical literature values are 50 and 10 μg/liter, respectively. The actual growth rate μ also depends on the concentrations of nitrogen (C_N) and phosphorus (C_p) in any box. Nutrient concentrations in the surface and bottom waters of the open sea must be known as they form a boundary condition.

The rate of change of oxygen concentration owing to algal production in a box can be stated in compatible units (see notation) as

$$\frac{dO}{dt} = 0.14\mu p \tag{8.17}$$

Algal respiration causes an oxygen loss which must be taken into account. Its rate R is mainly affected by temperature, and literature values suggest a range from 0 to 0.03 per day for temperatures up to $30^\circ C$ [37]. If the temperature in the lower boxes of a model is assumed as constant, the respiration rate R may also be assigned a constant value R_0. For the upper boxes, however, it may vary with the time of the year and can be formulated as before by

$$R = R_0\left(1 - a \cos\frac{2\pi t}{365}\right) \tag{8.18}$$

The oxygen consumption owing to respiration in which organic breakdown occurs is identical to the oxygen production when new algal matter is produced. Thus, every gram of organic matter destroyed will need 1.40 g of oxygen. Consequently, the rate of change of oxygen concentration in a box resulting from algal respiration can be stated as

$$\frac{dO}{dt} = 0.14 \ Rp \qquad\qquad\qquad (8.19)$$

8.10.4 Oxygen Balance

For any box in the model, an oxygen balance can be obtained if the various sources and sinks of oxygen applicable to that box are taken into consideration. Besides algal production and respiration, the other typical sources and sinks of oxygen are as formulated below.

Oxygen transfer at the air-water interface can be formulated in terms of the mass transfer coefficient K_L, the difference in actual oxygen concentration O from its saturation value O_{sat}, and the surface area A of the box in question divided by its volume. K_L values are likely to range from 10^{-1} to 10^{-3} m/day [37]. Thus, the rate of change of oxygen concentration with time due to surface uptake can be stated as

$$\frac{dO}{dt} = K_L(O_{sat} - O)\frac{A}{V} \qquad\qquad\qquad (8.20)$$

Field studies in bays and estuaries have shown actual re-aeration values to be much higher than those computed from standard equations [12]. This is attributed to the effect of winds, tides, ship traffic, and such other factors not included in the prediction equations which, therefore, tend to give conservative estimates of surface re-aeration. On the other hand, the presence of surface active agents in much polluted waters may tend to retard aeration.

Similarly, the rate of change of oxygen concentration in a box owing to downward flux can be formulated in terms of the difference in oxygen concentration between the upper and lower boxes, a vertical diffusion coefficient U_z, and the area per unit volume (also see Sec. 8.8.5 for a more elaborate approach) to give

$$\frac{dO}{dt} = U_z(O_{upper} - O_{lower})\frac{A}{V} \qquad\qquad\qquad (8.21)$$

Oxygen loss owing to the exertion of the BOD of the wastewater can be simply stated in terms of the mass of ultimate BOD of the wastewater S per unit time entering a box of volume V as follows (nitrification may be taken into account, if relevant):

$$\frac{dO}{dt} = \frac{S}{V} \tag{8.22}$$

Finally, the rate of change in oxygen concentration in a box resulting from various inflows and outflows can be stated in terms of the concerned flows Q, their respective oxygen concentrations O, and the box volume V. Thus, for the simple two-box model shown in Figure 8.15, the rate of change of oxygen in box 1 can be expressed as follows, using suffixes to identify the components and neglecting the oxygen brought in by the wastewater:

$$\frac{dO}{dt} = \frac{1}{V_1}\left[Q_{R1}(O_{R1} - O_1) + Q_{21}(O_2 - O_1) - Q_{12}(O_1 - O_2) - Q_{10}(O_1) \right.$$
$$\left. - Q_{20}(O_2) + Q_{01}(O_u) + Q_{02}(O_L) \right] \tag{8.23}$$

Thus, the <u>overall</u> oxygen balance for any box can now be formulated as the algebraic sum of the individual rates of oxygen change owing to each source or sink. For example, for box 1 the oxygen sources are algal production, surface uptake, and net flows, whereas the sinks are algal respiration, downward flux, and loss owing to an exertion of BOD. Hence, using the concerned equation numbers, we can state

$$\frac{dO(1)}{dt} = (8.17) + (8.20) + (8.23) - (8.19) - (8.21) - (8.22) \tag{8.24a}$$

For box 2 a similar formulation can be written except that now there will be no algal production (only respiration) and no surface uptake of oxygen. Adapting the individual equations to suit this box one can write

$$\frac{dO(2)}{dt} = (8.23) - (8.19) + (8.21) - (8.22) \tag{8.24b}$$

The output thus obtained from the pollution balance model will indicate the expected DO levels in each box from day to day.

8.10.5 Nutrients

The fate of the nutrients nitrogen and phosphorus can also be determined from the pollution balance model if their inputs into the system are known. In fact, it is necessary to know their concentrations in any box for determining the value of μ in Eq. (8.15) and thus solving Eq. (8.17) which is essential for finding the DO concentration in any box.

Phosphorus can be considered in two forms if desired: dissolved phosphorus and particulate phosphorus (algal biomass has been expressed as particulate phosphorus earlier). In both cases, the inflows into any box can be considered in a manner similar to the one seen earlier for oxygen [Eq. (8.23)] except that dissolved or particulate phosphorus values must now be used instead of oxygen, therefore Equation (8.23) must be adapted for phosphorus inflows.

[Equation (8.23) adapted for phosphorus inflows]   (8.23a)

Algal activity will have different effects on the dissolved and particulate fractions of phosphorus in a box. In the case of dissolved phosphorus, a reduction in its concentration will occur as a result of algal growth and an increase resulting from the destruction of organic matter in respiration. For particulate phosphorus the reverse will occur: an increase in concentration resulting from algal growth and a decrease resulting from respiration. Thus, phosphorus changes can be accounted for in terms of the algal growth and respiration rates discussed earlier [Eqs. (8.17) and (8.19)] as

Algal growth rate $= \mu p$ (8.17a)

Algal respiration rate $= Rp$ (8.19a)

Furthermore, two other mechanisms must be given some consideration. Dissolved phosphorus is likely to be affected in any of the lower boxes by exchange with bottom sediments, especially if substantial changes in pH and DO are expected to occur from pollution. This exchange can be accounted for by an equation of the following type incorporating a phosphate exchange coefficient γ to give the rate of change in concentration of dissolved phosphorus P_d as

$$\frac{d(P_d)}{dt} = \gamma \frac{A}{V}$$ (8.25)

Literature values for γ range widely from 0.27 to 500 mg P/m^2-day [21a] but, since our objective generally is to maintain sufficient DO in the bottom layers at which not much bottom exchange should occur, only a low value of the coefficient may be assumed until more data are available. In the upper boxes of the model, γ may be assumed to be zero.

Another mechanism to be considered is the sedimentation of particulate phosphorus (algae) likely to occur in a box. This mechanism accounts for the passing of algae from an upper box to a lower one; the assumption of a high sedimentation velocity implies a larger oxygen consumption resulting from respiration in the bottom waters. Sedimentation is accounted for in the model by assuming a suitable value of the settling velocity V_s (range: 0.1 to 7.5 m/day [37]). As a boundary condition, it may be assumed, in the interest of a conservative design, that all algal matter settling into the lower box is degraded in respiration and none enters the bottom sediments. A typical formulation for change in particulate phosphorus P_p is

$$\frac{d(P_p)}{dt} = V_s \frac{A}{V} p \tag{8.26}$$

Thus, in overall consideration, the rate of change of phosphorus concentration in any box may be stated using the relevant equation numbers as follows:

For dissolved phosphorus

$$\frac{d(P_d)_1}{dt} = 8.23(a) - 8.17(a) + 8.19(a) + 8.25 \tag{8.27}$$

$$\frac{d(P_d)_2}{dt} = 8.23(a) + 8.19(a) + 8.25 \tag{8.28}$$

For particulate phosphorus

$$\frac{d(P_p)_1}{dt} = 8.23(a) + 8.17(a) - 8.19(a) - 8.26 \tag{8.29}$$

$$\frac{d(P_p)_2}{dt} = 8.23(a) - 8.19(a) + 8.26 \tag{8.30}$$

Similar equations can be written for dissolved nitrogen concentration in any box after adapting Eq. (8.23) to include all inflows of dissolved nitrogen, and Eqs. (8.17) and (8.19) to include the value 7.2 instead of 0.14, since nitrogen has been assumed to constitute 7.2% by weight of algal matter. No denitrification term is included since design objectives require that the lower waters should always be aerobic. Hence, using the suffix b to denote the earlier equations adapted for nitrogen, we may state the rate of change of dissolved nitrogen N_d as

$$\frac{d(N_d)_1}{dt} = 8.23(b) - 8.17(b) + 8.19(b) \tag{8.31}$$

$$\frac{d(N_d)_2}{dt} = 8.23(b) + 8.19(b)$$ (8.32)

8.10.6 Calibration and Use of Models

In view of the various simplifications made to keep the pollution balance model manageable, and the various assumed values of constants and coefficients used, it is essential to validate the model by comparing actual field measurements of DO and nutrients with their values predicted by the model at present levels of pollutant inflows. Small discrepancies between observed and predicted values can often be removed by "fine tuning" the assumed data. After this calibration exercise, the model is ready for use in predicting the DO and nutrient levels likely to occur at other pollutant loads and conditions. Alternative strategies in waste disposal can also be examined with the help of the model. Inherent limitations in the use of such models must, of course, be kept in view.

As an illustration of the above principles, an interesting study of Izmit Bay (SWECO-BMB, Ref. 33), Istanbul, can be quoted. A four-box model was developed taking into account the geometry of the bay and its stratification pattern. A preliminary hydrodynamic study gave some idea of the water movements in the study area. Based on present-day pollutant loads (BOD, N, P, etc.), the DO levels predicted by the model were compared with observed values in the bay to facilitate calibration of the model.

Figure 8.16 shows the observed and predicted values of DO in the four boxes during a calibration exercise. The model appears to fit better to the condition in the lower boxes where DO is actually more critical. Some values of coefficients and constants used in this study are also shown below:

μ_{max}^o = 0.2 per day

R_0 = 0.02 per day

K_N and K_p = 50 and 10 µg/liter, respectively

γ = 1 mg P/m^2-day

V_s = 0.10 m/day

K_L = 0.50 m/day

U_z = 10^{-2} m/day

a = 0.3

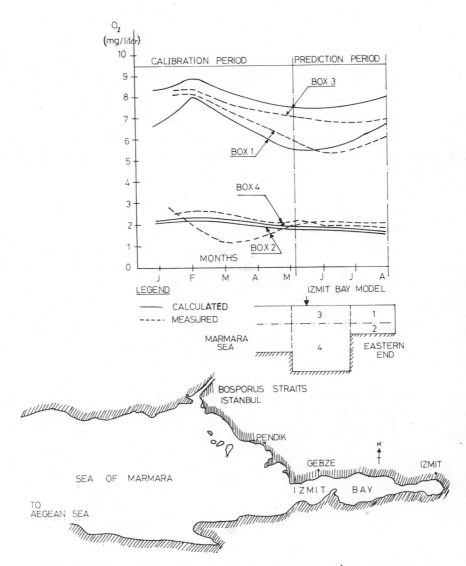

Figure 8.16 The pollution model for Izmit Bay (adapted from the SWECO-BMB Report on Izmit Sewate Master Plan Report, 1976).

8.10.7 Estimation of Overall Nutrient Buildup

Example 8.6 illustrates a relatively simple and practical method for estimating the likely buildup of phosphorus in a bay, in a gross sense, not

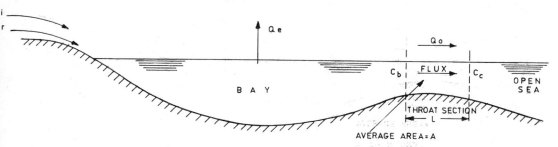

Figure 8.17 The concentration buildup in a bay.

boxwise, by preparing an overall materials balance of all inflows and outflows from the system.

Referring to Figure 8.17, a mass balance for water in the bay is given
by

$$Q_i + Q_r = Q_e + Q_o \tag{8.33}$$

where

Q_i = net inflows of wastewater with phosphate concentration C_i

Q_r = net inflow of natural runoff (including direct rainfall on bay)
with phosphate concentration Cr

Q_e = evaporation loss

Q_o = net outflow from bay with phosphate concentration C_o (net outflow
may be negative if $Q_e > Q_i + Q_r$)

Similarly, a mass balance for phosphorus, as phosphates, can be obtained if the actual concentrations just upstream and downstream at the
throat section between the bay and the open sea are measured (see Figure
8.17) and uniformly mixed conditions are assumed in the bay and the open
sea. Thus, we get

$$Q_i C_i + Q_r C_r = Q_o C_o + F + \frac{d(PO_4)}{dt} + R \tag{8.34}$$

in which the terms on the left-hand side denote all forms of inflow (the
term $Q_r C_r = 0$ in the dry season). On the right-hand side of the equation
we have

$Q_o C_c$ = phosphate carried out of the bay by net outflow

F = phosphate flux leaving as a result of diffusion, etc.,
owing to a concentration gradient across the throat

$d(PO_4)/dt$ = rate of change of phosphate storage in the bay (equals
zero at steady state)

R   = phosphate removed from the system by various removal
                        mechanisms (can be assumed to be zero under test condi-
                        tions provided the phosphorus measurements are made when
                        the algal uptake of phosphorus is negligible in the bay)

The flux F can be estimated as follows from site measurements of phosphate concentrations since

$$F = K''A(C_b - C_c)$$

(8.35)

where

K'' = transfer coefficient, length/time*

A   = average cross-sectional area of flow, length^2

C_b and C_c = phosphate concentrations at the bay end and sea end,
                   respectively, of the throat section

Hence, for steady-state conditions, Eq. (8.34) becomes

$$Q_iC_i + Q_rC_r = Q_oC_o + K''A(C_b - C_c)$$

(8.36)

in which C_o can be assumed to be equal to C_c in the net outflow.

Example 8.6   The inflow-outflow data for a eutrophic bay in a monsoon
climate were obtained as indicated in Table 8.12. The phosphate concentration on either side of the throat section was measured and gave C_b = 0.21
mg/liter and C_c = 0.10 mg/liter. The average cross-sectional area at the
throat section was found to be 20,000 m^2 and its length 2.0 km. Estimate
the average value of the dispersion coefficient D (assume the phosphate
concentration in the runoff equals 0.1 mg/liter and in wastewater it averages 28.0 mg/liter. From Eq. (8.32),

$$Q_rC_r + Q_iC_i = Q_oC_o + K''A(C_b - C_c)$$

$$(351,000)(0.1) + (141,000)(28) = (228,000)(0.10) + (K'' \text{ m/day})(20,000)$$
$$\times (0.21 - 0.10)$$

K'' = 1800 m/day

Thus, the dispersion coefficient D = $K''L$ = (1800 x 2000 m)/24 = 150,000
m^2/hr. This is a somewhat low value signifying that the bay does not have
much mass transport from it. It behaves more like a lake.

Example 8.7   From the data given in Example 8.6, estimate the likely
phosphate concentration in the bay if the phosphate concentration in inflow

---

*The transfer coefficient K'' is related to the eddy dispersion coefficient
D (or ε) by $K''L$ = D, where L = length of the throat section.

Table 8.12 Monthwise Inflow-outflow Inventory for a Bay (See Example 8.6)

Month	Inflow Q_r (runoff + direct, 10^6 m³)	Inflow Q_i (wastewater, 10^6 m³)	Total inflow ($Q_i + Q_r$, 10^6 m³)	Evaporation losses (Q_e, 10^6 m³)	Net outflow to sea (10^6 m³)
Jan	-	4.0	4.0	3.0	1.0
Feb	-	4.0	4.0	3.5	0.5
Mar	-	4.3	4.3	6.8	-2.5
Apr	-	5.0	5.3	12.2	-6.9
May	2.0	5.1	7.1	15.2	-8.1
June	21.0	4.7	25.7	10.7	15.0
July	43.5	4.0	47.5	8.9	38.6
Aug	34.7	4.0	38.7	8.1	30.6
Sept	22.0	4.0	26.0	7.8	18.2
Oct	5.0	4.3	9.3	10.0	-0.7
Nov	-	4.0	4.0	6.2	-2.2
Dec	-	4.0	4.0	4.1	-0.1
Annual total:	128.2	51.4	179.6	96.5	+83.1
Average per day:	0.351	0.141	0.492	0.264	0.228

is reduced to an average 36% of the present-day value. Assume all other conditions including K'' remain the same.

Again from Eq. (8.32),

$$Q_r C_r + Q_i C_i = Q_o C_o + K'' A(C_b - C_c)$$

$$(351,000)(0.1) + (141,000)(0.36 \times 28) = (228,000)(0.10)$$
$$+ (1800)(20,000)(C_b - 0.10)$$

or, $C_b = 0.14$ mg PO_4/liter in the bay at steady state.

8.11 DISCHARGE FROM A ROUND JET IN A HOMOGENEOUS LIQUID

Mixing by jet diffusion takes place when one fluid is introduced into another fluid by a jet. The discharge of sewage to seawater is a case in point where the two fluids have different densities. Sewage is 2.5 to 3%

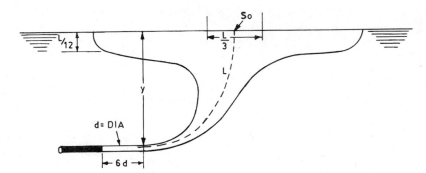

Figure 8.18 A definition sketch for a plume rising from a round jet in a homogeneous liquid.

lighter than seawater and, thus, tends to rise up by buoyancy, giving what is called a "plume"* as shown in Figure 8.18. A sort of aspirator action occurs which helps to draw in the surrounding seawater and dilute the sewage.

In their early studies introducing buoyant jets horizontally into a homogeneous liquid, Rawn and Palmer [11] observed that, if the length of the plume along the centerline was L, the other dimensions of the plume were approximately as shown in Figure 8.18.

However, the major interest is in the dilution effected at the top of a rising column. This dilution has been found by several workers to be a function of the Froude number N_{Fr} and of the depth/diameter ratio Y/d. The Reynolds number has no effect since it is high (in the range of 10^5 to 10^6 for outfalls).

Froude's number is the square root of the ratio of inertial forces to gravitational forces, which for a cube of length D can be written as

$$N_{Fr} = \sqrt{\frac{\rho D^2 U^2}{\rho g D^3}} = \frac{U}{\sqrt{gD}} \qquad (8.37)$$

where, in consistent units,

ρ = density of fluid

U = velocity

g = acceleration due to gravity

---

*A typical discharge from an outfall has both momentum and buoyancy at the source. The initial momentum is generally small and is soon diffused, while buoyancy is continually effective during its rise. Thus, it behaves more like a plume than a momentum jet. Some investigators refer to it as "forced plume," or "buoyant jet."

In a homogeneous medium with ρ_a = ambient water density, ρ_0 = wastewater density at source, and U_0 = jet velocity issuing from outfall of diameter d, we can rewrite

$$N_{Fr} = \frac{U_0}{\sqrt{g'd}} = \frac{U_0}{\sqrt{g[(\rho_a - \rho_0)/\rho_a]d}} \qquad (8.38)$$

where $g' = g[(\rho_a - \rho_0)/\rho_a]$ to account for buoyancy. Furthermore, if the discharge through an outfall is Q, issuing through a jet area $= \pi d^2/4$, the Froude number can be computed from Eq. (8.38) as follows:

$$N_{Fr} = \frac{Q}{(\pi d^2/4)\sqrt{g[(\rho_a - \rho_0)/\rho_a]d}} \qquad (8.39)$$

Where N_{Fr} is less than 1.0, gravity predominates and wastewater rises almost vertically, but where N_{Fr} is greater than 1.0, the jet issues with increasing momentum and the energy is utilized in turbulent mixing.

Figure 8.19 shows some experimentally observed relationships obtained between N_{Fr}, Y/d, and the dilution S_0 obtained at the centerline of the rising plume owing to aspirator effect [14,17]. These curves are applicable to horizontally issuing round buoyant jets after the flow is established (i.e., beyond a distance of about 6d for round jets; see Figure 8.19.

In these curves, therefore, Brooks [17] defines

S_0 = centerline (or minimum) dilution in the buoyant jet, relative to the concentration on the centerline at the end of the zone of flow establishment

Y = vertical distance from the center of the jet (at the end of the flow establishment) to the level at which S_0 is measured

d = diameter of the jet at the source (at the vena contracta if there is jet contraction)

The results are based on assumed values of the entrainment coefficient (0.082 for round jets) and a spreading ratio of 1.16. Recent studies have shown that the entrainment coefficient is not constant but is dependent on the local Froude number; the error due to this is not considered large [17].

Since S_0 refers to the end of the zone of flow establishment, it must be multiplied by 1.15 (for a spreading ratio of 1.16) to give dilution with reference to the initial discharge. Furthermore, to obtain the average dilution at the top of the plume S_0', the value of S_0 should be multiplied by 2.0 (assuming Gaussian distribution). Thus, values of S_0 obtained from Figure 8.19 for round jets should be doubled to estimate average dilutions.

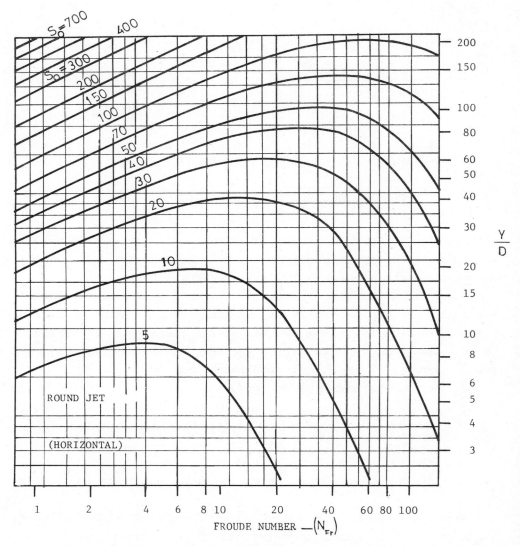

Figure 8.19 The centerline dilution of round, buoyant jets discharging horizontally into stagnant, uniform environments. To get the centerline dilution relative to the nozzle, multiply S_o by 1.15 to adjust for the zone of flow establishment; multiply by 2 to obtain an average (from Ref. 14).

For the sake of accuracy, the Froude number, which is generally calculated using the wastewater density at the source, should be multiplied by 1.07 to obtain the relevant value N_{Fr}' at the end of the zone of flow establishment, before referring to Figure 8.19. However, the difference is scarcely significant in this case. Similar curves are obtained for jets issuing an any inclination to the horizon or vertically [17].

It is evident from all these curves that, for a given depth and discharge, the value of N_{Fr} increases as the diameter is reduced. Thus, the dilution obtained also increases. The figures are useful in estimating the effects of alternative strategies in the design of outfalls to seas (and to lakes for heated discharges), as illustrated below.

Example 8.8 Estimate the value of the Froude number and the likely dilution at the top of a plume emerging at 1.57 m^3/sec from an open-ended outfall of 1 m diameter with a velocity of 2 m/sec and rising to a height of 25 m above the outfall. Assume ρ_a = 1.020 and ρ_0 = 1.001. From Eq. (8.38),

$$N_{Fr} = \frac{U_o}{\sqrt{g[(\rho_a - \rho_0)/\rho_a]d}} = \frac{2}{\sqrt{9.81[(1.020 - 1.001)/1.020](1 \text{ m})}} = 4.63$$

$$N_{Fr}' = 1.07 \times N_{Fr} \cong 5.0$$

Referring to Figure 8.19 with N_{Fr}' = 5.0 and Y/d = (25 + 0.5)/1.0, we get a minimum centerline dilution S_o of about 15 and the average dilution of twice that, namely, about 30.

If, for the data given in Example 8.8, it is desired to ensure a minimum centerline dilution of, for example, 30 instead of 15, what possible alternatives are open for consideration? We could examine the following four alternatives:

1. Extend the outfall farther out into the sea until Y = 45 m. This will give the required dilution of 30, but, depending on the topography, it may involve extending the length by several hundred meters and thus prove expensive.

2. If Y is retained at 25 m, the required dilution of 30 can be theoretically obtained by increasing the Froude number to as high a value as 85. This would imply increased pumping and energy costs to give the required velocity, which would be abnormally high and create other problems as well.

3. Incorporate a "jet pump" in the outfall to increase dilution. This has not yet been demonstrated as feasible in full-scale outfalls; laboratory tests have shown that the initial dilution could be increased up to 190% of that for the normal outfall [15].

4. Instead of discharging only from the end of the outfall, provide "n" number of openings, spaced such that their plumes rise without overlapping. In this manner the values of both N_{Fr} and Y/d can be increased as desired (since diameter d_1 of ports will be smaller than the diameter of pipe) without affecting the outfall length or overall pumping head.

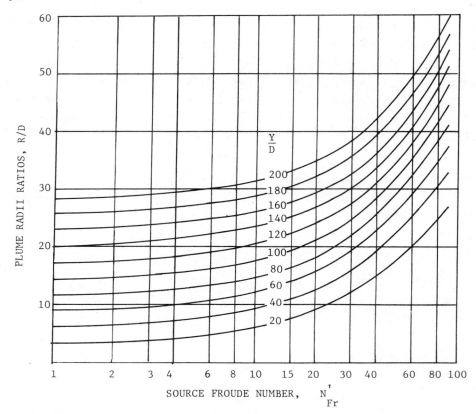

Figure 8.20 Radii ratios for round jets at the top of the plume from a horizontal discharge within a nonstratified environment (from Ref. 35).

For example, if n = 3, the discharge at each port is $Q/3$, i.e., $1.57/3$ m³/sec. If a velocity of, for instance, 3.5 m/sec is desired from each port, the diameter d_1 should now be 0.44 m; this gives $N_{Fr} = 12.2$ and $Y/d_1 = (25 + 0.5)/0.44 = 58$, which gives a dilution of nearly 30. The advantage of providing multiport discharge in securing higher dilutions is thus evident. Multiport arrangement is possible by providing several ports in the side of a pipe. If these ports are sufficiently far apart, their plumes do not interfere with one another and the dilutions can be estimated by the methods just described. If the ports are so close as to give what is called a "slot discharge" and constitute a "line source," estimation of dilution is done as discussed in the next section.

For Figure 8.19 to be applicable, it is necessary to ensure that rising plumes do not significantly interfere or overlap. For this purpose,

the width of the rising plume at its maximum height of rise must be estimated and the adjacent orifices spaced accordingly. As a rule of thumb,
the width w of the plume may be taken as one third of the height of rise Y,
namely, w ≅ Y/3.

However, the plume width can be better approximated from Figure 8.20
prepared by Russel Ludwig [35] which gives a relationship between N_{Fr}, the
depth-to-diameter ratio Y/d and the plume radii ratio r/d based on the work
of Fan and Brooks [14-17] for horizontal jet discharges into homogeneous
liquids.

From the data given in Example 8.8, the plume width at the top can be
estimated by reference to Figure 8.20 with N_{Fr} = 4.63 and Y/d = 25.5/1.0
to get r/d = 5, i.e., r = 5 x 1.0 m or w = 2 x 5 = 10 m.

8.12 DISCHARGE FROM A LINE DIFFUSER IN A HOMOGENEOUS LIQUID

If ports are spaced closely enough so that there is extensive overlapping in their rising plumes, they constitute a "line source" and produce
a flow pattern similar to that from a long horizontal slot (Figure 8.21).
Thus, when designing a line diffuser, its dilution characteristics are correlated to the discharge per unit length of diffuser rather than to the individual port discharges. For the same volume of discharge and buoyancy
conditions, line diffusers are generally capable of giving higher dilutions
than are single plumes. To this end, closely spaced jets which do interfere
are often preferred.

For most practical cases involving line diffusers, the angle of discharge is not a variable and the minimum centerline dilution S_o in the
rising plume can be approximated from the following [17]:

$$S_o = \frac{0.38(g')^{1/3}Y}{q^{2/3}} \qquad\qquad (8.40)$$

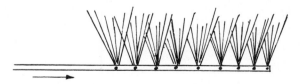

Figure 8.21 The discharge from a line diffuser (horizontal slot
discharge) giving overlapping plumes.

in which q = discharge per unit length; the other terms have been defined earlier. Brooks states that since the spreading ratio is now nearly equal to 1.0 no correction is necessary for the zone of flow establishment. The average dilution across the plume is, thus,

$$S_o' = \sqrt{2}\, S_o \tag{8.41}$$

Example 8.9 Consider the data given in Example 8.8 and assume a "line-diffuser" of 25 ports spaced 7 m apart. Compute the likely dilution.

Diffuser length = (25 - 1)7 = 168 m

$$q = \frac{1.57 \text{ m}^3/\text{sec}}{168 \text{ m}} = 0.00935 \text{ m}^3/\text{m-sec}$$

$$g' = g\left(\frac{\rho_a - \rho_0}{\rho_a}\right) = 9.81\left(\frac{1.020 - 1.001}{1.020}\right) = 0.1862$$

From Eq. (8.40),

$$S_o = \frac{(0.38)(0.1862)^{1/3}(25.5)}{(0.00935)^{2/3}} = 127$$

Thus it is seen that by incorporating a diffuser section in the earlier outfall, the minimum centerline dilution is increased from 15 to 127; the average dilution across the plume is now $\sqrt{2} \times 127 = 180$

With both round jets and line-diffusers discharging a lighter fluid into a heavier and homogeneous one, the lighter mixture tends to rise all the way up to the surface. Thus, in homogeneous environments the plumes invariably tend to surface. The behavior of a plume in a stratified environment, however, is different. This is discussed in the next section.

8.13 TRAPPING LEVEL IN A STRATIFIED LIQUID

When wastewater is discharged to a stratified liquid or one in which the density varies with depth, it mixes with the bottom water and rises up until the density of the mixture equals that of the seawater. Thus the mixture stays "trapped" at that level and cannot rise any farther (Figure 8.22). Evidently, the better the mixing, the lower is the trapping level reached, since the trapping level depends on the wastewater density, the mixing conditions, and the seawater density profile.

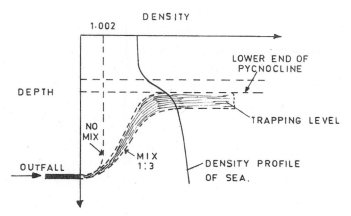

Figure 8.22 The trapping level within a stratified liquid.

When wastewater and seawater mix, the resulting density is estimated from the proportion in which they are mixed. For example, if 1 part of wastewater of density 1.002 mixes with 3 parts of seawater of density 1.033, the resulting density of the mixture ρ_m is given by

$$\rho_m = \frac{(1 \times 1.002) + (3 \times 1.033)}{3 + 1} = 1.025$$

and the mixture will rise only up to the level at which seawater is also of the same density. In general, we could state

$$\rho_m = \frac{\rho_0 + (S - 1)\rho_a}{S} \qquad (8.42)$$

where

ρ_0 = wastewater density

ρ_a = ambient seawater density at bottom

S = diluted volume of mixture

In Eq. (8.42) we could substitute values of the density parameter σ, if desired, instead of densities.

Trapping a rising plume in seawater is similar to the trapping due to atmospheric inversions in air pollution. However, unlike air pollution, trapping a wastewater plume is considered desirable wherever feasible, especially in the bathing season since the outfall site presents an esthetically more pleasing appearance than possible with surfacing plumes. Of course, the prevalence of a natural sea current at the trapping level then becomes a sine qua non for mass transport of the wastes out of the site.

Where only surface currents (e.g., wind-induced surface currents) are available or where the DO is already quite low in deeper waters, it may be advantageous to allow the plume to surface.

To ensure trapping of a plume at a desired subsurface level (for example, at the pycnocline), every effort is made to increase the dilution ratio. The greater the dilution obtained in the lower layers, the closer is the density of the mixture to that of the bottom seawater and the lower the trapping level. Thus, the provision of diffusers with multiports is preferred to outfalls with only a single end-discharge. In some climates the sea may exhibit only weak stratification and trapping may be difficult even with increased dilution ratios.

8.14 ESTIMATION OF TRAPPING LEVEL

8.14.1 For Round Jet Discharge

Brooks [17] gives the following equations which are useful in estimating the trapping level as well as the dilution for round jets discharging in stratified waters. For the sake of simplicity he assumes two conditions: the density profile of the sea has a linear variation with depth, and the rising plumes are discrete (no interference). The assumption of a linear density profile is often valid and not much error is introduced thereby, but if the density variation with depth is quite erratic, stepwise calculations can be made with the help of computer programs developed for the purpose [3,17]. Hansen [29] has described another method for stepwise calculations based on Cedervall's equations.

The maximum height of rise Y_{max} from the discharge level is given by Brooks [17] as

$$Y_{max} = 3.98(Qg_d')^{1/4}\left(-\frac{g}{\rho_1} \times \frac{d\rho}{dy}\right)^{-3/8} \tag{8.43a}$$

in which Q is the discharge per port, or full discharge if there is only one port, i.e., $Q = U_o(\pi d^2/4)$; and $g_d' = [(\rho_a - \rho_0)/\rho_a]g$ in which the terms have been defined before; $d\rho/dy$ = rate of change of ambient density with depth.

The plume centerline dilution is given by

$$S_o = 0.071(g_d')^{1/3}(Y_{max})^{5/3}(Q)^{-2/3} \tag{8.43b}$$

and the average dilution by

$$S_o' = \sqrt{2}\, S_o \qquad\qquad (8.43c)$$

Recently, in an outfall design for Bodrum, Turkey, Sarikaya [27] has shown that identical results were obtained using two different methods: one given by Brooks as above and the other by Hansen [29] based on Cedervall's equations. The density profile was reasonably linear.

Example 8.10   The density gradient at an outfall site is estimated to be uniformly 2.8×10^{-5} g/cm^3-m depth and the seawater density is 1.0278 at the proposed discharge depth of 36 m. If the wastewater discharge Q per orifice is 0.0123 m^3/sec and its density 0.999, find the maximum height of rise and the likely dilution assuming there is no interference between adjacent plumes.

1. From Eq. (8.43a),

$$Y_{max} = 3.98\left[\, 0.012\left(\frac{1.0278 - 0.999}{1.0278}\right)9.81\,\right]^{1/4}\left(-\frac{9.81}{1.0278}\times 2.8\right.$$
$$\left.\times 10^{-5}\right)^{-3/8}$$

$$= 20.87 \text{ m from the discharge level}$$
$$= 36 - 20.87 = 15.13 \text{ m from seawater surface}$$

2. From Eqs. (8.43b) and (8.43c),

$$S_o - 0.071\left(\frac{1.0278 - 0.999}{1.0278}\times 9.81\right)^{1/3}(20.87)^{5/3}(0.0123)^{-2/3} = 137$$

$$S_o' = 1.74(137) = 238$$

The density of the mixture has to be sufficiently greater than that of the overlying water to overcome the residual inertia of the rising plume. Furthermore, the edges of a rising plume become diluted more rapidly than the central core; consequently, the peripheral flow is stopped at a lower level, while the central core tends to rise higher [13]. Hansen [3] refers to two levels of plume rise: a "MAXI" level where jet density equals ambient water density, and a "MAXO" level (higher than the previous one) reached as a result of kinetic energy possessed in jet elements.

For Eqs. (8.43a, -b, and -c) to be valid, it is necessary to ensure that rising plumes do not significantly interfere or overlap. For this purpose, Hansen [29] has proposed the use of Cedervall's equation for determining the concentration profile in a plume assuming Gaussian distribution:

$$\frac{C}{C_m} = \exp\left(-\mu\, k\, \frac{r^2}{y^2}\right) \tag{8.43d}$$

where

C_m = maximum concentration along plume axis

C = concentration at distance r from axis

r = distance from plume axis (maximum value of r of concern in design equals half the distance between two adjacent ports)

y = height of rise from ports

μ and k = constants (with recommended values of $\mu = 1.0$ and $k = 80$)

Using the value of $Y_{max} = 21$ m, for instance, calculated in Example 8.10, if the orifices are spaced 8 m apart (i.e., r = 4 m), substitution in Eq. (8.43d) gives $C = 0.054 C_m$, namely, about 5% of the maximum concentration will be present at the edges of the plume. If this much overlap is tolerable, the orifices will not need to be spaced farther apart.

8.14.2   For Line Diffuser Discharge

For a line diffuser discharging in a stratified liquid, Brooks [17] and others have developed relations from theoretical considerations to enable the estimation of the height to which a plume would rise and the dilution it would give. His work has been done again on the assumption of a linear density profile in order to obtain generalized solutions. A direct computer solution can be found for irregular profiles normally occurring in seas or they can be approximated as linear ones with acceptable results, considering the fact that the results of calculations for stratified environments are probably no more accurate than ± 20%. Quantitative confirmation has so far been obtained only for round jets in laboratory tanks. Equations for line diffusers have not yet been validated quantitatively either in laboratory tanks or in the field [17]. Nonetheless, Brooks' recently modified equations are useful for design approximations and are given below:

1. Maximum height of rise of plume:

$$Y_{max} = 2.5(qg')\left(-\frac{g}{\rho_a} \times \frac{d\rho}{dy}\right)^{-1/2} \tag{8.44}$$

2. Centerline dilution at top of plume:

$$S_o = 0.089(qg')^{2/3} q^{-1}\left(-\frac{g}{\rho_a} \times \frac{d\rho}{dy}\right)^{-1/2} \tag{8.45}$$

3. Average dilution across plume:

$$S_o' = \sqrt{2}\,S_o \qquad\qquad\qquad\qquad (8.46)$$

in which $-(d\rho/dy)$ = rate of change of ambient density with respect to increasing depth.

When the density gradient is strong, the plume rises less and dilution
is reduced. Density gradients vary over time and consequently the dilutions
obtained also vary. As stated earlier, if stratification data are known,
the likely dilutions can be expressed statistically. Similarly, currents
have an effect on the trapping levels. The stronger the currents, the
deeper is the trapping level than that predicted by the above equations.

The above equations are valid for cases where the ports are provided
on both sides of an outfall pipe and not all on one side (see Example 8.11).

8.15 ESTIMATION OF INITIAL DILUTION

The plume dilution estimation methods presented thus far have assumed
a calm sea with no currents at all. Generally, currents do prevail and,
in fact, they are most desirable as it is only through currents that gross
exchanges of water occur. In the absence of any current, the continuous
release of wastewater into a stagnant seawater mass would only lead to a
remixing of the mixture and gradual buildup of high concentrations. Thus,
currents, however light, are essential for successful sea disposal.

The extent over which the current is effective depends on the magnitude
of the waste field generated, i.e., its width and depth. The dilution afforded by the seawater sweeping past this field can be simply stated as

$$S_d = \frac{m^3/\text{sec of seawater}}{m^3/\text{sec of wastewater}} = \frac{Ubh}{Q} \qquad\qquad (8.47)$$

where

   S_d = dilution resulting from displacement
   U = current velocity, m/sec
   b = width of field, m
   h = depth of field, m
   Q = wastewater discharge, m³/sec

The width b of the field is the one normal to the direction of current.
For a diffuser discharge it is approximately equal to the projected length

of the diffuser section normal to the current. For a single round jet it
is the width of the plume at the top (estimated from Figure 8.20 or from
Eq. (8.43d). For a set of several round jets it is the sum of the adjacent
plume widths.

The depth h of the waste field is difficult to estimate since it is
itself affected by the current velocity; the greater the velocity, the less
may be the value of h. In a quiet body of water it may be L/12 (Figure
8.13). However, with both surfacing and submerged fields a certain amount
of vertical mixing occurs as the field moves downcurrent, which increases
the effective value of h. According to Rawn et al. [13], values of 1/4 or
1/3 of the ocean depth are not considered unreasonable along the California
coastline. The density stratification pattern, no doubt affects the vertical spread of the field. Thus, a surface field may not be able to penetrate
below the thermocline, or a submerged one may not be able to rise above it
except at considerable distances downcurrent. Hansen [29] gives a method
for estimating the depth of the sewage field from the MAXI and MIDI depths.
The sewage field depth can also be estimated from Eqs. (8.48 and (8.50)
given by Brooks in connection with the reduction in dilution owing to the
so-called "cloud effect" at low current speeds.

In fact, when current speeds are very low, the waste field may hang like
a thick cloud over the outfall and reduce the effective height of rise of the
plume through uncontaminated seawater. This further reduces the value of S_o
as explained below, requiring a sort of correction factor to be applied to
take into account this "cloud effect" in estimating the initial dilution.

Brooks' method [17] of estimating the correction factor is based on
the reasoning that, for all buoyant plumes rising in stratified or unstrat-
ified environments, the dilution is approximately linear with height y.
Thus, with reference to Figure 8.23:

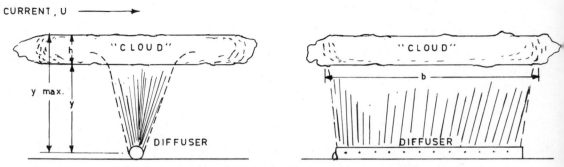

Figure 8.23 The reduction in dilution resulting from the "cloud
effect" in very slow currents (from Ref. 17).

$$\frac{S_y}{S_o} = \frac{y}{y_{max}} \qquad (8.48)$$

where

S_y = centerline dilution at height y (i.e., at bottom of waste field). This is the reduced dilution due to cloud effect. The plume continues to rise within the field itself but does not increase the net dilution.

y = height of rise through uncontaminated seawater.

The average dilution in the field at height y is $\sqrt{2}$ times the centerline concentration value, or

$$S_y{}' = \sqrt{2}\, S_y \qquad (8.49)$$

From Eq. (8.47) we could write

$$S_y{}'Q = Ubh = Ub(Y_{max} - y) \qquad (8.50)$$

Hence, substituting from Eq. (8.49) and dividing throughout by S_o, we get

$$\frac{S_y}{S_o} = \left[\frac{UbY_{max}}{\sqrt{2}\, QS_o}\right]\left(1 - \frac{y}{Y_{max}}\right) \qquad (8.51)$$

Let $p = (\sqrt{2}\, QS_o)/UbY_{max}$ in Eq. (8.51), and rewrite to give

$$\frac{S_y}{S_o} = \frac{1}{p}\left(1 - \frac{S_y}{S_o}\right)$$

or

$$S_y = S_o\left(\frac{1}{1 + p}\right) \qquad (8.52)$$

Again, the average dilution at level y may be obtained as $\sqrt{2}\, S_y$.

Thus, the term $1/(1 + p)$ constitutes a sort of correction factor to reduce the estimated plume dilution S_o and give a lower value S_y occurring as a result of blockage by the "cloud." For example, if $p = 1.0$, the effective dilution $S_y = S_o/2$. Brooks does not recommend the use of the above method of estimation if p is about equal to or greater than 2.0.

Under given site conditions for a proposed outfall design, the value of p can be estimated at the low current speeds likely to prevail in the critical season. The correction factor can then be computed and applied to the previously estimated value of S_o to find the centerline dilution S_y

at the bottom of the waste field. The average dilution at that level is
then obtained [Eq. (8.49)] and substituted in Eq. (8.50) to find the depth
of the sewage field h.

To find the <u>initial</u> dilution S_1, a trial and error solution may be
necessary, adjusting the value of b in order to have the value of the displacement dilution S_d (at the design current speed) approach as close as
possible the effective dilution S_y or S_o owing to plume rise. Where current
speeds are favorable, the initial dilution S_1 could well he higher than S_o.

The dilution resulting from displacement by currents may be considerable in some cases.* However, the reliability of occurrence of currents
must be taken into consideration and corresponding dilutions expected can
be expressed in statistical terms for summer and winter conditions. The
initial dilution for design purposes may then be taken as the one statistically likely to be equalled or exceeded a desired percentage of the time
(e.g., 80% of the time).

For specific discharge sites, graphical plots can also be developed
as shown in Figure 8.24, giving correlations between the current speeds

---

*At very high current speeds, a "blown plume" tends to occur and a modified
set of equations may need to be used to obtain a better estimation of the
dilution [8].

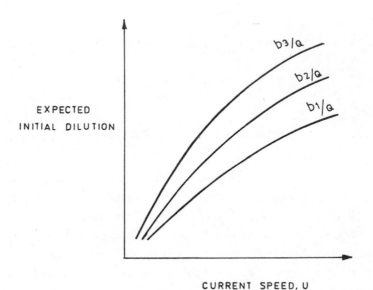

Figure 8.24 The relationship between current speed and expected initial dilution for different values of diffuser length b.

and initial dilutions obtainable for different values of diffuser length per unit discharge (b/Q). Example 8.11 illustrates how the initial dilution can be estimated from given data.

8.16 ESTIMATION OF DILUTION OWING TO DISPERSION

Lateral dispersion of the waste field occurs in horizontal travel after a plume surfaces, giving additional dilution. Factors affecting the turbulent dispersion coefficient ϵ and methods for its estimation were discussed in Sec. 8.8.4, together with some typical values in seas depending on their "scale" of phenomenon (Figure 8.13).

In order to estimate the dilution resulting from lateral dispersion, Brooks [18] considers three alternatives concerning the value of ϵ as the waste field travels horizontally with ever-increasing width b: either ϵ remains constant (conservative assumption) or it increases linearly with b, or it follows the "4/3 law." Depending upon how ϵ changes, he estimates the corresponding values of b. These are tabulated below as ratios introducing a dimensionless coefficient β.

ϵ/ϵ_0	1	b/b_0	$(b/b_0)^{4/3}$
b/b_0	$[1 + (2\beta x/b_0)]^{1/2}$	$1 + (\beta x/b_0)$	$\{1 + [(2/3)\beta x/b_0]\}^{3/2}$

in which

$$\beta = \frac{12\epsilon_0}{Ub_0} \tag{8.53}$$

where

   ϵ_0 = dispersion coefficient initially (when width equals b_0), cm^2/sec

   b_0 = initial width of field, cm (equal to projected length of diffuser in direction of alongshore currents. For onshore currents, b_0 may be assumed to be about 10 m or so).

   U = current velocity, cm/sec

   x = distance from outfall, cm

For each of the three cases, the dilution C_0/C can be estimated using "erf" (standard error function) values from mathematical tables:

For $\epsilon/\epsilon_0 = 1$,

$$\frac{c_0}{c} = \frac{1}{\text{erf}\sqrt{0.75/\beta\,(x/b_0)}} \qquad (8.54)$$

For $\epsilon/\epsilon_0 = b/b_0$,

$$\frac{c_0}{c} = \frac{1}{\text{erf}\sqrt{1.5/[1 + \beta\,(x/b_0)]^2 - 1}} \qquad (8.55)$$

For $\epsilon/\epsilon_0 = (b/b_0)^{4/3}$,

$$\frac{c_0}{c} = \frac{1}{\text{erf}\sqrt{1.5/[1 + 2/3(\beta)(x/b_0)]^3 - 1}} \qquad (8.56)$$

A graphical solution is given in Figure 8.25, from which the dilution factor can be readily estimated at any distance x from an outfall under given conditions. For example, an outfall has a diffuser section of 300 m normal to the direction of a current of 15 cm^2/sec, the initial value ϵ_0 is about 10^4 cm^2/sec [found by reference to Figure 8.13 or using Eq. (8.23).

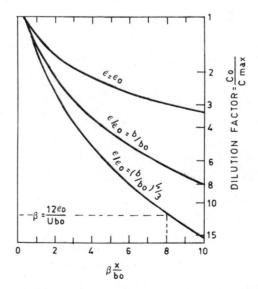

Figure 8.25 The dilution factors resulting from dispersion in horizontal travel (adapted from Ref. 9).

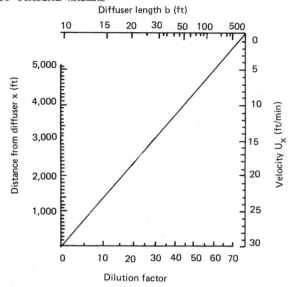

Figure 8.26 A nomograph for the estimation of dilution resulting from eddy diffusion in the horizontal travel of a waste field; based on the Four-Thirds law (adapted from Ref. 38).

Thus, $\beta = (12)(10^4)/(15)(300 \times 10^2) = 0.25$. If the dilution factor is to be estimated at a distance, for example, 30 times the diffuser length, $\beta(x/b_0) = 0.25 \times 30 = 7.5$. With this value of 7.5, we enter Figure 8.25 from which the dilution factor corresponding to the three different conditions can be read off as 3, 6, and 11.

Figure 8.26 gives a simple nomograph which can be used for estimating dilution in horizontal travel based only on the Four-Thirds law [38].

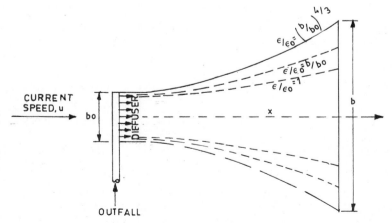

Figure 8.27 The extent of plume fanning in horizontal travel (adapted from Ref. 9).

Figure 8.27 illustrates how fanning out of the waste field occurs under the three different assumed conditions. The assumption that ϵ_0 remains constant even though the waste field spreads out gives, indeed, a conservative estimate of dilution in most cases. The choice, therefore, generally lies between the remaining two assumptions depending upon supporting evidence, if any, obtained at site [9].

The overall dilution obtained is the <u>product</u> of the initial dilution and the dispersion dilution (see illustrative Example 8.11).

8.17 A SIMPLIFIED PROCEDURE FOR THE PRELIMINARY DESIGN OF OUTFALLS

Based on little available data, engineers are often called upon to prepare, for initial consideration, very preliminary designs of outfalls to serve a community or industry. Some thumb rules can be proposed, especially for disposal of domestic sewage, although oversimplification can be dangerous in dealing with so complex a subject as disposal to sea. The designs so prepared can be useful only in assessing very approximate budgetary requirements and in comparing different waste disposal strategies.

A simplified design procedure is given below and can be followed in preparing preliminary designs of outfalls, not only for small and medium size communities but also for industrial discharges. A quick idea of the outfall size and the likely dilutions can be obtained with a minimum of site investigations. This should be followed by more detailed studies wherever the magnitude or nature of the wastewater so demands.

A tentative outfall site is first selected, generally depending on topography and the layout of the existing or proposed sewage system. Thereafter, the following data, which are considered the minimum necessary for preparing a preliminary design, must be obtained together with any data already available with the Navy, Oceanographic Institutes, etc.:

1. Wastewater characteristics (the average flow, its fluctuations, and its quality data).
2. Soundings along the proposed outfall alignment to obtain the ground profile.
3. Density profile, if possible, or at least the depth of the pycocline at the proposed site during the critical season (generally the summer).
4. Current velocities and directions at the surface and at middepth (or pycnocline depth) at the proposed site during the critical season. Supplementary information from float tracking may be useful.

Thus, it will be seen that current measurements are, generally, the major item of field work involved. Determination of T_{90} values for coliforms and the eddy diffusivity coefficient are not considered essential in all cases.

The simplifying assumptions on which a preliminary design can be made are outlined below step by step (also see Figure 8.28).

1. a. Estimate the outfall pipe diameter that would give a velocity of about 1.0 m/sec during average flow (see Sec. 8.18).

    b. Provide an outfall with a diffuser to get a maximum dilution and a likely "trapping" of the plume below or at the pycnocline. Thus, provide a diffuser length of about 2 m per 1000 m^3/day average wastewater flow. If sea currents are weak, provide even longer diffusers, for example, 3 m or more per 1000 m^3/day flow.

    c. Keep the discharge depth at a minimum of 20 m (or if impossible, extend the outfall up to 2 km) [30a].

    d. Align the outfall in such a manner that the diffuser is perpendicular to the direction of the prevailing current as far as possible. This is particularly useful when currents parallel to the shore are not very strong. A Y-shaped diffuser is advantageous for increasing the dilution when the currents are in the shoreward direction.

    e. Provide a minimum port diameter of 10 cm and a total port area approximately equal to the pipe area.

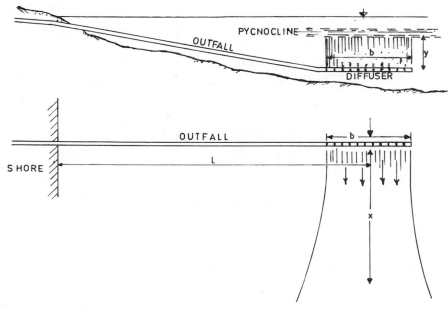

Figure 8.28 A design for small outfalls (see Example 8.11).

2. Assume that the plume will rise to a lower level of the pycnocline. (This assumption may not always be valid, especially if density strat-ification is weak or absent; more accurate formulae may be used, if desired.)

3. Select a middepth current velocity U, m/sec, that is likely to occur during the critical season. An average current speed at the outfall of 0.06 m/sec or more is desirable. At very low speeds, the rising plume is not readily displaced and tends to hang over the outfall like a cloud, thus reducing the initial dilution.

4. Estimate the initial dilution by a simple mass balance to give

$$D_1 = \frac{m^3/\text{sec seawater}}{m^3/\text{sec wastewater}} = \frac{Uby}{Q} \tag{8.57}$$

where

D_1 = initial dilution
U = current velocity in a direction perpendicular to diffuser, m/sec
b = diffuser length, m
y = height of plume rise, equal to half the distance from the diffuser to a lower level of the pycnocline, m (safe assump-tion applicable to small outfalls)
Q = average wastewater discharge, m^3/sec

As a rule of thumb, try to ensure that the initial dilution is about 100 in touristic and bathing areas and about 50 in other areas.

5. Estimate the dilution D_2 in horizontal travel based on a suitable value of the eddy diffusivity coefficient (Figure 8.25) and use Figure 8.26 assuming that the 4/3 law applies.

6. Estimate the coliform reduction factor D_3 based on a selected value of T_{90} from Figure 8.1, the assumed current velocity, and the distance of travel from the discharge point to about 100 m from the shoreline [Eq. (8.8)].

7. The concentration of all conservative substances in the wastewater will be reduced by a factor equal to $D_1 \times D_2$. Coliform reduction will occur by a factor equal to $D_1 \times D_2 \times D_3$.

8. Accumulation of heavy metals, radioactive isotopes, and nondegradable organics in fish and other seafoods can be estimated if the concentra-tions of these substances in seawater are first estimated and suitable "concentration factors" applied for the concerned species. However, often little data concerning local species may be available.

Example 8.11 Prepare a preliminary design for an outfall to serve a coastal town using the following data:

Population = 50,000 persons

Sewage flow = 200 liters per person per day (infiltration negligible)

Approximate depth of discharge = 20 m at 800 m from shoreline

Pycnocline depth = 9 m from surface

Middepth current speed (80 percentile) = 0.06 m/sec at right angles to the diffuser, and 0.05 m/sec in shoreward direction.

Beach coliform standard: 80 percentile not to exceed 1000 per 100 ml

Bacterial dieoff T_{90} = 2 hr

Raw waste coliforms = 2.25×10^7 per 100 ml

Velocity in outfall to be 1 m/sec at average flow

Solution:

1. Wastewater flow = 50,000 x 200 x $(1/10^3)$ = 10,000 m³/day = 0.1157 m³ per sec.

2. Pipe area = $\dfrac{0.1157 \text{ m}^3/\text{sec}}{1 \text{ m/sec}}$. Hence, pipe diameter = 38 cm or nearest.

3. Diffuser length: Providing 3 m length per 1000 m³/day flow, we need a total diffuser length b = 30 m.

4. Initial dilution D_1' with current at right angles to diffuser, from Eq. (8.47),

$$D_1' = \frac{(0.06 \text{ m/sec})(30 \text{ m})(0.5)(20 - 9 \text{ m})}{0.1157 \text{ m}^3/\text{sec}} = 85.5$$

5. Initial dilution D_1 with directly shoreward current:

$$D_1 = \frac{(0.05 \text{ m/sec})(10 \text{ m})(0.5)(20 - 9 \text{ m})}{0.1157 \text{ m}^3/\text{sec}} = 24$$

Note: Plume width is now assumed as 10 m only.

6. Dilution D_2 in shoreward travel to beach: Assume beach water quality standards have to be met at 100 m from shoreline, namely at 700 m distance from outfall. From Figure 8.26,

$$D_2 = 15$$

7. Overall dilution factor for conservative substances reaching beach:

$$D_1 \times D_2 = (24)(15) = 360$$

8. Bacterial reduction:

Dieoff rate K = $\dfrac{2.3}{T_{90}} = \dfrac{2.3}{2.0}$ = 1.15 per hr

Travel time to shore = $\dfrac{700 \text{ m}}{0.05 \text{ m/sec}}$ = 3.89 hr

Reduction factor $D_3 = 1/e^{-Kt}$ = 87.7

Overall reduction factor = $D_1 \times D_2 \times D_3$ = (24)(15)(87.7) = 31,572

Expected coliform concentration at beach = $\dfrac{2.25 \times 10^7}{31,572}$

$$= 712 \text{ per } 100 \text{ ml}$$

9. Diffuser ports: Use port diameter = 10 cm (area 78.5 cm²)

Total port area = pipe area. Number of ports = $\dfrac{1157 \text{ cm}^2}{78.5 \text{ cm}^2}$ = 8

Hence,

$$\text{Spacing} = \frac{30 \text{ m}}{8 - 1} = 4.28 \text{ m}$$

8.18 DESIGN OF OUTFALL SYSTEMS

Having decided upon the length of an outfall, type of discharge (single jet or line source), and other basic aspects in order to secure the desired dilution, one can take a closer look at the engineering design of the system as a whole. The essential considerations in such a system design include

1. Preliminary treatment and pumping (headworks)
2. Hydraulic design of outfall including diffuser
3. Construction and maintenance
4. Cost estimation

All other things being equal, the orientation of the outfall which gives optimum dilution is preferred. An alignment normal to the direction of the current is preferred although this is not essential. Also, generally, a straight length of diffuser is considered to be as efficient as a Y-shaped one [17].

8.18.1 Preliminary Treatment and Pumping (Headworks)

The considerations affecting the degree of pretreatment were discussed in Sec. 8.3. However, even where raw sewage is proposed to be discharged to the sea, some form of simple preliminary treatment such as screening and grit removal is considered desirable.

Pumping may be accomplished in one or two stages depending on heads and discharges. Flows greater than, for instance, twice the average may be discharged directly to the shoreline depending on economic considerations, extent of pollution by overflows, and other such factors. Sewer infiltration must be taken into account if substantial. Figure 8.29 gives a typical schematic diagram showing pumping arrangement and illustrating a few terms commonly used in connection with outfalls.

The total pumping head required for the high lift pumps is generally not more than 30 m, but depends on the following:

1. Friction losses (usually estimated on the assumption of n = 0.014 to 0.016 in Manning's formula or friction factor $f \cong 0.024$ for cement-lined outfalls).
2. Diffuser losses (can be calculated for a given diffuser, or estimated as a typical perforated pipe whose loss of head is 1/3 the loss corresponding to a regular pipe. As an approximation, diffuser loss can be taken as about 1 to 2 m or less).

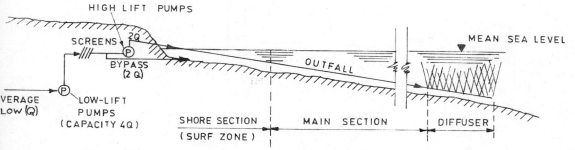

Figure 8.29 A typical schematic diagram of an outfall system.

3. Density head (equals the product of the average depth of water over a discharge point and $(\rho_s - \rho_0)/\rho_0$.

4. Tide difference between high tide and mean sea level.

5. Inlet losses (often on the order of 0.25 m or so).

Other hydraulic considerations must also be given due regard. Generally, the design has to be checked for water hammer, for which purposes a more conservative value of n may be used (for instance, 0.012) than above. Provision of a surge tower may be necessary.

8.18.2 Sizing of Outfall Including Diffuser

Velocities

The outfall diameter for carrying the average flow Q should be optimized, taking into account the capital costs of the pipe and the energy costs for pumping. For settled or secondary treated wastewater, the minimum velocity at design flow is restricted at about 0.5 m/sec, while for raw wastes (screened and degritted) the minimum is limited to about 0.70 m/sec. Velocities above 1 m/sec at peak flows tend to scour out settled material. Much higher velocities have been designed to facilitate operation and maintenance.

Along the diffuser, the pipe may need to be stepped down in diameter one or more times to avoid very low velocities as the flow progressively leaves the pipe (see Figure 8.30).

Arrangement and Sizing of Ports

A rule of thumb used in estimating the length of the diffuser section is to provide 1 to 2 m of diffuser length for every 1000 m^3/day average

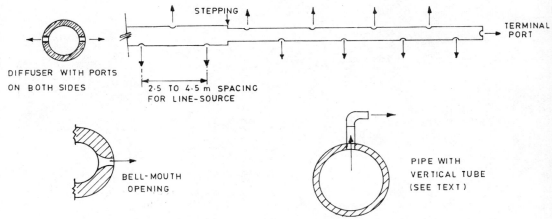

Figure 8.30 The layout of ports in a diffuser.

flow. This first approximation is then revised to suit local requirements of dilution.

To ensure that all ports flow full, the total area of ports should be less than the pipe area (preferably about 0.5 to 0.8 times the pipe area). Ports are generally circular, kept 0.15 to 0.25 m diameter (minimum 0.1 m diameter), and bell-mouthed inside for smooth discharge. Froude numbers larger than 1.0 must be ensured at all flow conditions. Jet velocities of about 3 m/sec are considered satisfactory. A diffuser designed for carrying a much larger flow in the future may need to be operated with a few ports blanked off in the early years.

Port spacing depends on whether a line source is desired or not. Where plumes are desired to rise without overlapping, the ports are spaced more than $Y/3$ apart as the plume cores are expected to widen that much in rising (Figure 8.20). If a line source is preferred, the ports must be spaced closer than above. They are then generally spaced 2.5 to 4.5 m apart. The trend is to use small diameter ports with spacing 20 to 30 times the port diameter.

The ports are generally made to discharge horizontally to benefit from their higher diluting capabilities. They are located at horizontal middepth on both sides of the pipe in zigzag fashion (Figure 8.30). Spacing limits then apply to adjacent ports to avoid interference.

Ports are not made to face upward for fear of sand and silt entry, especially where wastewater flow is intermittent. In the case of pipes which are buried or which are likely to be buried under silt or sand, an additional tube may be provided at each port as shown in Figure 8.30.

The terminal port diameter is kept 1.5 times the diameter of the other ports. In fact, it is preferable to provide it at invert level in a bulk-head which can be removed for cleaning.

Hydraulic Design of Diffuser Section

The aim in preparing a good hydraulic design is to see that the flow is divided as equally as possible between the ports. Once the port diameter and spacing are selected, the hydraulic calculations need to be started from the farthest point backwards. In general, the rate of discharge from a port is given by

$$q = C_d a \sqrt{2gH} \qquad\qquad (8.58)$$

where

q = discharge, m^3/sec

a = orifice area, m^2

H = total head, m (which is the sum of velocity head $v^2/2g$ and and static head ΔP)

C_d = coefficient of discharge

C_d is not a constant, but depends on how great a part of the total energy is kinetic energy <u>after</u> passage of the port. If ports are small ($< 1/10$ of pipe diameter) and pipe velocity is V,

$$C_d = 0.975 \left(1 - \frac{V^2}{2gH} \right)^{3/8} \qquad \text{for a smooth bell-mouth port}$$

$$C_d = 0.63 - 0.58 \frac{V^2}{2gH} \qquad \text{for a sharp-edged port}$$

The design procedure [13] is given below with reference to the notations used in Figure 8.31. Starting with the farthest port, with port diameter = d, find H_1 to give a required discharge q_1. For this port, assume

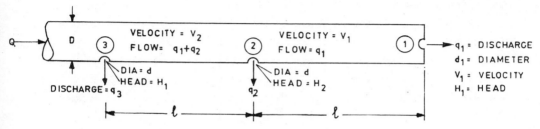

Figure 8.31 A definition sketch for diffuser design (see text for explanation).

C_{d1} = 0.91 or 0.61, depending on the type of port. Now,

$$v_1 = \frac{q_1}{\pi d_1^{\,2}/4} \quad\text{and}\quad V_1 = \frac{q_1}{\pi D^2/4}$$

Proceeding to port 2, there is an increase in total head due to

1. Friction loss,

$$h_f = f\frac{\ell}{D}\,\frac{V_1^{\,2}}{2g}$$

2. Rising bottom of the sea. The static head difference between ports 1 and 2 adjusted for density is

$$h_y = (Y_1 - Y_2)\left(\frac{\rho_s - \rho_0}{\rho_0}\right)$$

Thus, $H_2 = H_1 + h_f + h_y$. Knowing H_2 and V_1, the value of C_d applicable to port 2 (i.e., C_{d2}) can be found using the above formulae and q_2 can also be found from Eq. (8.58):

$$q_2 = C_{d2}a\sqrt{2gH_2}$$

Since H_2 is slightly greater than H_1, the value of q_2 will also be somewhat greater than that of q_1.

The pipe velocity $V_2 = (q_2 + q_1)/(\pi D^2/4)$. In a similar fashion, the discharge from other upstream ports is found step by step, checking finally that $\Sigma q = Q$. It will be evident that where ports of uniform diameter are provided, the discharge from each succeeding upstream port will be slightly larger. This can often be tolerated up to some maximum difference of, for instance, 10 to 20% between the discharge of the first and last ports. In long diffusers, the orifice diameter may need to be changed at intervals so as not to exceed the tolerable difference. Ideally, there should be no difference in discharge between the orifices. This would require the diameter to be varied from orifice to orifice.

Furthermore, if the diffuser is laid on a sloping sea bottom, it is impossible to achieve uniform distribution at _all_ flow rates. Brooks suggests that it is advisable to aim at a fairly uniform distribution at low or medium flow conditions. If the flows in the early years of an outfall are expected to be much less than the design flows, a few ports may just be blocked off.

Example 8.12 also illustrates a diffuser design. A slightly different approach is followed in order to provide a larger diameter orifice in the end bulkhead than in the other orifices. Computer programs can be readily developed for diffuser designs.

8.18.3 Construction and Maintenance

Outfall construction is generally a specialized type of work and not possible to describe in any great detail here. Construction techniques depend upon the nature of the sea bed, the depth at which the outfall is to be laid, the diameter and type of pipe to be laid, the extent of anchorage needed, and other such factors. • Techniques are also affected by climatic conditions, length of working season, availability of skilled labor, and equipment.

Some of the construction methods used include the following [39-41]:

1. Pipe by pipe
2. Float and drop
3. Bottom tow
4. Lay barge
5. Reel-off barge
6. On-site extrusion

In the pipe-by-pipe method, the pipes are laid on the sea bed or in a trench one by one and divers are sent down to join them up. Thus, mechanical joints (flanges or push-in type) are needed. The bottom soil should also be such that a trench can be maintained while work is in progress. Speed of work depends on weather and sea conditions. Under favorable conditions this may be the cheapest method of all and requires a minimum of construction equipment.

The float-and-drop method is, in fact, an extension of the one just described above. Instead of laying pipe by pipe, longer sections, each consisting of a few pipe lengths (or the entire length) may be floated out, positioned over the previously excavated trench, and lowered into it with the sections again jointed together by divers. It is difficult to prevent some bending of pipes during floating and laying and, hence, the pipe material selected should be capable of withstanding the expected stresses. Favorable conditions for adoption of these construction methods are a shallow and calm sea and a reasonably long working season.

The bottom-tow method is relatively more sophisticated and involves pulling a string of pipes into a preformed trench. Several pipes at a time are prepared together with their protective coating, aligned at the shore in the required direction, welded together to form a section, protective coatings finished over the joints, and then pulled into the trench with the help of a powerful winch mounted on a boat until the tail of the section arrives at the waterline. The next section is then assembled, welded to the previous one, and pulled in farther until, section by section, the whole length is pulled into position. Steel pipes are required owing to stresses generated in laying. The pipes are suitably coated, as discussed below, with reinforced concrete often being preferred for outside coating since its weight helps to get the desired amount of negative buoyancy. The bottom-tow method is often faster and preferred for nearshore work such as in outfalls, although more specialized equipment is essential. An important advantage is that navigation is unaffected, divers' work is minimum, and construction progress is less affected by sea conditions.

The lay-barge method is relatively the most sophisticated of all methods. All pipe jointing, coating, and laying is done from a large barge instead of on land. The method is, therefore, economical only for very long lengths as used in the oil industry and is presently used for pipes up to 125 cm diameter only [39]. The pipe trench may be precut or formed by jetting at high pressure.

With plastic pipes (flexible and without joints), either the pipe of required length can be mounted on a large diameter reel placed horizontally on a barge and unwound as the barge moves away from the shoreline as in the reel-off barge method (limited to 25 cm diameter pipes), or the pipe itself may be produced by on-site extrusion to give a single, jointless pipe of required length and diameter (up to 1 m diameter) from extrusion equipment temporarily erected at the shoreline. In either case, anchoring is needed to counteract buoyancy. Construction can be rapid and corrosion protection unnecessary. Jointed pipes are also used.

Generally, for diameters about 1 m or larger, reinforced cement concrete (RCC) pipes are preferred. Their heavy weight helps (it may be desirable to have a negative buoyancy of 5 to 10 kg/m length depending on currents, winch capacity, etc.) and they can be designed for specific foundation conditions. They are protected from inside with coal tar. Extended bell-type joints with neoprene rubber double ring gaskets can provide flexibility. Pipes of about 1 m or less in diameter may be fabricated

using bituminous coated steel with internal cement or epoxy coal tar and
an external reinforced concrete cover. High density polyethylene pipes
with butt fusion welding have also been used in recent times for increasingly larger diameter outfalls. In all cases, sufficient anchorage is
needed in the form of concrete blocks to hold down the pipes. Alternatively, where geological conditions permit, the outfall may be created
just by tunneling through rock at shallow depth.

In the surf zone, pipelines are generally buried so as not to be disturbed by wave action (see Figure 8.32). Beyond this zone they may be laid
only partly buried as shown in the figure unless obstruction to navigation
is feared. The design must take into account instabilities caused by erosion, shifting sands, currents and wave action, possible foundation fail-
ures, and earthquake forces. Thus, a variety of local conditions must be
taken into account and extensive site surveys must be conducted if necessary.

Figure 8.32 shows alternative ways in which outfalls may be laid.
Burial of the pipeline in the surf zone may entail dredging a trench in the
natural sea bed until a water depth about 7 to 10 m or so is reached. The
pipeline may then be laid either on a bedding or directly on the trench
floor and covered over to a depth of about 2 m over the pipe top. Sheet
piling may be provided on either side to ensure that the covering material
and pipe are not disturbed or eroded away by currents. Alternatively, where
rock is available near the surface, the pipeline may be anchored to it using
reinforced concrete blocks.

The crown elevation of the land portion of the outfall pipe is kept
above the mean sea level by not more than the minimum differential head
between the pipe inside and outside. Also, to prevent air from entering
the outfall, the portion between pump station and shore is kept low enough
to ensure that the line will remain full of water at zero flow. A manhole
(with an air-release valve near it) is provided in the shore section. This
manhole is generally extended about 1.5 to 2.0 m above sea level to enable
it to be opened without spillage owing to hydrostatic pressure differentials.

Often the surf section is more difficult to lay since water depth is
not enough for floating equipment to be used and work has to be done from
trestles or gantries placed on a temporary jetty. Thus, this section may
need to be costed differently from the rest of the outfall.

Beyond the surf zone, namely, in the main and diffuser sections, the
pipe is generally laid only partly buried as stated earlier. A shallow
trench may be made and a bedding made in it for the pipe. After the pipe

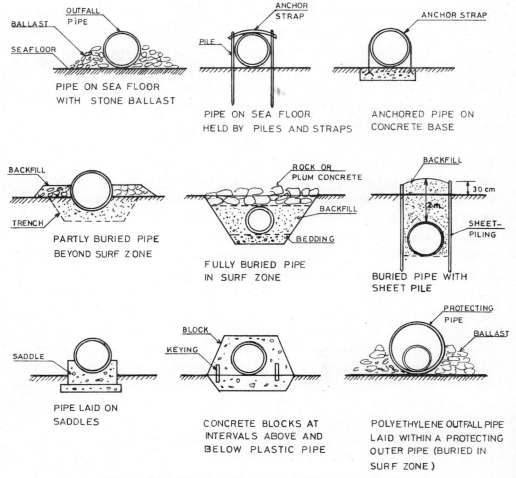

Figure 8.32 Alternative ways of laying outfalls.

is laid, it may be covered over (using excavated material) to either half or full depth as desired (see Figure 8.32). Of course, the diffuser section is only half buried when jets are to discharge from the side.

Trench cutting methods involve the use of either mechanical or hydraulic dredges. Among mechanical dredges, three types are used: the dipper,

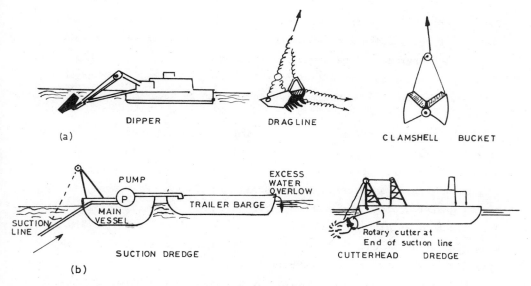

Figure 8.33 Different types of dredge for trench cutting: (a) mechanical dredges, and (b) hydraulic dredges.

the dragline, and the clamshell bucket. Hydraulic dredges include the plain suction and cutterhead types (see Figure 8.33).

The dipper is like an excavating shovel mounted on a boat. It is useful for removing hard materials after blasting is done. However, its use is generally limited to calm, protected waters. The dragline is essentially an open bucket with teeth which can be lowered into position over the trench and dragged by cables from a ship, or from the shoreline. The dredged materials are barged away. The dragline is simple, relatively low in capital cost, and can be used in deep waters. The depth of cut, however, is not easy to control. The clamshell bucket has jaws which open and close. It can bite into several tons of material at a time, but is mainly suitable in softer sea beds. It can work at a variety of depths (20-60 m). The dredged material may be either placed beside the trench for later backfilling or barged away to another site.

A suction dredge, being of the hydraulic type, is good for work with soft materials and can be used at depths of up to 20 m. It is preferred for use in seas with heavy navigational traffic because of its relatively fast speed of work. The suction line is lowered from the main vessel, and the material pumped up is either transferred to a trailer barge or held in the main vessel itself for later discharge elsewhere by bottom dumping.

The cutterhead dredge is similar in principle but has a rotary cutter at the end of the suction line enabling it to cut into a variety of soils (soft soils, clay, hardpan, limestone, and even soft basalt). Thus, this type of dredge is considered more efficient and versatile. Small units can work in water as shallow as 1 m, while larger units can be used in waters up to 60 m deep [40].

Acceptance tests for construction and performance need to be specified in the contracts. It is desirable to blind-flange the main pipe and test it hydraulically before proceeding with the diffuser section. A check for alignment (kinks) and burial depth is also necessary. Of course, all materials, gaskets, etc., are checked before use and, where pipes are welded, a 100% check of welding is necessary. A similar check for the pumping station must also be included. Finally, the outfall position should be clearly notified to the navigational authorities.

Performance tests may sometimes be demanded by a control agency. These may require the discharger to be responsible for satisfactory statistical data on E. coli observed at the shoreline, and on the number of occasions when surfacing of plume occurs compared to predictions.

Maintenance of an outfall essentially consists of occasional cleaning for which a suitable provision must be made. This may be particularly necessary where raw, unsettled wastewater is discharged. Cleaning can be done mechanically using devices similar to those used in cleaning sewers or water mains. Some cleaning can also be achieved by pumping seawater through the outfall at a rate high enough to flush out materials at a high velocity. Maintenance of pumping and preliminary treatment facilities, however, may constitute the major items of work and operating costs in a sea disposal system.

Shellfish may enter a diffuser and clog it unless ports are blocked off during prolonged stoppage of pumping.

Occasionally, a dragging anchor from a ship may get entangled with the outfall pipe and cause damage. Cracked pipes may result in much leakage, and even significantly reduced dilutions compared to design values. More serious problems may be caused by liquefaction of bottom support material (for example, due to earthquakes or strong waves) and failure of unsupported pipe. Erosion of soil may also lead to pipe failure. Wave forces during hurricanes have been known to move pipes bodily [40].

8.18.4 Costs

As costs can vary substantially from one site to another, only a rough indication can be given for outfalls of different diameters. The average life of an outfall is on the order of about 60 years, while that of pumps and other equipment is about 15 years. Annual operating and maintenance costs may be around 0.10 to 0.30% of construction costs, excluding repayment of a loan which may, in fact, be the major item of annual costs. In Spain, power requirement is reported to average about 0.05 kWh/m^3 flow [30a].

Example 8.12 A petrochemical complex proposes to discharge its wastewaters to the sea through an outfall provided with multiport diffusers. From the data given below, make a preliminary design for a discharge at point C (Figure 8.34) and

1. Find the overall dilution obtainable at 3 km from the outfall in the direction of the currents
2. Specify if any pretreatment is necessary
3. Prepare the hydraulic design for the diffuser up to 50 m from the endpoint
4. Estimate the pumping head

Data

1. Wastewater Characteristics

    Average flow = 3 m^3/sec (low = 2.0 m^3/sec; peak = 3.5 m^3/sec)

    CN^- = 10 mg/liter (allowable in seawater = 0.01 mg/liter at 3 km downstream

    Phenol = 20 mg/liter (allowable in seawater = 0.002 mg/liter at 3 km downstream)

    Oil and grease (will be removed using "best possible practice")

    E. coli = 0.3 x 10^8/100 ml (due to municipal sewage at 20% total flow)

    ρ_0 = 1.000

2. Site Characteristics

    The proposed site and its profile are shown in Figure 8.34. Discharging the waste at a point closer than C is not considered desirable and topography indicates that beyond point E the outfall cost will rise sharply.

3. Other Survey Data

    a. Temperature, salinity, and density distributions in the critical season are given in Figure 8.34.

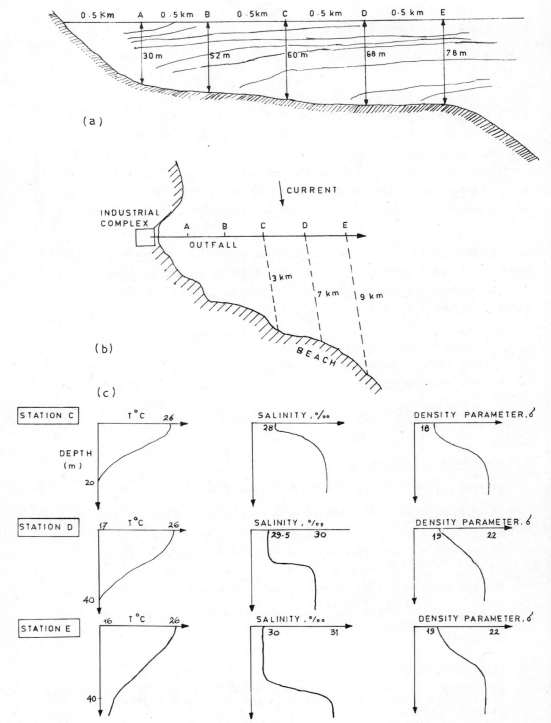

Figure 8.34 Some survey data pertaining to a proposed outfall site; (a) topography and density profile along proposed alignment, (b) site plan, and (c) details at different stations (see Example 8.12).

416

b. Current speed for design purposes = 7.5 cm/sec in a direction normal to the outfall, at trapping level. (No onshore currents parallel to outfall alignment.)

c. E. coli dieoff rate constant K_b = 0.676/hr (t_{90} = 3.4 hr).

d. Difference in elevation between high and mean sea level = 0.30 m.

e. Eddy dispersion coefficient may be taken in accordance with Figure 8.13 and assumed to follow the 4/3 law.

SOLUTION

Outfall diameter

As wastewater may be only screened and degritted, provide a minimum velocity = 0.80 m/sec at low flow.

$$\text{Pipe area} = \frac{2.0 \text{ m}^3/\text{sec}}{0.80 \text{ m/sec}} = 2.5 \text{ m}^2$$

Diameter D = 1.78 m

$$\text{Velocity at peak flow} = \frac{3.5 \text{ m}^3/\text{sec}}{2.5 \text{ m}^2} = 1.4 \text{ m/sec (acceptable)}$$

Length of diffuser section

As a trial, provide 1.5 m length/1000 m³-day. Hence,

$$\text{Diffuser length needed} = \frac{3.0 \text{ m}^3/\text{sec} \times 86,400 \times 1.5}{1000} = 388 \text{ m}$$

Take it as 396 m, such that 100 ports at 4.0 m apart will form the diffuser section. There will be 8.0 m between consecutive ports on each side of the diffuser since the ports will be staggered.

$$\text{Discharge per unit length } q = \frac{3.0 \text{ m}^3/\text{sec}}{396 \text{ m}} = 0.0075 \text{ m}^2/\text{sec}$$

Height of plume rise

At point C,

Discharge depth = 60 m

ρ_a = 1.0218 ρ_0 = 1.000

Thus,

$$g' = g\left(\frac{\rho_a - \rho_0}{\rho_a}\right) = 9.81\left(\frac{1.0218 - 1.000}{1.0218}\right) = 0.209 \text{ m/sec}^2$$

and the rate of change of ambient density with depth below the pycnocline (since submerged plume is likely) is

$$- \frac{d\rho}{dy} = \frac{1.0215 - 1.0218}{60 - 30} = 10^{-5} \text{ per meter}$$

From Eq. (8.44),

$$Y_{max} = \frac{2.5(qg')^{1/3}}{\left(- \dfrac{g}{\rho_a} \dfrac{d\rho}{dy} \right)^{1/2}}$$

Hence,

$$Y_{max} = \frac{2.5(0.0075 \times 0.209)^{1/3}}{\left[\dfrac{9.81}{1.0218} (10^{-5}) \right]^{1/2}} = 29.7 \text{ m from discharge level}$$

The plume is submerged below the pycnocline.

Dilution in vertical rise of plume

From Eq. (8.45),

$$s_o = \frac{0.89(qg')^{2/3}}{q\left(- \dfrac{g}{\rho_a} \dfrac{d\rho}{dy} \right)^{1/2}} = 160 \text{ (at centerline top)}$$

or

$$S_o = \frac{0.89(0.0075 \times 0.209)^{2/3}}{0.0075\left(- \dfrac{9.81}{1.0218} 10^{-5} \right)^{1/2}}$$

and

$$S_o' = \sqrt{2}\, S_o = 226 \qquad \text{(average dilution)}$$

Likely reduction in dilution S_o due to "cloud effect

From Eq. (8.52),

$$S_y = S_o\left(\frac{1}{1 + p} \right) \qquad \text{where} \qquad p = \frac{2QS_o}{UbY_{max}}$$

Now

$$p = \frac{(2)(3 \text{ m}^3/\text{sec})(160)}{(0.075 \text{ m/sec})(396)(29.7 \text{ m})} = 0.76$$

Hence,

$$S_y = 160\left(\frac{1}{1 + 0.76} \right) = 91 \qquad \text{(centerline)}$$

and the corresponding value of average dilution = $\sqrt{2} \times 91 = 128$.

Depth of plume (cloud)

From Eq. 8.50,

$$S_y' Q = Ubh \qquad \text{where} \qquad h = Y_{max} - y$$

Thus,

$$h = \frac{(128)(3 \ m^3/sec)}{(0.075 \ m/sec)(396 \ m)} = 12.92 \ m$$

Dilution due to displacement by current

From Eq. (8.47),

$$S_d = \frac{Ubh}{Q}$$

or $\quad S_d = \dfrac{0.075 \ m/sec \times 396 \ m \times 12.92 \ m}{3.0 \ m^3/sec} = 130$

This value is practically equal to the value of 128 estimated above after accounting for the "cloud effect." Hence, adopt 130 as the value of initial dilution. (Note: Other diffuser lengths can be tried to see if the initial dilution can be optimized.)

Dilution due to transverse dispersion

From Eq. (8.9),

$$\epsilon_0 = 0.01(b)^{4/3} = 0.01(396 \times 10^2)^{4/3} = 13,675 \ cm^2/sec$$

$$\beta = \frac{12\epsilon_0}{Ub_0} = \frac{12 \times 13,675}{7.5 \times (396 \times 10^2)} = 0.547$$

At a distance of 3.0 km from the outfall,

$$\frac{\beta x}{b_0} = \frac{0.547 \times 3000}{396 \ m} = 4.1$$

Thus from Figure 8.25, the dilution factor = 5.5.

Overall dilution for conservative substances

Overall = (130)(5.5) = 715

CN^- and Phenol Concentrations

Assuming CN^- and phenol behave like conservative substances, their concentrations will be reduced by dilution. Thus, at 3 km from the outfall in the direction of the current, their concentrations will be

$$CN^- = 10 \text{ mg/liter} \times \frac{1}{715} = 0.014 \text{ mg/liter}$$

$$\text{Phenol} = 20 \text{ mg/liter} \times \frac{1}{715} = 0.028 \text{ mg/liter}$$

While the CN^- value is fairly close to the specified water quality requirement (0.01), the phenol value is not. The specified limit for phenol is 0.002 mg/liter; thus the allowable concentration in the wastewater discharge is 0.002 x 715 = 1.43 mg/liter. Hence the percent removal necessary = (20 - 1.43)/20 = 93%. This calls for a fairly high degree of phenol removal, which may or may not be acceptable (the distance of 3 km from the outfall is arbitrarily fixed in this case for meeting standards). Fortunately, no currents prevail in the direction of the shore near the outfall. It may be desirable to extend the outfall farther and ensure more dilution rather than to treat the wastewater, depending on relative costs.

E. coli Concentrations

The time of travel in a critical season from the outfall to the beach 3 km away in the direction of the current is

$$t = \frac{3000 \text{ m}}{0.075 \text{ m/sec}} = 11.1 \text{ hr}$$

$$\frac{C_t}{C_0} = e^{-K_b t} = e^{-0.676(11.1)} = 5.51 \times 10^{-5}$$

Hence, the original concentration of $0.3 \times 10^8/100$ ml will now become

$$(0.3 \times 10^8)\left(\frac{1}{715}\right)(5.51 \times 10^{-4}) = 23 \text{ per 100 ml}$$

This is well below the requirements of beach standards.

Hydraulic Design of Diffuser

Low flow is chosen for design purposes: Pipe diameter = 1.78 m, number of ports = 100; average port discharge at low flow = (2.0 m³/sec)/100 = 0.02 m³/sec.

Port No. 1 (farthest port in diffuser)

Assume the total head in the main flow at the farthest port as $N_1 = 0.1$ m. To avoid deposition at the bulkhead region, choose $q_1 = 4$ x average per port = 4 x 0.02 = 0.08 m³/sec. Assume $C_d = 0.91$ (bell-mouth ports).

$$a_1 = \frac{q_1}{C_{d1}\sqrt{2gH_1}} = \frac{0.08}{0.91\sqrt{2 \times 9.81 \times 0.1}} = 0.063 \text{ m}^2$$

$$d_1 = \sqrt{\frac{4(0.063)}{\pi}} = 0.283 \text{ m}$$

Checking for Froude number we get

$$N_{Fr} = \frac{q_1}{a_1\sqrt{\left(\frac{\rho_a - \rho_0}{\rho_a}\right)gd_1}} = \frac{0.08}{0.063\sqrt{\left(\frac{1.0218 - 1.000}{1.000}\right)9.81 \times 0.283}} = 5.19$$

This acceptable as it is > 1.0.

Velocity in pipe $V_1 = \dfrac{0.08 \text{ m}^3/\text{sec}}{2.5 \text{ m}^2} = 0.032 \text{ m/sec}$

(This can be increased if the pipe diameter is reduced in a later part of the diffuser.)

Velocity head $\dfrac{V_1^2}{2g} = \dfrac{(0.032)^2}{2 \times 9.81} = 5 \times 10^{-5} \text{ m}$

Friction factor $f = 0.024$ (for instance)

Distance ℓ to next port $= 4.0 \text{ m}$

Head loss $h_{f1} = f\dfrac{\ell}{D}\left(\dfrac{V_1^2}{2g}\right) = \dfrac{0.024 \times 4.0 \times (5 \times 10^{-5})}{1.78}$

$$\cong 0.00 \text{ (negligible)}$$

Decrease in depth $\Delta y = 4.0 \text{ m}\left(\dfrac{68 \text{ m} - 60 \text{ m}}{500 \text{ m}}\right) = 0.064 \text{ m}$

Head $= h_y = \left(\dfrac{\rho_a - \rho_0}{\rho_0}\right)\Delta y = \dfrac{21.80}{1000} \times 0.064 = 0.0014 \text{ m}$

Total head $H_2 = H_1 + h_f + h_y = 0.10 + 0.0014 = 0.1014 \text{ m}$

$\dfrac{V_1^2}{2gH_2} = \dfrac{5 \times 10^{-5}}{0.1014} \cong 0.0005$ (hence no change in C_d value)

Port No. 2

Choose $a_2 = \dfrac{a_1}{4} = \dfrac{0.0628}{4} = 0.0157 \text{ m}^2$

$d_2 = 0.141 \text{ m}$ (no need to check N_{Fr} again)

Now,

$$q_2 = C_{d2}a_2\sqrt{2gH_2} = 0.91 \times 0.0157\sqrt{2 \times 9.81 \times 0.1014} = 0.0201 \text{ m}^3/\text{sec}$$

$$V_2 = \frac{0.0201 + 0.08 \ m^3/sec}{2.5 \ m^2} = 0.04 \ m/sec$$

Progressing similarly as for port 1 we get

$$h_{f2} \cong 0.00 \ (\text{negligible}), \quad h_y = 0.0014$$

$$H_3 = 0.1014 + 0.0014 = 0.1028 \ m$$

Again,

$$\frac{V_2^2}{2gH_3} \cong 0.0005 \ (\text{no change in } C_d \text{ value})$$

Ports 3 to 100

Proceed as above. See Table 8.12 for calculated values for 14 ports from the end. It will be observed from column 9 that the discharge increases gradually from port to port. The diameter of the ports may be varied after a difference of, for instance 10% or more is reached in the discharge.

Estimation of Pumping Head

Peak friction head in outfall = $h_f = \dfrac{fLV^2}{2gD}$

$$= \frac{(0.024)(1500 \ m)(1.4 \ m/sec)^2}{(2)(981)(1.78 \ m)}$$

$$= 2.02 \ m$$

Head due to outfall elevation = $\dfrac{21.80}{1000} \times 60 \ m = 1.3 \ m$

Peak friction head in diffuser = $h_{fd} = \dfrac{1}{3}\left(\dfrac{fLV^2}{2gD}\right)$

$$= \frac{1}{3}\left[\frac{(0.024)(396)(1.4)^2}{(2)(9.81)(1.78)}\right] = 0.18 \ m$$

Head $h_y = \dfrac{21.80}{1000} \times 396 \ m \times \left(\dfrac{68 - 60}{500}\right) = 0.14 \ m$

Head at port no. 1 assumed above = 0.10 m

Head owing to tide difference = 0.30 m

Head owing to inlet and other minor losses = 0.10 m (for example)

Total pumping head = 2.02 + 1.3 + 0.18 + 0.14 + 0.10 + 0.30 + 0.10

$$= 4.14 \ m$$

Note: This is not a complete design but an illustrative one giving some of the typical computations involved. As stated earlier, the length of

Table 8.13 Diffuser Design (Example 8.12)

(1) Port number	(2) Distance from end (m)	(3) Port diameter d_n (m)	(4) Port area a_n (m²)	(5) Pipe diameter D (m)	(6) Total head E_n (m) [from (18)]	(7) Ratio: (12)/(6)	(8) Discharge coefficient C_d	(9) Port discharge j_n (m³/sec) [(3)(4)$\sqrt{2g(6)}$]	(10) Increment in pipe velocity Δy (m/sec) [(9)/A]
1	0.0	0.883	0.0628	1.48	0.10	0	0.91	0.08	0.032
2	4.0	0.141	0.0157	1.48	0.1014	0	0.91	0.0201	0.008
3	8.0	0.141	0.0157	1.48	0.1028	0	0.91	0.0208	0.0081
4	13.0	0.141	0.0157	1.48	0.1042	0	0.91	0.0205	0.0082
5	16.0	0.141	0.0157	1.48	0.1056	0.0015	0.91	0.0206	0.0082
6	20.0	0.141	0.0157	1.48	0.1070	0.002	0.91	0.0207	0.0083
7	24.0	0.141	0.0157	1.48	0.1084	0.0025	0.91	0.0209	0.0083
8	28.0	0.141	0.0157	1.48	0.1098	0.0031	0.91	0.0210	0.0084
9	22.0	0.141	0.0157	1.48	0.1112	0.0036	0.91	0.0211	0.0085
10	26.0	0.141	0.0157	1.48	0.1126	0.0042	0.91	0.0212	0.0086
11	40.0	0.141	0.0157	1.48	0.1140	0.0045	0.91	0.0214	0.0086
12	44.0	0.141	0.0157	1.48	0.1154	0.0057	0.91	0.0215	0.0086
13	48.0	0.141	0.0157	1.48	0.1168	0.0066	0.91	0.0215	0.0086
14	52.00	0.141	0.0157	1.48	0.1182	0.0076	0.91	0.0217	0.0087

Note: Columns are numbered in parenthesis to explain derivation of values in individual columns.

Table 8.13   (Continued)

(11) Pipe velocity V_n (m/sec)	(12) Velocity head (m)	(13) Friction factor f	(14) Distance to next port L_n (m)	(15) Head loss h_m (m) [(12)(13)(14)/(5)]	(16) Increase in depth (m)	(17) Density head (m)	(18) Total head E_n (m) [(6) + (15) + (17)]	(19) Froude number
0.032	0.00086	0.024	4.0	0.000	0.064	0.0014	0.1014	5.19
0.040	0.00008	0.024	4.0	0.000	0.064	0.0014	0.1028	7.36
0.048	0.00012	0.024	4.0	0.000	0.064	0.0014	0.1042	>1
0.055	0.00016	0.024	4.0	0.000	0.064	0.0014	0.1056	>1
0.064	0.00021	0.024	4.0	0.000	0.064	0.0014	0.1070	>1
0.072	0.00027	0.024	4.0	0.000	0.064	0.0014	0.1084	>1
0.080	0.00033	0.024	4.0	0.000	0.064	0.0014	0.1098	>1
0.088	0.00040	0.024	4.0	0.000	0.064	0.0014	0.1420	>1
0.096	0.00047	0.024	4.0	0.000	0.064	0.0014	0.01126	>1
0.105	0.00055	0.024	4.0	0.000	0.064	0.0014	0.1140	>1
0.114	0.00064	0.024	4.0	0.000	0.064	0.0014	0.1154	>1
0.123	0.00077	0.024	4.0	0.000	0.064	0.0014	0.1168	>1
0.132	0.00088	0.024	4.0	0.000	0.064	0.0014	0.1182	>1
0.141	0.00101	0.024	4.0	0.000	0.064	0.0014	0.1197	>1

Note: Columns are numbered in parenthesis to explain derivation of values in individual columns.
Port numbers for this page are as given on the preceding page.

the outfall may need to be increased to obtain better dilution characteristics.

REFERENCES

1. A. T. Ippen, Estuary and Coastline Hydrodynamics, McGraw-Hill, N.Y., 1966.

2. H. Sverdrup, M. Johnson, and R. Fleming, The Oceans, Prentice Hall, N.J., 1963.

3. The control of pollution in coastal waters, WHO Interregional Training Course, Denmark, 1970.

4. Health hazards of the human environment, WHO, Geneva, 1972.

5. G. Fair, J. Geyer, and D. Okun, Water and Wastewater Engineering, Vol. 2, John Wiley and Sons, N.Y., 1968.

6. Mathematical and hydraulic modelling of estuarine pollution, Water Pl. Res. Tech. Paper 13, H.M. Stationary Office, London, 1972.

7. R. Koh and L. Fan, Mathematical models for prediction of temperature distribution due to discharge of heated water into large bodies of water, EPA-WQO Contract No. 14-12-570, U.S. Government Printing Office, 1970.

8. J. Buhler, Ph.D. thesis, University of California, Civil Eng. Dept., 1974.

9. N. Brooks, Waste Disposal in Maine Environment, Proceedings, International Conference, Pergamon Press, N.Y., 1959.

10. E. Burgess, Airphoto analysis of ocean outfall dispersion, Program 16070 ENA for EPA-USA, School of Eng., Oregon State University, 1971.

11. An investigation of the efficacy of submarine outfall disposal of sewage and sludge, Publication No. 14, State Water Poll. Cont. Bd., Sacramento, California.

12. G. Kramer, Predicting reaeration coefficient for pollution estuary, Jour. ASCE, 100, 1974.

13. A. Rawn, E. Bowerman, and N. Brooks, Diffusers for disposal of sewage in sea water, Proc. ASCE, 86, 1960.

14. L. Fan and N. Brooks, Numerical solutions of turbulent buoyant jet problems, Report KH-R-18, California Inst. Tech., Pasadena, 1969.

15. Y. Argaman, M. Vajda, and N. Galil, Use of jet pumps in marine waste disposal, Jour. ASCE, 101, 1975.

16. Hansen, International Report No. 15, Danish Isotops Centre, 1967.

17. N. Brooks, Dispersion of hydrologic and coastal environments, Report prepared for U.S. EPA, Grant No. 16070 DGY.

18. International Conference on Waste Disposal in Marine Environment, Proceedings, Pergamon Press, N.Y., 1959.

19. M. Oguz, Optimisation programme for ocean outfall design, Master's thesis, Environmental Engineering Department, Middle East Tech. University, Ankara, Turkey.

20. B. Ketchum, Eutrophication of estuaries, in Eutrophication: Causes, Consequences, Correctives, Nat. Acad. Sci., Washington, 1969.

21. R. Mitchell, Ed., Water Pollution Microbiology, Wiley-Interscience, 1972.

21a. Stumm and Lackie, 5th International Conference, IAWPR, Paper II-26, 1970.

22. W. Bascom, Instruments for studying ocean pollution, Jour. ASCE, Env. Eng. Div., 103, 1977.

23. Coastal Pollution Control, WHO/DANIDA Training Course, Denmark, Vol. I and II, 1976.

24. A. L. H. Gameson, Ed., Discharge to the Sea from Sewage Outfalls, Water Res. Centre International Symposium, Pergamon Press, 1975.

25. Marine Pollution and Marine Waste Disposal, Proceedings 2nd Int. Congress, Sanremo, Italy, Pergamon Press, 1975.

26. T. Solemez, Sea Disposal of wastes from small and medium-sized communities, WHO/METU/ITU Short Course, Oct. 3-14, 1977, Ankara, Turkey.

27. N. Kor and H. Sarikaya, Determination of design parameters for marine waste disposal in Turkish coasts, WHO/METU/ITU Short Course, Oct. 3-14, 1977, Ankara, Turkey.

28. E. Ulug, A case study on the design parameters of marine waste disposal, WHO/METU/ITU Short Course, Oct. 3-14, 1977, Ankara, Turkey.

29. J. A. Hansen, Sea disposal of wastes from small and medium-sized communities, WHO/METU/ITU Short Course, Oct. 3-14, 1977, Ankara, Turkey.

30. D. Mackay, Conduct and practice of marine surveys for pollution control, WHO/METU/ITU Short Course, Oct. 3-14, 1977, Ankara, Turkey.

30a. J. Fernando, Criteria for marine waste disposal in Spain, WHO/METU/ITU Short Course, Oct. 3-14, 1977, Ankara, Turkey.

31. P. Harremoes, In-situ determination of microbial disappearance, WHO/METU/ITU Short Course, Oct. 3-14, 1977, Ankara, Turkey. (Also, Water Res., 4, 1970.)

32. Master plan and feasibility study on water supply and sewerage for greater Instanbul, Turkey, DAMOC Consortium Report for WHO/UNDP, 1972.

33. SWECO-BMB report of Iller Bankasi on Izmit sewerage master plan report, 1976.

34. K. Curi, Factors affecting determination of T_{90} for bacterial die-off in coastal waters, Doctoral thesis, Bogazici University, Istanbul, Turkey.

35. L. Russel, The planning and design of ocean outfall systems, PAHO Symposium, Buenos Aires, Argentina, 1976.

36. H. Hydan, Water exchange in two layer stratified waters, Jour. Hydraulic Div., ASCE, 1974.

37. J. Newbold and J. Ligget, Oxygen depletion model for Cayuga Lake, Jour. ASCE, Env. Eng. Div., 1974.

38. Wastewater Engineering, Metcalf and Eddy Inc.

39. W. G. G. Snook, Design and construction of submarine pipelines and diffusers.

40. R. A. Grace, *Marine Outfall Systems*, Prentice Hall, 1978.

41. Rules for Design, Construction and Inspection of Submarine Pipelines and Pipeline Risers, Det Norske Veritas, Hovik, Norway, 1976.

Chapter 9

DISPOSAL OF WASTES ON LAND

9.1 INTRODUCTION

Interest in disposal of wastes on land has not been confined to countries with arid and warm climates. Even in the developed countries, increasing interest is now being taken in land disposal as an ecologically
satisfactory solution under certain conditions. Land application of wastewater leads to groundwater recharge and/or evapotranspiration. Land application can be achieved as follows:

1. Irrigation of agricultural crops
2. Infiltration through specially constructed recharge systems (e.g.,
    shallow basins for groundwater recharge)
3. Evaporation ponds with or without infiltration

Some of the essential features of these systems are discussed below, and
particular attention is paid to irrigational systems as they have a wider
scope for application especially in the developing countries.

9.2 WASTEWATER IRRIGATION

9.2.1 Essential Features

The irrigational use of treated effluent and sludge must be given the
fullest consideration in planning any waste disposal scheme. There is
sound logic in returning the solids back to land and reusing the wastewater
wherever feasible.

From the agricultural point of view, treated wastewater and sludge
could be used effectively provided certain quality constraints are met.
Their use has advantages over plain water, such as the following:

1. Presence of fertilizing constituents
2. Favorable soil conditioning properties

Land disposal becomes a successful method of treatment achieving primary, secondary, and tertiary treatment in a single operation and in many instances is capable of giving returns in the form of crops and reusable water. However, the main disadvantages associated with their use are

1. Contamination and pollution hazards likely in regard to agricultural workers and crop consumers

2. Possible chemical effects of the wastewater or sludge on the soil and in groundwater

3. Generally large land requirements

4. Seasonal nature of irrigation demands

Wastewaters* can be conveyed to irrigation fields in open channels or pipes designed to avoid the creation of stagnant pools and solids deposition en route. There they may be spread on the land either by spraying from special nozzles or by releasing from the channels to flow over the land by gravity. The runoff may flow to collecting channels at the lower end of the fields or percolate through the soil and be collected in underdrains laid for the purpose. Some typical arrangements are shown in Figure 9.1. Quantities of treated effluents expected from different domestic and industrial sources are indicated in Chapter 5.

A part of the gross quantity of wastewater is "lost" en route to the field crops which, therefore, receive a reduced or net quantity. Of this net quantity reaching the crops, some evaporates (evapotranspiration), while the remainder runs off the surface. In systems where the field is diked all around and flooded, the water percolates through the soil and is collected in underdrains. It is then subjected to greater "treatment" in the soil than is possible with surface runoff systems. Spray irrigation is very efficient but suffers from some handicaps as discussed later.

Physical activity can lead to a change in the soil composition and its size distribution through either deposition of new material or erosion of the existing soil. Biological activity can degrade organic matter into stabilized end products, while chemical activity in the soil can be very complex. Long-term changes in soil structure and quality can occur from irrigational use of wastewater. Figure 9.2 shows in brief the various inputs and outputs involved in an agricultural system.

---

*For the purpose of this chapter, the term "wastewaters" will include liquid sludges also.

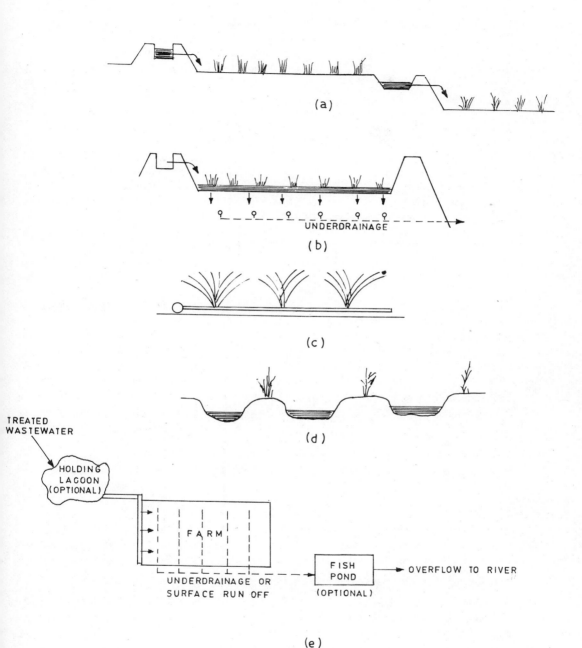

Figure 9.1 Some typical irrigation arrangements; (a) surface irrigation over a contoured field, (b) flood irrigation (underdrainage optional),
(c) spray irrigation (using movable pipe network), (d) ridge and furrow
irrigation, and (e) a typical layout for a wastewater irrigation system.

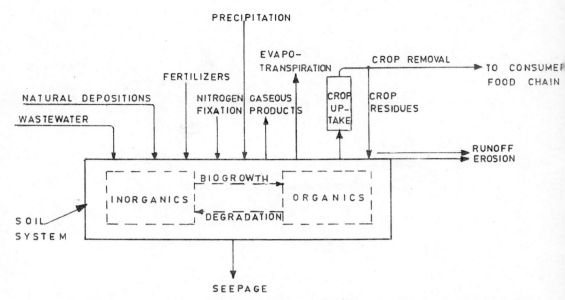

Figure 9.2 Some major inputs and outputs involved in an agricultural system.

Soil and water, in turn, affect plant growth and plant constituents. Certain fertilizing substances and "growth factors" contained in wastewaters can increase crop yields. Phytotoxic substances can destroy crops, while certain persistent substances, heavy metals, etc., accumulated in the soil may be picked up by the plants and passed back to animals and man through the food chain, thus causing health and ecological effects.

Health can also be affected by careless handling or inadequate pre-treatment of wastewaters, leading to exposure of the farm workers to parasites, direct contamination of food crops, especially those likely to be eaten raw, and groundwater pollution.

Owing to the large variety of factors involved, the area of land that can be irrigated by a given volume of water varies widely and must be estimated by the methods outlined further in this chapter. Also, careful attention has to be paid to the need for providing alternative means for disposal (or storage) of the wastewater when not required for farming purposes.

As a rough indication, the range in which irrigation land requirement lies is 5 to 40 hectares per 1000 m^3 daily flow (see Table 9.23).

A fairly large variety of crops have been grown on wastewater disposal farms depending on climate and local agricultural practices. From the point of view of health hazards and wastewater pretreatment requirements, some information on the choice of crops is given later.

Generally, with well-treated and disinfected effluents, crop selection is "unrestricted" while, with raw or partially treated effluents, the crop selection is "restricted" to certain varieties as shown in Table 9.26.

The chemical characteristics of the wastewater must meet some of the usual requirements discussed in Sec. 9.2.13. A few examples of water quality standards are given in Table 9.25. Chemical characteristics have particular relevance when dealing with industrial wastewaters.

Wastewater irrigation possibilities must be evaluated on the basis of costs and benefits using usual economic and environmental criteria. Irrigational use may be justified on one or more of the following grounds, provided enough land is available and the quality of the wastewater (either raw, after treatment, or after dilution with other water) is suitable for irrigational use at the proposed site:

1. The relatively arid nature of the climate and the lack of plentiful freshwater resources as alternative to wastewater, in which case the need for water conservation by reuse would overcome the usual economic considerations.

2. Land disposal in preference to direct disposal to a watercourse in order to avoid or minimize water pollution and related problems. This may be attractive when land disposal involves little or no pretreatment and when land disposal is itself a step in the needed overall treatment (for example, to remove nutrients, etc., as in tertiary treatment), or when the quality, year-round availability, and nearness of the wastewater to the irrigable land make its use favorable.

3. Governmental or local policy in favor of land disposal to encourage farming and food production, especially in the developing countries.

Irrigational use of wastewater is not to be regarded as just a primitive and inferior method of getting rid of sewage or sludge. Properly designed and operated, irrigation can be a convenient and low cost method, comparable with tertiary treatment and capable of satisfying environmental criteria. Furthermore, the cost of wastewater disposal can be at least partly offset by the sale value of the treated effluent supplied to the farm. For obvious reasons land disposal has had a much wider application in warm and arid regions, although there have been some notable examples in the temperate climates where increasing consideration is being given to it at the present time as a means of tertiary treatment [2].

In Germany, sewage has been utilized for agricultural purposes since around 1880. In 1962 in the Federal Republic of Germany, about 6000 hectares

of land were irrigated with sewage from nearly 70 plants; 4000 hectares by
irrigation channels and 2000 hectares by spraying. About 1.2×10^8 to 3×10^8 m^3 of sewage are used annually for land irrigation [1]. The methods
employed are damming up of the sewage, irrigation on inclined fields, and
ridge and furrow irrigation. With these methods, however, expensive drainage systems are required and new systems of this type are, therefore, not
being constructed. Spray irrigation systems using fixed or mobile piping
and sprayers are preferred because soil drainage is not necessary with them
[1].

The oldest sewage farm in India was established around 1895, and even
today in India one of the principal modes of final disposal of wastewater
is on land. The climatic conditions, the absence of dilution flows in most
streams during the summer, and the need to grow more food make irrigation
attractive. It is preferred that wastewaters from large cities be used in
this manner. Presently there are over 132 farms covering nearly 12,000
hectares and utilizing over 5×10^8 m^3 of sewage annually.

Examples are being increasingly reported from several countries of
the use of liquid sludge for irrigation; the sludge sources are primary
settlement, anaerobic digestion, or aerobic stabilization of sewage and
various industrial wastes. In several instances studies have been initiated
to see the possible detrimental effects on soil, crops, groundwater from
heavy metals, nitrates, etc.

A large number of treated, and sometimes untreated, wastewaters from
industries have been disposed by land application. Some of the industrial
wastes for which land disposal is favorable are from the food processing
industries such as:

Canneries
Milk dairies
Sugar factories
Breweries and distilleries
Beverage factories

Other industries where land disposal has also been considered or
adopted are:

Fertilizer manufacture
Pulp and paper
Tanneries

Yeast manufacture

Slaughterhouses

About 1300 industrial plants, including more than 300 canneries, in the U.S. utilize land irrigation for wastewater disposal. Besides this, some 950 municipalities, largely in the arid western parts of the U.S., also use infiltration or crop irrigation to treat and dispose their sanitary sewage [2].

It is interesting to see that in the U.S. currently much interest has been revived in land disposal of wastes as an economical method of achieving a high degree of treatment. Large cities such as Chicago are seriously exploring land disposal as an environmentally superior solution to the problem of their enormous quantities of liquid sludge produced daily at their sewage treatment plants [2]. To better sustain this renewed interest, more scientific design and management methods are being developed.

Depending on the method of irrigation followed on a farm, the runoff or underdrainage can (where levels and land permit) be impounded finally in a pond for the culture of fish in order to get still more benefit out of the wastewater. In some instances fish ponds precede the farms. The use of wastewater in fish ponds can also present certain hazards and demand better design and management.

Town planners could give greater consideration to the potential for land treatment and earmark locations suitable for sewage irrigation in the vicinity of their towns apart from aesthetic, topogarphical, soil, and other considerations. The large areas required for land treatment could help town planners in controlling urban sprawl and in siting certain industries or facilities (like refuse disposal) which must be located away from residential areas. Eroded and agriculturally poor lands could be developed gradually by the use of sewage sludges. The lands in all cases could be considered for the development of grasslands, croplands, or forest land. Thus, both land and water resources would be augmented.

A difficulty in this regard has been the possibility of odor problems and mosquito nuisance in the vicinity of such sites. This difficulty is likely to occur in any poorly managed farm or waste treatment plant where ponding of wastewater occurs, where the units are overloaded, or where other difficulties exist. This aspect is very important from the public relations point of view and cannot be overemphasized.

Governmental and other agencies could formulate their financial policies in such a manner as to encourage land disposal where it constitutes an environmentally sounder solution.

Data on the costs of wastewater irrigation systems are not so readily available. Land disposal of sludge has proved to be cheaper at times than other modes of sludge treatment and disposal, even when land application involves piping the sludge several kilometers away. Some studies in the U.S. have indicated that the cost of land application of liquid sludge is not much greater than ocean disposal and lagooning, and generally more economical than incineration or drying for sale as fertilizer [49].

9.2.2 Selection of Cropping Pattern

Selection of crops to be grown on a farm must be geared to the quality of the wastewater when selecting from among the likely crops that can be normally grown in a given area. Crops grown on wastewater farms have included the following:

Forage crops*
> Fodder grasses such as Bermuda grass, rye grass, tall wheatgrass, Reed Canary grass, alta fescue, para-grass, and alfalfa

Field crops†
> Corn, maize, wheat, barley oats, rye, rice, pulses, millets, sugarcane, sugar beets, cotton, flax, essential oil-bearing plants (citronella, mentha, etc.), and tobacco

Vegetables
> Tomato, potato, lettuce, red beet‡, artichokes, broccoli, spinach, soybeans, beans, cabbage, cauliflower, okra, and clover

Fruits
> Citrus fruits, stone fruits, berries, strawberries, grapes, and bananas

Others
> Various trees, woodlands, ornamental plants, and flowers

Wastewater irrigated farms in the vicinity of cities have a ready market for their crops and, naturally, cash crops such as vegetables, fruits, etc., are preferred. Stricter control of the effluent quality is

---

*Crops used as feed for stock.

†Crops such as those listed and including forage crops grown over a large area, but excluding fruits, vegetables, and ornamental plants.

‡Red beet, like lettuce, is a metal accumulator.

needed, therefore, to avoid health hazards as detailed later in this chapter. Farms adjoining food and beverage industries prefer to grow crops
which can be readily used in the industry.

Where municipal or domestic wastewater is used raw or partially treated,
crop selection is restricted to fodder grasses or any of the field crops
which will not be eaten raw. Herein lies a potential conflict between the
desire to maximize profits and the need to safeguard health. Regulations
in this regard exist in some countries. The chemical quality of the wastewater in the context of the local soil and climate may also affect the
choice of crops. For example, fruits are generally more sensitive to chemical quality, salinity, etc. Crops with a relatively high nutrient demand
may be more compatible for growing on sewage farms. Some examples are
Reed Canary grass, corn, alfalfa, soybeans, and red clover (see Sec. 9.2.9).

The selection of crops, therefore, generally depends on the agricultural practices and market demands prevailing in a given area, and on the
local climate, type of soil, and quality of wastewater. The mechanized
farming practices likely to be followed and the ready availability of
fertilizers, pesticides, etc., in the vicinity of cities may also need to
be taken into account in crop selection.

Another aspect to be considered is the cropping pattern, namely, what
percent of land is to be covered by what crop in each season. Since different crops have different watering requirements, the selection of the
cropping pattern must take into account the availability of wastewater,
the growing season of each crop, and the need for crop rotation to maintain soil fertility and permeability. (For example, inclusion of legumes,
pulses, etc., may be desirable.) In colder, wetter climates a combination
of agricultural and woodland irrigation may be preferred.

9.2.3 Estimating Crop Water Requirements

Several methods are used by irrigation engineers to estimate the water
requirements of different crops. Local estimates, wherever available,
should be considered. It is not within the scope of this book to go into
agricultural details but, as an illustration, one or two well-known methods
are described below. Plant cultivation in arid areas is reported to require
the quantities of water per year as indicated in Table 9.1 depending on
climate, soil, and method of irrigation [45,48].

Table 9.1 Water Requirement of Crops in the Growing Season

| Crop | Water requirement (total mm[a] in growing season) | |
	California	Hyderabad, India
Wheat	366-760	375
Barley	364-700	360
Citrus	500-600	-
Cotton	500-600	1070
Beet	700-900	-
Grasses	550-975	-
Lucern (alfalfa)	800-900	-
Sugar cane	400-950	2400
Rice	-	1060
Potatoes	-	680
Tobacco	-	1000
Willow, maple	1200-1500	-

[a]In irrigation practice, the water requirement of crops, precipitation, evaporation, etc., are often stated in terms of mm (1000 mm = 1 m^3 per m^2 area).
Source: Refs. 4, 45, and 48.

In the U.S., with its different climatic regions, Pound and Crites [50] surveyed wastewater irrigation facilities and found that land application rates generally varied from 12.7 to 101.6 mm per week of the crop growing season depending on the nature of the crop, soil, and climate. Total water requirement can be estimated if the land application rate and duration of growing season are known.

9.2.4 Use of Evapotranspiration Data

Consumptive use of water by plants, often termed evapotranspiration, includes any loss of water by evaporation from the soil surface and transpiration by plants. Consumptive use varies with the climatic conditions (sunlight, temperature, humidity, and wind) and ground coverage of the crop. The latter determines the amount of sunlight received by the leaves.

The more complete the ground coverage, the higher the consumptive use with all types of crops.

Records from 30 different plant species completely shading the ground show the consumptive use to be more or less the same. Hence, under the same climatic conditions, if the consumptive use of a crop with complete coverage is known, it can be estimated for other crops, roughly, if the percent coverage is known [46].

A method of estimating crop water requirement is to apply a suitable reduction factor α to the local pan evaporation rate E_p in order to find the rate of evapotranspiration E_t:

$$E_t = \alpha E_p \qquad\qquad\qquad\qquad (9.1)$$

The value of α ranges from 0.2 to 0.8 depending on the time of the year and on the growth stage of the plant. It generally increases as the plant matures. Some typical values are given in Table 9.2.

The values of mean evaporation from free surfaces also vary extensively from area to area. For example, in Central Europe evaporation from free surfaces is estimated to average about 60 mm/month, while in regions such as India and the Sudan it exceeds 330 mm/month. The type of pan and its placement can, naturally, affect the result. The U.S. Weather Bureau Class A pan is widely used in the U.S. and other countries.

Both for seasonal and perennial crops, the actual irrigation requirements can be estimated on a month-to-month basis by first estimating the crop water requirement E_t for a given month as described above and subtract-

Table 9.2 Values of Coefficient α to Estimate Plant Evapotranspiration from Open Water Evaporation

Plant	Value of α			
	Spring	Summer	Autumn	Winter
Fodder	0.4-0.6	0.8		
Alfalfa	0.7	0.9	0.7	0.6
Barley, wheat	0.8		0.4	0.6
Tomato		0.6		
Orchards	0.5	0.7	0.4	0.2

Source: Refs. 4, 46, and 48.

ing from it the estimated effective precipitation P in that month, calculating as explained in the next section.

Other methods for the estimation of water requirements can be found
in irrigation handbooks [4,48] (also see Sec. 9.2.16).

9.2.5 Effective Precipitation

Precipitation includes both rain and snow. Data are generally readily
available for most localities and isohyetal maps showing lines of equal
precipitation are also available. The annual precipitation in different
regions is shown below:

Climate	Annual rainfall (mm)
Desert	Under 120
Arid	120-250
Semiarid	250-500
Moderately humid	300-1000
Humid	1000-2000
Very humid	Over 2000

Only the growing season precipitation is of value and, furthermore,
the effectiveness of the precipitation from an agricultural point of view
has to be considered. Thus, rainfall less than 10 mm per month is sometimes
neglected whereas heavy rainfall is only partly effective, the rest merely
contributing to runoff. Recommendations of the U.S. Bureau of Reclamation
for arid and semiarid areas can be found in many standard texts on irrigation. A simplification of the estimation procedure is given in Table 9.3
for determining the effective precipitation P over a given time and area.

It may be assumed that the effective precipitation during one month in
excess of evapotranspiration percolates beyond the root zone and is not
available to meet the evapotranspiration need of the following month. In
this manner, the net irrigation requirement I is found on a month-to-month
basis as

$$I = E_t - P \tag{9.2}$$

Table 9.3 Estimation of Effective Precipitation
(U.S. Bureau of Reclamation)

Precipitation increments (mm/month)	Percent effective precipitation in stated increment[a]
0-20	95
20-50	90
50-100	20
100-150	30
> 150	0

[a]For example, if rainfall in a given month is 150 mm, the effective rainfall = 20(0.95) + 30(0.90) + 50(0.7) + 50(0.3) = 96 mm.

9.2.6 Method of Irrigation

There are several methods of introducing the irrigation water on a field; some of the factors affecting their choice are discussed below. The method should be chosen to suit the crops to be irrigated, the lay of the land, its slope, nature of soil, its holding capacity, permeability, etc. [46]. The method should also be related to the quality of the wastewater. Essentially, two types of systems are used: surface or gravity systems, and sprinkler or pressure systems (Figure 9.1).

Border or Strip Irrigation

The border or strip method of irrigation involves the division of a field into a number of strips, generally 5 to 15 m wide and 100 to 300 m long, separated by low dikes or border ridges. The water given from one end advances as a sheet down the strip, allowing it to enter the soil as the sheet advances. This method is suitable for a wide range of soil textures, although generally not for fine textured soils with low intake rates. It is often used with relatively large irrigation streams to cultivate hay pasture and grain crops. Land slopes may be up to 3%. In using wastewater, the design of the system should provide for ease of access and operation of valves, gates, etc., without having to dip in the wastewater or wet soil.

Basin Irrigation

In the basin method of irrigation used in many parts of the world, water is applied to level plots surrounded by dikes, and may be used on a

wide variety of soil textures and crops. It can also be used with fine
textured soils with low permeability rates. It promotes deep percolation*
of water and thus helps leaching of salts. The basin may be of any shape
and some are as large as 10 ha for growing rice, although they are often
4 m² or more. In an FAO-sponsored study in Iran, it was found that a basin
size of 25 m x 10 m gave the highest water distribution efficiency in growing sugar beets on a clay loam soil [47]. They are periodically filled
with water to the required depth to replenish the soil moisture. In areas
of heavy rainfall, surface drainage facilities may be necessary. Crops
grown include hay, pastures, grains, and many row crops. The use of this
method is generally restricted to smooth, level lands or to contour basins.

Furrow Irrigation

     Furrow irrigation is possible with various soil textures and land
slopes, and with either large or small streams of water. In fact, this
system is used extensively by farmers. In tilling the land, the soil moved
to make the furrows elevates the ridges between. Many crops and vegetables
are grown in this manner such as the row crops potatoes, corn, cotton,
fruit trees, grapes, etc. The spacing of the furrows is determined by the
spacing of the row crops.

     Unnecessary water losses can occur from deep percolation if the furrows
are too long. Overirrigation of the upper end of the field can occur by the
time the lower end is sufficiently irrigated. Large flows causing erosion
should also be prevented. Furrow irrigation has often been used with sewage
especially with raw or primary settled sewage.

Corrugation Irrigation

     This is similar to furrow irrigation, except that the furrows are more
closely spaced for use with certain small grains, hay, and perennial crops.

Contour Channel Irrigation

     A sloping field can be irrigated from a channel running along its
upper contour to allow the water to flow over the field until it is inter-
cepted by another channel running along a lower contour. The channels or
ditches should be spaced close enough and other measures should be taken
to ensure uniform distribution of water. The method is well suited to
sloping and rolling land for the cultivation of grasses and certain trees.

---

*Percolation beyond the root zone.

A similar method is used for the treatment of wastewater where the soil has poor percolation characteristics.

Sprinkler Irrigation

In sprinkler irrigation, the soil is wetted in much the same way as rain. It can give uniform distribution of water without much deep percolation and is particularly adapted to light applications of water for shallow-rooted crops and where annual water requirement is low. On porous soils such as sands, irrigation by other methods may give excessive losses by rapid percolation.

Sprinkler systems, however, are generally higher in initial and operating costs. They can be laid in the field as portable units (so that the land is not crisscrossed by dikes and furrows, thus enabling easy use of tractors and other machinery) but this handling may not be desirable on heavy soils or with wastewater. Large, slowly revolving sprinkler arms, 250 to 400 m long, have been used recently in the U.S. with nozzles facing downward to reduce aerosol effect and eliminate human handling of wastewater [2]. Pressures of 2 to 4 kg/cm^2 are most common in the conventional systems, although higher pressures up to 10 kg/cm^2 are used. Plastic pipes exposed to sunlight may deteriorate in 1 to 5 years, while most other components would have a useful life of from 7 to 15 years.

Higher evaporation, but lower percolation, losses occur with sprinklers than, for example, with furrow irrigation because sprinklers wet the entire soil surface as well as the leaves of the plants. Evaporation from soil surface is confined to a relatively shallow depth, not much below the upper 20 cm. With light irrigations wetting only the surface layer, just enough water can be added to the soil to meet evaporation requirements. With heavier irrigations, soils may be wetted to an appreciable depth with resultant percolation, and moisture loss through evaporation may be relatively small in comparison to the amount of water applied. Hence, sprinklers can be more efficient in usage of water, especially with light irrigations. Sprinklers are also suitable for a rolling topography.

Mechanical clogging of the system, handling of leaks, and spraying of pathogens if any from wastewater directly onto the plants or fruits make it desirable to confine the use of sprinklers to only well-treated wastewaters. Similarly, the use of sprinklers can encourage direct foliar absorption of chlorides, sodium, and certain heavy metals, thus escaping the complexing capacity of the soil (see Sec. 9.2.12). Sprinkler sprays tend to wash off pesticides and other chemicals applied to the plants. Under certain climatic conditions, some plant diseases may be encouraged.

9.2.7 Gross Irrigation Requirement

The gross irrigation requirement depends on the type of irrigation system used because a considerable portion of the incoming water is lost in transmission and distribution through percolation, evaporation, surface runoff, and inadequate farm management. Losses also depend on the type of soil and mode of water distribution used. Higher efficiencies are attainable with sprinkler systems than with channel or basin types. Glass houses exhibit the highest efficiencies owing to more careful design and operation. The gross quantity required at the farm, stated in terms of efficiency e is

$$I \text{ gross} = \frac{(E_t - P)}{e} \tag{9.3}$$

Water losses are invariably high where overirrigation is practiced, where distribution furrows are too long, or where the field is not properly leveled and ponding occurs.

A few typical values of efficiency e adopted in designing farm irrigation layouts are given in Table 9.4. The values given exclude losses in long canals and storage reservoirs.

In fact, overall efficiency of any irrigation system would be still further reduced as it would include all conveyance losses in canals, distributors, and storage. In the case of wastewater irrigation projects, such losses would be low owing to the shorter lengths generally involved.

Table 9.4 Efficiency of Irrigation Systems

Irrigation system	Approximate Efficiency[a] (percent)
Glass houses	75-80
Sprinkler type	65-75
Gravity type (basin or channels)	42-65

[a]Assuming artificial leaching is not required.

Source: Ref. 4.

Channel losses, for example, are estimated in the range of 0.1 to 0.7 m^3/m^2-day of channel area, depending on the type of soil and channel lining [4]. Where wastewater has to be stored in holding lagoons when not being used, the overall losses must also take lagoon evaporation and seepage into account, unless the lagoon bottom and sides are specially treated to avoid or minimize seepage.

9.2.8 Quality of Wastewaters Used for Irrigation

The physical, chemical, and biological quality of wastewaters (and liquid sludge) intended to be used for irrigation must be examined carefully in view of their long-term effects on crops and soils. Besides suspended solids, there can be a variety of substances in solution in the wastewater or sludge. All dissolved substances are referred to as "total dissolved solids" (TDS) in the wastewater; they include sodium, calcium, magnesium, carbonates, bicarbonates, etc., and nutrients and heavy metals in solution.

Some of the dissolved solids can participate in exchange reactions with soil colloids, depending on the exchange capacity of the soil, and thereby affect soil quality in the course of time. These ions are abundant under natural soil conditions and, in moderate concentrations, pose no problems in irrigation waters. Furthermore, with adequate flushing water, most are readily transported to the groundwater table. Difficulties with salt and sodium affected soils are caused by either excessively high salt application or extremely low rainfall with a consequently high evapotranspiration rate resulting in salt accumulation in the soil.

Crop yields may reduce substantially if irrigation over a prolonged period of time leads to high salt level in the soil. The soil saturation extract, therefore, can be correlated to the crop yield. One must distinguish between irrigation water quality and the soil saturation extract. The salt concentration in the saturation extract of a soil can be 2 to 10 times higher than that of the applied irrigation water in the absence of salt contribution from the groundwater. The concentrating effect in the soil is explained later in Sec. 9.2.18.

The qualities of domestic and certain industrial wastes have been discussed earlier in Chapter 5. Any irrigation water, including treated wastewater effluent or sludge, must be evaluated in terms of certain traditional criteria which have come to be developed over the years from good irrigation practice. The aspects which attract special attention are the following:

1. Nutrients available in wastewaters and nutritional requirements of crops

2. Organic load in wastewater

3. Pathogens and their transmission

4. Toxic substances and factors affecting their uptake by plants and onward transmission up the food chain

5. Traditional criteria for irrigation waters

   a. Total dissolved solids (TDS) or electrical conductivity (EC)

   b. Sodium content, sodium absorption ratio (SAR), exchangeable sodium percent (ESP), and percent sodium

   c. Bicarbonate or residual sodium carbonate (RSC)

   d. Boron

6. Existing soil quality and likely buildup of salts in future

7. Land treatment efficiency and groundwater pollution

It should be noted that irrigation water has no inherent quality and can be evaluated only in the specific local conditions of soil, climate, crops, and irrigation practice. A wastewater which may be satisfactory under one set of conditions may be unsatisfactory under a different set of conditions.

9.2.9 Nutrients Available in Wastewaters and
        Nutritional Requirement of Crops

Besides oxygen, hydrogen, and carbon, which are freely available, the chemical elements for plant growth include the following:

1. Nitrogen, phosphorus, potassium, sulfur, calcium, magnesium, and iron. These are generally required in relatively large quantities and, therefore, are called macronutrients. They are obtained from the soil. Nitrogen promotes leaf and stem growth, while phosphate stimulates root growth and hastens ripening. Potash also stimulates growth and is needed for the formation of chlorophyll.

2. Manganese, boron, zinc, copper, molybdenum, and chloride. These are needed in minute quantities in the soil (some as low as 0.01 part per million) and are called micronutrients.

Certain plants may not require all the above, whereas certain others may require elements not listed above such as selenium, cobalt, silicon, etc. Vitamins, hormones, and what are generally called "growth factors" also play an important part in plant growth.

Wastewaters and sludges contain many of these fertilizing elements. However, only their nitrogen, phosphorus, and potassium contents are generally measured when comparing them with artificial fertilizers. In fact, wastewaters can provide some of the micronutrients and "growth factors" not available in commercial fertilizers. Several controlled studies made on different crops have shown that plots irrigated with wastewater gave significantly higher yields than control plots using commercial fertilizers with N, P, and K in equivalent amounts as in the wastewater [4,5,5b,32].

The uptake of nutrients, nitrogen, and phosphorus by a few crops is shown in Table 9.5. When a crop is harvested, it implies removing that much nitrogen and phosphorus from the system. The use of grasses would enable larger uptakes of nitrogen and phosphorus per unit time and area as shown in the table, and their use would be more favorable with wastewaters rich in these nutrients. Warm weather would also give larger yields of these grasses and thus accomplish greater nutrient removal from the system than in colder climates.

Forage crops are often considered suitable for wastewater irrigation owing to their high nutrient uptakes and long growing seasons. From among the nutrients in wastewater applied on land, phosphorus and potassium are readily removed in their passage through the soil, but nitrogen is not. Hence, where wastewater is applied on land as a tertiary treatment step, it is essential to select carefully the crops and cropping pattern to ensure optimum removal of nitrogen at all times.

Table 9.5 Crop Uptake of Nutrients

Crop	Uptake (kg/ha-year)	
	Nitrogen as N	Phosphorus as P
Coastal Bermuda grass	538-672	39
Reed Canary grass	253	40
Alfalfa	174-246	18-24
Corn	174	28
Red clover	134	13
Soybeans	111-127	16-20
Wheat	69-85	13-16

Source: Ref. 2.

Solids in wastewater and sludges are often good soil builders or con-
ditioners. Like humus, they help to improve the water holding capacity of
soils and the growth of soil bacteria, which in turn help to make nitrogen
available to plants. They also contribute to soil fertility and erosion
control.

Secondary effluent from a trickling filter plant was used to grow
pearl millet (which ranks high among cereal grains in its requirement of
nutrients) in order to study its response under wastewater irrigation.
Field experiments were conducted planting pearl millet, Tiftlate variety
[Pennisetum typhoides (Burm), Stapf and E. C. Hubbard], on fine sand with
the water table 12 m below. Different plots were irrigated at different
rates varying from 51 to 200 mm per week so that they received correspondingly varying nutrient applications from the wastewater. Measurements were
made to determine what percentage of the applied nutrients are recovered in
the harvested crops [41].

Figure 9.3 shows that as the wastewater application rate increased
the total crop uptake increased, but the recovery efficiency for nitrogen
decreased. Crop uptakes generally exhibited an asymptotic approach toward
a maximum value for all nutrients measured, except Fe. The uptake could
be fitted by the empirical function

$$\frac{Y_{max} - Y}{Y_{max} - Y_{min}} = e^{-X/X'} \tag{9.4}$$

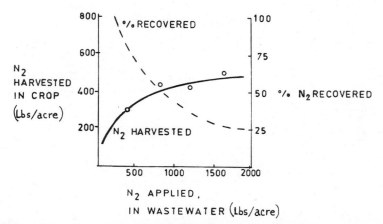

Figure 9.3 The observed relationship between nitrogen application
and recovery rates (adapted from Ref. 4.1).

where

Y   = nutrient content in crop, kg/ha

Y_{max} = maximum nutrient content in crop, kg/ha

Y_{min} = minimum nutrient content in crop, kg/ha (at zero application of nutrient)

X   = applied nutrient, kg/ha

X'   = fertility response coefficient, kg/ha

Y/X   = percent recovery

For nitrogen, the minimum content in pearl millet was 70 kg/ha, whereas the maximum was 538 kg/ha. The value of X' was found to be 515 kg per ha. The efficiency of recovery in the typical operating range was 50 to 75%. At a very high rate of nutrient application on land (around 2000 kg/ha), only about 25% of the applied nutrient was accounted for by the crop [41]. The rest partly denitrified in the soil and partly leached out as nitrates. Hence, the uptake of nutrients is a function of their availability, but maximum uptake is not affected by excess availability.

The nutrient application rates required for several types of crops are known from general agricultural practice, and chemical fertilizers are often used along with plain irrigation water. For example, the nutrient requirements of a few crops are given in Table 9.6.

When wastewater is applied instead of plain water, the need for chemical fertilizers is proportionately reduced. Alternatively, in the case of some industrial wastewaters where nutrient content is higher than necessary for the crops in question, some scope exists for dilution with plain water. Dilution may also be practiced for overcoming other possible objections to the chemical quality of the wastewater for irrigation purposes.

Table 9.6 Nutrient Requirements of Some Crops

Crop	Application rate of nutrients in chemical fertilizers (kg/ha)		
	N	P	K
Artichokes	45-135	15-65	0-110
Truck crops	65-200	25-110	0-160
Essential oil bearing plants (citronella and mentha)	86-130		

Source: Refs. 17 and 46.

In some instances, one of the nutrients may need to be supplemented, whereas the others may be present in sufficient quantity in the wastewater. For example, in the cultivation of millet it has been suggested that fortification of the wastewater effluent from secondary treatment by potassium may be desirable to maintain proper crop vigor [41].

In a 4 year study with high yielding varieties of wheat irrigated with raw domestic sewage (BOD mostly around 160 mg/liter diluted with plain water in the ratio of 1:1 and supplemented with N, P, and K to reach recommended application doses for wheat, the yield was found to be nearly 80% higher than that for plain wellwater fortified with required amounts of N, P, and K to the same levels [5b]. This would indicate, as stated earlier, that wastewater furnished something more than just N, P, and K to the crops. When liquid sludge is applied on land, the application rate for some high nitrogen sludge may have to be limited from the nitrogen point of view as discussed later with regard to ground water pollution caused by excess nitrogen conversion to nitrates. Alternatively, the nitrogen rich sludge can be diluted with freshwater or treated effluent wherever possible and applied over a greater area of land.

Item 9 in Table 9.7 presents the relative distribution of nutrients between the liquid phase and the solid phase of anaerobically digested sludges. Digestion converts some part of the organic nitrogen to ammonia which remains in the liquid phase. Hence, more total nitrogen is applied to land when liquid sludge is used instead of dried sludge. The case is similar with potash.

Example 9.1 If the sludge waste from an oxidation ditch is 35 g dry wt/person-day, roughly estimate the N, P_2O_5, and K_2O present in it per 1000 persons annually. Assuming the sludge volume (after thickening to 4% solids) is 0.875 liter/person-day, state the nutrient concentrations as mg/liter. Also estimate the nutrients in the treated effluent, assuming suitable values from Table 9.7.

1. In sludge:

$\quad$ N $\quad$ = 0.06 x 35 g/capita-day x 1000 x 365 = 766.5 kg/year (= 2400 mg /liter)

$\quad$ P_2O_5 = 0.04 x 35 g/capita-day x 1000 x 365 = 511 $\quad$ kg/year (= 1600 mg /liter)

$\quad$ K_2O $\quad$ = 0.013 x 35 g/capita-day x 1000 x 365 = 166 $\quad$ kg/year (= 520 mg /liter)

Table 9.7 Nutrient Content of Some Wastes

Item	Average nutrient content			Ref.
	Nitrogen as N	Phosphorus as P_2O_5	Potassium as K_2O	
Nutrients in raw sewage				
1. Urine and feces, g/capita-day	7.9	3.3	2.5	6
2. Raw domestic sewage, g/capita-day	6-12	4-18	2-6	7
3. Raw domestic sewage, mg/liter	20-40	6-20	17-36	8
4. Raw domestic sewage, mg/liter (average of 8 Indian cities)	48	10.6 (as PO_4)	21 (as K)	5a
Nutrients in treated effluents				
5. Secondary effluent, domestic, mg/liter (average of 33 U.S. cities)	13[a]	8 (as P)[b]	36	12
Nutrients in sewage sludge				
6. Raw primary sludge, % dry wt	1.5-4.0	0.8-2.8	0-1.0	8
7. Activated sludge (undigested)	5.0-6.0	4.0-7.0	0.3-0.5	-
8. Anaerobically digested sludge, % dry wt	1.6-6.0	1.5-4.0	0-3.0	8-10
9. Anaerobically digested sludge, % dry wt				
a. Total dry sludge	5.0	3.0 (as P)	0.5	2
b. Liquid fraction only	1.0	Slight	0.46	2
10. Aerobically stabilized milk sludge from oxidation ditch	6.1	5.5	Nil	39
11. Aerobically stabilized domestic sludge, % dry wt (oxidation ditch over a wide range of solids retention time)	5-8	3-5	0.8-1.8	11
Nutrients in animal wastes				
12. As percent of dry solids, cattle	0.3-1.3	0.15-0.5	0.13-0.92	36
13. As percent of dry solids, hogs	0.2-0.9	0.14-0.83	0.18-0.52	36
14. As percent of dry solids, hens	1.8-5.9	1.0-6.6	0.80-3.3	36
Other waste				
15. Protein waters from potato-starch plants, mg/ton of processed material	1.8	0.9 (as P)	3.6 (as K)	40

[a]Nitrogen removal in conventional secondary treatment averages $\approx 40\%$.

[b]Phosphorus removal in conventional secondary treatment averages 50%.

2. <u>In effluent</u> (assuming 50% of N and P_2O_5 and 10% of K_2O in raw sewage
    is removed in extended aeration treatment):

    N   = 8 g/capita-day x 0.5 x 1000 x 365 = 1460 kg/year

    P_2O_5 = 10 g/capita-day x 0.5 x 1000 x 365 = 1825 kg/year

    K_2O = 25 g/capita-day x 0.1 x 1000 x 365 = 912 kg/year

The above nutrient concentrations can also be expressed in terms of
kg/ha per cm of irrigant. For example, assuming the sewage flow per person
= 150 liters/day,

    N = 1460 kg/yr from 1000 persons = 26.7 mg/liter = 2.67 kg/ha-cm

In a case in India, the effluent from secondary treatment was estimated
to have the equivalent of $10* worth of commercial fertilizers in every 1000
m^3 of effluent [13]. This did not include the likely additional benefit
from increased crop yield referred to earlier.

The nutrient value of San Diego wastewater has been estimated at $14.60
per 1000 m^3 ($18 per acre-ft) when applied to agricultural use [14]. In
fact, the value of nutrients in municipal effluents of all U.S. cities is
estimated at $630 million per year at 1973 prices [2]. Any saving in arti-
ficial fertilizers that can be achieved in this manner should be very welcome in view of the general shortage of fertilizers in many countries and
the high amounts of energy required to manufacture them. The assured fertilizer content in the wastewater should be taken into account in fixing
its price for agricultural use.

The nitrogen and phosphorus contained in direct rainfall on an agricultural area can be estimated roughly from the typical concentrations
given in Chapter 5. Generally, nitrogen derived from rainfall varies between 1 and 20 kg/ha-year and is relatively insignificant compared with the
agriculturist's inputs.

Some nitrogen is also fixed by plants and microorganisms. This may
vary from about 12 to 35 kg/ha per year for grasses to as much as 225 kg/ha
per year for legumes under favorable conditions. Blue-green algae also
contribute to nitrogen fixation.

----

*1 U.S. $ = rupees 8/-.

9.2.10 Organic Load in Wastewaters

Although wastewaters may bring in welcome nutrients to the irrigation
system, they also bring in a load of organic matter in the form of dissolved
and suspended organics. Inorganics in dissolved and suspended form are also
present in varying concentrations.

The coarser particulates are filtered out at the soil surface, while
the finer ones tend to penetrate soil pores depending on the hydraulic
driving head available for this purpose. Most of the solids are held back
within the top 5 cm or so of the soil. The dissolved substances percolate
through, depending on the soil infiltration characteristics and other factors. Some flocculation and further removal of solids occur as the wastewater percolates. Organic substances undergo either aerobic or anaerobic
degradation depending on the oxygen availability, namely, on the soil aeration condition. For example, a lagoon or seepage pond would tend to develop
anaerobic conditions in the soil below it, while intermittently irrigated
soils with alternating wet and dry periods may tend to be aerobic.

The extent of biodegradation can be measured, for instance, in terms
of BOD removal as the wastewater passes through a depth of soil. It is
evident that the extent of biodegradation would depend upon whether the
conditions in the soil are aerobic or anaerobic, upon the mean residence
time in the soil layer, and upon the biodegradability of the waste itself.
A deep soil layer with a longer residence time would be likely to give lower
BOD values in the leachate.

The mean residence time in the soil layer depends on the wastewater
application rate (which is limited by soil permeability characteristics)
and the void space or pore volume of the soil (which again depends on the
type of soil) within the given layer.* The residence time is obtained by
dividing the total pore volume by the wastewater flow rate.

Organic removal with soil depth essentially follows plug-flow, first-
order removal kinetics. However, there are two routes by which organic
removal occurs. One route is the loss of organic matter through leaching.
This applies to all kinds of organics, degradable or nondegradable, provided

_____

*Can be determined in laboratory or assumed from available data (Sec. 9.2.15).

they are not prevented by the soil system from leaching out. As discussed in Sec. 9.2.18, the leaching of inorganic substances has also been shown to follow first-order kinetics [24]. The other route is bacterial degradation.

However, the two different removal mechanisms, leaching and degradation, both affect the residence time, thus giving an effective residence time in the soil. Consequently, the overall removal is governed by whichever is the faster of the two mechanisms. Organic matter tends to adhere to the soil particle surfaces, thus exhibiting slower leaching rates with the result that the degradation rate often seems to control the overall removal. Thus, irrigant application loads should also vary with the type of soils, with prevailing soil temperatures, and the nature of waste, as they all affect degradation rates. Excessive loadings can lead to soil anaerobicity, reduction in crop productivity, and odor nuisance.

In India, where domestic and municipal sewage and certain industrial organic wastes are often applied in the raw or untreated state to land, it is recommended that the BOD_5 be limited to 500 mg/liter as no particularly deleterious effects have been reported after many years of wastewater usage at this level. The crop watering practice often followed is to give about 20 cm of wastewater to the land every 8 to 10 days during the season. Sometimes the wastewater is diluted with one to three parts of supplementary water from rivers or wells. The BOD load using raw sewage on the land can be computed as follows:

$$\text{Wastewater applied per day over a hectare of land} = 0.2 \times 10{,}000 \text{ m}^2 \times \frac{1}{8}$$

$$= 250 \text{ m}^3/\text{ha-day}$$

$$\text{Hence, } BOD_5 \text{ applied to land} = 250 \times 10^3 \times 500 \times \frac{1}{10^6}$$

$$= 125 \text{ kg/ha-day}$$

Dilution would proportionately reduce the BOD_5 load. The load of 125 kg/ha-day (= 180 kg BOD ultimate/ha-day) is computed from recommended usage and is not necessarily the maximum applicable. Values range widely from 25 to 150 kg BOD_5/ha-day in India. A few values available in literature have been summarized together with the Indian values in Table 9.8.

The organic loadings for secondary municipal effluents given in Table 9.8 are not necessarily the upper limits of applicability. They are low because, with low BOD in treated effluents, the crop and soil watering limits control the rate of application. With several other wastes, very high organic loads are applied. Where concentrated liquid sludge is applied

Table 9.8 Typical Organic Loadings Used in Practice
for Land Application of Some Wastes

Waste	Organic loading (BOD$_5$ kg/ha-day)	Reference
Dextrose	1026	2
Pulp and paper	225	37
Horse manure	160	2
Raw domestic and industrial (India)	25-150	—
Secondary municipal effluent	2-5[a]	See text
Liquid digested sludge	2-90	Estimated from various data
Protein waters from potato starch plants	600[b]	40
Milk (without cheese)	12-125[a]	—
Canning wastes	100-2000	40

[a] In the case of several wastes, hydraulic load on the soil, rather than organic load, often controls the land requirement. In fact, in some Indian dairies with their own farms, the dung of animals is mixed with the milk waste and used for irrigating a high nutrient uptake grass, Brachiaria mutea staph (called para-grass), yielding 185,000 kg/ha.

[b] 2200 to 2400 ha per 100,000 tons potatoes processed. Generally, with this waste organic load is reported to control the land required.

directly to land, organic loadings may govern the design, although very little data are available in this regard.

The value of 225 kg/ha-day for pulp and paper mill wastes shown in Table 9.8 was derived as a result of a large number of experiments by the National Council of the Paper Industry for Air and Stream Improvement, Inc., U.S., using laboratory columns (lysimeters) filled to 50 cm depth with four representative soils: a sand loam, a silt loam, and two clay loams. A number of effluents were used including bleached and unbleached kraft, sulfite, insulating board-mill, etc. Based on these observations, with a correction for differences in soil temperatures existing between the field and laboratory, sustained BOD loadings of 225 kg/ha-day on soils of 37.5 to 50 cm depth could be expected to result in percolates of high quality, unimpaired growth of selected cover vegetation,* and no reduction in soil

---

*Alta fescue grass was used as it is a grass of high moisture resistance, reasonably high salt tolerance, and with an extensive root formation. Laboratory investigations proved it to be superior to Reed Canary grass.

permeability owing to BOD loading. The leachate was aerobic at all times
and gave BOD removal efficiencies of over 95% [37].

When the BOD loadings were 314 kg/ha-day, the BOD removal efficiencies
were still 60 to 84%, but the percolate was now septic, thus showing that
insufficient oxygen was gaining entry at the soil surface. The cover vegetation and soil permeability were also adversely affected after 8 months of
irrigation.

In general, results show that the more permeable the soil, the farther
the suspended and dissolved materials move into it. This results in a
thicker active zone, allowing a higher long-term acceptance rate [38].
Several studies have been conducted on septic tank seepage fields, but their
results cannot be applied directly to the design of an irrigation field
which has a crop cover and is periodically plowed.

9.2.11 Pathogens in Wastes and Their Transmission

Pathogenic organisms or parasites contained in wastewaters and sludges
used in irrigation could, under certain conditions, be directly transmitted
to those who work on the farms and to those who consume the products of
these farms. Groundwater contamination could also occur as discussed later.

Raw domestic sewage from any community usually contains the whole
range of pathogenic organisms found in the community. Table 5.1 gives the
range of values likely for different pathogens in domestic sewage. Industrial wastes could also contain pathogens, either from the processing steps
(for example, anthrax from animal handling) or from the sanitary sewage
from workers' toilets or residences adjoining the industry and sewered
together with the industrial wastes.

Organism dieoff occurs during treatment and in the watercourses into
which the wastewater is discharged. Table 9.9 gives some of the expected
removals in waste treatment. Generally, coliform removal rates also serve
as indicators of the removal rates for other organisms, although the rate
of dieoff is not the same in all cases. For example, helminth eggs with
their higher settling rate may be removed faster than coliforms, whereas
viruses in general are less effectively removed in waste treatment and
chlorination.

Furthermore, it is important to bear in mind that all treatment processes are not operated at optimum efficiency at all times. There are

Table 9.9 Removal of Organisms in Wastewater Treatment[a]

Organism	Nature of treatment	Overall percent removal	Reference
1. Coliforms	Primary sedimentation	30-40	18
	Primary and secondary bio-treatment	90-99	18
	Oxidation ditch (extended aeration)	91-97	14
	Primary and secondary and 15 to 20 mg/liter chlorine for contact periods of 1-2 hr	~99.9	18
	Waste stabilization ponds	90-99.99[b]	19,34
	Maturation ponds	98.7-99.9	34,20
	Aeration lagoons	90-95	
2. Viruses	Primary settlement	34-66	17a
	Trickling filter	~40	33
	Activated sludge	91-98	17a,33
	Oxidation ditch (extended aeration)	~99[c]	17
	Oxidation pond (single celled)	96-97	17a
3. Salmonellae	Oxidation ditch (extended aeration)	~91[c]	17
	Waste stabilization ponds, multicelled, series operation	99.6-100	19
4. Others:			
Fecal streptococci	Waste stabilization ponds, multicelled, series operation	99.99	19
Protozoan cysts and helminthic eggs	Primary settlement	30-60	14

[a]Bacterial removal kinetics have been discussed in Chapter 4.

[b]Ponds with two or three cells in series consistently give high removals in warm climates (see Chapter 14).

[c]No correlation was observed between the removal of pathogens and the solids retention time in the ditch (> 6 days).

breakdowns and stoppages caused by a variety of factors, during which time effluent quality can sharply deteriorate. Hence, certain precautions and adherence to good farming practices are essential.

Studies have been made on the survival of coliforms and pathogens in soils and on crops irrigated with wastewater. The viability of these organisms varies from several days to a few months depending on the type of organism, the climatic conditions (temperature), soil moisture, and the amount of protection provided by the crops themselves. In all cases, air, sunshine, and soil have been found to rapidly reduce their numbers (Table 9.10).

The many variables affecting the survival rate of different organisms make it difficult to apply available data to existing or new farms. In India, weekly samples of a variety of vegetables from five sewage farms were examined and only a few were found positive for salmonella and parasites during summer months, but almost all vegetables were positive for coliforms and fecal streptococci at all times [17a]. Other studies have also shown that coliform bacteria are generally present on crops grown both with and without sewage irrigation, thus raising doubts about the utility of coliform counts as far as farm crops are concerned.

Experiments with sewage irrigated crops in Germany were carried out to study the coliform survival rate on the clover plant spray irrigated with settled sewage. Examinations showed that all clover leaves cut immediately after spraying were infected with coliform bacteria (Figure 9.4). Over a 14-day period, the rate of infection decreased to 8% (curve D in Figure 9.4) and reached that of nonirrigated clover (curve F). Closely

Table 9.10 Survival of Pathogens on Some Food Crops

Crop	\multicolumn{5}{c}{Survival time (days)}
	Coliforms	Salmonella sp.	Shigella	Salmonella typhosa	Cholera vibrio
Fodder	12-34	12->42	<2		
Leaf vegetables and tomatoes	35	1-40 7[a]	2-7	31	2-29
Root crops		10-53			
Orchard crops		<3	6		

[a]Based on limited data from NEERI, Nagpur, India [14], in a relatively warm climate. Other data in table are from studies in the U.S. and Europe.

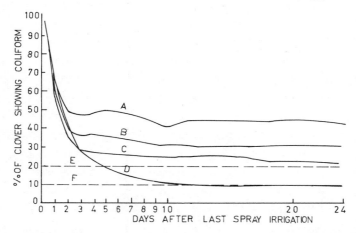

Figure 9.4 Coliform survival on crops under different conditions (adapted from Ref. 1).

spaced clover, however, showed a higher rate (30%, curve B). Higher rates were also observed where the clover had been irrigated repeatedly (curves A and C). Heavy rainfall did not wash away the bacteria, but, in fact, by secondary infection, increased the rates for all clover plants irrigated by sewage as well as those not irrigated. Intensive sunshine, however, rapidly decreased the rates. The investigation showed that an interval of 14 days after the last spray irrigation is generally safe, assuming that salmonellae from the sewage also decrease at the same or higher rate than coliforms [1].

In order to obviate the possibility of pathogen transmission, several farms have restricted themselves to the growth of grasses and fodder crops. Some efforts have been made to explore the scope of growing other more profitable crops, such as citronella and mentha, for example, from which essential oils can be extracted. The commercial value of these oils can make such cultivation attractive, while at the same time hygienically safe [5]. Pyrethrum (whose oil is useful for mosquito control in many developing countries) can also be grown similarly.

An outbreak of cholera in Jerusalem, Israel, in 1970 was investigated and it was concluded that vegetables irrigated with raw sewage (contrary to Ministry of Health regulations) were the most likely pathway for the secondary dissemination of the disease after its importation into the city. Cholera organisms were detected in raw sewage samples from all sections of

the city during the outbreak. They were also detected in the soil that
had been irrigated with raw sewage and on vegetables grown in it, some of
which were found in the markets [18]. Several other outbreaks of typhoid
and other diseases have been reported in the past [1,18].

Studies made in the U.S. and elsewhere have shown no evidence that bacteria, amoeba, or helminth eggs penetrate the surface of vegetables or cause
internal contamination. However, contamination can evidently occur from the
broken skin surface of vegetables and fruits; hence, the recommendation to
avoid growing on sewage farms vegetables and fruits which may be eaten raw.

Generally, proper disinfection of the treated wastewater is recommended before its unrestricted use for irrigation. Different countries
require different standards as shown in Table 9.26. In the absence of
standards, one rationale recommended is to bring down the coliform concentration to the levels generally found in nearby rivers.

The health of sewage farm workers has also attracted much attention,
and numerous studies have shown that proper care in operation is essential.
Similar to the case of workers handling solid wastes, there are increased
occupational hazards involved in handling wastewater and crops at sewage
farms.

Hookworm and other enteric infections are particularly to be guarded
against on sewage farms. In India, the rate of hookworm infection has been
found to be significantly higher among workers on sewage farms than among
the farming population in general [21]. This is attributed to the local
custom of walking barefoot on the sewage farms and thus providing an opportunity for the hookworm larvae to enter the human body. Wearing shoes and
gloves and observing reasonable standards of hygiene are recommended for
their effectiveness.

An examination of Berlin irrigation field workers has not shown them
to have a rate of infectious diseases or of worm infestation higher than
the rest of the population [1]. A follow-up study of the health of workers
at sewage treatment plants in the U.S. has not revealed any excessive risk
of disease in this group [22]. Hence, proper care and precautions must be
taken in operating sewage farms.

A few essential precautions that must be taken in designing sewage
farms are discussed later. Sewage farming must be operated in such a manner
as to be both safe and profitable.

9.2.12 Toxic Substances in Wastes and Factors
        Affecting Their Uptake or Removal

    In designing irrigation systems, concern must be directed toward those
substances contained in wastewater that may become concentrated in soil,
plants, animals feeding on such plants, and, ultimately, man. Regular applications of treated wastewater effluent or sludge over long periods of
time can lead to objectionable accumulations.

    The principal elements of concern are heavy metals such as cadmium,
mercury, lead, chromium, nickel, zinc, copper, etc. Boron, occurring naturally in soils and in wastewaters, can also be toxic to plant growth when
in high concentration. Copper, nickel, zinc, cadmium, and boron can be
toxic to plants (phytotoxins) in concentrations found in some treated wastewater effluent. The other elements are generally present in very minute
amounts, but some may accumulate in the food chain [15].

    Some of the factors affecting plant uptake of heavy metals are the
following:

1. Levels of toxic elements in the wastewater or sludge and their charac-
    teristics

2. Species of plants grown, their age, condition, and rooting depth. Some
    plants are known to be metal accumulators.

3. Background concentration of toxic elements in the soil and their dis-
    tribution

4. Ability of soil chemical constituents to convert toxic elements to non-
    available chemical compounds. This ability is in turn affected by the
    nature of toxic element and the type of soil, for example:

    a. pH of the soil solution

    b. Organic and clay content and type in the soil

    c. Phosphate level in the soil

    d. Cation exchange capacity (CEC) of the soil and the exchangeable
        sodium percent (ESP) in the soil

    e. Adsorption and precipitation

    The above factors determine the deleterious effects of heavy metals
and other substances on plants and soils. For example, in soils with high
pH, most heavy metals occur in precipitated, unavailable forms. Similarly,
the phosphorus uptake by plants is also retarded at high pH. Thus, calcare-
ous solids are generally less amenable to manifestations of heavy metal
toxicity than are the neutral or acidic soils. It has been suggested that

the principal factor in retention of heavy metals is sorption on hydrous oxides of manganese and iron [49]. Problems, particularly with heavy metals, are possible where soil pH is below the range of 6.0 to 4.9, depending upon the soil and crop. This is generally unlikely because farmers contrive to keep their soil on the alkaline side, as liming, for example, gets better crop yields.

Organic matter, in the form of complex molecules of humus, is capable of chelating and fixing heavy metals, thus making them unavailable to plants. Thus, soils rich in organic matter can hold back heavy metals from the plants over short periods of time, until eventually saturation or breakdown of organic matter releases the accumulated elements into the soil solution.

Clay type and content determine the amount of exchangeable bases such as Ca^{2+}, Mg^{2+}, K^+, and Na^+ which influence the CEC of the soil. This capacity is an indication of the capacity of the soil to chemically "tie up" heavy metal ions. Unfortunately, the more desirable soils for wastewater irrigation are less clayey in order to have higher permeability.

The uptake rate of any element is dependent on the nature of the plant species and its age, the nature of the element itself, and the "competition" it meets from other substances present in the soil. The combination in which the elements occur in the soil also affects the damage caused. For example, certain toxic element availabilities can be reduced in some instances by the available phosphate level.

The mode of distributing the wastewater can also have an effect on the accumulation of toxic elements in plants. Spray irrigation is undesirable from this point of view, since the toxic ions contained in the wastewater escape the complexing capacity of the soil, to a large extent, and tend to be absorbed directly through the leaves.

More serious consideration is being given at the present time to the fate of cadmium, copper, and zinc in the food chain. The other heavy metals do not seem to accumulate in edible portions of crops and are essentially phytotoxins (some substances in trace amounts may, in fact, be advantageous for crop growth). Zinc is more readily absorbed than most other heavy metals. The presence of copper somewhat inhibits zinc transport through the plant.

There is some uncertainty at the present time concerning safe levels of many elements accumulating in plants. Little is known concerning the long-range effects of toxic elements applied to agricultural lands through the continuous use of wastewater and sludges. Sewage farms have been in operation in many parts of the world for a number of years, but criteria for establishing and quantifying metabolic damages and yield reductions resulting from various substances present in the wastewater have not yet been developed. Effective analytical methods for detecting and measuring small concentrations of heavy metals in plants and soils have only recently been applied. A further complicating factor is the combined and possibly synergistic effect on health of a mixture of toxic and nontoxic chemicals.

Trace metals and toxic substances are of particular significance in dealing with industrial wastewaters and sludges. Their presence in micro-quantities may not affect biological waste treatment plants, but may be of consequence later owing to accumulation in soils and plants. Table 9.11 gives data from a Japanese sewage treatment plant, showing the relative concentrations of a few heavy metals in raw sewage, treated effluent, and in partially dried sludge. The corresponding concentrations in most cases in sludge are one to three orders of magnitude larger than in raw sewage [14]. Recommended limits for potentially toxic trace elements in irriga-tion waters are not available at present. For comparison, concentrations of some elements in large rivers of North America are given in Table 9.12, as their waters are being used for irrigation.

Table 9.11 Concentration of Some Heavy Metals in Sewage and Sludge at Ukima Treatment Plant, Tokyo, 1970

| | Concentration (mg/liter) | | |
Item	Raw sewage influent	Treated secondary effluent	Partially dried sludge (25% solids)
Fe	4.4	4.7	4,650
Cu	1.75	1.09	744
Cr	4.0	2.6	242
Zn	4.6	3.9	884
Cd	0.10	0.05	42
Hg	0.57	0.52	16.2
Pb	19.2	8.4	970
Total solids	1,947	1,697	252,000
Volatile solids	678	465	129,000

Source: Ref. 14.

Table 9.12 Minor Elements in Large Rivers of North America

Element	Median (μg/liter)	Range (μg/liter)
Ag	0.09	0-0.94
Al	238	12-2550
B	10	1.4-58
Ba	45	9-152
Co	0	0-5.8
Cr	5.8	0.72-84
Cu	5.3	0.83-105
Fe	300	31-1670
Li	1.1	0.075-37
Mn	20	0-185
Ni	10	0-71
Pb	4	0-55
Sr	60	6.3-802
Ti	8.6	0-107
V	0	0-6.7
Zn	0	0-215

Source: Ref. 4.

Generally, in sludge digestion many of the heavy metals are precipitated as sulfides and are, therefore, more likely to be found in the solid phase. Dried sludges from industrial cities have been used for a long time for agricultural purposes with no adverse reports on plants or health. But the potential for damage does exist (Table 9.13).

University of Illinois agronomists studying the application of Chicago sludge on land concluded that the land in question would have the ability to accept accumulations of the heavy metals in Chicago's sludge for at least 100 years without any adverse effects on crop productivity or usefulness [2].

Since time immemorial, rivers have been the major source of irrigation waters, and some indication about heavy metal concentrations can be obtained from their quality (Table 9.12). It is, however, true that in recent times the rivers have been receiving increased pollutional loads and, hence, experience with irrigational use of more polluted waters over prolonged periods

Table 9.13 Composition of Fixed Solids in Sludge from Domestic Sewage

Constituent	Composition of fixed solids in anaerobically digested primary sludge (percent dry basis)
Aluminum (Al_2O_3)	4.3
Arsenic (As)	Trace
Barium (BaO)	0.10
Calcium (CaO)	5.70
Chromium (Cr_2O_3)	-
Cobalt (CoO)	-
Copper (CuO)	0.05
Iron (Fe_2O_3)	6.0
Lead (PbO)	0.20
Magnesium (MgO)	0.60
Manganese (MnO)	0.04
Nickel (NiO)	-
Phosphorus (P_2O_5)	1.50
Potassium (K_2O)	0.50
Silicon (SiO_2)	27.60
Sodium (Na_2O)	1.50
Sulfur (SO_3)	2.50
Titanium (T_iO_2)	0.10
Zinc (ZnO)	0.04

Source: Ref. 16.

is not available. For example, in recent times the concentration of detergents (ABS) has been increasing. Several studies have shown the undesirable effect of ABS on crop yields. Several rivers in Europe and the U.S. sampled between 1962 and 1967 have shown anionic detergent concentrations averaging from 0.1 to 0.3 mg/liter [9]. Some irrigation water standards limit their concentration to less than 3 mg/liter as anionic detergents, whereas some drinking water standards allow up to 1 mg/liter.

Example 9.2 If cadmium contained in treated effluent and liquid sludge is 0.05 and 8 mg/liter, respectively, estimate the amount that would be applied to land in kg/cm of application, over 1 hectare of land.

1 cm over 1 ha $= 0.01 \times 10{,}000 = 100 \text{ m}^3$

Cd in effluent $= 0.05 \text{ mg/liter} \times 100 \times \dfrac{10^3}{10^6} = 5 \times 10^{-3} \text{ kg/ha-cm}$

Cd in liquid sludge $= 0.8 \text{ kg/ha-cm}$

Substances such as chlorides have long been recognized for their deleterious effects on crops, and recommended limits have been included specifically in the standards of some countries for irrigation waters (see Table
9.23). The entry of chlorides, sulfates, and salts in general from the
infiltration of brackish water into sewers can ruin the utility of an otherwise perfectly usable city wastewater and should be guarded against as they
are not removed in conventional waste treatment processes.

Specific industrial wastewaters which may be used directly for irrigation could bring in certain toxic substances and heavy metals of concern.
For example, plating wastes containing zinc or chromium or other metals
may be mixed with the rest of the wastes of a factory that would otherwise
be dilute and disposable by irrigation. Wastes from nitrogenous fertilizer
factories may be disposable on land after dilution to control total dissolved
solids, provided arsenic wastes if any are segregated for separate disposal.
Mining and metal refining industries may generate dilute wastes containing
copper or other metallic compounds.

Generally, straight chemical industry wastes, which are likely to contain acids, alkalis, heavy metals, and such other substances, are not likely
to be considered for land disposal. The problem arises in the case of those
industries which deal with organic products like leather tanning, cotton
textiles, pulp and paper, etc., whose manufacture contains some metals or
compounds of concern in irrigation. Leather tanneries are often located
adjacent to slaughterhouses and may have a common disposal facility and
tanneries may discharge chromium, for example. The case may be similar for
pulp and paper mill wastes, which may contain mercury compounds from the
fungicides used in manufacture.

Food industry wastes are generally most suitable for discharge by
land irrigation, although there is growing concern about the fate of certain nondegradable preservatives, pesticides, etc., found in small quantities in such wastes.

9.2.13 Traditional Criteria for Irrigation Waters

A substantial amount of information is available concerning the permissible quality of freshwaters for irrigational use. This knowledge is
equally applicable to the evaluation of wastewater quality. For pulp and
paper mill wastes, a specific study was made to demonstrate the striking
similarity in soil response to mill effluent and a synthetic effluent prepared by adding similar salt concentrations to freshwater [37]. Similar
results could be expected with other wastewaters. Hence, some of the traditional quality criteria used for irrigation waters are discussed below.

1. TDS, often reported in terms of EC
2. Sodium content, expressed in terms of SAR, ESP, and percent sodium
3. Bicarbonate or RSC
4. Boron
5. Chlorides
6. Other ions
7. pH

As the salts contained in the irrigation water can accumulate over a
period of time in the soil, it is often the salt concentration in the soil
which is of greater interest than that in the irrigation water. The soil
concentration is determined by collecting soil samples and extracting a
solution either by squeezing out on a special press or by treatment with
alcohol. For methods of analysis, avoidance of interference in testing,
correction for negative adsorption, etc., many reference works are available.
The irrigation water standards of some countries are given in Table 9.25.

Total Dissolved Solids

This is a very widely used criterion for judging the quality of irrigation waters. It is a measure of the salinity of the water and is expressed
as mg/liter of TDS or as EC in μmhos/cm. The latter is the preferred manner of measurement and expression.* High salinity in the irrigation water

---

*Electrical conductivity can be approximately related to the total dissolved
solids concentration with the empirical formula TDS (mg/liter) $\simeq$ 0.64 EC
(μmhos/cm). This is only an approximation as the real relationship is a
function of the chemical composition of the given water. Calibration curves
can be prepared for specific waters.

Table 9.14 Classification of U.S. Irrigation Waters

Salinity of water	Uses	EC (μmhos/cm, 25°)
Low	For most crops on most soils	< 250
Medium	Suitable in most instances with moderate drainage	250-750
High	For salt tolerant plants on adequately drained soils	750-2250
Very high	May be used in special circumstances with very tolerant plants and excess leaching	2250-5000

Source: Refs. 42 and 48.

can affect the osmotic pressure of the soil solution and can produce critical conditions with regard to soil water availability for the plant, thus affecting plant growth and even survival. Waters in the U.S. are commonly classified as given in Table 9.14. In the USSR, irrigation waters are evaluated as shown in Table 9.15.

Highly saline groundwaters have been used since time immemorial in North Africa and the Middle East (with salts up to 7000 mg/liter) for the irrigation of sandy oases without causing salinization or deterioration of the soils. Similarly, in Israel, water with 2300 mg/liter total salt (including 100 mg/liter chlorides) have been used to irrigate tolerant crops

Table 9.15 Evaluation of Irrigation Waters in the USSR

Salt content (g/liter)	Approximate EC (μmhos/cm, 25°)	Evaluation
0.2-0.5	300-750	Water of the best quality
1.0-2.0	1,500-3,000	Water causing salinity and alkalinity hazard
3.0-7.0	4,500-10,500	Water could be used for irrigation only with leaching and perfect drainage

Source: Ref. 48.

on sandy soils [48]. Leaching and easy drainage are essential in such cases.

Treated wastewater effluent from mainly domestic sewage will exhibit EC values generally around 1000 μmhos/cm (TDS ≃ 650 mg/liter) for the relatively more dilute sewages as one would expect on the average in the U.S. In the developing countries, with less water usage, the EC values for treated effluent would be in the range of 1000 to 2000 μmhos/cm (TDS ≃ 650 to 1300 mg/liter), or even more, especially if the original water supplies are high in dissolved salts.

The national average range of mineral pickup of TDS by domestic usage in the U.S. is reported to be between 100 and 300 mg/liter (EC ≃ 150 to 250 μmhos/cm) above the background value in the water supply [12] (see Table 5.2).

Values of EC are not readily available for treated industrial wastewaters but can be estimated roughly from TDS values, or measured directly as EC.

The permissible salinity in irrigation water depends on the type of crops proposed to be cultivated and their tolerance to salinity built up in the soil. Buildup of salinity in the soil, in turn, depends on the soil drainage and reference is made to Sec. 9.2.18 where the question of buildup of salinity in the soil is discussed more fully. For most crops, the conductivity of the soil saturation extract (not of irrigation water) has to be limited, as shown in Table 9.19, from about 4000 for fruits to 16,000 to 18,000 μmhos/cm for fodder grasses. The permissible salinity is also affected by the sodium content. Figure 9.5 gives a diagram which is useful in classifying various waters based on the recommendations of the U.S. Salinity Laboratory (1954).

Another procedure for evaluating the salinity of irrigation waters is based on what is called "effective salinity." Effective salinity includes all dissolved salts in irrigation water except calcium sulfate and calcium and magnesium bicarbonates. It is assumed that these salts precipitate and so do not contribute to soil salinity. The suggested classification is given in Table 9.16.

Sodium Content

The presence of sodium in the soil in exchangeable form can cause adverse effects even at low concentrations. Exchangeable sodium tends to make a moist soil impermeable to air and water and, on drying, the soil

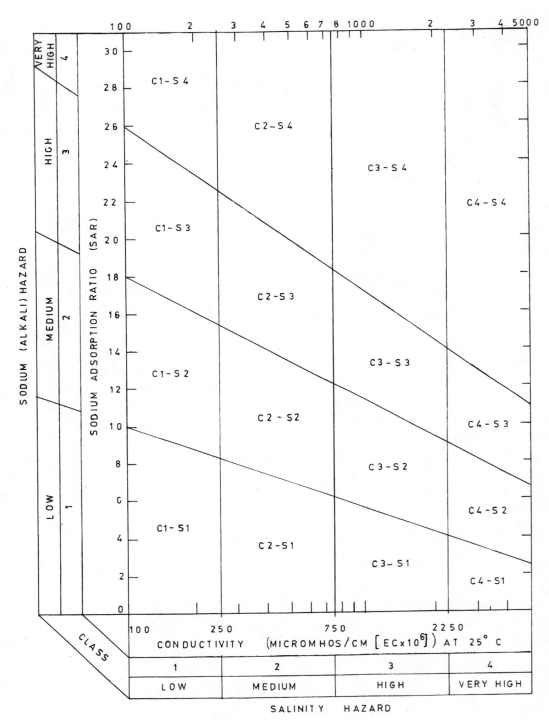

Figure 9.5 A diagram for the classification of irrigation waters (from the U.S. Salinity Laboratory).

Table 9.16 A Tentative Classification for Effective Salinity
of Irrigation Waters

Soil condition	Permissible range of "effective salinity" (meq/liter)[a]
Little or no leaching of the soil can be expected	3-5
Restricted leaching; slow drainage	5-10
Open soils; easy percolation	7-15

[a]1000 μmhos/cm at 20° C ≃ 10 meq/liter.
Source: Ref. 4.

can become hard and difficult to till. Even interference with germination
and seedling emergence can occur.

Several methods have been used to evaluate the sodium hazard [48]. A
reliable index used in the U.S. and some other countries to estimate the
tendency of irrigation water to form exchangeable sodium in the soil is the
SAR which is a calculated value.

$$SAR = \frac{Na^+}{\sqrt{\dfrac{Ca^{2+} + Mg^{2+}}{2}}} \qquad (9.5a)$$

in which all concentrations are expressed as meq/liter. Table 9.17 gives
a few typical values.

In fact, the sodium hazard also depends on the TDS concentration, and
for a given value of SAR, the hazard increases as the EC value increases.
The graphical classification given by the U.S. Salinity Laboratory (1954)

Table 9.17 Some Typical Values of SAR

Sodium hazard to soil	SAR value in irrigation water	Approximate ESP
Low	0-10[a]	0-12
Medium	10-18	12-20
High	18-26	20-27
Very high	Above 26	Above 27

[a]Certain fruit crops and woody plants may show damaging sodium accumulations at about SAR 4 to 8.
Source: Ref. 23.

is often followed (Figure 9.5) to determine the quality of water for irrigation purposes.

The SAR value of the original drinking water can be expected to rise by about 3 to 5 meq/liter in the treated wastewater, which is mainly domestic in character. However, actual values can be estimated for any wastewater from the above expression for SAR.

In Figure 9.6, the values of ESP corresponding to different SAR values are given. The ESP of soil is the fraction of negatively charged adsorption sites in soil occupied by sodium ions. The correlation between ESP and SAR is given by the empirical equation:

$$ESP = \frac{100(-0.0126 + 0.01475SAR)}{1 + (-0.0126 + 0.01475SAR)} \tag{9.5b}$$

If more than 10 to 20% of the cation exchange capacity of the soil is taken up by sodium, the physical condition of the soil deteriorates. The finer the soil, the lower the value at which soil structure deteriorates. Soils with high ESP are "dispersed" and are unsuitable for crop production owing to low permeability to water and air.

Nonsaline "sodic" soils (high in ESP but low in salinity) may show nutritional disturbances to crops at about 10% ESP values. Sodium-sensitive crops and fruit crops may be affected at even 4 to 5% ESP values.

SAR varies as the square root of the total salt content of the water when cationic proportions are constant. Thus a ninefold increase of total salts in a poorly draining soil would increase the SAR threefold above the value of the incoming water.

SAR values for bleached kraft mill effluents are reported to average around 4.6, whereas effluents from unbleached kraft mills may be as high as 7.6. Values less than 8.0 are considered tolerable for permeable soils, whereas lower values are desirable for clayey soils [37].

Another parameter concerning sodium is "percent sodium"; it is defined as the percentage of the sodium content of water in the total cation content. In fact, previously, water quality was evaluated on the basis of its sodium percentage alone. This is calculated as follows:

$$\text{Percent sodium as } Na^+ = \frac{100Na^+}{Na^+ + Ca^{2+} + Mg^{2+} + K^+} \tag{9.6}$$

where all values are expressed as meq/liter.*

---

*Calcium   (Ca^{2+}), 1 mg/liter = 0.049 meq/liter   Carbonate   (CO_3^{2-}), 1 mg/liter = 0.0333 meq/liter
Magnesium (Mg^{2+}), 1 mg/liter = 0.082 meq/liter   Chloride   (Cl^-), 1 mg/liter = 0.0282 meq/liter
Sodium   (Na^+), 1 mg/liter = 0.043 meq/liter   Bicarbonate (HCO_3^-), 1 mg/liter = 0.0164 meq/liter
Potassium (K^+), 1 mg/liter = 0.025 meq/liter

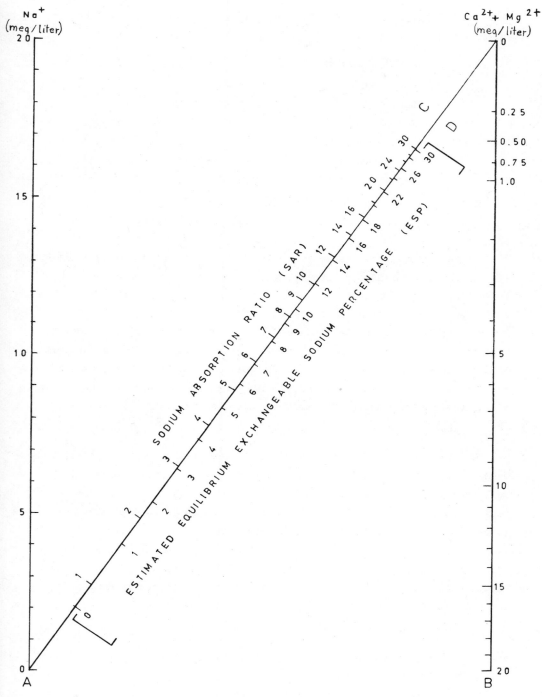

Figure 9.6 A nomogram for determining the SAR value of irrigation water and for estimating the corresponding ESP value of a soil that is at equilibrium with the water (from the U.S. Salinity Laboratory).

Generally, the values given below are recommended for irrigation waters [23]:

Water condition	Percent sodium
Very good	20
Good	20-40
Medium	40-60
May not be acceptable	60-80
Not acceptable	Over 80

The likely increases in these four ions in domestic sewage over their original concentrations in drinking water supply can be estimated from Table 5.2.

Certain industrial wastewaters exceed the above recommended values for SAR, or percent sodium, or both and cannot be considered for continuous irrigation without precautions. Cotton textile wastes are often high in sodium owing to the presence of caustic soda from certain kiering and dyeing operations. Corrective measures include the addition of Ca^{2+} ions to improve the ratio of Na^+ and Ca^{2+} ions. Calcium can be added in the form of periodic applications of lime to the soil, or the treated wastewater could be passed over gypsum ($CaSO_4$) contact beds prior to discharge on land. Effluents from pulp and paper mills may also be high in Na^{2+}.

Bicarbonate or Residual Sodium Carbonate Classification

A large amount of bicarbonate tends to precipitate out the calcium, as $CaCO_3$, from water and soil. Magnesium enters the exchange complex of the soil, replacing the precipitated calcium. Ordinarily, magnesium does not replace calcium to any great extent but, if calcium is precipitated as it is released, the reaction proceeds to completion.

As calcium and magnesium are lost from the soil water, the relative portion of sodium is increased, with an attendant increase in sodium hazard discussed in the preceding section. This hazard is generally evaluated in terms of "residual sodium carbonate" (RSC), defined as

$$RSC = (CO_3^{2-} + HCO_3^-) - (Ca^{2+} + Mg^{2+})$$ (9.7a)

where all ion concentrations are expressed as meq/liter.

Some typical recommended values for irrigation waters are given below [23]:

Water quality	RSC (meq/liter)
Safe	Less than 1.25
Marginal	1.35-2.5
Unsuitable	Over 2.5

With an increase in "leaching fraction" (Sec. 9.2.18), the percent of bicarbonate that precipitates decreases. The use of marginal waters can be corrected by the use of gypsum, for example, as stated in the previous section.

Boron

Although boron is necessary in small quantities for plant growth, it is one of the important phytotoxic substances because it can occur in toxic concentrations in natural irrigation waters. Very few other substances occur in toxic concentrations in natural waters and, hence, it is necessary to check for boron in assessing irrigation water quality. Industrial sources involving borax mining, refining, and usage may discharge considerable quantities in wastewaters.

In areas where boron occurs naturally in excess in the soil or in the irrigation water, boron-tolerant crops may grow satisfactorily. The relative boron tolerance of a few crops is shown below. More exhaustive lists are available in several standard texts on irrigation.

Crops	Limiting boron concentration in irrigation water [23] (mg/liter)
Fruits	0.3-1.0
Vegetables, wheat, barley, corn, cotton, potato	1.0-2.0
Sugar beet, onion, carrot, lettuce, palm trees, alfalfa, etc.	2.0-4.0

Assuming boron in the soil acts as other soluble salts in the soil, the boron content of the soil saturation extract will build up to a steady-state value in the course of time as explained in Sec. 9.2.18. A concentration of boron below 0.7 mg/liter in the soil extract is considered safe for all plants.

Chlorides

Chlorides are not adsorbed on a soil complex and have no effect on
the physical properties of soil per se. However, woody plants including
most fruit trees and ornamental shrubs are particularly sensitive to sodium
and chloride ions, exhibiting characteristic leaf burns when leaves accumulate about 0.25 and 0.5% on a dry weight basis of these elements, respectively [42]. Among these plants, however, there are wide variations in the
rate of chloride accumulation and, hence, tolerances of different crops
also vary widely as shown below.

Crop	Limit of tolerance to Cl⁻ in soil extract (meq/liter)
Citrus fruits	20-30
Stone fruits	14-50
Avocado	10-16
Grapes	20-50
Berries	10-20

The domestic usage of water adds 4 to 8 g/person of chlorides daily
to the quantity present in the water supply. Depending upon the water
usage, this adds about 30 to 50 mg/liter Cl^- in wastewater, or about 0.85
to 1.4 meq/liter, which may further concentrate about two to ten times in
the soil extract. Hence, chlorides in domestic sewage generally do not
cause much of a problem for most fruit trees. However, the original water
supply itself may be high in chlorides, there may be infiltration of brackish or saline water into sewers, or certain industrial wastes may bring in
chlorides in objectionable quantities and impair the irrigation value of
the wastewater.

If the upper limit of 50 meq Cl^-/liter of soil saturation extract for
the more tolerant fruit trees is not to be exceeded, and if the salt buildup in the soil is about five times the value in the irrigation water, the
latter should not have chlorides in excess of 10 meq/liter, or approximately
330 mg/liter. An excess concentration might be permissible if the soil was
sandy with very good drainage characteristics, or if the climate was wet,
thus encouraging greater leaching out of the salt from the soil, or if
chloride-sensitive crops were avoided. Some irrigation water standards,
as in India, permit chloride concentrations even up to 600 mg/liter (16.92

meq/liter) under favorable conditions. For citrus fruits in Israel, irrigation waters with chlorides up to 15 meq/liter are considered to be low or of no risk on sandy and loamy soils, and up to 7.5 meq/liter on clayey soils. Since chlorides do not affect the soil, they have not been included in some water classification systems.

Incidentally, the use of sprinklers that wet the foliage may demand lower values of chlorides and sodium owing to possible direct foliar absorption and resulting leaf injury in fruit trees, citrus plants, etc., which are sensitive to them [42].

Other Ions

Magnesium

It has been suggested [48] that the magnesium content of irrigation water can affect soils unfavorably if the following ratio exceeds a value of 50:

$$\text{Magnesium ratio} = \frac{Mg^{2+} \times 100}{Ca^{2+} + Mg^{2+}} \qquad (9.7b)$$

However, there is limited experience available in this regard.

Sulfates

The term "potential salinity" of water has been suggested in terms of $Cl^- + \frac{1}{2}SO_4^{2-}$ in meq/liter, and the following values have been proposed [48]:

Desirable soil permeability	$Cl^- + \frac{1}{2}SO_4^{2-}$ (meq/liter)
Good	5-20
Medium	3-15
Low	3-7

pH Values

Except in the case of certain industrial wastes, pH values are seldom low enough to be a problem since the soil itself has buffer capacity. Values as low as pH 2.8 have been reported as a result of drainage from coal mines. Continuous addition of acidic wastes would, however, decrease the pH of the soil and alter the equilibrium conditions in the soil moisture, thus possibly rendering certain metal ions, such as iron, aluminum, and manganese, soluble and facilitating their uptake by plants. The addition of lime to the soil can help to avoid or retard damage to soils and crops.

The study referred to earlier regarding the irrigational use of pulp and paper mill wastes suggests restricting the effluent pH between 6.5 and 9.0 to avoid deleterious effects on the soil and crops [37]. It should be feasible to apply this to most other wastewaters as well.

9.2.14 Soil Quality

Before designing a large irrigation system, it is necessary to have some data on existing soil quality in the potentially valuable areas for irrigation, not only to plan the system better, but also to be able to evaluate possible damage to soil quality over a long enough period of time. Several chemical processes are at work in the soil when it is used for land treatment; the most important ones are ion exchange, adsorption, precipitation, and chemical alteration.

As soil data can be interpreted better in the context of climatic, topographical, and geomorphological data in general, it would be helpful to examine the latter along with the soil.

After preliminary investigations of the type and thickness of the soil layer, soil samples are obtained generally from within the top 40 to 50 cm, taking usual precautions in practice. For example, two types of samples are obtained: disturbed and undisturbed, and tested for the following:

Tests on undisturbed samples:

1. Moisture content
2. Bulk density
3. Permeability

Tests on disturbed samples

1. Texture (% of sand, silt, clay, etc.)
2. Percent $CaCO_3$
3. Organic matter (organic carbon and organic nitrogen C/N ratio, phosphorus content, and potassium content)
4. pH of soil saturation extract
5. EC of soil saturation extract
6. CEC of soil, and cation composition of soil extract (Ca^{2+}, Mg^{2+}, Na^+, and K^+)
7. Anion composition (HCO_3^-, CO_3^{2-}, SO_4^{2-}, and Cl^-)
8. Boron

The effects of some of the above characteristics have already been discussed earlier with regard to trace metal uptake. The purpose of carrying out the above tests is to get a better insight into the soil quality. The type and number of tests to be performed depend on the extent of the system to be designed and on the quality of the wastewater proposed to be used.

An idea of the fertility of the soil is obtained from its pH, organic carbon and N, P, and K concentrations, some indication of which is given in Table 9.18 [4]. Organic carbon and organic nitrogen are on the order of 50 and 5%, respectively, of the organic matter in well-humified soils, giving a C/N ratio of about 10. Organic matter itself constitutes a small fraction of the total soil mass.

Interpretation of soil data with regard to the ability of the soil to support the desired type of crops under irrigation is the work of agricultural specialists who should be associated with not only the soil but all other aspects of any wastewater irrigation project.

In the absence of locally available data, soil infiltration tests need to be conducted at the site to establish the permeability of the soil, and prolonged tests over 3 to 4 months may give a reasonable idea. Field irrigation rates must be kept below the values obtained in prolonged infiltration tests.

Example 9.3 A soil contains 1.0% organic matter, 0.025% P, and has a carbon/nitrogen ratio of 10:1. If the soil bulk density is 1.2 g/cm^3, estimate the nitrogen and phosphorus contents in the top 30 cm of the soil as kg/ha, and the N/P ratio of the soil.

Table 9.18 Presence of Carbon, Nitrogen, Phosphorus, and Potassium in Soils

Description	Percent in soil				C/N
	Organic C	N	P_2O_5	K[a]	
Low	0.5-1.0	< 0.10	0.025-0.05	< 0.5	10-12
Moderate	1.0-1.5	0.10-0.15	0.05-0.075	0.5-1.0	12-20
High	Over 1.5	Over 0.15	0.075-0.13	1.0-2.0	Over 20

[a]Meq K/100 g.
Source: Refs. 4 and 48.

Weight of soil = 10^8 cm^2 x 30 cm x $\frac{1.2}{10^3}$ = 3.6 x 10^6 kg/ha

Weight of organic matter = 0.01 x 3.6 x 10^6 = 3.6 x 10^4 kg/ha

If carbon is assumed to be 50% of the organic matter, nitrogen will be equal to 10% of the carbon as C/N = 10:1. Thus,

Carbon   = 0.5(3.6 x 10^4) = 18,000 kg/ha

Nitrogen   = 0.1(18,000) = 1,800 kg/ha

Phosphorus = 0.00025 x 3.6 x 10^6 = 900 kg/ha

N/P ratio = $\frac{1800}{900}$ = 2:1

The soil has a relatively higher amount of P compared with N than is necessary for plant growth to occur. Generally, biological growth requires an N/P ratio at about 16:1. Hence, nitrogen is the limiting nutrient in this case.

Land Classification and Topography

Many different kinds of land classification schemes and systems are in use in the world. Lands are often classified according to their capability for cultivation without deterioration over a long period of time. Lands are also classified according to their suitability for irrigation. As irrigation is generally expensive, such a classification also takes into account economic criteria. The land classification standards followed by the U.S. Bureau of Reclamation are given in Table 9.19 as a typical example of how lands are classified. The various practices followed in other countries are summarized elsewhere [48].

Table 9.19 Typical Classification of Land

Land class	Definition
I	Soils suitable for irrigation, without any problems.
II	Soils suitable for irrigation with some small problems such as slope, texture, structure, depth, and rock clearing.
III	Soils suitable for irrigation but having more extensive problems concerning slope, texture, structure, depth, and rock clearing.
IV	Soils suitable for irrigation after special economic and engineering measures are taken.

Table 9.19 (Continued)

Land class	Definition
V	Soils temporarily not suitable for irrigation. Require further economic and engineering study for determination.
VI	Soils not suitable for irrigation

Source: U.S. Bureau of Reclamation.

In this connection, the general topography of the land also needs to be noted. Flat lands may have slopes of 0 to 2%, whereas the steeper ones may slope 2 to 10% up or down. Drainage in the general area may or may not be well defined, and areas likely to be flooded at high flows need to be demarcated. Groundwater depth and the nature of strata up to that depth are of primary interest. A depth of agricultural soil less than 30 cm would limit cultivation to crops with shallow roots, while from the point of view of pathogen removal and general filtration, a depth of 1.5 m in good, unfissured soil would be desirable.

9.2.15 Soil Texture and Structure

The sizes and types of particles making up a soil determine its texture. These particles range in size from clay to fine gravel as shown below:

Clay	0.002 mm and smaller
Silt	0.002-0.05 mm
Sand	0.05-2.0 mm
Gravel	Larger than 2.0 mm

Sandy soils are classed as coarse-textured or light soils; they are gritty to the touch and crumble easily. Clay soils are fine textured and tend to be sticky and heavy. Most soils lie somewhere in between and are called "loams." They are mixtures of sand, silt, and clay, but their texture is governed by their clay content. A soil with more than 35% clay behaves, for all practical purposes, like a clay, while one with less than 10% clay is just like sand. A "loamy sand" has 10-15% clay; a "medium loam" has 15-23% clay; while a "clay loam" has 23-35% clay.

Table 9.20 Typical Physical Properties of Some Soils

Soil texture	Infiltration rate,[a] typical rank (cm/day)	Approximate total pore space (g/cm^3)	Approximate bulk density (g/cm^3)	Approximate moisture content (%)		
				At field capacity	At permanent wilting	Total available[b] (dry wt. basis)
Silty sand	120 (60-600)	38	1.65	6-12	2-6	4-6
Sandy loam	60 (30-180)	43	1.50	10-18	4-8	6-10
Loam	30 (18-50)	47	1.40	18-26	8-12	10-14
Clay loam	18 (6-36)	49	1.35	23-31	11-15	12-16
Silty clay	6 (0.6-12)	51	1.30	27-35	13-17	14-18
Clay	<0.6 (0.3-2.5)	53	1.25	31-39	15-19	16-20

[a]These are generally fixed or long-term infiltration rates. Rates vary greatly with soil structure and other factors, even beyond the ranges shown in parentheses.

[b]Readily available fraction may be 70-80% of stated values.

Note: Infiltration rate of 1 cm/day = 100 m^3/ha-day.

Texture has an important influence on drainage characteristics, and thus on the amount of moisture that can be stored in a soil (see Table 9.20). This helps to determine the frequency and duration of irrigations and how soon a soil can be worked after a rainfall. Sandy soils retain less moisture than clayey soils, often have little organic matter and nutrients present in them, are liable to become acidic more quickly, and need more manure and irrigant applications.

The depth and texture of the subsoil is also important as it determines the depth to which plant roots can penetrate and the moisture that can be held in reserve. Poor drainage tends to cause anaerobic (reducing) conditions under which the normally rich brown ferric oxides in the soil are converted to grey-colored ferrous compounds. The depth at which this occurs indicates how deep the soil is useful for rooting.

Recommended methods for soil sampling may be used. The use of a soil auger is convenient; it can be screwed into the ground 15 cm at a

time until a depth of 90 cm is reached. The cores thus obtained are examined for texture, color, and drainage quality.

Soil classification is often done according to the diagram shown in
Figure 9.7, in which, for example, a soil containing 13% clay, 41% silt,
and 46% sand would be classified as a "loamy sand."

Soil structure is determined by the variation in grain sizes. Uniform-
grained soils are more porous, whereas those in which sizes vary greatly
are more densely packed. Soil structure is also affected by agricultural
practices. Pore space in irrigated soils often ranges between 25 and 55%
(Table 9.20). A reduction in pore space by 5 to 10% would be considerable
and could seriously impede plant growth by obstructing the movement of air
and water.

Irrigation practices can affect grain size distribution and, thus, the
soil structure. For example, rotation of crops can help to improve or
maintain the soil structure. An unfavorable concentration of sodium salts
in the soil can cause the deterioration of soil structure and the loss of
porosity by a breakdown or dispersion of clay aggregates.

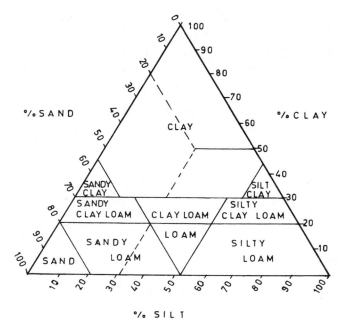

Figure 9.7 A soil classification diagram.

The real specific gravity of a soil is the specific gravity of a single soil particle, with values ranging from 2.65 for inorganic soil to 1.20 for organic soil. However, owing to air entrained in a soil mass, the volume weight or bulk density is less than the above values and is termed apparent specific gravity. Hence, both texture and structure affect apparent specific gravity.

9.2.16 Soil Moisture

In all irrigation methods where water is applied to the surface of land, it enters the soil within a short time, generally ranging from a few minutes to a day or so, and is "stored" in the soil for later use by plants. For this reason, irrigation is generally intermittent, spread over an hour or so per day, or alternatively, as long as required to flood the field to 15 to 20 cm depth once in a week or so. The object of irrigation is to get water into the soil where it can be stored. Thus, the rate of entry of water into soil under field conditions, called the intake rate, is of importance.

Gravitational water is that part of the moisture which will move out of the soil if favorable drainage is provided. Capillary water is that part which exists in the pore space of the soil and is retained against the force of gravity. Hygroscopic water is only on the surface of the soil grains and is not capable of movement by the action of gravity or capillary forces. When gravitational water has drained away, the remaining moisture content of soil is called the "field capacity."

A plant will wilt when it is no longer able to extract sufficient moisture from the soil. The soil moisture content at which the plant permanently wilts is called the "permanent wilting point" or the wilting coefficient. The difference in moisture content of the soil between field capacity and permanent wilting is called the available moisture (it is mainly the capillary water) and, in fact, only 75% of this is relatively readily available to a plant.

A fine-textured soil retains more water than a coarse-textured soil. At field capacity, for example, a cubic meter of clay soil will hold about 400 liters of water, a loam soil about 270 liters, and a sandy soil only about 135 liters. The addition of organic matter as manure not only provides nutrients, but has a generally favorable influence on soil structure and permeability; its effects, however, on moisture availability in the soil may be practically negligible. The available moisture in a few typical soil types is given in Table 9.20.

The time taken after rainfall or irrigation to reach the field capacity depends on the rapidity with which the gravitational water drains out; this ranges from 1 to 2 days for clays and loams. Thereafter, the soil moisture content continues to decrease and, unless further rainfall or irrigation occurs, the permanent wilting point is eventually reached. According to one school of thought, plants are capable of using water equally readily over the range of soil moisture content between field capacity and the permanent wilting point. There is another view that, under certain conditions of soil and climate, water is not equally available over the whole range and, to ensure maximum crop yields, the soil moisture content should not be less than 1/4 to 1/2 of the field capacity [48]. This affects the irrigation practice followed.

Soil samples are collected in airtight containers to prevent any loss of moisture during transport to the laboratory. The moist samples are weighed, dried in an oven at 105° C, and reweighed. The difference in weight divided by dry weight of the soil gives the percent of moisture on a dry weight basis. Several other methods for field determinations are also available [4].

The effect of irrigation on the soil moisture content can be estimated from the following equation whose use is illustrated in Example 9.4.

$$I_e = \frac{P_f - P_i}{100} S_s D_r \qquad\qquad (9.8)$$

where

I_e = effective infiltrated irrigation water, mm (= field capacity)

P_f = final moisture content of soil, percent by weight, after irrigation

P_i = initial moisture content of soil, percent by weight, before irrigation

S_s = apparent specific gravity of soil, g/cm^3

D_r = depth of soil to the root zone to be wetted, mm

Example 9.4 For a clay-loam soil which is irrigated periodically, the moisture content before irrigation is 19% by weight. If the effective infiltrated irrigation water is 1000 m^3 per hectare, estimate the soil moisture content after irrigation. Assume the soil apparent specific gravity

= 1.35, and the depth to which the soil is wetted = 90 cm. What should be the frequency of irrigation if the evapotranspiration rate = 250 mm/month?

$$\text{Infiltrated water } I_e = \frac{1,000 \text{ m}^3}{10,000 \text{ m}^2} = 100 \text{ mm}$$

From Eq. (9.8):

$$100 \text{ mm} = \frac{P_f - 19\%}{100} \times 1.35 \times 900 \text{ mm}$$

Thus

$$P_f = 27.3\%$$

$$\text{Required irrigation frequency} = \frac{100 \text{ mm}}{250 \text{ mm/month}} = 12 \text{ days}$$

The purpose of periodic irrigation is to restore the field capacity of the soil for moisture, and a sufficient quantity must be used for that purpose. Generally, an excess quantity is used so that the full field capacity is restored at each irrigation; the balance leaches down below the root zone. The frequency of irrigation depends on the climate and the nature of the crops as reflected by the evapotranspiration rate E_t. Table 9.20 gives the range of moisture values for different types of soils at field capacity and at the permanent wilting point. As stated earlier, the minimum desirable moisture content may be somewhere between the field capacity and the wilting point, depending on the soil, crops, and other factors.

9.2.17 Infiltration, Intake, and Permeability

The time rate at which water will percolate into soil, or the rate of infiltration, is important for irrigation. It is influenced by soil properties and by the moisture gradient. The infiltration rate applies to a level surface covered with water. When the soil is furrowed, the term intake rate is preferred since it denotes infiltration under a particular configuration of the soil surface, although both terms are sometimes used interchangeably. Infiltration and intake rates are expressed as mm/hr or cm/day for a given soil.

The permeability of a soil is defined as the velocity of flow caused by a unit hydraulic gradient. Permeability is a velocity term, length

divided by time, and varies greatly from soil to soil. Also, the term
permeability is used for designating flow through soils in any direction.
Usually, soils are nonhomogeneous and anisotropic, meaning that the permeability not only varies from point to point but the permeability in one
direction may be different from the value in another direction. This is
most likely in alluvial soils and is discussed further in Sec. 9.3. Knowledge of permeability is essential in irrigation studies and in the design
of soil drainage systems. It can be measured in the field or in the laboratory by the use of suitable permeameters or by water draw-down tests. (The
latter method is discussed in Sec. 9.3.)

A permeameter used in the field is essentially a hollow tube of 30 to
60 cm diameter, pushed into the soil to a depth of about 60 cm with another
60 cm or so protruding out of the soil. The length of the tube is, thus,
1.0 to 1.2 m. It is desirable to have the tube made of thin steel with a
sharp edge to facilitate driving it in without disturbing the soil too much.

The permeameter can be used at a constant head by providing for either
an overflow arrangement or a ballfloat valve as shown in Figure 9.8. The
head may be kept at about 15 to 20 cm above the soil to approximate irrigation conditions. The length of travel through the soil is L, the distance
to which the tube has penetrated. The head causing a flow is H_L, from the

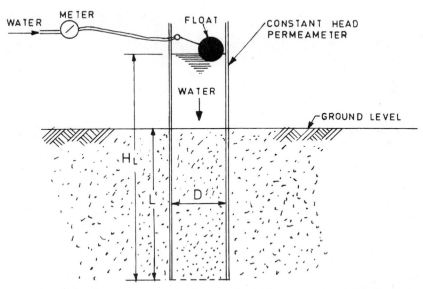

Figure 9.8 An arrangement for conducting a constant head permeameter test.

bottom edge of the tube to the constant water level. For a given diameter
D, the area of the tube is A, and the quantity of water Q passed through
the tube in a given time is measured either by noting a drop in the level
of the tube (for small flows only), by noting a drop in the level in an
adjoining tank, or by metering the inflow when it is large. Freshwater
available at the site is used, and the tube may be covered to prevent evaporation. There may be three to five such tubes set up in close proximity
and used simultaneously.

From the Darcy-Weisback equation, if the friction factor f is known,
the velocity V can be found from

$$H_L = f \frac{L}{D} \frac{V^2}{2g} \qquad (9.9)$$

For soils, this can be simplified because of negligible kinetic energy
and written in the following form:

$$V = k \frac{H_L}{L} \qquad (9.10)$$

where k = soil permeability (distance/time). It is a combined measure of
soil and fluid properties and is, therefore, affected by soil characteristics
as well as temperature and viscosity. But,

$$Q = AV \qquad (9.11)$$

and hence,

$$k = \frac{QL}{AH_L} \qquad (9.12)$$

Example 9.5 If the flow of water through a soil of 30 cm depth is
9.5 liters in 20 minutes at a head loss H_L of 70 cm, and the tube area is
1000 cm^2, estimate the permeability k and the infiltration rate.

From Eq. (9.12), the permeability

$$k = \frac{(9.5 \times 10^{-3})(30)}{(1000 \times 10^{-4})(70 \times 10^{-2})(20/60 \times 24)} = 293 \text{ cm/day}$$

The infiltration rate can be calculated as

$$\frac{Q}{A} = \frac{(9.5 \times 10^{-3})100}{(1000 \times 10^{-4})(20/60 \times 24)} = 684 \text{ cm/day}$$

For irrigation purposes, our primary interest is in intake rates. The
intake rate obtained at the start of the test is generally high and falls

off rapidly with time. Hence, the test must be continued over a long enough period of time to allow the percolation to level off. The set of tubes erected at a site could be run continuously for several days, or even for a month or more, to obtain percolation rates at fairly saturated conditions in the soil. Some other methods used involve spreading water along furrows or basins, or spraying it in order to simulate the system proposed to be used in full-scale irrigation. Where furrows are used, the quantity of percolated water in a given time is divided by the entire soil surface (length of furrow times spacing between furrows) to obtain the average intake rate. A few typical physical properties of some soils are given in Table 9.20.

The accumulation of inorganics from biodegradation, as well as those in the wastewater, tends to form a new layer of soil at the interface. Part of this is removed by leaching and surface runoff during wet periods; plowing of the field mixes the remaining with the original soil. Where liquid sludge is used for irrigation, deep plowing of the field would be necessary between plantings.

The cover crop plays an important role in affecting the percolation rate in the field. In laboratory lysimeter tests using pulp and paper mill wastes to irrigate alta fescue grass which has an extensive root formation, it was found that the presence of a good cover crop gave a twentyfold increase in the percolation rate on dense silt loam soil, whereas death or impairment of the cover crop consistently gave reduced rates [27]. Spray irrigation on bare soil has been reported to compact the top soil and reduce its permeability. Economical land use would demand that the irrigation system be designed with a proper crop cover in view.

However, in the absence of locally available data, soil infiltration tests as described above need to be conducted at various spots over a site to establish the infiltration rate of the soil. Actual irrigation rates must be kept below the values obtained from prolonged tests.

It would not be out of place here to mention that the continued use of poor quality wastewater which does not satisfy the various quality criteria discussed earlier may gradually deteriorate the soil and reduce its permeability. The gradual buildup of salt in the soil is discussed in the next section. A large amount of sodium in relation to other cations may cause a high ESP in the soil complex, which will have an adverse effect on the soil structure. Dispersion of clay particles can occur as a result of ion exchange, and difficulties may occur which, among other things, could reduce soil percolation rates. Practically every patch of irrigated soil, particularly in arid regions, is under the potential threat of salinization and alkalization and must be well controlled through water quality and soil drainage.

9.2.18 Buildup of Salts in Soil

The irrigated soil, with crops undergoing evapotranspiration, acts as a sort of concentrating mechanism leading to a gradual buildup of salts with time. This concentrating effect of the soil-crop system can be explained in terms of a simple model based on a mass balance of salt and water in the system.

With reference to Figure 9.9, for a unit area of soil at steady conditions, the water balance can be obtained by equating all incoming moisture to all outgoing moisture as follows:

$$I + P = E_t + L \qquad\qquad\qquad (9.13)$$

where

    I = applied irrigation water, volume/time

    P = effective precipitation, volume/time

    L = leachate or water loss below root zone, volume/time

    E_t = evapotranspiration, volume/time

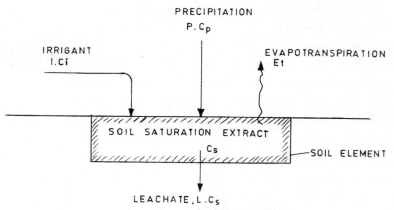

Figure 9.9 Salt buildup in soil.

A salt balance can be prepared by neglecting the amount of salts taken up by the plants and thus removed from the system when the plants are harvested. Also, it can be assumed that, over a small element of the soil, the concentration of salt built up in it is uniform throughout the element. It can be further assumed that the leachate is equally effective over the whole area and carries away the accumulated salt with which it comes into contact.

By neglecting plant uptake, the model gives more conservative estimates of salt buildup with time. However, all leachate is not equally effective in carrying away salt from the soil element with all types of soils, and to that extent the model underestimates the salt buildup.

As irrigation proceeds, the salt concentration starts to build up. But as buildup occurs, the mass lost per unit volume of leachate also increases until a steady-state condition is reached where the salt balance is given by

$$(IC_i) + (PC_p) = (LC_s) + (E_t C_E)$$

where

C_i = concentration of substance in irrigation water, mass/volume

C_s = concentration of substance in soil saturation extract, mass/volume

C_p = concentration of substance in precipitation

C_E = concentration of substance in evaporated moisture (also equal to zero as salts are left behind during evaporation)

Thus,

$$(IC_i) + (PC_p) = LC_s$$

The average incoming concentration $C_{i(av)} = (IC_i + PC_p)/(I + P)$ and $(I + P)C_{i(av)} = LC_s$. Thus,

$$C_s = \frac{C_{i(av)}}{[L/(I + P)]} = \frac{C_i}{(L/I)} \qquad \text{when} \qquad P = 0 \qquad (9.14)$$

Also, by combining with Eq. (9.13) we get

$$C_s = \frac{C_{i(av)}}{(I + P - E_t)/(I + P)} = \frac{C_i}{(I + P - E_t)/I} \qquad \text{when} \qquad P = 0 \qquad (9.15)$$

The term $C_s/C_{i(av)}$ is referred to as the salt buildup ratio.

From the resulting relationships* it is evident that the salt concentration in the soil saturation extract C_s is

1. A function of the original concentration in the irrigation water C_i
2. Inversely dependent on the ratio $L/(I + P)$, called the "<u>leaching fraction</u>" (= L/I when $P = 0$)
3. Increase with E_t (and decreases as I, P, and L increase)

The buildup of salt concentration in the soil is greater in the dry and hot areas where P is small and E_t is high. Conversely, in wet areas, both P and L increase, thus giving a low buildup of salts in the soil. In fact, with sufficient precipitation and leaching, the soil concentration may not be very different from that of the irrigation water, thus giving latitude for use of water with higher salt concentrations. Leaching fractions of 0.3 or 0.4 are rarely practical except with some well-drained, sandy soils. With more clayey soils they may more often be around 0.2 or even 0.1. Hence, the soil saturation extract C_s often ranges between 2 and 10 times the original concentration in the irrigating water.

Equations (9.14) and (9.15) also explain how it may be beneficial in certain circumstances to provide additional flow merely for leaching before a crop is planted or between two successive crops to flush out accumulated salts. Additional flow may even be given during crop growth as long as the crop can tolerate it and soil aeration is not adversely retarded. This may be particularly desirable with some industrial wastewaters high in salts.

In designing irrigation systems, an effort is made to restrict the concentration of dissolved salts in soil saturation extract C_s to the approximate values given in Table 9.21 to not reduce crop yields adversely. These values are only approximate and vary with the specific plant to be grown, and other factors affecting crop yield. More specific values are obtainable in irrigation texts, and reference should also be made to irrigation water standards. Furthermore, as salt concentrations build up in the soil, conditions affecting chemical equilibria may change, thus causing some substances to precipitate, and vice versa, when conditions change during rainfall, for example.

---

*The values of I, L, P, and E_t can also be stated as mm/year and the concentration terms C_i, C_s, C_p, and C_e as µmhos/cm, to match the usual mode in which irrigation data are expressed.

Table 9.21 Maximum Permissible Salinities in Soil Saturation Extract

| Crop | Maximum permissible values of EC (μmhos/cm) at 25° C in soil saturation extract | |
	North Africa [48]	U.S. [42]
Bermuda grass, tall wheat grass	-	18,000
Cotton	12,000	16,000
Wheat	14,000	14,000
Barley	17,000	10,000 (at seedling stage)
Sugar beets	-	8,000 (at germination)
Alfalfa	-	8,000
Tomato, broccoli, spinach	-	8,000
Corn, flax, potato	7,000	6,000
Rice	-	4,000[a] (at seedling and flowering stages)
Citrus and most stone fruits	-	4,000
Strawberries and other berries	-	2,000-4,000

[a]Draining the field prior to seed setting and flowering and reflooding with less saline water, where feasible, has been suggested since at other times rice can tolerate higher salinities. Rice fields can be flooded since they do not require aerated soils.

The utility of Eqs. (9.14) and (9.15) is actually limited by the fact that they are based on several simplifying assumptions which may not be justified in reality. In order to explain the buildup of salts in soils, more elaborate mathematical models could be developed, but their utility would again be questionable in many cases where, in utilizing those models, values of coefficients would ultimately have to be assumed and based on insufficient evidence. A typical result from the use of our simple model is the development of curves such as the ones shown in Figure 9.9, where a log-log plot is made [using Eq. (9.15)] of the applied irrigation water in mm/year against the permissible EC of the irrigation water for different conditions of leaching and different limits of EC in the soil saturation extract.

The curves shown in Figure 9.9 are only qualitative in nature. They indicate in general that the higher the salinity of the water the less the

quantity that can be used for irrigation unless leachate is more. For a
given soil in areas of greater rainfall and longer rainfall season, where
leachate is more, a more saline wastewater can be used without exceeding
the limit of conductivity EC_e in the soil extract. Also, any salinity
built up in the soil from irrigation in the previous season may be leached
out somewhat during the following rainfall season (Table 9.22).

Thus, to estimate salt buildup, it is necessary to estimate the effective leachate from the field. The leachate L can be estimated on the assumption that the available moisture content of the soil remains the same
at the beginning of each irrigation cycle, namely, as seen earlier:

$$L = (I + P - E_t) \qquad\qquad\qquad (9.16)$$

As an example, if applied irrigation water is 1000 mm/year with EC =
400 μmhos/cm, and if the leaching fraction is 0.10, namely, leachate = 100
mm/year, then from Figure 9.10, EC_e = 4000 μmhos/cm, giving a salt buildup
of 10 times the original concentration [$C_s = C_i/(L/I)$].

If the irrigation water is more saline with, for example, EC = 1000
μmhos/cm and leachate = 300 mm/year, the soil saturation concentration EC_e
can still be held at 4000 μmhos/cm as above, provided the irrigated water
quantity is 1200 mm/year. Hence, the extra quantity required for leaching
purposes = 1200 - 1000 = 200 mm/year. The total leachate is 300 mm/year
and the leaching fraction is 300/1200 = 0.25. The soil moisture content
at field capacity is assumed to be the same in both cases.

If the net requirement of irrigation water I (= E_t - P) is known, the
irrigator will generally apply about 10 to 20% more water than is necessary
to meet the consumptive use because of nonuniformity of application and
infiltration. This increases the leachate correspondingly, but this is not
the same as deliberate additional use of irrigation water specifically for
enhancing leaching in order to control salt buildup. The specific leaching
requirement can be estimated from Eq. (9.14) by substituting in it the desired value of soil concentration C_s' that is not to be exceeded and the
actual average concentration $C_{i(av)}$ in the irrigation water, and finding
the required quantity of additional water q to be infiltrated into the soil
to give the necessary leaching fraction. Thus,

$$C_s' = \frac{C_{i(av)}}{(L + q)/(I + P + q)} \qquad\qquad\qquad (9.17)$$

If the original leaching fraction $L/(I + P)$ is denoted by "p" for the
soil under consideration (generally in the range 0.1 to 0.3), and an

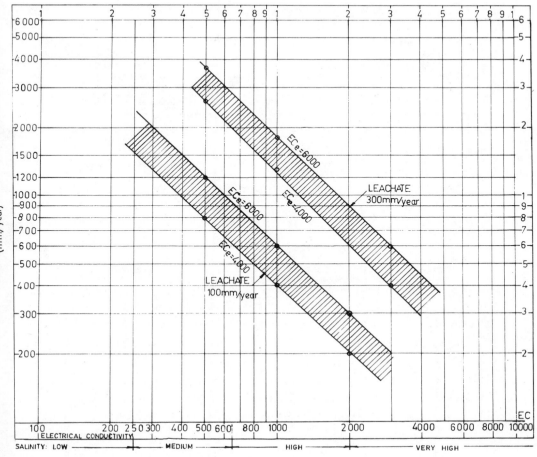

Figure 9.10 The permissible conductivity of irrigation water at
different irrigation and leaching rates. EC = permissible conductivity of
irrigation water in μmhos/cm. EC_e = electrical conductivity of the soil
saturation extract in μmhos/cm.

assumption implied here that all the extra quantity q will leach out readily
from the soil and not be retained in it,

$$C'_s = \frac{C_{i(av)}}{[p(I + P) + q]/(I + P + q)} \qquad (9.18)$$

Thus, the quantity I must be increased by the quantity q to hold the
salt concentration below C'_s.

Example 9.6 illustrates how an irrigation system can be designed knowing the quality of effluent available from a wastewater treatment plant.

Table 9.22 Leaching Requirement as Related to Electrical Conductivities
of Irrigation and Leachate Waters

Electrical conductivity of irrigation waters (μmhos/cm)	Leaching requirement[a] for the indicated maximum values for conductivity of leachate water at bottom of root zone (%)			
	4,000	8,000	12,000	16,000
250	6.2	3.1	2.1	1.6
750	18.8	9.4	6.2	4.7
2,250	56.2	28.1	18.8	14.1
5,000		62.5	41.7	31.2

[a]Fraction of the applied irrigation water that must be leached out.
Source: Ref. 42.

Increased infiltration is achievable by using excess water, provided
it will not cause harm in other ways. For example, the soil must be such
as to permit the higher infiltration rate desired. As stated earlier,
leaching fractions greater than 30 to 40% of the quantity applied is rarely
practical but may be achieved on some well-drained sandy soils. The crops
may find the water excessive. Prolonged irrigation can damage crops because of waterlogging and poor aeration. Thus, opportunities for inducing
extra leaching by excess water usage may be limited by such factors. Table
9.22 gives some estimates of leaching requirements in relation to the quality of irrigation water and crop tolerance [42].

It may be repeated here that the salt content of irrigation water together with rainfall plays a major role in the determination of the leaching requirement. A little careful designing of an irrigation system can
enable the use of various wastewaters of industrial and municipal origin.

One further aspect to be considered when studying salt buildup in soils
is to take into account the differences in the solubility of the salts.
Solubility depends on the nature of the salt, the solute temperature, and
the presence of other salts. For most salts, the solubility increases with
temperature; some salts increase more rapidly than others. In complex solutions, the solubility of most salts changes. Their presence in a solution
of salts having a common ion causes the solubility of these salts to drop.
For example, the presence of magnesium chloride can sharply decrease the
solubility of sodium chloride. In the case of mixtures of salts of dis-
similar ions, the solubility of the component having a lower level of solu-

bility increases because of dissociation of salts. An example of this is
the increase in solubility of gypsum in the presence of chlorides.

It is important to take the above effects into account when artificial
leaching of alkaline soils is planned. The order in which the salts will
be leached out will depend on their solubilities; the leaching is generally
more successful in the summer than winter. For a further discussion of
this subject, see Refs. 4 and 48.

The following empirical formulation has been suggested in the U.S. to
aid farmers in evaluating the effect of brackish water irrigation on soil
salinity in humid regions [43]:

$$EC_{e(f)} = EC_{e(i)} + \frac{nEC_{(iw)}}{2} \qquad\qquad (9.19)$$

where

EC = electrical conductivity and the subscripts f and i refer to
final and initial soil values and iw to irrigation water

n = number of irrigations (between rainfalls)

In studying the soil salinity increase through the use of different
qualities of irrigation water in Texas (annual precipitation = 400-600 mm),
it was observed that the use of water with a conductivity of 4360 μmhos/cm
at 25°C for irrigating gray silt and silt loam over a period of 15 to 20
years gave the following buildup at different depths [48]:

Depth (cm)	EC_e (μmhos/cm at 25°C)		
	Nonirrigated	Irrigated	Ratio
0-25	1915	5525	2.88
25-60	2190	5595	3.55
60-90	2770	5580	2.00
90-120	3285	5165	1.57

This shows that salt accumulation is maximum at the top and reduces progressively with depth in a sort of plug-flow fashion.

9.2.19 Land Treatment Efficiency and Groundwater Pollution

Possibilities for groundwater pollution exist and must be evaluated
in connection with any wastewater irrigation project. Pollution can be
either biological, chemical, or both, depending on the soil and wastewater

Table 9.23 Effectiveness of Land Disposal Techniques

Item	Approximate efficiency of removal (%)		
	Spray irrigation	Flood irrigation[a]	Rapid infiltration ponds
BOD	99	80	99
Suspended solids	99+	80	99
N	80-90	80	80
P	99	80	90
Heavy metals	99	10-30	95
Organic compounds	99	50	90
Viruses	99+	90	99+
Bacteria	99	90	99+
Total cations	0-75	0-50	0-75
Total anions	0-50	0-10	0-50

[a]Percentages are best estimates in light of limited information.
Note: The efficiencies of the three methods should not be compared with each other because each is affected by its own set of local conditions, application rates, soil types, etc.
Source: Ref. 28.

quality. The percolating wastewater is partially "purified" in its passage through the soil and ultimately returns by gravity flow to the subsurface aquifer and is mixed with natural groundwater.

An investigation of the effectiveness of different land disposal techniques carried out by the U.S. Army Corps of Engineers showed the relative efficiencies of removal of different constituents from the wastewater in its passage through the soil (see Table 9.23).

In order to understand the fate of the two principal nutrients, nitrogen and phosphorus, as they pass through the soil, it is necessary to recall the earlier discussion in Chapter 4 regarding the various biochemical transformations that can occur from one form of nitrogen to another. Natural soils may contain from a few hundred to a few thousand kilograms of N per hectare of area. The major amount of this nitrogen occurs in the organic form. Microbial action results in the transformation of some of this organic-N to the ammonia or NH_4^+ form; this is called ammoniation. In aerated soils and favorable conditions, nitrification occurs giving nitrates (NO_3^-);

this is called mineralization. Under reducing conditions in waterlogged or otherwise anaerobic soils, the nitrates or nitrites may be reduced to N_2 and be given off as a gas; this is denitrification.

The nitrate ions are completely soluble and move with the soil water. Thus, leachates contain nitrogen in the nitrate form (as well as in certain organic forms which are soluble). On the other hand, NH_4^+ ions enter into cation exchange relationships with soils and their concentration is, therefore, relatively low in leachates. In fact, the fate of nitrogen in the soil is very complex because of many factors affecting its form and transport; these include immobilization, adsorption, volatilization, cation-exchange, and convection in addition to the mineralization, denitrification, fixation, and plant uptake already discussed.

Phosphorus occurs in soils in both organic and inorganic forms, but neither form is very soluble. The phosphorus content of natural soils generally ranges from 0.01 to 0.15%. The phosphorus added as fertilizer or released by the decomposition of organic matter is generally converted to an inorganic, insoluble form such as iron or aluminum phosphate in acidic soils, or precipitated as calcium phosphate in the more commonly alkaline soils. Thus, soils display a high retention capacity for phosphorus, and the soluble phosphorus concentration in leachates is thus low, ranging between 0.01 and 0.10 mg/liter in most cases.

Owing to the various mechanisms of removal described above, the concentrations of nitrogen and phosphorus in surface runoff are different from those in leachates. Surface runoff has low concentrations of NH_4^+ and NO_3^- as they are highly soluble and, therefore, dissolved and carried into the soil by the initial rain before runoff occurs. Often the NO_3^- content of the surface runoffs have been observed to be as low as those in the rain itself. Leachates, on the other hand, contain all the NO_3^- washed into the soil, except that taken up by the plants in the root zone.

Phosphorus concentrations in surface runoff and soil leachate display reverse characteristics compared with nitrogen. As stated above, both organic and inorganic forms of phosphorus are only slightly soluble in water and, therefore, tend to precipitate in the topsoil. Thus, water percolating through the soil has a low concentration. However, surface runoff has a higher phosphorus concentration resulting from the soluble fraction and the fraction contained in the finer soil particles carried with it through erosion. Thus, from an agricultural field, the transport of the two nutrients, nitrogen and phosphorus, to a waterbody (e.g., a river) follows different routes: nitrogen travels mainly through ground flow, while phosphorus travels through surface flow. Of course, in terms of relative magnitudes, nitrogen generally exceeds phosphorus. Erosion adds to the phosphorus loss from a soil, but an irrigated field with a proper crop cover is less likely to suffer erosion than is a barren one.

Nitrogen and phosphorus removals measured in the irrigation system at Pennsylvania State University show that the wastewater is relieved of most of its phosphorus by the soil, and of most of its nitrogen by the crop (Table 9.24) [2].

All soils are effective at removing P, K, Ca, and Mg from the wastewater percolating through them, but the degree of effectiveness depends on the soil texture. Coarse-textured soils are good with P and Ca, but not with K and Mg. None are too effective at removing nitrogen except through the growth of crops. Without crops, the nitrogen may only leach through as nitrate; hence proper crop management is essential to ensure nitrogen removal. The likely contents of nitrogen and phosphorus in surface waters and irrigation return flows are given in Chapter 5.

Water and its dissolved constituents do not necessarily move together in their downward journey through a layer of soil. A certain amount of intermixing and redistribution occurs owing to differences in transport rates brought about by diffusion, attraction, exchange, adsorption, and other phenomena within the soil layer. Plug-flow formulae are, therefore, not strictly applicable. Better predictions of nutrient distribution can be made if water-flux estimates can be combined with mixing or diffusivity considerations.

Table 9.24   Nutrient Removal in Different Parts of an Irrigation System

Mode of nutrient removal	Percent removal	
	N	P
By crops	83	27
By leaching	23	1
By soil retention	-6	72
Total	100	100

Source: Ref. 2.

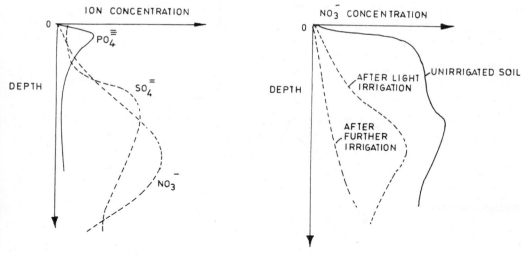

Figure 9.11 Probable soil concentration profiles.

Several investigators have attempted to develop equations to predict the concentration of ions such as Cl^-, NO_3^-, and SO_4^{2-} at different depths in soils mainly for understanding soil fertility in the root zone. As most of these studies are for specific soil conditions, it will suffice to give here a qualitative idea of how distribution generally occurs in soils. Figure 9.11 illustrates typical concentration profiles likely to be obtained in soils [20].

Generally, in experimental studies using soil lysimeters, a solution containing the ions is added to the surface of a moist soil followed by additional water without any salt. For nitrate studies, well-aerated soils are necessary to avoid any loss of nitrogen by denitrification. As would be expected, a "spread" is observed owing to dispersion and mixing as the solution moves through the soil. The NO_3^- ion, which is relatively most free to move, gives a typical concentration profile as shown in Figure 9.11. The soil partially adsorbs SO_4^{2-}, whose movement is thus slowed and more skewed with depth. The PO_4^{3-} is confined mainly to the top layer.

Figure 9.11 shows the progressive downward displacement of the NO_3^- profile as the extent of irrigation is increased. The case is similar after heavy rainfall percolation.

Owing to their health significance, our interest lies particularly in the amount of nitrate in groundwaters. It is evident that, as further downward displacement through the soil occurs, the nutrients are more widely

distributed and maximum concentrations are reduced. There is continued
water and nutrient redistribution with a downward flow.

In a recent study it has been found that, depending on the method of
application, 22 to 36% of the nitrogen in liquid sewage sludge applied on
soil is lost principally as nitrogen gas from "aerobic" soils within an
18-week period [26]. Furthermore, organic nitrogen contained in the soil
itself undergoes bacterial degradation or mineralization, which yields inorganic nitrates capable of being leached out. The mineralization rate is
a function of temperature, among other things; the effect of temperature
is given by the following equation [25]:

$$m_t = m_o e^{\left[\dfrac{0.5E(T_t - T_o)}{T_t T_o} \right]} \tag{9.20}$$

where

m_t = mineralization rate per month at temperature T_t

m_o = mineralization rate per month at temperature T_o

E = energy of activation, cal/mole

T_o = reference temperature, $^\circ K$

T_t = average soil temperature, $^\circ K$

In this study, two field measurements with different soils gave m_o =
0.0015 per month and E = 11,000 for a silt loam soil, and m_o = 0.00092 per
month and E = 9,500 for a gravelly loam soil. This implies an average
mineralization rate of approximately 1% per year. In tropical climates,
a higher rate may be expected.

For a soil organic nitrogen content of 4500 kg/ha in the above study,
the inorganic nitrogen loss by percolation through the soil without any
addition of wastes or fertilizers was estimated at about 1% per annum (42
kg/ha annually), whereas the corresponding annual seepage was 227 mm. This
gave an estimated average concentration of 13 mg/liter in the seepage water
leaching the top 40 cm of soil. Such a concentration was comparable to observed values in tile drainage water from cropped areas [27]. However,
infiltration rates are not homogeneous in time or space.

This would indicate that a potential for nitrogen contamination does
exist, although actual values would depend on the background levels in the
groundwater, the dilution obtained, and the fact that all nitrogen leached
from the soil would not necessarily reach groundwater; some ammonium may
be fixed to soil particles and some nitrate may be denitrified in anaerobic

subsoil regions. WHO drinking water standards for nitrogen recommend that
the nitrate concentration should be less than 45 mg/liter where possible.

In the temperate-humid regions of North America, soils are reported
to contain 400 to 10,000 kg/ha of nitrogen within the top 40 cm or so of
the soil (average = 3400 kg/ha). As a rough estimate, if it is assumed that
the organic nitrogen content of a top soil averages, for example, 3400 kg/ha
and its mineralization rate averages 1% per year, a nitrogen replenishment
rate of 34 kg/ha would be necessary from natural or artificial sources to
sustain the original level. Furthermore, if a crop is grown, such as a
grass, needing, for example, 250 kg of N/ha-year (see Table 9.5), the minimum nitrogen addition to the soil should be 250 + 34 kg N/ha-year. The
efficiency of nutrient uptake by plants is typically 50 to 70% only (Sec.
9.2.9) and, thus, actual nutrient addition would need to be in the range
of 500 kg N/ha-year. The unused portion of nitrogen addition would partly
sustain the soil level and partly be lost in different ways described above.
The excessive addition of nitrogen would increase losses, including leaching of the nitrogen mineralized to nitrates.

In describing the leaching of a nondegradable soluble ion such as
chloride from a porous medium, a formulation of the phenomenon has been
proposed and tested on the assumption that the solute and solution remain
uniformly distributed through the soil surface layer, and reduction follows
first-order kinetics [24] and plug-flow conditions.

$$N_t = N_o e^{-KL} \tag{9.21}$$

where

L = amount of water leaching through the soil (mm) between the
two observations

K = constant (with matching units) (mm^{-1}) depending on type of
soil, nature of substance, etc.

N_o = initial content in layer

N_t = content in layer, after volume L has leached through

In another study [25] in which Eq. (9.21) was applied to the leaching
of nitrates from two corn growing sites near New York, with level, well-
drained soils, it was found that the values of K were 0.11 per cm for a
gravelly loam and 0.087 per cm for a silt loam soil. From this, one can
estimate that the amount of leachate required to wash away, for instance,
90% of a given content from a layer is about 200 to 250 mm.

In wastewater irrigation, however, repeated applications of various dissolved salts occur at the soil surface. These salts, in passing through the soil, may enter the soil complexing system and be retained there until saturation capacity is reached, after which elimination from the system may occur. At steady state, the mass of incoming salt must equal the mass going out in the leachate.

If repeated dosing of a substance is taking place, and if elimination is assumed to follow first-order kinetics, an analogy could be drawn from the phenomenon of "total body burden" and accumulation of the substance in soil could be described by the following equation based on Eq. (4.3) in Chapter 4.

$$A = \frac{a}{K'}\left[1 - e^{-K'(LC_s)} \right] \tag{9.22}$$

where

A = mass of substance accumulated in soil in a given time

a = mass of substance dosed on soil

= $(IC_i + PC_p)$ where the terms have already been defined previously

L = volume of leachate in given time

C_s = concentration of substance in leachate, mass/vol

K' = fraction of accumulated mass eliminated in leachate volume in given time. First-order assumption implies that a fixed fraction of "A" is eliminated per unit volume of leachate. K' depends on various factors such as the nature of the substance, the type of soil, etc.

As irrigation and leaching proceed, the maximum accumulation at steady state is obtained from Eq. (9.22) as

$$A_{max} \rightarrow \frac{a}{K'} \tag{9.23}$$

or

$$A_{max} = \frac{(IC_i + PC_p)}{K'} \tag{9.24}$$

Hence, the maximum accumulation in the soil is a function of the load "a" and the elimination rate constant K'. The smaller the value of K', the smaller should be the corresponding load "a" on the soil for a given value of A_{max}. A clayey soil would be expected to have a lower value of K' than would a sandy soil. Accumulation of salts would be less in wet areas where leachate is high.

The major interest, however, lies in the maximum concentration reached in the soil at saturation or steady-state conditions when, from Eqs. (9.14) and (9.23) we get

$$C_s = \frac{(IC_i + PC_p)}{L} \qquad (9.25)$$

or

$$C_s = \frac{A_{max}K'}{L} \qquad (9.26)$$

Furthermore, when irrigation is stopped and the substance is no longer dosed on the soil, natural rainfall (or artificially added freshwater) would continue to leach it out until a new equilibrium is established. This leaching out after further dosing is stopped can be approximated as a typical first-order dieoff equation similar to Eq. (9.21).

$$A_t = A_o e^{-K'(LC_s)} \qquad (9.27)$$

where A_o and A_t = the mass of substance present in the soil at the start, and at time t, respectively.

The equations just given could be applied to estimate the accumulation and leaching of various substances such as nitrates, dissolved salts, trace and heavy metals, etc., provided values of K' or K are known for the specific conditions. Such data are not available at the present time, nor is the effect of different factors on their values clearly known.

The fate of degradable substances such as sewage organic matter, organic nitrogen, or radioactive materials would be governed by two dieoff mechanisms of the first-order type: one owing to leaching and the other owing to bacterial or physical degradation of the substance. Again, drawing an analogy from the effective elimination time for radioactive substances from an organ [Eq. (4.6)], one can say, in general, that the effective residence time for a degradable substance in soil is related to the leaching time and degradation time as follows:

$$T_e = \frac{T_L T_d}{T_L + T_d} \qquad (9.28)$$

where

T_e = effective half-life of degradable substances in the soil, days

T_L = time for half the substance to be removed through leaching only, days (leaching half-life)

T_d = time for half the substance to be removed by degradation mechanism only, days (degradation half-life)

From Eq. (9.28) it is evident that the larger of the two values, T_L or T_d, will tend to cancel itself out if it is much larger than the other. For example, if T_L = 2 days and T_d = 200 days, the effective residence time in the soil T_e = 1.98 days. But if T_L and T_d are 50 and 200 days, respectively, then T_e = 40 days. In general, more slowly degrading substances would show higher concentrations in a given leachate than rapidly degrading substances would, assuming there were no differences in their leaching characteristics. A degradable substance which is not rapidly (or easily) leached out would be found in low concentrations in the leachate. Radioactive materials of long half-life would have their accumulation in the soil controlled by their leaching time.

A potential problem with groundwater pollution is the dissolved solids leaching through the soil. The efficiency of removal of total anions is low as seen from Table 9.23, and may affect groundwater qualities downstream. Chlorides, sulfates, etc. from the wastewater could percolate through the soil and travel long distances, thus causing pollution of wellwater supplies in the vicinity. Certain industrial wastewaters, such as distillery effluents, tannery effluents, etc., could contain large concentrations of chlorides which could pollute underground water supplies. Chemical pollution can travel long distances. Color also could persist for long distances in spite of much natural "treatment" in the ground, as would nondegradable organics and hard detergents.

The persistence of nondegradable insecticides and pesticides is a problem of relatively recent origin, but is not confined to wastewater irrigation systems.

In a study on groundwater pollution below a garbage dump in South Bavaria, the groundwater was sampled through bore holes to measure chemical quality and compared with similar studies at another dump in south Hesse, Germany. In Hesse, with a groundwater flow of 0.5 to 1.0 m/day, the pollution spread with an oval-shaped front for not more than 650 m, whereas in Bavaria, with a groundwater flow of 5 to 10 m/day, pollution spread in the form of a narrow, elongated tongue extending for more than 3000 m [31]. The velocity of groundwater movement varies widely in nature, ranging from as little as 0.3 m per year to several kilometers per month. In deep layers, stratification may exist and, thus, it may not always be correct to assume that nutrients and other materials contained in the invading waters will be uniformly distributed in the resident groundwaters. Mathematical models and techniques used in studying flow through porous media must be applied with care to investigations of the effects of the invading waters on the resident groundwaters.

Bacterial contamination, on the other hand, is generally readily filtered out in the soil, although it could travel considerable distances through fissured strata and gravelly soils.

Generally, a minimum depth of 1.0 to 1.20 m of good soft soil is sufficient to filter out pathogenic organisms. However, human viruses have been detected in groundwaters at 6 m depth [29]. There are also several reports of high bacterial and coliform contamination of wellwaters in the vicinity of sewage farms using raw sewage in India; however, open dug wells in India and other countries have almost always shown high coliform concentrations, irrespective of their siting [14]. In dune sands in Europe, complete removal of E. coli has been observed within 5 m travel.

Some information on the problems of groundwater pollution can also be obtained from numerous studies concerning leachates from garbage dumps. In one study in Germany concerning leachate from a dump where garbage was mixed with sewage sludge, it was shown that E. coli and salmonellae were considerably more abundant than in the case of another dump with garbage alone. However, during passage through compact soil with no possibility of short-circuiting, all salmonellae were removed in the first 20 cm, and E. coli and sulfate-reducing bacteria were no longer detected at a depth of 2 m [30].

9.2.20 Irrigation Land Requirements

From the foregoing discussion it will be evident that the land requirement for irrigation with a given volume of wastewater depends on a variety of factors such as:

1. Cropping pattern and water requirement of crops
2. Climate
3. Quality of wastewater
4. Type of soil
5. Efficiency of irrigation

Different cropping patterns over a given area yield different annual irrigation requirements for that area (Sec. 9.1.3). Either area may be limiting, the quantity of wastewater may be limiting, or the wastewater quality, especially in the case of industrial wastes, may also set the limit to the amount of watering than can be done on a unit area without deleterious effect on the soil. In the latter case, a cropping pattern can be found to match the limited quantity of the available wastewater of given quality. If fresh dilution water is available, the wastewater could be diluted to a desired quality and either spread over a greater area or used for a different set of crops needing more water.

Where land application is being done essentially as a mode of tertiary treatment, the need to optimize nitrogen removal may necessitate spreading the farm over a larger area to reduce the wastewater application rate on the land, minimize the "slippage" of nitrogen through the soil as nitrate, and maximize its uptake by plants.

Irrigation land requirements must also take into account the fact that the operation of a farm is seasonal, with periods when the crop is harvested, and the soil kept fallow, then tilled and replanted. This demands a flexibility of operation which generally does not suit the wastewater treatment plant discharge which is at a more or less constant rate throughout the year, unless the wastewater is also of a seasonal nature as it could be, for example, from a food processing factory (canning, sugar, etc.).

Hence, flexibility demands that either more land be kept available than strictly necessary for the farm to be able to divert the wastewater onto the extra land for grass cultivation when not required on the main cropland, or arrangements be made for holding the wastewater in a lagoon until required, or for discharging the surplus treated wastewater to a natural watercourse. This additional land requirement may be on the order of 10 to 100% of the main cropland, depending on the size of the farm, its mode of operation, and whether the wastewater or sludge is to be held in a lagoon or used on land when not required at the main farm (also see the next section).

In view of the large number of variables involved, the area that can be irrigated by wastewater varies widely. Approximate area requirements vary from 5 to 10 hectares per 1000 m^3 daily flow in warmer climates, up to 40 hectares or even more in less favorable climates (see Table 9.25). The U.S. survey of Pound and Crites [50] showed that during the crop growing season, land requirements ranged from 7 to 55 ha/1000 m^3-day.

Table 9.25 Typical Land Requirements for Wastewater Irrigation

Wastewater	Climate	Overall land requirement[a] (ha/1000 m^3-day)
Domestic sewage	Warm	5-10
	Temperate	10-40
Food industry wastes (milk, breweries, sugar, canning, etc.) and pulp and paper mill wastes	Warm to temperate	5-25

[a]Land requirement varies widely, depending on climate, type of soil, type of crops proposed to be grown, mode of irrigation, and the need to have additional land for use when water is not needed on the main farm during harvesting.

Where digested sludge has been applied on land, the loading in terms of dry suspended solids in kilograms per hectare per day has varied considerably from 14 to 70, and in some cases has gone as high as 700 and even more. For a solids load of 70 kg/ha-day, with an average moisture content of 95%, the liquid sludge application rate amounts to 1.4 m^3/ha-day [49]. This would require about 700 ha for a liquid sludge flow of 1000 m^3/day. If the solids loading is 700 kg/ha-day, the corresponding land requirement will be only 70 ha for 1000 m^3/day sludge flow. The wide variation found in practice is, no doubt, partly a result of differences in local conditions, but also reflects to some extent the need for more rational design criteria to be developed.

9.2.21 Precautions in Operating Wastewater Irrigation Systems

In the context of previous discussions, some guidelines such as the following could be drawn up for the design and operation of wastewater irrigation systems.

1. Wastewater quality should be in relation to the nature of the soil and the type of crops proposed to be grown on the farm. In this connection, Table 9.26 gives a brief summary of existing standards in some countries for the use of wastewater in agriculture. Toxic substances, hard detergents, and the like are best prevented at their source as their removal in treatment may be very expensive or almost impossible.

Table 9.26 Quality Limits for Waters Used for Irrigation
in Some Countries

| Item | Los Angeles, California | | India |
	Agriculture	Parks	
Total dissolved solids, mg/liter	2100	1500	2100
Electrical conductivity, μmhos/cm	-	-	3000 at 25° C
Chlorides, mg/liter	355	250	600
Sulfates, mg/liter	-	250	1000
Boron, mg/liter	2.0	2.0	2.0
Percent sodium	-	-	60
SAR, meq/liter	10	8	-
Residual sodium carbonate, meq/liter	2.5		-

Note: Irrigation water quality can be evaluated only in the context of local factors such as type of soil, its drainage characteristics, climate, salt tolerance of crops, and irrigation management as discussed in previous sections of this chapter.

Some attempts at sewage irrigation have failed because of soil problems caused by waterlogging and from excessive buildup of salts as discussed in previous sections of this chapter.

2. It is desirable to maintain a proper crop cover on the disposal site. It prevents soil erosion, improves soil porosity, and also helps in the uptake of nitrogen and its removal by harvesting, thus making less nitrogen available for conversion to nitrate and eventual seepage into groundwater.

3. The selection of crop cover depends on climatic and soil conditions, its water requirements, tolerance to buildup of salts in the soil, agricultural markets, farming practice and such other factors as discussed earlier. For industrial wastewaters and sludges, the possible ill effects of toxic substances and heavy metals must be kept in view (alkaline and highly organic soils may then be preferable).

To minimize hazards to the health of those who consume the farm products, several countries have formulated some essential requirements, some of which are summarized in Table 9.27. Some of them require chlorination after a good degree of treatment before the wastewater is suitable for unrestricted use on farms; with less than this quality, only a restricted variety of crops may be permitted. Another requirement sometimes imposed from the viewpoint of health safety is delayed harvesting, wherever feasible, to allow 2 to 4 weeks for any pathogens possibly present on the crops to die before harvesting or consumption.

Table 9.27 A Summary of Existing Requirements in Some Countries for Controlling the Agricultural Use of Wastewater

Crop	California	Israel	South Africa	Federal Republic of Germany
Orchards and vineyards	Primary effluent; no spray irrigation; no use of dropped fruit	Secondary effluent	Primary effluent heavily chlorinated where possible. No spray irrigation	No spray irrigation in the vicinity
Fodder fiber crops and seed crops	Primary effluent; surface or spray irrigation	Secondary effluent, but irrigation of seed crops for producing edible vegetables not permitted	Tertiary effluent	Pretreatment with screening and settling tanks. For spray irrigation, biological treatment and chlorination are necessary
Crops for human consumption that will be processed to kill pathogens	For surface irrigation primary effluent may be used. For spray irrigation, disinfected secondary effluent only (no more than 23 coliform organisms/100 ml)	Vegetables for consumption not to be irrigated with renovated wastewater without disinfection (less than 1000 coliforms per 100 ml in 80% of samples)	Tertiary effluent	Irrigation only up to 4 weeks before harvesting. Potatoes and cereals: irrigation through flowering stage only
Crops for human consumption in a raw state	For surface irrigation no more than 2.2 coliforms/100 ml. For spray irrigation, disinfected filtered wastewater with turbidity of 10 units permitted provided it has been treated by coagulation	Not to be irrigated with renovated wastewater unless they consist of fruits which are peeled before eating		Potatoes and cereals: irrigation through flowering stage only

Source: Ref. 18.

4. The method of irrigation is also important. The safest method is subsoil irrigation (it also prevents flies and odors); contamination danger is maximum with spray irrigation. Sprinklers directly apply sewage pathogens or toxic substances to the plants, thus bypassing the capacity of the soil to "filter" them out. However, spray irrigation has other advantages as seen earlier. Grazing of milch cows or buffaloes on the farm is to be carefully controlled to avoid the contamination of udders. The efficient usage of irrigation water is also dependent on the method of irrigation as discussed earlier. Where sludge is used, deep plowing of soil between applications helps to aerate the soil and restore its permeability.

5. Poor farming practices can lead to nuisance conditions from odors, flies, and mosquitoes, leading to closure of farms or consideration of other alternative methods of disposal to take care of future increases in wastes from a community.

6. Farm workmen must be protected from health hazards by observance of all personal hygiene rules. For example, insistence on wearing boots and gloves while working in the fields may be necessary in the developing countries in warmer regions. Regular medical checkups and vaccinations of personnel are also advisable. Education of the farmer and adoption of correct practices are generally more feasible on larger farm units.

It should be mentioned that the engineering design of a sewage farm should ensure that safe handling is built into the system from the start. Ease of access and operation of channels, etc., without having to come into contact with sewage or wet soil would help. Careful provision of such facilities would set them apart from conventional farms that use freshwater for irrigation.

7. Satisfactory alternative arrangements must also be provided for the disposal of wastewater during nonirrigating periods. As discussed earlier, a farm does not need irrigation water all the time, nor at a uniform rate. This poses a problem which must be fully considered at the design stage. In fact, one of the major differences between the primitive wastewater irrigation systems and the present-day engineered systems is the manner in which the above problem is handled. In a properly engineered system, the wastewater, whenever surplus to the irrigation requirements, can be stored or discharged elsewhere without causing any environmental hazards. The kind of backup or stand-by facilities sometimes provided include
   a. Holding lagoons for off-season storage
   b. Additional land (agricultural or forest)
   c. Discharge to rivers (with or without additional treatment)
   d. Discharge to sea (during a nonbathing season)
   e. Groundwater recharge (surface or subsurface)

The need for a balancing or holding lagoon is felt when a fixed area of land is to be irrigated at a varying rate of crop demand. The reader is referred to Example 9.6 in which the manner of determining lagoon capacity is discussed. Lagoon land requirements can be quite substantial as storage may be needed for about 1 to 4 months of wastewater flow.

However, balancing devices do not provide a constant detention time to the inflow. By their very nature, they provide a long detention time for some periods and a very short detention time at other periods when the lagoons are at a low level. At some time of the year the lagoon may even be practically empty. This is important to keep in mind as the incidental treatment afforded by the lagoon (bacterial dieoff, further BOD removal, denitrification, etc.) is variable with the time of year. Also for this reason, the use of the holding lagoon as a fish pond, as a treatment device, or as an aerated facultative lagoon demands that it be so constructed as to ensure the desired minimum lagoon volume at all times beyond the balancing requirements. Aeration could be achieved by the provision of floating aerators in the lagoon.

In tropical climates with wet (monsoon) periods limited to 3 or 4 months, and fairly steady irrigation demands for the rest of the year, holding lagoons of large capacity are not required since, during wet periods, river flows are often large and sufficient to enable wastewater disposal by dilution. Between croppings, however, when irrigation is not required, the wastewater flows need to be held in short-period balancing lagoons or diverted to additional croplands used by rotation. The latter mode is often preferred where feasible.

The feasibility of spraying wastewater on land, even during cold winters, has been demonstrated in the case of the Nittany Valley, Pennsylvania, where treated wastewater is spread on forest tracts during the winter, taking advantage of the porous forest floor and using equipment that would not clog by freezing.

Discharge of surplus wastewater to river or sea, with or without treatment, requires that various factors be examined as discussed in Chapters 7 and 8. Timing of the wastewater discharge to high flood flows in rivers or to nonbathing season in the sea would, in many cases, favor reduced costs of treatment before discharge.

Land treatment may itself help to polish a secondary effluent and give high removals of nitrogen, phosphorus, BOD, ABS, and microorganisms, which may justify its use as a regular tertiary treatment step before discharge to a river or coastal water that is to be protected from eutrophication. Here, crop management may take on an added significance as an important factor in ensuring sustained high removals of nitrogen.

Groundwater recharge by surface or subsurface methods as a backup device for an irrigation system can be considered where soil permeabilities at the surface or in the subsurface aquifer would permit economical design. Its major objective would be to ensure that, one way or another, either by irrigation or by deliberate recharge, the groundwater is replenished. This may be important for preventing salt water intrusion and protecting downstream supplies.

8. Success of an irrigation system can be jeopardized by poor surface or underdrainage from the fields. Wastewater application must be held within the limits of soil permeability. Depending on the depth to the groundwater table, the type of soil in the intervening layers, the infiltration rates, and other factors, some artificial underdrains may be needed to protect the soil from waterlogging and loss of natural aeration. It may also be necessary in order to reduce groundwater pollution potential. This may add substantially to the cost of preparing the irrigation fields, but may be essential in some cases.

9. Present and potential use of groundwater must be carefully considered. Percolation from holding lagoons, canals, etc., may need to be diverted

to collecting wells and pumped back to the lagoon or to farm in order to avoid groundwater pollution where such pollution is feared to have detrimental consequences.

10. Provision must be made to enable monitoring of land-treated effluent or return irrigation flows. This can be done through tile drains, collecting wells, and other structures to which the return flow can find its way. Monitoring for nitrates has a special significance with regard to groundwater pollution. There is no reason to dispense with the usual sampling for various constituents that are normally analyzed in conventional waste treatment plants. Needless to say, flow measurement facilities such as Parshall flumes must be provided.

11. From the agricultural point of view, a few experimental plots located within the farm could help to conduct studies for maximizing crop yields through the use of various techniques and fertilization levels.

12. The design of farms using wastewater, particularly in the vicinity of larger towns in developing countries, must take into account the fact that farming practices near urban areas are likely to be less primitive than in the deep rural areas. The farmers are likely to demand and use more mechanized methods of farming, distributing the wastewater, controlling pests and insecticides, etc. For example, the use of movable spray irrigation devices facilitates the use of tractors without the ditches and dikes required for flood irrigation. Also, as stated earlier, sustained high removal of nitrogen can be assured only through proper crop management.

13. Some organizational or institutional changes may be necessary in order to control and promote wastewater irrigation practice on a sound basis. Scientifically designed and carefully operated and monitored, wastewater irrigation can be accomplished successfully with maximum ecological compatibility with nature. The environmental impact of any wastewater irrigation scheme can be assessed as for any other project.

9.2.22 Design of a Wastewater Irrigation System

In order to illustrate the concepts discussed thus far, an example has been solved below, for assumed data.

Example 9.6 Design an irrigation farm in order to utilize the unchlorinated effluent from a proposed wastewater treatment plant using a facultative aerated lagoon (BOD efficiency is 85%) treating sewage from a population of 70,000 persons. Sewage flow rate is expected to be 150 liters per person daily. The water, soil, and climatic data are given below.

Water Quality

TDS = 370 mg/liter (EC = 580 μmhos/cm), pH = 7.80

Total alkalinity = 195 mg/liter as HCO_3^-

Total hardness = 105 mg/liter as $CaCO_3$

Na^+ = 89.8 mg/liter, K^+ = 9.5 mg/liter, Ca^{2+} = 48.6 mg/liter, Mg^{2+} = 23.2 mg/liter

SO_4^{2-} = 41.4 mg/liter, Cl^- = 37.0 mg/liter

Boron = 0.30 mg/liter

Climatic Data (35° N latitude)

	J	F	M	A	M	J	J	A	S	O	N	D
Average temperature (°C)	5	8	12	15	20	25	30	31	27	21	11	6
Precipitation (mm/month)	29	23	27	23	12	2	-	-	2	11	22	25
Pan evaporation (mm/month)	27	44	74	104	152	278	424	412	298	157	55	28

Soil Data

The following predominant soil characteristics at the proposed site are as follows:

CEC = 30 meq/liter (Ca^{2+} = 24, Mg^{2+} = 4.5, K^+ = 0.5, Na^+ = 1.5)

EC_e = 400 μmhos/cm

pH = 7.8

Percent C = 1.17, C/N = 13.0

Bulk density = 1.3 g/cm³

Texture: sand (> 50 μ) = 7.5%, silt (2.50 μ) = 34.8%, clay (< 2 μ) = 43%, $CaCO_3$ = 13.9%, and stones = 4.8%

Infiltration rate (from field tests) = 5.0 cm/day at steady saturation condition

Moisture content by weight = 30% at field capacity, 15% at wilting

In the absence of a sewage system in the town, the likely quality of treated sewage can be estimated only by making suitable assumptions. One method is to be guided by values given in Table 5.2 showing the average increment in different mineral constituents owing to domestic water usage. These constituents are not much affected by treatment. Hence, the quality of treated wastewater is roughly estimated as follows:

Total wastewater flow = 70,000 x 150 liters/capita-day = 10,500 m³/day

Effluent BOD_5 = (1 - 0.85)54 g/capita-day = 8.1 g/capita-day (= 53 mg per liter)

Since sewage flow per capita is not high, we may use the highest value from the range shown for each mineral constituent in Table 5.2. Thus,

TDS = 370 + 300 (for example) = 670 mg/liter

Approximate EC $\simeq$ 1050 μmhos/cm

Total alkalinity = 195 + 150 = 345 mg/liter as HCO_3^- (= 5.66 meq/liter)

Na^+ = 89.8 + 70 = 159.8 mg/liter as Na^+ (= 6.84 meq/liter)

K^+ = 9.5 + 15 = 24.5 mg/liter as K^+ (= 0.613 meq/liter)

Ca^{2+} = 42.6 + 40 × (40/100) = 58.6 mg/liter as Ca^{2+} (= 2.87 meq/liter)

Mg^{2+} = 23.2 + 40 × (24/100) = 32.8 mg/liter as Mg^{2+} (= 2.69 meq/liter)

pH will be slightly less than original value of 7.80.

Check for Wastewater Quality as Irrigant (Sec. 9.2.13)

1. EC value indicates somewhat high salinity

2. pH value is not objectionable

3. $\displaystyle SAR = \frac{Na^+}{\sqrt{\dfrac{Ca^{2+} + Mg^{2+}}{2}}} = \frac{6.84}{\sqrt{\dfrac{2.87 + 2.69}{2}}} = 4.10$ meq/liter (acceptable)

4. $\displaystyle \text{Percent sodium} = \frac{100\,(Na^+)}{(Na^+) + (Ca^{2+}) + (Mg^{2+}) + (K^+)}$

   $\displaystyle = \frac{100(6.84)}{6.84 + 2.87 + 2.69 + 0.613}$

   $= 52.6$ ("medium" range)

   However, in terms of SAR value, the sodium hazard is "low." With reference to Figure 9.5, the wastewater can be classified as "$C_3 - S_1$," namely, satisfactory for almost all soils.

5. $RSC = (CO_3^{2-} + HCO_3^-) - (Ca^{2+} + Mg^{2+})$

   $= (0 + 5.66) - (2.87 + 2.69)$

   $= 0.10$ meq/liter (safe)

6. Boron concentration of 0.3 mg/liter in irrigation water will be tolerable for most plants. Assuming boron in the soil will build up as other soluble salts would, over a period of time, the steady-state concentration would be about 5 times more than the original concentration in the irrigating water as shown below.

7. It is assumed that, in view of the essentially domestic character of the wastewater, no toxic or hazardous concentration of trace metals and other substances are involved. The ABS concentration is also assumed to be at an unobjectionable level (bacteriological quality may be controlled by chlorination, if desired).

Soil Data

The soil data do not reveal any strikingly objectionable feature. It can be classified as "silty-clay" by reference to the diagram in Figure 9.7. The pH of the soil is only slightly alkaline and the electrical conductivity of the soil extract indicates a relatively low salinity. Both pH and EC_e will increase somewhat as irrigation proceeds. An increase in pH cannot be

predicted, but an increase in EC_e can be guessed at as shown further below.
The clay content of the soil is fairly high. This affects its CEC value
and permeability. In this context, the presence of $CaCO_3$ should be beneficial. However, a detailed interpretation of soil data should be made
only with the assistance of soil specialists and agronomists.

Selection of Cropping Pattern

Selection of the cropping pattern depends on local agricultural practices and the constraints imposed by the soil and wastewater quality. In
view of the unchlorinated effluent from the proposed treatment plant, assume
that only perennial fodder grass will be grown. Its irrigation water requirements are estimated in Table 9.28.

Method of Irrigation

Basin irrigation with open distribution channels should be used.
(Spray irrigation could have been used, if desired.)

Irrigation Frequency and Infiltration Rate

The moisture content of the soil is 30% by weight at field capacity
and 15% at the wilting point. Assume that it is not desirable to allow the
moisture content to fall below 22% at any time between irrigations. Hence,
from Eq. (9.8):

$$I_e = \frac{P_f - P_i}{100} S_s D_r$$

$$= \frac{30 - 22}{100} (1.3)(900) \approx 100 \text{ mm}$$

At the July maximum evapotranspiration rate of 382 mm per month (Table
9.28), the above quantity of 100 mm will be consumed in $(100/382)30 = 7.8$
days. Hence, in July a given plot of field should be irrigated every 7 to
8 days at 100 mm depth of wastewater per application. In fact, since irrigation efficiency for this type of system is likely to be only about 50%,
the wastewater required should be estimated at 200 mm every 7 to 8 days.
About 70 mm may be lost en route and in uneven application, and only 130 mm
would be applied to the basin, giving a leachate of about 30 mm.

Field tests for infiltration have given a saturation condition rate of
5.0 cm/day. Hence, the above quantity of 130 mm will require a maximum of
3 days to infiltrate into the soil. In fact, the time required will be
somewhat less than 3 days since initial infiltration will be at a higher
rate than the saturation condition rate.

Table 9.28 Calculated Irrigation Requirements for Data Given in Example 9.6

(1) Month	(2) Precipitation (mm/mo.)	(3) Effective precipitation P (mm/mo.)	(4) Pan evaporation E_p (mm/mo.)	(5) Assumed factor for fodder grass (α)	(6) Evapotranspiration E_t (= αE_p) (mm/mo.)	Net monthly irrigation requirement, I = (E_t - P)		(9) Gross irrigation requirements at 50% efficiency[a] (m³/ha-mo.)	(10) Gross requirements for 100 ha farm (cumulative m³)[b]
						(7) mm/mo.	(8) m³/ha-mo.		
Jan	29	27	27	0.6	16	–	–	–	–
Feb	23	22	44	0.6	26	4	40	80	8,000
Mar	27	25	74	0.7	52	27	270	540	62,000
Apr	23	22	104	0.7	73	51	510	1,020	164,000
May	12	11	152	0.7	106	95	950	1,900	354,000
June	2	2	278	0.7	222	220	2,200	4,400	794,000
July	–	–	424	0.9	382	382	3,820	7,640	1,558,000
Aug	–	–	412	0.9	371	371	3,710	7,420	2,300,000
Sep	2	2	298	0.8	238	236	2,360	4,720	2,772,000
Oct	11	10	157	0.7	110	100	1,000	2,000	2,972,000
Nov	22	21	55	0.6	33	12	120	240	2,996,000
Dec	25	23.5	28	0.6	17	–	–	–	–
Total	176 mm/year	166 mm/year			1,646 mm/year	1,498 mm/year	14,980 m³/ha-year	29,960 m³/ha-year	2,996,000 m³/year

[a] Assuming basin irrigation system with open channel distribution.

[b] See Figure 9.11.

At other times of the year the evapotranspiration rate will be less and, correspondingly, the required frequency of irrigation will be reduced (for example, the frequency in June will be once in 13.5 days, since the evapotranspiration rate will be only 222 mm/month) and, thus, more land will be needed to utilize the given wastewater or the latter will have to be stored in a holding lagoon as shown below.

Holding Lagoon and Farmland Requirement

Various possibilities present themselves for design of the irrigation system and must be evaluated in the light of the local conditions such as availability of agricultural land, dilution flow in adjoining watercourses, and other factors affecting costs and benefits, as discussed previously.

The simplest alternative would be to size the farm on the basis of the highest estimated requirement of irrigation, which in the given case occurs during June and July each year. At that time, the gross irrigation requirement is 7640 m³/ha-month or 255 m³/ha-day. A wastewater flow at 10,500 m³/day can irrigate a maximum of 10,500/255 = 41 hectares of net croplands. In the adjoining months of May and August, the flow will be more than necessary for 41 hectares and nearly 40% of the flow will need to be discharged to the nearest watercourse. Progressively larger quantities will need to be discharged in other months and practically the full flow will have to go to the watercourse in December, January, and February. If good stream flows coupled with low temperatures do not create an objectionable condition, and meet other disposal objectives, the arrangements may be the simplest and stream disposal and land disposal could complement each other. Alternatively, the area of land irrigated may be varied to suit the available wastewater flow if this is possible.

If, on the other hand, it is desired to use all the wastewater flow for irrigation over a fixed area of land, some form of flow balancing would be required in a holding lagoon. As a trial solution, if it is proposed to irrigate 100 ha, the cumulative irrigation requirements from month to month would be as shown in column 10 of Table 9.28, totaling 2,996,000 m³/year. The wastewater inflow would be at the calculated rate of 10,500 m³/day, or 315,000 m³/month (assumed at a steady rate although there would be some seasonal differences) and would total 3,780,000 m³ in a year. Thus, the wastewater utilized would be only about 80% of the total available. This would be fairly realistic since storage would entail losses by way of net evaporation, seepage, etc., from the lagoon. Lagoon losses can be estimated better after a trial size is first estimated.

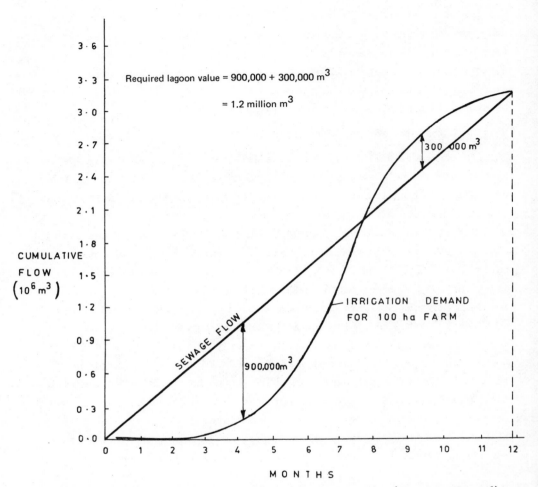

Figure 9.12 An estimation of holding lagoon capacity (see Example 9.6).

The capacity of the lagoon can be estimated from a cumulative plot of
the incoming and outgoing flows, as shown in Figure 9.12. A minimum volume
of 1.2 million m³ is required in this case. The siting of the lagoon would
be facilitated by the availability of a natural depression; if not, the
depth of excavation would be governed by the type of soil, the depth to
groundwater table, etc. Assuming an average depth of 5 m, the holding lagoon would have an average water spread of 24 ha.

The lagoon will receive direct precipitation and suffer evaporation
loss. Net loss from the lagoon will be equal to 2053 - 176 = 1877 mm/year,
i.e., 450,500 m³/year from a 24 ha area. This will be accommodated within

the 20% losses allowed for above, provided seepage is not much. The choice
of a lagoon site will be important. Incidentally, a lagoon of this capacity
would give a theoretical detention time of 114 days, or 3.8 months, to the
wastewater flow when the lagoon is full, but very little detention when the
lagoon is at low level. Hence, the incidental treatment afforded by the
lagoon would be variable with time (see Sec. 9.2.22 also).

Check for Salt Buildup in Soil

On an annual basis, assume that the water effectively applied to the
irrigation basins is 30% over the net requirement, and the excess quantity
leaches through. Thus,

$I_{applied}$ = 1498 x 1.3 = 1947 mm/year
Leachate L = 1498 x 0.3 = 450 mm/year
Precipitation P = 166 mm/year

Hence, from Eq. (9.14),

$$\text{Buildup ratio } \frac{C_s}{C_{i(av)}} = \frac{1}{L/(I_{app} + P)} = \frac{1}{450/(1947 + 166)} = 4.8$$

Thus, electrical conductivity in the soil will be likely to reach the following value at steady-state condition (assuming $C_{i(av)} \cong C_i$):

$$EC_e \cong 4.8 \times 1050 \cong 5040 \text{ } \mu\text{mhos/cm}$$

which is acceptable for grass cultivation. However, this estimation of
salt buildup is only approximate as it is based on several simplifying
assumptions.

Check for BOD Load on Field

The maximum gross irrigation volume is used in July; 7640 m³/ha-month
(= 250 m³/ha-day).

$$\text{Ultimate BOD load} = \frac{53 \text{ mg/liter}}{0.7 \text{ (for example)}} \times 250 \times \frac{10^3}{10^6} = 19.0 \text{ kg/ha-day}$$

This is the maximum load likely and is quite acceptable. At other times
of the year, the load will be less.

Nutrient Value in Wastewater

Assuming typical per capita values of N, P, and K in raw sewage, and
their likely removal in treatment, we get, as shown in Example 9.2 earlier:

$$N = 1460 \text{ kg/1000 persons} \times 70 = 102.2 \text{ tons/year}$$
$$P_2O_5 = 1825 \times 70 = 127.7 \text{ tons/year}$$
$$K_2O = 912 \times 70 = 63.8 \text{ tons/year}$$

Land Requirement

A 100 ha farm signifies 9.52 ha per 1000 m^3/day flow (see Table 9.25). However, total land = lagoon + farm = 24 + 100 + 124 ha = 140 ha gross (for example) = 140 ha/70,000 persons = 20 m^2/person.

9.3 GROUNDWATER RECHARGE SYSTEMS

The principal aim of groundwater recharge systems is to dispose of wastewater in such a manner as to avoid pollution of streams, lakes, or coastal waters and, at the same time, secure groundwater replenishment and possible reuse.* Such systems, if properly designed and operated, can yield reusable water safely, economically, and in an aesthetically unobjectionable manner. Crop cultivation with its attendant evapotranspiration may be avoided in order to maximize groundwater recharge although crops should improve soil surface permeability and nitrogen removal.

The availability of suitable sandy or gravelly soils with good infiltration characteristics is essential for the success of such a project. A number of shallow recharge basins are provided in parallel. Each basin may be a long rectangle a few hundred meters long, filled with wastewater to a depth of about 20 to 30 cm, and operated intermittently, that is, a few days wet followed by a few days dry, for reasons given below. Various other methods of recharge are also used. The physical, chemical, and biological quality of the wastewater has to be compatible with the characteristics of the soil and the aquifer into which recharge occurs. Suspended solids, organic and inorganic, algae, precipitated deposits, etc., can affect the continued infiltration rates and reclaimed water quality. The latter is also affected by the distance of travel within the aquifer. Some pretreatment of the wastewater before recharge is generally required.

Experiences reported by Bouwer [51] in operating an experimental recharge system in Phoenix, Arizona showed that, where secondary sewage

---

*Groundwater recharge systems must be differentiated from "deep well" injection systems in which the aim is wastewater disposal to a deep aquifer of poor water quality (e.g., brackish water) with no possibility for consumptive use.

effluent is used, the pollution of groundwater by nitrates can be controlled by intermittent operation of the basins, in which sequences of long inundation periods (14 days wet, 7 days dry) yielded about 90% removal of nitrogen, whereas, with sequences of short inundation periods (2 days wet, 3 days dry) all the nitrogen in the effluent was converted to nitrate in the renovated water. Apparently, the longer wet period with consequent anaerobicity encouraged denitrification; the dry periods aerate the soil and favor aerobic degradation of organics, and denitrification.

Suspended solids, as in secondary effluents, have been found not to clog soils at application rates of 6.6 cm/day, whereas clogging occurring at application rates as high as 32.8 cm/day has been controlled by intermittent operation of the recharge beds to provide alternating wet and dry periods [2].

The facility with which most soils can remove phosphorus and microorganisms has already been discussed in Sec. 9.2.19. The performance characteristics of rapid infiltration ponds have been given in Table 9.23.

Proper design of a groundwater recharge system necessitates the determination of its hydraulic properties. It will be recalled from Sec. 9.2.17 that the hydraulic conductivity of a soil may be markedly different in the horizontal direction from its value in the vertical direction (anisotropic), especially if it has layers of different types of soil. Thus, for example, if alternate layers of sandy and gravelly beds of equal thickness exist, with, for example, K values of 4 m/day and 150 m/day, respectively, the hydraulic conductivity in the horizontal direction K_h is the arithmetic mean of the two values (i.e., 77 m/day), whereas the conductivity in the vertical direction K_v is the harmonic mean of the two values (i.e., 13.2 m per day). The ratio K_h/K_v equals 5.83. Furthermore, each layer may be anisotropic in itself.

Bouwer [51] explains how field tests may be performed to estimate the values of K_h and K_v for a soil by simulation on an electrical resistance network analog using the knowledge of the water levels in two adjoining observation wells corresponding to given infiltration rates in experimental recharge basins. In his Phoenix, Arizona studies, the ratio K_h/K_v was found to be about 16.0. Diagrams showing equipotentials and streamlines are then prepared for any average recharge rate. The shape of the groundwater "mound" formed below the recharge basin is shown in Figure 9.13 with the vertical scale exaggerated.

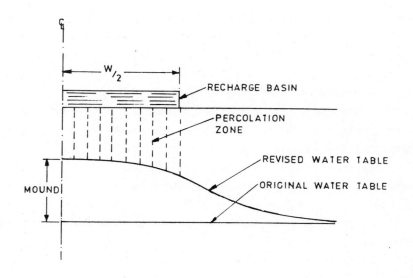

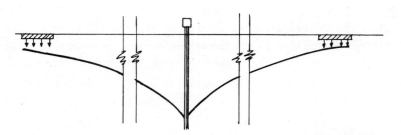

Figure 9.13 (a) A groundwater mound below a recharge basin; the vertical scale is exaggerated (adapted from Ref. 51). (b) A section through two parallel recharge strips with the withdrawal of groundwater by wells midway between the strips.

Care must be taken in designing a system so that the mound beneath the recharge area does not "back up" to the soil surface, with a resulting reduction in infiltration rate. A high mound would also make it difficult to maintain a sufficient aerobic percolation zone (of "at least several feet") and permit rapid drainage of the soil profile during the dry-up periods. It should be noted that the saturated zone has a capillary fringe extending above it by about 0.3 to 1 m, depending upon the type of soil.

In a water reclamation project, one more element has to be incorporated in the above, namely, a water extraction system. This may be in the form of wells, or infiltration drains and galleries of sufficient number and length. The distance of underground travel (on the order of several hundred meters) and the underground detention time (on the order of several months) must be kept as high as practicable to ensure the desired water quality at all times.

The extracted water can be conveniently used for a variety of purposes, such as recreational, industrial, agricultural, and can even be used as drinking water after suitable treatment. Indirect reuse of wastewater in this manner normally does not meet with objections on aesthetic grounds, as the wastewater loses its identity upon infiltration into the soil. A few water reclamation systems of this type have been installed in the U.S., Israel, and elsewhere. In Europe, recharge systems have been used especially with the polluted waters of the Rhine and the Ruhr rivers to make them suitable for drinking water supplies.

In many developing countries, groundwater recharge occurs, but in an incidental manner arising out of crop irrigation with wastewaters. These are "low rate" systems requiring much land, but providing food and employment for the people. Direct recharge systems in permeable soils are of the "high rate" type; they require less land (0.3-1.0 ha/1000 m^3-day) and operating labor.

REFERENCES

1. H. Kruse, Some present-day sewage treatment methods in use in Fed. Rep. of Germany, Bull. WHO, 26, 1962.

2. G. Kneeland, Jr., Land treatment of municipal sewage, Jour. ASCE, Civil Eng., 1973.

3. J. Penman, Natural evaporation from open water, bare soil and grass, Royal Society, London, Ser. A 193, 1948.

4. Irrigation of Agricultural Lands, American Society of Agronomy, U.S., 1967.

5. CPHERI (NEERI) Nagpur, India, Tech. Dig. No. 3, 1970.

5a. Short Communication, Ind. Jour. Env. Health, CPHERI, 14, 1972.

5b. CPHERI (NEERI) Nagpur, India, Tech. Dig. No. 47, 1975.

6. G. Fair, J. Geyer, and D. Okun, Water and Wastewater Engineering, John Wiley & Sons, N.Y., 1968.

7. S. Arceivala, Simple Waste Treatment Methods, Middle East Tech. Univ., Ankara, 1973.

8. Wastewater Engineering, Metcalf & Eddy, Inc., McGraw-Hill, N.Y., 1972.

9. Report of the special commission on pollution of surface waters, International Water Supply Association, Vienna Conference, Vienna, Austria, 1969.

10. F. Pöpel, Sludge Digestion and Disposal, 3rd Edition, Univ. of Stuttgart, Germany, 1967.

11. B. Handa, NEERI, NAGPUR, personal communication.

12. Advanced Waste Treatment Programme, USA.

13. S. Arceivala, Reuse of water in India, in Water Renovation and Reuse, H. Shuval, Ed., Academic Press, N.Y., 1976.

14. Personal communication.

15. The hazards to health of persistent substances in water, Report of Working Group, WHO, 1972.

16. Utilization of sewage sludge as fertilizer: Manual of practice, No. 2, FSWA, U.S., 1946.

17. CPHERI (NEERI) Nagpur, India, Tech. Digest No. 38, 1973.

17a. CPHERI, Nagpur, India, private communication.

18. Reuse of effluents: Methods of wastewater treatment and health safeguards, WHO Tech. Rep., Series, No. 517, 1973.

19. S. Arceivala, J. Laxminarayana, Algarsamy, and Shastri, Design, Construction and Operation of Waste Stabilisation Ponds in India, CPHERI, Nagpur, 1973.

20. Agricultural drainage and eutrophication, in Eutrophication: Causes, Consequences and Remedies, Nat. Acad. Sci., Washington, D.C., 1967.

21. CPHERI (NEERI) Nagpur, India, Tech. Digest No.

22. F. Dixon and L. McCabe, Health aspects of wastewater treatment, Jour. WPCF, 36, 1964.

23. Water quality criteria, U.S. Public Health Service, 1974.

24. W. Gardener, Movement of nitrogen in soil, in Soil Nitrogen, American Society of Agronomy, 1965.

25. D. Harth, Optimal control of nitrogen losses from land disposal areas, Jour. ASCE, Env. Eng. Div., 99, 1973.

26.

27. Nutrients from tile drainage systems, California Dept. of Natural Resources, U.S. EPA, 1971.

28. Assessment of the effectiveness and effects of land disposal methodologies of wastewater management, U.S. Army Corps and Engineers Report 72-1.

29. L. Dove, Recycling allows zero wastewater discharge, Jour., Civil Eng., 1975.

30. Muller, Bacteriological investigations on garbage seepage waters, Gewasserschutz, Wasser, Abwasser, 10, 1973 (in German).

31. H. Exler, The extent of groundwater pollution below a garbage dump, Gewasserschutz, Wasser, Abwasser, 10, 1973 (in German).

32. A. Aleti, R. Meier, and G. Agarwal, Wastewater reuse for large cities: An experimental study at Kanpur, India, Univ. of California, Berkeley, SERL Rep. No. 71-9, 1971.

33. N. Clerke and P. Kabler, Human enteric viruses in sewage, Health Laboratory Science, 1, 1964.

34. E. Gloyna, Waste stabilisation ponds, WHO Monograph Series No. 60, 1971.

35. J. Lamb, Guidelines for control of industrial wastes No. 5 (pulp and paper manufacturing wastes), WHO Document, WHO/WD/72.11, 1972.

36. S. Barnes, Some aspects of the utilisation and disposal of farm wastes, Jour. Inst. Sewage Purif., 1964.

37. R. Blosser and E. Owens, Irrigation and land disposal of pulp mill effluents, Water Sew. Works Jour., 111, 1964.

38. K. Healy and R. Laak, Site evaluation and design of seepage fields, Jour. ASCE, Env. Eng. Div., 100, 1974.

39. H. Scheltraga, Economic Treatment of Dairy Waste in Netherlands, SCS/70 Symposium, Washington, D.C., 1970.

40. Koziorowski and Kucharski, Industrial Waste Disposal, Pergamon Press, 1972.

41. A. Overman, Effluent irrigation of pearl millet, Jour. ASCE, Env. Eng. Div., 101, 1975.

42. L. Bernstein, Quantitative assessment of irrigation water quality, in Water Quality Criteria, ASTM, STP 416, Amer. Soc. Test Mat., 1967.

43. J. Lunin and M. Gallatin, Brackish water for irrigation in humid regions, USDA, ARS-41-29, 1960.

44. Approaches to land classification, UN, FAO Soils Bulletin 22, 1974.

45. Water and the environment, in Irrigation and Drainage, Paper 8, UN, FAO, Rome, 1971.

46. Irrigation practice and water management, Irrigation and Drainage, Paper 1, UN, FAO, Rome, 1971.

47. Design criteria for basin irrigation systems, Irrigation and Drainage, Paper 3, UN, FAO, Rome, 1971.

48. Irrigation, Drainage and Salinity, an International Source Book, FAO/UNESCO, 1973.

49. B. Ewing and R. Dick, Disposal of sludge on land, in Water Quality Improvement by Physical and Chemical Processes, Univ. of Texas Press, Austin, Texas, 1970.

50. Pound and Crites, Jour. ASCE, Civil Eng., 1973.

51. H. Bouwer, Groundwater recharge design for renovating wastewater, Jour. ASCE, 96, 1970.

52. A. Koenia and D. Loucks, Management model for waste water disposal on land, Jour. ASCE, 103, 1977.

Part 2

WASTEWATER TREATMENT

Chapter 10

WASTE TREATMENT METHODS

10.1 INTRODUCTION

In every country, and particularly in the developing countries, there
are many competing demands on the limited amount of funds available for
development. Wastewater treatment, although important from public health,
ecological, aesthetic, and other points of view, is generally apt to be given
a low priority. Thus, within the limited funds available, the designer is
called upon to select a method of waste treatment which will be capable of
meeting the environmental quality objectives and giving the right degree of
treatment required before discharge to fresh or coastal waters or to land.

From among the treatment methods qualifying for consideration, the
designer must then select those which have a bearing on local conditions
such as climate, availability of land, equipment, power, and perhaps more
important in some cases, the availability of skilled personnel and facilities for operation and maintenance of the plant once it is installed. Finally, if he is left with more than one type of process to choose from, he
must make an economic evaluation to be able to recommend the method of
treatment which will satisfy all the essential technological-economic criteria.

The approach just outlined is so basic that it must be followed in
every case to be able to identify what is now called "appropriate technology," a phrase which has replaced the earlier, less accurate but simpler
one, "low-cost methods of treatment." Instances of inappropriate transfer
of technology from the developed to the developing countries are not want-
ing when a global look is taken. This has sometimes been due to overselling,
sometimes due to a desire on the part of the developing countries themselves
to be identified with high technology, "going first class" so to speak, and
sometimes due to lack of data, or a misunderstanding of the ability of some
of the simpler methods involving ponds, ditches, and lagoons to give reliable and high quality treatment.

Over the years, it has become increasingly apparent that simple, low cost waste treatment does not necessarily mean low quality treatment. Without sacrificing quality, ways and means of reducing the costs and complexities of waste treatment have been developed and should be welcomed by all designers who would like to stretch their "dollar" of public funds as far as possible to meet the rising expectations of the people and build plants which will work under the various constructional and operational difficulties to which anyone from the developing countries will readily testify. That method which is technologically simplest, economically competitive, and operationally capable of meeting effluent quality requirements must invariably be preferred if a project is to be viable, whether in a developed or developing country.

The objective of this book is to focus on a few methods of treatment, such as the following listed in increasing order of mechanization, for treating both domestic and industrial wastes.

1. Waste stabilization ponds with or without anaerobic ponds preceding them.
2. Mechanically aerated lagoons and combinations of lagoons and ponds
3. Pasveer and carrousel-type oxidation ditches and other extended aeration processes
4. Activated sludge process and some of its modifications; trickling filters and rotating discs

Any one of these methods, along with land disposal of effluent should give, in most cases, a high degree of removal. Item 4 above may seem out of place in a book which emphasizes simpler methods of treatment. However, it should not be so when it is stated that the aim is to look for "appropriate technology," and the activated sludge process or one of its modifications may be the most appropriate method for an industry or subdivision where land is expensive or unavailable, and where good operational facilities may exist. The methods listed above are by no means the only ones. There are other biological and even physicochemical methods which may be more appropriate to consider in some cases.

Some urban centers are growing so rapidly that they are doubling in population every 12 to 15 years. World urban growth in the period 1970-1990 is expected to increase urban population by nearly 240% as a world average [1a]. With severe "pockets" of pollution thus being formed, treatment plant performance criteria will need to be tightened further. Already, in several instances, performance criteria have widened to include not only removal of biochemical oxygen demand (BOD) but also microorganisms and nutrients from domestic and municipal wastewaters. Are the simpler methods of treatment capable of meeting these criteria? In most cases, yes. Ditches, lagoons, and ponds followed by land irrigation, for example, can often give a combination which would be unbeatable even by a highly sophisticated tertiary treatment plant. Furthermore, lagoons and ponds have the added advantage of flexibility in expansion of the plant, relocation to another site, changing over to a more advanced process, etc., as discussed in the appropriate chapters.

But, first, let us briefly examine the essential steps in plant design which must be followed no matter whether the plant is to serve a community or an industry.

10.2 WASTEWATER CHARACTERISTICS*

It is important not only to know the values of the usual parameters such as flow Q, BOD_5, BOD_u, chemical oxygen demand (COD), suspended solids (volatile and fixed), total solids, temperature, pH, alkalinity, total Kjeldahl nitrogen (TKN), PO_4-P, and such other items, but also whether any toxic substances are likely to be present and, if so, in what concentrations and for what durations of time. It is also equally important to know the likely magnitude of fluctuations, and with a sufficient number of samples it may be possible to express flows, BOD, etc., as 50 percentile, 90 percentile, and 99 percentile values.

Generally, the hydraulic design of a plant has to be sufficient to meet 99 percentile values, whereas organic loading is often based on 50 percentile values, although higher values have been used sometimes in the interest of a consistently high quality effluent. Sludge handling facilities are also based on 50 percentile values, but final settling tanks are preferably sized on peak flow rates.

In the case of municipal wastewaters, the typical range of per capita values discussed below can be used along with expected flows to estimate BOD, TKN, solids, etc. As is well known, plant designs are sometimes based on little, if any, reliable field data, and success is then a hit or miss proposition.

Raw domestic wastewater characteristics are shown in Table 10.1. The range of values given is also typical for municipal wastewaters which are

_____

*This section repeats the information presented in Section 5.1.

Table 10.1 Domestic Wastewater Characteristics

Item	Range of values contributed in wastes (g/capita-day)
BOD_5	45-54
COD (dichromate)	1.6 to 1.9 x BOD_5
TOC	0.5 to 1.0 x BOD_5
Total solids	170-220
Suspended solids	70-145
Dissolved solids	50-150
Grit (inorganic, 0.2 mm and above)	5-15
Grease	10-30
Alkalinity, as $CaCO_3$	20-30
Chlorides	4-8
Nitrogen, total, as N	5-12
Organic nitrogen	~0.4 x total N
Free ammonia	~0.6 x total N
Nitrite nitrogen	-
Nitrate nitrogen	-
Phosphorus, total, as P	0.8-4.0
Organic P	~0.3 x total P
Inorganic (ortho- and polyphosphates)	~0.7 x total P
Potassium, as K_2O	2.0-4.0
Microorganisms present in wastewater	(per 100 ml wastewater)
Total bacteria	10^9-10^{10}
Coliforms	10^6-10^9
Fecal streptococci	10^5-10^6
Salmonella typhosa	10^1-10^4
Protozoan cysts	Up to 10^3
Helminthic eggs	Up to 10^3
Virus (plaque forming units)	10^2-10^4

Note: Values given in the table represent approximate daily per capita contributions. To these must be added the contents in the original water supply. Per capita daily water usage ranges from 80 to 300 liters in most sewered communities, although consumption in some U.S. and other communities is even higher than 500 liters. The pH of wastewater generally ranges from 6.8 to 8.0, again depending on raw water quality.
Source: Ref. 1.

predominantly domestic in character. The wastes are mainly organic in nature, containing carbon, nitrogen, and phosphorus, among others, with relatively high concentrations of microorganisms. They are readily putrescible,
and biological degradation of organic matter proceeds even as the wastes
flow through sewers. This alters some characteristics as time passes. Values given in Table 10.1 refer to relatively fresh wastes.

Furthermore, all values given in Table 10.1 are stated in terms of
grams/capita to facilitate their use for design purposes, and for comparison between different communities. Values given in terms of mg/liter are
obscured by the fact that water usage varies markedly between communities,
especially between those in developed and developing countries.

Values of BOD generally average around 54 g per person per day where
the sewage collection system is reasonably efficient. In some developing
countries, the BOD values may be only 30 to 40 g per person per day, as all
the sewage produced may not enter the sewer system. Where combined sewers
are used, the BOD values may be about 40% higher, namely, 77 g per person
per day. In the case of offices, factories, schools, etc., where there is
part-time occupancy, the BOD values are generally taken as half of 54 g per
person per day, or less. For restaurants and cafeterias, each meal served
may be taken to contribute one-fourth of 54 g of BOD. Theaters and cinemas
may be taken to contribute one-sixth of 54 g of BOD per seat. Hotels and
hospitals, on the other hand, may contribute as much as 1.5 to 2.5 times
more than the usual 54 g of BOD per person per day.

Since interest in many countries often lies in irrigational use of
wastewater, it is interesting to know the mineral pick-up from domestic
usage of water. Table 10.2 gives some typical values reported as national
averages in the U.S. [2], while Table 10.3 gives some values reported from
Israel [3]. A detailed discussion of the nature of wastes from industries
is beyond the scope of this book.

Table 10.2 Mineral Pick-up from Domestic Usage of Water in the U.S.

Item	Pick-up from one cycle of domestic water usage (mg/liter)
Chlorides as Cl	20-50
Sulfates as SO_4	15-30
Nitrates as NO_3	20-40
Phosphates as PO_4	20-40

Table 10.2 (Continued)

Item	Pick-up from one cycle of domestic water usage (mg/liter)
Sodium as Na	40-70
Potassium as K	7-15
Calcium as CaCO$_3$	15-40
Magnesium as Mg	
Boron as B	0.1-0.4
Total dissolved solids	100-300
Total alkalinity as CaCO$_3$	100-150

Source: Ref. 2.

Table 10.3 Mineral Pick-up in Sewage from Towns
and Agricultural Settlements in Israel

Item	Municipal Sewage[a]	Agricultural Settlements[b]
Nitrogen as N, g/capita-day	5.18	7.0
Potassium as K, g/capita-day	2.12	3.22
Phosphorus as P, g/capita-day	0.68	1.23
Chlorides as Cl, g/capita-day	0.54	14.65
Boron as B, g/capita-day	0.04	0.06
Sodium as Na, g/capita-day	0.60	14.75
Total hardness as CaCO$_3$, g/capita-day	2.50	6.25
Total dissolved solids, g/capita-day	40	78
Electrical conductivity, μmhos/cm	600	470
Sodium absorption ratio, meq/liter	2	1.5

[a]Average of 62 municipalities surveyed (2.1 million total population). Sewage flow at 100 liters/capita-day.

[b]Includes sewage and livestock water; total flow at 250 liters/capita-day. Population surveyed was 96,880 in 267 settlements.
Source: Ref. 3.

10.3 DESIGN REQUIREMENTS

For sewage treatment plant design, the shorter the period of design used, the better it is. Short periods would be 7 to 10 years for the treatment units per se, but for the outfall sewer, main pumping station, and land requirement for the treatment plant, longer periods are advisable and often their sizing is done so as to be sufficient at full development in the future. If treatment units are designed for unduly longer periods, they may be quite oversized in their early life, and can sometimes be a disaster, especially in a warm climate, where septic conditions can be readily induced in settling tanks and other unaerated units. Oversized aeration equipment can only lead to waste of electrical energy, and may even adversely affect performance under certain conditions. However, oversizing does not lead to any particular problems with facultative type aerated lagoons and oxidation ponds.

10.3.1 Plant Performance

Next, the performance requirements of the process must be fixed especially with regard to BOD removal efficiency, need for nutrients removal, desirability of nitrification, and required bacterial removal efficiency. Some consideration also needs to be given to the strictness with which the performance criteria must be met under varying conditions of climate and wastewater flow. This aspect may also have a bearing on the choice of the treatment process.

Knowing the projected population to be served, the quantities and characteristics of the wastewater, and the permissible concentration of different constituents that can be discharged to fresh or coastal waters or to land for irrigation, one determines the efficiency of treatment necessary. This estimation is relatively easier where clear-cut effluent standards are available. Where receiving water quality standards are available, the effluent quality requirements can be derived from the former from a knowledge of the likely available dilution ratios. Where no standards at all exist, the designer would guide himself by the fact that the available dilution capacity in the watercourse (without disturbing the legitimate uses to which the watercourse is presently put) should be used in the interest of economy, with provision for stepping up the degree of

treatment at a later date when necessary. He must then estimate the presently available dilution capacity using methods described elsewhere.

Alternatively, wherever feasible, the designer may wish to explore the
possibility of discharging to land for irrigation, taking advantage of the
fact that effluent quality requirements for irrigational use are often
easier to meet than those for discharging to freshwater, especially where
the wastewater is mainly domestic in character. In some cases, the designer
may even find a combination of two methods of disposal more favorable to
use. For example, he may wish to discharge to land only during the summer
season when river flows are likely to be at minimum values and coastal
bathing beaches in full use; thereafter, in nonirrigating seasons, the effluent may be discharged to a river or coastal waters.

Thus, one can determine the approximate degree of treatment required
in terms of BOD removal and/or any other suitable parameter.

10.4   CHOICE OF TREATMENT METHOD

The major factors which affect the choice of a treatment method are
the following:

1.   Overall cost (capital and operating)
2.   Performance capability
3.   Operating characteristics
4.   Equipment availability
5.   Manpower availability

Items (4) and (5) are more in the nature of constraints which may be
crucial in some developing countries. The appropriate technology to use
in a given situation would be one which would take into account these constraints and produce a plant at the lowest overall costs that is capable
of meeting the performance requirements in the most dependable manner.

Table 10.4 gives the efficiency of BOD removal that can be expected
in the case of predominantly domestic sewages treated by a few different
methods. This table can be used to make a preliminary selection of the
processes likely to meet efficiency requirements. Additional criteria
(e.g., nutrient and coliform removal) may have to be taken into account in
some cases [4].

Next, a quick idea of land requirements can be obtained from Table
10.4. Wide variations are to be seen in the values given. This is due to

Table 10.4 Characteristics of Some Methods for Domestic Sewage Treatment

Item	Extended aeration systems	Conventional activated sludge	Facultative aerated lagoons	Waste stabilization ponds	Land treatment	Rapid infiltration systems
Performance						
BOD removal, %	95-98	85-93	70-90	70-90	80-90	99
Nutrient removal, % N	15-30[a]	30-40[a]	-	40-50	80-90	80
% P	10-20[a]	30-45[a]	-	20-60	80-99	80
Coliform removal, %	60-90	60-90	60-96	60-99.9	90-99	99
Land requirement, m²/person[b]						
Warm	0.25-0.35	0.16-0.20	0.15-0.45	1.0-2.8	10-20	0.7-2.0
Temperate	0.35-0.65	0.20-0.40	0.45-1.0	3-12	20-80	in favorable soils (extra for pretreat)
Process power requirement (kWh/person-year)[c]	13-20	12-17	12-15	Nil	Nil	Nil
Sludge handling	No digestion; drying beds/mechanized devices	Digestion; drying beds/mechanized devices	Manual desludging once in 5 to 10 years	Nil	None	-
Equipment requirement (excluding screening and grit removal)	Aerators, recycle pumps, sludge scrapers (for large settlers)	Aerators, recycle pumps, scrapers, thickeners, digesters, driers, gas equipment	Aerators only	Nil	Sprinklers (optional)	Nil
Operational characteristics	Simpler than activated sludge (all aerobic)	Skilled operation required	Very simple	Simplest	Very simple	Very simple

Table 10.4 (Continued)

Item	Extended aeration systems	Conventional activated sludge	Facultative aerated lagoons	Waste stabilization ponds	Land treatment	Rapid infiltration systems
Effect of population size on unit cost	Relatively little	Considerable	Slight	Slight	Slight	Slight
Special features	High efficiency; nitrified effluent	-	Easy to enlarge or relocate if desired; combination of lagoon and pond can enable optimization for cost	Slight	20-100% of O and M cost offset by revenue from crops Some pretreatment may be required.	Groundwater recharge benefit occurs Some pretreatment may be required.

aAdditional nutrient removal can be achieved through special measures.

bBased on population equivalent = 54 g/person-day, 3 m depth of water in lagoons, and embankment slopes 2.0 horizontal:1.0 vertical.

cBased on aerator capacity of 2 kg O_2/kWh at standard conditions (20°C, zero DO, and plain water) and 0.75 of standard value delivered at field conditions.

the fact that temperature has a considerable effect on the performance of processes involving lagoons and ponds. Sludge dewatering methods also have a considerable influence on land requirements. The choice of a method (e.g., waste stabilization pond) may have to be ruled out from further consideration because of the unavailability of land in some cases. In preparing Table 10.4, effort has been made to distinguish between operation in colder and in warmer climates. The reader is also referred to each of the relevant chapters in the book concerning the different methods.

Finally the cost aspect is discussed in Sections 10.7 and 10.8.

10.5 AEROBIC BIOLOGICAL PROCESSES

Among the aerobic biological processes, the <u>activated sludge process</u> is widely used in waste treatment. Developed around 1914, the complex biochemical mechanism of the process has fascinated more research workers than any other waste treatment method and, particularly in the last 20 years or so, various modifications of the process have been developed.

Power requirement varies from 12 to 17 kWh/person-year to run the whole plant, whereas the land requirement varies from 0.16 to 0.4 m^2/person, depending on the climate, the mode of sludge dewatering (open drying beds or mechanical dewatering), and the population served. Some design criteria are summarized in Table 10.4 and in the flowsheet in Figure 10.1.

Operationally, the activated sludge process including sludge treatment demands skilled operation of a relatively high order and the largest manpower requirement per unit of population served. The presence of the sludge digester makes it necessary for the operator to understand the operation of both aerobic and anaerobic processes, and sludge gas utilization for power generation at the site often seems attractive, but again introduces a high degree of mechanization which is beyond the scope of facilities readily available in many developing countries. Artificial heating of digesters in cold climates and mechanical sludge dewatering systems similarly introduce a high degree of mechanization. In fact, it can be said that, in the activated sludge process, it is not so much the actual activated sludge aeration which causes problems as it is the treatment and disposal of the surplus sludge. Only in warmer climates can some simplification be introduced by having unheated digesters, with no gas utilization, and with sludge dewatering on open sand beds. Alternatively, aerobic digestion of sludge, which is simpler to operate, can be adopted, but needs somewhat more power.

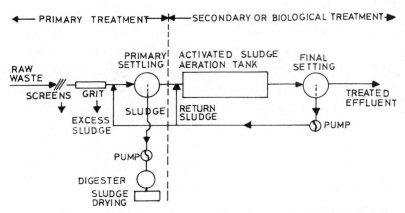

Figure 10.1 A flowsheet of a conventional activated sludge process.

It should also be evident that mechanical and electrical equipment components of an activated sludge plant are relatively expensive, and this aspect might go against its choice for a developing country that would need to import most of these items and their spare parts. Costwise, the activated sludge plant is also likely to be somewhat unfavorable, except perhaps for very large sizes. This can be seen in Figure 10.5 (also see Example 10.2).

Modifications of the activated sludge plant which have gained popularity in recent years are the <u>extended aeration process</u>, of which a typical application is to be found in the Pasveer and Carrousel type oxidation ditches widely used in Europe and elsewhere to serve even quite large populations, and in the "package" plants available in the U.S. for small installations. This process is simpler to construct and operate than the conventional activated sludge plant. Lengthening of the aeration time has earned it the name "extended aeration," and gives the advantage that the sludge is sufficiently mineralized and the excess quantity does not need any further treatment in a digester before dewatering. The operator now deals with only one type of process, namely, the aerobic type, and general operational control is easier. Equipment requirement is also simplified both in variety and complexity (see Figure 10.2).

However, this simplification is offset to some extent by the fact that more power is consumed in extended aeration systems since all organics are now stabilized aerobically only. Thus, the break even point between this process and the conventional one depends on the relative capital and operational costs of the two systems. The typical land and power requirements are summarized in Table 10.4.

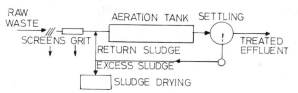

Figure   10.2 A flowsheet of an extended aeration process as used in
Pasveer-type oxidation ditches and other extended aeration systems.

A special advantage of this process is the fact that it is generally
capable of giving the highest BOD removal (95 to 98%) compared with any
other process. The effluent is also generally fully nitrified [5].

Construction costs for oxidation ditches are generally lower than those
for conventional activated sludge plants. In the Netherlands the cost per
capita for oxidation ditches is considered to be about 75% of the cost for
conventional plants [8]. In India, Arceivala et al. [7] have shown this
difference to be even higher, especially for small size plants. Figure
10.6 shows construction costs for oxidation ditches in the Netherlands and
in India; the differences are illustrative of the widely different climatic
conditions, construction specifications, and equipment costs. Instrumentation and automatic operation are held to a minimum in India as it should be
in a developing country.

At the other end of the spectrum of waste treatment processes is the
waste stabilization pond, which is the simplest method of treatment (Figure
10.3).

The various advantages and disadvantages stem from the fact that the
process is entirely natural, without any man-made accelerating devices
(e.g., mechanical aeration, heating, etc.) being used. Consequently, there
is practically no equipment requirement, no power consumed, and minimal attention is required for day to day operation. On the negative side, however,
is the fact that the land requirement of this process is the highest compared with any other. Land requirement is also considerably affected by
climatic conditions (Table 10.4). Thus, the method is competitive where
land is available, relatively inexpensive, and the climate is favorable [6].

Figure   10.3 A simple waste stabilization pond installation.

Figure   10.4   A flow sheet for mechanically aerated facultative lagoons.

Construction cost is least, compared with other methods, as shown in
Figure 10.7.   However, the land cost may be a crucial factor affecting the
overall capital cost.

The use of anaerobic ponds can be advantageous under certain conditions,
especially with high-strength wastes as discussed in Chapter 16.

Mechanically aerated lagoons fall between the above methods.   Facultative aerated lagoons (Figure 10.4) are similar to waste stabilization ponds,
except that oxygen is now supplied artificially through mechanical aeration.

With the advent of mechanical aerators of the vertical-axis type available either as fixed or floating units, facultative aerated lagoons have
rapidly increased in popularity since they combine simplicity of construction and operation with minimal mechanization.   In some developing countries,
the aerators could even be readily manufactured, if necessary, in standard
sizes as discussed more fully in Chapter 15.

Facultative aerated lagoons can give 70 to 90% BOD removal at detention
times ranging from 3 to 12 days depending on the climate.   Similar to stabilization ponds, these lagoons are also considerably affected in their
performance by temperature.   Thus, land requirements also vary considerably
with climate, ranging from 0.15 to 1.0 m^2/person.   Power requirements vary
from 12 to 15 kWh/person-year (see Table 10.4).   Owing to favorable land
and power requirements, aerated lagoons have the potential for becoming the
most favored form of treatment in both developing and developed countries.
Figure 10.8 shows typical costs for facultative aerated lagoons.

10.6   ENERGY CONSERVATION

Energy conservation and reduction methods now need to be given increasing attention when designing waste treatment plants.   This is again
needed equally in both developing and developed countries.   In this regard,
a twofold approach needs to be considered.   First, adopt every feasible
method to conserve energy without adding to the costs or complexities of

the treatment process. This implies a judicious selection of equipment and processes, and an emphasis on good engineering and architectural design without bringing in exotic technology. Second, consider only on a cost-benefit basis the adoption of more advanced devices, provided the mechanization so introduced is within the technological competence of the people . concerned. Thus, the latter approach would have to be restricted to more developed countries.

Among the methods which could be readily included in the first approach above are the following:

1. Select, as far as possible, the least energy-intensive processes capable of meeting effluent quality requirements.

   Table 10.4 shows that power requirements are nil for waste stabilization ponds, relatively low for facultative aerated lagoons, and highest for extended aeration systems. Aerobic digestion also generally needs more power than does anaerobic digestion. Thus, if a more energy-intensive process is selected, there should be strong justification for it, such as the need for very high BOD removal efficiency, or the need for nitrification, or the need for reliability in operation, etc. If, in an instance, a simple waste stabilization pond is found feasible, no further energy conservation effort is necessary and the method should be preferred.

   Often there is also considerable scope for combining two processes having different requirements (e.g., for power and land) in such a manner as to be able to optimize for overall costs, or for either power or land or some other requirement. Several examples of this type are possible (Sec. 15.7). Some other considerations have also been discussed in Chapters 14 to 17.

   Again, within a selected process it may be possible to reduce the energy requirement in certain ways. For example, an interesting method for reducing energy requirements in extended aeration and activated sludge processes is to benefit from the oxygen release resulting from denitrification (see Sec. 14.6).

   Similarly, in the case of mechanically aerated lagoons, energy requirements can perhaps be reduced in warmer climates by constructing deeper units as discussed in Sec. 15.3. Better attention to mixing kinetics, heat conservation, etc., can also help reduce lagoon sizes and power requirements as discussed in Sec. 11.18 and in Chapter 15. More research is needed in this regard, especially in warmer countries.

2. Within a given process use, wherever possible, the least energy consuming equipment and construction techniques.

   For example, screw pumps are reported to generally use less energy than conventional pumps. Mechanical aeration systems similarly require less energy than do pneumatic systems for a given oxygen input rate in lagoons.

   Benefits must be secured from natural land contours to avoid pumping as far as possible. In cold climates, the proper choice of building material and construction techniques becomes important for heat conservation; so does good architectural design, in any climate. Various ways for bringing about savings in energy requirements can be found without adding to the costs or complexities of the treatment process.

Among the more mechanized methods of energy conservation could be listed the following:

a. Supplementation of conventional energy sources by wind and solar energy wherever possible. Windmills have been used since time immemorial, and could perhaps be extended to waste treatment plants to run pumps, aeration rotors, rotating discs, etc., without introducing much mechanization.

Solar energy collection devices are becoming available, as are other devices, on an experimental basis for use in space heating. Other uses (e.g., digester heating) must be explored further before such devices can be used on a wider scale.

b. Adoption of more advanced heat recovery devices. Among these, sanitary engineers are already familiar with the use of methane from digester gas for heat or power generation. The system introduces a considerable extent of mechanization which may not be desirable at all times, and the adoption of such a device should, therefore, be carefully evaluated.

Waste heat recovery from the wastewater itself through the use of an electric heat pump has been conceived. Again, its use must be carefully evaluated before adoption, especially by developing countries.

Wilke and Fuller [13] described the design of a proposed plant requiring minimum external energy sources for wastewater treatment at Wilton, Maine (population 4200). This is achieved through a combination of good engineering and architectural design for heat devices, waste heat recovery from the effluent by electric heat pump, and choice of low-energy requiring processes and equipment, besides the usual recovery of heat energy from methane gas. The operation of this plant, when built, will be watched with great interest.

At this stage, it may not be out of place to mention that, in some countries, it may be considered more appropriate to set aside some energy conservation thoughts in favor of greater encouragement for locally manufactured items, even though not of optimum efficiency in energy consumption. For example, locally made aerators for lagoons may be preferred over imported units, however efficient the latter may be in oxygen transfer characteristics. A certain amount of "deliberate diseconomy" may not be a bad thing for a developing country.

10.7 COST OF WASTE TREATMENT

The costs of various wastewater treatment methods vary widely depending on the nature of the waste, the process, the climate, the design criteria, the site conditions and local cost of labor, materials, land, and power. Nonetheless, costs must be estimated reasonably well in advance to be able to (1) choose the least-cost alternative from among those which are technically capable of performing the desired function, (2) raise adequate finance for implementing the scheme, and (3) determine appropriate wastewater treatment rates for revenue purposes for the costing of a product or for tax purposes.

It is generally difficult to obtain accurate cost figures owing to the widely differing conditions just stated, even for domestic sewage of relatively uniform character; local data must be evaluated to generate unit costs as far as possible. Plant costs are often stated as capital costs and annual costs which are made up of various components as listed below.

Capital Costs

These include all initial costs incurred up to plant start-up, such as:

1. Construction costs (including equipment supply and erection)
2. Land purchase, right-of-way, etc.
3. Engineering design and supervision fees, legal and any other fees
4. Interest charges on loans during construction period

Annual Costs

After start-up of the plant, the following annual costs must be incurred:

1. Interest charges on capital borrowings (loans)
2. Amortization of loans
3. Depreciation of plant
4. Insurance of plant
5. Operation and maintenance of plant (including minor repairs)
6. Tax, royalties, and profits

Items 1 to 4 in annual costs may be regarded as "fixed" charges since they generally have to be incurred whether the plant works or not. Item 6 may not be relevant to public utility concerns as they do not pay tax to the government nor operate for profit, although in some instances even public utilities are required to take these items into account when costing a product or service as if they were private undertaking competing in the capital market.

Generally, for domestic and municipal wastes, published data are available concerning plant construction costs and plant operation and maintenance costs, both of which are related to the plant inflow or the population served. It is interesting to observe that, in almost all such studies, a log-log relationship has given good fit to the data, and equations used have been of the form

$$C = a(X)^b \qquad\qquad (10.1)$$

where

C   = cost per person or per unit flow

X   = population or flow

a and b = constants

It is evident that the constant "a" reflects the cost when X is equal to a "unit" of population (e.g., 1000 persons) or a "unit" of flow (e.g., mgd or 1000 m^3/day), and the exponent "b" reflects the economy of scale. In general, "b" has a range of -0.3 to -0.6 in most wastewater treatment plants and lift stations. A value of -0.6 indicates that the increase in plant cost is relatively small for a large increase in the size of a plant. The value of "b" for trunk sewers is reported to be even higher than -0.6, indicating that the cost is not greatly affected by diameter, presumably since laying costs predominate. A similar situation should be expected in the case of sea outfalls.

In the case of treatment plants, however, the construction cost is considerably affected by the size of the plant, and it is well known that, for the conventional activated sludge plants and trickling filter plants, the costs rise sharply as the size of the plant reduces, since "b" values are around -0.3. Waste stabilization ponds, on the other hand, may show "b" values around -0.5 to -0.6, indicating that there is relatively little economy of scale and, thus, a lower period of design can be used. Ponds need not be built for a large future population but, rather, expanded from

time to time as the population increases. Aerated lagoon costs generally
vary in a manner similar to stabilization pond costs, whereas oxidation
ditches may show a cost variation with size between that of lagoons and
conventional activated sludge plants, and are perhaps more like the latter.

As cost data collected from different countries in different years can
only be of relative value for illustration purposes, equations developed
in each case are not given here. However, typical log-log plots are illustrated in Figures 10.5, 10.6, 10.7, and 10.8 for conventional activated
sludge plants, oxidation ditches, waste stabilization ponds, and facultative
aerated lagoons, respectively. These may be useful to the reader to prepare
only preliminary, order-of-magnitude estimates for construction, operation,
and maintenance of these plants in the absence of local cost data.

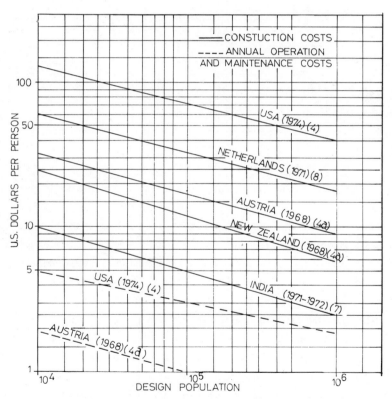

Figure 10.5 Construction and annual O and M costs versus the population served by complete waste treatment plants using an activated sludge
process (references are given in parentheses). Note: the daily average
flow per person is 380 liters in the U.S. and 135 liters in India; the
daily BOD$_5$ per person is 77 g in the U.S. and 54 g in India. The monetary
conversion is as follows: $1 U.S. = 3 Dutch Fl. = 19 Australian Sh. = 7.5
Indian Rs.

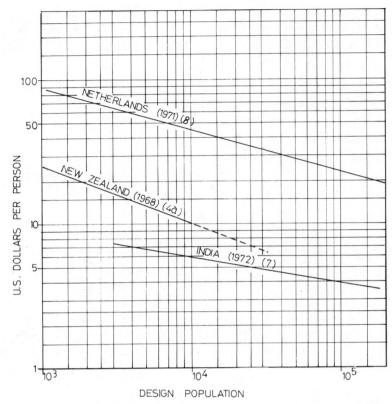

Figure 10.6 Construction costs versus the population served for Pasveer and carrousel oxidation ditches (references are given in parentheses).

Particularly in the case of aerated lagoons and stabilization ponds, it should be remembered that the design and, therefore, their costs, are considerably affected by temperature. This aspect is unfortunately "averaged out" in the generalized types of curves given in Figures 10.7 and 10.8. Also, with all types of plants, costs are substantially affected by the "specifications" imposed on the designer and by the extent of instrumentation and automatic operation built into the plant. The cost curves shown for India represent a situation in which the latter is kept to an absolute minimum, and where all equipment needed for constructing any type of waste treatment plant is manufactured fully within the country.

The size range over which a given method of treatment is economically competitive compared with other methods can be determined as explained in Sec. 10.8 (see Examples 10.2 and 10.3). Several studies on costs have been reported in the literature [1,4a,8-11].

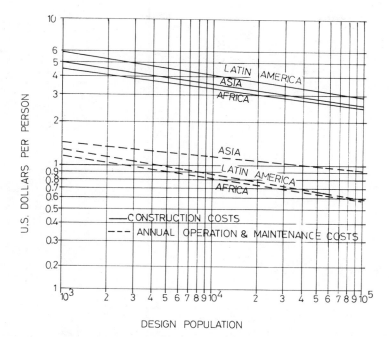

DESIGN POPULATION

Figure 10.7 Construction and operation and maintenance (O and M) costs versus the design population served by waste stabilization ponds. The data were obtained by Reid et al. [12] through a questionnaire survey; the BOD_5 = 77 g/person-day and all costs are updated to 1975 values.

10.8 DETERMINATION OF ANNUAL COSTS

Since it "costs" money to borrow money and since borrowings must gradually be returned over a period of time, all costs can be stated in terms of annual costs for the sake of uniformity in determining the total costs, including operation and maintenance of a plant. The major factor often affecting the total annual costs is the annual interest rate charged by the funding agency and the interest that can be obtained by deposits made by the user.

Without going into details, the various sources from which funds can be obtained are as follows: (1) public bonds issues, (2) national banks and lending agencies, (3) government, (4) potential beneficiaries of a project, (5) internal borrowings (e.g., replowing of company profits), and (6) international and bilateral agencies (e.g., World Bank, IDA, Asian Development Bank, etc.).

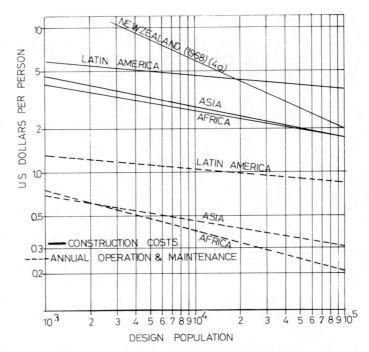

Figure 10.8 Construction and O and M costs versus the design population served by facultative aerated lagoons. The data were obtained by Reid
et al. [12] through a questionnaire survey; the BOD_5 = 77 g/person-day and
all costs are updated to 1975 values.

10.8.1 Interest Charges

Depending upon the funding agency, interest rates can vary substantially. Some government agencies advance loans for waste treatment purposes
at 5 to 6% interest; some even give a part of the amount as an outright
grant to municipalities in order to encourage pollution control activities.
Other agencies may charge around 8% or even more. The World Bank is reported (1979) to charge 7.25% on the disbursed part of the loan and 0.75%
on the undisbursed part of the total amount earmarked by it for a project
[4a]. The IDA, on the other hand, charges no interest, and only a 0.75%
service charge annually. No repayment is required in the first 10 years
and, thereafter, time up to 40 years is given for repayment. These are
indeed what are called "soft terms" and only some developing countries can
qualify for them. IDA loans are made to the governments themselves. In
most instances, loans have to be obtained under "harder" terms, especially

by industries wishing to install waste treatment plants for either pollution control or reuse of water or both.

10.8.2 Loan Amortization

The gradual repayment of a loan is necessary to amortize one's debts. Ideally, the amortization period should be spread over the average lifespan of the assets. But banks and industries may require return in a shorter period. Two alternatives for repayment of loan are generally available.

For public bodies constructing water supply and sewage projects for which bonds are issued by them in the capital market, repayment of bonds at maturity can be refinanced by issuing new bonds. Thus, only the interest on loan has to be continually paid. When the loan becomes due, fresh bonds will again be issued. This is a valid procedure for public bodies always likely to command response in the capital market and helps to spread the burden equally between the present generation and the next.

For other public bodies, and for industries which must borrow funds from the government or national and international funding agencies, not only must the interest charge be paid, but also the loan fully amortized in an agreed period. This method is generally preferred in costing a product or computing profits, etc. Once the loan is amortized, the amounts may continue to be set aside annually as they help to build up reserves for future expansion. The amount to be set aside annually for amortization of a loan is given by

$$A = \frac{Pi(1 + i)^n}{(1 + i)^n - 1} \qquad\qquad (10.2)$$

where

A = annual payment or annuity

P = present amount (loan)

i = rate of interest per year

n = number of years in which the loan is to be returned

10.8.3 Depreciation of Assets

Besides interest charges and loan amortization, another item of fixed charge which must be created is depreciation, to enable replacement of an

Table 10.5 Estimated Life and Depreciation of a Plant

Item	Estimated life (years)	Annual depreciation as percent of capital cost (straight line method)
1. Waste treatment plant (average applied to whole plant)	20	5
2. Lift stations (machinery)	10	10
3. Major civil structures (dams, trunk sewers, outfalls)	50	2
4. For preliminary estimates		
All equipment and piping	10	10
Masonry buildings	33	3
Frame buildings	20-25	4-5

item after its normal "life." Owing to wear and tear and obsolescence with time, different items of equipment, machinery, and civil works have different economic lifespans, after which it would be cheaper to replace the items. Typical values are shown in Table 10.5. Sometimes, funds set aside for depreciation are also used in case of very major repairs, additions, or alterations, but not for day to day maintenance, for which funds must come from the operation and maintenance account.

Depreciation allowance can be calculated in different ways, some of which are as follows:

1. Sinking fund method
2. Present worth of future returns
3. Fixed percentage of depreciated value
4. Straight line method
5. Production units method

Only the first two methods take into account the interest earned by the annual installments set aside for depreciation. Thus, the percent of capital cost set aside in the earlier years is higher in the latter three methods compared with the former.

The sinking fund method is generally considered suitable for public works. The annual amount A required to be set aside to produce a future sum F is given by

$$A = \frac{Fi}{(1 + i)^n - 1}$$ (10.3)

For public works, the future sum F desired to be accumulated over a period of n years at an annual rate of interest i may be taken as the present capital cost of the item, although it is appreciated that, owing to inflation, the same item will cost more when the time comes to replace it. Thus, strictly speaking, F should be the future value of present worth estimated from

$$F = P(1 + i)^n \qquad\qquad (10.4)$$

in which i and n may be assigned different numerical values from those in Eq. (10.3). If i and n have the same values in Eqs. (10.3) and (10.4), the two equations can be combined to give Eq. (10.2) seen earlier.

In the straight line method, which is often preferred for use with industrial waste treatment plants, the amount set aside each year is a fixed percentage of the capital cost. This is shown in Table 10.5. Other methods are also followed. It is important to know what the national tax authorities permit private industries to charge as depreciation in computing taxable income.

10.8.4 Operation and Maintenance (O and M)

The various items included under O and M are (1) staff, (2) chemicals, (3) electricity and fuel, (4) transport, rentals and other direct costs, (5) maintenance and repairs, and (6) overheads. These costs depend considerably on the type of waste treatment selected, local labor and energy costs, availability of spares, etc.

Maintenance costs under item (5) include all the minor repairs required from time to time in any plant. They are generally estimated as a percentage of the capital cost of the item of equipment or civil works as shown in Table 10.6.

Table 10.6 Estimation of Costs for Routine Maintenance and Repairs

Item	Percentage of capital cost
Buildings	2-5
General equipment and machinery	2-10
Instruments	25

10.8.5 Tax, Insurance, Royalties, and Profits

As stated earlier, most of these items are inapplicable to public works except for costing purposes. But for industrial wastewater plants these items may need to be taken into consideration.

Insurance on equipment and buildings is generally at 1% of the capital cost to cover against fire.

A few examples are solved below to illustrate how costing is done, and how two processes with different capital and running costs can be compared. Cost models can be developed either for computer or for simple graphical optimization. An example of graphical optimization is given in Sec. 15.7.

Example 10.1 Find the likely cost of treatment per kg of BOD_5 which will be received at a proposed activated sludge process plant for treating sewage from 100,000 persons. At the proposed site, the construction, operation, and maintenance costs as given for Austria in Figure 10.5 are applicable today after a 6% per year adjustment for inflation and other factors.

Assume the interest charge on loan = 8% per year, land cost = $10 per m^2, annual depreciation allowances fetch 5% interest only (other assumptions are stated by item below):

Itemized Costs	U.S.$
1. Plant construction at $17 per person (1968 costs, Figure 10.5)[a]	1,700,000
Add 6% per year to update to 1978 level	1,020,000
Construction total	2,720,000
2. Land cost at 0.22 m^2/person (Table 10.4) and $10/$m^2$	220,000
3. Engineering and other fees at assumed rate of 6% of item 1	163,200
4. Interest charges during construction at an average 8% of item 1	217,600
Total capital	3,320,800

Annual Costs

5. Interest charges at 8% on total capital $3,320,800 (no amortization; loan replacement by issuing new bonds)[a]	265,664

6. Depreciation on item 1, average life 20 years,
    sinking fund attaining 5% interest [Eq. (10.3)][b]   82,000

7. Operational and maintenance cost at $1.60 per person
    per year (from Figure 10.5 after updating to 1978
    level)   160,000

8. Tax, insurance, royalties, and profits   nil

                                    Total annual costs   507,664

---

[a]Alternatively, if amortization is also required, in 20 equal installments, the annual amount calculated from Eq. (10.2) will be equal to
$332,080.

[b]Other methods for estimating depreciation may be preferred, especially
by an industry.

Hence, on the various assumed bases as detailed above, the total annual
cost per person (1978 basis) = $507,664/100,000 persons = $5.08.

Assuming BOD_5 = 54 g/person-day = 19.71 kg/person-year, we get cost
per BOD = $0.26 per kg BOD_5 received at plant.

Example 10.2 A cost comparison has to be made between two alternatives, namely, activated sludge process and extended aeration (oxidation
ditch) for serving a population of 60,000 persons for which the following
data are available:

Item	Activated sludge	Extended aeration (oxidation ditch)
1. All construction costs (including engineering fees), per person	$8.0	$6.25
2. Power consumption, kWh/person-year	14	20
3. Land required, ha	1.8	1.8
4. Construction time, months	12	9

Assume: Power cost = $0.03/kWh; loan return in 20 equal installments
with 8% interest; annual depreciation by sinking fund invested at 5% interest; average plant life = 20 years; annual plant maintenance (excluding
power) = 5% of plant cost for both types of plant; land cost = $5.00/m^2.

The estimated capital and annual cost are tabulated below for each
process.

Cost item	Activated sludge ($)	Extended aeration ($)
Capital Costs		
1. Construction and engineering costs	480,000	375,000
2. Land	90,000	90,000
3. Interest charges at 8% of items 1 and 2 during construction	45,600	37,200
Total capital	615,600	502,200
Annual Costs		
4. Annual payment for capital recovery in 20 equal installments at 8% interest[a]	62,717	51,163
5. Depreciation on item 1 by sinking fund method, 20 years life, 5% interest [Eq. 10.3)]	14,400	11,250
6. Power cost at $0.03/kWh	25,200	36,000
7. Other operation and maintenance cost at 5% of item 1	24,000	18,750
Total annual	126,317	117,163
Total annual cost per person	2.10	1.95

[a]From [Eq. (10.2)] using n = 20 and i = 0.08, the "capital recovery factor" is found to be 0.1019. Thus, annual payments equal 0.1019 x P, where P = total capital cost of the plant.

Under stated financial terms and the construction and energy costs, the extended aeration process (oxidation ditch) is slightly cheaper than the activated sludge process.

Example 10.3 Assuming that enough land is available, compute the land cost in U.S.$/$m^2$ at which an oxidation pond will just break even in annual cost with an oxidation ditch for 60,000 persons using data given in Example 10.2. Assume: the oxidation pond construction and engineering costs (including interest charges during the construction period) to be $2.5 per person; land required = 4 m^2/person; and pond operation and maintenance cost = $0.5. The pond will be replaced in 20 years by a more mechanized technology requiring less land, but neglect the resale value of the surplus land.

Total construction cost of oxidation pond = $2.5 x 60,000 = $150,000

Land required = 4 x 60,000 = 240,000 m^2 (24 ha)

Annual costs for pond	U.S.$
1. Annual payment for capital recovery in 20 equal installments at 8% interest on construction cost only	15,280
2. Depreciation on construction cost, sinking fund method, 20 years life, 5% interest [Eq. (10.3)]	4,500
3. O and M cost at $0.5/person-year	30,000
	49,780 (without land)

The annual cost of oxidation ditch process (Example 10.2) = $117,163. Hence, the difference between the two annual costs is the maximum amount available per year for meeting the interest charges and loan repayment for the capital cost of the land, i.e., 117,163 - 49,780 = $67,383 per year.

At the loan condition stated at item 1 above, the maximum capital that can be borrowed for land purchase is found from Eq. (10.2) to be $673,830. Thus, the break even land cost = $673,830/240,000 m^2 = $2.8 per m^2.

REFERENCES

1. S. Arceivala, Simple Waste Treatment Methods, Middle East Tech. Univ., Ankara, Turkey, 1973.

1a. World health statistics report, Vol. 29, 1976.

2. Metcalf and Eddy, Inc., Wastewater Engineering, McGraw-Hill, N.Y., 1972.

3. A. Feinmesser, Advances in water pollution, International Association on Water Pollution Research, 1970 Conference, 1.

4. Klemetson and Grenney, Physical and economic parameters for planning regional wastewater treatment systems, Jour. WPCF, 48, 1976.

4a. D. Okun and G. Ponghis, Community waste water disposal, WHO Monograph, Geneva, 1975.

5. A. Pasveer, Oxidation ditches, Ind. Jour. Env. Health, Nagpur, India, 1963.

6. E. Gloyna, Waste stabilization ponds, WHO Monograph, 1971.

7. S. Arceivala, Bhalerao, and Algarsami, Cost Estimates for Various Sewage Treatment Processes in India, symposium, NEERI, Nagpur, India, 1969.

8. J. Zeper and A. DeMan, Large oxidation ditch carrousel, 5th Congress, International Association on Water Pollution Research, San Francisco, California, 1970.

9. S. Selcuk, Masters thesis, Middle East Tech. Univ., Ankara, Turkey, 1973.

10. K. Saffet, Masters thesis, Middle East Tech. Univ., Ankara, Turkey, 1973.

11. W. W. Eckenfelder, Jr., Water Quality Engineering for Practicing Engineers, Barnes and Noble, Inc., N.Y., 1970.

12. G. Reid and M. Muiga, <u>Mathematical Model for Predicting Wastewater Disposal Systems in Developing Countries</u>, University of Oklahoma, 1976.

13. D. Wilke and D. Fuller, Highly energy efficient Wilton wastewater treatment plant, <u>Jour. ASCE, Civil Eng.</u>, 1976.

Chapter 11

PRINCIPLES OF REACTOR DESIGN

11.1 PRINCIPLES OF REACTOR DESIGN

In order to design various systems for wastewater treatment and disposal, it is necessary to know the extent of changes occurring in the composition and concentration of materials in a reactor, and the rate of such
changes. Changes may be due to reactions between materials, biological
activity, decay with time, and mass transport.

Materials balance relationships can be established for specific components of open systems, which at steady-state conditions can be generalized
as:

Input ± change in system = output

The mechanism of mass transport under quiescent conditions may be
molecular diffusion, depending on the concentration gradient and other
factors as given by Fick's first law. Convective or advective mass transport is characteristic of mixed systems, and involves eddy dispersion as
well as molecular diffusion; the eddy coefficient is several orders of
magnitude higher than the diffusion coefficient in most systems. Other
mass transport considerations involve the transport of material at an interface as, for example, the transfer of oxygen from air to water.

Besides mass transport, changes within a system can be brought about
by various biological and/or chemical reactions. Their kinetics are discussed below followed by a discussion of the mixing phenomenon in reactors.

11.2 TYPES OF REACTIONS

Many reactions occurring in waste treatment are slow, and kinetic
considerations are, therefore, important. The general equation relating
the rate of change of concentration with time to the concentration of the
reacting substance can be expressed as

$$\frac{dC_A}{dt} = \pm\ KC_A{}^n \tag{11.1}$$

where

C_A = concentration of reacting substance A

K = reaction rate constant, per unit time

n = order of the reaction (n = 1 signifies first order, n = 2 signifies second order, and so on)

Major factors which affect the K values include the following:

1. Temperature
2. Presence of catalyst
3. Presence of any toxic substances
4. Availability of nutrients and "growth factors"
5. Other environmental conditions

11.2.1 Zero-order Reactions

Zero-order reactions (n = 0) are independent of the concentration C_A and, hence, their rate dC_A/dt is constant as shown in Figure 11.1.

$$\frac{dC_A}{dt} = K \tag{11.2}$$

Certain catalytic reactions occur in this fashion. Some first-order biochemical reactions may also seem to behave like zero-order ones (see below).

11.2.2 First-order Reactions

First-order reactions (n = 1) are those where the rate of change of concentration of a substance A is proportional to the first power of the concentration. Thus,

$$\frac{dC_A}{dt} = K(C_A) \tag{11.3}$$

Referring to Figure 11.1 it will be seen that the rate diminishes with time. A true first-order reaction is one in which a single substance [e.g., H_2O_2 or $Ca(OCl)_2$] is decomposing. Aeration and heat dissipation also proceed in a similar manner.

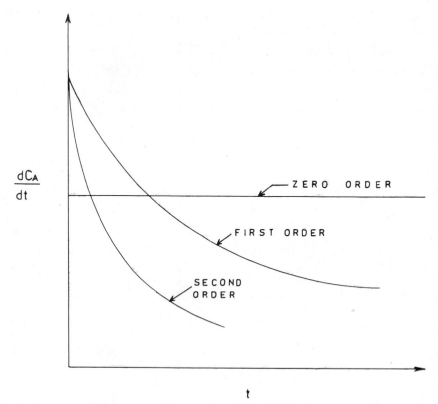

Figure 11.1 The change in removal rate (dC_A/dt) with time t for different order reactions (adapted from Ref. 2).

Biological stabilization of organic matter in batch systems is a typical example of a pseudo first-order reaction. Although several items are involved, such as dissolved oxygen concentration, number of organisms, and organic matter concentration, the rate is proportional to the concentration of a single item (organic matter in this case) provided the other items are in relative abundance. If the organic matter (substrate concentration) is also maintained within a narrow range (as in a continuous flowing, complete-mixing reactor), then the rate is practically constant and the reaction behaves like a pseudo zero-order type of reaction. Some biological treatment systems do behave in this manner.

There are various complex processes whose <u>overall</u> rates are approximately first order in nature. Several single substances by themselves exhibit zero-order rates, but complex substrates (e.g., sewage and industrial wastes), which may contain several such substances, may show a high rate of overall removal at the start when all their component substances are being removed (consumed) simultaneously. Later, the overall rate may be slower since only the relatively more difficult substances now remain. Thus, the overall rate may seem like a typical first-order rate.

Equation 11.3 may be integrated within the limits of concentration C_1 and C_2 at time t_1 and t_2, respectively, to give

$$\int_{C_1}^{C_2} \frac{dC_A}{C_A} = K \int_{t_1}^{t_2} dt$$

or

$$\ln \left(\frac{C_1}{C_2} \right) = K(t_2 - t_1)$$

or, if the concentration is C_0 at the start ($t = 0$), then the concentration C_t at any time t is

$$C_t = C_0 e^{-Kt} \tag{11.4}$$

For logarithms to the base 10, Eq. (11.4) becomes

$$C_t = C_0 10^{-kt} \tag{11.5}$$

where $K = 2.3k$, and

$$k = \left(\frac{\log (C_0/C_t)}{t} \right) \tag{11.6}$$

The units of k and K are per unit time.*

A graphical plot of $\log (C_0/C_t)$ against t should give a straight line whose slope equals k. In fact, obtaining a straight line is taken as a test for concluding that the reaction follows first-order kinetics (see Figure 11.2 and Example 11.1).

---

*Generally, in this book, capital letter K is used to denote its derivation from natural or Napierian logarithms, while the small k refers to logs to the base 10.

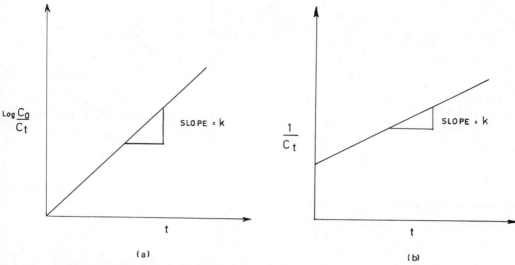

Figure 11.2 Typical curves obtained with (a) first- and (b) second-order reactions.

11.2.3 Second-order Reactions

<u>Second-order reactions</u> (n = 2) proceed at a rate proportional to the second power of the concentration, namely,

$$\frac{dC_A}{dt} = K(C_A)^2 \tag{11.7}$$

As shown in Figure 11.1, small changes in the concentration can affect the rate considerably. For second-order reactions, a graphical plot of $1/C_A$ against t should give a straight line whose slope gives K (see Figure 11.2) since the integration of Eq. (11.7) gives

$$K = \left(\frac{1/C_2 - 1/C_1}{t_2 - t_1} \right) \tag{11.8}$$

The units of K are now $(\text{time})^{-1}(\text{concentration})^{-1}$. Demographers have noted that population growth rates fit second-order kinetics. The world population has been growing at a second-order rate since early times!

11.2.4 Other Reactions

Other types of reactions include (1) higher than second order, (2) fractional order, (3) sequential reactions in which one substance is removed first and the other afterward, and (4) chain or train reactions in which various intermediate steps occur (e.g., $NH_3 \rightarrow NO_2^- \rightarrow NO^-$).

Example 11.1 The following results were obtained for the concentration of a pollutant in an industrial waste at different periods of plain aeration in the laboratory. Check whether the data fit first-order or second-order kinetics.

Aeration time t, hr: 0   0.5   1.0   1.5   2.0

Pollutant concentration C_t, mg/liter: 610   420   320   260   220

The problem may be solved graphically, or analytically, as shown below:

1. For first order, $k = \dfrac{\log (C_0/C_t)}{t}$ per hour

   t = 0   0.5   1.0   1.5   2.0

   k = -   0.324   0.28   0.246   0.22

   The k value is not constant; hence, the reaction is not following first-order kinetics.

2. For second order, $K = \dfrac{(1/C_t) - (1/C_0)}{t}$ per hour per mg/liter

   t   = 0   0.5   1.0   1.5   2.0

   $K \times 10^3$ = 1   1.5   1.48   1.47   1.48

   The K value is reasonably constant and, hence, second-order formulation appears to fit these data better.

11.3 FLOW PATTERNS IN REACTORS

All tanks, lagoons, and ponds used for waste treatment can be referred to as "reactors." Their flow patterns depend on their mixing conditions, which depend on the shape of the reactor, energy input per unit volume, the size or scale of the unit, and other factors. Flow patterns affect the time of exposure to treatment and substrate distribution in the reactor. Typical flow and mixing patterns include

1. Batch reactors
2. Continuous flow reactors
    a. Ideal plug flow
    b. Ideal completely mixed flow
    c. Nonideal dispersed flow
    d. Series or parallel combinations of the above

11.4 BATCH REACTORS

Batch reactors are closed systems having no continuous flow. But the contents are generally well mixed to ensure that no temperature or concentration gradients exist. All elements are exposed to treatment for the same length of time for which the substrate is held in the reactor. Hence, they are like ideal plug-flow reactors (see below).

Batch reactors are often used in laboratory studies. Care must be exercised in extrapolating their results to other types of reactors. A biochemical oxygen demand (BOD) bottle can also be regarded as a batch reactor (see Figure 11.4a).

11.5 TRACER TESTS FOR CONTINUOUS-FLOW REACTORS

Except for batch reactors, all other types of reactors are used on a continuous-flow basis. Therefore, tracer tests help to distinguish between the different types and to characterize them.

The tracers used may be dyes, chemicals, or radioactive isotopes which are injected at the inlet end of the reactor either as a sudden slug (pulse input) or on a continuous basis (step input). The effluent is monitored for the concentration of the tracer by taking samples at regular time intervals (every 15 minutes, for example) from which "time-concentration" curves are obtained. These curves can be expressed in dimensionless units as explained below. When the tracer is added as a slug, the curves obtained are called "C-diagrams" (see Figures 11.3a and 11.3b).

In C-diagrams, the axes are made dimensionless by plotting C/C_0 against t/t_0, in which

    C = actual concentration in effluent sample

    C_0 = concentration if the slug dose of tracer was mixed uniformly with the entire reactor volume

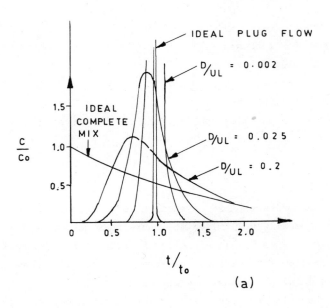

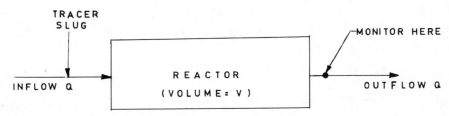

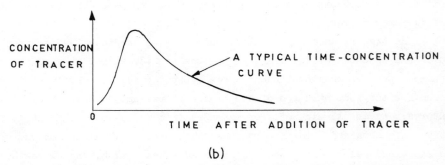

Figure 11.3 (a) Some typical C-diagrams for different types of reactors; (b) a typical tracer test.

t = actual time of sampling after addition of the tracer

t_0 = theoretical detention time in reactor = reactor volume V/flow rate Q

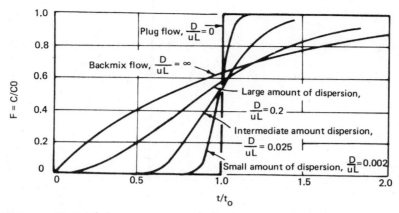

Figure 11.3 (c) F curves for various extents of dispersion as predicted by the dispersion model for step or continuous input of tracer.

The C-diagrams give the "age" distribution at the exit. It is evident that different reactors must show different results. With this background, we may now see how different types of reactors respond to tracer tests in terms of the dimensionless term D/UL which is generally used for characterizing the different reactors. This term is also called the "dispersion number" of the reactor and is further discussed later.

The F-diagram relates to step or continuous tracer input and gives the fraction of the coming concentration that appears in the effluent. Thus, the dimensionless ratio of concentrations C/C_O shown by the ordinate now consists of

C = actual concentration of tracer in effluent

C_O = actual concentration of tracer continuously entering the reactor (it being assumed that no tracer was initially present in the reactor)

F-diagrams give the existing age distribution for continuous tracer inputs (see Figure 11.3c). Again, different reactors give different F-diagrams.

11.6 IDEAL PLUG FLOW

Ideal plug flow, also called piston flow, is one in which every element of flow leaves the reactor in the same order in which it entered; there is no overtaking or falling behind, no intermixing or dispersion. Hence, every element is exposed to treatment for the same length of time (as in a batch reactor), namely, the theoretical detention time t_O (see Figure 11.4b).

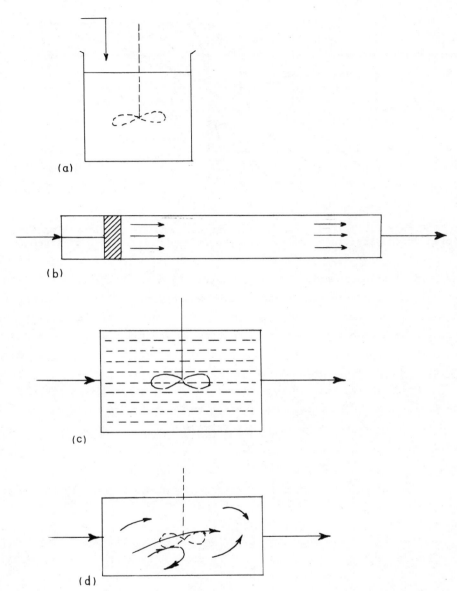

Figure 11.4 Different types of reactors: (a) batch-type reactor;
(b) plug-flow reactor (piston flow); (c) completely mixed reactor; (d) dis-
persed-flow reactor (see text for explanation).

The slug of inert tracer added instantaneously at the inlet end flows
through the tank uniformly and appears all at once in the effluent at time
$t = t_0$, or $t/t_0 = 1.0$ (see Figure 11.3b). The concentration C is high since

the slug of tracer does not mix with the tank contents, theoretically
speaking. Thus, the ratio C/C_0 is high at $t/t_0 = 1.0$. At other values of
t/t_0, less or greater than 1.0, the value of C/C_0 is zero. Notice that the
ideal nature of the reactor enables the tracer slug to be discharged sud-
denly, not gradually. In practice, no reactor follows ideal plug-flow
conditions because some intermixing always occurs.

The F-diagram in Figure 11.3c shows the response of the ideal plug-flow
reactor to a step or continuous input of inert tracer, which appears as a
continuous output of the same concentration, again starting all at once
after a time lag equal to the theoretical detention time (i.e., $C/C_0 = 1.0$
at $t/t_0 = 1.0$). Thus, an ideal plug-flow reactor receiving a substance
either as a slug or as a continuous input provides a detention time to it
equal to the theoretical time. This information can now be used to study
the effect of exposing degradable substances, such as those contained in
wastewater, to treatment in such reactors.

Biodegradable substances reduce in concentration during their passage
through the reactor owing to biochemical activity. As seen earlier, substrate removal for a first-order equation is given by

$$S = S_0 e^{-Kt} \qquad\qquad (11.9)$$

where

S = substrate concentration in effluent

S_0 = substrate concentration in influent

t = treatment time in reactor (equal to theoretical detention time
t_0 for ideal plug-flow reactors)

K = substrate removal rate, time^{-1}

All through the reactor, K is constant, but the concentration of the
degradable substrate decreases gradually as the flow progresses through the
reactor. Thus, at the head end of the reactor, the substrate concentration
is high and the removal is also correspondingly high for first and higher
order reactions. As the flow arrives at the tail end, the substrate concentration is lower and, hence, the removal is also lower. This does happen
in the case of long, rectangular tanks or lagoons used in wastewater treatment. On an overall basis, plug-flow reactors are found to be more efficient
than other types of reactors for first and higher order reactions. For a
zero-order reaction, the removal rate is theoretically constant from the
beginning to the end of the reactor.

11.7   IDEAL COMPLETELY MIXED FLOW

Ideal completely mixed flow is one in which all inflowing elements are instantly and thoroughly dispersed in the reactor so that the contents are perfectly homogeneous at all points in the reactor. Consequently, the concentration of the effluent from the reactor is the same as that in the reactor (see Figure 11.4c).

The tracer slug added in this case is, theoretically, completely mixed with the tank contents as soon as it enters. Thus, at the start the concentration $C = C_0$, or $C/C_0 = 1.0$ at $t/t_0 = 0$. Thereafter, as t/t_0 increases, the concentration C decreases because the tracer is gradually washed out from the reactor and, therefore, C/C_0 also decreases (see Figure 11.3b).

If the tracer input is uniformly continued over a period of time (F-diagram), there will be some input and some output throughout the time, unlike the plug flow case seen earlier where the output occurred only after time t_0. Hence, for an inert tracer, a mass balance around the reactor gives

Rate of inflow - rate of outflow = rate of change in reactor

$$QC_0 - QC = \frac{dc}{dt} V$$

Integration gives the effluent concentration C as

$$C = C_0(1 - e)^{-(t/t_0)} \qquad\qquad (11.10a)$$

where all terms have been defined before.

From the F-diagram shown in Figure 11.3c it will be seen that, when $t/t_0 = 1.0$, the corresponding value of the ratio $C/C_0 = 0.632$, and it takes a long time before the value of C/C_0 reaches 1.0. For example:

t/t_0:   0.1   0.5   1.0   2.3   4.6

C/C_0:   0.015   0.4   0.632   0.90   0.99

Furthermore, if the incoming tracer is stopped after the concentration has reached a certain value, for example, C_m in the reactor, the tracer will be gradually purged out and the falling part of the curve will be similar to the one shown in Figure 11.3b (C-diagram). The effluent concentration C will now be similarly obtained as

$$C = C_m e^{-t'/t_0} \qquad\qquad (11.10b)$$

where t' = time elapsed <u>after</u> stoppage of the tracer. Again, the time-concentration ratios can be illustrated as below:

t'/t_0: 0.1   0.5   1.0   3.3   4.6

C/C_m: 0.9   0.6   0.368   0.10   0.01

Thus, for $C = 0.01C_m$ (99% flushout from the system), the time required is 4.6 times the theoretical detention time in the reactor. Applications of Eqs. (11.10a) and (11.10b) are further discussed in Sec. 11.16.

For degradable substances following first-order kinetics, we get, as before

$$\frac{ds}{dt} = -KS$$

Again, a mass balance around the reactor at steady-state conditions gives:

Rate of input + rate of change in reactor = rate of output

$$QS_0 - KSV = QS \qquad\qquad (11.11)$$

where, in compatible units, we have

V   = reactor volume
Q   = inflow rate
K   = removal rate constant for given process ($time^{-1}$)
S_0 and S = incoming and outgoing substrate concentrations, respectively

Equation (11.11) can be rewritten in the following well-known forms for complete-mixing systems, enabling an estimation of the effluent substrate concentration under these conditions as

$$S = \frac{S_0}{1 + K(V/Q)} \qquad\qquad (11.12a)$$

or

$$S = \frac{S_0}{1 + Kt} \qquad\qquad (11.12b)$$

The derivation and application of these equations are further discussed in Chapter 12.

Square or round tanks with a high degree of agitation as used in wastewater treatment often approach completely mixed conditions. Even the long, rectangular tanks used in activated sludge plants can approach completely mixed conditions provided the feed is given and collected all along the tank length instead of feeding at the head end and collecting at the tail end.

Completely mixed flow regimes are particularly useful to adopt where uniform concentrations throughout the reactor are desirable for one reason or another. For example, with fluctuations in inflows, concentrations, or both, completely mixed reactors would be likely to perform better. Similarly, where toxic inflows are expected, completely mixed systems would be likely to perform better as they would immediately disperse any incoming material throughout the tank and dilute it down, hopefully, to a safe level so as not to disrupt biological activity (see Sec. 11.16 also).

11.8 DISPERSED FLOW

Dispersed flow is one in which each element of incoming flow resides in the reactor for a different length of time (Figure 11.4). It is also called intermixing or arbitrary flow and lies between the two boundary cases seen earlier, namely, ideal plug flow and complete mixing. Thus, dispersed flow is a nonideal case and can be used in practice to describe the flow conditions in most reactors.

Using Fick's second law, but replacing his diffusion coefficient by a dispersion coefficient D, the change in concentration is given as

$$\frac{\partial C}{\partial t} = D \frac{\partial^2 C}{\partial X^2} \tag{11.13}$$

where

C = concentration of substance, mass/volume

D = axial or longitudinal dispersion coefficient, L^2/t

X = distance in direction of main flow, L

Taking a steady-state differential materials balance at any cross section in a reactor, Levenspiel [1] has given the following general equation for any reactant following nth-order kinetics:

$$D \frac{\partial^2 C}{\partial X^2} - U \frac{\partial C}{\partial X} - KC^n = 0 \tag{11.14}$$

where

      U = mean flow velocity along reactor, L/t

      K = removal rate constant per unit time

      n = order of the reaction

For a conservative tracer K = 0, the last term is dropped and solution of the equation gives the family of curves shown in the C-diagram in Figure 11.3b for different values of D/UL. It will be recalled that D/UL is called the "dispersion number," which characterizes the mixing conditions in a reactor. It is affected by all the factors which affect its component terms, D, U, and L.

By definition, dispersion is absent in ideal plug flow and, therefore, $D = 0$ and $D/UL = 0$. Conversely, dispersion is infinite in ideal complete mixing and $D = \infty$ and $D/UL = \infty$. For various nonideal or dispersed flow conditions, the value of D/UL lies anywhere between 0 and ∞. From the C-diagrams given in Figure 11.3b it will be seen that as D/UL increases, the spread increases, the peak concentration reduces in height, and the peak occurs earlier, namely, at $t/t_0 < 1$.

It is interesting to observe from Table 11.1 that the aerated lagoons, oxidation ponds, and oxidation ditches with which we are most concerned in this book are widely varying in their mixing conditions, practically from one end of the spectrum to another. This aspect must be given more consideration in their design than has been done hitherto. For this reason, the subject of mixing kinetics is being dealt with at some length in this chapter.

Among the factors affecting dispersion in waste treatment units may be listed the following [3]:

1. The scale of the mixing phenomenon
2. The geometry of the unit
3. The power input per unit volume (mechanical or pneumatic)
4. Type and disposition of inlets and outlets
5. Inflow velocity and its fluctuations
6. Density and temperature differences between inflow and reactor contents
7. Reynolds number (which, in turn, is affected by some of the factors already listed)

Temperature variations between $12°$ C and $20°$ C have been observed to have no significant effect on the dispersion condition in tanks [5]. Strong

Table 11.1 Typical D/UL Values for Different Waste Treatment Units

Treatment unit	D/UL[a] (likely range)
Rectangular sedimentation tanks	0.2-2.0
Activated sludge aeration tanks	
Long, plug-flow type	0.1-1.0
Complete-mixing type	3.0-4.0 and over
Waste stabilization ponds	
Multiple cells in series	0.1-1.0
Long, rectangular ponds	
Single ponds	1.0-4.0 and over
Mechanically aerated lagoons	
Rectangular, long	0.2-1.0
Square shaped	3.0-4.0 and over
Pasveer and carrousel-type oxidation ditches	3.0-4.0 and over

[a]Dispersion number can also be expressed in terms of the detention time t of a unit. Since $U = L/t$ where L = length of travel path in reactor, we can write

$$\frac{D}{UL} = \frac{Dt}{L^2} \qquad\qquad (11.15a)$$

Source: Adapted from Ref. 2.

wind can distort the time-concentration curve, pushing the peak more leftward, implying short-circuiting.

For the above-stated reasons, the same unit may show different D/UL values on different occasions if flow, wind, and other conditions vary substantially. In this regard, units constructed as cells in series should show greater consistency in their overall D/UL values.

The actual C-diagram obtained by conducting a tracer test on a reactor can be compared with the theoretical curves of known D/UL values to find the curve that most closely fits the experimental curve. This comparison can be done by comparing their variances or their peak times as discussed later.

In practice, the typical range of D/UL values observed in different types of treatment reactors lies between 0.1 and 4.0. Of course, some reactors may show values less than 0.1 and some may show values much greater than 4.0 (even 100). None, however, are ideal plug flow or ideally complete mixing; all lie somewhere in between. Table 11.1 gives some typical D/UL values given by Arceivala [2]. When a unit shows a D/UL value of about 0.2 or less, it can be said to be approaching plug flow, whereas at 3.0 to 4.0 it can be considered well mixed or approaching complete mixing.

11.9 SUBSTRATE REMOVAL IN DISPERSED FLOW

Equation 11.14 was solved for first-order kinetics by Wehner and Wilhem [6] in 1956 with regard to chemical engineering processes, but is only now finding increasing application in waste treatment process design. For reactions other than first order, only a numerical solution is possible. Their equation is

$$\frac{S}{S_0} = \frac{4ae^{d/2}}{(1 + a)^2 e^{a/2d} - (1 - a)^2 e^{-a/2d}} \qquad (11.15b)$$

where

a = $\sqrt{1 + 4Ktd}$ (dimensionless)

d = $D/UL = Dt/L^2$ (dimensionless)

L = length of axial travel path

t = detention time, theoretical (= V/Q)

K = substrate removal rate constant, time^{-1}

S_0 and S = initial and final substrate concentration, respectively (mass/vol)

As seen earlier, D/UL values may vary widely from 0 to ∞. When D/UL approaches zero (plug flow), Eq. (11.15b) above gives practically the same prediction of effluent concentration S as is given by Eq. (11.9) seen earlier. Similarly, when D/UL approaches infinity (complete mixing), Eq. (11.15b) gives practically the same prediction of S as is given in Eq. (11.12b) for complete mixing systems. The advantage in using Eq. (11.15b), however, is that it is universally applicable over the whole range of D/UL values, while Eqs. (11.9) and (11.12b) are only specifically applicable to the two ideal boundary conditions.

Use of the Wehner-Wilhem equation is facilitated by the use of charts. Figure 11.5 shows a graphical plot of the dimensionless product Kt against percent remaining (S/S_0) and percent removed ($1 - S/S_0$). The family of curves shown pertain to different values of D/UL from 0 to ∞, and are seen to flare out from the point of origin.

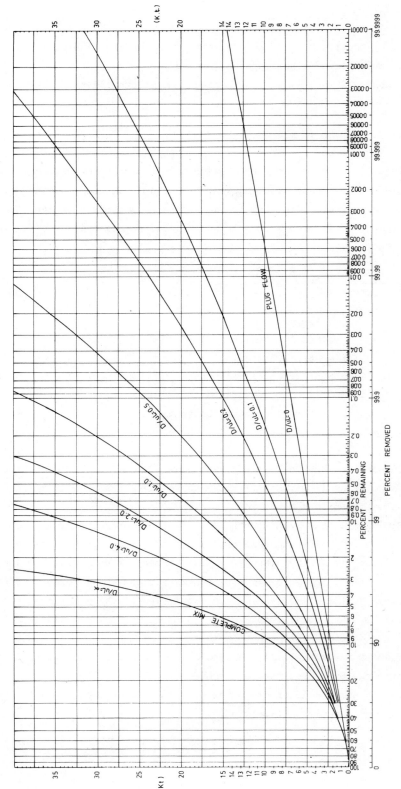

Figure 11.5 Substrate removal efficiency determination using the Wehner-Wilhem equation.

It is evident from Figure 11.5 that (1) a completely mixed or even a
reasonably well-mixed system (D/UL > 4.0) is incapable of giving more than
90 to 97% efficiency for Kt values less than 20, (2) for a given Kt value,
plug flow-type systems always tend to give a higher efficiency than do
well-mixed systems, (3) very high efficiencies (greater than 99%) as are
often required in the case of bacterial removal in oxidation ponds, for
example, can be reached only if the system approaches plug-flow conditions.
Thus, it is seen that D/UL is as important a design parameter as is the
substrate removal rate K. Figure 11.6 is a plot similar to Figure 11.5,
but is useful for low values of Kt.

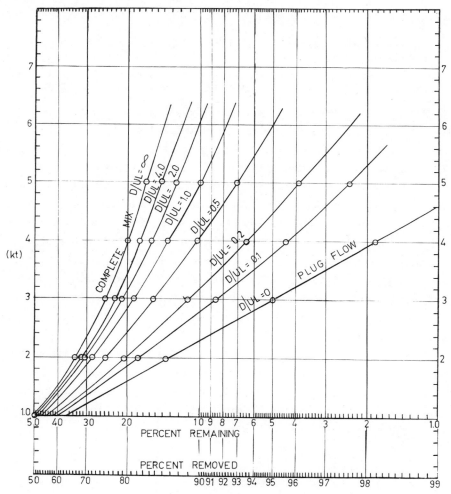

Figure 11.6 Determination of substrate removal efficiency using
the Wehner-Wilhem equation.

Example 11.2 Estimate the process efficiency (percent removed) if a
wastewater is treated for t = 6 hours and the substrate removal rate K = 0.2
per hour at reactor conditions when D/UL = 0, 0.2, 4.0, and ∞.

Entering Figure 11.6 at Kt = (0.2)(6) = 1.2, we get

D/UL: 0   0.2   4.0   ∞

Efficiency (%):   70   65   56   55

11.10 ESTIMATION OF DISPERSION NUMBER, D/UL

The dispersion number D/UL can be estimated in either of the following
two ways:

1. By conducting a tracer test on an existing unit or a suitably scaled-
    down model.
2. By using empirical methods which are less accurate but useful in pre-
    dicting the likely dispersion number before a unit is constructed.

Tracer test results give C-diagrams or time-concentration curves from
which the variance of the curve can be calculated and D/UL found [1] as
shown in Example 11.3 where the necessary equations are also given. Generally, the time-concentration curves show long "tails," which cannot be
neglected as they considerably affect the calculation of variance. Therefore, the sampling must continue for a long enough period of time (about
four to even ten times the theoretical detention time of the unit). This
may not be feasible, in which case the tracer concentration beyond twice
the theoretical time can be estimated from a semilog plot of the data,
where the concentration generally decreases in a linear fashion, which makes
extrapolation easier. If the time-concentration curve is cut off prematurely, a very serious error in the estimation of D/UL can result as illustrated in Example 11.3.

To get over this difficulty, Murphy and Timpany [5] have developed
another method of estimating D/UL from the ratio of the time at which peak
occurs t_p to the theoretical time t_0. This method is also illustrated in
Example 11.3, where the necessary equations are given. The method makes
field work as well as computations easier.

Example 11.3 Estimate the dispersion number D/UL for a small tertiary
pond of 2.2 days theoretical detention time for which the tracer test results

are given below for a period of 9 hr only. Use both variance and peak-time techniques.

| Observations | | Variance computations | | |
Time t (hr)	Concentration C (mg/liter)	t^2	tc	t^2c
1	23	1	23	23
2	23.5	4	47	94
3	25	9	75	225
4	26.5	16	106	424
5	28	25	140	700
6	31	36	186	1116
7	27	49	189	1323
8	26	64	208	1664
9	25.4	81	228.6	2057.4
Σ	258.4	285	1202.6	7626.4

1. Variance Technique [1]

   a. Mean time $\bar{t} = \dfrac{\Sigma\, tc}{\Sigma\, c}$

   $$= \frac{1202.6}{258.4} = 4.654$$

   b. If the standard deviation of the time-concentration curve is σ_t, the variance is σ_t^2 where

   $$\sigma_t^2 = \frac{\Sigma\, t^2 c}{\Sigma\, c} - \bar{t}^2$$

   $$= \frac{7626.4}{258.4} - (4.654)^2 = 7.89$$

   c. Variance in terms of dimensionless time (C-curve)

   $$\sigma^2 = \frac{\sigma_t^2}{\bar{t}^2}$$

   $$= \frac{7.89}{(4.654)^2} = 0.364$$

Using the following relationship for closed vessels of finite length

$$\sigma^2 = 2\left(\frac{D}{UL}\right) - 2\left(\frac{D}{UL}\right)^2 (1 - e^{-UL/D}) \qquad (11.16)$$

the dispersion number D/UL is obtained by trial and error as 0.24.

2. <u>Peak-time Technique</u> [5]

The following ratio is first computed:

$$\frac{t_p}{t_0} = \frac{\text{peak time}}{\text{theoretical detention time}} = \frac{6 \text{ hr}}{2.2 \text{ days} \times 24} = 0.114$$

The value of D/UL is then computed from one of the following two equations:

$$\frac{D}{UL} = 4.027(10)^{-2.09(t_p/t_0)} \qquad \text{when} \qquad 0.3 < \left(\frac{t_p}{t_0}\right) < 0.8 \quad (11.17)$$

or

$$\frac{D}{UL} = 0.2\left(\frac{t_p}{t_0}\right)^{-1.34} \qquad \text{when} \qquad 0.03 < \left(\frac{t_p}{t_0}\right) < 0.3 \quad (11.18)$$

Hence, in the above case, $D/UL = 0.2(0.114)^{-1.34} = 3.67$.

Note: Here, the variance technique gives erroneous results since the time-concentration curve is prematurely stopped after only 9 hr, without allowing the full "tail" to be developed.

In recent years, empirical correlations between dispersion and certain parameters have been attempted in order to enable greater use of the dispersion number concept in reactor design without recourse to field tests or model studies every time.

For diffused-air aeration tanks of the type used in activated sludge plants, Murphy [4] has reported the following correlation:

$$D = 3.118(W^2)(q_A) \tag{11.19}$$

where

    W = tank width, ft

    q_A = air flow rate per unit tank volume, standard $ft^3/1000 \ ft^3/min$

This equation was derived from studies on tanks up to about 9 m (30 ft) in width, and air supply rates ranging from less than 10 to more than 100 scfm per 1000 ft^3.

Arceivala [3] has pooled tracer study data from various sources covering mechanically aerated lagoons, oxidation ponds, rectangular settling tanks, and such other units, to arrive at an empirical correlation between dispersion D and the width W of a unit. Essentially, his results show that D varies as W^2 for small size units of less than 10 m, whereas the variation is linear for larger widths (W ≥ 30 m). Furthermore, in each case he found that units possessing baffles which lengthened the flow path (around-the-end type baffles) invariably gave much higher D values than those without baffles but having the same width. This is because bends in flow increase dispersion.

The empirical correlations tentatively obtained can be expressed as follows where D is in m^2/hr and W in m:

1. <u>For lagoons and ponds of widths larger than 30 m</u>

    a. With baffles, $D = 33W$ (11.20)

    b. Without baffles, $D = 16.7W$ (11.21)

2. <u>For lagoons and ponds of widths less than 10 m</u>

    a. With baffles, $D = 11W^2$ (11.22)

    b. Without baffles, $D = 2W^2$ (11.23)

As further data become available, these equations may need revision to improve their prediction accuracy. Meanwhile they may be useful in making rough approximations, and may be supplemented by actual model studies where desired.

Among the factors affecting dispersion, it may be mentioned that on three large scale lagoons studied by Murphy and Wilson [7] variation in power input from 0.47 to 2.29 $kW/1000\ m^3$ lagoon volume showed no significant effect on D and D/UL values. Thus, the mixing imparted by the aerators apparently did not affect the overall value of D, which was perhaps already large owing to large lagoon widths (76 m and 106 m). However, enough data are not in hand at present to be able to state that aerated lagoons of lesser widths will also similarly be unaffected by power input.

In designing a lagoon or pond to give a desired value of D/UL, a trial width first needs to be selected. Generally, the depth is already decided upon based on other considerations. Thus, for the given flow and detention time, the length is computed and, using one of the equations just given, D is then computed. It is now possible to find D/UL or its equivalent expression, Dt/L^2 (see Example 11.4). Inserting a baffle in the unit helps to quickly reduce the D/UL value since lengthwise baffles reduce the width of flow and increase both the velocity and length of the flow path.

Presently, a relationship is not available to enable one to predict the <u>overall</u> value of D/UL when two or more cells are placed in series such that the flow passes from one distinct cell to another. Preliminary studies

show that a sharp decrease in the overall value of D/UL occurs when a group
of cells are placed in series, although the D/UL value of each individual
cell may not, relatively, be so low. As an approximation, it can be said
that

Cells in series	Likely overall D/UL value
2	0.2-0.7
4	0.1 or less

However, it is not essential to know the overall D/UL values. For
cases involving cells in series, each cell may be designed individually
using the specific value of D/UL applicable to that cell. In this manner,
each cell in the series arrangement can be so designed as to promote well-
mixed or plug-flow conditions as desired. The cells may also be unequal in
size if it is beneficial for the process. It is not possible to benefit
from this flexibility in design when the equal, completely mixed cells-in-
series models described below are used.

For example, where shock loads are likely to be received, and it is
desired to provide a cells-in-series system, the first one or two cells may
be made as both large and well-mixed types, whereas the remaining cells may
be smaller in size and their geometry so adopted as to promote plug-flow
conditions. Thus, the designer can manipulate the geometry of each individual cell in the series arrangement as desired.

Where considerations of shock load are not involved, as in ordinary
domestic sewage treatment, a more economical design can generally be prepared if the dispersion model is used rather than the ideal complete mixing
model so often used.

Example 11.4 A mechanically aerated lagoon provides 5 days' detention time to a wastewater flow of 10,000 m^3/day. If its depth is to be
restricted to 3 m, estimate the lagoon dimensions so that the dispersion
number D/UL will be 0.5 or less.

Lagoon volume = 10,000 x 5 = 50,000 m^3

Lagoon area   = $\frac{50,000}{3}$ = 16,667 m^2

As a trial, assume L/W = 4/1. This gives L = 258 m and W = 64.5 m. Equation (11.21) gives D = 16.7(W) = 16.7 x 64.5 = 1115 m^2/hr.

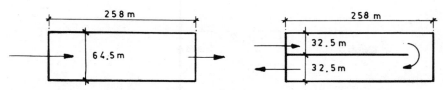

Figure 11.7 A lagoon arrangement (see Example 11.4).

Thus,

$$\frac{D}{UL} = \frac{Dt}{L^2} = \frac{1115 \times (5 \times 24)}{(258)^2} = 2.0$$

This value is high, and lagoon dimensions need to be readjusted to increase the length/width ratio. This objective can also be achieved by inserting one baffle in the lengthwise direction as shown in Figure 11.7 so that W = 32.5 m and L ≅ 510 m. Equation (11.20) must now be used to estimate D. Thus,

$$D = 33W = 33 \times 32.5 = 1073 \ m^2/hr$$

and

$$\frac{Dt}{L^2} = \frac{1073 \times (5 \times 24)}{(510)^2} = 0.495 \ (\text{acceptable})$$

11.11 EQUAL CELLS IN SERIES

Another model often used in reactor design is the equal cells-in-series model. It is possible to use it with first-order or any order reactions. In this model it is assumed that there are "n" number of equal cells in series where each cell in itself is completely mixed and has detention time t' and volume v (see Figure 11.8a). Thus,

Σ v = total volume of cells in series

$t' = \dfrac{\Sigma v}{nQ}$ = detention time per cell.

Since each cell is assumed to be completely mixing, Eq. (11.12) can be used, with first-order kinetics, to give the output from the first cell as

$$S_1 = \frac{S_0}{(1 + Kt')}$$

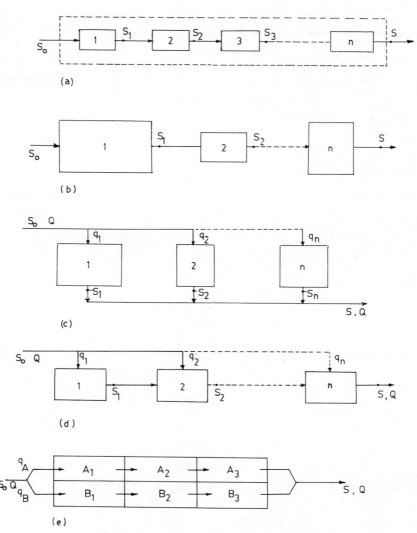

Figure 11.8 Different types of arrangements for reactor cells. (a) Equal cells in series; (b) unequal cells in series; (c) equal or unequal cells in parallel; (d) cells in series with incremental feed; (e) series-parallel arrangement.

The output from the first cell becomes input for the second cell; hence,

$$S_2 = \frac{S_1}{(1 + Kt')} = \frac{S_0}{(1 + Kt')(1 + Kt')}$$

Generalizing in this manner for "n" cells we get

$$S = \frac{S_0}{(1 + Kt')^n} \qquad (11.24)$$

This can also be written as

$$S = \frac{S_0}{[1 + K(\Sigma v/nQ)]^n} \qquad (11.25)$$

For a given efficiency, if n is increased ($n \to \infty$), the required volume (Σv) approaches that required for a plug-flow system with first-order kinetics, namely, Eq. (11.24) reduces to Eq. (11.9). On the other hand, if n = 1, Eq. (11.24) is the same as Eq. (11.12b). The required total volume for any number of cells in series is indicated in Table 11.2 from which it is seen, as in the case of the dispersed flow model, that the higher the efficiency required, the more advantageous it is to have plug flow. It is also seen from the table that, in most cases, a sharp drop in total reactor volume occurs if two to four cells in series are used instead of just one. Having still more cells does reduce the volume requirement further, but not so

Table 11.2 Relationship Between Number of Cells, Their Total Volume, and Efficiency

Number of completely mixed cells in series	Total relative volume of reactors[a] required to attain efficiency levels of			
	85%	90%	95%	99%
1	5.67	9.0	19.0	49.0
2	3.18	4.32	6.96	12.14
4	2.48	3.10	4.48	6.64
6	2.22	2.82	3.90	5.50
8	2.16	2.64	3.60	5.04
-	-	-	-	-
-	-	-	-	-
-	-	-	-	-
-	-	-	-	-
∞ (plug flow)	1.90	2.30	3.0	3.91

[a]Values given in the table are the values of the dimensionless product Kt at stated conditions (see Example 11.5).

sharply as before. It may be well to remember that in the case of large lagoons and ponds each cell may not really be completely mixing and it may be more realistic to apply the dispersed flow model individually to each cell in order to estimate the efficiency, as stated in the previous section.

   Example 11.5 Estimate the relative volume of reactors that will give 90% efficiency with $n = 1$, $n = 2$, and $n = \infty$.

1. When $n = 1$,

$$\frac{S}{S_0} = 0.1 = \frac{1}{1 + K(V/Q)} \quad \text{or} \quad K(V/Q) = 9.0$$

2. When $n = 2$,

$$\frac{S}{S_0} = 0.1 = \frac{1}{[1 + K(V/nQ)]^n} = \frac{1}{[1 + (KV/2Q)]^2}$$

Hence,

$$K\left(\frac{V}{Q}\right) = 4.32$$

3. When $n = \infty$ (plug flow),

$$\frac{S}{S_0} = 0.1 = e^{-K(V/Q)}$$

Hence,

$$K\left(\frac{V}{Q}\right) = 2.30$$

11.11.1 Unequal Cells in Series

   Sometimes, unequal cells in series need to be provided for certain reasons. Examples include oxidation ponds where one cell may be larger than another cell for topographical reasons, or where the first cell may be deliberately kept larger than the second cell to avoid anaerobicity owing to an overload of the first cell or an upset owing to shock load (Figure 11.8b).

   For two completely mixed cells in series, Marais [8] has shown that unequal sizes give less overall efficiency than do equal ones. Thus, whenever possible, cells should be equal in size for maximizing efficiency.

If total detention time is fixed, for example, t, and both cells are first assumed equal, with time t' in each cell, we get from Eq. (11.24)

$$S_2 = \frac{S_0}{(1 + Kt')^2} \tag{11.26}$$

Now, if one cell is made $t' + \Delta t$ and the other $t' - \Delta t$ to keep total time constant, we get

$$S_2' = \frac{S_0}{(1 + Kt')^2 - K^2 \Delta t} \tag{11.27}$$

An inspection of Eq. (11.27) shows that S_2' will be higher than S_2, namely, the efficiency will be lower with unequal cells. Of course, as stated earlier, other considerations may lead to a choice of unequal cells.

Since the cells may not be really completely mixed, their design can be better made using the dispersed flow model with the appropriate value of D/UL applicable to each of the unequal-sized cells.

11.12 CELLS IN PARALLEL

Cells may be provided in parallel as shown in Figure 11.8c, if desired. This is sometimes done with ponds and lagoons. The following aspects may then be kept in view:

1. The cells can be either equal or unequal in size as they operate independently of each other.
2. Even if the cells are unequal in size, they can be operated to give equal detention time by adjusting the individual feed rates.
3. Each cell can be designed individually, using the dispersed flow model and an appropriate value of D/UL for that cell. The D/UL values may vary from cell to cell [3].
4. If each cell is assumed to be completely mixed, the same result is obtained as if the system were a single cell of equivalent total volume [8]. Thus,

$$S = \frac{S_0}{1 + K(\Sigma v/Q)} \tag{11.28}$$

5. For a given total volume, the substrate removal efficiency is lower for cells in parallel than for cells in series. However, a parallel arrangement may be preferred sometimes for operational reasons (e.g., flexibility of operation, shutdown of one cell without disturbing other cells during repairs, sludge removal, etc.).

11.13 CELLS IN SERIES WITH INCREMENTAL FEED

A series arrangement of cells may come to be provided as shown in Figure 11.8d. Sometimes, in order to relieve overload on the first cell of a series of cells, a part of the inflow may be diverted to the second or third cells, thus forming a series arrangement with incremental feed. Such arrangements have been used in the case of ponds and lagoons. They have also been used in trickling filters and conventional activated sludge systems where incremental feed is desired.

In a series arrangement with "n" number of equal or unequal volume cells fed incrementally as shown in Figure 11.8d, all fractions of flow do not receive the same exposure to treatment. The first increment of feed receives treatment in all cells; the second increment is treated in n - 1 cells; the third in n - 2 cells, and so on. If the cells are of equal volume v, Marais [8] has shown that

$$S = \frac{S_0}{1 + K(nv/Q)} \tag{11.29}$$

which makes it the same in efficiency as a single completely mixed cell of equivalent volume. Thus, the benefit of series arrangement is lost.

If the cells are of unequal volume, other methods can be used to estimate efficiency. Furthermore, the dispersed flow model can also be usefully applied whether the cells are of equal or unequal volume provided the appropriate value of D/UL is used for each cell individually and a stepwise computation is made of substrate removal.

Example 11.6 Three cells, each 16,000 m^3, are provided in series to treat a wastewater flow of 10,000 m^3/day. If K is 0.7 per day, estimate the efficiency using the equal cells-in-series model. Also estimate the efficiency if the cells are connected in series with an incremental feeding arrangement. (Assume K is equally applicable to both cases and the cells are completely mixed.)

1. As cells in series

From Eq. (11.25):

$$\frac{S}{S_0} = \frac{1}{\left[1 + K\left(\dfrac{\Sigma v}{nQ}\right)\right]^n} = \frac{1}{\left[1 + 0.7\left(\dfrac{16,000 \times 3}{3 \times 10,000}\right)\right]^3} = 0.10$$

and efficiency = (1 - 0.10)100 = 90%.

2. As cells in series with incremental feeding, Eq. (11.29) gives

$$\frac{S}{S_0} = \frac{1}{1 + K(nv/Q)} = \frac{1}{1 + (0.7)(3)(16,000/10,000)}$$

$$= 0.23 \text{ (or efficiency = 77\% only)}$$

11.14 SERIES-PARALLEL ARRANGEMENT

In the case of large lagoons and ponds, it is possible and often advantageous to make a series-parallel arrangement of cells as shown in Figure 11.8e. Such an arrangement permits one to benefit from the normally
higher efficiencies obtainable with cells in series while also having the
benefit of the flexibility in operation possible with cells in parallel.

An added advantage of this arrangement is that the individual cells
can be stepped in level to suit falling ground contours both lengthwise
and sideways.

The estimation of substrate removal efficiency is done using Eq. (11.24)
or applying the dispersed flow model with the appropriate value of D/UL for
each individual cell.

11.15 CELLS IN SERIES WITH RECYCLING

The introduction of recycling in an existing set of cells in series
is sometimes necessary. Its effect on treatment is often twofold: first,
the hydraulic flow rate through the concerned cells is increased with a
consequent reduction in detention time in the cells and a change in the
value of the dispersion number D/UL; second, recycling may change the value
of the substrate removal rate K itself by affecting the solids concentration
in the system, seeding, etc. One factor may offset the other to some extent
and, therefore, it is not easy to evaluate the overall effect of recycling.
The effect of recycling is discussed more fully in Chapter 12.

11.16 EFFECT OF SHOCK LOADS

Diurnal fluctuations of flow and BOD in municipal plants are well
known. Generally, plants are capable of handling these fluctuations; however, industrial wastewater flows may sometimes show very high fluctuations

in flows, concentrations, or both. This is particularly true where certain operations are carried out on a batch basis, giving sudden discharges for short periods (slugs).

The effect of a slug discharge on a plant depends on (1) the duration of the shock load, (2) the nature of the shock or material (e.g., acidic, toxic, organic, hydraulic, or thermal shock, and (3) the capacity of the plant for withstanding shock load.

The reader is referred once again to the C-diagrams presented earlier (Figure 11.3) from which it will be seen that complete mixing gives the lowest peak (C/C_0 = 1.0). The peak concentration is higher for all other cases; the highest is for plug flow. A similar situation occurs even if the slug continues to be received for some time. The system fails if microbes are adversely affected at the peak concentration reached in the reactor. It is, therefore, evident that, where shock loads are likely to come, the system is best protected by making it completely mixed.

It should also be evident that, in relation to a given inflow, larger sized units are less likely to be upset than are smaller ones in high-rate systems, since the incoming slug is more diluted in a larger unit. Shock load considerations are particularly important when cells in series are proposed to be used. The first one or two cells in the series may be upset owing to their relatively smaller size compared with a single cell system of equivalent total volume.

Similarly, in the case of a step input occurring over a certain period of time (F-diagram), Figure 11.3 shows how the different reactors behave. The complete mix reactor shows a gradual increase in concentration with time, whereas the others all show higher concentrations than the former just after $t/t_0 > 1.0$. Thus, it is evident that the maximum concentration reached in an ideal complete mix reactor will be lower than in any other type of reactor until $t \gg t_0$. This aspect may be crucial to the success or failure of a system subjected to shock loads.

For this and other reasons, the trend in recent times has been to design activated sludge plants as complete mixing systems. Equalization tanks are also designed as complete mixing systems wherever feasible.

It was seen earlier that, for inert nondegradable substances, Eq. (11.10a) holds for the ascending part of the curve and Eq. (11.10b) for the descending or purging portion after the substance has stopped entering the complete-mixing reactor. Uses of these equations are illustrated in the examples given below. Furthermore, Example 11.9 illustrates the inherent vulnerability of smaller size units to given shock loads. Completely

mixed extended aeration systems, for example, with their longer detention
times, are often better able to withstand shock loads than are the smaller
high-rate units.

Several studies have been made on the effects of shock loads to determine the transient response of the activated sludge system [10]. Interest
has been shown not only for organic or toxic shock loads but also for hydraulic shock loads. In a recent study on hydraulic shock loads, two types
of situations were studied. In one case, the hydraulic loads were varied,
keeping the organic substrate concentration constant in the influent. In
the other case, the hydraulic loads were varied, keeping the total daily
organic load constant. In both cases it was observed that, for completely
mixed activated sludge reactors designed for operation with a mean hydraulic
residence time of 8 hr, the system could accommodate, without serious disruption of treatment efficiency, hydraulic shocks consisting of step increases in the flow rate up to 100%. Decreases in the flow rate even greater
than 100% could be accommodated. Based on their study, the researchers
recommended that, in the interest of providing more steady and reliable
performance with regard to substrate removal efficiency, an activated sludge
system be afforded protection (by insertion of, for example, an equalization
or surge basin) against a change in flow rate greater than 100% [10].

The effects of sudden changes in temperature on biological systems
have also been investigated. Two types of problems can occur with shock
temperatures: reduced oxygen solubility and an effect on bacterial activity. A control unit run at 35° C was used and the oxygen uptake rate,
solids production, etc., were studied using a Warburg respirometer. Shock
temperatures did not significantly inhibit the biological system until 45° C.
A shock temperature of 50° C required an acclimation period of 5 hr. No
inhibition was noticed when the temperature was dropped to 20° C [11].

Eckenfelder [9] states that the provision of an equalization tank
prior to biological treatment may be desirable in the interest of obtaining
a more uniform treatment efficiency when the BOD load (based on 4-hr com-
posites) is likely to exceed a ratio of 3:1.

When a treatment system has a cells-in-series arrangement and receives
a shock load, the resulting effect on the final effluent depends on the
biodegradability or otherwise of the incoming slug of material, and on the
number of cells in series. For a given total volume, the greater the number of cells in a series, the more the system behaves like a plug-flow one,
and nondegradable material tends to pass through from cell to cell giving
a peak concentration in the final effluent after a time somewhat less than

the theoretical detention time of the system. The fewer the number of cells in series, the lower is the peak concentration in the effluent; this may be a desirable arrangement when the nondegradable material is also toxic.

For slugs of degradable material, the effect on the effluent quality tends to be just the reverse, provided the degradation follows first-order or higher order kinetics. For a given total volume, the more the cells in a series, the greater is the time of exposure to treatment and, thus, the lower is the effluent concentration. Of course, in saying this it is presumed that the other components of the system are designed properly. For example, the first cell which receives the brunt of the organic load should have sufficient aeration capacity if it is not to turn anaerobic. A judicious choice of the number of cells in series has to be made, depending upon the nature of the incoming material, the extent of fluctuation, the total volume of cells in the system, the rate of degradation, and other factors.

Example 11.7 Estimate the maximum concentration of a conservative pollutant reached in two reactors, one an intermixing type with $D/UL = 0.025$ and the other an ideal complete-mixing type, each of 600 m^3 volume and providing a 6-hr theoretical detention time to a flow of 1000 m^3 per hr, when the pollutant (1) enters as an instantaneous slug of 600 kg and (2) enters at a steady rate of 100 mg/liter over 12 hr.

1. Instantaneous Input

C_0 = concentration if mixed with entire tank volume

$$= \frac{600 \text{ kg} \times 10^6}{600 \text{ m}^3 \times 10^3} = 100 \text{ mg/liter}$$

From the C-diagram (Figure 11.3) we get

 a. For intermixing ($D/UL = 0.025$), maximum $C = 1.8C_0 = 180$ mg/liter.
 b. Complete mix condition, $C = C_0 = 100$ mg/liter, maximum initially and thereafter decreasing gradually.

2. Continuous Input

$C_0 = 100$ mg/liter

From the F-diagram (Figure 11.3), read for $t/t_0 = 12$ hr/6 hr = 2.0.

 a. For intermixing ($D/UL = 0.025$), $C/C_0 = 1.0$ for $t/t_0 > 1.5$. Hence, $C = 100$ mg/liter is reached at about 9 hr and remains so until discontinued at 12 hr.

b. For complete mixing, $C/C_0 = 0.864$ at $t/t_0 = 2.0$. Hence, effluent concentration C = 86.4 mg/liter at 12 hr. The concentration will then decrease gradually.

Example 11.8 For the data given in Example 11.7, estimate the time required for the effluent from the complete-mixing reactor to show a concentration of only 1 mg/liter through gradual purging out of the pollutant after having reached the maximum value of 86.4 mg/liter.

From Eq. (11.10b),

$$C = C_m(e^{-t'/t_0}) \quad \text{or} \quad \frac{1.0}{86.4} = e^{-t'/6}$$

Hence, $t' = 14.7$ hr after the incoming pollutant is discontinued.

Example 11.9 If 100 mg/liter of a conservative toxic substance is received for 30 min at the inlet of an activated sludge aeration tank with a complete-mixing regime, estimate the maximum concentration reached in the unit when detention time in the tank is (1) 18 hr (as may be possible in an extended aeration plant), (2) 6 hr (as in conventional activated sludge), and (3) 2 hr (contact stabilization).

From Eq. (11.10),

$$C = C_0(1 - e^{-t/t_0})$$

For (1),

$$t/t_0 = \frac{30 \text{ min}}{18 \times 60 \text{ min}} = 0.027$$

Thus,

$$C/C_0 = 0.028 \quad \text{or} \quad C = 0.028 \times 100 = 2.8 \text{ mg/liter}$$

Similarly, for (2),

C = 8.0 mg/liter

and for (3),

C = 22.2 mg/liter

11.17 ESTIMATION OF WASTEWATER TEMPERATURE IN LARGE REACTORS

In order to apply the kind of temperature correction equation given in Chapter 12 [Eq. (12.30)], it is necessary to estimate the likely temper-

ature in a lagoon, pond, or other large reactor from a knowledge of the
incoming wastewater temperature and the ambient conditions.

Generally speaking, large reactors such as lagoons and ponds gain heat
from the incoming wastewater and from solar radiation and bacterial activity in the reactor. Often, the incoming wastewater is the major source of
heat gain, whereas solar radiation is a relatively minor one, and bacterial
activity is even more so. Thus, heat gain is essentially proportional to
the inflow rate and the temperature difference between influent and reactor
conditions. Heat losses from the pond occur simultaneously, largely through
convection and radiation and, to a lesser extent, through evaporation.
These losses are proportional to the surface area exposed and the ambient
conditions. Neglecting the minor losses and gains, we may concentrate on
the major ones and balance them, since the total heat gain must equal the
total heat loss at steady-state or equilibrium conditions in the reactor.
Thus, total heat gain = total heat loss, or

$$Q(T_i - T_e) = fA(T_w - T_a) \tag{11.30}$$

where

Q = wastewater inflow rate, m^3/day

T_i = influent temperature, $°C$

T_e = effluent temperature, $°C$

T_w = average wastewater temperature in reactor, $°C$

T_a = air temperature, $°C$

A = surface area of reactor, m^2

f = exchange coefficient, m/day

If complete mixing conditions are assumed, $T_e = T_w$ and the above
equation can be written as

$$\frac{A}{Q} = \frac{t}{d} = \frac{T_i - T_w}{f(T_w - T_a)} \tag{11.31}$$

where

t = detention time (days)

d = depth of reactor (m)

The above coefficient f takes into account several variables, such as wind,
humidity, turbulence (thus, net surface area exposed), and other factors
which affect heat exchange from the reactor. The minimum value of the heat
exchange coefficient is 0.325 m/day. Hence, values of f somewhat larger

Table 11.3 Typical Values of Exchange Coefficient f

Unit	Coefficient f $(m/day)^a$	Reference
Aerated lagoons	0.49 ± 0.06	9
Waste stabilization ponds	0.40	2,12

[a]Multiply by 24.59×10^{-6} to give million gal/day/ft^2.

than this value would be likely to be observed in the field. Some values reported in the literature are given in Table 11.3. The value of f is obtained by plotting the data graphically as shown in Figure 11.9. Equation 11.31 given earlier is applicable to complete mixing conditions.

In a Canadian study, a linear temperature decrease was found along the length of a plug flow type of lagoon. In this case, the mean lagoon temperature T_w could be calculated as the arithmetic average of the influent and effluent temperatures.

As a rough rule of thumb, temperatures in waste stabilization ponds can be assumed as 5°C higher than the winter air temperatures. This has been an approximate observation in both temperate and warm climates. In temperate climates the influent wastewater temperatures depend on the use of domestic hot water systems and other factors, which counter the low ambient

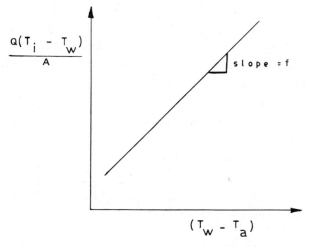

Figure 11.9 The determination of the exchange coefficient f.

temperatures. For the relatively small size reactors used in activated
sludge treatment, reactor temperatures are often practically equal to those
of the influent wastewater.

The design of holding ponds for thermal pollution control, especially
in the case of hot industrial discharges which need to be held in a lagoon
for some length of time to enable the temperature to be dropped below a
desirable value, can also be made using Eq. (11.31).

Example 11.10 A lagoon is 3.3 m deep and provides 10 days' detention
to an industrial wastewater entering at $55^\circ C$. If the mean ambient temperature in the given season is $10^\circ C$, estimate the lagoon temperature, assuming
complete mixing conditions and f = 0.5 m/day.

$$\frac{t}{d} = \frac{T_i - T_w}{f(T_w - T_a)} = \frac{10 \text{ days}}{3.3 \text{ m}} = \frac{55^\circ C - T_w}{(T_w - 10^\circ C)0.5 \text{ m/day}}$$

Hence, $T_w = 28^\circ C$.

11.18 FACTORS AFFECTING CHOICE OF REACTORS

From the foregoing discussions it will be seen that reactors can be
provided in different ways in order to emphasize desired mixing and other
conditions. Particularly, in the case of large aerated lagoons and oxidation ponds, the designer must decide whether he wishes to emphasize plug
flow conditions or complete mixing conditions. Plug flow conditions are
approached either by providing a long and narrow unit, by providing around-
the-end type baffles, or by providing cells in series. In each case, the
idea is to reduce the value of D/UL. Conversely, complete mixing conditions
are approached by providing more squarish geometry, more mixing, etc., to
increase the value of D/UL.

Proper choice of reactors is important for the success of a treatment
project. The aim of any designer should be to maximize the substrate removal efficiency at minimum capital and running costs and to build into the
system better operating characteristics and reliability. The factors affecting reactor choice can be grouped and listed briefly as follows since
they have already been discussed earlier:

1. Nature of the waste
    a. Reaction order. Zero-order reactions perform equally well in
        plug flow or complete mixing. First and higher order reactions

perform better in plug flow, giving higher efficiencies. Very high efficiencies (for example, in bacterial removal) are possible only with plug flow.

b.   Reaction rate constant K. Wastes with very low K values perform more or less the same in plug flow and complete mixing. Wastes with higher K values, and first-order kinetics, are best removed in plug flow.

2. Process optimization

a.   Functionalize wherever possible. For example, environmental requirements of acid-formers are different from those of methane fermenters. They can, therefore, be handled better in successive steps. Deep lagoons and ponds can be more assured of having an anaerobic layer lower down where gasification can occur and BOD can be removed without expensive aeration. Similarly, compare Inhoff tanks with septic tanks.

b.   Sequential removal of different substrates is better achieved in plug flow reactors and cells in series. Nitrification occurs after the carbonaceous demand is satisfied. Nitrifiers have a slower growth rate; hence, plug flow conditions help in achieving better nitrification.

c.   Heat conservation to help increase waste treatability (K value). In treating hot industrial wastes, especially in cold climates, two or more cells in series (plug flow) give a better overall removal efficiency than does one large cell of equivalent volume in which the incoming temperature will drop more quickly.

d.   Load distribution. This may be necessary to avoid overload on the first cell of a series of cells, which may reduce the K value. Unequal cells in series or recirculation may be necessary.

3. Other factors

a.   Operational flexibility. Provision of cells in parallel is often done for the sake of flexibility since one cell can be put out of commission for repairs or sludge removal, etc., without affecting the operation of the other cells. The sizing of cells may also be based on flexibility and other considerations such as equipment availability.

b.   Site conditions. In the case of large lagoons and ponds, site conditions may dictate the size and relative location of cells in series or in series-parallel. Unequal cells in series also sometimes may be necessitated purely from site considerations.

REFERENCES

1. O. Levenspiel, Chemical Reaction Engineering, John Wiley and Sons, N.Y., 1962.

2. S. Arceivala, Use of Dispersed Flow Model in Designing Waste Treatment Units, NATO Advanced Institute, Istanbul, Noordhoff Int. Pub., Netherlands, 1976.

3. S. Arceivala, Predicting Dispersion in Aerated Lagoons and Oxidation Ponds, in press.

4. K. Murphy, Flow patterns in waste treatment, in <u>Biological Waste Treat-</u>
    <u>ment</u>, R. Canale, Ed., Wiley-Interscience, N.Y., 1971.

5. K. Murphy and P. Timpany, <u>Proc. ASCE</u>, <u>93</u>, 1967.

6. J. Wehner and R. Wilhem, <u>Chemical Eng. Sc.</u>, <u>6</u>, 1956.

7. K. Murphy and A. Wilson, Characterization of mixing in aerated lagoons,
    <u>Jour. ASCE, Env. Eng. Div.</u>, <u>100</u>, 1974.

8. G. Marais, Fecal bacterial kinetics in stabilization ponds, <u>Jour. ASCE,</u>
    <u>Env. Eng. Div.</u>, <u>100</u>, 1974.

9. W. W. Eckenfelder, <u>Water Quality Engineering for Practicing Engineers</u>,
    Barnes and Noble, Inc., N.Y., 1970.

10. T. George and A. Gaudy, Response of completely mixed systems to hy-
    draulic shock loads, <u>Jour. ASCE, Env. Eng. Div.</u>, <u>99</u>, 1973.

11. J. Carter and W. Barry, Shock temperature effects, <u>Jour. ASCE, Env.</u>
    <u>Eng. Div.</u>, <u>101</u>, 1975.

12. S. Selcuk., M.S. thesis, Middle East Tech. Univ., Ankara, Turkey, 1974.

Chapter 12

PRINCIPLES OF BIOLOGICAL TREATMENT

12.1 PRINCIPLES OF BIOLOGICAL TREATMENT

The basic principles of biological growth accompanying substrate removal are discussed in this chapter, and their application to actual plant
design is discussed in succeeding chapters. A unified approach is desirable
for the development of kinetic models.

Organic substrates support heterotrophic microorganisms which use the
organic carbon contained in the substrate partly as a source of energy and
partly as a nutrient for new cell growth. This is shown schematically in
Figure 12.1. The organic carbon supplies both the energy and the carbon
needed for the building of protoplasm. Other nutrients such as nitrogen
and phosphorus are also necessary. In aerobic biological growth, a sufficient amount of oxygen is also essential. A part of the new cells produced
is destroyed by endogenous respiration* until, finally, a nondegradable
cellular residue remains.

The microbial cells produced are often depicted by the empirical formula $C_5H_7O_2N$. These cells may remain as <u>suspended growths</u> in the biological
reactor as in the case of activated sludge systems or form <u>attached growths</u>
on surfaces (media) available to them in the reactor as in the case of
trickling filters. The basic principles of biological treatment apply
equally well to suspended and attached growths.

In engineering terms, aerobic biological degradation can be simply
depicted as follows using the term "sludge" instead of microbial solids:

Waste + sludge + air → surplus sludge + end-products

Figure 12.2 depicts, in essence, what happens in any biological waste treat-

---

*Endogenous respiration or autooxidation involves the breakdown (oxidation)
of organics in the cell itself in order to produce energy for the maintenance of the remaining cell mass. Thus, there is continuously a fractional
decrease in cell mass with time. (In humans, this kind of decrease becomes
evident during starvation!)

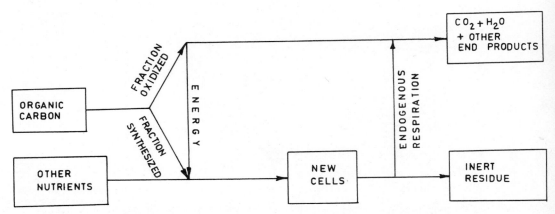

Figure   12.1   The biological degradation of organics.

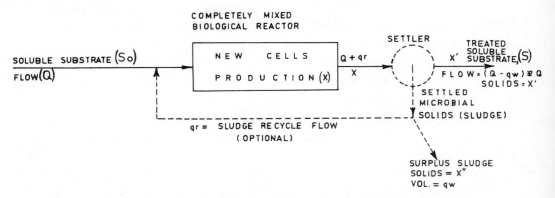

Figure   12.2   A typical biological treatment system.

ment process. In order to remove organic matter which is in soluble form
in the incoming wastewater, it must be converted into a settleable form by
producing new biological growths, suspended or attached, which will readily
flocculate to form settleable sludge. If the settling is efficient, the
final effluent will contain only that fraction of the soluble substrate
which is not converted into microbial solids. The settled sludge can be
withdrawn from the system and disposed of separately. The sludge can also
be returned to the biological reactor if desired. In fact, often, especially in the case of suspended growth systems, wherever possible, sludge
recycling is preferred for reasons we shall see later. In any event, some
surplus sludge is produced which must be removed from the system and disposed of separately.

12.2 CELL YIELD

The relation between new cell production and soluble substrate consumption can be stated as

$$\frac{dx}{dt} = Y\left(\frac{dS_r}{dt}\right) \tag{12.1}$$

where

x = microorganisms, mass or concentration

S_r = substrate removed (consumed), mass or concentration
    $[= (S_0 - S)/t]$

t = time

Y = true growth yield coefficient (mass of microbial cells
    produced per unit mass of substrate utilized)

The value of Y is virtually constant for a wide variety of substrates treated aerobically. This is explained by the fact that the synthesis of protoplasm involves so many transformations that the overall energy requirements are virtually the same regardless of the substrate being metabolized [1]. Synthesis is also shown to be proportional to oxidation, and for soluble substrates the relation is [2]:

$Y = 0.39$ g microbial solids/g COD removed

  $= 0.57$ g microbial solids/g BOD_5 removed

Anaerobic breakdown of organic substrates releases less energy and, therefore, the value of the yield coefficient Y is also less. Chemoautotrophs (for example, nitrifiers) do not draw their energy requirements from organic carbon but from the oxidation of organic compounds and, thus, they also show lower values of Y compared with those for heterotrophs. A few typical values of Y are given in Table 12.1.

In practice, X is generally measured as volatile suspended solids (VSS). The substrate may be measured in terms of COD, BOD_u, BOD_5, total organic carbon (TOC), etc., but should be clearly stated since in each case a different numerical value of Y is obtained. Furthermore, the value of Y is based on the soluble substrate removed. If solids are present in the inflow as in full-scale plants, they will lead to larger estimates of Y compared with well-controlled laboratory experiments using only soluble substrates.

Table 12.1 Typical Values of Constants at 20°C

Wastewater	Growth coefficients and constants				Substrate removal rate constants		Overall removal rate constant K_a per day for process				Basis	Type of study	Author and reference
	Y (mass/mass)	K_d (day^{-1})	μ_{max} (day^{-1})	K_s (mg/liter)	k (day^{-1})	k'[c] (day^{-1})(mg/liter)$^{-1}$	Activated sludge	Extended aeration	Facultative aerated lagoon	Oxidation pond			
Aerobic													
Domestic[a]	0.4	0.09	~3.2	60.0	8.0						COD		Sherrard and Lawrence [23]
Domestic[b]	0.73	0.075				0.017-0.03					BOD5		Eckenfelder [8]
Domestic	0.372	0.098		45.5	8.35						BOD5		Hashimoto and Fujita [5]
Domestic[b]	0.56	0.035	0.331	16.8		0.0257					BODu	Field	Handa [9]
Domestic[b]	-	-	-	-	-	-	13-30	20-30	0.6-0.8	0.1-0.13	BOD5	Field	Arceivala [10]
Skim milk	0.48	0.045		100	5.1						BOD5	Lab	Gram [11]
Chem./petrochem.[b]	0.31-0.72	0.05-0.18				0.003-0.018					BOD5	Lab	Eckenfelder [8]
Textile composite[b]	0.72	0.10	0.89	52	1.24						BOD5	Lab	Samsunlu [12]
Synthetic sewage (58°C)	0.34	0.48	5.2	740	15.38	0.0207					COD	Lab	Surucu [13]
Anaerobic													
Synthetic milk	0.37	0.07		24.3	0.38						COD	Lab	Gates [14]
Packing house (35°C)	0.76	0.17		5.5	0.32						BOD5	Lab	Gates [14]

[a] The values are typical of wastewaters treated aerobically [23].

[b] Values include the effect of influent suspended solids.

[c] $k'S_o = k''$ ranges around 7 to 8 per day for domestic sewage.

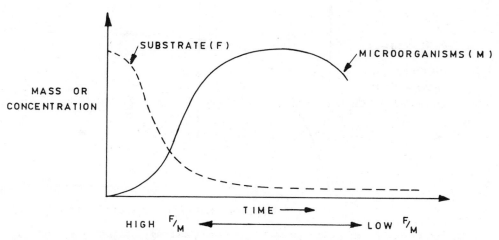

Figure   12.3 Microbial growth and substrate removal in batch culturing.

12.3 MICROBIAL GROWTH RATES

Microbial growth with time in batch culturing follows the typical curve
given in Figure 12.3, progressively passing through the lag phase, followed
by an exponential growth phase, a constant growth phase, and finally a declining phase. Since microorganisms multiply as a result of binary fission,
their growth is a function of their numbers present at any given instant.
Thus, it is necessary to study the microorganism growth rate per unit of
microorganisms in order to evaluate the effect of substrate concentration
and other environmental factors (e.g., temperature) on growth rates. Hence,
the term "specific growth rate" is used, where its observed or actual value
is given by

$$\mu = \frac{dx/dt}{x} \qquad\qquad (12.2)$$

The manner in which μ changes in the constant rate and declining growth
rate regions was given in the form of an empirical equation that Monod [3]
developed in 1942 from pure culture studies. According to this equation,
μ varies in accordance with a hyperbolic function to account for a lower
growth rate when there is nutrient limitation. Figure 12.4 shows an increase in specific growth rate μ as the substrate (limiting nutrient) concentration increases until, after a while, it levels off at the maximum
possible value of μ, namely, μ_{max} at the reactor conditions. Monod's empirical function gives

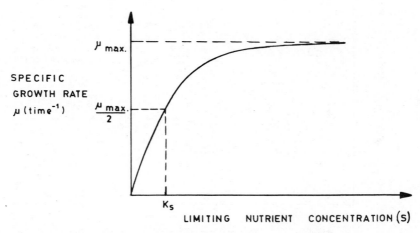

Figure 12.4 Specific growth rate and limiting nutrient concentration [3].

$$\mu = \mu_{max} \left(\frac{S}{K_s + S} \right) \tag{12.3}$$

where

μ_{max} = maximum specific growth rate, $time^{-1}$

S = concentration of rate limiting nutrient or substrate (e.g., organic carbon, nitrogen, phosphorus, etc.)

K_s = concentration of rate limiting nutrient at which $\mu = \mu_{max}/2$ (also called "saturation concentration")

Wuhrmann [4] reported K_s = 0.2 mg/liter for a mixed culture activated sludge growing on glucose. If the actual glucose concentration in the substrate is 10 mg/liter, Eq. (12.3) gives that $\mu \cong \mu_{max}$. On the other hand, for domestic sewage, Hashimoto and Fujita give K_s = 45.5 mg/liter BOD_5. If the substrate concentration is again 10 mg/liter, Eq. (12.3) gives μ = 0.18 μ_{max}. Thus, when K_s values are relatively high, the specific growth rate may be in the lower region of the curve and affected by substrate concentration similar to a first-order equation. But when K_s values are low relative to the substrate concentration, the specific growth rate may be at or near the maximum value and not much affected by the concentration as in a typical zero-order reaction. Table 12.1 gives some values of K_s observed for some wastes.

12.4 NET SOLIDS PRODUCTION

Combining Eqs. (12.2) and (12.3) gives

$$\frac{dx}{dt} = \mu_{max} \left(\frac{S}{K_s + S} \right) X \qquad (12.4)$$

However, in waste treatment, where cells generally have long residence times, losses in cell mass owing to endogenous respiration must be taken into account to arrive at the <u>net</u> growth rates. Thus, Eqs. (12.2) and (12.4) can be written as follows to give net rates, as

$$\frac{dx}{dt} = \mu x - K_d X \qquad (12.5)$$

and

$$\frac{dx}{dt} = \mu_{max} \left(\frac{S}{K_s + S} \right) X - K_d X \qquad (12.6)$$

in which K_d = cell decay rate (time^{-1}) due to endogenous respiration. It is the fraction of cells destroyed per unit time. A few typical values of K_d observed with some wastewaters are given in Table 12.1.

For finite conditions, Eq. (12.5) can be rewritten, substituting from Eq. (12.1) to give

$$\frac{\text{Net solids produced}}{\text{time}} = \frac{\Delta X}{\Delta t} = Y \left(\frac{S_0 - S}{t} \right) - K_d X \qquad (12.7)$$

12.5 HYDRAULIC DETENTION AND CELL RESIDENCE TIMES

By definition, the hydraulic detention time t is given by

$$t = \frac{V}{Q} = \frac{\text{volume of water in system}}{\text{volume leaving system/time}} \qquad (12.8)$$

Similarly, the mean cell residence time Θ_c or the solids retention time (SRT) is given by

$$\Theta_c = \text{SRT} = \frac{\text{weight of solids in system}}{\text{weight leaving system/time}} \qquad (12.9)$$

Some suspended growth systems are operated on a flow-through basis
with no recycling (e.g., most oxidation ponds and aerated lagoons), whereas
others are operated with recycling (e.g., activated sludge or extended
aeration). This makes a considerable difference in the relative values of
t and Θ_c as shown below:

1. Completely mixed reactor, no recycling (Figure 12.2). From Eq. (12.8),
    $t = V/Q$; from Eq. (12.9), $\Theta_c = XV/XQ$; thus,

$$\Theta_c = t \tag{12.10}$$

2. Completely mixed reactor, with recycling (Figure 12.2). Again, from
    Eq. (12.8), $t = V/Q$, but, from Eq. (12.9),

$$\Theta_c = \frac{XV}{(Q - qw)x' + q_w X''} \tag{12.11}$$

Generally, the numerator > the denominator and, therefore, $\Theta_c > t$.

When a reactor is operated for a while, a "steady-state" condition
is reached at which the solids produced per unit time equal the solids
leaving per unit time. This is true whether there is recycling or no recycling.

If there is no recycling, the more solids produced in the reactor,
the more are washed out until the solids production equals the solids leaving. If there is recycling, the surplus sludge withdrawal is so controlled
by the operator that, again, solids production equals solids leaving so that
the solids concentration in the reactor is held relatively constant. Thus,
at steady state, we can also write the mean cell residence time as

$$\Theta_c = \frac{\text{weight of solids in system}}{\text{net weight of solids produced/time}} \tag{12.12}$$

$$= \frac{XV}{(\Delta x/\Delta t)} \quad \text{if X is in concentration units} \tag{12.13a}$$

$$= \frac{X}{(\Delta x/\Delta t)} \quad \text{if X is in mass units} \tag{12.13b}$$

Equation (12.13) is the reciprocal of Eq. (12.2) on a finite time basis.
Thus,

$$\Theta_c = \frac{1}{\mu} \tag{12.13c}$$

in which μ is the observed or actual specific growth rate per unit time.

In order to get a feel for the range of values observed in full scale
treatment plants, it may be mentioned that typical values of Θ_c range from
3 to 10 days for activated sludge systems and from 10 to 30 days or more for
extended aeration. By comparison, the t values range from less than 6 hr for
activated sludge to 12 hr or more for extended aeration (see Table 14.1).

12.6 CELL WASHOUT TIME

The time during which a cell remains in a treatment system (Θ_c) must be greater than its doubling time or else the cell would be "washed out" of the system before it has had a chance to multiply, and the process would fail. Thus, it is essential to know the doubling time of a cell at the critical (winter) temperature.

$$\text{Doubling time } t_d = \frac{0.693}{\text{observed specific growth rate } \mu}$$

1. For suspended growth reactors without recycling, $\Theta_c = t$ should be equal to or greater than t_d. In the design of oxidation ponds and flow-through type aerated lagoons where there is no recycling, it is essential to make sure the minimum detention time in a unit is not less than the algal or bacterial doubling time at the critical operating temperature. In designing such units as cells in series, the minimum cell size is controlled by this aspect (see Sec. 16.5).

2. For suspended growth reactors with recycling ($\Theta_c > t$), the recycling rate can be adjusted to keep $\Theta_c \geq t_d$, while the hydraulic detention time t can be kept to the minimum, consistent with the other considerations discussed later (e.g., MLSS concentration and fluctuations in inflow). Thus, recycling of sludge becomes a way of increasing Θ_c without necessarily increasing t.

In attached growth systems, the hydraulic detention time t may be very small indeed, while the biological growths attached to the media provide relatively high residence time θ_c, and recycling is not essential, though it may be useful for other reasons.

12.7 FOOD/MICROORGANISM RATIO

A term often used in waste treatment is the so-called "food-to-micro-organism" ratio (F/M). For a continuous flowing system in which S_0 and S are the initial and final substrate masses, respectively,

$$F = \frac{S_0 - S}{t} \quad \text{and} \quad M = \text{microorganism mass, } X$$

Then

$$\frac{F}{M} = \frac{S_0 - S}{Xt} \tag{12.14}$$

This ratio is often stated as kg BOD_5 removed per day per kg of volatile suspended solids in mixed liquor (MLVSS). Several values are reported in the literature, some in terms of BOD_5, BOD_u, TOC and COD, some in terms

of MLVSS, and some in terms of MLSS (mixed liquor suspended solids, which
include both volatile and fixed). Some also give $F/M = S_0/Xt$. Thus, it
is necessary to be careful in comparing literature values (see Table 14.1).

This concept of F/M is not applied to attached growth systems since
the mass of biogrowths is difficult to determine in them.

Example 12.1 Estimate the net weight of solids produced in an activated sludge system in which influent BOD_5 is reduced from 250 to 30 mg
per liter in a flow of 4000 m^3/day. The aeration tank volume is 700 m^3,
and MLVSS 3000 mg/liter. Assume Y = 0.5 and K_d = 0.09 per day. Also compute Θ_c and t, and F/M.

Substrate removed $= (250 - 30)(4000) \times \dfrac{10^3}{10^6} = 880$ kg/day

MLVSS in system $= (3000)(700) \times \dfrac{10^3}{10^6} = 2100$ kg

Net solids production rate = Y (substrate removal rate) - K_d (MLVSS)
$$= 0.5(880) - 0.09(2100)$$
$$= 251 \text{ kg/day}$$

Net solids production/BOD_5 removed $= \dfrac{251}{880} = 0.285$

$\Theta_c = \dfrac{\text{solids in system}}{\text{net solids produced/day}} = \dfrac{2100}{251} = 8.36$ days

$t = \dfrac{700}{4000} = 0.175$ days = 4.2 hr

$\dfrac{F}{M} = \dfrac{880}{2100} = 0.42$ kg BOD_5 per day/kg MLVSS

12.8 SOLIDS CONCENTRATION IN SUSPENDED GROWTH REACTORS

Particularly in the design of activated sludge and extended aeration
systems, it is necessary to know what the concentration of microbial solids
will be in a reactor under given conditions. Since in these processes the
reactors used are often the complete mixing type, the following derivation
is given:

A mass balance around a completely mixed reactor gives, at steady-state
conditions:

Net rate of solids production in system = net rate at which solids leave system

Thus,

$$Y\left(\frac{dS_r}{dt}\right) - K_d X = \frac{\Delta x}{\Delta t}$$

or

$$\frac{\Delta x/\Delta t}{X} = \frac{Y(dS_r/dt)}{X} - K_d \qquad\qquad (12.15)$$

or

$$\frac{1}{\Theta_c} = Y\left(\frac{S_0 - S}{Xt}\right) - K_d \qquad\qquad (12.16)$$

or

$$\frac{1}{\Theta_c} = Y\left(\frac{F}{M}\right) - K_d \qquad\qquad (12.17)$$

Equation (12.16) can be rewritten to give, either in mass or concentration units,

$$X = \frac{Y(S_0 - S)}{1 + K_d\Theta_c} \left(\frac{\Theta_c}{t}\right) \qquad\qquad (12.18)$$

Also, substituting $t = V/Q$ gives

$$XV = \frac{Y\Theta_c Q(S_0 - S)}{1 + K_d\Theta_c} \qquad\qquad (12.19)$$

Thus, for a given wastewater, the volume V of the reactor (aeration tank) can be found if the flow Q and the BOD proposed to be removed $(S_0 - S)$ is known, the constants Y and K_d are known, a suitable value of Θ_c is assumed as discussed later, and the maximum concentration of microbial solids X to be allowed to build up in the reactor is fixed. Once the various parameters are fixed, the actual concentration X will be directly proportional to the BOD removed, namely, $S_0 - S$. Application of this equation to practical design is shown later.

Equations (12.15) to (12.19) derived above are equally applicable whether or not there is recycling, because the basic assumption (namely, solids production = solids leaving at steady state) remains the same in both cases.

However, if there is no recycling, $t = \Theta_c$ and Eq. (12.18) merely simplifies further to give

$$X = \frac{Y(S_0 - S)}{1 + K_d t} \tag{12.20}$$

Equation (12.20) is often useful in the design of aerobic flow-through type lagoons as long as the lagoon is assumed to be a complete-mixing reactor.

Marais [24] has recently shown that Eq. (12.18) needs to be modified to take into account the fact that, in practice, all volatile solids in a system are not "active." Some are merely the residue after endogenous respiration. If this endogenous residue = X_e, and "active solids" = X_a, then the MLVSS value is given by

$$MLVSS = X_a + X_e$$

About 20% of the active, degradable solids end up as endogenous residue in the system. Now, where recycling is practiced, it can be shown from the definition of Θ_c that

$$\frac{X_e}{0.2 K_d X_a} = \Theta_c$$

and

$$MLVSS = X_a (1 + 0.2 K_d \Theta_c)$$

Thus, Eq. (12.18) can be extended to account for this buildup by incorporating the above equation to give

$$MLVSS = \left(\frac{Y(S_0 - S)}{1 + K_d \Theta_c} \frac{\Theta_c}{t} \right) (1 + 0.2 K_d \Theta_c) \tag{12.21}$$

Marais has also indicated a method for the estimation of Y and K_d using Eq. (12.21). However, most values of Y and K_d found in the literature so far (and listed in Table 12.1) have been evaluated using Eq. (12.18).

12.9 SUBSTRATE REMOVAL IN SUSPENDED GROWTH REACTORS

Equations (12.18), (12.19), and (12.20) can be used for estimating the final effluent concentration S, provided other terms are known. S can also be determined as follows in terms of Monod's constants. From Eq. (12.6) we can write

$$\frac{\Delta x/\Delta t}{x} = \frac{1}{\Theta_c} = \mu_{max}\left(\frac{S}{K_s + S}\right) - K_d$$

From which we can rearrange to give

$$S = \frac{K_s[(1/\Theta_c) + K_d]}{\mu_{max} - [(1/\Theta_c) + K_d]} \tag{12.22}$$

Thus, it will be observed that, for a given wastewater in completely mixed reactors, K_s, K_d, and μ_{max} are constants and the value of S only depends on the value of Θ_c adopted by the designer. It is independent of the X values since X will adjust itself to the amount of substrate removed.

In terms of another parameter, k' (discussed later), the final substrate concentration in the case of completely mixed reactors is given by

$$S = \frac{1}{k'Y}\left(\frac{1}{\Theta_c} + K_d\right) \tag{12.23}$$

Here again, for a given wastewater, S only depends on Θ_c.

12.10 APPROXIMATION OF SUBSTRATE REMOVAL

Under certain conditions, substrate removal in suspended growth systems can be approximated as follows. From Eq. (12.1) the substrate removal rate is directly proportional to the rate at which new cells are produced. Thus, substituting from Eq. (12.1) to Eq. (12.4) gives

$$Y\left(\frac{dS_r}{dt}\right) = \mu_{max}\left(\frac{S}{K_s + S}\right)X \tag{12.24}$$

or

$$\frac{dS_r}{dt} = \frac{\mu_{max}}{Y}\left(\frac{S}{K_s + S}\right)X \tag{12.25}$$

Two boundary conditions may now be considered.

Case 1: When $S \gg K_s$, Eq. (12.25) simplifies to give

$$\frac{dS_r}{dt} \simeq \left(\frac{\mu_{max}}{Y}\right)X$$

or

$$\frac{dS_r}{dt} \cong kX \tag{12.26}$$

where $k = \mu_{max}/Y$ = maximum substrate utilization rate per unit time per unit mass of microorganisms (time^{-1}). In words, this signifies that the substrate concentration is so high that it does not affect the removal rate, which now depends only on the microorganism concentration X. Thus, substrate removal is zero order with respect to substrate concentration, but first order with respect to organism concentration. Referring to Figure 12.4, the specific growth rate is at its maximum (μ_{max}). Substrates which display low values of K_s fall in this category.

Case 2: When $S \ll K_s$, Eq. (12.25) can be written as

$$\frac{dS_r}{dt} = \left(\frac{\mu_{max}}{YK_s} \right) SX$$

or

$$\frac{dS_r}{dt} = k'SX \tag{12.27a}$$

where $k' = \mu_{max}/YK_s$ = specific substrate utilization rate (time^{-1}) (mg/liter^{-1}).

Many wastewaters including domestic sewage often fall in this category. Treatment requirements dictate that the final effluent BOD concentration (soluble) be kept between 10 and 20 mg/liter. Thus, when completely mixed reactors are used, their concentrations are the same as that required in the effluent. These concentrations may be lower than the K_s values and, hence, their removal kinetics may fall in this category. Under such conditions, k' values are readily determined from graphical plots of $(S_0 - S)/Xt$ against S, which tend to give straight lines whose slopes equal k' (see Figure 12.5). Different wastewaters give different slopes, reflecting their treatability under given conditions. Table 12.1 gives a few typical values of k'.

The advantage in using this approximation is that only one parameter, k', is required for design purposes, whereas in the Monod case, three constants, K_s, K_d, and μ_{max}, are required to be known. Of course, both approaches should give the same answers if the constants used are correct.

Recently, Eckenfelder [6] and others have shown that actual substrate removal data give a better fit when the initial substrate concentration is also taken into account by plotting $S_0[(S_0 - S)/Xt]$ against S (see Figure

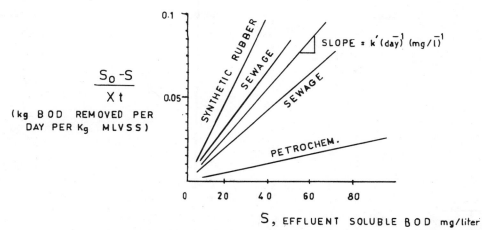

Figure 12.5 BOD removal characteristics of various organic wastes in terms of K' (from Ref. 8).

12.6). The slope of the line then gives the value of the treatability rate k'' per unit time, where

$$k'' = k'S_0 \qquad\qquad (12.27b)$$

For treating domestic sewage, Eckenfelder states that k'' values range between 7.0 and 8.0 per day in U.S. experience. The use of the rate constant k'' implies that, at a constant organic loading, the effluent soluble BOD is directly proportional to the influent BOD in a plant.

12.11 OVERALL SUBSTRATE REMOVAL RATES

The k' values are given per unit time per unit solids and, therefore, in their use the designer has first to decide what the microbial solids concentration X in mg/liter will be in the reactor as a result of recycling. Thus, K_a, the <u>overall</u> substrate removal rate per unit time at the solids concentration prevailing in the reactor, is

$$K_a = k'X \qquad\qquad (12.28)$$

and, from Eq. (12.27a),

$$\frac{dS_r}{dt} = K_a S \qquad\qquad (12.29)$$

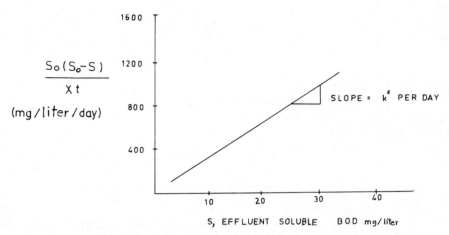

$\dfrac{S_0(S_0-S)}{Xt}$

(mg/liter/day)

Figure 12.6 BOD removal characteristics of various organic wastes in terms of K'' (from Ref. 6).

Equation (12.29) indicates that substrate removal is first order with respect to the substrate concentration S in the reactor. Consequently, the substrate removal efficiency for completely mixed reactors can be estimated from the well-known equation seen earlier [Eq. (11.12)]:

$$S = \frac{S_0}{1 + K_a t} \qquad (12.30)$$

which can also be stated as

$$S = \frac{S_0}{1 + (k'x)t} \qquad (12.31)$$

in which S_0 and S are the initial and final substrate concentrations, respectively, and t is the hydraulic detention time.

The dispersed flow model for first-order kinetics given by Wehner and Wilhem [Eq. (11.15)] can be conveniently used provided the <u>overall</u> substrate removal rate constant K_a per unit time for the process in question is known. Typical values of K_a are included in Table 12.1.

12.12 ESTIMATION OF K VALUES FROM FIELD DATA

The various K values can be determined from the actual operating data of a treatment unit. The information required is the initial and final substrate concentration (BOD, for instance), the theoretical detention time t in the unit, and the mixed liquor volatile solids concentration and temperature. In lab-scale units, mixing conditions are generally well defined.

However, a word of caution is necessary at this stage with regard to full-scale units, where the actual mixing conditions may not be anywhere near the ideal complete mixing condition assumed if Eqs. (12.30) or (12.31) are used. This would give K_a values specific for the reactor in question, which cannot be generalized and applied to other reactors where mixing conditions may be different. Thus, it is advisable (especially when dealing with large lagoons and ponds) to simultaneously perform a tracer test for determining the actual mixing conditions in terms of the dispersion number D/UL discussed in Chapter 11. If D/UL values lie between 0.1 and 4.0, it may be advisable to use the dispersed flow model [Eq. (11.15)] in order to determine more accurately the K_a value. If D/UL ≥ 4.0, not much error will be made if the ideal complete-mixing equations are used. If D/UL ≤ 0.1, the ideal plug flow equations may be used, if desired, instead of the dispersed flow equation.

It is unfortunate that a large number of K values reported in the literature for lagoons and ponds have been given without any reference at all to the actual mixing conditions in the unit.

12.13 TEMPERATURE EFFECTS

Temperature effects on treatment kinetics are generally accounted for by the well-known equation relating the K value at any temperature to the K value at 20°C by

$$K_{T °C} = K_{20 °C} (\Theta)^{T-20} \tag{12.32}$$

where Θ is a parameter depending on the nature of the process, as shown in Table 12.2. Experimental determination of Θ requires that at least three bench-scale units be run in parallel at different temperatures.

Eckenfelder [6] states that in domestic sewage, BOD is mainly in suspended and colloidal form so that removal on bioflocs is largely physical and relatively independent of temperature; but in the case of soluble industrial wastes, temperature effects may be higher (Θ = 1.06 to 1.10). Also, with the activated sludge process, at higher loadings the floc is more dispersed and each organism is more directly affected by temperature. Generally, in case of lagoons and ponds, the less the solids concentration in the unit, the higher the value of Θ.

Table 12.2 Effect of Temperature on Process Kinetics

Process	θ
1. Activated sludge treating domestic sewage	
a. Less than 0.6 kg BOD/kg MLSS	1.0-1.01
b. Over 0.6 kg BOD/kg MLSS	1.01-1.04
2. Trickling filters	1.035
3. Mechanically aerated lagoons	
a. Aerobic	1.035
b. Facultative	1.035-1.09
c. Extended aeration type	1.01-1.03
4. Waste stabilization ponds	1.035

Between 10 and 30°C, μ_{max} and K_d increase with temperature. K_s decreases slightly from 10 to 20°C, then increases substantially up to 30°C. Y increases from 10 to 20°C, but decreases thereafter. Thus, according to Muck and Grady [7], the temperature effect on the substrate removal rate k' depends on the combined effect of μ_{max}, K_s, and Y. Similarly, the effect on net solids production X depends on the combined effect on K_d and Y.

12.14 SOLUBLE AND TOTAL BOD OF EFFLUENT

It will be recalled that in deriving the substrate removal equations seen earlier, only soluble substrates were assumed to be received in a plant; leftover soluble substrates went out in the effluent. However, solids separation processes are never ideal and some proportion of the solids also leave along with the effluent. These solids are partly fixed and partly volatile. The volatile solids, therefore, add to the effluent BOD and reduce the overall removal efficiency.

The microbial solids (VSS) produced as a result of substrate consumption in the reactor are themselves only about 77% biodegradable; the remaining 23% is "inert" over the usual time period involved in waste treatment. The oxygen demand of biodegradable matter will be seen from Eq. (12.36) to be

$$D = 1.42 \text{ mg BOD}_u/\text{mg biodegradable solids}$$
$$= 1.0 \text{ mg BOD}_5/\text{mg biodegradable solids}$$
$$= 0.77 \text{ mg BOD}_5/\text{mg VSS}$$

Recycling of sludge leads to an accumulation of the organic but inert fraction, leading to a decrease in the degradable fraction of VSS in the mixed liquor. Any inorganic solids coming in the influent also tend to accumulate through recycling. Thus, as the mean cell residence time Θ_c increases, the BOD per unit VSS decreases and the ratio of VSS/SS also decreases.

Figures 12.11(i) and 12.11(j) show the effect of Θ_c on the biodegradable fraction in the mixed liquor, and on the BOD per unit suspended solids, respectively, as given by Eckenfelder [6].

Adams and Eckenfelder [15] give the biodegradable fraction in the mixed liquor VSS as follows:

$$f_b = \frac{YS_r + K_dX - \sqrt{(YS_r + K_dX)^2 - 4K_dX(0.77YS_r)}}{2K_dX} \qquad (12.33)$$

in which X = average MLVSS in the system, kg, and S_r = substrate removed, kg/day. (Y and K_d have been defined earlier.)

For example, if at Θ_c = 8.36 days (Example 12.1), and S_r = 880 kg/day, X = 2100 kg/day, Y = 0.5, and K_d = 0.09/day. Equation (12.33) gives f_b = 0.675. Thus, only 67.5% of the VSS are biodegradable, and the oxygen requirement can be computed from the above as 0.675 mg BOD_5/mg VSS. Furthermore, if the ratio of VSS/SS is, for example, 0.6, the oxygen demand becomes 0.6 x 0.675 = 0.4 mg BOD_5/mg SS.

Typical ratios observed in practice are given in Tables 14.1 and 15.1.

The total BOD of the effluent is the sum of the soluble BOD and the BOD resulting from the escape of volatile solids, and can be written as

Total BOD_5 in effluent, mg/liter	=	Soluble BOD_5 in effluent S, mg/liter	+	BOD_5 due to mg VSS/liter in effluent	(12.34)

12.15 OXYGEN REQUIREMENTS

In aerobic biological treatment, oxygen is required for the various demands briefly listed below:

1. Oxidation of the fraction of organic carbon which is oxidized in order to provide energy for synthesis (see Figure 12.1).
2. Endogenous respiration of cells in the system.
3. Nitrification (if conditions are favorable).

12.15.1 Carbonaceous Demand

In waste treatment it is always the ultimate BOD of the organic matter that has to be satisfied.

$$\frac{BOD_u}{BOD_5} = \frac{1}{1 - 10^{-k5}} \tag{12.35}$$

If k = 0.1 per day at 20°C, as is commonly used in BOD bottle tests, $BOD_u/BOD_5 = 1/0.68 = 1.47$. Furthermore, if the BOD_u of microbial solids is taken as equal to its COD, then

$$C_5H_7O_2N + 5O_2 + H^+ = 5CO_2 + 2H_2O + NH_4^+ + energy$$
$$(113) \quad\quad (160)$$

or

$$\frac{kg\ O_2}{kg\ cells} = \frac{160}{113} = 1.42 \tag{12.36}$$

In suspended growth systems the oxygen required in biological treatment for meeting the carbonaceous demand must be equal to the BOD_u satisfied in the system. A part of the microbial solids produced is withdrawn from the system in the form of sludge, thus, only the ultimate oxygen demand of the remaining fraction must be taken into account as shown below:

$$\begin{array}{ccccc} O_2\ required, & = & BOD_u\ removed\ in & - & (1.42)net\ solids \\ kg/day & & system,\ kg/day & & withdrawn,\ kg/day \end{array}$$

At steady state, the net solids withdrawn per day (partly as surplus sludge and partly as loss of solids in effluent) must equal net solids produced per day. The latter can be estimated from Eq. (12.12) as XV/Θ_c. Furthermore, if the BOD concentrations are measured on a BOD_5 basis, they must be converted to BOD_u to give

$$O_2\ required,\ kg/day = Q(S_0 - S) - 1.42\left(\frac{XV}{\Theta_c}\right)$$

Substituting from Eq. (12.19), we get

$$O_2\ required,\ kg/day = Q(S_0 - S)\left(1 - \frac{1.42Y}{1 + K_d\Theta_c}\right) \tag{12.37}$$

in which Q = flow per day and S_0 and S are the initial and final BOD <u>ultimate</u> concentrations, respectively, to give kg/day, Y is the yield coefficient, K_d = endogenous decay rate constant, per day, and Θ_c = mean cell residence time, days.

The foregoing derivation can also be given in a slightly different manner since the net solids produced per day can be written as

$$Y(BOD_u \text{ removed/day}) - K_d XV$$

Thus,

$$\begin{aligned}
O_2 \text{ required} &= BOD_u \text{ removed} - 1.42\left[Y(BOD_u \text{ removed} - K_d XV) \right] \\
\text{per day} & \qquad \text{per day} \qquad\qquad\qquad\qquad \text{per day} \\
&= (1 - 1.42Y)BOD_u \text{ removed} + 1.42K_d XV \\
& \qquad\qquad\qquad\qquad \text{per day}
\end{aligned}$$

or

$$O_2 \text{ required} = a'(BOD_u \text{ removed}) + b'(XV) \qquad\qquad (12.38)$$
$$\text{per day} \qquad\qquad \text{per day}$$

in which

$$a' = 1 - 1.42Y \qquad \text{and } b' = 1.42K_d \qquad\qquad (12.39)$$

Thus, if the constants Y and K_d are known on a BOD_u basis, they can be readily used to find the values of a' and b'.

Typical oxygen requirements for treating domestic sewage in activated sludge and extended aeration are indicated in Table 14.1.

Example 12.2 Estimate the oxygen required per day for satisfying the carbonaceous demand if a wastewater is treated at (1) Θ_c = 4 days and (2) Θ_c = 20 days. Assume Y = 0.36 and K_d = 0.08 per day (BOD_u basis). State the result in terms of BOD_u and BOD_5 removed per day assuming BOD_u/BOD_5 = 1.47.

1. For Θ_c = 4 days

$$\begin{aligned}
O_2/\text{day} &= Q(S_0 - S)\left[1 - \frac{1.42 \times 0.36}{1 + (0.08 \times 4.0)} \right] \\
&= 0.61 \, (BOD_u \text{ removed}) \\
&= 0.9 \, (BOD_5 \text{ removed})
\end{aligned}$$

2. For Θ_c = 20 days

$$\begin{aligned}
O_2/\text{day} &= Q(S_0 - S)\left[1 - \frac{1.42 \times 0.36}{1 + (0.08 \times 20)} \right] \\
&= 0.8 \, (BOD_u \text{ removed}) \\
&= 1.18 \, (BOD_5 \text{ removed})
\end{aligned}$$

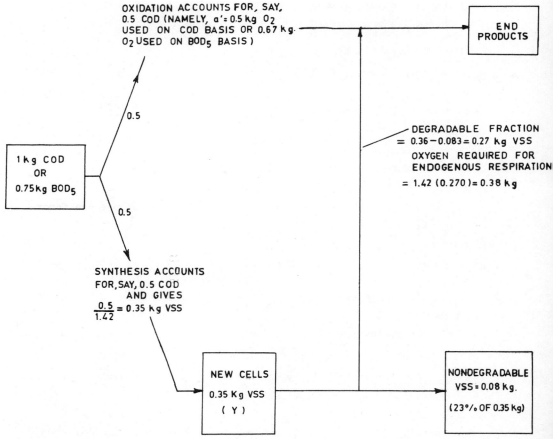

Figure 12.7 A typical materials balance for oxygen and organics. Aerobic removal of 1 kg COD gives 0.08 kg nondegradable residue of solids and the total O_2 required = 0.5 + 0.38 = 0.88 kg/kg COD removed. The following data are assumed: Y = 0.35 kg VSS/kg COD removed; nondegradable fraction = 23%; oxygen required = 1.42 kg O_2/kg VSS; a' = 1 - 1.42Y on COD or BOD_u basis.

Note: As Θ_c increases, the oxygen required also increases owing to a greater endogenous loss, until finally a nondegradable residue is left (see Figure 12.7).

12.15.2 Oxygen Demand for Nitrification

Section 12.20.1 gives the conditions favorable for the occurrence of nitrification in biological treatment, and the oxygen required for it.

Thus, the oxygen demand for nitrification may be computed from the total Kjeldahl nitrogen (TKN) oxidized as:

$$O_2 \text{ required/day} = 4.33(\text{TKN oxidized in treatment}) \qquad (12.40)$$

in which TKN oxidized = incoming TKN - TKN in effluent and excess sludge withdrawn.

The total oxygen demand of the treatment process is the sum of demands for nitrification and carbonaceous oxidation seen in the previous section. The carbonaceous demand is always satisfied first, and the nitrogenous demand afterward, provided conditions are favorable for nitrification and the aeration capacity of the unit is adequate. If the aeration capacity is not sufficient, nitrification will not occur even if other conditions are favorable. The nitrogen removed by way of excess sludge is reduced as the value of Θ_c increases.

Computation of the total oxygen demand for a plant is illustrated in Examples 14.1 and 14.2; some typical oxygen uptake rates for activated sludge and extended aeration plants are given in Table 14.1.

12.16 SLUDGE QUALITY AND RECYCLING

If it is desired to keep $\Theta_c > t$, sludge recycling is essential. The quantity of sludge which has to be recycled from the final settling tank back to the aeration tank depends upon how compactly the sludge is settled and how much solids concentration is desired in the aeration tank. Thus, the quantity depends first on its quality (settleability).

Sludge quality also has an effect on its further handling. The surplus sludge withdrawn from the system may need additional stabilization (as in the conventional activated sludge process) before dewatering and disposal. If the surplus sludge is already in a well-stabilized condition (as in the extended aeration process), further disposal is simplified.

From both stability and settleability points of view, the mean cell residence time Θ_c has a considerable effect. At low values of Θ_c (i.e., high F/M), bacterial growth tends to be dispersed rather than flocculated. This gives poor settling characteristics. If the aeration capacity is also not adequate and dissolved oxygen in the aeration tank is less than 0.5 mg per liter, filamentous organisms, which have a relatively high surface area per volume ratio, tend to outgrow the bacterial population and further impair the settling characteristics [25,27]. Poor settling is referred to as "sludge bulking."

As Θ_c increases beyond 2 to 3 days, and aeration is sufficient, the settling characteristics progressively improve since the biogrowth readily flocculates and settles. For this reason, F/M loadings are generally held within the range shown in Table 14.1. Values of Θ_c greater than 10 to 20 days are provided where it is desired that the sludge be sufficiently mineralized for dewatering without any further treatment on digesters.

In laboratory studies, Grau et al. [16] have shown that cells-in-series type completely mixed reactors produce a denser sludge than do single-celled completely mixed reactors. This is due to the plug flow characteristics of the former arrangement, with greater organic loading in the initial cells producing more flocculable bacteria. Excessively high loadings on the initial cells may, of course, reduce the benefits.

White et al. [37] have similarly shown a correlation between the dispersion number (D/UL) of the aeration tank and the settleability of the sludge. Better settleability was obtained at low dispersion numbers.

The settling properties of sludge are assessed in terms of the "sludge volume index" (SVI) which is a useful parameter in process control and is readily determined by any operator. The SVI is defined as the volume in ml occupied by 1 g of sludge solids after settling for 30 min. The sample is taken from the mixed liquor in the aeration tank and settled in a laboratory cylinder. It is a standardized way of estimating the solids concentration likely to be present in the settled sludge recycled from the settling tank since

$$\frac{10^6}{SVI} = \text{solids concentration, mg/liter, in settled sludge} \qquad (12.41)$$
$$\text{in laboratory cylinder}$$

A bulking or poorly settling sludge will show a higher value of SVI since its solids concentration will be low and it will occupy a greater volume in the test cylinder. Typical characteristics are

SVI	Corresponding solids concentration in settled sludge (mg/liter)	Settling characteristics
50	20,000	Excellent
50-100	20,000-10,000	Good
100-150	10,000-6,700	Satisfactory
150	6,700	Increasing "bulking" of sludge

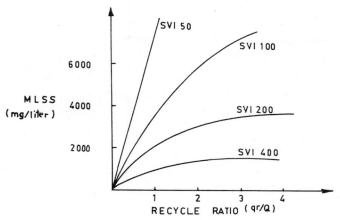

Figure 12.8 The relationship between SVI, MLSS, and the recycle ratio.

The actual solids concentration in the return flows often averages around 8,000 to 12,000 mg/liter. Referring to Figure 12.2, a materials balance for solids entering and leaving the reactor gives

$$(Q + q_r)MLSS = q_r \left(\frac{10^6}{SVI} \right)$$

or

$$MLSS = \frac{10^6 (q_r/Q)}{SVI[1 + (q_r/Q)]} \qquad (12.42)$$

where q_r/Q = recycle ratio R.

Hence, a relationship exists between SVI, MLSS, and the recycle ratio q_r/Q. This is shown graphically in Figure 12.8. In practice, certain recycling ratios are essential if the desired MLSS concentrations are to be maintained in the aeration tank. The lower the value of SVI, the lower is the required recycling ratio for a given MLSS concentration. If sludge bulking increases SVI, the recycling ratio must be correspondingly increased to maintain the same MLSS concentration. Here, the operator may find that recycling cannot be increased endlessly as it is limited by pumping and other plant capacities. The recycling ratio can also be estimated for completely mixed reactors from

$$R = \frac{MLSS}{SS \text{ in return flow} - MLSS} \qquad (12.43)$$

or

$$R = \frac{MLVSS}{VSS \text{ in return flow} - MLVSS} \qquad (12.44)$$

Recycling ratios generally range from 0.25 to 1.0 in order to maintain the MLSS concentration in aeration tanks, which is generally limited to 2000 to 2500 mg/liter in plug flow type tanks and 4000 to 5000 mg/liter in completely mixed tanks. The upper limit of MLSS is generally tied in with two other aspects of plant design: the final settling tank capacity for settling an inflow with high solids concentration and the oxygenation capacity of the aeration system in the presence of a very high solids concentration. These aspects are discussed further in Chapters 13 and 17. In terms of the discussion in this chapter, the higher the MLSS (and, therefore, MLVSS) concentration in a system, the higher is the substrate removal efficiency.

While sludge settleability is important with all biological processes involving settling and recycling, the extent of sludge stabilization (mineralization) is important in dewatering and disposal. This question has been discussed in Chapter 17. Raw sludges characteristically show higher values of specific resistance associated with poor drainage characteristics. Well-treated sludges show relatively low values of specific resistance.

12.17 IMPORTANCE OF "Θ_c" AND "t" IN RELATION TO SHOCK LOADS

Shock loads disturb steady-state conditions in reactors. These shock loads may be either qualitative or quantitative in nature and affect treatment performance during the transient state.

As seen earlier in Section 11.15, quantitative shock loads (e.g., toxic substances in influent) are better withstood by complete-mixing reactors, in which the incoming load is rapidly and uniformly distributed and, thus, diluted. It is obvious, therefore, that, in complete mixing systems, the larger the value of the hydraulic detention time t, the better the ability to buffer shock loads. Where t is reduced much for reasons of economy, the process is increasingly vulnerable to disturbances during the transient state.

The mean cell resistance time Θ_c appears to play a more dominant role with regard to quantitative shock loads. Modeling studies have shown that the mass of substrate escaping during a transient period is highly dependent upon the steady-state Θ_c, but almost completely independent of t [28]. Increasing t does not much increase the buffering capacity of the system against quantitative shock loads.

12.18 NUTRIENT REQUIREMENTS

The question of nitrogen and phosphorus utilization must be considered from two viewpoints: (1) minimal requirements, and (2) the extent of removal in biological treatment.

To ensure maximum removal of carbon (and, therefore, BOD) from the substrate, carbon must be made the limiting nutrient, while N and P (and other nutrients) must be present in at least the minimum required proportion for cell growth.

Domestic sewage contains more than a sufficient amount of N and P, compared with carbon, but some industrial wastes are deficient in them (e.g., cotton textiles and pulp and paper), and the artificial addition of nutrients may be essential for biological treatment at optimum rates.

New cells synthesized from the degradation of organic matter have been shown to contain, at maximum level, 12.3% N and 2.6% P on a dry weight basis (C:N:P = 106:16:1). After endogenous respiration, the nondegradable residue contains only 7% N and 1% P; the rest is released back into the substrate [15]. Nutrients contained in the surplus sludge withdrawn from the system, and in the microorganisms leaving with the effluent, are, of course, unavailable for recycling. Thus, depending upon the age of the cells (Θ_c) in the system, the N and P requirements vary. The higher the value of Θ_c, the greater is the recycling and, hence, the less the nutrient requirements. Table 12.3 indicates some typical values.

Theoretically, the nutrient requirements can be approximated from the above stated percentages in cells, assuming that only 77% of the cells produced are biodegradable [15]. Thus, the requirements are

$$N, \text{ kg/day} = \frac{0.123 f_b \, \Delta X}{0.77} + \frac{0.07(0.77 - f_b) \, \Delta X}{0.77} \qquad (12.45)$$

Table 12.3 Minimal Nutrient Requirements

Process	Θ_c (days)	BOD_5 removed	Proportional nutrient requirements	
			N	P
Activated sludge	3-10	100	5.0	1.0
Extended aeration	20-40	100	0.825	0.165

$$P, \text{ kg/day} = \frac{0.026f_b \, \Delta X}{0.77} + \frac{0.01(0.77 - f_b) \, \Delta X}{0.77} \qquad (12.46)$$

where

ΔX = net microbial solids produced in treatment, kg/day

f_b = biodegradable fraction of MLVSS

Nutrient requirements in the conventional activated sludge process are higher than those of the extended aeration process since, in the latter, there is more recycling and less sludge withdrawal. In facultative aerated lagoons and oxidation ponds, no sludge is removed continually from the unit and some N and P are released back from the accumulated sludge undergoing anaerobic decomposition in the lower layers. Some nitrogen loss resulting from denitrification may occur in such units. Some phosphorus precipitation may also occur in the pond bottom at sufficiently high pH values, and some N and P may also be lost with the bacterial and algal solids escaping out with the effluent. Thus, nutrient requirements are affected by the nature of the treatment process.

12.19 PHOSPHORUS REMOVAL

Phosphorus removal in biological treatment is due mainly to the removal of surplus microbial solids from the system. At steady state, the surplus microbial solids removed equal those produced per unit time. Thus, it is necessary to estimate the net volatile solids produced, and to know the phosphorus content of the solids (which depends on cell age as seen in Sec. 12.18). The method of computation is illustrated in Example 12.3. The ratio of C:P in the waste also affects the removal efficiency since carbon is generally the limiting nutrient.

Thus, it is observed that phosphorus removal from a given wastewater is also a function of the mean cell residence time Θ_c. As the value of Θ_c increases, the phosphorus removal rate decreases, since less surplus sludge is removed from such a system. Thus, extended aeration should give less phosphorus removal than conventional activated sludge.

Conversely, when treating phosphorus-deficient wastewaters, the extended aeration process should require less artificial phosphorus addition than the activated sludge process.

Example 12.3 A wastewater discharge of 2000 m^3/day with BOD_u = 400 mg/liter shows a BOD removal efficiency of 90% in activated sludge treatment at Θ_c = 5 days. Estimate the phosphorus removal; assume the net solids production is 0.28 kg per kg BOD_u removed.

BOD_u incoming to aeration = 400 x 2000 x $10^3/10^6$ = 800 kg/day

Net solids production rate = 0.28(0.9 x 800) = 202 kg/day

Phosphorus removal at 2.6% content = 5.26 kg/day

Thus,

$$\frac{Phosphorus\ removed}{BOD_u\ removed} = \frac{5.26}{720} = 0.0073\ (or\ 0.73\%)$$

or

$$\frac{Phosphorus\ removed}{BOD_5\ removed} = 0.73\%\ x\ 1.47 = 1.0\%$$

The same result can be obtained using Eq. (12.46) with f_b = 0.77.

Note: If the waste was deficient in phosphorus, artificial addition would be necessary at the same rate, namely, approximately 1% or 1 kg P/100 kg BOD_5 removed (see Table 12.3).

12.20 NITROGEN REMOVAL (NITRIFICATION-DENITRIFICATION)

The per capita contribution of nitrogen in sewage generally ranges between 6 and 12 g/day. Raw sewage generally has about 60% of this nitrogen as free ammonia (NH_3-N) and about 40% as organic nitrogen (proteins, amino acids, urea, etc.). The latter are also decomposed by bacteria to form ammonia; thus, one should consider the organic nitrogen as well as the ammonia nitrogen in sewage, namely the total Kjeldahl nitrogen (TKN). Depending on the extent of water used in a community, raw sewages often show TKN values ranging from 30 to 50 mg/liter.

The ammonia may be biologically oxidized to nitrate nitrogen; this oxidation is called nitrification. Under certain conditions, the nitrate formed may be reduced biologically to give nitrogen gas; this is called denitrification. In some treatment processes, only nitrification may occur, whereas in others the nitrification may be followed by denitrification.

Some nitrogen removal (20-30%) also occurs as a result of the removal of surplus sludge from a system. Thus, as seen in the previous section concerning phosphorus removal, there will be a greater nitrogen removal when Θ_c is low (activated sludge) than when Θ_c is high. But at a high Θ_c, nitrification will be complete. If this nitrification is followed by denitrification, quite high removals of nitrogen (70-90%) can be achieved.

12.20.1 Nitrification

Nitrification of ammonia is a two-step process involving two different types of nitrifying microorganisms:

$$2NH_3 \longrightarrow 2NH_4^+ \text{ in water} \tag{12.47a}$$

$$2NH_4^+ + 3O_2 \xrightarrow{\text{Nitrosomonas}} 2NO_2^- + 2H_2O + 4H^- \tag{12.47b}$$

$$2NO_2^- + O_2 \xrightarrow{\text{Nitrobacter}} 2NO_3^- \tag{12.47c}$$

Thus, stoichiometrically, 4.6 kg of oxygen are required for every kg of nitrogen. Experimentally determined values are less (3.9 to 4.33) because some nitrogen is used in synthesis of bacterial cells and some oxygen is obtained from carbon dioxide and bicarbonates present in the water [15, 17].

The nitrifying organisms are strict, aerobic autotrophs. They use ammonia oxidation as the energy-yielding metabolic reaction. They are sensitive to their environment, and even under favorable conditions they are slower growing than the usual BOD-removing heterotrophs.

Nitrification can be achieved in suspended growth systems as well as in fixed-film reactors. The various factors affecting nitrification have been summarized by Wuhrmann [4], Eckenfelder [8], Downing [18], and others [31,34]. They are discussed briefly below.

In terms of Monod kinetics, the <u>growth</u> rate of nitrifiers can be described by

$$\mu = \mu_{max}\left[\frac{(NH_3\text{-}N)}{K_{s(amm)} + (NH_3\text{-}N)} \right] \tag{12.48}$$

where

μ and μ_{max} = the specific and maximum growth rates per unit time, respectively

$K_{s(amm)}$ = saturation constant, mg/liter for ammonia

NH_3N = ammonia concentration expressed as nitrogen, mg/liter

Typical values of $K_{s(amm)}$ range from 0.5 to 1.0 mg NH_3-N/liter, while μ_{max} at 20°C approximates 0.5/day for <u>Nitrosomonas</u>, and 0.8/day for <u>Nitrobacter</u> (see Table 12.4). Thus, the two-step nitrification process is controlled by the slower growing of the two microorganisms, namely, the <u>Nitrosomonas</u>. (By way of contrast, the growth rate of heterotrophs is about 50/day.)

Ammonia concentration in the reactor also affects the growth rate μ. At concentrations around 5 mg NH_3-N/liter or more, it is seen from Eq. (12.48) that μ is practically equal to μ_{max}, and the reaction proceeds like a zero-order one. In fact, the rate remains practically zero order until the ammonia concentration in the reactor reduces to about 1.5 to 2 mg/liter, below which the nitrification rate is reported to drop off rapidly [32]. Thus, 100% nitrification is difficult to achieve in practice.

The growth rate of nitrifiers is further affected by temperature, oxygen concentration in the reactor, its pH, and presence of toxic substances, etc.

Temperature effects on growth rate (long-term, not shock effects) are estimated from

$$\mu_{max(t)} = \mu_{max(20°C)}\theta^{(t-20)} \tag{12.49}$$

in which θ varies from 1.08 to 1.12 between the temperature range 5 to 20°C. The value of 1.12 applies to suspended growth systems where biological oxidation and nitrification are combined in one reactor. The value of 1.08 applies to attached growth systems with only nitrification. Consequently, temperature has a considerable effect on growth rate (see Figure 12.11d).

Nitrification has been reported to occur between 5 and 45°C, but the optimum temperature lies somewhere between 25 and 32°C.

The optimum pH range for nitrification is 7.8 to 9.2 for <u>Nitrosomonas</u> and 8.5 to 9.2 for <u>Nitrobacter</u>.

The effect of oxygen concentration in the reactor can also be expressed in the form of a Monod equation [Eq. (12.48)] in which oxygen concentrations are substituted for ammonia. The saturation concentration K_s for oxygen is around 1 mg O_2/liter. Thus, the oxygen concentration in the reactor should be 3 to 4 mg/liter to avoid oxygen limitation. Nitrification occurs even at lower oxygen levels, but the growth rate of nitrifiers is slower.

Table 12.4 Summary of Parameters for Nitrification-Denitrification Systems

Parameter	Nitrification		Denitrification	
	Nitrosomonas	Nitrobacter	Methanol	Internal carbon
μ_{max}(20°C), per day	0.5	0.8	0.5	0.2
K_s, mg N/liter	0.5	1.0	<1.0 (attached growth)	-
K_s, mg O_2/liter	1.0	1.0	-	-
Yield constant Y, g/g	0.08	0.03	0.6-0.8	-
Reaction kinetics				
Suspended growths	Monod	Monod	Zero order	Zero order
Attached growths	Monod	Monod	Monod	-
Optimum pH	7.8-9.2	8.5-9.2	7.5-9.2	7.5-9.2
Nitrification rates, mg NH_3-N/g VSS/hr				
Suspended, separate systems, 20°C	9-13			
Suspended, combined systems, 20°C	2-5			
Denitrification rates, mg NO_3-N/g VSS/hr				
Suspended growths, 20°C	-	-	10	{0.4 (endogenous) / 3.0 (raw sewage)}
Temperature correction θ and range				
Attached, separate	1.08 (5-25°C)		1.05 (5-20°C)	
Suspended, separate	-		1.12 (10-25°C)	
Suspended, combined	1.12 (5-20°C)		-	{1.15 (5-20°C) raw sewage / 1.20 (15-25°C) endogenous}

632

The presence of toxic substances even in small concentrations can inhibit nitrifiers. Among these may be listed organic-sulfur compounds, phenols, cyanides (each capable of giving 75% inhibition at 10^{-5} to 10^{-6} moles), chromium, nickel, and zinc (at about 0.25 mg/liter each). The observation that nitrification generally occurs after satisfaction of the carbonaceous demand has been explained by the "toxic" effect of carbon on nitrification, but is perhaps due more to the much slower growth rate of nitrifiers compared to the heterotrophs.

Sludge age (or mean cell residence time) Θ_c in a treatment plant must be sufficiently high to take into account the actual growth rate of nitrifiers at field conditions. Unless the cell residence time is somewhat greater than the cell doubling time, cell washout would occur. Thus,

$$\Theta_c > \frac{1}{\mu} \tag{12.13b}$$

where μ is the growth rate for Nitrosomonas at the worst operating temperature and other conditions. For industrial wastes, laboratory studies may be needed. For domestic sewage without any specially inhibiting factors, the following values may be useful to consider:

Mixed liquor temperature (°C)	Θ_c for complete nitrification (days)
5	12
10	9.5
15	6.5
20	3.5

Activated sludge plants in cold climates designed for Θ_c less than 10 days may show relatively poorer nitrification in winter. Extended aeration plants generally designed for Θ_c greater than 10 days may show nitrification uniformly over all seasons. Even activated sludge plants in warm climates may show nitrification uniformly, provided sufficient aeration capacity is available to meet the total oxygen demand and maintain required high dissolved oxygen levels in mixed liquors.

Contact time between incoming ammonia and the biomass in the mixed liquor must also be sufficient, besides appropriate sludge age, for successful nitrification.

The rate at which ammonia is converted to nitrates is a function of the mass of nitrifiers in the system and the ammonia removal rate per unit mass of nitrifiers and time. McCarty [20] has shown this rate to be about 100 mg NH_3-N/g NVSS-hr.

The mass of nitrifiers (NVSS) in a system can be estimated from the yield coefficient of 0.08 g/g NH_3-N removed (see Table 12.4), while the mass of usual heterotrophs can be estimated from 0.5 g/g BOD removed. Thus, the percentage of NVSS in total VSS in mixed liquors is quite small and varies from one plant to another. Consequently, nitrification rates also vary. The ratio NVSS/VSS varies from about 1 to 2% for systems where bio-oxidation and nitrification are combined in one reactor to about 5 to 10% for systems with separate reactors for biooxidation and nitrification and separate sludge recycle arrangements. Thus, actual nitrification rates in treatment plants range as follows until, as stated, low ammonia concentrations are reached in the reactors.

System	Nitrification rates (20°C) (mg NH_3-N/g VSS-hr)
Suspended, separate	9-13
Suspended, combined	2-5

In some cases, if influent ammonia concentration is high, it may become necessary to increase the reactor detention time even though the sludge age may be adequate.

Nitrification rates in fixed-film reactors have been reported to vary from 0.07 to 0.15 kg NH_3-N/m^3-day for air and from 0.12 to 0.42 kg NH_3-N per m^3-day for pure oxygen. However, experimental data from fixed-film reactors have been shown to fit better a Michaelis-Menten type of equation such as

$$V = V_{max}\left[\frac{NH_3\text{-}N}{K_{(amm)} + NH_3\text{-}N}\right] \tag{12.50}$$

where

V and V_{max} = the specific and maximum nitrification rates, respectively, expressed as mass of ammonia-nitrogen nitrified per unit time and surface area of media

$K_{(amm)}$ = the saturation constant

NH_3-N = ammonia concentration in inflow

Practical methods of achieving nitrification are discussed in Sec. 14.6.

Example   12.4   If a complete mixing reactor with sludge recycle is operated with Θ_c = 9 days, estimate the ammonia nitrogen concentration in the effluent assuming μ_{max} = 0.285/day at reactor temperature and K_s = 0.4 mg/liter as NH_3-N.

$$\mu = \frac{1}{\theta} = \mu_{max}\left[\frac{NH_3\text{-}N}{K_s + NH_3\text{-}N}\right]$$

or

$$\frac{1}{9} = 0.285/\text{day}\left[\frac{NH_3\text{-}N}{0.4 \text{ mg/liter} + NH_3\text{-}N}\right]$$

Hence, NH_3-N = 0.25 mg/liter.

Example   12.5   From the data given in Example 12.1, estimate whether complete nitrification is likely to occur at a mixed liquor temperature of 10°C. Influent NH_3-N = 30 mg/liter.

1. From Example 12.1, Θ_c = 8.36 days. From Sec. 12.20.1, the sludge age needs to be increased to 9.5 days at least.

2. Also, average inflow = 400 m³/day = 16.67 m³/hr, or average NH_3-N inflow = (16.67 m³/hr)(30 mg/liter)(10^3) = 0.5 kg/hr. Assume nitrification rate = 3 mg NH_3-N/g VSS-hr (combined system). From Example 12.1, VSS = 2100 kg, or nitrification capacity = (2100 x 10^3)(3) = 6.3 kg/hr. Hence, contact time is ample, but sludge age is not enough for 10°C.

12.20.2 Denitrification

Nitrates formed earlier may be reduced biochemically to give nitrogen gas which is released to the atmosphere, thus removing nitrogen from the wastewater. But this mode of removal requires that nitrification must precede denitrification since ammonia cannot be denitrified directly.

Denitrification involves microorganisms like _Pseudomonas_, _Micrococcus_, _Acinetobacter_, etc., all of which are facultative in character and abundant in sewage. They work over a wider range of environmental conditions than do the nitrifiers, but need _anoxic_* conditions during which they reduce

---

*_Anoxic_ conditions occur when dissolved oxygen is exhausted and microorganisms draw their oxygen requirement by the breakdown (reduction) of nitrates. _Anaerobic_ conditions begin when nitrates are also exhausted and sulfide liberation begins.

nitrates to obtain their oxygen. Nitrates serve as hydrogen acceptors in
the respiration of heterotrophs in the absence of dissolved oxygen.

Anoxic conditions refer to the immediate microenvironment of the bacteria (namely, within the biofloc or film) where redox potentials may be
about -200 mV, although dissolved oxygen may be present in the bulk liquid.

A carbon source is also essential for denitrification. This may be in
the form of carbon internally available in sewage (either in the raw sewage
or from endogenous respiration) or artificially added in the form of methanol (CH_3OH) or as any carbon-rich, nitrogen-deficient industrial wastewater
(e.g., brewery, molasses, nitrocellulose, and other chemical industry or
food industry waste). Factors in choice include costs, availability, and
reaction rates.

<u>With methanol</u> as electron donor, McCarty et al. [20] give

$$1.08CH_3OH + NO_3^- + H^+ \xrightarrow{\text{Denitrific}} 0.065C_5H_7O_2N + 0.47N_2 + 2.44H_2O + 0.76CO_2$$

If the wastewater has nitrates (nitrites and some dissolved oxygen
also), the methanol requirement is

$$\text{Methanol, mg/liter} = 2.47(NO_3\text{-N, mg/liter}) + 1.53(NO_2\text{-N, mg/liter}) + 0.87(DO\text{, mg/liter}) \qquad (12.51a)$$

<u>With internally available carbon</u> as electron donor,

$$C_5H_7NO_2 + 4NO_3^- \xrightarrow{\text{Denitrific}} 5CO_2 + 2N_2 + NH_3 + 4OH^- \qquad (12.51b)$$

Including assimilation, the consumption is about 3 mg $C_5H_7NO_2$/mg NO_3-N
or about 4.5 mg BOD_5/mg NO_3-N.

Since most community wastewaters have a higher ratio of BOD to nitrogen
than the one just mentioned, the use of internally available carbon as electron donor becomes an attractive and economical method for bringing about
denitrification.

Denitrification can be achieved in suspended growth systems as well as
in fixed-film reactors. But, an added advantage in the case of suspended
growth systems is that the oxygen released from breakdown of nitrates can
be made readily available for biological oxidation of organic matter in
the same suspension.

Oxygen release by breakdown of nitrates can be as follows:

$$2NO_3^- + 2H^+ \longrightarrow N_2 + 2.5O_2 + H_2O \qquad (12.52)$$

Thus, each kg of N_2 released is accompanied by 2.86 kg of O_2. From Eq. (12.47) we saw that each molecule of N_2 needs 4 molecules of O_2; now we see that denitrification releases back 2.5 molecules. Thus, theoretically, 62.5% (2.5/4.0) of the oxygen used is released back in denitrification. Advantage is taken of this in plant design, as discussed in Chapter 14.

Among the various factors which affect denitrification, the following may be briefly examined:

1. In terms of Monod kinetics the maximum, specific growth rate of denitrifiers, $\mu_{max(20°C)}$ = 0.2/day for internal carbon and 0.5/day for methanol (Table 12.4). Temperature effect on growth rate is quite substantial and is accounted for by the Θ_c values.

2. The effect of pH on denitrification needs further study, but the optimum pH range is believed to be 7.5 to 9.2.

3. The effects of toxic substances also need further study, but are believed to be similar to those affecting BOD removal because many microorganisms are facultative and responsible for BOD removal and nitrification.

4. Contact time required to bring about denitrification is affected by the type of carbon source. The denitrification rate is directly proportional to the demand for oxygen to meet the respiration rate of the heterotrophs. Consequently, it is fastest when methanol is used, less so with raw sewage, and least with endogenous respiration (see Table 12.4). The rates are generally zero order but drop off a great deal after about 2 mg/liter NO_3-N is left in the reactor [32]. Consequently, 100% denitrification is difficult.

Figure 12.9 gives a relation between the denitrification rate and the oxygen uptake rate owing to respiration. This relationship can be used

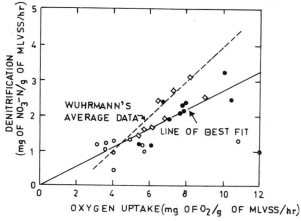

Figure 12.9 The correlation between the rates of denitrification and oxygen uptake (adapted from Ref. 8).

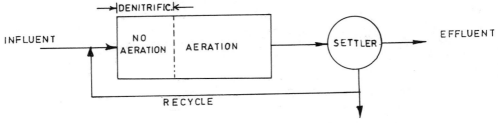

Figure 12.10 Typical flowsheets for denitrification in an activated sludge process.

for the design of denitrification basins as illustrated in Example 14.4. Proper choice of contact time and denitrification rate is also discussed further in Sec. 14.6.

Denitrification rates in fixed-film reactors have been studied mainly with methanol usage. The ratio of methanol/NO_3-N has to be somewhat larger than 3, although stoichiometrically 2.5 should be enough. The process is mostly zero order, but tends to become half order with thick biofilms, owing to diffusional resistance [33].

Practical methods of achieving denitrification in activated sludge systems (Figure 12.10) are discussed in Sec. 14.7. Nitrogen removal in aerated lagoons and waste stabilization ponds are discussed elsewhere.

12.21   CHOICE BETWEEN PROCESS LOADING PARAMETERS F/M AND Θ_c

In the activated sludge system and its modifications, process loading has often been stated in the past as BOD load per unit time per unit volume of aeration tank. This did not take into account solids concentration and, therefore, came to be replaced by the F/M ratio which gave the BOD (substrate) removal rate per unit microbial solids per unit time.

However, as stated in Sec. 12.7, some confusion has occurred owing to the different modes of expressing F/M. Furthermore, as Lawrence and McCarty

[22] point out, all volatile solids in the system are not active microbial solids. Some are just nondegradable organic residues accumulating in the system because of recycling. The active microbial solids may be only 20 to 50% of the total volatile solids in some cases. Thus, the parameter Θ_c is recommended for use as it does not suffer from these difficulties.

The mean cell residence time Θ_c is only a ratio of the solids in any system to the solids leaving it. As long as the solids are assumed to be homogeneous, it does not matter what fraction is active and what is inert. The ratio will be the same whether applied to MLSS, MLVSS, or active MLVSS. Hence, Θ_c lends itself better to process design and control. The value of Θ_c is related to that of F/M by Eq. (12.17).

It is not possible to use the concept of Θ_c with facultative-type aerated lagoons and oxidation ponds since the solids in the system are not determinable, nor are they homogeneous as settlement occurs in facultative-type units. Their process loadings are, therefore, simply given on an areal, volumetric, or time basis. With other types of aerated lagoons, the concept of Θ_c can be used. With trickling filters and rotating discs, a modified form of the Θ_c concept can be used since solids in the system are proportional to the surface areas available for biogrowths.

12.22   EFFECT OF Θ_c ON PERFORMANCE

The factors involved in choosing an appropriate value of Θ_c when designing the activated sludge process are summarized below in brief and are discussed with Figure 12.11 showing the effect of Θ_c on performance. The curves shown are merely indicative of the general trend. Their exact delineation would depend on the relative values of the kinetic constants for the particular wastewater.

Activated sludge plants are generally designed to keep Θ_c values between 4 and 10 days, whereas for extended aeration plants the values range from 10 to 20 days or even more. Thus, the variation of Θ_c is over a wide range, and some rational basis is provided through the following discussion to facilitate an appropriate choice of Θ_c values in design. Illustrative examples are given in the next chapter.

The succeeding chapters follow up on the concepts stated in the captions to Figures 12.11(a) through 12.11(l), which show the effect of Θ_c on performance.

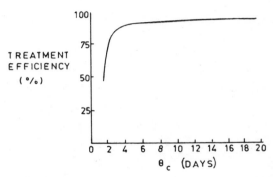

Figure 12.11 (a) Shows how efficiency increases rapidly once Θ_c is greater than the cell washout time. There is some, but not much, improvement in efficiency thereafter. The cell washout time depends on the specific growth rate at the critical (winter) operating temperature, and fixes the minimum value of Θ_c for successful operation.

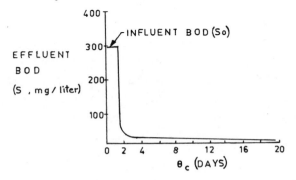

Figure 12.11 (b) Also shows that the effluent quality is improved only after Θ_c exceeds cell washout time. Thereafter, the effluent soluble BOD is not much improved. The BOD of any volatile solids carried over in the effluent will, of course, decrease as Θ_c increases. Thus, as shown in the figure, the total effluent BOD reduces as Θ_c increases.

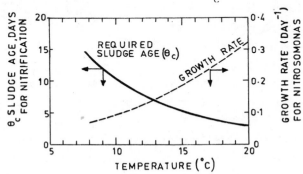

Figure 12.11 (c) Shows the required value of Θ_c to ensure nitrification in activated sludge plants. Where nitrification is desired, the Θ_c value should equal or exceed the required value at the winter operating temperatures [8,18]. Nitrification is more likely to occur at all times of the year in extended aeration plants.

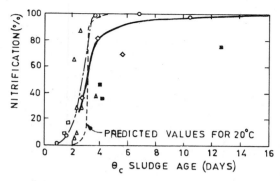

Figure 12.11 (d) Shows similar data on nitrification. More or less complete nitrification can be expected beyond Θ_c values of about 3 days at 20° C. Data from various sources have been pooled to give the expected percent nitrification [8]. Nitrogen removal as such takes place when denitrification occurs. Also, some nitrogen is withdrawn from the system in the form of excess sludge. The latter reduces as Θ_c increases.

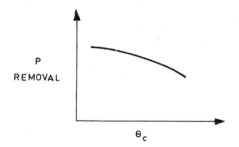

Figure 12.11 (e) Gives the phosphorus removal efficiency at different Θ_c values. The removal efficiency decreases as Θ_c increases for a given ratio of C:P in the waste. The phosphorus requirement in the case of nutrient deficient wastes reduces as Θ_c increases.

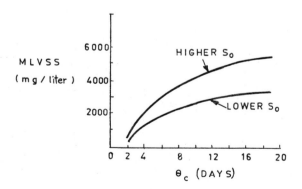

Figure 12.11 (f) Shows that the MLVSS concentration X in the aeration tank is considerably affected by the initial substrate concentration S_0 at any given value of Θ_c [Eq. (12.18)]. As long as Θ_c remains unchanged, X increases as S_0 increases. This keeps the F/M ratio constant and the efficiency of removal remains unaffected.

641

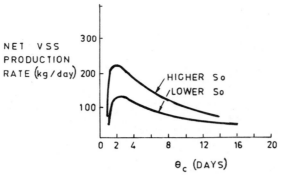

Figure 12.11 (g) Shows the variation in the net solids production rate at different Θ_c values. The rate is maximum at about 2 days, as not much substrate removal occurs, whereas at more than 2 days there is progressively more time for endogenous decay [23].

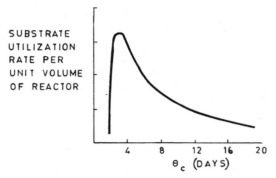

Figure 12.11 (h) Shows that the substrate utilization rate per unit volume of reactor varies with Θ_c in much the same manner as the VSS rate in (g). Recycling helps to build up the solids concentration X in the reactor, thus increasing the overall substrate removal rate K_a, since $K_a = k'X$ [Eq. (12.28)]. However, as the value of Θ_c increases, there is more endogenous decay and the "active" fraction of X tends to decrease [26].

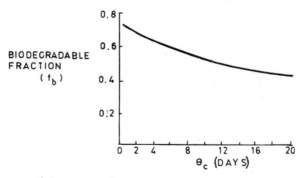

Figure 12.11 (i) Shows how the degradable fraction in the MLVSS decreases as Θ_c increases. Even in freshly produced cells ($\Theta_c = 0$), the biodegradable fraction is only 0.77 of the cell mass, and decreases with age [Eq. (12.33)] as the organic but inert fraction accumulates. Thus, all MLVSS are not "active" microorganisms.

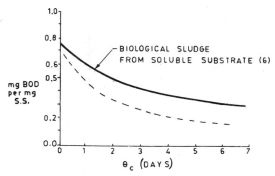

Figure 12.11 (j) Shows how well stabilized the solids are at different Θ_c values. As Θ_c increases, the biodegradable fraction decreases, as seen in Figure 12.11(b), and the BOD of the suspended solids diminishes. In systems where inflow has inert suspended solids, the latter also tend to accumulate because of recycling, thus giving a reduced BOD per unit weight of solids at a given value of Θ_c (dotted curve).

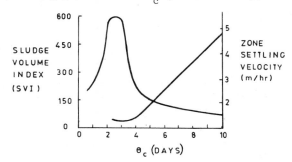

Figure 12.11 (k) Shows that sludge settleability (SVI) is unsatisfactory at Θ_c values of about 2 to 3 days, but improves thereafter as the sludge solids are more mineralized. Temperature and nature of microbial growths also affect their settleability. Good sludge settleability is an important criterion in the choice of a Θ_c value for the design of plants.

This figure also shows that the zone settling velocity of the sludge steadily improves beyond Θ_c of 4 days.

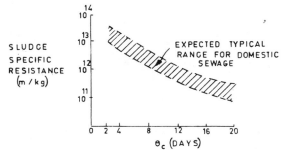

Figure 12.11 (1) Gives an indication of the effect of Θ_c on the specific resistance of the sludge to draining. As Θ_c increases, the drainage characteristics of sludge improve (see Chapter 17). Specific resistance is an additional parameter affecting choice of a Θ_c value in the case of extended aeration and other systems where the surplus sludge is dewatered directly without any digestion.

12.23 CHARACTERISTICS OF ATTACHED GROWTH SYSTEMS

The basic principles of biological treatment enunciated previously
for suspended growth systems are equally applicable to attached growths in
what are called fixed-film reactors. But the design of fixed-film reactors
must take into account several additional factors such as the following:

1. Diffusion of the substrate and nutrients into the biofilm, and of the
    end products out of the biofilm

2. Intrinsic reaction rates in biofilms

3. Growth and sloughing of biofilms, and their effects on diffusion and
    penetration of substances into the biofilm and on reaction rates

4. Hydrodynamic properties of the reactor in terms of effective area
    available for attached growths, contact time in passage of flow through
    the reactor, etc.

These factors affect all fixed-film reactors, be they trickling filters,
rotating discs, submerged, or fluidized media beds.

Processes occurring in fixed films are said to be reaction-rate limited
if the substrate and nutrients have diffused fully through the biofilm, and
bacterial kinetics control the process. This is the maximum rate of reaction and akin to that seen in suspended growth systems. But, when penetration of any substrate or nutrient into the fixed film is only partial, the
process may be said·to be diffusion-rate limited.

The combined effect of diffusion and reaction on removal rates in
biofilms can be expressed by

$$\frac{\delta^2 S_f}{\delta x^2} = \frac{K_{f(zero)}}{D} \left(\frac{S_f}{K_s + S_f} \right) \tag{12.53}$$

where

$\quad S_f \qquad$ = concentration of limiting substrate in biofilm

$\quad x \qquad\quad$ = any distance in biofilm from interface with liquid

$\quad K_{f(zero)}$ = zero-order reaction rate constant, in film

$\quad D \qquad\quad$ = diffusion coefficient

$\quad K_s \qquad$ = Monod's saturation constant

Equation (12.53) is cumbersome but can be solved on a computer. But,
within the practical range of operation, it has been found that similar
results can be obtained by using a much simpler half-order approximation
of the general type shown below [33-35].

$$V_a = K_{a(half)}(S^{1/2})$$ (12.54)

where

V_a = removal rate, per unit area and time

$K_{a(half)}$ = half-order reaction rate constant, on areal basis

S = bulk concentration of substrate at interface

Kornegay and Andrews [35], experimenting with glucose oxidation in biofilms, showed that the removal rate V_a increases up to a maximum value as the biofilm thickens up to about 65 μ (since the relatively thin film is fully penetrated), but as thickness increases further the removal rate remains constant (since diffusion-rate limitation occurs and the thicker film is not penetrated farther).

It is this diffusion-rate limitation in biofilms that gives rise to a half-order situation. Most experiments have been done with well-defined substrates such as glucose [35], nitrates [33], etc. with low values of K_s and removal rates shown to be zero order in terms of biofilm concentrations, which become half order in terms of bulk concentrations. Harremoes [34] presents evidence from the operation of rotating discs to show this might also be the case with complex substrates like sewage in which BOD removal in fixed films might be in general a zero-order process which behaves like a half-order one owing to diffusion limitation.

For this reason, treatment kinetics of fixed-film reactors are often studied by determining the removal rates V_a at different bulk concentrations S as just stated. If observed over a wide enough range of S values, one may obtain curves of the type shown in Figure 12.16. These curves resemble those of Monod but are not quite the same. They can be fitted by a general equation of the type

$$V_a = V_{a(max)} \left[\frac{S}{K_m + S} \right]$$ (12.55)

An experimental method for determining the kinetic constants in this equation has been illustrated in Sec. 12.4. It should be noted that the substrate concentration K_m, at which $V_a = 0.5\ V_{a(max)}$, has a numerically different value from that of K_s obtained from Monod's growth kinetics.

Attempts have been made to apply Monod kinetics to fixed-film reactors but difficulties have been experienced in determining the mass of attached growths and distinguishing between their active and inert fractions.

Diffusion phenomena in fixed films can also help explain how denitrification can occur in them. A biofilm may not be aerobic to its full thickness. Only its outermost zone may be aerobic. The oxygen diffusion gradient from the liquid-biofilm interface may diminish rapidly as the substrate is oxidized. Farther inward, there may now be an anoxic zone in the biofilm in which substrate oxidation may result from the breakdown of nitrates, which diffuse into this zone along with some substrate, thus giving denitrification. Finally, the innermost zone may become anaerobic when nitrates are exhausted. Consequently, at steady-state conditions, three distinct zones may be present along the biofilm thickness so that simultaneous nitrification-denitrification can occur in a reactor.

12.24 DETERMINATION OF TREATABILITY AND KINETIC CONSTANTS

Laboratory studies are required to determine the "treatability" of a wastewater, namely, its substrate removal rate constants such as k' or k'' seen earlier, or K_s and μ_{max}, along with other kinetic constants such as Y and K_d if the activated sludge process or any of its modifications are proposed to be used (see Example 12.4). But if facultative-type aerated lagoons or oxidation ponds are being considered, only the overall substrate removal rate constant K_a needs to be determined (see Sec. 12.24.2). Often, in dealing with typical municipal sewages, laboratory experiments are not essential and the values of the constants are assumed from literature data (Table 12.1).

Sometimes, in the case of certain industrial wastes, laboratory data may need to be supplemented by pilot plant studies in the field in order to evaluate more specifically any operating problems likely to occur, such as (1) toxicity effects, (2) temperature effects, (3) foaming problems, and (4) sludge settling and separation problems. Based on these studies, the need for any pretreatment facilities, flow or concentration equalization units, etc., may be determined. The choice of reactor type may also be assisted by such studies.

In dealing with certain industrial establishments where the product mix manufactured changes from time to time (e.g., chemicals and pharmaceuticals), the nature of the wastewater may vary substantially, and it may be possible to express K values statistically as "likely to equal or exceed a certain value 50, 90, or any other percent of the time," provided sufficiently long sampling results are available.

The nutrient content of the wastewater may sometimes need to be adjusted by artificial addition of nitrogen, phosphorus, or both in order to bring the C, N, and P content into balance for proper biological activity.

12.24.1 Continuous Flow Studies

The above kinetic coefficients are often determined from bench-scale continuous flow studies in an attempt to simulate actual conditions in the prototype. The waste which is to be studied is fed continuously to a reactor (see Figure 12.12) in which "seed" sludge (from a well-operated activated sludge plant) is under aeration. This may have to be done for a few days to ensure that the activated sludge has the opportunity to be acclimatized to the particular waste which is to be studied. COD measurements will show a gradual improvement in removal efficiency as the system gets acclimatized and a stable COD concentration is reached. In the absence of "seed" sludge, several days of aeration may be needed, using a not too dilute feed.

If the pH of the waste is not within 6.0 to 8.0, it may have to be adjusted to bring it within this limit. The mixed liquor pH needs to be measured daily as a check. If the nutrient level is not adequate, it may have to be supplemented by addition of suitable chemicals. Any of the readily available compounds of nitrogen and phosphorus may be used; the quantity added must be sufficient to ensure that the ratio of BOD:N:P is

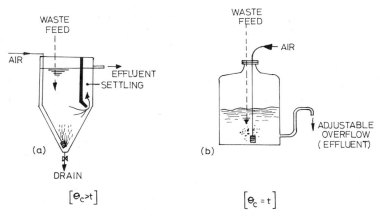

Figure 12.12 Laboratory studies using continuous flow reactors; (a) with settling compartment, (b) flow-through type.

maintained at 100:5:1 (Sec. 12.18). This may be necessary when attempting to treat certain industrial wastes.

For laboratory studies, use of soluble wastes gives good results in accordance with theory, and feeding the waste to the laboratory unit can be easily done. High concentrations of suspended solids and greases or oil may need prior removal. The waste must be fed to the aeration chamber at a constant rate. This can be done by setting up a drum or carboy to hold the day's requirements and providing a positive displacement, peristaltic, or other pump or a simple constant head feed device.

The experiments should be performed under as uniform temperature conditions as practicable. Aeration in the laboratory is generally sufficient to give nearly complete mixing conditions in the small reactors used.

The reactor shown in Figure 12.12a has a settling compartment and, thus, resembles a typical reactor with a recycling arrangement enabling the gradual buildup of solids in the system to give desired values of Θ_c. Another kind of reactor that can be used is shown in Figure 12.12b; it can be set up quite easily with ordinary laboratory glassware. It resembles a flow-through type reactor, as the mixed liquor is displaced out by the incoming waste. There is no arrangement for preventing the solids from flowing out with the effluent, and thus, Θ_c is always equal to t in these reactors. The position of the adjustable overflow arm fixes the volume under aeration.

Generally, it is advantageous to operate about five or more such reactors in parallel with different detention times and/or flow rates in each to give varying values of Θ_c from less than 1 day to 15 to 20 days. The values of the coefficients are determined graphically and, hence, the need to obtain sufficient plotting points with different conditions.

The substrate that is to be studied is, generally, first measured for COD or BOD_u. The quantity of "seed" activated sludge to be put, initially, into the flow-through type reactors (Figure 12.12b) is estimated as follows assuming a yield coefficient of 0.5:

$$X, \text{ mg/liter} \cong 0.5(BOD_u, \text{ mg/liter}) \qquad\qquad (12.53)$$

Thereafter, as the reactors run, some new microbial solids are produced and some solids are displaced out in the effluent. There is no need for any additional withdrawal of solids or mixed liquor. For the reactors with settling compartments (Figure 12.12a), initially, sufficient seed activated sludge is added to give X equal to about 1500 mg/liter or so. This value is typical of mixed liquor from activated sludge plants which can be used without dilution. As these reactors run, the new microbial solids produced are retained in the unit, thus giving a buildup of solids. A part must be withdrawn manually as sludge to keep a more or less steady-state condition in the reactor.

Withdrawal of solids from the unit is generally done by pipetting off some of the mixed liquor a few times a day. If, for example, v ml of mixed liquor are to be pipetted 4 times a day so that solids removed equal the solids produced per day, we can estimate the value of v to maintain, for instance, 1500 mg/liter in the reactor:

$$\left(\frac{4 \times v, \; ml}{10^3} \right) (1500 \; mg/liter) = 0.5(BOD_u, \; mg/liter)Q \; liters/day$$

or

$$v, \; ml = \frac{0.5(BOD_u)(Q)(10^3)}{4 \times 1500} \tag{12.54}$$

This relationship helps to estimate the quantity to be withdrawn from the reactor when flow Q and BOD_u are known.

Considerations similar to those outlined above enter into the conduct of pilot plant studies since they are also of the continuous flow type, but are on a larger scale.

For both bench-scale and pilot plant studies, the analysis to be performed during the period of experimentation includes the following. Tests may be performed two to four times a week after steady-state conditions are reached.

1. BOD, COD, or both in influent and treated effluent. TOC measurements may be made, alternatively, if desired. It is generally necessary to test filtered samples.

2. Volatile suspended solids in effluent. In flow-through type reactors, VSS in the effluent equals VSS under aeration. In reactors with settling compartments, VSS in the mixed liquor must be measured separately.

Other data to be recorded must include the flow rate, reactor volume, and wastewater temperature.

The following example illustrates how the experimental information is analyzed to find the values of the various coefficients.

Example 12.4 The data from five continuous flow reactors of the type permitting recycling, operated in a laboratory study on waste treatability, are given in Table 12.5. Compute graphically the values of the constants

Table 12.5 Observed and Computed Data in Laboratory Studies for Waste Treatability

| | Observed data | | | | | | | Computed values | | | |
Reactor	Feed Q (liters/day)	Feed S_0 (mg/liter)	Effluent S (mg/liter)	t (hr)	MLVSS X (mg/liter)	MLVSS X (g)	Solids produced X/Δt (g/day)	$\dfrac{S_0 - S}{Xt}$ (day^{-1})	Specific solids growth rate (ΔX/Δt)/X (day^{-1})	Θ_c (days)	$\dfrac{1}{S}$ (mg/liter)$^{-1}$
1	7.8	310	7.5	4.6	2680	4.1	0.910	0.575[a]	0.222	4.5[b]	0.133
2	9.0	310	13.2	4.0	1510	2.25	1.185	1.028	0.526	1.90	0.075
3	9.0	310	18.0	4.0	1100	1.63	1.210	1.61	0.740	1.35	0.056
4	9.0	310	31	4.0	850	1.27	1.175	1.97	0.925	1.08	0.032
5	9.0	310	40.5	4.0	805	1.20	1.150	2.02	0.961	1.04	0.025

[a] $\dfrac{S_0 - S}{Xt} = \dfrac{F}{M} = \dfrac{(310 - 7.5)(7.8)}{(4.1 \times 10^3)} = 0.575$ mg/mg-day

[b] $\Theta_c = \dfrac{4.1}{0.91} = 4.5$ days.

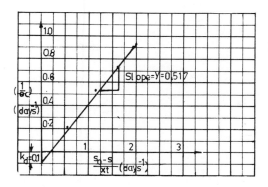

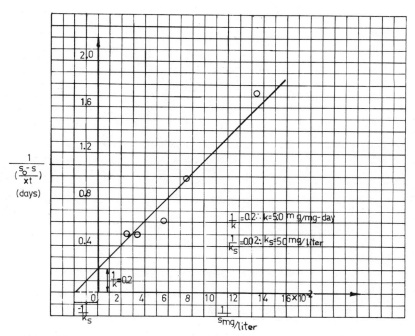

Figure   12.13   Graphic plots of data from Example 12.4.

Y, K_d, k, and K_s for the waste.   The computations are also tabulated in Table 12.5.

From a linearized plot of $1/\Theta_c$ against $(S_0 - S)/Xt$ based on Eq. (12.16), we get Y = 0.517 from the slope of the line and K_d = 0.1 from the intercept (see Figure 12.13a).

Similarly, the values of k and K_s can be found as follows.   Equation (12.25) can be rewritten in terms of k to give

$$\frac{dS_r}{dt} = \frac{kS}{K_s + S} \tag{12.55}$$

Substituting from $dS_r/dt = (S_0 - S)/t$ and taking the reciprocal gives the Lineweaver-Burke form of the equation which is linear

$$\frac{1}{(S_0 - S)/Xt} = \frac{K_s}{k}\left(\frac{1}{S}\right) + \frac{1}{k} \tag{12.56}$$

From a plot of this equation (see Figure 12.13b) we get $k = 5.0$ mg/mg-day, and $K_s = 50$ mg/liter.

For $S \ll K_s$, a plot of $(S_0 - S)/Xt$ against S gives, from the slope of the line, $k' = 0.1$ (mg/liter)$^{-1}$(day)$^{-1}$.

12.24.2 Batch Tests

Instead of continuous flow bench-scale or pilot plant studies to evaluate the waste treatability constants, it is possible to conduct batch tests for the same purpose. Batch tests are similar to ideal plug flow conditions with $D/UL = 0$ and, thus, the overall substrate removal rates K determined from such tests are readily applicable to other mixing conditions in full-scale reactors through the use of the Wehner-Wilhem equation [Eq. (11.15b) provided first-order kinetics are involved.

The manner in which batch tests are conducted and results interpreted is described by Bhatla et al. [29]. Essentially, the method first involves the growth of an acclimatized biological culture in the laboratory using samples of the wastewater under study. This is done in a manner similar to the one just described in the preceding section. Thereafter, the following steps are carried out:

1. Feed to the culture of microorganisms is stopped and aeration is continued until only endogenous reactions exist. The culture is then allowed to settle. The supernatant is removed.

2. Four reactors are filled with volumes of the settled microbial solids such that the concentrations of MLVSS will cover the extremes of probable operating loading conditions. For example, add 0.5, 1.0, 2.0, and 4.0 liters, respectively, to the four reactors.

3. The previously removed supernatant is then added to each reactor to make a total volume of 4.0 liters.

4. The wastewater in question is then added in the amount of 4 liters (or less in the case of high BOD) to each reactor and aerated.

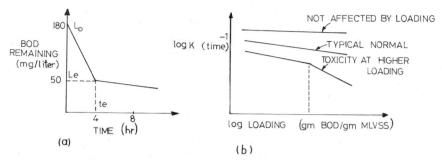

Figure 12.14 The substrate removal rate K/time in batch testing at (a) given loading and (b) different loadings (adapted from Ref. 29).

5. Immediately following the addition of the wastewater, each reactor is sampled after 1, 15, 30, and 45 min, 1, 1.5, 2.0, 4.0, 8.0, 12.0, and 24 hr, or other time intervals as desired and tested for BOD.

6. Oxygen uptake rate measurements may also be made if desired.

When the wastewater is added to the previously aerated MLVSS in a reactor, the BOD is first removed by transfer of organic substrate to the surface of the solids and then followed by stabilization. Generally, for most substances which are readily biodegradable, stabilization reactions occur almost as rapidly as BOD transfer. The BOD removal rate is thus obtained from a plot of the BOD remaining versus time, as shown in Figure 12.14a. Initially, BOD removal is more rapid owing to oxidation, synthesis, and bioprecipitation. When the BOD has been substantially removed, there remains a residual BOD owing to endogenous reactions. Thus, the values of BOD, L_e, and the time t_e at the equilibrium point where the slope of the line sharply changes are noted. If the initial BOD is L_0 (mg/liter), the substrate removal rate at the test temperature is given as [29]:

$$K \ (\text{time}^{-1}) = \frac{L_0 - L_e}{L_e t_e} \qquad\qquad (12.57)$$

For example, for the data shown in the figure for a wastewater exhibiting conventional activated sludge kinetics:

$$K = \frac{180 \text{ mg/liter} - 50 \text{ mg/liter}}{50 \text{ mg/liter} \times 4 \text{ hr}} = 0.65 \text{ per hr}$$

Since the ratio of BOD applied to MLVSS is different in each of the above four test reactors, four different values of K will be obtained. These will plot as shown in Figure 12.14b.

The BOD transfer rate is directly proportional to the mass of micro-organisms MLVSS in a reactor. Thus, theoretically, the slope of the line in Figure 12.14b should be exactly 45° if the mass of organisms used in calculating the loading ratio represents the transfer area. But, within a batch experiment, the concentration of microorganisms increases as the substrate is converted to cell material. Since the initial concentration of MLVSS is used in calculating the loading ratio F/M, the observed slope is less than 45° as shown in the figure. At higher values of F/M for a given BOD, the MLVSS is low and the production of new cell material is proportionately high. Thus, the slope is flatter than 45°.

For design purposes, the appropriate value of K must be selected from Figure 12.14b, depending upon the value of F/M at which the prototype is likely to be operated. A temperature correction may be applied if necessary.

This method helps to bring out the effect of the process loading rate F/M on the removal rate K per unit time. For some wastes, the removal rate sharply decreases at higher loadings owing to toxicity, and a lower loading (dilution) is called for. The treatability of different industrial wastes can also be compared by this method.

12.24.3 Trickling Filter Modeling

Trickling filter modeling in laboratory or pilot-scale work involves setting up a filter column of about 2-m height filled with stones or other media exactly of the type to be used in a full-scale plant. Dosing rate should also be in a practical range. Effluent samples are withdrawn from different depths, and for each hydraulic load applied to the filter the percent BOD remaining at each depth is found.

From such data, the value of coefficient n and of the reaction rate constant K_f' in Eq. (14.9) can be computed as shown in Figure 12.15.

Trickling filter modeling can also be done based on the use of Eq. (12.54) by noting the removal rate V_a per unit area of media per unit time at different substrate concentrations S. But such results are specific for the hydraulic loading rate applied, and therefore the experiment must be repeated to obtain results at other hydraulic rates.

DATA

DEPTH (m)	Percent BOD REMAINING AT STATED HYDRAULIC LOADS, Q (liter / min - m^2)			
	20	40	60	80
1·0	50	70	75	82
1·5	40	50	60	60
2·0	25	30	40	50
2·5	15	20	30	40

COMPUTATION

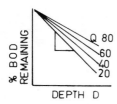

1. PLOT Percent BOD REMAINING (LOG SCALE) AGAINST DEPTH D (ORDINARY SCALE) AT DIFFERENT HYDRAULIC LOADS, Q. COMPUTE SLOPE OF EACH LINE.

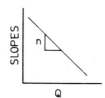

2. PLOT HYDRAULIC LOADS (LOG SCALE) AGAINST THE CORRESPONDING SLOPES COMPUTED ABOVE (LOG SCALE) DETERMINE SLOPE OF LINE. THIS GIVES "n".

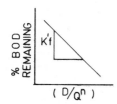

3 FINALLY, PLOT Percent BOD REMAINING (LOG SCALE) AGAINST THE CORRESPONDING TERM (D/Q^n) (ORDINARY SCALE) SLOPE OF THIS LINE EQUALS $K_f' = 0.62 \, M^2 \cdot (min)^{1/2}/l^{1/2}$

INSERTING IN EQUATION 14.9 WE GET FROM LABORATORY RESULTS

$$S = S_0 \cdot e^{-0.62 \, D/Q^{0.5}}$$

Figure 12.15 Determination of trickling filter equation from laboratory data.

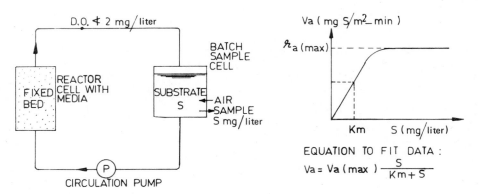

Figure 12.16 Experimental determination of removal kinetics for fixed-film submerged reactors. Data obtained [t (min), S (mg/liter), and V (mg S removed/m²-min)] are fitted using Lineweaver-Burke linearization.

12.24.4 Modeling of Submerged Beds

Experimental data from submerged-type, fixed-film reactors have been shown to fit better a Michaelis-Menten type of equation such as the one shown in Eq. (12.50), in which removal rates are measured instead of the growth rates of Monod. One method used is described here [36].

A differential batch reactor cell of about 2 liters is used of the type shown in Figure 12.16. It is filled with 3-mm diameter glass beads (or other medium whose exact surface area is known). The sample batch of known substrate concentration is circulated through it at a velocity high enough so that there is no appreciable gradient of substrate concentration between the inlet and outlet of the reactor cell. For an aerobic system, the sample cell should be aerated so that DO of not less than 2 mg/liter is observed after passage through the reactor cell.

After operating the closed system for a few minutes, a small sample is withdrawn from the sample cell and checked for substrate concentration. This step is repeated several times and the removal per unit time and area of medium (mg substrate/m²-min) obtained at each experiment is plotted against the corresponding substrate concentration to give a curve as shown in Figure 12.16. Then, using Eq. (12.55), the removal rate V_a corresponding to the actual concentration S desired in the effluent is calculated.

REFERENCES

1. R. McKinney, Microbiology for Sanitary Engineers, McGraw-Hill, N.Y., 1962.

2. J. Servizi and R. Bogan, Free energy as a parameter in biological treatment, Proc. ASCE, 89, 1963.

3. J. Monod, Recherche sur la Croissance des Cultures Bacteriennes, Herman et Cie., Paris, 1942.

4. K. Wuhrmann, Grossversuche Verlag, Zurich, Switzerland, 1964.

5. S. Hashimoto and M. Fujita, Kinetic studies on optimal operation of activated sludge, Jour. Water Waste, Japan, 17, 1975.

6. W. W. Eckenfelder, Jr., Principles of Biological Treatment, NATO Advanced Study Institute, Istanbul, Turkey, Noordhoff Int. Pub., Leiden, Netherlands, 1976.

7. R. Much and L. Grady, Temperature effects on microbial growth, Jour. ASCE, Env. Eng. Div., 100, 1974.

8. W. W. Eckenfelder, Jr., Water Quality Engineering for Practicing Engineers, Barnes and Noble, Inc., N.Y., 1970.

9. B. Handa, Ph.D. Thesis, Nagpur University, Nagpur, India, 1972.

10. S. Arceivala, Use of Dispersed Flow Model in Designing Waste Treatment "Units", NATO Advanced Institute, Istanbul, Turkey, Noordhoff Int. Pub., Leiden, Netherlands, 1976.

11. A. Gram, Reaction kinetics of aerobic process, San. Eng. Res. Lab. No. 2, IER Series 90, Univ. of Calif., Berkeley, California, 1956.

12. A. Samsunlu, Activated sludge kinetics for an industrial waste, Egredir Univ. Eng. Sc. Fak. Publication 10, Izmir, Turkey, 1975 (in Turkish).

13. G. Surucu et al., Thermophilic treatment of high strength wastewaters, in Biotechnology and Bioengineering, Vol. XVII, John Wiley & Sons, 1975.

14. W. Gates et al., A rational model for the anaerobic contact process, Jour. WPCF, 39, 1967.

15. C. Adams and W. Eckenfelder, Process Design Techniques for Industrial Waste Treatment, Env. Press, Nashville, Tennessee, 1974.

16. Grau et al., Kinetics of substrate removal by activated sludge, Water Research, 1977.

17. D. Liebich, Influence of Nitrification on Oxygen Balance of Streams, Technical University, Berlin, Federal Republic of Germany, 1972.

18. A. Downing, Factors to be considered in the design of activated sludge plants, in Advances in Water Quality Improvement, Vol. I, Univ. of Texas Press, Austin, Texas, 1966.

19. A. Lawrence and C. Brown, Design and control of nitrifying activated sludge systems, Jour. WPCF, 48, 1976.

20. McCarty, Beck, and St. Amant, Biological denitrification by addition of organic materials, Proceedings of the 24th Industrial Wastes Conference, Purdue Univ., W. Lafayette, Indiana, 1969.

21. R. Michael and V. Jewell, Optimization of denitrification process, Jour. ASCE, Env. Eng. Div., 101, 1975.

22. A. Lawrence and P. McCarty, Unified basis for biological treatment design and operation, Jour. ASCE, San. Eng. Div., 96, 1970.

23. J. Sherrard and A. Lawrence, Design and operation model of activated sludge, Jour. ASCE, Env. Eng. Div., 99, 1973.

24. I. Essen, Second-stage BOD using a stochastic model, Symposium on Environmental Engineering Problems (Turkish-German), Ege. Univ., Pub. 10, 1975.

25. J. Bisogni and A. Lawrence, Relations between solids retention time and settling characteristics of activated sludge, Water Research, 5, 1971.

26. J. Andrews, Kinetic models of waste treatment, in Biological Waste Treatment (R. Canade, Ed.), Wiley-Interscience, 1971.

27. A. Pasveer, A case of filamentous activated sludge, Jour. WPCF, 41, 1969.

28. R. Roper et al., Discussion of Ref. 23, in Jour. ASCE, 100, 1974.

29. M. Bhatla, V. Stack, and R. Weston, Design of waste water treatment plants from lab data, Jour. WPCF, 38, 1966.

30. M. Degyansky and J. Sherrard, Effects of nitrification in the activated sludge process, Water and Sewage Works Jour., 124, 1977.

31. H. Painter, Microbial transformation of inorganic nitrogen, IAWPR Conference on Nitrogen as a Water Pollutant, Proceedings, Vol. 1, Copenhagen, Denmark, 1975.

32. B. Heide, Combined N and P removal in a low-loaded activated sludge system, IAWPR Conference on Nitrogen as a Water Pollutant, Proceedings, Vol. 3, Copenhagen, Denmark, 1975.

33. P. Harremoes, The significance of pore diffusion to filter denitrification, IAWPR Conference on Nitrogen as a Water Pollutant, Proceedings, Vol. 3, Copenhagen, Denmark, 1975.

34. P. Harremoes, Half-order reactions in biofilms and filter kinetics, Department of Sanitary Engineering, Technical University, Denmark, Dec. 1975 (Preprint).

35. B. Kornegay and J. Andrews, Characteristics and Kinetics of Biological Fixed Film Reactors, Department of Environmental Systems Engineering, Clemson University, 1969.

36. I. Sekoulov, University of Stuttgart, Personal Communication, 1979.

37. White, Tomlinson, and Chambers, The effect of plant configuration on sludge bulking, IAWPR Workshop on Treatment of Wastewaters in Large Plants, Vienna, Austria, Sept. 1979.

Chapter 13

PRINCIPLES OF AERATION

13.1 INTRODUCTION

Improvements made in recent years in the efficient design of mechanical
aerators have led to their increasing use in waste treatment plants. Today,
mechanical aerators are often preferred over pneumatic devices using com-
pressed air, and are available in a variety of sizes and designs. They are
used in activated sludge aeration, aerobic digestion, and other aerobic
processes. They are particularly preferred for use in oxidation ditches
and aerated lagoons. Thus, in this chapter, attention will be focused on
mechanical aerators.

13.2 THEORY OF AERATION

Aeration is a gas-liquid mass transfer process in which the driving
force in the gas phase is the partial pressure of the gas P_g in accordance
with Henry's law, whereas in the liquid phase it is the concentration gradient C_s - C, in which C_s is the saturation concentration at the gas-liquid
interface and C is the concentration in the body of the liquid. Oxygen
molecules from the gas are first transferred at the interface, which may
be assumed to be of nearly molecular thickness, enabling it to rapidly
reach saturation condition. Thereafter, the oxygen penetrates downward by
diffusion and convection. The resistance to penetration in the gas and
liquid phases can be considered to be in series, and of the two, the controlling rate is that which offers the greater resistance.

For slightly soluble gases such as O_2 and CO_2, as in the case of most
waste treatment processes, it is the liquid phase which governs the transfer process. The concentration gradient is, therefore, important, and the
mass transfer per unit time can be expressed in terms of rate of change of
concentration as follows:

$$\frac{dc}{dt} = K_L a(C_s - c) \tag{13.1}$$

where

K_L = liquid film coefficient (L/t)

a = interfacial area for transfer per unit volume of liquid (1/L)

C_s = saturation concentration of gas in a given liquid at a given temperature (M/L^3)

C = actual concentration of gas in a given liquid at a given temperature (M/L^3)

Under quiescent conditions, the surface film is stationary and molecular diffusion controls. Thus, $K_L = D_L/L$ where D_L = molecular diffusion coefficient (L^2/t) and L = depth of film (length). Under turbulent conditions, however, the surface film is continually renewed so that liquid with concentration C replaces liquid with concentration C_s at the interface. Danckwerts [1], Dobbins [2], and others have shown the liquid film coefficient to be affected in such cases not only by the molecular diffusion coefficient but also by the rate at which the surface film is renewed, namely, $K_L = \sqrt{D_L - r}$ in which r = frequency of renewal of film (1/t). Thus, the reaeration rate is affected by the degree of turbulence.

Again, with increasing turbulence and spray, the value of the parameter "a" increases as finer and finer droplets are formed, thus increasing the overall gas transfer rate. But in practice, it is not possible to measure this area, and hence the overall parameter $K_L a$ is determined by experimentation as illustrated in Example 13.1.

The integrated form of Eq. (13.1) gives, for concentrations C_1 and C_2 measured at time t_1 and t_2,

$$K_L a = \frac{2.303 \log[(C_s - C_1)/(C_s - C_2)]}{t_2 - t_1} \tag{13.2}$$

Thus, $K_L a$ can be estimated analytically using Eq. (13.2) or graphically from the slope of the line obtained when logarithms of C_s - C values are plotted against time t on semilog paper. The units of $K_L a$ are time^{-1}. The value so obtained will be specific to the conditions of the experiment as discussed further below.

13.3 FACTORS AFFECTING AERATION

The major factors which affect the efficiency of oxygen transfer are listed and discussed below. Since the performance of aerators in both model tests and prototypes is likely to be considerably affected by these factors, their testing must be done under some arbitrarily standardized conditions or the results obtained under other conditions must be suitably adjusted for the standardized ones (Sec. 13.4):

1. Temperature
2. Dissolved oxygen concentration in the liquid under aeration
3. Characteristics of the wastewater
4. Characteristics of the aerator

13.3.1 Temperature

Temperature has a considerable effect on the oxygen transfer rates since aeration involves gas transfer at the interface followed by diffusion within the liquid phase. Temperature also tends to affect the air bubble size to some extent. As temperature increases, the values of K_L increase in accordance with

$$K_{L(T)} = K_{L(20^\circ C)} \Theta^{T-20} \tag{13.3}$$

Eckenfelder [4] and Ford [5] report that Θ values vary from 1.016 to 1.037 depending on the aeration system, with values generally ranging from 1.020 to 1.028 for bubble aeration systems and from 1.024 to 1.028 for mechanical aeration systems. Aerator performance is generally rated at $20^\circ C$ ($10^\circ C$ in some countries) by manufacturers, and must be adjusted to the estimated temperature in the treatment unit. Winter conditions are critical in this regard, and the temperature likely to prevail in the treatment unit can be estimated using Eq. (12.33). Adjustment of aerator performance is then done using Eq. (13.5) given below.

Temperature also affects the solubility of dissolved oxygen and thus the dissolved oxygen gradient C_s - C, discussed in the next section.

13.3.2 Dissolved Oxygen Concentration

The saturation concentrations C_s for dissolved oxygen in distilled water exposed to water-saturated air at different temperatures are given in Table 13.1. These values are also applicable for our purposes to most tap waters in which salt concentrations are within drinking water limits [3].

Table 13.1 Dissolved Oxygen Saturation Concentration
in Distilled Water at Different Temperatures

Temperature (°C)	DO Saturation C_s (mg/liter)	Temperature (°C)	DO Saturation C_s (mg/liter)
0	14.6	20	9.2
1	14.2	21	9.0
2	13.8	22	8.8
3	13.5	23	8.7
4	13.1	24	8.5
5	12.8	25	8.4
6	12.5	26	8.2
7	12.2	27	8.1
8	11.9	28	7.9
9	11.6	29	7.8
10	11.3	30	7.6
11	11.1	31	7.5
12	10.8	32	7.4
13	10.6	33	7.3
14	10.4	34	7.2
15	10.2	35	7.1
16	10.0	36	7.0
17	9.7	37	6.9
18	9.5	38	6.8
19	9.4	40	6.6

However they are often not valid for wastewaters which have various substances in solution. The saturation concentration for wastewater C_{sw} may
be 90 to 98% of the C_s value for distilled water and may need to be determined by aerating a treated effluent sample overnight. A correction for
atmospheric pressure may be needed where it is expected to vary from the
assumed value of 760 mm Hg.

The performance of aerators is generally rated by manufacturers using
plain tap water at 20°C (760 mm Hg) and with no dissolved oxygen in the
test tank. The oxygen transfer efficiency thus needs to be adjusted for
the actual concentration of dissolved oxygen C_L likely to exist in the
full-scale unit under operating conditions. The greater the dissolved

oxygen present in the aeration tank, the less is the oxygen transfer efficiency of the aerator. For conventional activated sludge plants where nitrification is not intended, dissolved oxygen values in the aeration tank would range between 0.5 and 1.0 mg/liter, whereas if nitrification is desired, the dissolved oxygen values would more nearly have to average around 1.5 to 2.0 mg/liter. The latter range would also generally apply to extended aeration units where nitrification is to be ensured. Adjustment of the aerator oxygenation capacity from the manufacturer's rated conditions to field conditions is done in the ratio of $(C_{sw} - C_L)/C_{s(20^\circ C)}$.

13.3.3 Wastewater Characteristics

The oxygen transfer rate is less in wastewater compared to distilled or tap water for which the aerator is rated by the manufacturer. Thus, a correction factor must be applied depending on the characteristics of the wastewater. The oxygen transfer rate for wastewater is generally less because of the presence of dissolved solids, organics, surface-active agents, etc. present in the waste. The latter tend to concentrate at the interface, change the surface tension, and create a sort of barrier to diffusion. This lowers the value of K_L.

Small concentrations of surface-active agents reduce surface tension and therefore the value of K_L. As concentrations increase, however, the size of bubbles tends to decrease, thus increasing A/V or the value of "a." Consequently, in bubble aeration the value of the parameter $K_L a$ first tends to decrease and then to increase with increasing concentration of surface-active agents [4]. Generally, as dissolved organic material is removed in a biological process, the value of $K_L a$ for wastewater tends to increase and approach that for plain water.

The wastewater characteristics are accounted for by determining a parameter α, which is a ratio of the values for plain water and wastewater.

$$\alpha = \frac{K_L a \text{ for wastewater}}{K_L a \text{ for plain water}} \qquad (13.4)$$

This ratio is determined by experimentation, first using plain water and then using the wastewater in question under identical aeration conditions. Example 13.1 illustrates the method.

Values of α range from 0.4 to 0.98, depending on the nature of the waste and its extent of treatment. Values tend to approach unity as substances affecting the transfer of oxygen are removed in the biological process. Values of α are considerably affected by the viscosity of the wastewater. At very high solids concentration ($> 2\%$) the viscosity may increase and α may decrease [11]. This may be of special significance in some industrial and animal wastes and in the design of aerobic digesters. Values of α are also affected by the mixing intensity since the latter affects the rate of renewal of the film at the interface [5]. For this reason, a laboratory determination of α values must be regarded as only an approximation of the field values since it is not easy to have identical mixing turbulence in the model and the prototype.

Example 13.1 A sample of wastewater showed a value of C_{sw} = 9 mg/liter when aerated overnight at 15.5°C. About 10 liters of sample were then deaerated using Na_2SO_3* and $CaCl_2$†, and the aerator was switched on. Samples were then withdrawn at 5-min intervals and checked for DO, using a DO probe and double-checked with the Winkler method. From the results tabulated below, estimate K_La per hour. Compute α if a similar test with plain water has given K_La = 2.8 per hr at 15.5°C.

Time (min)	Observed DO concentration C (mg/liter)	C_{sw} - C (mg/liter)
5	0.5	8.5
10	1.5	7.5
15	3.0	6.0
20	4.0	5.0
25	5.1	3.9
30	6.0	3.0

Figure 13.1 shows a semilog plot of the data from which K_La is computed as the slope of the line [Eq. (13.2)]. Thus,

---

*Stoichiometrically, at 8 mg/liter of Na_2SO_3 per 1 mg/liter dissolved oxygen in the sample (20 to 30% excess generally used).
†At about 0.5 mg/liter of cobalt chloride per mg/liter dissolved oxygen in sample (not to exceed 1.5 mg/liter).

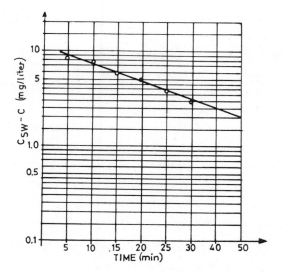

Figure 13.1 The determination of $K_L a$.

$$K_L a = \frac{2.3 \ \log(8.9/3.01) \times 60}{30 \ min - 5 \ min} = 2.53 \ per \ hr \ at \ 15.5^{\circ} C$$

$$\alpha = \frac{2.53 \ per \ hr \ at \ 15.5^{\circ} C}{2.8 \ per \ hr \ at \ 15.5^{\circ} C} = 0.9$$

13.4 OXYGENATION CAPACITY UNDER FIELD CONDITIONS

From the foregoing discussion it is evident that the rated oxygenation capacity of an aerator given at "standard" conditions is necessary to compensate for the actual conditions likely to obtain in the field so as to take into account variations in temperature, presence of DO in the tank, and reduced oxygen transfer rate for wastewater owing to the substances present in it or to high solids concentration.

The oxygen transfer rate under field conditions N_f is related to the oxygen transfer rate under standard conditions N_O as follows:

$$N_O = \frac{N_f}{[(c_{sw} - c_L)/c_{s(20^{\circ} C)}](\alpha)(\theta)^{T-20}} \qquad (13.5)$$

Often, in practice, the value of N_f may be only 60 to 80% of the N_O value

at standard conditions, and if this compensation is neglected, the design
may become inadequate.

Winter is critical from the point of view of oxygen transfer characteristics, whereas summer is critical from the point of view of the oxygen
demand itself since greater BOD removal and nitrification are likely in the
summer. Thus, it is desirable to estimate the required oxygen transfer rate
at both winter and summer conditions and to provide the larger of the two
values.

Any given aerator will generally display its highest oxygenation capacity in plain water, less in sewage, and still less in aerating certain
industrial and animal wastes, especially with certain substances in solution
and at high solids concentrations, as stated earlier.

Example 13.2 By what percentage should the "standard" oxygenation
capacity of a mechanical aerator be decreased in order to compensate for
field conditions in which $T = 10^\circ C$, $\alpha = 0.9$, $C_{sw} = 10.17$, and $C_L = 2$ mg per
liter?

From Eq. (13.5),

$$N_f = \left[\left(\frac{10.17 - 2.0}{9.2} \right) (0.9)(1.024)^{10-20} \right] N_0^{-1}$$
$$= 0.63\ N_0$$

Thus, under field conditions the aerator may be expected to give only
63% of its rated output.

13.5 MECHANICAL AERATION SYSTEMS

Mechanical aerators can be grouped as follows:

1. Vertical axis type
    a. Radial flow, slow speed
    b. Axial flow, high speed
2. Horizontal axis type; brush and cage rotors

Table 13.2 gives a summary of the characteristics, uses, and typical details of different types of mechanical aerators. The units are also illustrated in Figures 13.2, 13.3, and 13.4.

Turbulence and liquid spray are caused by the motion of an aerator.
Fine spray creates a relatively large surface area, thus entraining much

Table 13.2 Typical Characteristics and Uses of Mechanical Aerators

Aerator type	Characteristics	Oxygenation capacity[a] (kg O_2/kWh)	Components and available sizes	Suitable application	Remarks
Radial flow, slow speed	Like a low-head, high-volume pump. Air entrainment results from an induced hydraulic jump. Output speed generally 30-60 rpm (even 100 rpm).	2.0-2.3 (Efficiency drops with size.)	TEFC[b] motor: 3-150 hp; impeller: 0.9-3.7 m; speed reducer with service factor 1.5-2.0. Units usually fixed on bridges or platforms. Float-mounted units also available.	Activated sludge and extended aeration tanks/ditches. Contact stabilization. Aerobic digestion. Aerated lagoons and large aeration basins of 0.9 to 5.5 m depth (deeper basins need draft tubes).	Higher initial cost. Speed reducer problems likely. Correct aerator submergence essential in the case of fixed units.
Axial flow, high speed	Like a low-head, high-volume pump. Most oxygen transfer occurs because of spray, and some because of turbulence. Output speed generally 900-1440 rpm.	1.5-2.0	TEFC[b] motor, 1-150 hp. Propeller with riser tube and deflector (no speed reducer). Usually mounted on floats of fiberglass or other noncorrodible material.	Generally same as for radial-flow, slow-speed aerators. Basin depths 0.9-4.6 m, depending on aerator size.	Lower initial cost. Simple to install and operate. Floating units adjust to varying water level in basin.
Brush and cage rotors	Various types available in which rotation is around horizontal axis, creating spray contact and air entrainment. Output speed is generally 30-60 rpm.	1.5-2.0	TEFC[b] motor, 3-50 hp; speed reducer, chain drive. Rotor mounted on hollow tube shafting. Intermediate supports with bearings needed depending on length.	Smaller plants using activated sludge or extended aeration tanks/ditches, of 0.9-2.5 m depths.	Easy to fabricate locally. Problems may be caused by long shaftings and intermediate bearings. Correct rotor submergence is essential.

[a]At standard conditions (20°C, 760 mm Hg, zero DO in basin, and tap water) and optimum depth of immersion of rotating element. Power is based on gross input to motor (see Sec. 13.7).

[b]TEFC: totally enclosed, fan cooled.

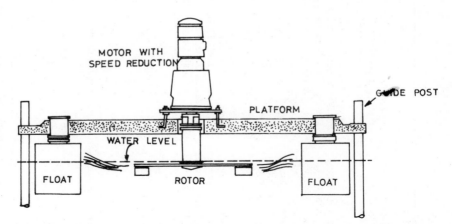

Figure 13.2 Detail of a typical floating low-speed vertical-axis aerator (several types of rotor designs are available).

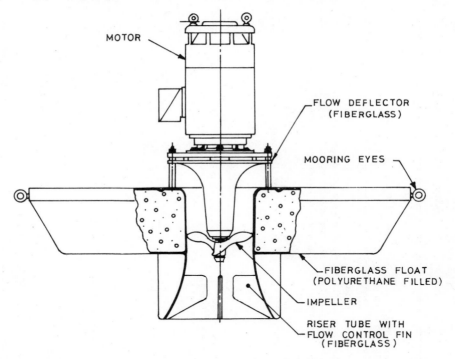

Figure 13.3 Detail of a floating high-speed vertical-axis aerator.

air. As the spray falls back on the liquid, air in the form of small bubbles enters and circulates below the surface. The aerator also sets up a circulatory motion in the entire unit, provided the power level (power input per unit volume) is adequate.

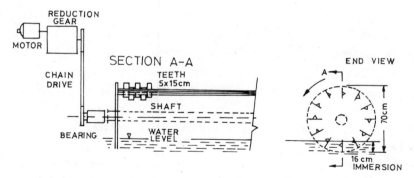

Figure 13.4 A cage rotor (horizontal shaft).

Turbulence increases the gas transfer rate by its effect on $K_L a$ as discussed earlier. For surface aerators the effect on $K_L a$ has been shown to be nearly proportional to the square of the power level [6].

Generally, submergence or depth of immersion of an aerator in waste-water has a very significant effect on its oxygenation capacity and power requirement. At the correct submergence (as recommended by the manufacturer and at which the unit is rated) the performance will be optimum. There will be good turbulence and entrainment of air relative to its power consumption. If the submergence is increased, the unit tends to act as a liquid mixer rather than aerator, and power consumption increases without any commensurate benefit in oxygenation capacity. On the other hand, if submergence approaches zero, only some surface spray is formed in the vicinity of the aerator but without any effective turbulence. The power consumption decreases, but so does the oxygenation capacity. Thus, care has to be exercised when erecting an aerator to ensure its optimum depth of immersion as recommended by the manufacturer. Most aerators are quite sensitive to the depth of immersion and their rated capacity is severely reduced with incorrect immersion.

Where fixed aerators are used (mounted on columns or stilts), it is essential that the liquid level in the unit be kept constant to ensure the required degree of submergence of the aerator blades. This implies a need for watertight conditions in the unit and requires care in the design to avoid heading up at peak flows.

This difficulty is obviated when floating aerators are used since they adjust themselves to the level in the basin. Their use is further discussed in Sec. 15.7. Vertical-axis aerators, either fixed or floating, are often preferred as they can be used with both deep and wide basins, unlike the horizontal-shaft units which are somewhat restricted in this regard. Proper design is essential to avoid surge and vortex formation, which tend to reduce the efficiency of oxygenation. Baffles attached to the columns on which the vertical-axis aerator is mounted may need to be provided to break up vortex formation. It may also be desirable to provide concrete pads on the floor of an earthen basin directly below the aerator where units over 75 hp are used.

Icing problems may occur in cold climates and some protection should be provided in the form of cover plates or a housing over the aeration units.

13.6 OXYGENATION AND MIXING

An aerator is often called upon to perform two functions:

1. Oxygenation of the wastewater
2. Mixing of basin contents to prevent any settlement of biological flocs and other suspended solids

Both functions have to be performed by aerators in the case of certain processes such as activated sludge, extended aeration, aerobic flow through lagoons, etc., and the power requirements must be determined for both purposes, using the higher of the two values. Computation methods are illustrated in Chapters 14 and 15. Where oxygenation and mixing are required, the minimum power level should be kept equal to or higher than 2.75 kW per 1000 m^3 basin volume so that sufficient circulatory velocities are created and all solids are kept in suspension [7]. These results are based on observations in several full-scale aerated lagoons.

Where mixing is not essential as in the case of facultative-type aerated lagoons, the minimum power level required to ensure the availability of dissolved oxygen in all parts of the basin is 0.75 kW per 1000 m^3 basin volume [7].

Where several aerators are to be installed in a large basin, their spacing is kept so that their circles of influence do not overlap. Recommendations in this regard generally depend on the horsepower of each unit and are available from the manufacturers. Large aerators may be spaced 60 to 80 m apart. The basin depth should also be compatible with the type and size of the aerator proposed to be used, otherwise oxygenation and mixing

may not be adequate at all depths. Some indication of the permissible range
of basin depths is given in Table 13.2

13.7 OXYGENATION EFFICIENCY

Generally speaking, radial-flow slow-speed aerators have a relatively
higher oxygen transfer efficiency than do axial-flow high-speed aerators,
and the brush and cage type rotors used in practice. Table 13.2 gives
the typical range of oxygenation efficiencies, generally expressed in terms
of kg O_2 per kWh or per hp-hr at the "standard" conditions noted earlier
and at the optimum immersion depth specified by the manufacturer. The gross
horsepower is the brake horsepower or "nameplate horsepower," although oxygenation efficiency can also be expressed in terms of the actual power delivered by the unit to the water (net power). The purchaser should be clear
about the basis on which efficiency is stated:

Power delivered, watts = (line voltage)(line amperage)(power
                              factor)(motor and gear efficiencies) (13.6)

The efficiency of a good geared-drive arrangement may be about 90%
after a sufficient period of "running in."

Single units ranging from 1 hp to as much as 150 or even 200 hp with
rotating element diameters as large as 3 to 4 m are available. Most vertical-axis radial-flow slow-speed aerators are capable of delivering 2 to 2.3
kg O_2/kWh at standard conditions. The high-speed axial-flow units and the
brush or cage type rotors generally deliver between 1.5 and 2.0 kg O_2/kWh.
Actual values at field conditions are considerably less in all cases for
reasons discussed earlier.

The cost of aerators depends on the type of unit, horsepower, type of
mounting, and availability within the country. For larger plants, vertical-
axis aerators are generally cheaper than the horizontal-axis units. Among
vertical-axis types, the high-speed units are generally cheaper and easier
to install and operate. The low-speed units are generally more expensive,
but are preferred for larger installations owing to their higher efficiency.

Figure 13.5 shows the total costs versus the plant design capacity
for various devices in the U.S. [10].

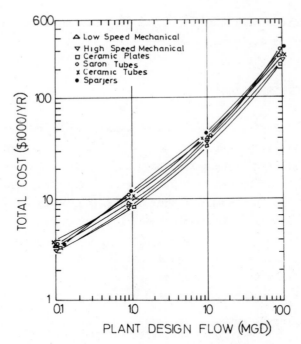

Figure 13.5 Total costs versus plant design capacity for various aeration devices in the U.S.

13.8 FIELD TESTING OF AERATION SYSTEMS

Field testing of aeration systems under actual operating conditions may be necessary to verify the manufacturer's claims or the parameters assumed by designing engineers. The method generally followed is first to determine the value of $K_L a$ under field conditions using tap water or wastewater undergoing biological treatment, and then to find the oxygenation capacity (OC) as follows, assuming complete mixing:

$$OC = (K_L a)_{20^\circ C} (C_{s(20^\circ C)} - C)(\text{basin volume}) \qquad (13.7)$$

in which C = 0 at "standard" conditions. If the $K_L a$ value is determined in the field using any one of the methods stated below, the result is converted from the test temperature to the 20° C condition using an agreed value of Θ in Equation (13.3). For example, let us assume $K_L a$ = 3.0 per hr at 20° C, C_s = 9.2 mg/liter at 20° C for tap water, and basin volume = 2000 m³. The oxygenation capacity is then obtained from Eq. (13.7) as:

$$OC = (3.0 \text{ per hr})(9.2 \text{ mg/liter})(2000 \times 10^3 \text{ liters})\left(\frac{1}{10^6}\right)$$

$$= 55.2 \text{ kg } O_2/\text{hr}$$

This value may be divided by the brake horsepower of the unit to obtain its gross oxygen transfer efficiency in kg O_2/hp-hr or kg O_2/kWh. Equation (13.6) may be used if the result is desired in the form of actual power delivered to the water, namely, shaft power or net power.

Generally, $K_L a$ can be determined in the field by conducting a large-scale experiment on the same general lines as illustrated earlier in Example 13.1. In fact, this method is generally considered as a standard one. However, it should be remembered that the $K_L a$ value is itself a function of the power level as discussed earlier and thus the field test cannot necessarily be compared with the equipment rating, which may have been done at an entirely different power level. Similarly, tank geometry may also affect the result [5].

Other methods used are to measure the oxygen uptake rate at various points in the basin while wastewater treatment by the activated sludge or other biological process is in progress at (1) steady-state conditions and (2) nonsteady-state conditions. These methods are described more fully elsewhere [5,8,9].

Sweeris and Trietsch [8] describe in detail the field testing of "Simcar" vertical-axis-type low-speed aerators in a Carrousel ditch in the Netherlands using plain water. Their results gave an oxygen transfer efficiency varying from 2.0 to 2.59 kg O_2/gross kWh at different depths of immersion. The test temperature averaged around $9°$ C.

REFERENCES

1. P. Danckwerts, Significance of liquid film coefficients in gas absorption, Ind. Eng. Chem., 93, 1951.

2. W. Dobbins, The nature of the oxygen transfer coefficient in aeration systems, in Biological Treatment of Sewage and Industrial Wastes (MeCaba and Eckenfelder, Eds.), Reinhold, 1956.

3. Standard Methods for the Examination of Water & Waste Water, U.S. PHS, AWWA, APHA, 1965.

4. W. W. Eckenfelder, Water Quality Engineering for Practicing Engineers, Barnes and Noble, Inc., N.Y., 1970.

5. D. L. Ford, Oxygen transfer and aeration equipment selection, in <u>Bio-logical Treatment</u>, NATO Advanced Study Institute, Istanbul, Turkey, Noordhoff Int. Pub., Leiden, Netherlands, 1976.

6. Cooney and Wang, Aeration, in <u>Biological Waste Treatment</u>, R. Canale, Ed., Wiley-Interscience, N.Y., 1971.

7. C. Adams and W. W. Eckenfelder, Jr., Eds., <u>Process Design Techniques for Industrial Waste Treatment</u>, Enviro Press, Nashville, Tennessee, 1974.

8. S. Sweeris and R. Trietsch, Determination of the oxygenation capacity in carrousel plants, <u>H$_2$O</u>, No. 5, 1974.

9. R. Conway and G. Kumke, Field techniques for evaluating aerators, <u>Jour. ASCE, San. Eng. Div.</u>, 1966.

10. D. L. Gibbon, Ed., <u>Aeration of Activated Sludge in Sewage Treatment</u>, Pergamon Press, 1974.

11. D. Baker, R. Loehr, and A. Anthonisen, Oxygen transfer at high solids concentration, <u>Jour. ASCE, Env. Eng. Div.</u>, <u>101</u>, 1975.

Chapter 14

SOME BIOLOGICAL TREATMENT METHODS

14.1 SUSPENDED AND ATTACHED GROWTH SYSTEMS

In this chapter we will discuss some practical applications of the
various concepts presented earlier concerning reactor design and biological
treatment. Methods commonly used can be divided into suspended and attached
growth systems. These divisions are based on how the biological flocs or
films exist in the reactors. It should be re-emphasized that the general
principles of biological treatment are equally applicable in both cases,
although there are certain differences brought about by the habitat of the
growths.

Examples of <u>suspended growth systems</u> are the following:

Activated sludge and its modifications
Aerobic and anaerobic sludge digesters
Aerated lagoons
Waste stabilization ponds

The activated sludge process, which was first developed in the early
part of this century, is very popular, and various modifications have been
developed to improve performance or operational characteristics.

Tapered and step-aeration processes started the gradual transition
from the more plug-flow-type (conventional activated sludge units) to the
more complete-mixing-type units used presently. Other modifications have
included longer aeration times — extended aeration systems — of which the
Pasveer-type oxidation ditch is one of the most popular, and the contact
stabilization process. Aerobic digestion is also of interest.

<u>Attached growth systems</u> occur naturally at any water-solid interface
as in soils, lakes, river bottoms, etc. Their engineered forms are exem-
plified by some of the following:

Trickling (or percolating) filters
Rotating discs

675

Submerged media beds (downflow, upflow, and fluidized beds)

Land treatment and infiltration systems

Although in recent times a more fundamental approach to biofilm kinetics in fixed film reactors has been followed, their design has remained largely empirical.

In the following pages some essentials of practical plant design are covered for some of the methods just listed. Lagoons, oxidation ponds, and land disposal systems are discussed in their respective chapters.

14.2 ACTIVATED SLUDGE

In both conventional and complete mixing activated sludge systems, the essential elements are the same. The raw waste undergoes screening, grit removal, and primary settling before aeration. The mixed liquor from the aeration tank is settled to give a clear supernatant which may be disinfected and discharged or treated further depending on the end use. The sludge is withdrawn from the final settling tank. Its major part is recycled, while a small fraction is wasted in order to keep the system at steady state. The fraction wasted is generally thickened and digested aerobically or anaerobically before dewatering.

The essential difference between the two types of system lies in the mixing conditions in their aeration tanks (see Figure 14.1). The choice between a more plug flow or a more completely mixed pattern must depend upon the factors discussed in Sec. 11.18. In the interest of better operational stability, the present trend is to prefer the complete mixing type wherever feasible, especially with industrial wastes where both the magnitude of flow fluctuations and possibilities of toxic inflows are greater. With municipal sewage, two or three complete mixing cells in series would be more desirable for obtaining better sludge settling characteristics (Sec. 12.16).

Typical design criteria used with municipal wastewaters are included in Table 14.1. Some differences in design criteria are noticed between "warm" and "temperate" climates. A precise definition of climates is difficult, but for practical purposes of design in this book, "warm" is taken to imply those climates where the contents of the reactor (whether activated sludge, aerated lagoon, or oxidation pond) are around 18 to 20°C or higher all through the year, and "temperate" refers to conditions where reactor contents are around 9 to 10°C or so during the winter. In "cold" climates, the reactor contents may be around 3 to 7°C or lower in the winter.

Table 14.1 Summary of Typical Design Criteria for Treatment of Domestic Sewage in Activated Sludge and Extended Aeration Systems

Parameter	Conventional and complete mixing activated sludge systems	Extended aeration (including Pasveer and Carrousel-type ditches)
Mean cell residence time Θ_c, days[a]		
Warm	4-5	10-20
Temperate	5-10	20-30
Cold	10-15	Over 30
F/M ratio, kg BOD_5/kg MLVSS-day[a]		
Warm	0.6-0.9	0.2-0.25
Temperate	0.4-0.6	0.1-0.2
Cold	0.2-0.4	0.1 or less
Hydraulic detention time t, hr	3-8[b]	12-36
MLSS, mg/liter	2000-5000[b]	4000-5000
VSS/SS	0.7-0.85	0.5-0.8
Biodegradable fraction of VSS, f_b	0.55-0.7	0.4-0.65
BOD_5 of MLVSS, mg/mg MLVSS	0.55-0.7	0.4-0.65
Suspended solids in effluent, mg/liter	10-30	10-30
Substrate removal rate constants	See Table 12.1	See Table 12.1
Cell growth constants Y and K_d	See Table 12.1	See Table 12.1
Temperature activity correction	See Table 12.2	See Table 12.2
Recirculation pump capacity	0.5-0.75 Q	0.75-1.0 Q
Net VSS production, kg/kg BOD_u removed	0.3-0.5	0.25-0.4
Surplus sludge, g VSS/person-day	22-35	18-30
Surplus sludge, g SS/person-day	30-40	26-40
Separate sludge digestion	Needed	Generally not needed
Oxygen requirements, kg/kg BOD_u removed		
Without nitrification	0.6-0.65	0.8-0.85
With nitrification	0.8-1.2	1.0-1.3
Dissolved oxygen in aeration tank, mg/liter	0.5-2.0[a]	1.0-2.0[a]
Nutrient requirements	See Table 12.3	See Table 12.3
Sludge drying bed area	See Table 17.2	See Table 17.2
Power required in aeration, kWh/person-year	8-17[a]	15-19[a]
Expected removal efficiency		
Percent BOD	85-93	95-98
Percent coliforms	90-95	90-98
kg P removed/100 kg BOD_u removed	0.4-1.0	0.1-0.4
kg N removed/100 kg BOD_u removed	4-5	2-4

[a]Also depends on nitrification requirements.

[b]Higher values in given range used with complete mixing systems.

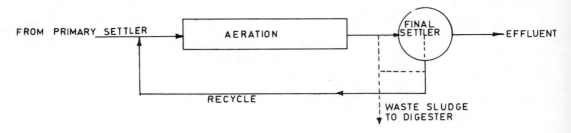

(a)

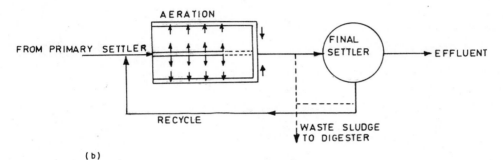

(b)

Figure 14.1 Typical flowsheets for (a) conventional and (b) completely mixing activated sludge systems.

14.2.1 Aeration Tank

The first step in design is to choose a suitable value of Θ_c (or F/M). This choice depends on the expected winter temperature of mixed liquor, the type of reactor, the expected settling characteristics of the sludge, the efficiency of BOD removal, and the nitrification required. The reader is referred to Sec. 12.22, which gives guidance in choosing Θ_c values. The choice generally lies between a Θ_c of about 5 days in warmer climates to 10 days in temperate ones where nitrification is desired along with good BOD removal, and complete mixing systems are employed for treating domestic or similar wastewater. Reactor temperatures are more or less equal to those of the influent wastewater since hydraulic detention times are relatively small.

The second step is to select two interrelated parameters, the hydraulic detention time t and the MLSS concentration in mg/liter. For typical domestic sewage, the value of t is kept the minimum possible (to reduce aeration tank volume) without exceeding an MLSS value of 2000-3000 mg/liter if a conventional plug flow type aeration system is to be provided, or 3000-5000 mg/liter for complete mixing types. Generally, concentrations higher than 6000 mg/liter are not kept owing to the effect on the settling tank design (see below) and possibly on the oxygen transfer capacity of the aeration device. Hence, the upper limit is fixed. In estimating the total mass of solids in the system as a whole, the solids contained in the final settling tank and recirculation may be added to the solids contained in the aeration tank if desired.

However, in treating wastewaters which are highly fluctuating in volume or likely to contain occasional slugs of toxic substances, the aeration tank volume should preferably be increased over the minimum, allowing the MLSS concentration to be proportionately reduced. The tank geometry should also be adopted so as to promote as nearly complete mixing conditions as possible. Thus, for such a wastewater, a detention time t of 6 to 8 hr may be more desirable to adopt than the commonly used minimum of 3 to 4 hr (see Sec. 11.15).

Once the required volume of the aeration tank is determined using Eq. (12.19), the geometric configuration of the tank can be fixed in such a manner as to give a low value of the dispersion number ($D/UL \le 0.2$) for conventional plug flow designs and higher values ($D/UL \ge 4.0$) for the so-called completely mixed units. This can be done using the methods given in Sec. 11.10.

14.2.2 Sludge Settling and Recycling

Recycling ratios upward of 0.5 are generally preferred as being able to maintain the desired MLSS concentrations in completely mixed reactors under varying conditions of sludge settleability (SVI) as discussed in the previous chapter. Recirculation pump numbers and total capacity should be designed for flexibility in operation.

Settling tank design is now beginning to receive more attention since it is realized that traditional designs based simply on overflow rate considerations are not always adequate, and the solids loading of such units must also be taken into account as "hindered settling" often occurs in them. Thus, design criteria must be properly selected as discussed in Chapter 17

using batch flux techniques to be able to size the final settling tanks adequately to transmit the required solids flux to the tank bottom. In computing the normal overflow rate of the tank, the recirculation flow is
neglected as it is withdrawn from the tank bottom as "underflow" and therefore does not constitute part of the overflow. The trend in the U.S. is
to design final settling tanks for the peak flows likely to be received at
a plant. Winter conditions can considerably affect settling characteristics
of sludge. For industrial wastes, it may be desirable to develop specific
experimental data for design purposes.

14.2.3 Surplus Sludge Removal and Disposal

The surplus sludge solids which must be removed from the system can be
taken from the underflow of the settling tank, namely, a part of the recirculated flow can be diverted out of the system. This is easy to accomplish,
but the actual mass of solids withdrawn depends on the settled sludge density, which may vary somewhat from time to time. Hence, some preference is
shown for withdrawing a small fraction of the mixed liquor just before entering the settling tank. In this manner, it is possible to keep Θ_c fairly
constant if the flow withdrawn is always kept as a constant fraction of the
total flow of mixed liquor since the mass of solids withdrawn is then also
a constant fraction of the mass in the system. Automatic control of Θ_c is
also possible in this manner.

However, the surplus sludge is now too dilute to be sent directly to
the digester, and may be thickened first in a thickener or sent to the
primary settling tank to settle along with the primary sludge and then pass
on to the digester. Sludge drying beds or lagoons can be designed using
methods outlined in Chapter 17.

14.2.4 Oxygenation

Oxygen requirements are estimated for meeting the carbonaceous and
nitrification demands where nitrification is desired. Aeration power requirement is computed for both summer and winter conditions and the higher
of the two values is installed, as explained in Chapter 13. Typical values
are given in Table 14.1.

14.2.5 Nutrients

Where nutrient addition is required to make up deficiency, commercially available compounds of nitrogen and phosphorus are used; their dosages are computed using Eqs. (12.45) and (12.46).

Nutrient removal from wastewater can also be estimated as explained in Secs. 12.19 and 12.20. Denitrification system design and phosphorus removal through chemical precipitation can be done as detailed in Secs. 14.6 and 14.7.

Example 14.1 Design a complete mixing activated sludge system to serve 60,000 people that will give a final effluent that is nitrified and has a BOD_5 not exceeding 25 mg/liter. The following design data are available:

Sewage flow = 150 liters/person-day = 9000 m^3/day

BOD_5 = 54 g/person-day = 360 mg/liter; BOD_u = 1.47 BOD_5

TKN = 8 g/person-day = 53 mg/liter;   P = 2 g/person-day
= 13.3 mg/liter

Nutrient addition is not necessary. .

Winter temperature in aeration tank = 18° C

Y = 0.6, K_d = 0.07 per day,   k' = 0.038 $(mg/liter)^{-1}(hr)^{-1}$ at 18° C

(all BOD_5 basis)

Assume 30% raw BOD_5 is removed in primary sedimentation, and BOD_5 going to aeration is therefore 252 mg/liter (0.7 x 360 mg/liter).

Selection of Θ_c, t, and MLSS

Considering the operating temperature and the desire to have nitrification and good sludge settling characteristics, adopt Θ_c = 5 days. As there is no special fear of toxic inflows, the hydraulic detention time t may be kept between 3 and 4 hr, and MLSS = 4000 mg/liter (complete mixing).

Effluent BOD_5 and efficiency

From Eq. (12.23),

$$S = \frac{1}{k'Y}\left(\frac{1}{\Theta_c} + K_d\right)$$

$$= \frac{1}{(0.038)(0.6)} \left(\frac{1}{5} + 0.07 \right) = 12 \text{ mg/liter}$$

Assume SS in effluent = 20 mg/liter and VSS/SS = 0.8. If degradable fraction of VSS = 0.7 (check later), BOD_5 of VSS in effluent = (0.7)(0.8 × 20) = 11 mg/liter. Thus, total effluent BOD_5 = 12 + 11 = 23 mg/liter (acceptable).

$$\text{Soluble } BOD_5 \text{ removal efficiency} = \frac{252 - 12}{252} = 95\%$$

$$\text{Overall } BOD_5 \text{ removal efficiency} = \frac{252 - 23}{252} = 91\%$$

$$BOD_5 \text{ removed} = (252 - 12)(9000)\frac{10^3}{10^6} = 2160 \text{ kg/day}$$

Aeration tank

Neglecting solids in final settling tank, we can estimate from Eq. (12.19)

$$XV = \frac{Y\Theta_c Q(S_0 - S)}{1 + K_d \Theta_c} \qquad \text{where } X = 0.8(4000) = 3200 \text{ mg/liter}$$

or

$$(3200 \text{ mg/liter})V = \frac{(0.6)(5)(9000)(252 - 12)}{[1 + (0.07)(5)]}$$

and

Tank volume V = 1500 m^3

Check for detention time $t = \frac{1500 \text{ m}^3 \times 24}{9000 \text{ m}^3/\text{day}} = 4 \text{ hr}$

Check for F/M ratio $= \frac{(252 - 12)9000}{(3200)(1500)} = 0.45 \text{ kg } BOD_5/\text{kg MLVSS-day}$

Let the aeration tank be in the form of four square-shaped compartments operated in two parallel rows, each with two cells of, for example, 11 m × 11 m × 3.1 m deep to ensure well-mixed conditions. (Check for D/UL or Dt/L_2, using Eq. (11.22). This gives a value of about 10.0.)

Return sludge pumping

If SS concentration of return flow = 1% = 10,000 mg/liter, we get from Eq. (12.43) the recirculation ratio

$$R = \frac{MLSS}{(10,000) - MLSS} = 0.67$$

or

   Return flow = 0.67Q = 0.67 × 9000 = 6000 m³/day

To ensure the possibility for a greater return flow if SVI is low, provide
3 pumps of 3000 m³/day capacity each. (Normally, two pumps will be used
for recycling and one will remain as a standby. Thus, when necessary, the
recirculation ratio can be increased from 0.67 to 1.0.

Settling tank design

   (See Chapter 17, Example 17.2)

Surplus sludge production

   From Eq. (12.13a),

$$\text{Net VSS produced/day} = \frac{XV}{\Theta_c} = \frac{(3200)(1500)(10^3/10^6)}{(5)} = \frac{4800}{5}$$

$$= 960 \text{ kg/day}$$

or, in terms of SS,

$$\text{Net production/day} = \frac{960}{0.8} = 1200 \text{ kg/day}$$

1. If these SS are removed as underflow from the final settling tank with
   a solids concentration of 1%, and if the specific gravity of sludge is
   assumed as nearly 1.00,

$$\text{Liquid sludge to be removed/day} = 1200 \times \frac{100}{1.0} = 120,000 \text{ kg/day}$$

$$= 120 \text{ m}^3/\text{day (namely, about 1.3\%}$$
$$\text{of Q)}$$

   This quantity is small and the recirculation pumps can be used to pump
   this flow also, but to the primary settling tank or to a thickener.

2. Alternatively, a flow of 300 m³/day may be withdrawn directly from the
   mixed liquor (concentration 4000 mg/liter) so as to withdraw the same
   weight of SS from the system per day. This can also be pumped to a
   primary settling tank or thickener.

$$\text{Check VSS production/BOD removed} = \frac{960}{2160} = 0.45 \text{ kg/kg BOD}_5 \text{ removed}$$

$$\cong 0.30 \text{ kg/kg BOD}_u \text{ removed}$$

$$= 24 \text{ g/person-day}$$

$$\text{Check SS production/BOD removed} = \frac{1200}{2160} = 0.56 \text{ kg/kg BOD}_5 \text{ removed}$$

$$\cong 0.38 \text{ kg/kg BOD}_u \text{ removed}$$

$$= 30.3 \text{ g/person-day}$$

This does not include the solids removed in the primary settling tank.

Digester design

It is proposed to provide aerobic digestion of the mixed primary and excess activated sludge as shown in Sec. 14.4.

Sludge drying

Determination of sludge drying beds area is discussed in Chapter 17.

Biodegradable fraction of VSS

From Eq. (12.33),

$$f_b = \frac{(YS_r + K_dX) - \sqrt{(YS_r + K_dX)^2 - 4K_dX(0.77YS_r)}}{2K_dX}$$

or

$$f_b = \left\{ (0.6)(2160) + (0.07)(4800) \right.$$
$$\left. - \sqrt{[(0.6)(2160) + (0.07)(4800)]^2 - (4)(0.07)(4800)(0.77)(0.6)(2160)} \right\}$$
$$/[(2)(0.07)(4800)] = 0.72 \quad \text{(A value of 0.70 was assumed earlier in the BOD computation.)}$$

Phosphorus removal

From Eq. (12.46), or by using the method shown in Example 12.3:

$$P, \text{ kg/day} = \frac{0.026 f_b \, \Delta X}{0.77} + \frac{0.01(0.77 - f_b) \, \Delta X}{0.77}$$

$$= \frac{(0.026)(0.72)(960)}{0.77} + \frac{0.01(0.77 - 0.72)(960)}{0.77} = 24 \text{ kg/day}$$

$$\text{kg P/100 kg BOD}_u \text{ removed} = \frac{24}{(1.47)(2160)} = 0.76$$

This P is removed from the system along with surplus sludge.

Incoming P at 2 g/person-day = 120 kg/day in raw sewage

Assume 25% removal in primary sedimentation, namely, 30 kg/day. Thus, the removal efficiency in the activated sludge section = 24/(120 - 30) = 26.7% and overall removal = (30 + 24)/120 = 45%.

Nitrogen removal

Similarly, incoming total Kjeldahl nitrogen (TKN) at 8.0 g/person-day = 480 kg/day. Assume 30% (i.e., 144 kg/day) is removed in primary sedimentation and the balance of 336 kg/day is oxidized to nitrates. Nitrogen removed along with surplus sludge can be estimated from Eq. (12.45), or as about 10% of surplus solids withdrawn, i.e., 0.10 x 960 = 96 kg/day.

Thus,

$$\text{Total nitrogen removal} = \frac{144 + 96}{480} = 50\%$$

Oxygen requirement

1. For carbonaceous demand, using Eq. (12.37), oxygen required

$$O_2 \text{ kg/day} = (BOD_u \text{ removed}) - (BOD_u \text{ of solids leaving})$$

$$= 1.47(2160 \text{ kg/day}) - 1.42(960 \text{ kg/day})$$

$$= 1812 \text{ kg/day} = 75.5 \text{ kg/hr} = 0.57 \text{ kg } O_2/\text{kg } BOD_u \text{ removed}$$

2. For nitrification, using Eq. (12.40), oxygen required

$$O_2 \text{ kg/day} = 4.33 \text{ (TKN oxidized, kg/day)}$$

Neglecting the nitrogen contained in the final effluent and in the surplus sludge withdrawn, the maximum amount of nitrogen to be oxidized equals 336 kg/day as computed above.

$$O_2 \text{ kg/day} = 4.33 \times 336 = 1455 \text{ kg/day} = 60.6 \text{ kg/hr}$$

3. Total oxygen required = 75.5 + 60.6 = 136 kg/hr = 1.0 kg O_2/kg BOD_u removed

$$\text{Oxygen uptake rate per unit tank volume} = \frac{136 \times 10^6}{1500 \times 10^3}$$

$$= 90.6 \text{ mg } O_2/\text{hr/liter tank volume}$$

$$= 50 \text{ mg } O_2/\text{hr/liter (without nitrification)}$$

Power requirement

Assume oxygenation capacity of aerators at field conditions is only 70% of the capacity at standard conditions (Chapter 13) and mechanical aerators are capable of giving 2 kg O_2/kWh at standard conditions.

$$\text{Power required} = \frac{136 \text{ kg/hr}}{0.7(2 \text{ kg/kWh})} = 97 \text{ kW (130 hp)}$$

$$= \frac{97 \text{ kW} \times 24 \text{ hr/day} \times 365 \text{ days}}{60,000 \text{ persons}}$$

$$= 14.2 \text{ kWh/year/person}$$

14.3 EXTENDED AERATION

Extended aeration systems are simpler in construction and operation than are activated sludge systems, and are often entirely aerobic since the

primary settling of the wastewater and the anaerobic digestion of the sludge
are omitted. The wastewater is brought directly to the aeration basin after
screening and grit removal. Aeration is now done for an extended period of
time, thus mineralizing the sludge solids sufficiently so that they can be
dewatered without any digestion. This helps to improve efficiency and to
simplify the whole operation. But the power requirement may be higher than
for activated sludge systems; this can be minimized by encouraging denitrification concurrently with nitrification in the aeration tank.

Pasveer and carrousel-type oxidation ditches are essentially extended
aeration systems. First developed in Holland, by 1977 there were well over
2000 installations of Pasveer-type ditches in western Europe alone, with
about 100 carrousels existing and another 80 under construction. The largest
carrousel-type ditch serves a population equivalent of about 8,300,000 for
BASF, Germany [1]. Several units of both types treat various industrial
wastes. The so-called "package plants" popular in the U.S. and some other
countries for small installations are also often based on the extended
aeration principle. Mechanically aerated lagoons can also be designed on
the extended aeration principle if desired.

Pasveer ditches of the type shown in Figure 14.2a are often preferred
for small installations. The ditch is used on a discontinuous basis — as
an aeration tank for some hours and as a settling tank for a little while
during which the aeration and inflow are switched off and the biological
flocs are given a chance to settle. The clear supernatant is decanted off.
Then aeration is switched on and inflow is resumed. This intermittent
operation gives good BOD and nitrogen removal, but requires that the raw
waste be held in the pump sump when the aeration is stopped, a practice
which can lead to anaerobicity in warm climates.

Consequently, continuous inflow and ditch operation are generally preferred, especially if flows are large and the climate warm. Continuous
operation of the ditch can be provided in one of the following ways: (1)
by providing two discontinuous ditches in parallel so that the continuous
inflow can be diverted from one to the other as required, (2) by providing
an arm in the ditch which can be isolated for use as a settling compartment
without disturbing the rest of the ditch, where aeration goes on continuously (Figure 14.2b), or (3) by providing a separate settling tank (Figure
14.2c). The adoption of the first two alternatives often necessitates
installation of automatic time switches to start and stop aeration and
open and close certain inflow and outflow valves. This may not be feasible

in all developing countries, and the provision of a separate settling tank
may be preferred even for small installations. Pasveer-type ditches are
about 1.0 to 1.5 m deep and are generally aerated by means of horizontal-
axis mechanical rotors ("cage" rotors) which create sufficient velocities
(30 cm/sec or more) to keep all solids in suspension.

Carrousel ditches have been gaining in popularity as they have some
constructional advantages over the Pasveer type in the case of larger flows.
With the trend toward larger oxidation ditches, the adoption of vertical-
axis aerators enables the use of deeper ditches (about 2.5 to 5.0 m), thus
saving on land requirement. The width of the ditches is also not restricted
by the availability of cage rotors, which need to be provided with intermediate supports when shafts are long. Vertical-axis rotors have been shown
to be considerably cheaper than cage rotors for larger ditches [10].

Figure 14.2d shows a typical carrousel aeration unit. One or more
surface aerators of the required oxygenation capacity are mounted centrally
at one end of the ditch, and the flow circulates around a dividing wall
placed lengthwise. The depth of the ditch is related to the diameter of
the aerator. For example, a depth up to 5 m may be necessary with a large
aerator of about 4 m diameter. Flow velocities of 30 cm/sec or more in the
ditch are claimed [1]. Thus, all solids are kept in suspension. The
developers claim that the hydraulic flow pattern in the ditch enhances the
oxygen transfer characteristics of the aerators. Aeration is concentrated
at one point, rather than being distributed over the length of the ditch.
Thus, the mixed liquor is rich in oxygen as it flows out of the aeration
zone, but low in oxygen as it returns back to the zone. This enhances
oxygen transfer, which is proportional to the dissolved oxygen deficit
from saturation. Separate settling tanks are generally provided.

The configuration shown in Figure 14.2e is recommended as it incor-
porates nitrification and denitrification along with BOD removal. This
reduces some oxygen requirement and thus helps conserve energy. The
channel section where denitrification takes place is maintained with a
dissolved oxygen concentration of 0.5 mg/liter or less, and a part of the
influent is diverted to this section to serve as a carbon source for the
denitrification. The other sections of the ditch have dissolved oxygen
around 2 mg/liter and, with sufficient mean cell residence time, nitrification occurs (see Sec. 14.6).

Figure 15.3 shows an extended aeration lagoon. Here, the ditch is
replaced by a lagoon. The name "extended aeration lagoon" may be used in

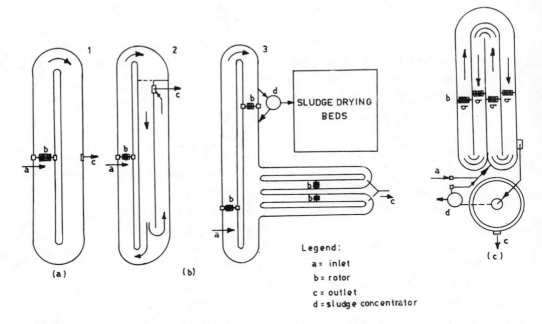

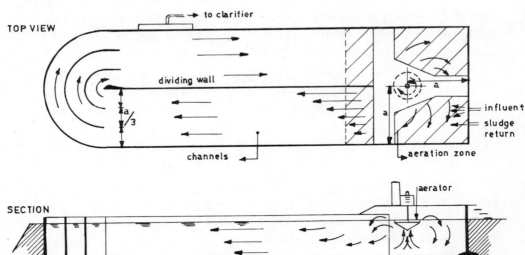

Figure 14.2 Different types of Pasveer oxidation ditches; (a) discontinuous operation, (b) continuous operation, (c) continuous operation
with separate settling tank, (d) a carrousel oxidation ditch (adapted from
Ref. 1), (e) a layout carrousel oxidation ditch with DO control for nitrification; although the receiving waterbody may not require a denitrified

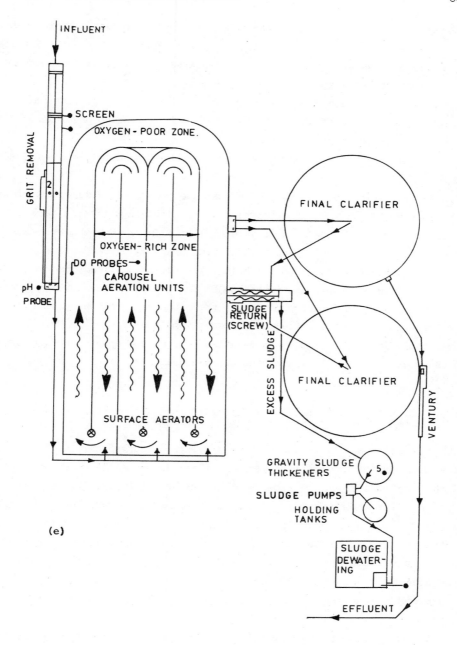

(e)

effluent, it is advisable to incorporate denitrification in the design of the plant as an energy conservation measure. Theoretically, 62.5% of the oxygen required for nitrification can be used for BOD removal by nitrifiers, thus reducing the power consumption for oxygenation (see text for explanation).

order to distinguish it from other types of aerated lagoons. The process is not affected in any manner if ditches are replaced by lagoons. Vertical-axis aerators are available for use in lagoons. In fact, the carrousel ditch may be regarded as a special case of the extended aeration lagoon. Land requirement is reduced compared with the Pasveer-type ditches. Intermittent operation and benefit from partial denitrification is also possible as discussed further in Chapter 15.

Package plants based on the extended aeration principle are available in several designs in the U.S. and elsewhere.

In U.S. practice the extended aeration system is considered economical for flows up to 1.0 mgd [2], but this is definitely not true in many other countries. In India, Arceivala et al. [3] have shown Pasveer-type ditches to be economical up to a population of 150,000, compared with conventional activated sludge or trickling filters. In Holland, the limit for carrousel ditches is now considered to be about 300,000 population equivalent [4]. For even larger populations the extended aeration systems are preferred in South Africa [19]. This question is discussed in Chapter 10, and Example 10.2 illustrates how relative capital and operating costs must be taken into account in making an economic analysis.

14.3.1 Selection of Θ_c and F/M in Extended Aeration Systems

As in the case of activated sludge design discussed earlier, the first step in designing any extended aeration plant is to select a suitable value of the mean cell residence time Θ_c and, consequently, F/M. The major factor affecting this choice is sludge stabilization and nitrification. The surplus sludge withdrawn from the system should be in such a sufficiently mineralized condition that it can be sent directly for dewatering without any further treatment. As seen in Sec. 12.22, the effluent quality is not much improved by extending the aeration time or Θ_c, but both the quantity of net volatile solids produced and their biodegradable fraction (BOD) are reduced.

In fact, theoretically speaking, in extended aeration systems, such a value of Θ_c can be adopted that all the volatile solids produced are destroyed by endogenous respiration to give an inert residue. As seen earlier (Figure 12.7), if aeration is continued for a sufficiently long time, the non-degradable residue is 23% of the new cells produced. Furthermore, if the substrate mass (BOD) removed per day is S_r, the VSS produced per day will be YS_r and the degradable solids produced per day will be $0.77YS_r$. If, as seen earlier, the mass of degradable solids in the system is f_bX and the fraction destroyed endogenously is K_d per day, then at steady state,

$$0.77YS_r = K_d f_b X \qquad (14.1)$$

or

$$X = \frac{0.77YS_r}{K_d f_b} \qquad (14.2)$$

Thus

$$\Theta_c = \frac{\text{VSS in system}}{\text{VSS produced/day}} = \frac{X}{YS_r} = \frac{0.77}{K_d f_b} \qquad (14.3)$$

Temperature affects both the constants K_d and f_b. The value of K_d also depends on the nature of the wastewater, whereas the fraction f_b depends on various factors embodied in Eq. (12.33) and generally varies from 0.3 to 0.5 for extended aeration systems (Table 14.1). For example, at 20° C, if $K_d = 0.09$ per day and $f_b = 0.5$, we get $\Theta_c = 22.2$ days. However, if $f_b = 0.4$, Θ_c must be increased to 27.7 days in order to completely destroy all degradable solids.

The above result can also be stated in the form of the food-to-microorganism ratio F/M since, from Eq. (12.14),

$$\frac{F}{M} = \frac{S_r}{X}$$

Substituting for S_r from above we get

$$\frac{F}{M} = \frac{K_d f_b}{0.77Y} \qquad (14.4)$$

If, in the example just seen, Y is taken as 0.58 (BOD$_5$ basis and 20° C), we get F/M = (0.09)(for example, 0.5)/(0.77)(0.58) = 0.1 kg BOD$_5$ removed per kg MLVSS-day.

Thus, Eqs. (14.3) and (14.4) provide a rational basis for the selection of Θ_c and F/M values for extended aeration systems where complete destruction of all degradable solids is desired on the premise that the remaining solids will be inert (mineralized) and will not need any further treatment in digesters before dewatering. The effect of temperature on the constants is difficult to take into account separately as discussed in Sec. 12.13, but their combined effect on process loading can be taken into account by selecting a suitable value of the temperature activity correction coefficient Θ from Table 12.1. Because of the low value of Θ, temperature does not have a large effect on the loading rate of extended aeration systems treating domestic sewage.

The values of Θ_c and F/M computed in the manner just described may be regarded as the limiting values for complete degradation. They are also generally more than sufficient for achieving full nitrification (Sec. 14.6.1). Lower values of Θ_c can also give nitrification and a sludge which can be dewatered equally readily without any odor nuisance in warmer climates. For example, Arceivala [5] has shown in full-scale sewage plants in India that very satisfactory effluents and sludge are obtained at F/M values as high as 0.22 kg BOD$_5$/kg MLVSS-day. Similarly, Handa [6], working on a pilot-scale oxidation ditch of 20,000 gal (90.8 m^3) volume treating domestic sewage, observed on the basis of sludge specific resistance that mean cell residence times as low as 10 days gave a very satisfactory performance when mixed liquor temperatures varied between 20 and 28° C. This corresponded to 0.2 kg BOD/kg MLVSS-day. BOD removals were around 97% in both the above cases.

For Dutch climatic conditions, Pasveer [7] suggests that small plants be provided with a volume of at least 250 liters/person on the basis of 54 g BOD$_5$/person and MLSS of 4000 mg/liter. Assuming VSS/SS = 0.65, an approximate F/M ratio of 0.083 kg BOD$_5$/kg MLVSS-day is implied. Similarly, Θ_c values implied in this criterion range between 25 and 30 days. For larger plants where better operational control is likely, the developers of the carrousel ditch [1] suggest higher loadings (lower Θ_c), which must be determined individually in each case so as not to affect sludge dewatering characteristics.

For subarctic conditions in Alaska, Coutts and Christianson [8] recommended that loadings be restricted to 0.08 kg BOD$_5$/kg MLSS-day (i.e., approximately 0.1 kg BOD/kg MLVSS) at mixed liquor temperatures continually below 7° C. Typical values of Θ_c and F/M under different climatic conditions are indicated in Table 14.1 for domestic sewage.

14.3.2 Aeration Tank Sizing

The next step in design is to select the MLSS concentration in the
aeration tank. Most extended aeration systems, especially the ditches and
package plants, are generally designed as complete mixing reactors, and
MLSS concentrations are kept between 3000 and 5000 mg/liter for reasons
stated earlier in the case of activated sludge. The size of the aeration
tank is then estimated either from the F/M ratio selected, or from Θ_c by
using Eq. (12.19). In estimating the total mass of solids in the system
as a whole, the solids in the final settling tank may also be taken into
account if desired, or omitted in the interest of a safer design.

The detention time in aeration tanks is always larger in the case of
extended aeration systems compared with activated sludge, owing to differences in loading rates. This becomes a distinct advantage when the inflows
are highly fluctuating in character or when they bring in toxic substances
as short period slugs. This aspect has been discussed in Sec. 11.15. For
the same reason, extended aeration systems show less variation in effluent
quality as a result of inflow variations. Typical detention times range
from 12 to 36 hr as shown in Table 14.1.

The geometric configuration of the aeration tank and the power input
per unit volume are kept so as to ensure as completely mixed a condition
as possible (D/UL ≥ 4.0), using methods given in Sec. 11.10. Both Pasveer
and carrousel-type ditches invariably show D/UL values greater than 4.0 in
tracer tests [1,5,6,8]. The total required aeration volume may also be
provided in the form of two or three aeration cells in series, each cell
in itself being a complete-mixing ditch or tank, with the series arrangement
helping to enhance efficiency by reducing the overall D/UL value. Such an
arrangement would, of course, not be desirable where toxic slugs are likely
to be received.

14.3.3 Sludge Handling

Recycling ratios between 0.75 and 1.0 are generally preferred to enable
maintining MLSS concentrations of 4000 mg/liter in aeration tanks under
varying conditions of sludge settleability (SVI).

Dissolved oxygen concentrations in the mixed liquor less than 0.5 mg per liter must be avoided as they tend to favor the growth of filamentous organisms and increase the sludge "bulking" (SVI). Increased aeration or reduced BOD load may restore the growth of floc-forming bacteria and reduce the SVI. Overaeration may lead to the formation of filamentous coli and again affect the SVI. Thus, variable recycling ratios are essential for the success of the process.

Design of the final settling tank must be based on batch flux studies to ensure that the expected solids flux can be handled in the "hindered settling" generally likely in such tanks. This is explained in Chapter 17.

Surplus sludge removal may be done from the settling tank, the underflow being recycled, or from the mixed liquor before settling as explained in Sec. 14.2.3. The latter has certain advantages and permits better control of the Θ_c value. The surplus sludge is generally too dilute for direct disposal to sludge drying beds, although this is often done in the case of small plants. But for larger ones, it may be desirable to send the surplus sludge first to a thickener, from which the supernatant can be returned back to aeration and the thickened underflow pumped to a drying bed. Alternatively, the dilute sludge may be sent directly to a lagoon wherever feasible. Design principles for sludge drying beds and sludge lagoons are given in Chapter 17.

The net VSS production rate in extended aeration systems ranges from 0.25 to 0.4 kg/kg BOD_u. To withdraw this much VSS, one must withdraw more SS since VSS constitutes only 50 to 80% of the SS.

Baars [9] states that 30 g dry weight of volatile and inert suspended solids per person must be removed daily from oxidation ditches in Holland. Pasveer points out [7] that this surplus sludge production of 30 g/person-day may increase to about 40 to 45 g SS/person-day in the winter. Zeper and DeMan [10] also give similar values from their experience in Holland. Coutts and Christianson [8] state from work in subarctic Alaskan conditions that surplus sludge production may reach 0.6 kg SS/kg BOD_5 applied (which is nearly equivalent to 32 g/person-day at 54 g BOD/person) at temperatures continually below 7° C.

Experience in India with mixed liquor temperatures ranging between 23 and 28° C and with higher loadings indicates a surplus sludge production of 18 to 20 g VSS/person-day (on 54 g BOD_5 p.e. basis) [6]. With a VSS/SS ratio of only about 50 to 60%, the surplus solids production should range between 26 and 40 g SS/person-day. Thus, it is interesting to observe that the solids production rate in extended aeration systems seems to vary with climate over only a relatively narrow range.

As some of the surplus solids escape with the effluent, the net quantity of surplus solids to be withdrawn is somewhat reduced in practice.

14.3.4 Oxygenation and Power Requirements

As in the case of activated sludge, the oxygenation requirement is estimated for meeting the carbonaceous demand using Eq. (12.37) and the nitrification demand using Eq. (12.40). Complete nitrification is generally likely to occur in extended aeration systems owing to the relatively larger Θ_c values used, especially in warmer temperatures, as long as sufficient oxygenation capacity is provided. This should be done where nitrification is desired.

Complete nitrification followed by partial denitrification can be achieved in the same unit if, as discussed earlier with regard to the ditch configuration shown in Figure 14.2e, a part of the ditch is so operated as to promote nitrification and another part to promote denitrification (see Sec. 14.6).

The advantage of such a system is that, even if a denitrified effluent is not essential, the system requires less aeration power since denitrification releases some oxygen. As stated in Sec. 12.20.2, theoretically, 62.5% of the oxygen used earlier in nitrification can be released by the reduction of nitrates in denitrification. Several workers have reported on this advantageous method of operation [11]. Oxygen requirements with and without nitrification are indicated in Table 14.1.

The aeration power requirement is computed for both summer and winter conditions and the higher of the two values is installed as explained in Chapter 13. The selection of aeration equipment is also discussed in Chapter 13. The power requirement generally varies from 13 to 20 kWh/person/year for domestic sewage, depending on the extent of nitrification and denitrification occurring in the system.

14.3.5 Nutrients

Nutrient addition, if required, may be computed as outlined for activated sludge systems as long as it is borne in mind that, generally, the nutrient requirement is less for extended aeration than it is for activated sludge (Sec. 12.18).

Nutrient removal from wastewater can be estimated as explained in Sec. 12.18. Nitrogen removal by nitrification, denitrification, and phosphorus removal by chemical precipitation are discussed in Sec. 14.6.

14.3.6 Construction of Oxidation Ditches

Construction of Pasveer-type oxidation ditches may be either in earthwork with sloping embankments or in brick or stone masonry with vertical walls. Whatever method is used, watertightness is essential to ensure the desired immersion of the cage rotor, Kessener brush, or vertical-axis aerator. Typical layouts are shown in Figure 14.2(a-e). The depth of the ditches is kept between 1.0 and 1.5 m, while their width is limited by the type and availability of the aeration rotors. The length of the ditches is not too important, but some effort is made to keep the perimeter of the walls to a minimum when two or more ditches are used side by side. It is now possible to construct deeper ditches (2.5 to 5.0 m) using vertical-axis rotors.

The design of the inlet and outlet from an aeration tank is relatively simple and several suggestions exist in the literature [19,20]. However, it is important to see that the outlet is sufficient to handle peak flows without undue heading up of liquid in the aeration tank, or else the aerator submergence will change and the oxygenation capacity will be affected. A weir of adequate length or a submerged pipe of adequate diameter may be used to connect the aeration tank to a separate settling tank.

Hopper-type settling tanks need to have hoppers with at least $60°$ slopes; hydrostatic removal of sludge (or underflow) should be on a continuous basis, especially in warmer climates. The sludge may be pumped back to the aeration tank inlet either by the use of screw pumps or an ordinary centrifugal-type sewage pump. Where possible, the sludge may be drained by gravity from the settling tank to the wet-well in a raw sewage pump house in order to be pumped back to the aeration tank.

Sludge drying is generally accomplished on open sand beds similar to the ones used in conventional treatment plants. If the beds are located sufficiently high, the underdrains can be sloped to convey the filtrate back to the ditch or to the effluent channel.

Where irrigation is proposed either with treated effluent or with wet sludge or both, the plant should be so located as to command the area to be irrigated by gravity or else additional pumping would be required.

The costs of construction and operation are discussed in Chapter 10.

The land requirement for oxidation ditches varies from 0.25 to 0.65 m^2 per capita and mainly depends upon the size of the aeration tank and drying beds.

Example 14.2 Using the same data given earlier in Example 14.1, design an extended aeration system to serve 60,000 people. Primary settling of the raw sewage is now to be omitted, and the surplus sludge should be fit for direct dewatering. Assume complete nitrification occurs, but neglect denitrification. The winter temperature in the aeration tank may be taken as $18°$ C as in Example 14.1.

Selection of Θ_c

From Eq. (14.3),

$$\Theta_c = \frac{0.77}{K_d f_b} = \frac{0.77}{(0.07)(0.63 \text{ assumed})} = 17 \text{ days}$$

The assumed value of f_b will be checked later, However, provide $\Theta_c = 20$ days.

Effluent BOD$_5$ and efficiency

From Eq. (12.21a),

$$S = \frac{1}{k'Y}\left(\frac{1}{\Theta_c} + K_d\right)$$

$$= \frac{1}{(0.038)(0.6)}\left(\frac{1}{20} + 0.07\right) = 5.0 \text{ mg/liter}$$

Assume SS in effluent = 20 mg/liter and VSS/SS = 0.7. If $f_b = 0.063$ (check later),

BOD$_5$ of VSS in effluent = $(0.63)(0.7 \times 20) \cong 9.0$ mg/liter

Thus, total effluent BOD$_5$ = 5.0 + 9.0 = 14.0 mg/liter (acceptable).

Influent BOD$_5$ = 360 mg/liter (no primary sedimentation)

Soluble BOD$_5$ removal efficiency = $\frac{360 - 5.0}{360}$ = 98.6%

Overall BOD$_5$ removal efficiency = $\frac{360 - 14.0}{360}$ = 96%

BOD$_5$ removed in aeration = $(360 - 5)(9000)(10^3/10^6)$ = 3195 kg/day

Aeration tank

Assume MLSS = 4000 mg/liter (complete mixing) and MLVSS = 0.7(4000) = 2800 mg/liter. Neglecting solids in final settling tank, we estimate from Eq. (12.19)

$$XV = \frac{Y\Theta_c Q(S_0 - S)}{1 + (K_d \Theta_c)}$$

or

$$2800V = \frac{(0.6)(20)(9000)(360 - 5.0)}{1 + (0.07)(20)}$$

Thus,

Tank volume V = 5700 m^3

Check for detention time

$$t = \frac{V}{Q} = \frac{5700 \times 24}{9000 \text{ m}^3/\text{day}} = 15.2 \text{ hr}$$

Check for F/M ratio

$$\text{F/M ratio} = \frac{(360 - 5)9000}{2800 \times 5700} = 0.2 \text{ kg BOD}_5/\text{kg MLVSS per day}$$

The aeration tank may be designed in the form of a set of Pasveer or carrousel-type ditches or as an extended aeration lagoon. The configuration can be selected on the basis of various considerations such as available equipment and its efficiency, available land, etc. The configuration chosen should be such as to promote complete mixing and prevent solids from settling in the aeration tank.

Return sludge pumping

1. Identical to Example 14.1. Recirculation ratio will be between 0.67 and 1.0, and 3 pumps of 3000 m^3/day capacity will be required.
2. Settling tank design will be made on the basis given in Example 17.2.

Surplus sludge production

From Eq. (12.12)

$$\text{Net VSS produced/day} = \frac{XV}{\Theta_c}$$

$$= \frac{(2800)(5700)(10^3/10^6)}{20} = \frac{15960}{20} = 798 \text{ kg/day}$$

or, in terms of SS, net production = 798/0.7 = 1140 kg/day.

1. If these SS are removed as underflow from the final settling tank with a solids concentration of 1%, the volume of sludge to be withdrawn is approximately 108 m^3/day. The same recirculation pump can also be used for this purpose, and the surplus sludge pumped either directly to drying beds or to a thickener before sending to drying beds.

2. Alternatively, a flow of 285 m^3/day may be withdrawn directly from the mixed liquor (concentration 4000 mg/liter) to withdraw the same weight of SS from the system daily. This flow would need to be pumped to a thickener before sending to drying beds.

$$\frac{\text{Check VSS production}}{\text{BOD removed}} = \frac{798}{3195} = 0.25 \text{ kg/kg BOD}_5 \text{ removed}$$

$$= 0.17 \text{ kg/kg BOD}_u \text{ removed}$$

$$= 14 \text{ g/person/day}$$

$$\frac{\text{Check SS production}}{\text{BOD removed}} = \frac{1140}{3195} = 0.36 \text{ kg/kg BOD}_5 \text{ removed}$$

$$= 0.24 \text{ kg/kg BOD}_u \text{ removed}$$

$$= 20 \text{ g/person-day}$$

(Both VSS and SS production values computed here theoretically are lower than the values likely in practice given in Table 14.1. This is due to the fact that all incoming BOD is not in soluble form; some solids are also present and some other factors can also affect it.)

Sludge drying or lagooning

See illustrative examples solved in Chapter 17.

Biodegradable fraction of VSS

From Eq. (12.33)

$$f_b = \frac{(YS_r + K_dX) - \sqrt{(YS_r + K_dX)^2 - 4K_dX(0.77YS_r)}}{2K_dX}$$

$$(YS_r + K_dX) = 0.6(3195) + 0.07(15960) = 3034$$

and

$$f_b = \frac{3034 - \sqrt{(3034)^2 - 4(0.07)(15,960)(0.77)(0.6)(3195)}}{2(0.07)(15,960)}$$

$$= 0.635 \text{ (retain value assumed earlier)}$$

Phosphorus removal

From Eq. (12.46)

$$P, \text{ kg/day} = \frac{0.026f_b\Delta X}{0.77} + \frac{0.01(0.77 - f_b)\Delta X}{0.77}$$

or

$$P = \frac{0.026(0.635)(798)}{0.77} + \frac{0.01(0.77 - 0.635)(798)}{0.77} = 18.5 \text{ kg/day}$$

$$\frac{\text{kg P removed}}{100 \text{ kg BOD}_u \text{ removed}} = \frac{18.5}{(1.47)(3195)} = 0.394$$

As there is no primary sedimentation, the total phosphorus removal is as given above. Influent P is 120 kg/day (Example 14.1); thus,

$$\text{Removal} = \frac{18.5}{120} = 15.4\%$$

Nitrogen removal

From Eq. (12.45), N removal = 90.7 kg/day = 1.93 kg/100 kg BOD$_u$ removed. Influent N = 480 kg/day (Example 14.1). Thus, removal = 18.9%. Besides this removal, some more N$_2$ is likely to be removed because of denitrification, which is neglected in this example (see Sec. 14.6).

Oxygen requirement

1. Using Eq. (12.37), the oxygen required for meeting the carbonaceous demand is computed as

    O$_2$, kg/day = BOD$_u$ removed - BOD$_u$ of solids leaving

    $$= 1.47(3195 \text{ kg/day}) - 1.42(798 \text{ kg/day})$$
    $$= 3564 \text{ kg/day} = 148.5 \text{ kg/hr}$$
    $$= 0.76 \text{ kg O}_2/\text{kg BOD}_u \text{ removed}$$

2. From Eq. (12.40), the oxygen required for nitrification is obtained (neglecting nitrogen removed in sludge, lost in effluent, and lost in denitrification) as a maximum upper limit of 4.33 (influent TKN/day), namely,

    O$_2$ for nitrification = 4.33(480 kg/day) = 2078 kg/day
    $$= 86.6 \text{ kg/hr}$$

    Total oxygen required = 148.5 + 86.6 = 235 kg/hr
    $$= 1.2 \text{ kg/kg BOD}_u \text{ removed}$$

    Oxygen uptake rate per unit tank volume = $\dfrac{235 \times 10^6}{5700 \times 10^3}$

    $$= 41.2 \text{ mg O}_2/\text{hr/liter tank volume}$$

Power requirement

Assuming the oxygenation capacity of aerators = 2 kg O$_2$/kWh at standard conditions, and only 70% capacity at field conditions,

$$\text{Power required} = \frac{235 \text{ kg/hr}}{0.7(2 \text{ kg } O_2/\text{kWh})} = 168 \text{ kW (225 hp)}$$

$$= \frac{168 \text{ kW} \times 24 \text{ hr/day} \times 365 \text{ days}}{60,000 \text{ persons}}$$

$$= 24.5 \text{ kWh/person/year}$$

This is the maximum power requirement assuming all incoming nitrogen has to be oxidized.

14.4 AEROBIC DIGESTION

The aerobic digestion of sludge achieves volatile solids stabilization similar to anaerobic digestion, but requires mechanical or pneumatic aeration as the process is aerobic. It is sometimes used in stabilizing surplus biological sludges generated in the activated sludge process or any of its modifications, and in trickling filtration (along with the primary sludge where settling is done prior to biological treatment). Its usage has been particularly popular with the smaller sized biological plants using the contact stabilization process and with certain "package" plants for domestic and industrial purposes, but there is no reason why it should not be considered even for larger sized plants if costs are favorable. A few typical flowsheets are illustrated in Figure 14.3.

Essentially, construction costs for aerobic digesters are lower, but operating costs are somewhat higher, than for anaerobic digesters. Among the operating advantages for aerobic units can be listed (1) production of a readily dewaterable sludge in less detention time than is generally required for anaerobic systems at comparable temperatures, (2) relatively odor-free operation, (3) lower BOD of supernatant, and (4) generally simpler operation with less chance of upsets owing to the presence of a more mixed and diverse ecological community. The fact that no methane gas is produced (which could partly offset energy requirements in the case of anaerobic units) should not be held against aerobic digesters, especially in developing countries where skilled operators and the necessary equipment may be lacking. Methane gas production and utilization can be a tricky business and may completely upset the order of priorities at a plant where the primary focus of limited resources should be on effluent quality and not on gas manufacture.

In aerobic digestion, microorganisms are in the "endogenous" phase, and a net decrease in microbial solids occurs with time. In this sense

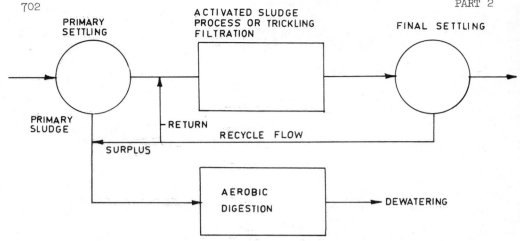

(a)

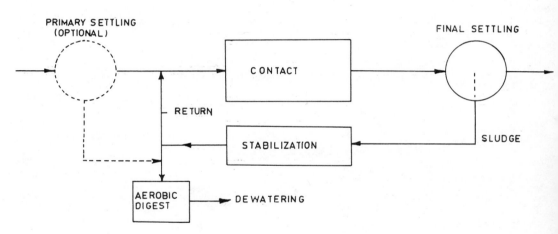

(b)

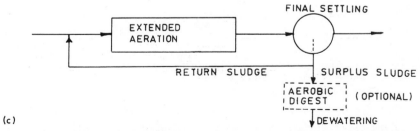

(c)

Figure 14.3 Some uses of aerobic digestion in waste treatment; (a) aerobic digestion of surplus sludge from activated sludge or trickling filters, (b) use of aerobic digestion in contact stabilization process, (c) use of aerobic digestion for further stabilization of surplus sludge from extended aeration if desired.

it can be said that aerobic digestion is similar to the extended aeration process, but is applied only to the sludge; many of the comments made earlier in Sec. 14.3 apply equally well here. Not much design and performance data are available directly from full-scale units, but they can be derived from comparable extended aeration plants which give readily disposable sludge. Most of the smaller sized aerobic digesters have been designed on an empirical basis as given in Table 14.2.

Generally, aerobic digesters are designed as flow-through-type reactors with no recycling. Thus, the mean cell residence time Θ_c is equal to the hydraulic detention time t. However, in the interest of reducing digester volume, various ways can be found of increasing "Θ_c" without having to increase "t," such as by introducing recycling or devices to hold back solids in the digester while allowing the supernatant to be displaced out by incoming sludge. Therefore, it is also more correct to talk in terms of "Θ_c" than in terms of "t."

The typical values of Θ_c indicated in Table 14.2 are at 20° C and need to be increased suitably at lower temperatures. Aerobic digesters are generally constructed as open unheated units and their winter temperatures must be estimated.

A recent trend is to develop design criteria based on laboratory-scale batch-type aerobic digesters, feeding the actual domestic or industrial sludge and operating at temperatures and solids concentrations around the anticipated values in the full-scale digesters. Batch digestion may be done for 25 to 30 days, withdrawing samples every few days to check for their VSS concentration. The results so obtained can be plotted as shown in Figure 14.4, from which the slope of the line would give K_v, the rate of

Table 14.2 Typical Design Basis for Aerobic Digesters
Treating Domestic Sludges

Item	Typical values
1. Mean cell residence time Θ_c, days (or hydraulic detention time, whichever controls) at 20° C	15-20 or more
2. Solids loading, kg volatile solids/m^3-day	1.6-3.2
3. Power level (surface aerators), kW/1000 m^3 digester volume	~20

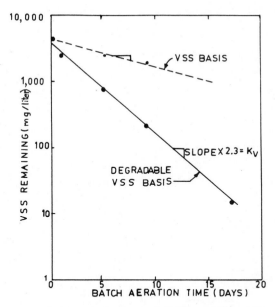

Figure 14.4 The correlation between volatile solids with detention time.

volatile solids destruction per unit time since, following first-order
kinetics in batch or ideal plug flow,

$$\frac{(VSS)_t}{(VSS)_0} = e^{-K_v t}$$ (14.5)

Thus, knowing the value of K_v, the required detention time can be
computed in order to reduce the VSS concentration from a given original
value to any desired final value. Generally, the percent reduction of VSS
aimed at varies from about 40 to 60%. It is evident that the required reduction in organic content should be correlated to the drainage characteristics of the final sludge, which may be dewatered either on open or on
covered sand beds or using vacuum filters or filter presses. The specific
resistance of sludge to dewatering and its correlation with its free drainage characteristics are discussed more fully in Chapter 17. Where mechanical methods of dewatering are used, a higher percent reduction in VSS may
be desirable than when drying beds are provided. Seasonal variations in
VSS reduction owing to temperature differences will also be observed in
existing units.

Typical values of K_v/day at $20°$ C range from about 0.03 to 0.05 for
mixed primary and activated sludges, and from about 0.05 to 0.07 for excess

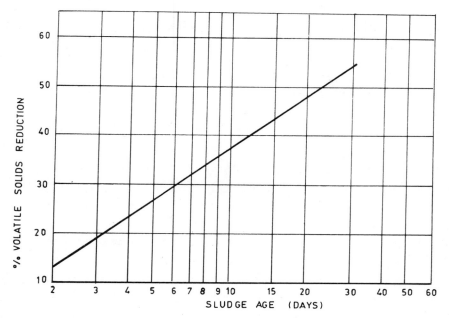

Figure 14.5 The correlation between volatile solids reduction and sludge age (from Ref. 12).

activated sludges alone. Thus, aerobic digesters treating excess activated sludges alone may be expected to give a higher reduction of volatile solids than those treating mixed sludges at the same detention time. Figure 14.5 shows the percent VSS reduction that can be expected at different mean cell residence times (Θ_c) for mixed primary and activated sludges, based on laboratory studies [12].

In designing full scale digesters, one can estimate the required detention time t in order to obtain a desired VSS reduction, on the assumption of complete mixing conditions in the digester, using Eq. (11.12b) which can be rewritten in the form

$$t = \frac{X_0 - X_e}{K_v X_e} \qquad\qquad (14.6)$$

where

X_0 = initial VSS concentration, or mass

X_e = VSS concentration or mass after digestion

K_v = VSS destruction rate constant, day^{-1}

t = required detention time, days

It would, however, appear to be more realistic to design aerobic digesters
of such long detention times as dispersed flow reactors rather than completely mixed ones. Referring to the discussion in Sec. 11.8, Eq. (11.15b)
or Figure 14.6 may be used with the above K_v values after selecting or
computing an appropriate value of the dispersion number D/UL as explained
in Sec. 11.7. Tracer tests to determine mixing conditions in full-scale
aerobic digesters have not yet been reported.

The oxygen requirements of aerobic digesters can be approximated either
from actual oxygen uptake tests on batch laboratory units or from theoretical computations using Eq. (12.36) and assuming that the nitrogen is also
fully oxidized to nitrate. The total oxygen requirement is about 2 kg per
kg of microbial cells oxidized.

The power requirement for oxygenation is thus computed knowing the
oxygen transfer capacity of aerators at the critical field conditions. The
power requirement must also be computed from the point of view of mixing
to keep all solids in suspension, and the higher of the two values used.
Recommended power levels range around 20 kW per 1000 m^3 digester volume
when the suspended solids concentration is 8000 to 12,000 mg/liter. At
high solids levels (greater than 20,000 mg/liter), Adams et al. [13] state
that up to 40 kW/1000 m^3 may be required to ensure that all solids are kept
in suspension. Most mechanical aerators in aerobic digesters require bottom
mixers for solids greater than 8000 mg/liter, especially if tanks deeper
than about 3.8 m are used [14].

Detention time in aerobic digesters is generally governed by the minimum value of K_v likely at winter temperatures in the digester. K_v values
at 20° C are adjusted to other temperatures using Eq. (12.32) with θ = 1.01
to 1.05. Oxygenation requirements are generally maximum at summer conditions when the destruction of volatile solids is likely to be highest.
Thus, both winter and summer operating temperatures must be known.

More recent ideas in design include the use of pure oxygen, higher
temperature operation, and blending of sludge in a mechanical blender to
improve its settleability and autooxidation rate [15].

Example 14.3 Using the data given earlier in Example 14.1, design
an aerobic digester to treat the mixed primary and surplus activated sludge
likely to be received from an activated sludge plant serving 60,000 people.
The following design data may be used:

Surplus SS produced/day = 1200 kg/day (from Example 14.1)

Surplus VSS produced/day = 960 kg/day (from Example 14.1).

Primary sludge produced/day = 60 g dry SS/person and 70% volatile

Sludge moisture content after thickening = 97% (assumed)

Sludge specific gravity = 1.00 (assumed)

K_v = 0.05 per day at 20°C. θ for temperature correction = 1.05

Desired VSS reduction = 45%

Minimum operating temperature = 15°C; maximum = 20°C

Solution

Primary sludge solids = 60 x 60,000 x 10^{-3} = 3600 kg/day

Surplus SS produced in activated sludge = 1200 kg/day

Total SS to digester = 3600 + 1200 = 4800 kg/day

Sludge volume at 3% solids = 4800/0.03 = 160 m^3/day

VSS concentration X_0 = [(0.7)(3600) + (960)]/(160)(10^{-3}) = 21,750 mg per liter

At 45% reduction of VSS,

Effluent concentration X_e = 11,960 mg/liter

K_v at 15°C = 0.05$(1.05)^{15-20}$ = 0.04 per day

Assuming complete mixing conditions in digester, we get from Eq. (14.6)

$$t = \frac{X_0 - X_e}{K_v X_e} = \frac{12,750 - 11,960}{(0.04)(11,960)} = 20.5 \text{ days}$$

Digester volume = 20.5 x 160 = 3280 m^3

Solids loading = $\frac{(0.7 \times 3600) + 960}{3280} = \frac{3480}{3280}$ = 1.1 kg/m^3-day

At 2 kg/kg VSS fully oxidized,

Oxygen required = (2)(0.45 x 3480 kg/day) = 3132 kg/day
                                          = 130.5 kg/hr

Assuming that at actual field conditions the mechanical aerators can give 70% of their standard oxygenation capacity of 2 kg O_2/kWh (Chapter 13),

Power required = $\frac{130.5}{(0.7)(2.0)}$ = 93.2 kW (125 hp)

Check for minimum power level = $\frac{93.2}{3.28}$ = 28.4 kW/1000 m^3

14.5   CONTACT STABILIZATION PROCESS

The contact stabilization process is yet another modification of the
activated sludge process in which advantage is taken of the fact that substrate removal, especially in the case of domestic sewage, occurs in two
stages. In the first stage (lasting 0.5 to 1.0 hr) the colloidal, finely
suspended and dissolved organics present in the sewage are rapidly adsorbed
on the activated sludge solids. In the second phase, the adsorbed organics
are separated and oxidized in 3 to 6 hr. In the processes seen so far,
both of these steps occurred in the same unit. In the contact stabilization
process, the two steps are separated or functionalized. The first step
occurs in what is called the "contact" tank, and the second one in the
"stabilization" tank. Thus, so to speak, the growth phase is separated
from the decay phase. A typical arrangement in shown in Figure 14.6.

This functionalization helps to reduce the overall volumetric requirement of tanks compared with the activated sludge system, although the oxygen
requirement remains practically the same because the organics to be oxidized
are the same in both cases. The surplus sludge is withdrawn for further

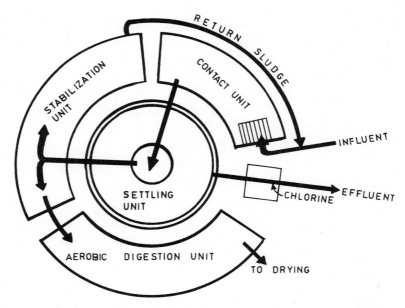

Figure 14.6 A typical "package" plant using the contact stabilization process with aerobic digestion of surplus sludge.

treatment in an aerobic or anaerobic digester to facilitate its further disposal. In efficiency of performance, the contact process is somewhat similar to the conventional activated sludge process. An MLSS concentration up to 3000 mg/liter is maintained in contact units, and even up to 10,000 mg/liter in stabilization units.

The contact stabilization process generally lends itself well to the treatment of domestic sewage. But, before deciding on its use for industrial wastes, it is desirable to ensure in laboratory studies that about 85% removal of BOD_5 occurs in about 15 min contact time with previously stabilized sludge. Unless BOD is sufficiently present in colloidal or suspended form, the contact step will not be very effective. Where toxic or highly fluctuating inflows are expected, it may not be advisable to use the contact process (Sec. 11.16).

Existing conventional activated sludge plants have sometimes been converted into contact stabilization plants capable of handling nearly double the original flow without any major additions to the overall tank capacities. This is possible because, for example, if the conventional plant receives a flow of Q m^3/hr and the aeration time required is, for example, 6 hr, the tank volume required is $6Q_c$. Now, for a contact stabilization plant, if the "contact" time required is 1 hr and the "stabilization" time is 4 hr, based on a return flow of, for example, 0.5Q, the total tank volume required is only (Q × 1 hr) + (0.5Q × 4 hr) = 3Q. Thus, double the flow can be accommodated by converting from the conventional to the contact process.

The contact stabilization process is also popular in the case of small sized "package" plants. In fact, most of the package plants are based on either the extended aeration system or the contact stabilization system. In the latter case, separate aerated compartments are generally provided: one for "contact," one for "stabilization" of the return sludge, and one for the aerobic digestion of the surplus sludge. As shown in Figure 14.6, a separate settling compartment is also necessary, as are various pumps and piping, etc., to transfer liquids from one compartment to the other. A very compact arrangement is possible in some "package" designs. However, in developing countries, especially in warm climates, the use of air and hydraulic lift devices, as well as diffused air systems incorporated in such plants, can become a source of maintenance troubles in operation and some redesigning may be called for to use conventional pumps and mechanical aerators.

14.6 NITRIFICATION-DENITRIFICATION SYSTEMS

A certain amount of nitrogen removal (20-30%) occurs in conventional
activated sludge and extended aeration systems. If additional removal is
desired in order to meet wastewater discharge standards, some process modification is necessary. Nitrogen removal ranging from 70 to over 90% can
be obtained by use of the nitrification-denitrification method in plants
based on activated sludge and other suspended growth systems. The activated sludge process has been used successfully with ammonia **concentrations**
as high as 500 mg NH_3-N/liter to achieve 90% removal.

Recalling our discussion in Secs. 12.20.1 and 12.20.2, biological denitrification requires prior nitrification of all ammonia and organic nitrogen in the incoming waste. Thus, an important requirement of plant design
is that **nitrification** should occur at all times, even under the worst operating conditions (e.g., winter conditions, slug discharges, etc.). For
this reason, the sludge age or mean cell residence time Θ_c should be carefully selected with a sufficient factor for safety. Extended aeration plants
with their higher Θ_c values are likely to be more uniform in nitrification
performance than activated sludge plants of low sludge age.

14.6.1 Nitrification

Typical flowsheets for achieving <u>nitrification</u> in suspended growth
systems are shown in Figure 14.7.

Figure 14.7a shows a <u>combined system</u> for biological oxidation (BOD
removal) and nitrification in a single tank. This is the favored method
of operation. It is less sensitive to load variations — owing to the larger
sized aeration tank — generally produces a smaller volume of surplus sludge
owing to higher values of Θ_c adopted, and there are some reports to indicate
that such sludges have better settling characteristics. The sludge mainly
has usual heterotrophs with only about 1 to 2% nitrifiers. Thus, nitrification per se gives little additional sludge.

In north European countries (e.g., Holland) such plants are designed
with the ratio of F/M = 0.05-0.10 kg BOD_5 per kg MLSS to have nitrification
all year round ($\sim 10^\circ$C). Higher F/M ratios may not show nitrification during some winter months. Choice of this ratio under any climatic conditions
should be guided by sludge age considerations outlined in Sec. 12.20.1.

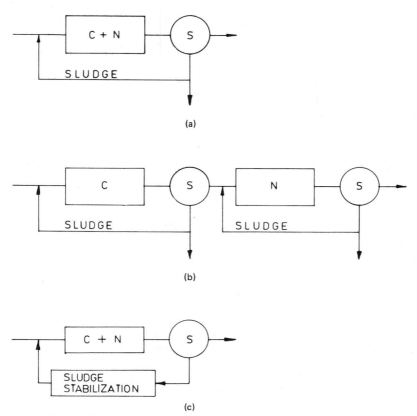

Figure 14.7 Typical flowsheets for nitrification systems; (a) suspended growth, combined system, (b) suspended growth, separate system, (c) suspended growth combined system with contact stabilization; C — biological oxidation of carbonaceous matter, N — nitrification, S — sedimentation.

The nitrification rate reported from Holland [16] for an extended aeration plant has been 1.6 mg NH_3-N/g VSS-hr at $10°C$.

Care should be taken to ensure that the oxygenation capacity of aeration tanks is sufficient to meet oxygen uptake due to carbonaceous demand and nitrification. Recycling of sludge must be rapid enough to prevent denitrification (and "rising" sludge) owing to anoxic conditions in the settling tank — a phenomenon often experienced in warm seasons.

Figure 14.7b shows a separate system in which biological oxidation and nitrification take place in two separate successive tanks, each having its separate sludge recycling arrangement. Consequently, the first tank can now be smaller in size since a higher F/M ratio can be used, but this makes the system somewhat more sensitive to load variations and also tends to produce more sludge for disposal. An additional settling tank is also necessary between the two aeration tanks to keep the two sludges separate. A principal advantage of this system is its higher efficiency of nitrification and its better performance when toxic substances are feared to be in the inflow. As long as these toxics are biodegradable, the first part of the plant will take care of them and the nitrifiers will be protected.

Figure 14.7c shows a typical contact stabilization system in which nitrification can occur along with biological oxidation in the contact step if the hydraulic detention time in this tank is adequate for nitrification which, as seen in Table 12.4, occurs only at about 2-5 mg NH_3-N/g VSS-hr at $20°C$ in combined, suspended growth systems. Thus, in the design of contact stabilization plants, not only should sludge age be adequate but also the contact time. This may make the choice of this method unattractive unless influent ammonia concentration is quite low.

14.6.2 Denitrification

Denitrification in suspended growth systems can be achieved using any one of the typical flowsheets shown in Figure 14.8.

The use of methanol requires a flowsheet of the type shown in Figure 14.8a. The wastewater has first to be nitrified using either a combined or separate system as just discussed. The nitrified waste is then dosed with methanol and allowed to denitrify in a stirred, anoxic tank. The rest of the steps include reaeration, settling, and sludge recycling as in usual waste treatment plants.

Methanol is relatively more expensive to use, but plant capital cost is somewhat reduced by the fact that the denitrification rate is quite rapid (about 10 mg NO_3-N/g VSS-hr at $20°C$), thus requiring a much smaller sized anoxic tank. Some operating difficulties may arise from the dosing rate of methanol, which in the case of a fluctuating load situation may become either too much or too little. Too much would introduce an unnecessary BOD in the effluent while too little would leave some nitrates undenitrified.

Generally speaking, the use of methanol or another artificial carbon source should be avoided as far as possible since they add to the cost of treatment. A more satisfactory arrangement would be to use the carbon contained in the waste itself to bring about denitrification. With methanol

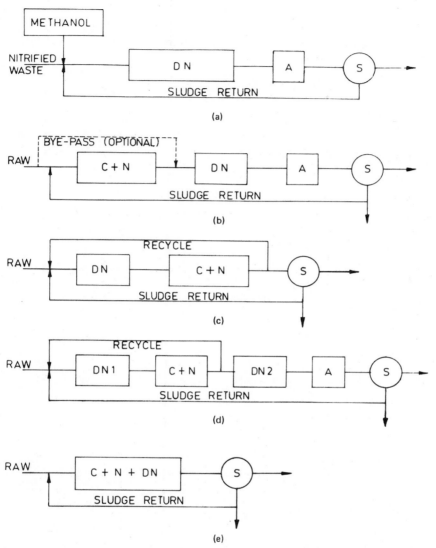

Figure 14.8 Typical flowsheets for denitrification systems; (a) separate denitrification of nitrified waste, (b) postdenitrification, (c) predenitrification, (d) "Bardenpho" arrangement, (e) simultaneous nitrification-denitrification; C - carbonaceous matter, N - nitrification, DN - denitrification (anoxic), S - sedimentation.

usage nitrification-denitrification may increase the cost of treatment by as much as 60% over that for conventional activated sludge [18]. The use of internally available carbon may increase plant costs by only about 10% over conventional activated sludge when nitrification-denitrification is achieved simultaneously in extended aeration systems and by about 30% in case of separate nitrification and denitrification systems.

The system shown in Figure 14.8b is similar to the one first proposed by Wuhrmann in which denitrification is allowed to take place in a separate anoxic tank following the nitrification step. The internally available carbon is used and no chemical is necessary.

However, the anoxic tank has to be of sufficient detention time for denitrification to occur which, as discussed in Sec. 12.20.2, has a slower rate (about 0.4-2 mg NO_3-N/g VSS-hr at 20°C) since the corresponding oxygen uptake rate of the mixed liquor is due mainly to endogenous respiration and thus low. The denitrification rate therefore in a way also depends on the F/M ratio used in the prior aeration tank [18]. Figure 12.9 can be used as illustrated in Example 14.5 to determine the likely denitrification rate in the anoxic tank from the oxygen uptake rate at a known temperature. The rate can then be adjusted for any other temperature using θ values shown in Table 12.4. The anoxic tank must be kept gently stirred (not aerated) to prevent solids settling in it.

Finally, the flow from the anoxic tank is desirable to reaerate for 10-15 min to drive off nitrogen gas bubbles and add oxygen prior to sedimentation. Sludge recycling from the settling tank should be effected rapidly to avoid its becoming anoxic once again and causing sludge lifting by nitrogen gas bubbles.

While the flowsheet is perfectly feasible and high removal of nitrogen can be achieved, the detention time required in the anoxic tank is rather high owing to the low endogenous respiration rate. Consequently, if desired, a portion of the raw waste may be bypassed to enter directly into the anoxic tank and thus contribute to an increased respiration rate. This arrangement helps to reduce the size of both the anoxic and aerobic tanks, but reduces overall nitrogen removal since some unnitrified ammonia in the bypassed waste cannot be denitrified.

Figure 14.8c shows a similar denitrification system but with the relative positions of the anoxic and aerobic tanks reversed. The stirred anoxic tank now receives the incoming raw waste directly. In its anoxic state, the oxygen requirement of the waste is met by the release of oxygen from nitrates in the recycled flow taken from the end of the nitrification tank. Primary settling of the raw waste may be omitted so as to bring more carbon into the anoxic tank. Owing to the presence of the raw waste, denitrification rates in the anoxic tank are higher (about 2-3 mg NO_3-N/g VSS-hr at

20° C). The method benefits from the potential that raw unaerated waste has for denitrification. Example 14.5 illustrates the design method.

This arrangement has some advantages and is thus often preferred to the one described previously, e.g., the size of the anoxic tank is reduced owing to faster denitrification rates; the overall oxygen requirement is reduced since oxygen released by nitrate reduction in the anoxic tank is useful in satisfying the oxygen demand of the incoming waste; and a separate short-detention reaeration tank is not necessary prior to settling. However, the requirement of recycling of nitrified liquor is quite high. This recycle has to be on the order of 2 to 3 times the raw waste inflow to achieve upward of 80% denitrification. Sludge recycling from the settling tank is also necessary.

Since some biological stabilization occurs in the anoxic tank, the size of the subsequent aerobic tank can be reduced somewhat if its detention time is not reduced below that necessary for completing nitrification at the expected rate at liquor temperature.

More complete nitrification-denitrification can be achieved by use of a flowsheet as shown in Figure 14.8d which gives the so-called Bardenpho method developed in South Africa [19]. It is similar to the previous two types from which it combines their features by having two anoxic tanks, one on either side of the aerobic tank. The first anoxic tank has the advantage of a higher denitrification rate for reasons just discussed. The percentage removal of nitrogen in it can be increased up to about 85% by adopting a suitably high rate of recycling of nitrified liquor, while the nitrates remaining in the liquor passing out of the tank can be denitrified further in a second anoxic tank through endogenous respiration, thus giving an overall removal upward of 90%. The liquor is finally aerated before settling and settled sludge recycled.

Heide [16], working with a similar flowsheet (but with lime addition prior to final settling to achieve phosphorus removal), has also reported good results (> 85% N removal at 10° C) from a plant serving 500 p.e. in Holland. Actual denitrification rates measured at different temperatures in both these plants [16,19] are summarized in Table 14.3.

Simultaneous nitrification and denitrification can also be achieved in a single, large enough tank (with sufficient sludge age) using a simple flowsheet of the type shown in Figures 14.8e and 14.2e. It results from having aerobic and anoxic zones within the same tank — a condition often

Table 14.3 Denitrification Rates [16,19]

Tank	Carbon source	Denitrification rate (mg NO_3-N/g VSS-hr)	
		South Africa [19]	Holland [16]
1st anoxic	Raw sewage	3.6 (20° C)	1.1 (10° C)
2nd anoxic	Endogenous respiration	1.3 (20° C)	0.5 (10° C)

occurring in extended aeration ditches where zones in the immediate vicinity of aerators are rich in oxygen and others away from aerators or deeper in the ditch are anoxic. The nitrifying and denitrifying bacteria in the biomass are not adversely affected by the rapidly alternating aerobic and anoxic environments so that wherever dissolved oxygen is available, nitrates are formed and wherever dissolved oxygen is absent, denitrification occurs. This phenomenon has also been demonstrated in laboratory-scale reactors subjected to alternating conditions [20]. Consequently, simultaneous nitrification-denitrification occurs continuously in such systems.

However, this simultaneous action remains largely uncontrolled in full-scale plants as it is dependent on the fluctuating influent loads in relation to a relatively constant aeration capacity. The relative proportion of anoxic zones changes from hour to hour and is highest when the plant receives peak loads [21]. Even under such fluctuating conditions about 50% nitrogen removals have often been reported [20,21].

A more controlled form of simultaneous action can be achieved by suitable placement and spacing of aerators, which is especially feasible with extended aeration systems in long ditch-like constructions, so that zones farthest from the aerators are more evenly anoxic at all times, and high nitrogen removals (> 80%) are assured [20]. Oxygenation requirements are also reduced.

In existing extended aeration plants built with evenly spaced aerators, much improvement in nitrogen removal efficiencies has been demonstrated by switching off a few aerators at the head end of the tank where the raw waste and return sludge are received. This has, in fact, helped to conserve electrical energy without affecting BOD removals by using the oxygen released from denitrification [19].

Recycle pumping is avoided if there is continuous circulation of the mixed liquor through the aerobic and anoxic zones. This is done anyway in the case of ditch-like constructions and should cause no extra cost. A return motion of the flow can also be established by a clever choice of type and placement of aerators [19]. In the case of extended aeration-type lagoons the objective of simultaneous nitrification-denitrification-sedimentation and recycle can be achieved by alternately switching on and off the aerators in the lagoon (see Chapter 15).

At the present time it would appear that the most promising approach to biological treatment is to use the extended aeration process in a relatively deep (4-5 m) ditch or lagoon-type configuration with some effort at process functionalizing to allow better operational control and higher treatment efficiency at minimum power cost. This can be achieved by adopting, first, a suitable sludge age Θ_c and, second, engineering the aeration system, the inflow, outflow, and sludge return so as to promote fairly well-defined aerobic and anoxic zones in the ditch in a sequence similar to the Bardenpho or other flowsheets described previously. Such arrangements would give not only high BOD removal ($> 95\%$) but also high nitrogen removal ($> 85\%$), often at no extra cost. In fact, there is even a reduced aeration cost by benefiting from oxygen released by denitrification. Generally speaking, these arrangements give an effluent of more uniform quality and less sensitivity to fluctuating inflows and loads while keeping the plant's equipment requirement and operation relatively simple.

One such arrangement is shown in Figure 14.2e based on the carrousel ditch system [20] and another in Figure 15.5 based on an intermittently operated lagoon which is eminently suitable for smaller flows.

Example 14.4 Design a complete-mixing activated sludge system with nitrification-denitrification to treat wastewater from a community of 60,000 persons based on the following data from Example 14.1:

Flow $Q = 9000$ m^3/day; BOD$_5$ = 360 mg/liter (raw)

Assume 30% BOD removal in primary settling and 90% in biological step.

TKN = 53 mg/liter (raw) = 40 mg/liter (for example) after settling

Winter temperature of mixed liquor = 10° C

Yield Y = 0.6

K$_d$ = 0.07/day (BOD$_5$ basis, 15° C)

MLSS = 4000 mg/liter; VSS/SS = 0.8

Assume organic N in effluent = 5 mg/liter

A. Nitrification Step

Adopt sludge age Θ_c = 10 days (Sec. 12.20.1). Provide combined bio-oxidation and nitrification in a single tank (Figure 14.7a).

Aeration tank

Raw settled BOD to aeration = 360(1 - 0.3) = 252 mg/liter

$$XV = \frac{YQ(S_0 - S)\Theta_c}{1 + K_d\Theta_c} \qquad \text{[Equation (12.20)]}$$

$$(0.8 \times 4000)V = \frac{(0.6)(9000)(252 \times 0.9)(10) \times 10^{-3}}{(1 + 0.07)(10)}$$

$V = 2250 \ m^3$

$$MLVSS = \frac{2250(0.8 \times 4000)}{10^3} = 7200 \ kg$$

$$\text{Detention time} = \frac{2250 \ m^3 \times 4}{9000 \ m^3/day} = 6 \ hr$$

F/M ratio

$$\frac{BOD_5/day}{MLVSS} = \frac{9000 \times 252 \times 10^{-3}}{7200} = 0.315 \ kg \ BOD_5/kg \ MLVSS\text{-}day$$

$$BOD \ removed/day = (9000 \ m^3/day)(252 \times 0.9) \times 10^{-3} = 1620 \ kg \ BOD_5/day$$

Excess solids produced/day

$$= Y(BOD_5) - K_d(MLVSS) = 0.6(1620 \ kg/day) - (0.07/day)(7200 \ kg)$$
$$= 468 \ kg/day$$

Oxygen requirement

For carbonaceous demand:

$$O_2 \ uptake/day = (BOD_u \ removed/day) - (BOD_u \ of \ solids \ withdrawn/day)$$
$$= 1.47(1620 \ kg/day) - (1.42 \times 468 \ kg)$$
$$= 1716 \ kg/day = 71.5 \ kg/hr$$

For nitrification:

$$Total \ N \ to \ be \ oxidized = (incoming \ N) - (N \ in \ effluent) - (N \ in \ sludge)$$
$$= [40 \ mg/liter \ (say) \times 9000 \times 10^{-3}] - [5 \ mg/liter$$
$$\times 9000 \times 10^{-3}] - [7\% \ (say) \times 468 \ kg/day]$$
$$= 282 \ kg \ NH_3\text{-}N/day = 11.75 \ kg/hr$$

$$O_2 \ uptake/day = 4.33(282) = 1222 \ kg/day = 51 \ kg/hr$$

Hence, total oxygen uptake rate = 71.5 + 51 = 122.5 kg/hr at field conditions.

$$\text{Check for nitrification rate} = \frac{11.75 \text{ kg/hr} \times 10^6}{7200 \text{ kg} \times 10^3}$$

$$= 1.63 \text{ mg } NH_3\text{-N/g VSS-hr (acceptable at } 10^\circ C)$$

B. Alternative Methods of Denitrification

1. **Aerobic tank followed by anoxic tank (Fig. 14.8b)**

Estimate denitrification rate using Figure 12.9 as follows:

$$O_2 \text{ for endogenous respiration} = 1.42 \times K_d = 1.42 \times 0.07 = 0.1/\text{day}$$

$$= 0.0041/\text{hr; namely, } 0.0041 \text{ g } O_2/\text{g VSS-hr,}$$
$$\text{or } 4.1 \text{ mg } O_2/\text{g VSS-hr}$$

At this oxygen uptake rate the corresponding denitrification rate = 1.1 mg NO_3-N/g VSS-hr. However, this rate corresponds to 15°C (since the value of K_d was at 15°C). Hence, corrected denitrification rate at 10°C (using θ = 1.15) = 0.5 mg NO_3-N/g VSS-hr.

Total N to be reduced = 282 kg NO_3-N/day = 11.75 kg/hr. Hence,

$$\text{VSS required in anoxic tank} = \frac{11.75 \text{ kg/hr} \times 10^6}{0.5 \times 10^3} = 23,500 \text{ kg}$$

Assume MLVSS concentration in anoxic tank is the same as in aerobic tank (i.e., 3200 mg/liter).

$$\text{Tank volume} = \frac{2350 \times 10^3}{3200} = 7343 \text{ m}^3$$

$$\text{Detention time} = \frac{7343}{9000} \times 24 = 19.58 \text{ hr}$$

(Slightly more time is needed if DO present in influent to anoxic tank is taken into account.)

2. **Anoxic tank followed by aerobic tank (Figure 14.8c)**

From the above, carbonaceous demand for O_2/day = 71.5 kg/hr. Assume that anoxic tank gives 8-hr detention (vol = 3000 m³) and MLVSS = 3200 mg/liter = 3200 × 3000 × 10^{-3} = 9,600 kg. Hence,

$$O_2 \text{ uptake rate} = \frac{71.5 \text{ kg/hr} \times 10^6}{9.600 \text{ kg} \times 10^3} = 7.5 \text{ mg } O_2/\text{g VSS-hr}$$

From Figure 12.9, the corresponding denitrific rate is about 2.2 mg NO_3-N/g VSS-hr. This is at 15°C. The corresponding denitrification

rate at $10°C$ (using $\theta = 1.15$) = 1.1 mg NO_3-N/g VSS-hr. Hence, the minimum VSS required in the anoxic tank to reduce 11.75 kg NO_3-N/hr

$$= \frac{11.75 \times 10^6}{1.1 \times 10^3} = 10,680 \text{ kg}$$

Thus, increase anoxic tank capacity to give 9-hr detention.

Recycling rate

Each kg of NO_3-N releases 2.86 kg O_2. O_2 required/hr for carbonaceous
demand in anoxic tank = 71.5 kg/hr. Hence, NO_3-N required = 25 kg/hr (but
inflow Q gives only 11.75 kg/hr). Therefore,

$$\text{Recycling rate} = \frac{25}{11.75} = 2.13 \text{ Q}$$

Aerobic tank

Since biological "oxidation" has already been occurring in the anoxic
tank, the size of the aerobic tank can be reduced, but the size should be
enough for nitrification to be complete. Hence, in the above case there is
not much value in reducing the aerobic tank since the nitrification rate
determined earlier is already high enough at $10°C$.

Power saving

Oxygenation requirement (and hence power) is reduced owing to release
of oxygen from nitrates. Saving can be estimated at about 20% in this case
(also see Example 14.5).

3. Use of methanol prior to anoxic tank
    (no bypass of sewage) (Figure 14.8a)

Methanol required, kg/day = 2.47 (NO_3-N, kg/day)
                          = 2.47(282 kg/day) = 696 kg/day (cost?)

Assume denitrification rate = 10 mg NO_3-N/g VSS-hr at $20°C$. Hence, at $10°C$
(using $\theta = 1.15$)

Rate = 2.5 mg NO_3-N/g VSS-hr

Total N to be reduced = 11.75 kg/hr

VSS required in anoxic tank = $\frac{11.75 \times 10^6}{2.5 \times 10^3}$ = 4.700 kg

Hence,

Volume of tank = $\frac{4700 \text{ kg} \times 10^6}{3200 \text{ mg/liter} \times 10^3}$ = 14.68 m³ i.e.,

$$\text{Detention time} = \frac{1468 \times 24}{9000} = 3.9 \text{ hr}$$

Final aeration tank

Detention time = 15 min. (Settling, sludge recycling, etc., may be designed following methods outlined earlier.)

Example 14.5 Recalling the data used in solving Example 14.2 concerning an extended aeration system for 60,000 persons, a ditch unit is to be provided with appropriate anoxic and aerobic zones for promoting nitrification-denitrification in an arrangement similar to the Bardenpho method. Estimate what fraction of total ditch volume may have to be allocated for each purpose.

Minimum temperature = $18°$ C
Sludge age Θ_c = 20 days
Inflow = 9000 m^3/day
BOD_5 influent = 360 mg/liter = 3240 kg/day
TKN = 53 mg/liter = 480 kg/day
Detention time in ditch = 15.2 hr
Total ditch volume = 5700 m^3
MLSS = 4000 mg/liter; VSS/SS = 0.5

Nitrification-denitrification rates

Based on observed rates (Table 14.3) which are duly adjusted for temperature and also based on use of Figure 12.9, we may assume the following:

First anoxic zone with raw sewage: 3 mg NO_3-N/g VSS-hr ($18°$ C)
Second anoxic zone (endogenous):   1.2 mg NO_3-N/g VSS-hr ($18°$ C)
Nitrification zone (Table 12.4):   4.0 mg NH_3-N/g VSS-hr ($18°$ C)

Total N to be removed

= (Incoming N) - (N in sludge) - (N in effluent)
= 480 kg/day - (say) 18% x 480 kg/day - (say) 7% x 480 kg/day
= 360 kg/day

First denitrification zone volume

Assume 85% of 360 kg N/day denitrified in the first zone, i.e., 12.75 kg/hr. Hence,

$$\text{VSS required} = \frac{12.75 \times 10^6}{3 \times 10^3} = 4250 \text{ kg VSS}$$

or

$$\text{Zone volume required} = \frac{4250 \times 10^3}{2000 \text{ mg/liter}} = 2125 \text{ m}^3$$

or

$$\text{Detention time} = \frac{2125}{9000} = 5.67 \text{ hr (i.e., 37\% of ditch volume)}$$

Second denitrification zone volume

Now to denitrify 15% of 360 kg N/day at only 1.2 mg NO_3-N/g VSS-hr we need 1875 kg VSS, i.e., zone volume = 937 m^3 (i.e., 10% of ditch volume).

Check if remaining volume enough for nitrification

$$\text{VSS required to nitrify 15 kg N/hr} = \frac{15 \times 10^6}{4.0 \times 10^3} = 3750 \text{ kg VSS}$$

$$\text{Zone volume required} = \frac{3750 \times 10^3}{2000 \text{ mg/liter}} = 1875 \text{ m}^3 \text{ (only 21\% of ditch}$$
$$\text{volume required;}$$
$$\text{hence, ample)}$$

Final aeration volume

At 10-15 min detention, one of the aerator pockets should suffice. The overall ditch volume is thus quite sufficient for both nitrification and denitrification to occur. The normal biological oxidation will proceed concurrently with the above within the same volume. Of course, volumetric sufficiency is one factor; the other is sludge age at the operating temperature, for successful treatment.

Power requirement

From Example 14.2,

OC required = carbonaceous + nitrogenous demands
                     = 148.5 + 86.6 = 235 kg/hr

Theoretically, oxygen released by denitrification is 62.5% of that used in nitrification. Hence,

Denitrification may release 62.5% × 86.6 = 54 kg/hr (maximum)

Power saving = 54/235 = 23% of total

14.7 PHOSPHORUS PRECIPITATION

Phosphorus precipitation is usually achieved by addition of chemicals like $Ca(OH)_2$, ferrous or ferric chloride, or alum, either in the primary or the final settling tank. Alum is more expensive and generates more hydroxide, which creates extra sludge, difficult to dewater. For extended aeration plants there is no primary settling; chemical addition has to be done in the final settling tank. The total phosphorus contained in the effluent depends on the chemical dose. Heide [16] recommends the use of Ca/P (wt/wt) $\geq$ 7.0 so as to be able to obtain a treated effluent with about 2 mg total P/liter with pH values between 8.4 and 8.8. An increase of approximately 50% in surplus sludge production may be expected as a result of lime dosing, but the sludge is reported to have good dewatering properties.

When using iron salts, a molar ratio of 1.0:1.4 Fe/P is reported to give 91 to 96% removal, respectively, of total phosphorus using $FeCl_2$ (from steel mill pickle liquor) dosed directly beneath the aerator [18].

Chemical addition prior to biological treatment is also feasible if a primary settling tank exists as in the case of the conventional activated sludge process. The dose requirement then increases, but chemical precipitation also improves organic removal, thus reducing BOD load on the biological treatment section of the plant.

14.8 TRICKLING FILTERS

A trickling or percolating filter is perhaps the earliest form of a packed-bed, fixed-film reactor used in waste treatment. Biological degradation occurs in a manner similar to that in the activated sludge process except that the filter is a three-phase system in which the biofilm (corresponding to MLSS in activated sludge) is fixed on the solid medium (stones or plastic). Stone media are placed in a bed about 2 m deep while the wastewater is dosed over it and air flows past it between the voids in the media bed. No aeration is necessary, but with some deep bed designs forced ventilation has been practiced.

Although the contact time between the percolating wastewater and the biofilm is on the order of a few minutes to about an hour, much BOD removal

is accomplished since it is transferred to the biofilm where oxidation and
synthesis of new cells occur and the end products are washed back into the
wastewater. Synthesis leads to growth of the biofilm thickness which eventually sloughs (peels) off and is washed out of the bed by the flowing
wastewater. The biofilm (also called zoogleal film) is rich in various heterotrophs, predominantly bacteria. The microbial composition varies with
depth of filter, the nature of the waste, and the season of the year. Fungi,
for example, may be observed more in the upper zones of a filter bed while
nitrifiers may be present more in the lower zones. A rich population of
grazers (protozoa, rotifers, crustacea, etc.) would also be present, feeding
on the microorganisms, with a natural balance maintained in the community
as a whole.

Historically, the trickling filter has been considered more rugged in
operation and easier to maintain than activated sludge plants. This is
partly because attached growth systems can have on the whole a much larger
mass of biological solids than the corresponding mass of MLSS in suspended
growth systems, and the growths are not easily destroyed or washed out of
the system by incoming slugs (fluctuations) in quantity and quality of the
wastewater. Along with domestic sewage, the presence of considerable concentrations of even toxic substances like cyanide, phenols, and formalde-
hyde, etc., can be tolerated and treated.

Trickling filters are classified into low- and high-rate filters depending on organic and hydraulic loads placed on them.

<u>Low-rate filters</u> are best used for serving small communities where
simple and sturdy operation is desired. They are easy to construct with
local materials and need no mechanical equipment except a simple tipping
trough or dosing siphon. These filters are particularly suitable for sloping sites since a loss of head of 2.5-3.0 m occurs between inlet and outlet.
Recirculation of flow is not practiced. Thus, pumping is not required.
Land requirement varies from 0.5 to 0.7 m^2/person. In their design, simple
empirical methods borne out of experience are used.

Low-rate filters are designed to keep organic loads between 0.15 and
0.3 kg BOD_5/m^3 stone volume. Hydraulic loads are small and average 1-2 m^3
per m^2-day. Hence, stone-medium volumes average from 0.17 to 0.33 m^3/person
(population equivalent 50 g BOD_5/person-day). Generally, these filters
receive sewage settled earlier in septic or Imhoff tanks and BOD removed in
them may be neglected in the interest of a safer design.

For proper functioning, hard, frost-resistant stones (5-7 cm in diameter) are generally used and laid to a depth of 1.5-2.0 m. Uniform and frequent dosing of wastewater is essential to keep the biofilms on the stones wet. But often intermittent dosing is necessary in the case of small communities. This is achieved by the use of a "tipping tray" of steel or fiberglass or a dosing siphon so as to dose at about 10-15 min intervals. To ensure clear flow and ventilation, underdrainage is provided by supporting the stone medium on open-jointed, half-round pipes of 15 cm in diameter on a floor sloping toward a collecting channel. Filters are generally left open at the top.

Generally, sloughing from such filters is only occasional. Sloughed sludge averages about 0.2 to 0.3 g/g BOD removed, namely, about 10-15 g/person-day. But the sloughed solids are quite mineralized and could be disposed of along with the effluent, especially if land application follows. On some occasions it may be desirable to allow the whole effluent to pass through a "polishing pond" of 2-5 days' detention time in which sloughed solids would also settle and be held for occasional manual removal. Trickling filter effluent could also be applied to a sand filter bed for fine polishing.

Possible systems incorporating low-rate trickling filters (LRTF) could be as follows:

In each case above, pumping can be completely avoided if the ground is adequately sloping.

In fact, if the land is quite hilly and rocky, the nonmechanized, low-rate trickling filter arrangement may become a viable alternative to the simple oxidation pond which may then be difficult and expensive to construct, and could merit strong consideration even for medium-sized communities perched on hilltops.

The energy conservation needs of today demand that the low-rate trickling filter be favorably considered where both hydraulic head and filter stone are available to provide a "natural" solution as far as possible.

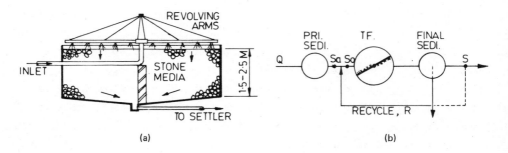

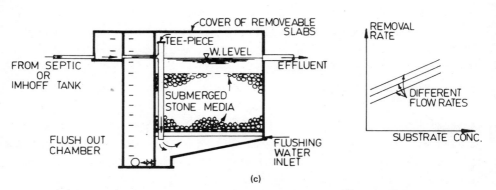

Figure 14.9 Trickling filter arrangements; (a) high-rate trickling filter, (b) flowsheet defining symbols used in Eq. (14.10), (c) submerged upflow filter.

High-rate filters [Figures 14.9 (a) and (b)] are preferred for larger flows since they involve more mechanization resulting from the need for pre- and postfiltration settling, continuous sludge withdrawal, recirculation pumping, etc. Their excess sludge also needs further treatment.

Over the last 60 years, they have alternated in popularity with the activated sludge process. Their design methods have also attracted much attention and research. No less than 14 methods of design have been proposed in the literature, but unfortunately most are empirical in nature [22]. A proper biofilm kinetic approach is only now emerging.

Biofilm kinetic approach Recent studies [30] have shown that as would be expected from biofilm kinetic considerations for attached growths (Sec. 12.22) in general a half-order reaction rate is observed with varying substrate concentrations at a constant flow rate through a trickling filter. At other flow rates, the reaction is still half order, but the value of the reaction rate constant changes (see Figure 14.9). This is because the hydrodynamics of a trickling filter is still quite complicated for modelling.

Removal rate is also a function of flow rate and the geometric configuration and resulting effective area of media. Until the biofilm kinetics and mass transport concepts are incorporated into one acceptable formulation, empirical design methods will need to be used.

   Empirical methods Among empirical methods, one that has been widely used in the United States and other countries over rather wide temperature ranges is the Tentative Method of Ten States, USA which, briefly, requires the following for single-stage filters [31]:

1. Raw settled domestic sewage BOD applied to filters should not exceed 1.2 kg BOD_5/m^3 filter volume-day.

2. Hydraulic load (including recirculation) should not exceed 30 m^3/m^2 filter surface-day.

3. Recirculation ratio (R/Q) should be such that BOD entering filter (including recirculation) is not more than three times the BOD expected in effluent. This implies that as long as the above conditions are satisfied efficiency is only a function of recirculation.

$$\text{Efficiency } E = \frac{(R/Q) + 1}{(R/Q) + 1.5}$$

   Eckenfelder's empirical method [32] is one which takes kinetic considerations into account and is found applicable to sewage and industrial wastes, stone and plastic media, and is also amenable to laboratory experimentation. It is based on the observation of Velz that BOD removal with depth of filter follows a first-order form owing to plug-flow-type conditions. Thus, in terms of biofilm mass X_v, contact time t with depth, the substrate in effluent is given by

$$S = S_0 e^{-K'(X_v)(t)} \tag{14.7}$$

   However, in trickling filters, X_v is a function of the surface area A of the medium, and the contact time is likewise a function of depth, medium characteristics, and the flow rate per unit area and time. Hence,

$$S = S_0 e^{-K'(A^m)(CD/Q^n)} \tag{14.8}$$

which can be simplified for a given type of filter medium, as follows:

$$S = S_0 e^{-K_f'(D/Q^n)} \tag{14.9}$$

in which

   S and S_0 = substrate (BOD) concentration, mg/liter

K'_f = apparent substrate removal rate constant (compatible units)

D   = filter depth, m

Q   = hydraulic load, m^3/m^2-day

c, m, n = constants

Evidently, K'_f depends on several factors resulting from its derivation such as surface area A of the medium, which can vary widely from one type to another (see Table 14.4), the effect of its geometry (coefficient m), and of its other characteristics (constant c), and of the type of flow (reflected by coefficient n which varies from 0.3 for turbulent flow to 0.67 for laminar flow and is about 0.5 in between). The nature of waste and its treatability also affect K'_f. Generally, the higher the initial BOD the better is the removal.

For this reason, the value of K'_f is specific for a given waste treated on a given filter of known characteristics, and cannot be extrapolated for use with other types of medium or flow. But it lends itself to laboratory and pilot plant experimentation, from which data can be generated for any type of waste or filter (see Sec. 12.23.3).

Temperature correction can be applied using θ = 1.035 in the range of $10°$ -25° C.

Plastic media, because they are expensive, are limited to industrial waste treatment generally involving strong organic wastes and deep filter towers. In most other cases stone media are used.

If recirculation is employed, as shown in Figure 14.9b, materials balance gives

$$QS_a + RS = S_0(Q + R)$$

Table 14.4 Types of Filter Media [24,32]

Media	Area (m^2/m^3)	Filter depth[a] (m)	Wet density (kg/m^3)
Stones		1-2.5	~ 2000
10 cm diameter	30		
5 cm diameter	90		
Plastics	90-240	Up to 12	< 500

[a]To ensure ease of natural ventilation.

in which

S_a = BOD, mg/liter, after primary settling

S_0 = BOD, mg/liter, after mixing with recirculated flow (namely, BOD entering filter)

S = effluent BOD, mg/liter, after final settling

Q and R = inflow and recirculated flow, respectively

Hence,

$$S_0 = \frac{S_a + (R/Q)S}{1 + (R/Q)} \qquad\qquad (14.10)$$

Substituting S_0 from Eq. (14.10) into Eq. (14.9) gives

$$\frac{S}{S_a} = \frac{e^{-K_f'(D/Q^n)}}{[1 + (R/Q)] - (R/Q)e^{-K_f'(D/Q^n)}} \qquad\qquad (14.11)$$

Recirculation is not essential per se for fixed-film reactors provided adequate reactor volume is provided to obtain desired BOD removal. In fact, recirculation tends to dilute the influent BOD to the filter. However, experience with high-rate trickling filters shows that recirculation helps to improve performance. Recirculation helps to "seed" the incoming flow and distribute the BOD load over the filter depth. It also helps to keep down nuisance from filter flies and ponding in the filter. But recirculation adds much to the operating costs and in some cases affects the size of the pre- or postfiltration settling tank. Generally, a recirculation ratio over 2.0 tends to become uneconomic.

Nitrification being a zero-order process at the usual ammonia concentrations in sewage (15-60 mg NH_3-N/liter), dilution of inflow by recirculation does not affect the nitrification rate. Nitrifying organisms are slow growing and confined to aerobic portions of biofilms. They are overpowered by fast-growing heterotrophs in the upper part of filters. Farther down, both BOD and biofilm growth decrease, while DO in the bulk flow increases, and thus nitrifiers now appear. About 50% nitrification is observed when trickling filter effluent averages about 15-25 mg/liter BOD. A 90% nitrification requires effluent BOD to be brought down to 5-15 mg/liter [22].

Excess sludge production from high-rate trickling filter operation is in the same general range as for other conventional activated sludge methods, namely, about 0.8 g/g BOD removed. However, trickling filter sludge generally has better settling characteristics than those of activated sludge and compacts more readily in final settling tanks to give lower moisture concentration (97-98%) by weight.

Example 14.6 illustrates the design methods just discussed.

Construction Stone media are most commonly preferred, with size ranging from 7 to 10 cm in diameter. Size variation in stones should be minimal to keep a high void ratio for ventilation. Filter depths are generally limited to 2.5 m with stone media.

Underdrains must also be designed to ensure free-flow conditions at all flows or else ventilation will be affected. Filter sidewalls could be constructed (like venetian windows) to be open and permit ventilation, so important in warm countries.

Dosing arms can revolve because of the hydraulic action of the issuing jets. The center turntable is the key to trouble-free operation. To accommodate revolving arms, circular filters are often built. This leads to some land wastage between successive units, but very satisfactory installations can be provided.

Land requirements for a trickling filter plant together with settling, sludge digestion, sludge drying, etc., may range from 0.3 to 0.5 m^2/person.

Example 14.6 Design a trickling filter using the empirical method of U.S. Ten States for the following data. Calculate the corresponding value of K_f' in Eckenfelder's equation.

Sewage flow = 5000 m^3/day

Raw settled BOD = 200 mg/liter = 1000 kg/day

Filter depth D = 1.8 m

Media = 7.5-10 cm in diameter

Recirculation as necessary

Efficiency desired - 85% (to give effluent BOD = 30 mg/liter)

(a) U.S. Ten States empirical method

1. Recirculation required to give 85% efficiency

$$E = 0.85 = \frac{1 + (R/Q)}{1.5 + (R/Q)} \quad \text{or} \quad R/Q = 1.83$$

2. Raw settled BOD applied $\not> 1.2$ kg/m^3-day. Hence,

$$\text{Filter volume} = \frac{1000 \text{ kg/day}}{1.2 \text{ kg/}m^3\text{-day}} = 833.3 \text{ }m^3$$

At depth = 1.8 m, filter diameter = 24.3 m.

3. Check for hydraulic load including recirculation

$$= \frac{5000 \text{ m}^3/\text{day} + 1.83(5000 \text{ m}^3/\text{day})}{[\pi \times (24)^2]/4} = 30 \text{ m}^3/\text{m}^2\text{-day (just enough)}$$

(b) Estimation of K_f'

From Eq. (14.10)

BOD to filter $S_0 = \dfrac{S_a + (R/Q)S}{1 + (R/Q)} = \dfrac{200 \text{ mg/liter} + 1.83(30)}{1 + 1.83}$

$$= 90 \text{ mg/liter}$$

Q load = 30 m^3/m^2-day

Using Eq. (14.9) and assuming n = 0.5,

$$\frac{S}{S_0} = e^{-K_f'(D/Q^n)} \qquad \text{or} \qquad \frac{30}{90} = e^{-K_f'[1.8 \text{ m}/(30 \text{ m/day})^{0.5}]}$$

Hence,

$$K_f' = 3.36 \text{ m}^{1/2}\text{-day}^{1/2}$$

14.9 SUBMERGED MEDIA BEDS

In the last few years attention has begun to be paid to the use of submerged media beds as opposed to trickling or percolating beds. In submerged contact beds the direction of wastewater flow may be downward or upward. If the flow is upward, the media may be fluidized or nonfluidized depending on the size of media and upflow velocity. The beds may be aerobic or anaerobic, and in some upflow systems pure oxygen is added at the bottom of the bed to supplement that already present (or put in by pre-aeration) in dissolved form in the wastewater.

Very interesting possibilities are opening up as a result of the use of these concepts. First, with upflow beds, media size can be much reduced (down to from 3- to 10-mm-size grains for nonfluidized and 0.5-mm for fluidized beds). This increases area available for biofilm growths and, therefore, their volumetric reaction rates. Loss of head is also much reduced with upflow types, and the flow itself tends to carry away the sloughed solids out of the bed, although some form of backwashing arrangement may be essential. With fluidized beds, proper solids-liquid serparation (biomass from media, and biomass and media from effluent) is essential and adds to cost and complexities. They have particular application in industrial waste treatment and in nitrification-denitrification of large ammonia concentrations.

One simple application of nonfluidized, upflow-type contact bed is to be found in the anaerobic treatment of domestic sewage after septic tank or other pretreatment. In this so-called "upflow filter" (Figure 14.9c) the medium used is stones of 1.8-2.5 cm in diameter laid to a depth of 1.0-2.0 m. Flow enters from the bottom and leaves from the top with a loss of head of only about 10-15 cm in small installations. In treating septic tank effluent, a stone bed volume of about 0.1 m^3/person would give an upflow face velocity of 1-2 m/day for a flow of 100-200 liters/person-day, respectively, giving about 70-80% BOD removal. A study in India [33] reports similar BOD removals at only about one-half the per capita stone volume just stated.

The effluent from such a unit is anaerobic and may need to be held in a polishing pond of short detention to aerate it. Some natural aeration would also occur as the effluent flows over ground to its final disposal point.

Biological treatment of wastes occurring in submerged or nonsubmerged downflow in natural soil has been discussed in Chapter 9.

14.10 ROTATING DISC SYSTEMS

In biological principles, rotating disc systems are similar to trickling filters. The discs (up to about 3 m in diameter) made of plastic mesh, PVC or polystyrene foam, or just plain asbestos provide the surface for attached biogrowths. The discs are partially submerged (about 30-40%) in a tank containing the wastewater and rotated in it at about 3-6 rpm (see Figure 14.10). Disc rotation helps to expose the biogrowths alternately to the substrate and to air and also keeps the wastewater tank aerated. An increase in rotational speed increases the dissolved oxygen in the tank where it is considered desirable to have a minimum of 1-2 mg/liter. The discs are also spaced at least 2-3 cm apart to prevent "bridging" between the growths.

The biofilm undergoes sloughing as in trickling filters and arrangements must be made for settling the sloughed solids and disposing them suitably as from any other biological process. For a given efficiency of BOD removal and/or nitrification, the rotating disc systems have reportedly

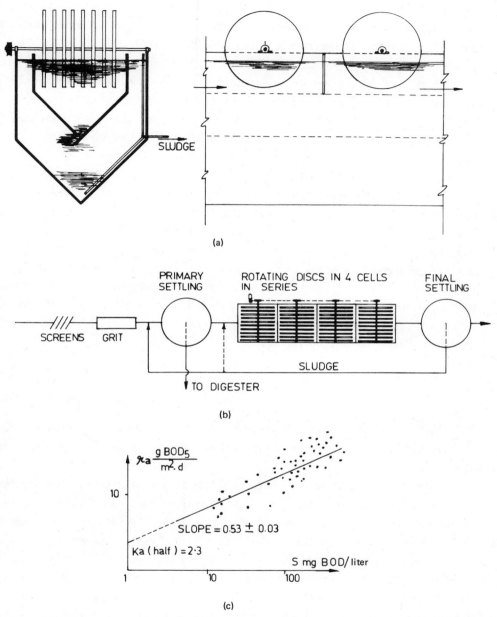

Figure 14.10 Rotating discs; (a) rotating discs installed in an Im- hoff tank, (b) rotating discs with separate settling and digestion, (c) BOD removal rate as a function of substrate concentration (adapted from Ref. 22).

required less power than needed for other comparable processes [26]. Better
performance has been obtained when the discs have been installed in three or
more compartments to promote plug-flow conditions [28]. Recycling of flow
is not essential.

Rotating disc systems are conveniently marketed as "package plants"
and are particularly popular for communities of less than 2000 persons.
Over 1500 units exist in Europe alone (1978). Larger sized installations
are also to be seen in the United States and France. The largest one of
its kind (93,000 m^3/day) designed for BOD, nitrogen, and phosphorus removal
is under construction at Orlando, Florida, and is expected to save energy
costs by about 30% compared to conventional processes [27b]. Efforts are
concentrated in developing lightweight but strong discs of inert material
having a high ratio of surface area to a given volume. The discs in each
compartment have to be supported on a shaft and bearings and be driven at
the required speed with the whole assembly being capable of withstanding
torque due to possible imbalance caused by a wet lower half and a dry upper
half resulting from prolonged idleness of disc. Disc rotation in one system
is made possible by releasing air from a header located below the disc and
attaching "cups" to disc periphery.

14.10.1 Sizing of Disc Units

Design of disc units involves estimation of surface areas required for
biofilm growth so as to bring about desired BOD removals (and nitrification
if desired).

Referring to Sec. 12.22, it will be recalled that with attached growths,
diffusion-rate limitation occurs which gives rise to a "half-order" regime
in terms of bulk concentrations. Using Popel's published data (1964) on
the performance of seven disc units placed in series and operated with
settled domestic sewage (210-620 mg BOD/liter) kept well mixed in each unit
to have uniform bulk concentration applied to the discs, Harremoes [22] has
shown that the BOD removal rate r_a per unit area and time exhibits a half-
order relation when plotted against the bulk concentration S applied to the
discs (see Figure 14.10). As the sewage was kept well mixed in each unit,
the value of S also denotes the effluent concentration from that unit.

From the figure, it is seen that the line of best fit has a slope of
0.53 ± 0.03 which Harremoes [22] regards as good evidence that BOD removal

in rotating discs is close to half order as would be expected from above for a pilot scale experiment with actual sewage. Thus,

$$BOD_{removed} = Q(S_0 - S) = (r_a)(A)$$

and from Figure 14.10,

$$r_a = K_{a(half)}(S)^{1/2}$$

or

$$Q(S_0 - S) = K_{a(half)}(S)^{1/2}A \tag{14.12a}$$

in which

$K_{a(half)}$ = half-order reaction rate constant, $g^{1/2}/m^{1/2}$-day
S_0 and S = initial and final substrate concentration, mg/liter or g/m^3
Q = flow rate, m^3/day
A = area of discs, m^2
r_a = BOD removal rate, $g\ BOD_5/m^2$-day

From Figure 14.10, $K_{a(half)}$ = 2.3 $g^{1/2}/m^{1/2}$-day over the temperature range of 18 to 20° C. Temperature correction may be applied using θ = 1.035 as in trickling filters, but recently (1976), Ellis and Banaga [23] have shown θ = 1.006 between 11 and 18° C and 1.002 between 18 and 27° C.

Equation (14.12a) can be rewritten as follows to find the effluent concentration S when other variables are known [22]:

$$S = \left[-\frac{K_{a(half)}A}{2Q} \pm \sqrt{\left(\frac{K_{a(half)}A}{2Q}\right)^2 + S_0} \right]^2 \tag{14.12b}$$

and

$$\text{Efficiency} = (S_0 - S)/S$$

From the equations just seen, the efficiency of treatment is found to be independent of detention time of flow (as in the case of trickling filters). Hydraulic loads used in the above experiments ranged from 13.2 to 155 liters/m^2-day. At the BOD loads used in the experiments, oxygen concentrations did not appear to have any significant effect but may be important for high BOD wastes.

Equation (14.12) corroborates rather well with actual experience in pilot and full-scale plants treating domestic sewage in varied conditions, summarized in Table 14.5.

Table 14.5 Performance of Some Rotating Disc Plants

Country	Plant	BOD applied (g/m^2-day)	Remarks	Ref.
England	Full-scale	6	Effluent BOD 20 mg per liter (for 95% of time). Nitrification occurs.	24
	Pilot	8	Nitrification successful (17-23° C)	28
Federal Republic of Germany	Full-scale	6-10	Over 90% BOD removal	24
		12-18	Loadings used in larger plants with slightly reduced efficiency	
India	Pilot (operated 10 hr/day)	25	85-89% BOD removal (temp. > 23° C)	25
United States	Full-scale	20[a]	85-94% BOD removal	26

[a]Estimated from data in Ref. 26.

It is interesting to compare the values given in Table 14.5 with those normally adopted for trickling filters. For example, low-rate trickling filters in England are loaded at 0.1-0.12 kg/m^3-day, and stones used average 5 cm in diameter (i.e., stone surface area = 90 m^2/m^3), thus giving a loading of 120 g/90 m^2-day, namely, 1.33 g/m^2-day. A rotating disc unit required to give the same performance can be loaded about 5 times higher [24].

Various other approaches to sizing of disc plants have been suggested in the literature [28] including some based on Monod kinetics [27a]. But, as discussed in Sec. 3.22, the Monod kinetic approach is not easy owing to difficulties experienced in determining the active mass of attached growths.

As stated in the previous section, it is desirable to install the discs in three or more compartments to promote plug flow and improve performance (Figure 14.10b).

14.10.2 Sludge Production

Plant designs are often based on an excess sludge production of about 0.6 kg/kg BOD applied. Recent studies have shown the production to be temperature dependent, affected by plant efficiency, and to be based on whether raw or settled waste was applied.

In one case [27a] at optimum conditions, sludge production was shown to decrease from 0.56 to 0.41 and even to 0.33 kg dry solids per kg BOD destroyed as temperature rose from $11° C$ to $18° C$ and $27° C$, respectively. Results from Taiwan [29] have shown that at about 85% BOD removal efficiency the excess solids production averaged about 0.55 kg SS/kg BOD removed from raw sewage application, and about 0.42 kg SS/kg BOD removed from settled sewage. Sludge production was lower than from conventional activated sludge systems.

Excess sludge has also been shown to exhibit good settling characteristics. Zone settling occurs, and settling is complete within 15 min [29].

Further treatment and disposal of the excess sludge may require digestion, dewatering, and/or land application. For small plants the excess sludge can be held in a tank or lagoon (with lime addition if desired) for periodic carting away for off-site disposal generally by land application or along with refuse land filling. In-situ sludge digestion can be provided if discs are installed directly in an Imhoff-type tank (Figure 14.10a). For larger plants, separate sludge digestion and/or dewatering (with or without chemical addition) may be necessary. It should be noted that in comparing rotating disc-type plants with other methods the cost and difficulties of sludge handling and disposal should be given proper consideration.

14.10.3 Power and Other Requirements

Power requirements for mechanically operated disc units of smaller sizes have been reported to vary widely from 6 to 12 kWh/person-year for about 80% efficiency, and upward of 16 kWh/person-year for higher efficiencies coupled with nitrification. Additional power may be needed for sludge disposal.

Power requirements for larger plants are reported to be not much different from those for activated sludge [24], while initial investment may often be higher than for activated sludge.

Land requirements may be quite small (< 0.1 m^2/person), especially where off-site sludge disposal is practiced. Loss of head through the disc units is quite low, about 3 cm or so [26], making them very suitable for flat terrains. The disc units may be covered over, if desired, and a forced ventilation system provided to take care of possible odors.

Example 14.7 Prepare preliminary design for a rotating disc-type installation to serve 1000 persons. Assume 80% BOD removal at an organic load of 20 g BOD/m²-day and 3-m-diameter discs, spaced 5 cm apart on centers.

Flow and BOD

At 54 g BOD/person-day and 200 liter flow/person-day the influent BOD = 54,000 g/day = 270 mg/liter, and Q = 200 m³/day.

Disc unit

$$\text{Disc area required} = \frac{54,000 \text{ g/day}}{20 \text{ g/m}^2\text{-day}} = 2700 \text{ m}^2$$

$$\text{Area per disc of 3-m diameter} = 2 \text{ sides} \times \frac{\pi d^2}{4} = 14 \text{ m}^2/\text{disc}$$

Number of discs required = 2700/14 = 195 (say)

At 5 cm apart, c/c, the tank for housing the discs will have:

Length = 195 × 5 = 9.75 m (say 10 m)

Width = 3-m diameter + 0.2-m clearance = 3.2 m

Depth = 2 m average (say)

Tank volume = 64 m³

Tank area = 32 m² (net) = 0.04 m² gross (approx.)/person

$$\text{Hydraulic load on discs} = \frac{200 \text{ m}^3/\text{day} \times 10^3}{2700 \text{ m}^2} = 74 \text{ liters/m}^2\text{-day}$$

$$\text{Surface load on tank} = \frac{200 \text{ m}^3/\text{day}}{32 \text{ m}^2} = 6.25 \text{ m}^3/\text{m}^2\text{-day}$$

Efficiency check

Use Eq. (14.12b) to estimate effluent BOD and verify if efficiency of removal assumed in data is reasonable. (Assume $K_{a(half)}$ = 2.3.)

$$S = \left[-\frac{2.3(2700 \text{ m}^2)}{2(200 \text{ m}^3/\text{day})} \pm \sqrt{\left(\frac{2.3(2700)}{2(200)}\right)^2 + 270 \text{ mg/liter}} \right]^2 = 49 \text{ mg per liter}$$

$$\text{Efficiency} = \frac{270 - 49}{270} = 81\% \text{ (compares with assumed value)}$$

Excess sludge

At 0.6 kg/kg BOD removed, sludge = 0.6[200(270 - 49) × 10⁻³] = 26.5 kg/day; sludge volume at 1% solids = 2.7 m³/day.

REFERENCES

1. DHV Consulting Engineers, Carrousel Treatment Plants, Dwars, Hecderik en Verhey BV, Amersfoort, Netherlands.

2. Metcalf and Eddy, Inc., Wastewater Engineering, McGraw-Hill, N.Y., 1972.

3. S. Arceivala, B. Bhalerao, and S. Algersamy, Cost estimates for various sewage treatment processes in India, CPHERI Symposium, Nagpur, India, 1969.

4. Personal Communication.

5. S. Arceivala, Design and performance of extended aeration plants; reports for Associated Industrial Consultants, Bombay, India, 1970, 1971.

6. B. Handa, Ph.D. thesis, Nagpur University, India, 1974.

7. A. Pasveer, Oxidation ditches, Ind. Jour. Env. Health, Nagpur, India, 1963.

8. H. Coutts and C. Christianson, Extended A.S.T.C.C., Nat. Env. Res. Center, Office Res. Dev., U.S. EPA, Corvallis, Oregon, 1974.

9. J. Baars, The Use of Oxidation Ditches for Treatment of Sewage, T.N.O., Netherlands, 1960.

10. J. Zeper and A. DeMan, Large Oxidation Ditch, Carrousel, Fifth Congress, International Association on Water Pollution Research, San Francisco, California, 1970.

11. Int. Assoc. W.P.R., conference on N_2 as a pollutant, Copenhagen, 1975.

12. J. Walker and D. Dreier, Aerobic Digestion of Sewage Solids, Thirty-fifth Annual Session, Georgia Water & Pollution Control Association, 1966.

13. C. Adams and W. W. Eckenfelder, Jr., Eds., Process Design Techniques for Industrial Waste Treatment, Enviro Press, Nashville, Tennessee, 1974.

14. Roy F. Weston, Inc., Upgrading Existing Wastewater Treatment Plants, EPA tech. transfer contract. 14.12.933, 1971.

15. S. Bokil and J. Bewtra, Influence of mechanical blending on A.D.W.A.S., in Advances in Water Pollution Research, 6th International Conference, Jerusalem, Israel, 1972.

16. B. A. Heide, Combined N and P removal in an A.S.S. operating on oxidation principles, in Nitrogen as a Water Pollutant, IAWPR Conference, Denmark, 1975.

17. C. Adams and W. Eckenfelder, Jr., Process Design Techniques for Industrial Waste Treatment, Enviro Press, Nashville, Tennessee, 1979.

18. C. Christianson and P. Harremoes, A literature review of biological denitrification of sewage, International Association on Water Pollution Research Conference on Nitrogen as a Water Pollutant, Proceedings, Vol. 3, Copenhagen, 1975.

19. J. Barnard, Biological denitrification, Jour. Inst. Wat. Poll. Con., 72 (6), 1973.

20. Van der Geest and Witvoet, Nitrification and denitrification in carrousel systems, International Association on Water Pollution Research Conference on Nitrogen as a Water Pollutant, Proceedings, Vol. 3, Copenhagen, 1975.

21. N. Matsche, Removal of nitrogen in an activated sludge plant with mammoth rotor aeration, International Association on Water Pollution Research Conference on Nitrogen as a Water Pollutant, Proceedings, Vol. 3, Copenhagen, 1975.

22. P. Harremoes, Half-Order Reactions in Biofilms and Filter Kinetics, Dept. San. Eng. Tech. University, Denmark, 1975 (preprint).

23. K. Ellis and S. Banaga, Rotating disc units at different temperatures, Jour. Inst. WPC, 75, 1976.

24. Water Research Centre, UK, Tech. Report No. 93, 1978.

25. Nat. Env. Eng. Res. Inst. (India) Tech. Digest No. 60, April, 1978.

26. E. Borchardt, Chapter on rotating discs, in Biological Waste Treatment, P. Canale, Ed., Wiley-Interscience, N.Y., 1971.

27a. . Clark et al., Jour. WPCF, 1978.

27b. G. Dallaire, United States Largest Rotating Biological Contactor Plant, Civil Engineering Magazine, A.S.C.E., Jan. 1979.

28. A. Dereli, A comparison of rotating disc and extended aeration performance, Ph.D. thesis, University of Newcastle upon Tyne, 1979.

29. W. Liu and C. Ouyang, Characteristics of RBD-Sludge (Taiwan), presented at Bogazici University Symposium, Istanbul, Turkey, 1979.

30. B. Atkinson and A. Ali, Wetted Area, Slime Thickness, and Liquid Phase Mass Transfer in Trickling Filters, Dept. Chem. Eng., University College, Swansea, 1976 (preprint).

31. Great Lakes-Upper Mississippi River Board of State Sanitary Engineers, Recommended Standards for Sewage Works, 1971.

32. W. W. Eckenfelder, Water Quality Engineering for Practicing Engineers, Barnes and Noble, N. Y., 1971.

33. Nat. Env. Eng. Res. Inst. (India), Tech. Digest No. 61, July 1978.

Chapter 15

MECHANICALLY AERATED LAGOONS

15.1 INTRODUCTION

Mechanically aerated lagoons are earthen basins, generally 2.5 to 5.0 m deep, provided with mechanical aerators installed on floats or fixed columns. Raw sewage is fed from one end into the lagoon (after screening) and leaves from the other end after a desired period of aeration. Such lagoons are much smaller in size (less than 10 to 20%) compared with waste stabilization ponds, partly because lagoons are deeper and partly because the detention time needed for stabilization is less than for natural algal ponds.

Mechanical equipment is now needed in the form of aerators. Recent advances made in the design of mechanical aerators have resulted in higher oxygen inputs per unit power consumed and have made them the preferred method of aeration for use in lagoons. In a few instances pneumatic methods of aeration have also been used. Electrical power is required at the site for running the aerators and providing general electrification.

Construction features are simple and similar in many ways to those seen in the case of waste stabilization ponds discussed in Chapter 16. Aerated lagoons permit a considerable amount of flexibility in design, spanning from the simple facultative type units at one end to the more efficient and compact ones employing recycling of solids at the other end, while at all times maintaining their relative simplicity in construction and operation. Consequently, they find useful applications in both developing and developed countries. Their design can be so manipulated as to optimize either for land or power requirement, or for overall costs, as desired, by adopting combinations of different types of aerated lagoons and algal ponds.

Aerated lagoons also have the potential for becoming "semipackaged plants." Aerators can be manufactured in standard sizes (5, 10, 25 hp, etc.), and the necessary ones can be ordered and readily installed in an earthen basin of required volume. Thus, some part of the work would be done at the site on a "do-it-yourself" basis, and the factory assembly and imports kept to a minimum [1].

Various examples exist of the successful use of aerated lagoons for the treatment of domestic, municipal, and industrial wastewaters [2,3,4]. Extensive use of aerated lagoons has been made by the pulp and paper industry, by food processing, petrochemical, and various other industries [5,6, 7]. There are several instances of overloaded algal ponds having been successfully converted into mechanically aerated ponds. McKinney and Benjes [3] give detailed operational data evaluating two such aerated lagoons treating domestic sewage. Bartsch and Randall [4] describe the operation of aerated lagoons treating textile wastes. Adams and Eckenfelder [7] have shown how design parameters can be developed for petrochemical waste.

15.2 TYPES OF AERATED LAGOONS

Depending upon the power input per unit lagoon volume, and the provision or otherwise of a recirculation arrangement, the solids in the system will either settle, flow through, or build up. Thus, based on the way solids are handled, three types of aerated lagoons are distinguished [1]:

1. Facultative
2. Aerobic flow through
3. Aerobic with solids recycling

The different ways of handling the solids have a substantial effect on efficiency, power requirement, detention time, sludge disposal, etc., and the design methods have to take these differences into account, although the basic principles of biological treatment apply equally well to all the three types. Some of their essential characteristics are given in Table 15.1.

15.2.1 Facultative Aerated Lagoons

In facultative aerated lagoons, the power input per unit volume is sufficient for diffusing the required amount of oxygen into the liquid, but not sufficient for maintaining all the solids in suspension. Consequently, some of the suspended solids entering the lagoon and some of the new solids produced in the lagoon as a result of substrate removal tend to settle down

Table 15.1 Some Typical Characteristics of Different Types of Aerated Lagoons Treating Domestic Sewage

Characteristic	Facultative type	Aerobic flow-through type	Aerobic with solids recycling (extended aeration type)
1. Solids control	None (some settle, some flow out with effluent)	Partial (solids cannot settle; they must flow out with effluent)	Full control (solids neither settle nor flow out. They are recycled with controlled withdrawal of surplus solids.)
2. Suspended solids concentration in lagoon, mg/liter	50-150	100-350	4000-5000
3. VSS/SS, percent	60-80	70-80	50-80
4. Mean cell residence time θ_c, days	High (because of settlement)	Generally < 5	Warm: 10-20 Temperate: 20-30 Cold: over 30
5. Overall BOD removal rate K_L per day, 20°C			
(a) Filtered samples	0.6-0.8	1-1.5	20-30
(b) Unfiltered samples	0.3-0.5	0.4-0.5 [12]	20-25
6. Temperature coefficient θ	1.035	1.035	1.01-1.03
7. Detention time, days	3-12	Generally < 5	0.5-2.0
8. BOD removal efficiency, percent	70-80	50-60	95-98
9. Nitrification	None	None	Likely under favorable conditions
10. Coliform removal, percent	60-99 [a]	60-90	60-90

Table 15.1 (Continued)

Characteristics	Facultative type	Aerobic flow-through type	Aerobic with solids recycling (extended aeration type)
11. Lagoon depth[b]	2.5-5.0	2.5-5.0	2.5-5.0
12. Land requirements, m²/person:			
Warm	0.15-0.45	0.10-0.35	0.13-0.25[c]
Temperate	0.45-0.90	0.35-0.70	0.25-0.55[c]
13. Power requirement, kW/person-year	12-15[d]	12-14	13-20 (Depends on nitrification and denitrification)
hp/1000 m³	2.0-2.5[d]	2.0-2.5	2.0-3.0
14. Minimum power level, kW/1000 m³ lagoon volume	0.75 (to ensure uniform diffusion of oxygen into lagoon)	2.75 (to ensure that all solids are kept in suspension)	2.75 (to ensure that all solids are kept in suspension)
15. Sludge	Accumulates in lagoon. Manual removal after some years	No accumulation; solids go out with effluent	Surplus sludge is withdrawn continuously (daily) and disposed of suitably
16. Outlet arrangement	Effluent flows over a weir	Partially or fully submerged pipe outlet	Weir or pipe (see Figures 15.3, 15.4, and 15.5)

[a] High efficiency is feasible only if two or more cells are provided in series, and temperatures are favorable.

[b] Lagoon depth should be compatible with the type of aerator used.

[c] Including sludge drying.

[d] Power requirement is reduced if anaerobic gasification is dependable (see text).

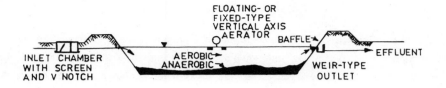

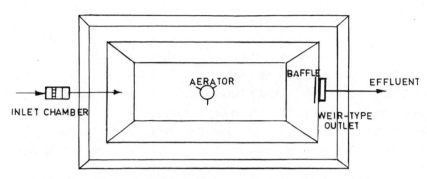

Figure 15.1 A mechanically aerated facultative lagoon.

and undergo anaerobic decomposition at the bottom. The activity in such a lagoon is, therefore, partly aerobic and partly anaerobic, which gives it the name "facultative." They are also sometimes referred to simply as "aerated lagoons" (see Figure 15.1). Such lagoons are similar to the facultative algal ponds used for waste stabilization, except that the oxygen is now derived from mechanical aeration instead of algal photosynthesis, and the stabilization rate is somewhat faster so that less detention time, and consequently less land, is required for comparable treatment.

Facultative aerated lagoons are being used successfully in the treatment of sewage and many different types of industrial wastes. They are simple to operate and give BOD removals of 70 to 90% in treating domestic sewages. Like all lagoons, they are sensitive to temperature and adequate detention times must be provided depending on the climate. Other characteristics are detailed in Table 15.1.

15.2.2 Aerobic Flow-through-type Lagoons

Aerobic flow-through-type lagoons are those where the power level is
high enough not only to diffuse enough oxygen into the liquid but also to
keep all solids in suspension as in an activated sludge aeration tank. No
settlement of solids occurs. In these lagoons the wastewater leaves along
with the solids under aeration (Figure 15.2). Thus, the efficiency of BOD
removal attained in these lagoons is not very high (about 50 to 60%) since
much solids are likely to be present in the effluent. Additional treatment
is necessary if better BOD and solids removal is desired. In fact, these
lagoons are generally followed by facultative stabilization ponds or aerated
lagoons, or are designed with the intention of converting them eventually
to aerobic lagoons with solids recycling. As the entire lagoon contents
are aerobic, these lagoons are also sometimes referred to simply as "aerobic lagoons" to distinguish then from "aerated lagoons," which may be only
partly aerobic as just seen. The power requirement of aerobic lagoons is
somewhat high when seen in relation to the BOD removed (Table 15.1).

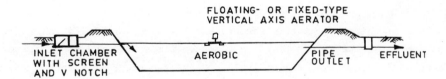

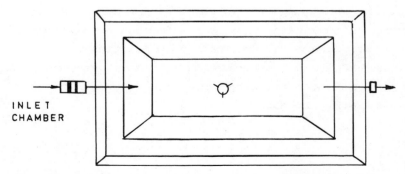

Figure 15.2 A mechanically aerated flow-through lagoon.

15.2.3 Aerobic Lagoons with Solids Recycling

Aerobic lagoons with solids recycling are very similar to activated sludge or extended aeration plants. The power input level is sufficient to meet the oxygen requirements and keep all solids in suspension. But the solids concentration in the lagoon is now quite high, since the system is designed to prevent the solids from escaping with the effluent by the incorporation of some form of solids settling and recycling.

In order to preserve the simplicity of operation afforded by such lagoons, they are mostly of the extended aeration type with direct disposal of surplus sludge to drying beds or sludge lagoons (see Figures 15.3, 15.4, and 15.5). Carrousel-type oxidation ditches are of this type, and could also be referred to as "extended aeration-type lagoons." Land requirements are less for extended aeration-type lagoons than for the shallower Pasveer-

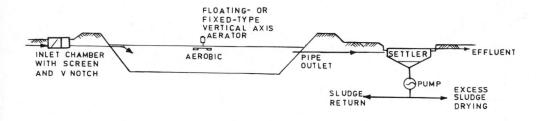

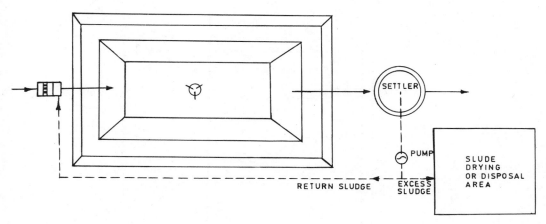

Figure 15.3 An extended aeration lagoon with separate settling and sludge recirculation.

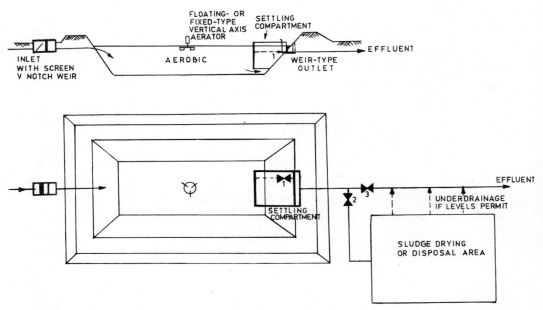

Figure 15.4 An extended aeration lagoon with a settling compartment within the lagoon. Note: Normally, valve 3 is open and valves 1 and 2 are closed. When valve 1 is opened, the settling compartment is short-circuited and mixed liquor can be withdrawn from the lagoon; this can then be sent to the disposal area by closing valve 3 and opening valve 2.

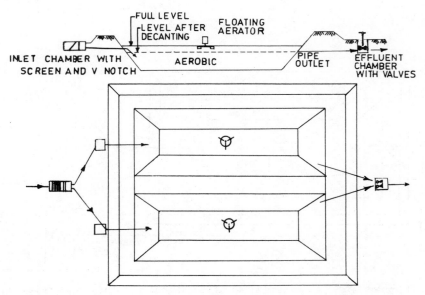

Figure 15.5 An extended aeration lagoon with two cells for intermittent operation.

type oxidation ditches, and the former are thus preferred for larger populations.

The efficiency of BOD removal in this type of lagoon can be as high as
95-98%, and nitrification can also be achieved. This is discussed more
fully below, and some typical characteristics are included in Table 15.1.

15.3 DESIGN OF FACULTATIVE LAGOONS

The general approach to the design of facultative-type aerated lagoons
is similar to that followed later in Chapter 16 for waste stabilization
ponds, except for the mode of oxygen supply.

15.3.1 Substrate Removal Rate

For the removal of organic matter, which generally follows first-order
kinetics, the removal rate can be recapitulated from Chapter 12 [Equation
(12.27a)] as

$$\frac{dS}{dt} = -k'XS = K_L S \tag{15.1}$$

where

S = substrate concentration, mass/volume

k' = the specific substrate removal rate, $(t)^{-1}(mass/volume)^{-1}$,
  $[k' = \mu_{max}/K_s Y$ when $S \ll K_s$ (see Sec. 12.10)]

X = volatile solids in the system, mass/volume

t = time of exposure to treatment

K_L = overall substrate removal rate, $time^{-1}$

The exact concentration of volatile solids X in the system is difficult
to estimate for facultative lagoons since (mixing) power levels are not high
and the settlement of solids can occur (see Sec. 15.3.5). Hence, it is
possible to follow two approaches: either the values of k' and X are estimated separately or they are merged into one single value K_L, as shown
above.

Values of K_L can be determined either from laboratory studies, as outlined in Sec. 12.24, or from field measurements of substrate removal under
known conditions. A few typical values of the substrate removal rate constant K_L per day are listed in Table 15.1 for domestic sewage and other

wastes. Values of k' range from 0.01 to 0.03 $\text{day}^{-1}\text{-(mg/liter)}^{-1}$ and
values of K_L range from 0.6 to 0.8 day^{-1} at $20°C$ for domestic sewage [9].
The latter values have been estimated from field studies on facultative
aerated lagoons after taking the actual mixing conditions into account. A
value of K_L equal to about 0.25 per day has been reported for pulp and
paper mill waste treatment lagoons [8].

15.3.2 Temperature Effects

As discussed in Chapters 12 and 13, temperature has a significant effect on the performance of aerated lagoons. The lagoon temperature is the
result of various factors affecting it, such as the incoming wastewater
temperature, the ambient air temperature, and the heat exchange coefficient
under given climatic and lagoon aeration conditions.

The lagoon temperature is generally estimated using Eq. (11.31) and
values of the heat exchange coefficient f listed in Table 11.3 for different
types of unit. A typical value of f for aerated lagoons is 0.49 m/day.

From the point of view of the optimum detention time required in a
lagoon to achieve a desired efficiency of performance, the minimum winter
temperature is the critical one for design. The substrate removal rate
known at one temperature can be estimated at any other temperature using,
from Chapter 12 [Eq. (12.30)]

$$K_{L(T°C)} = K_{L(20°C)} \theta^{T-20} \qquad\qquad (12.30)$$

Values of θ have been reported in the literature to range from 1.035 [7] to
1.097 [4]. For domestic sewage, a value of $\theta = 1.035$ may be used.

An interesting field study by Bartsch and Randall [4] on an existing
aerated lagoon treating textile wastes has shown that a lagoon temperature
of $14.4°C$ ($58°F$) was critical for the BOD removal efficiency of the system.
Temperatures somewhat greater than $14.4°C$ did not appear to have any significant effect on effluent BOD but, when temperatures dropped below $14.4°C$
in the lagoon, the effluent BOD increased sharply. From their field data,
they found the value of θ averaged was as high as 1.097 for the textile
waste.

15.3.3 Lagoon Efficiency

The ideal complete mixing model has generally been used by Eckenfelder [2] and others [5,7] to estimate the substrate removal efficiency in facultative lagoons:

For complete mixing single-celled lagoons

$$S = \frac{S_0}{1 + K_L t} \tag{15.2}$$

For complete mixing equal cells in series

$$S = \frac{S_0}{(1 + K_L t')^n} \tag{15.3}$$

where

  n   = number of equal-sized cells in series
  t'   = detention time per cell
  S_0, S = initial and final substrate concentrations in mg/liter or kg/day, respectively

The term K_L in Eqs. (15.2) and (15.3) can be replaced by the term k'X if desired.

For intermixing or dispersed flow

Where the configuration of the lagoon is not likely to promote well-mixed conditions, it may be more appropriate to use other models such as the dispersed flow model as advocated by Murphy [8], Arceivala [9], and others. Because of either some process considerations or the lay of the land, it may be preferable to have a relatively long and anrrow path of flow, thus promoting more plug-flow-type conditions. In fact, in most cases, the substrate removal and oxygenation requirement can be better approximated by use of the dispersed flow model discussed in Chapter 11, since lagoons are neither ideally completely mixed or ideally plug flow type. This dispersed flow model has been given by Wehner and Wilhem for first-order kinetics as follows:

$$S = S_0 \left[\frac{4a e^{1/2d}}{(1 + a)^2 e^{a/2d} - (1 - a)^2 e^{-a/2d}} \right] \tag{15.4}$$

where

$\quad a = \sqrt{1 + 4K_L td}$, dimensionless

$\quad d = D/UL$ = dispersion number, dimensionless

$\quad D$ = axial dispersion coefficient, L^2/t

$\quad U$ = velocity of flow through the lagoon, L/t

$\quad L$ = characteristic length of travel

The reader is referred to Sec. 11.8 for a fuller discussion on this subject and to Table 11.1 for a few typical values of D/UL observed in different treatment reactors, which may be recapitulated briefly as:

Aerated lagoon	Likely value of D/UL
Rectangular, long	0.2-1.0
Squarish to rectangular	2.0-4.0 and over

In order to use the dispersion number D/UL as a design parameter, either model or prototype studies may be performed using suitable tracers, or the values of D/UL estimated using analytical methods. The latter require that the value of the dispersion coefficient D be known for the type of lagoon configuration or size, after which the value of D/UL can be easily computed knowing the flow-through velocity U and the length L of the lagoon.

The value of D for aerated lagoons can be approximated using the empirical equation given by Eq. (11.20) if the width of flow is known. For a lagoon of any width larger than 30 m, this equation is:

$\quad D, m^2/hr = 33 \times$ lagoon width, m $\qquad\qquad\qquad$ (15.5)

Murphy and Wilson [8] studied three large aerated lagoons of different configurations in Canada. Two lagoons were long rectangular ones whose flow paths were approximately 76 m wide x 1200 m long x 4 m deep (average) and 106 m wide x 1350 m long x 4 m deep. The third lagoon was squarish, being 273 m wide x 334 m long x 4 m deep. Using rhodamine-WT as a tracer, the values of the dispersion number D/UL were determined from the time-concentration curves as explained in Sec. 11.10. As was expected, the square-shaped lagoon gave higher values of D/UL (~ 4), while the other two had values ranging mostly between 0.3 and 1.0.

Since the flow-through velocity U is equal to the flow rate Q divided by the cross-sectional area of the lagoon, we can rewrite the dispersion number in terms of the detention time t as

$$\frac{D}{UL} = \frac{Dt}{L^2} \tag{15.6}$$

Thus, from a graphical plot of D/UL versus t/L^2 as shown in Figure 15.6, one can find the value of D as the slope of the line. In the study just described, the average value of D was found to be 2881 m^2/hr (31,000 ft^2/hr) for the three lagoons in question.

Since dispersion is related to the scale of the phenomenon (Sec. 11.8), it is clear that the value of D can be used for designing other lagoons which are more or less of the same scale of size. For aerated lagoons of other sizes, for example, those with much smaller or larger widths, the likely value of D can be estimated from Eqs. (11.20) to (11.23). Model studies may be carried out if the conditions so justify.

An interesting observation made by Murphy and Wilson in their study just described was that no significant correlation could be established between D or D/UL and the levels of power input per unit volume in the range of 0.47 to 2.29 kW/1000 m^3 lagoon volume studied by them. The power input was varied by shutting off some of the aerators. A possible explanation for this observation, apart from the masking effect of likely experimental errors, is the relatively large width of the lagoons at which the value of

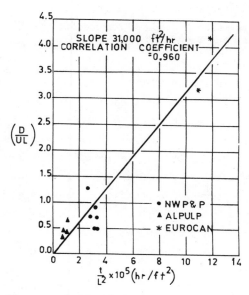

Figure 15.6 The correlation between dispersion number and geometrical parameters for three aerated lagoons (from Ref. 8).

D, already large, was not significantly increased by surface aeration in any overall manner. As stated in Sec. 11.10, data presently available do not enable one to state at what widths the effect of surface aeration would be insignificant.

From Example 11.4 it is seen that the geometry of the lagoon makes a considerable difference to the value of the dispersion number D/UL. In one case, it is approaching plug flow (D/UL = 0.49), while in the other case it is approaching well-mixed conditions (D/UL = 2.29). Thus, the two lagoons are likely to display different efficiencies of BOD removal, coliforms, etc., and also possibly show some differences in the microbial species present in them, although both would afford the same theoretical detention time to a given wastewater.

If a lagoon is so dimensioned that well-mixed (D/UL $\geq$ 4.0) conditions are promoted, not much error is made whether the ideal complete mixing model, or the dispersed flow model with an appropriate value of D/UL, is used. But, such is not the case when the value of D/UL is low.

It is desirable to keep as low a value of D/UL as possible when (1) removal follows first-order or higher order kinetics, (2) high efficiencies of removal are desired as in the case of coliform removals, for example, where sometimes efficiencies of 99 to 99.99% or even more may be desired, and (3) the process is benefitted by approaching plug flow conditions as, for example, when nitrification is desired or heat is to be conserved. In such cases, an attempt is made to keep D/UL around 0.2 or less. Economy in lagoon size is readily achievable in this manner in treating domestic and similar wastewaters.

However, under certain specific conditions, plug flow may be contraindicated. For example, if the wastewater is industrial in character and likely to receive slug discharges of toxic substances from time to time, it may be advantageous to opt for well-mixed conditions and adopt such dimensions for the lagoon that D/UL will be around 4.0 or more. Thus, the basis for selection of D/UL values should be process requirements.

Once having selected the value of D/UL, the designer has to play with the geometry or configuration of the units to achieve the desired value. As explained in Chapter 11, low values of D/UL require that D be kept as low as possible by adopting a narrow width of lagoon or by providing around-the-end-type baffles. These measures also generally have the effect of increasing the values of U and L and, consequently, help considerably in reducing the overall value of D/UL. Provision of cells in series also has

the same overall effect. Exactly opposite steps are necessary to follow
when high values of D/UL are desired.

15.3.4 Lagoon Depth

The depth of facultative aerated lagoons is generally kept between
2.5 and 5.0 m. The factors involved in the choice of depth are (1) the
type of soil and constructional problems likely, (2) compatibility with the
type of aeration equipment used, as there is a limitation to the maximum
depth with different types of aerators, and (3) the need to have a certain
minimum depth above the anaerobic zone to assist in oxidizing any H_2S gas
released from the lower layers, thus preventing the formation of odorous
conditions. Generally, about 1 m depth is sufficient for this purpose in
temperate climates, and up to 2 m in warmer climates or where many sulfides
are likely to be produced.

Deeper lagoons are preferred where it is desired to ensure an anaerobic
zone devoid of all oxygen to promote the growth of methane-fermenters where
temperatures permit (this is discussed further later).

15.3.5 Solids in Suspension

The extent of solids in suspension in lagoons naturally depends on the
size and nature of the solids and the mixing power input per unit lagoon
volume. Generally, in the case of facultative lagoons, a minimum aeration
power input of 0.75 kW/1000 m^3 lagoon volume is recommended (see Chapter
13). At this power input the solids in suspension may range between 30 and
150 mg/liter, although when treating domestic sewage the concentration may
well be around 40 to 60 mg/liter.

At higher power inputs, the solids concentration in suspension is
higher. Adams and Eckenfelder [7] suggest a linear relationship between
the solids concentration and the power input as given in Table 15.2 based
on observations in existing aerated lagoons.

When biological growth is very finely dispersed, as may occur at low
organic loadings, all solids may be maintained in suspension even at low
power input levels (e.g., pulp and paper mill aerated lagoons loaded at
less than 1800 kg/ha-day). Observations on 12 aerated lagoons treating
pulp and paper mill wastes showed that power levels of 2.75 kW/1000 m^3 were

Table 15.2 Likely Solids Concentration in Suspension at Different Power Levels in Facultative Lagoons Treating Domestic Sewage

Power level (kW/1000 m^3 lagoon volume)	Suspended solids in lagoon (mg/liter)
0.75	50
1.75	175
2.75	300

Source: adapted from Ref. 7.

sufficient to keep all solids in suspension (total solids varied from 400 to 500 mg/liter) [7,11].

15.3.6 Anaerobic Activity in Facultative Lagoons

As indicated earlier, a portion of the incoming suspended solids, and the new solids produced in the lagoon as a result of substrate removal, tend to settle down and enter the anaerobic zone in the lagoon. Under anaerobic conditions, a heterogeneous group of facultative and anaerobic bacteria collectively referred to as "acid formers" convert carbohydrates, proteins, and fats into fatty acids by hydrolysis and fermentation. Thereafter, if conditions are favorable, another group of bacteria known as "methane formers" convert these substances into carbon dioxide and methane.

The first stage conversion by acid formers merely changes the form of the organic matter without any substantial BOD reduction. It is only in the second stage that BOD reduction occurs as organic carbon is converted to CO_2 and CH_4. The methane-producing bacteria are strictly anaerobic organisms and, therefore, are not likely to grow in the presence of oxygen, which may be present even in the lower layers of well-aerated lagoons. These bacteria are also sensitive to temperature and little gasification occurs below 15°C (also see Sec. 14.9.1). Thus, BOD removal by anaerobic activity in facultative lagoons can occur only when the combination of sufficiently low power input (resulting in lower layers being devoid of oxygen) and a relatively warm temperature, is favorable for gasification. Such a condition is most likely to occur in warmer climates and where lagoons are relatively deep and highly loaded. A number of small-sized aerators may possibly be used to diffuse the oxygen uniformly in the upper layers of the lagoon without disturbing the anaerobic activity in the lower layers. Also, ways and means must be found for increasing the anaerobic removal rate by increasing contact surfaces and mass of attached growths in the lower layers without creating clogging problems. This is a fruitful subject for research as any success in reducing the aeration power requirement should be welcome.

In colder climates, and where lagoons are relatively lightly loaded, well-aerated, or shallow, gasification, and therefore BOD removal in the anaerobic zone, cannot be relied upon and is best neglected. No doubt, liquefaction by acid formers (which are not all strictly anaerobes) occurs in such lagoons as long as the temperatures are favorable, but the gasification step is inhibited by the presence of oxygen. Even liquefaction may not occur at very cold temperatures ($< 5°C$) and the settled solids may lie undegraded on the lagoon bottom until favorable conditions are reestablished.

Liquefaction results in a feedback of soluble organic matter (partially or wholly devoid of oxygen and perhaps containing H_2S and other oxidizable substances) to the upper aerobic layer. The BOD is then satisfied aerobically. The feedback BOD is estimated by Marais at about 0.4 g/g of VSS deposited in the lagoon [10].

For the reasons just discussed, the organic load on facultative lagoons must be established taking into account the effect of anaerobic activity.

If it is assumed in Eqs. (15.2), (15.3), and (15.4) that the term S_0 refers to the soluble BOD_u, then its value may be estimated from that of the total incoming BOD_u (soluble + solids) as follows:

1. No liquefaction, no gasification (very cold climates):

$$S_0 = 0.7 \text{ to } 1.0 \text{ (total incoming } BOD_u) \qquad (12.7a)$$

This is based on the assumption that a portion of the suspended volatile solids settle and remain undegraded on the lagoon bottom during the cold winter temperatures. Thus, the effective BOD to be removed is reduced during this period. The undegraded solids, of course, eventually degrade when the temperature rises after the winter, leading to a temporary deterioration of the effluent quality in some cases.

2. Only liquefaction, no gasification:

$$S_0 = 1.0 \text{ (total incoming } BOD_u) \qquad (12.7b)$$

This is due to liquefaction of settled volatile solids, giving a soluble organic feedback to the upper layers so that total incoming BOD_u has to be satisfied aerobically. Temperature and/or the presence of dissolved oxygen inhibit gasification.

3. <u>Both liquefaction and gasification occur</u>:

$$S_0 = 0.4 \text{ to } 0.7 \text{ (total incoming BOD}_u) \qquad\qquad (12.7c)$$

In warm climates and in relatively highly loaded deep lagoons operated at low power levels, anaerobic activity in the lower layers may possibly give about 30 to 60% removal of the total incoming BOD_u, thus leaving the balance for aerobic removal.

The maximum value of S_0 is given under case (2) above. This condition may occur during the summer for lagoons in colder climates, and in the winter for lagoons in warmer climates. Often, as a safe assumption, lagoons are designed for this maximum value of S_0.

15.3.7 Detention Time

The optimum detention time required for achieving the desired degree of treatment can now be estimated both for summer and winter conditions from a knowledge of the values of S_0 and k' or K_L at the corresponding temperatures, and the value of the dispersion number D/UL as explained earlier.

For domestic sewage treatment, the detention time required in facultative aerated lagoons may range from 3 to 6 days in warmer climates and up to 12 days or even more in colder climates (see Table 15.1).

15.3.8 Oxygen Requirements

It is evident from the foregoing discussion that the oxygen requirements of facultative aerated lagoons would be equal to the value of the soluble BOD_u (S_0) to be satisfied aerobically. Thus oxygen requirements are affected by the extent of dependable anaerobic activity occurring in the lower layers of the lagoons.

The value of the soluble BOD_u removed in the system (i.e., $S_0 - S$) is found using any of the three possible cases outlined in the previous section, depending on the temperature, to give the value of S_0, and subtracting from it the soluble BOD_u in the final effluent at that temperature.

Since nitrification does not generally occur in facultative aerated lagoons, there is no additional oxygen demand on this account.

15.3.9   Aeration Power and Power Level

The power required for running aerators is calculated from the oxygenation requirements, taking into account the fact that aerators are normally
rated by the manufacturers for operation at standard conditions (20° C, no DO,
tap water) and must therefore be adjusted for likely field conditions at the
crucial time of the year which, for aeration purposes, is generally the summer.

Generally, the oxygenation achieved at field conditions is only 60-80%
of that at standard conditions. A method has been given in Chapter 13 for
estimating the oxygenation capacity at different conditions (Eq. 13.5).

For treating municipal wastewater to obtain about 85% removal of BOD
in facultative aerated lagoons, the power requirement ranges as shown in
Table 15.1 subject to the proviso that the minimum power level required to
ensure availability of dissolved oxygen in all parts of the lagoon is 0.75
kW per 1000 m^3 lagoon volume [7].

Several aerators are provided to make up the total power requirement.
A minimum of two aerators are provided in smaller lagoons. The aerators
are often spaced uniformly, apportioning the total lagoon surface into a
number of squares coinciding with the number of aerators anticipated. Murphy
and Wilson [8] suggest the use of Eq. (15.4) to compute the BOD profile along
the lagoon length, and to space the aerators accordingly.

15.3.10   Nutrient Requirements

Nutrient addition may be necessary in the case of certain industrial
wastes. When so needed, the nutrient requirements can be estimated using
methods outlined in Sec. 12.18.

15.3.11   Performance

Facultative aerated lagoons can give 70-90% BOD removal from readily
degradable wastes such as domestic sewage. The total effluent BOD is made
up of the soluble fraction S and the fraction due to solids. Estimation of

the latter can be done using the procedure given in Sec. 12.14. The concentration of suspended solids in the effluent may be the same as that in the lagoon, or somewhat less if the outlet is baffled. About 50-80% of these solids may be assumed as volatile.

The BOD_5 corresponding to the volatile solids may be assumed as about 0.5 mg/mg VSS in the effluent. Owing to the relatively long detention time provided in aerated lagoons, the BOD_5 is not likely to be much more than this value. Thus, as an approximation,

Final effluent BOD_5, mg/liter = S, mg/liter + 0.5 (VSS mg/liter) (15.8)

Nitrification generally does not occur in facultative aerated lagoons. Some nitrogen removal, however, can occur as a result of solids settlement and some, as yet unexplained, mechanisms in the lower layers of lagoons under certain conditions.

Coliform removal generally ranges from 90 to 99% in summer and 60% or less in winter in single-celled lagoons. Higher removals can occur as stated earlier in lagoons operated with two or more cells in series or where other steps are taken to ensure very low values of D/UL. Design estimates can be prepared on the same basis as done for waste stabilization ponds (Chapter 16) using the coliform die off rate constant K_b as 1.0 to 1.2 per day at $20°C$, and converting to other temperatures using $\theta = 1.19$. Preliminary data would seem to indicate that K_b values in aerated lagoons are not much different from those observed in oxidation ponds.

15.3.12 Sludge Accumulation

A certain amount of net accumulation of sludge occurs with time, depending on the rate at which new sludge solids are deposited and the rate at which they are degraded. The degradation rate is estimated to vary from 40 to 60% per year in $20-30°C$ climates [7] although not much data are available in this regard. At any given time, the wet volume of accumulated sludge is estimated from the accumulated solids (undegraded organics + inert residue after degradation + inorganic settleables, in influent) and the likely sludge moisture content (about 85 to 96%), depending on the accumulation interval. This is explained in greater detail in Chapter 17.

As an approximation, the sludge accumulation in treating domestic sewage may be taken as 0.03 to 0.08 m^3/person-year, as in waste stabilization ponds, if the cleaning interval is around 5 years.

McKinney and Benjes [3] took field samples of settled sludge using a 3/4-in. core sampler driven manually through the sludge layer into the clay bottom. The clay plugged the bottom of the sampler, preventing the loss of sample. (See Chapter 16 for a similar method used in oxidation ponds.) The specific gravities of the bottom samples and the percentage of volatile solids varied considerably with distance from the rotor; the volatile fraction increased with distance.

As shown by Arceivala [1], prior removal of grit would reduce the accumulation of sludge by only about 5% or even less. Manual removal of sludge may be needed after some years so as not to decrease the effective lagoon volume. The dried sludge constitutes a welcome soil conditioner for farmers.

Example 15.1 Design a facultative-type aerated lagoon utilizing the same data as given in Example 16.4 for a town of 70,000 persons, namely:

Raw sewage flow = 12,400 m^3/day

Raw BOD_5 = 300 mg/liter

Coliforms = 10^6/100 ml

Final BOD_5 to be not more than 70 mg/liter (irrigational use)

Temperature: ambient = 1.7° C in January, 23° C in summer

Wastewater = 20° C

Assume K_L = 0.6 per day at 20° C and K_b = 1.2 per day at 20° C

Complete mixing conditions not essential

Assume aerators can give 2 kg O_2/kWh at standard conditions and only 75% of this input is likely at field conditions in any season

Lagoon detention time and size

As a trial, assume detention time t = 8 days. Thus,

Lagoon volume = 12,400 m^3/day x 8 days = 99,200 m^3

Since complete mixing conditions are not considered essential, we may design for a low value of D/UL. Again, as a trial, let the average lagoon dimensions be 62 m wide x 400 m long x 4 m deep, with a longitudinal baffle (see Figure 15.7) so that the effective width is 31 m and length 800 m.

Estimation of dispersion number

From Eq. (15.5), dispersion coefficient D = 33 x width. Hence,

$$\frac{D}{UL} = \frac{D_t}{L^2} = \frac{(33 \times 31 \text{ m})(8 \text{ days} \times 24)}{(800 \text{ m})^2} = 0.29 = 0.5 \text{ (for example)}$$

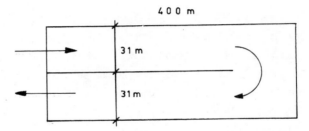

Figure 15.7 A lagoon arrangement (see Example 15.1).

Estimation of lagoon temperature T_L in winter
_____

From Eq. (11.31),

$$\frac{t}{h} = \frac{T_i - T_L}{f(T_L - T_a)} \qquad \text{where} \qquad f = 0.49 \text{ m/day for aerated lagoons}$$

$$\frac{8 \text{ days}}{4 \text{ m}} = \frac{20^\circ C - T_L}{(0.49 \text{ m/day})(T_L - 1.7^\circ C)} \qquad \text{or} \qquad T_L = 10^\circ C$$

Estimation of BOD removal rate K_L in winter
_____

From data, K_L at $20^\circ C$ = 0.6 per day, and θ = 1.035. Then,

$$K_{L(10^\circ C)} = K_{L(20^\circ C)} \theta^{T-20} = (0.6)(1.035)^{10-20} = 0.42 \text{ per day}$$

Estimation of lagoon efficiency
_____

1. In winter

$$K_L t = (0.42)(8) = 3.36$$

Enter Figure 11.5 (Wehner-Wilhem equation) at K_t = 3.36 and D/UL = 0.5, and read "percent removal" as 87% of the initial soluble BOD S_0, mg per liter.

Since the lagoon temperature is about $10^\circ C$, we may expect lique-faction of settled solids, but no gasification. Hence, S_0 = 1.0 x total incoming BOD_u = (incoming BOD_5) x 1.4 (for example) = 300 x 1.4 = 420 mg/liter, and soluble BOD_u in effluent = 420(1 - 0.87) = 54.6 mg/liter or soluble BOD_5 in effluent = 39 mg/liter. Likely concentration of SS in effluent = 100 mg/liter (see power level below). Likely VSS concen-tration in effluent = (for example) 60 mg/liter. Thus,

Total effluent BOD_5 = soluble + solids BOD at, e.g., 0.5 mg/mg VSS

$$= 39 + (0.5)(60) = 69 \text{ mg/liter}$$

Overall efficiency $= \dfrac{300 - 57}{300} = 77\%$

Note: If the longitudinal baffle is omitted, but the overall lagoon dimensions are retained as before, the value of D/UL will increase to 2.86, while the lagoon temperature T_L and BOD removal rate K_L will be unaffected. The "percent removal" of BOD will then be 81% only, and the total effluent BOD_5 will be about 87 mg/liter (overall efficiency = 71%). Thus, the provision of a baffle is necessary. If desired, the calculations may be repeated assuming a detention time of only 7 days and other suitable strategies to ensure low values of D/UL, so that total effluent BOD_5 is not greater than 70 mg/liter.

2. In summer:

Under summer conditions, the lagoon temperature is likely to be equal to or greater than 20° C. At 20° C, K_L x t days = 0.6 x 8 = 4.8. Assuming the same value of D/UL as before, namely, 0.5, we obtain from Figure 11.5 the "percent removal" as 92.5%. At higher temperatures, the percent removal will be even higher. Estimation of the initial soluble BOD_u S_0, mg/liter, under summer conditions can be made in either of the following two ways:

a. As a safe assumption, take S_0 = 1.0 x total incoming BOD_u, since gasification may be inhibited by the presence of dissolved oxygen in the lower layers although in terms of temperature, the conditions are favorable.

b. Take S_0 = 0.4 to 0.7 (total incoming BOD_u) if the lagoon is deep and gasification can be depended upon to reduce a portion of the incoming BOD.

Designing on the basis of assumption a above, S_0 = 420 mg/liter and

Soluble effluent S = 420 x (1 - 0.925)

$$= 30 \text{ mg/liter as } BOD_u = 21 \text{ mg/liter as } BOD_5$$

and Total effluent BOD_5 = soluble + solids BOD

$$= 21 + 0.5(60) = 51 \text{ mg/liter}$$

Overall efficiency $= \dfrac{300 - 51}{300} = 83$

Thus, it is seen that the facultative aerated lagoon of 8 days' detention time will be capable of giving a final effluent BOD_5 ranging from 51 mg/liter in the summer to 69 mg/liter in the winter. Without the provision of the longitudinal baffle, the range will be from 68 to 87 mg/liter.

Coliform die off

Assume the die off rate constant for coliforms $K_b = 1.2$ per day at $20°C$ as given in the data, and the temperature conversion constant $\theta = 1.19$. Then, at winter temperature of $10°C$ in the lagoon,

$$K_{b(10°C)} = K_{b(20°C)}\theta^{T-20} = (1.2)(1.19)^{10-20} = 0.21 \text{ per day}$$

For 8 days' detention time and $D/UL = 0.5$, we get from Figure 11.5 the following removal efficiencies in summer and winter:

Lagoon temperature	K_b (day^{-1})	$K_b t$	Percent removal
$10°C$	0.21	1.68	70.0
$20°C$	1.20	9.60	98.2

Hence, the coliform removal efficiency will fluctuate between 70% and >98.2%, depending on the season. Chlorination of the effluent may be required to meet irrigation standards. Alternatively, a maturation pond may be provided (Chapter 16).

Power requirement

The oxygenation capacity should be enough to take care of all the BOD_u removed aerobically in the lagoon.

1. In winter:

All BOD_u is removed aerobically. Overall efficiency = 77%.

$$O_2 \text{ required per day} = 0.77 \left(420 \text{ mg/liter} \times 12,400 \times \frac{10^3}{10^6}\right)$$

$$= 4010 \text{ kg/day (167 kg/hr) at field conditions}$$

Aerator capacity at field conditions = $0.75(2 \text{ kg } O_2/kWh)$. Hence,

$$\text{Power requirement} = \frac{167 \text{ kg/hr}}{(0.75)(2 \text{ kg } O_2/kWh)} = 111 \text{ kW (150 hp)}$$

Power level

Power input per unit volume = 111 kW/99,200 m^3 = 1.118 kW/1000 m^3. This power level is acceptable as it is higher than the minimum necessary. Also, at this power level the SS concentration is likely to be about 100 mg/liter in the lagoon.

2. **In summer:**

Generally, the oxygen requirement is higher in summer. However, if anaerobic gasification can be depended upon to reduce some portion of the BOD, then the aeration capacity can be reduced proportionately. This can be done by shutting off some of the aerators during the warm months.

However, as a safe assumption, provide enough capacity to remove all BOD$_u$ aerobically even in the summer. Since overall efficiency of removal is 83% in summer,

$$O_2 \text{ required} = 0.83 \left(420 \text{ mg/liter} \times 12,400 \times \frac{10^3}{10^6} \right)$$

$$= 4323 \text{ kg/day} = 180 \text{ kg/hr}$$

$$\text{Power required} = \frac{180}{(0.75)(2)} = 120 \text{ kW} \ (= 162 \text{ hp})$$

Thus, provide a total capacity of 120 kW (162 hp) and cut off some aerators from operation when not necessary. Assuming the average power consumption is 120 kW, the per capita annual power comsumption is estimated as

$$\frac{120 \text{ kW} \times 24 \text{ hr} \times 365 \text{ days}}{70,000 \text{ persons}} = 15 \text{ kW/person-year}$$

Sludge accumulation

If the cleaning interval is 5 years and accumulation occurs at 0.05 m^3/person-year, on an average,

$$\text{Sludge volume} \qquad = 0.05 \times 70,000 \times 5 \text{ years} = 17,500 \text{ m}^3$$

$$\text{Extra depth required} = \frac{17,500 \text{ m}^3}{\text{lagoon bottom area}} = 1 \text{ m, approximately}$$

Thus, overall liquid depth = 5 m. The design may be revised if soil or other conditions do not permit this depth.

Land requirement

Net lagoon area = 62 m x 400 m = 24,800 m^2

Add 30 to 35% more area to accommodate embankments and other requirements. Thus,

Gross area = 31,000 m^2 = 3.1 ha

Note: Gross area required = 33,480 m^2/70,000 persons = 0.478, or roughly
0.5 m^2/person. It will be seen that, for the provision of oxidation ponds
for the same population under identical temperature and other conditions,
the gross land requirement is estimated to be as much as 3.7 m^2/person (see
Example 16.4).

15.4 DESIGN OF AEROBIC FLOW-THROUGH TYPE LAGOONS

As stated earlier, aerobic flow-through type lagoons are entirely
aerobic in nature, with their power input high enough to keep all solids
in suspension. In this sense, they are similar to the aeration tanks used
in activated sludge plants except that solids are not recycled. Their design is based on the principles of biological treatment presented in Chapter 12 and recapitulated briefly below.

Since there is only flow through and no recycling of solids in the
system, the hydraulic detention time t is equal to the mean cell residence
time Θ_c in the lagoon. Thus,

$$t = \Theta_c \tag{15.9}$$

Theoretically, the mixing pattern in these lagoons can range from
reasonably well-mixed conditions to those approaching plug flow, depending
on the configuration of the lagoon and other factors outlined earlier.
Generally, however, their designs have been based on the use of the ideal
completely mixed model [2], although attempts have also been made to use
other models. As long as the geometry of the lagoon is such as to promote
well-mixed conditions, not much error is made by applying the ideal complete
mixing model.

15.4.1 Substrate Removal and Solids Concentration

If the complete mixing model is used, a mass balance for the microbial
solids present in the lagoon at equilibrium conditions and no recycling can

be prepared as explained in Sec. 12.8 [Eq. (12.20)] to give

$$X = \frac{X(S_0 - S)}{1 + K_d \Theta_c} \qquad (15.10)$$

where

X = mass or concentration of microbial solids present in the lagoon at equilibrium conditions and no recycling

Y = true growth yield constant

K_d = decay rate for microbial solids, time^{-1}, in endogenous respiration

Θ_c = mean cell residence time or detention time t in flow-through systems

S_0 and S = initial and final mass or concentration of substrate removed, respectively

Similarly, if the complete mixing model is used, the substrate removal can be estimated using Eq. (15.2) given earlier in this chapter and replacing K_L by $k'X$ to give

$$S = \frac{S_0}{1 + k'Xt} \qquad (15.11)$$

in which k' = specific substrate removal rate $(t^{-1})(\text{mass/vol})^{-1} = \mu_{max}/K_s Y$ when $S \ll K_s$ (Sec. 12.10).

Combining Eqs. (15.10) and (15.11), we obtain

$$S = \frac{1 + K_d t}{Y k' t} \qquad (15.12a)$$

It will be observed from Eq. (15.12a) that the final effluent BOD is independent of the incoming BOD S_0.

This is because, as S_0 increases (within a small range), the microbial solids concentration X also increases and, hence, the overall substrate removal rate $k'X$ correspondingly increases so that the final concentration remains the same. In fact, the value of S is dependent only on the values of Y, K_d, and k' for the particular waste, and the detention time t.

The reader is referred to Chapter 12 for laboratory methods of determining the various constants such as $k'Y$ and K_d from continuous flow tests. The value of K_L may also be estimated from batch tests described therein (Sec. 13.16.2). A few typical values of the constants are given in Table 15.1.

Values of k' for municipal wastewater generally range from 0.01 to
0.03 $(mg/liter)^{-1}(day^{-1})$ at $20^{\circ}C$, and can be converted to other temperatures
using $\theta = 1.035$ in Eq. (12.32). Overall K_L values are shown in Table 15.1.

15.4.2 Detention Time

Assuming ideal complete mixing conditions, the required detention time
for achieving any desired BOD removal can be estimated from Eq. (15.12),
provided the appropriate values of the constants k', Y, and K_d are used
after adjusting for the critical temperature. It must be mentioned that
slight changes in the values of the constants can have a considerable effect
on the lagoon size, and the constants need to be evaluated carefully, preferably based on laboratory field data.

Where k' values are determined from field data, care should be taken
to see that the actual mixing conditions in the lagoon are such as to justify the use of the ideal complete mixing model. This can be ascertained from
knowledge of the dispersion number D/UL for the lagoon, which should be well
above a value of 4.0 to minimize error in the estimation of a k' value if
using the ideal complete mixing model.

The detention time required in treating municipal wastewaters to give
about 50% BOD removal ranges from 2 to 3 days in favorable climates and up
to 5 days or more in colder climates.

The minimum detention time permissible is governed by cell washout time
(Sec. 12.6) and is given by

$$t_{min} = \frac{1}{Y(k'X) - K_d} \qquad\qquad (15.12b)$$

For example, for the data given in Example 15.2, the minimum detention time
permissible $t_m = 1.48$ days.

15.4.3 Solids Concentration

The solids concentration in aerobic, flow-through type lagoons depends
on the incoming concentration of solids, if any, and the solids produced as
a result of substrate removal in the lagoon. The latter are estimated from
Eq. (15.10) and, for a given wastewater, they are directly proportional to
the extent of substrate removal (i.e., $S_0 - S$) and inversely to the detention

time in the lagoon. Thus, the total amount of volatile solids present in an aerobic flow-through lagoon can be estimated as

$$\text{Total VSS, mg/liter} = \frac{X_i}{1 + K_d t} + \frac{Y(S_O - S)}{1 + K_d t} \qquad (15.13)$$

in which X_i = influent biological VSS in mg/liter, and all other terms have been defined earlier.

The total VSS present in an aerobic lagoon may constitute about 60 to 80% of the total SS present in the lagoon. As a rough estimate, the VSS concentration may be taken as 30 to 50% of the incoming BOD_5 concentration.

The SS concentration in such lagoons often varies from 100 to 350 mg per liter depending on the nature of the waste, the lagoon loading, the efficiency, and power input per unit volume. Thus, it is seen that, generally, the solids concentration X in aerobic flow-through lagoons is higher than that in aerated lagoons and, hence, for a given wastewater treatability k′, the overall removal rate k′X will be correspondingly higher in an aerobic lagoon.

However, the overall removal may not, in reality, be so high since the presence of these same solids in the final effluent adds to the final BOD value.

The BOD_5 of the SS present in the effluent can be estimated from the consideration outlined in Sec. 12.13. The value may range from 0.5 to 0.8 mg BOD_5/mg VSS depending on the mean cell residence time. Thus,

Final effluent BOD_5, mg/liter = S, mg/liter + 0.5 to 0.8(VSS, mg/liter)

$$(15.14)$$

15.4.4 Oxygen Requirements

The oxygen supply has to be adequate to meet the demand for biochemical oxidation of carbonaceous matter. As in the case of aerated lagoons, nitrification does not occur in aerobic lagoons.

The carbonaceous demand can be estimated using the same basic approach as presented in Sec. 12.15. In summary, the oxygen requirement is estimated as

$$\begin{bmatrix} O_2 \text{ required} \\ \text{(per day)} \end{bmatrix} = \begin{bmatrix} \text{soluble } BOD_u \text{ removed} \\ \text{(per day)} \end{bmatrix} - \begin{bmatrix} BOD_u \text{ of solids leaving} \\ \text{the system (per day)} \end{bmatrix}$$

$$(15.15)$$

Alternatively, the requirement can be stated, using Eq. (12.38), as follows:

$$\left[\begin{array}{c} O_2 \text{ required} \\ \text{(per day)} \end{array}\right] = a'\left[\begin{array}{c} \text{soluble BOD}_u \text{ removed} \\ \text{(per day)} \end{array}\right] + b'(XV) \qquad (15.16)$$

where all the terms have been defined earlier.

15.4.5 Aeration Power and Power Level

Exactly the same procedure is used in estimating the power requirement of aerobic lagoons as the one given earlier for facultative aerated lagoons. The only difference is that the minimum power level has to be kept equal to or higher than 2.75 kW per 1000 m^3 lagoon volume to ensure that all solids will be stirred up and kept in suspension throughout the lagoon. This is important to observe; otherwise the lagoon will operate more like a facultative aerated lagoon than an aerobic flow-through lagoon.

This requirement of minimum power level often places a constraint on the detention time that can be economically provided in such lagoons. Economy in power consumption dictates that the maximum detention time be such that the power requirement based on BOD removal considerations and that based on mixing considerations coincide as far as possible.

15.4.6 Other Characteristics

Among some of the other characteristics of aerobic lagoons, the following may be mentioned:

1. The maximum lagoon depth should be selected in such a manner as to be compatible with the aerator manufacturer's recommendations, or else all solids will not be kept in suspension.
2. No sludge accumulation occurs with time since all solids are kept in suspension and tend to flow out with the effluent.
3. Nutrient addition, when necessary, may be done using the methods outlined in Sec. 12.17 to estimate the requirements.
4. Neither nitrification nor denitrification normally occurs.
5. Coliform removal rates are identical to those expected in facultative lagoons.
6. The overall BOD removal is low (around 50 to 60%) considering the power input per unit volume. As stated earlier, this is due to the high concentration of solids in the effluent. Thus, the aerobic lagoon is often

followed by further treatment to remove the solids, with or without recycling. This is further discussed later.

Example 15.2 For the wastewater (flow = 12,400 m³/day; BOD_5 = 300
mg/liter) and other conditions detailed earlier in Example 15.1, design an
aerobic flow-through-type lagoon to serve a town of 70,000 persons, using
the ideal complete-mixing model. Since the lagoon is proposed to be followed
by another treatment unit, its size is proposed to be restricted to give a
detention time of only 3 days. The following additional data may be used:
k' = 0.015 per day at 20° C; Y = 0.5; K_d = 0.07 per day (all on BOD_5 basis).

Lagoon size

Lagoon volume = 12,400 m³/day x 3 days = 37,200 m³

Let lagoon depth = 4 m (check compatibility with aerator manufacturer's
requirements)

Lagoon area = $\dfrac{37,200 \text{ m}^3}{4 \text{ m}}$ = 9300 m²

To promote conditions approaching complete mixing, choose a squarish
shape, again checking its compatibility with the aerator manufacturer's
recommendations. Let the lagoon size be 93 m wide x 100 m long x 4 m deep.

Lagoon temperature in winter

From Eq. (11.31),

$$\frac{t}{h} = \frac{T_i - T_L}{f(T_L - T_a)} \qquad \text{where} \qquad f = 0.49 \text{ m/day for aerated lagoons}$$

or

$$\frac{3 \text{ days}}{4 \text{ m}} = \frac{20° \text{C} - T_L}{(0.49 \text{ m/day})(T_L - 1.7° \text{C})} \qquad \text{and} \qquad T_L = 15° \text{C}$$

BOD removal rate in winter

$$k'_{15° C} = k'_{20° C}\theta^{15-20} \qquad \text{where} \qquad \theta = 1.035$$

Thus,

k' at 15° C = $(0.015)(1.035)^{-5}$ = 0.0126 per day

Effluent BOD in winter

From Eq. (15.12),

$$S = \frac{1 + K_d t}{Yk't} = \frac{1 + (0.07)(3)}{(0.5)(0.0126)(3)}$$

$$= 64 \text{ mg/liter (soluble BOD}_5)$$

From Eq. (15.10), VSS in the lagoon will be

$$X = \frac{Y(S_0 - S)}{1 + K_d t} = \frac{0.5(300 - 64)}{1 + (0.07)(3)} = 97.5 \text{ mg/liter}$$

Assume BOD_5 of VSS in effluent = 0.8 mg/mg VSS.

Final effluent BOD_5 = soluble + BOD of solids

$$= 64 \text{ mg/liter} + 0.8(97.5 \text{ mg/liter})$$

$$= 142 \text{ mg/liter}$$

Overall efficiency $= \dfrac{300 - 142}{300} = 52.6\%$

Effluent BOD in summer

The lagoon temperature in the summer can be computed as $20.8°$ C. The corresponding BOD removal rate k' will be 0.0154 per day. Thus, the effluent soluble BOD_5 = 52.4 mg/liter, X = 102 mg/liter, and final effluent BOD_5 = 134 mg/liter. Overall efficiency = 55.3%.

Oxygen requirement in summer

$$O_2/\text{day} = (\text{soluble BOD}_u \text{ removed/day}) - (\text{BOD}_u \text{ of solids leaving/day})$$

$$\text{kg } O_2/\text{day} = \frac{\text{kg BOD}_5 \text{ removed/day}}{0.68} - 142(\text{kg solids leaving/day})$$

$$= \frac{12,400 \text{ m}^3/\text{day} \times (300 - 52.4) \times (10^3/10^6)}{0.68}$$

$$- 1.42 \left[12,400 \times 102 \times \left(\frac{10^3}{10^6} \right) \right]$$

$$= 3250 = 135.4 \text{ kg/hr at field conditions}$$

Power requirement

On the basis that aerators can give 2 kg O_2/kWh at standard conditions, and the oxygen transfer efficiency is only 75% at field conditions in the summer,

$$\text{Power required} = \frac{135.4 \text{ kg/hr}}{(0.75)(2.0)} = 90 \text{ kW (121 hp)}$$

Check for power level

Power input per unit lagoon volume = $\dfrac{90 \text{ kW}}{37,200 \text{ m}^3}$ = 2.42 kW per 1000 m^3

This is less than the minimum required. Hence, provide at least 2.75 kW per 1000 m^3, or 102 kW (138 hp) for this lagoon. The winter requirement will also be met by this minimum.

Power consumption

$\dfrac{102 \text{ kW} \times 24 \text{ hr} \times 365 \text{ days}}{70,000 \text{ persons}}$ = 12.76 kW/person-year

Note: In order to compare the performance of aerobic flow-through and facultative aerated lagoons at the same detention time, the former could also be designed for a detention time of 8 days (as was done for the latter in Example 15.16). At 8 days time, in winter, the aerobic lagoon can be shown to give a final effluent BOD$_5$ = 125 mg/liter (overall efficiency = 58.4%).

Although the efficiency has not increased much over the value of 52.6% corresponding to 3 days detention time, the power requirement now increases considerably since, on the minimum basis, it has to be = 2.75 kW × (12,400 m^3 × 8 days)/1000 = 273 kW = 34 kW/person-year.

This power consumption would be clearly uneconomical, considering the relatively low BOD removal. Thus, there is a constraint on the selection of detention time as discussed earlier. At longer detention times, it may be advantageous to design it as a facultative lagoon and secure good removal efficiencies. At small detention times, high efficiencies can be obtained only with solids recycling.

15.5 DESIGN OF AEROBIC LAGOONS WITH SOLIDS RECYCLING

As stated earlier, the design principles followed in the case of aerobic lagoons with solids recycling are the same as those followed in the case of the activated sludge process, and particularly its modification such as the extended aeration process which is most readily adaptable to lagoons.

The microbial solids produced as a result of substrate removal are not allowed to flow out with the effluent but are recycled or retained within the system by using either one of the following methods:

1. Provision of a separate settling tank after the lagoon and recirculation of the settled solids (Figure 15.3). This method is commonly followed in plants of all sizes. The Pasveer and carrousel type "ditches" can be provided in this manner.

2. Incorporation of a settling compartment in the lagoon itself to hold back the solids and allow only the relatively clear effluent to flow out (Figure 15.4). This is a possible method for small plants. The design of the settling compartment is similar to that for an upflow-type settling tank. When surplus sludge is to be removed from such a lagoon, the settling compartment can be bypassed by opening a valve in a pipe connecting the aeration side to the outlet weir as shown in the figure.

3. Intermittent operation of the lagoon. This is another possible method suitable for small plants (see Figure 15.5). After some hours of aeration, the aerators may be switched off and the inflow of raw sewage stopped or diverted to another compartment. The first compartment then undergoes a short period of quiescent settling (45-60 minutes) after which the relatively clear liquid is decanted off the top while the solids are retained in the unit. Raw sewage inflow is now resumed and the aerators are switched on for another round of aeration. The outlet is simply in the form of a pipe located at the lowest level to be reached when decanting the settled effluent from the lagoon. The pipe may also be provided, if desired, with a floating arm in the lagoon to ensure uniform decanting from the top layer only. The outlet pipe is closed by a valve when aeration is in progress. Intermittent operation has been used in the case of some small Pasveer-type ditches, and lagoons serving small communities [1,13].

Whenever solids are not allowed to escape with the effluent, the lagoons quickly succeed in building up a solids concentration to be able to operate either as conventional activated sludge systems or as extended aeration systems, depending on the mean cell residence time Θ_c in the system, where

$$\theta_c = \frac{\text{mass of microbial solids in system}}{\text{mass of microbial solids leaving per unit time}} \qquad (15.17)$$

From Chapter 12 it will be recalled that the process efficiency is considerably affected by the value of Θ_c. In fact, various other operating characteristics are also dependent upon the value of Θ_c and, thus, it should be selected with care.

Using the ideal complete-mixing model, the solids concentration in lagoons with recycling can be estimated as shown in Sec. 12.7 [Eq. (12.18)] as follows:

$$Y = \left[\frac{Y(S_0 - S)}{1 + K_d \Theta_c} \right] \left[\frac{\Theta_c}{t} \right] \qquad (15.18)$$

The first term on the right-hand side of the equation is the same as the one in Eq. (15.10) for lagoons with no recycling. The second term, i.e., $\theta_{c/t}$, shows the "multiplier" effect that recycling has on the solids concentration in the system. For example, if $\Theta_c = 20$ days, while $t = 0.75$ days, the factor $\Theta_c = 26.7$, signifying that the solids concentration in the lagoon is 26.7 times that in a flow-through-type lagoon of the same size.

At the high solids concentration carried in the system, the overall rate of substrate removal K_a per unit time is also high since $K_a = k'X$, and higher efficiencies of substrate removal can be expected than in the other types of lagoons seen previously. A few typical values of the constants are given in Tables 12.2 and 15.1. Values of K_a per day for BOD removal in domestic sewage often range from 20 to 30 at $20°C$. Also, because of the higher concentration of solids carried in the system, the effect of temperature on process efficiency is less and generally varies from 1.01 to 1.03.

The quantity of settled flow which must be recirculated can be determined using the method outlined in Chapter 12. This, in turn, determines the mass of solids which must be withdrawn from the system as surplus sludge and disposed of separately.

Some of the typical design criteria followed in the case of aerobic lagoons with solids recycling, and working on the extended aeration principle, are given in Table 15.1 with regard to municipal wastewater treatment. The ideal complete-mixing model is often used in determining the detention time and substrate removal. In the case of large lagoons where mixing conditions may deviate substantially from the ideal, it may be advantageous to use the dispersed flow model as illustrated earlier for facultative lagoons.

The overall BOD removal efficiency is of a high order, generally greater than 95% for municipal wastewater, as is to be expected with extended aeration systems. Nitrification can well occur if sufficient aeration is provided and other conditions (e.g., temperature) are favorable for it to occur.

The aeration system then has to be adequate to satisfy the oxygen demand of both carbonaceous and nitrogenous materials. The power requirement varies from about 13 to 20 kW/person-year, depending on whether nitrification and denitrification occur or not. The power input per unit volume in such lagoons calculated from BOD considerations is generally higher than that required to keep all the solids in suspension. Thus, the former controls.

The design of the settling tank and recirculation system has to be made with care, as detailed in Chapter 14 for all activated sludge systems. Operating problems may sometimes be caused as a result of poor settling or

gasification and lifting of settled sludge, for which typical causes and
remedies have been discussed. These difficulties are generally not experienced when the lagoons are operated as intermittent systems described earlier. In fact, a good amount of denitrification also occurs in intermittent
type lagoons, which aids in giving better nitrogen removals and less oxygenation requirements. Goronszy [13] gives operating data from an intermittent
lagoon serving 4000 persons in Australia.

To maintain the desired concentration of solids in the lagoon, surplus
solids must be withdrawn as sludge from the system. This can best be done
from the settled sludge just before its recirculation, as its solids concentration is the highest compared with any other location in the system
and the volume to be withdrawn is the least. The sludge so withdrawn is
still around 1% solids concentration by weight, and can be thickened further,
if desired, to about 3% solids in a thickener of relatively simple design,
before it is directly placed on a sand bed for open drying. A sludge lagoon
could also be provided, if desired, instead of drying beds. The design of
drying beds and sludge lagoons can be prepared as detailed in Chapter 17.

Typical computations for designing extended aeration systems have been
illustrated in Example 14.2.

15.6 FACTORS AFFECTING CHOICE OF LAGOONS AND THEIR LAYOUTS

The important factors affecting choice of lagoons are their power and
land requirements, vis-à-vis their performance and operating characteristics.
Generally, the choice lies between the facultative-type aerated lagoon and
the aerobic lagoon with solids recycling or intermittent operation. The
flow-through-type lagoon is often considered in stagewise development leading eventually to the recycling of solids or is followed by other treatment
to give a better quality of effluent.

Lagoons with solids recycling or intermittent operation are mostly of
the extended aeration type in the interest of simplicity and the desire to
avoid sludge digestion, which would be necessary with conventional activated
sludge systems. Extended aeration-type lagoons, although relatively easy to
operate, do require a greater degree of attention compared with the facultative lagoons which are, in fact, the simplest to operate as their extent of
mechanization is truly minimal. This may be an important factor in many
developing countries.

In terms of performance, the extended aeration lagoons, no doubt, give the best quality effluent. Nitrification and denitrification obtained in them may also be a special advantage in some cases. Consistency of performance over varying temperatures may be counted as another advantage. Consequently, intermittently operated lagoons deserve greater consideration than they have been given so far.

With regard to power and land requirements, the two types of lagoons have opposite characteristics: the extended aeration lagoons require more power but much less land; the facultative lagoons require more land but less power.

Thus, extended aeration-type lagoons may be favored for industrial or suburban developments where land is expensive or where better quality effluent is desired. Larger municipalities may also wish to consider them up to certain population sizes, beyond which conventional activated sludge may be cheaper.

Facultative lagoons may be favored for the large number of situations under which the oxidation pond may not be acceptable owing to its high land requirement, and other methods such as activated sludge may not be as desirable either owing to their technological requirements or simply because a higher quality effluent is not essential.

An advantage that all types of aerated lagoons have is the relative ease with which the existing lagoon volume or aeration capacity or both can be added to as populations increase or as better efficiency is desired. Lagoons present a considerable scope for progressive development. An aerated lagoon could also be built in the normal way but operated in the initial years as a simple stabilization pond by omitting the use of aerators and restricting the liquid depth, if desired. Later, aerators may be introduced and the unit operated in the normal manner to cater to a larger population. This may be of special value where the population is expected to increase very rapidly, as is often the case with certain industrial townships, tourism developments, camps, etc.

Lagoons also have the advantage of being relatively easily relocated at another site if necessary.

15.7 COMBINATIONS OF AERATED LAGOONS AND OXIDATION PONDS

At this stage, it may be interesting to consider the various possibilities which exist for developing different layouts combining either the dif-

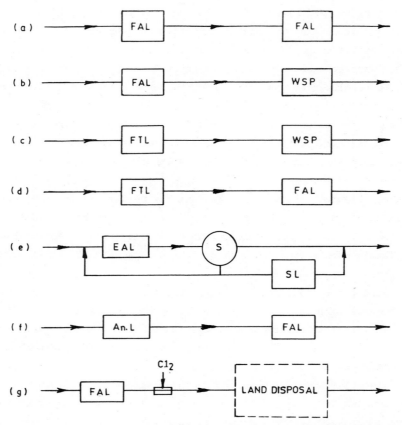

Figure 15.8 A few typical layouts combining different types of lagoons and ponds; FAL — facultative aerated lagoons, FTL — flow-through lagoons, EAL — extended aeration lagoons, WSP — waste stabilization pond, S — settling tank, SL — sludge lagoon, An.L — anaerobic lagoon.

ferent types of lagoons or lagoons and waste stabilization ponds to meet desired objectives. The objectives could be either to optimize land and power requirements, enhance simplicity of operation, promote treatability (e.g., by conserving heat), or some such characteristic. Figure 15.8 illustrates a few possible alternative layouts.

For example, Figure 15.8a shows how two (or more) facultative aerated lagoons could be provided as equal or unequal cells in series to enhance the overall removal of BOD, coliforms, etc., benefitting from mixing kinetics as explained earlier. Other reasons for such layouts could be the desire to conserve heat in the incoming waste to benefit from a higher temperature and faster reaction rate in the first cell (see Example 15.3), or to meet certain process requirements such as those discussed in Sec. 11.18.

Eckenfelder et al. [10] point out that the benefits of multiple lagoons for heat conservation may be marginal in warm climates and with relatively low-strength wastes.

The layout shown in Figure 15.8b is of considerable value in many areas where the combination of a facultative aerated lagoon followed by a stabilization pond can help the designer to optimize between land and power requirements, depending upon their relative costs, without much affecting the operating simplicity (see Example 15.4).

Figure 15.8c shows a flow-through-type lagoon followed by a waste stabilization pond. This combination can be designed to be convertible to the next combination shown in Figure 15.8d at some time in the future when the incoming load increases. Again, in such combinations, the designer has the opportunity to apportion the substrate removal between the two types of lagoons in such a manner as to optimize either for land, power, or overall costs.

Figure 15.8e gives a combination in which an extended aeration lagoon is used along with lagooning of the surplus sludge. The sludge lagoon can be considered as an alternative to other methods for handling the sludge. Settling and recycle are unnecessary if the aerobic lagoon is used on an intermittent basis.

Figure 15.8f gives an interesting possibility for the use of lagoons in warmer countries where anaerobic degradation (including gasification and consequent removal of BOD) can be depended upon at all times. The load thus placed on the facultative aerated lagoon is reduced to the extent that the anaerobic pretreatment is effective. This arrangement is functional in its approach since it separates the aerobic from the anaerobic activity.

Land disposal of the effluent from any lagoon would be worthy of consideration where land is available and the climate favorable. The benefits of securing further treatment on land include removal of nitrogen and phosphorus. Further removal of BOD and coliforms, as explained in Chapter 9, should not be overlooked.

Example 15.3 An industrial wastewater flow of 10,000 m^3/day is aerated in a two-celled facultative-type aerated lagoon operated in series and having 5 days' detention time in each cell. The depth in each cell is 3.3 m. If the wastewater inflow is at 55°C and the ambient temperature 10°C, estimate the temperature in each cell. Assume f = 0.49 m/day.

Volume of each cell = 10,000 x 5 = 50,000 m^3

$$\text{Area} = \frac{50{,}000}{3.3} \cong 15{,}000 \text{ m}^2$$

From Eq. (11.31),

$$\frac{t}{h} = \frac{T_i - T_L}{f(T_L - T_a)}$$

Thus, for the first cell:

$$\frac{5 \text{ days}}{3.3 \text{ m}} = \frac{(55°\text{C} - T_L)}{0.49(T_L - 10°\text{C})} \qquad \text{or} \qquad T_L = 35°\text{C}$$

Similarly, for the second cell, $T_L = 25°$ C.

If the lagoon had only one cell of the same total detention time, i.e., 10 days, T_L would have been equal to $27.5°$ C.

Note: The two cells in series need not be of equal size. They may be of unequal size, and various combinations could be tried to find the optimum combination which will require a minimum total detention time for obtaining a desired quality effluent. The results of three or four different trials can be plotted as shown in Figure 15.9 to obtain the needed detention time in the first cell by which the total detention time in the combination will be minimum.

Wherever possible, it may be advantageous, especially when dealing with high-temperature wastes, to keep the first cell of small size such that lagoon conditions will favor greater anaerobic activity, including gasification, so that the oxygenation power requirement is minimized.

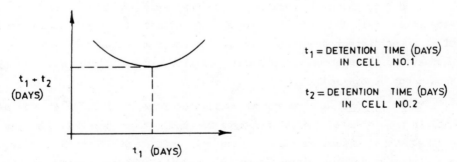

Figure 15.9 Optimizing for detention time in a lagoon with two cells in series treating hot wastewater (adapted from Ref. 10).

Example 15.4 Design a wastewater treatment system for a population
of 70,000, using a combination of a facultative aerated lagoon followed by
a waste stabilization pond to give the minimum total cost. Use the data
furnished earlier in Example 15.1 for facultative aerated lagoon design and
in Example 16.4 for stabilization ponds.

This calls for several trial designs to be prepared and costed to be
able graphically to arrive at the optimum conbination. In each trial, the
detention time in the aerated lagoon is assumed, and the detention time required in the stabilization pond is computed. Then the power and land costs
are computed along with construction and equipment costs as shown in Figure
15.10.

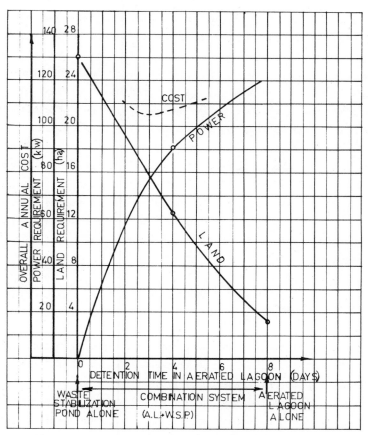

Figure 15.10 A determination of the optimum combination of facultative
aerated lagoon and stabilization pond (data from Example 15.4).

Using the results obtained in Examples 15.1 and 16.4, all for the population of 70,000 under identical conditions, we may summarize as follows:

Unit	Overall detention time (days)	Requirements	
		Land (ha)	Power (kW)
1. Stabilization pond alone	27	26	0
2. Combination system with 4 days' detention in facultative aerated lagoon	14.5	12.5	90
3. Facultative aerated lagoon alone	8	3.1	120

Figure 15.10 shows the above results graphically, along with a typical curve for the annual cost which must be arrived at after taking into account all capital and operating costs, including interest on capital.

If land costs are low, a stabilization pond alone will invariably be cheaper than either an aerated lagoon or combination system. As land costs increase, some combination of lagoon and pond will be optimum. At very high land costs, the aerated lagoon alone will be the optimum solution. Power costs will similarly affect the optimum point.

Generally, the combination which is likely to be the optimum is the one in which the detention time in the facultative aerated lagoon is small enough so that (1) the power input required per unit volume from BOD considerations is close enough to the minimum value of 0.75 kW/1000 m^3 and (2) the resulting temperature in the lagoon is around 13 to 15°C or higher in the winter.

15.8 CONSTRUCTION FEATURES

As stated earlier, the general construction features are similar to those seen in Chapter 16 for oxidation ponds. A few additional requirements are involved owing to the presence of aerators.

Aerobic as well as facultative lagoons are generally 2.5 to 5 m in liquid depth and are built in earthwork with slopes partly or fully pitched in stone or concrete. The inlet and outlet are located on opposite banks.

Where fixed aerators are used (mounted on columns or stilts) it is essential that the liquid level in the lagoon be maintained constant to ensure the required degree of submergence of the aerator blades. Most of the aerators available are quite sensitive to the depth of submergence and their rated oxygenation capacity can be severely reduced if the specified submergence is not maintained. This implies watertight conditions in the lagoon, which may sometimes be difficult to have in the early life of a lagoon. The discharge of effluent can be controlled over a weir, fixed or adjustable to any desired level, but of sufficient length to avoid much heading up at peak flows, which would change the aerator submergence.

To avoid percolation, a lagoon has to be located in relatively impervious soil or be suitably lined or constructed of masonry or concrete. The latter would be quite expensive owing to the large volume involved. Compaction of the soil to Proctor density is recommended. Control of percolation is also to be considered from the point of view of groundwater pollution.

Where floating aerators are used, percolation does not affect aeration since the aerator submergence always remains the same. Hence, the lagoon can be constructed easily and inexpensively in earthwork and provided with stone pitching if necessary. This is the main advantage of using floating aerators. Fabrication of aerators from noncorrodible material is, of course, necessary to ensure durability. Electric cable may be carried overhead to the aerator from the banks of the lagoon. The steel ropes used to anchor the aerator to the side banks can also be used to carry the cable. For repairs or maintenance, the aerator can be pulled or dragged in water to a corner of the lagoon where a small loop or arm can be provided to "wet dock" the aerator and enable it to be lifted up for inspection [1].

Other precautions in installing aerators depend on the size and type of aerator and the climatic conditions, and must be checked with the aerator manufacturer. In severely cold climates, special precautions need to be taken to keep the aerators running despite ice formations, and during sudden thawing, which may cause imbalance resulting in damage to the gear boxes. These may need the provision of special cover plates [11].

In the case of large aerators (over 75 hp) it is advisable to provide a concrete pad directly below them on the lagoon bottom, or to provide a shallow tube. The aeration capacity of all types of aerators tends to be reduced by the formation of a vortex which is generally controlled by providing "baffles" attached to the aerator support pillars.

Sometimes, it may be advisable to leave room for the later installation of additional aerators either to meet expected increases in incoming BOD or to increase the suspended solids concentration in the lagoon to desired values.

Aerated lagoons are often rectangular in shape, although they could be built to conform to natural topographical features. However, it appears that the power levels required for mixing are increased if the length-to-width ratio exceeds 2:1 [10]. In the case of uneven shape, the number and disposition of aerators should be such as to ensure uniform oxygenation.

The usual principles of good civil engineering practice would apply to the design and construction of aerated lagoons, particularly with regard to embankment slopes, compaction of earthwork, pitching, flood protection, etc. Where the groundwater table is relatively high, the lagoon level must be raised accordingly. Where the irrigation of adjoining fields is desired, the lagoon level may be kept so as to permit gravity flow to the fields.

The provision of a preliminary screening facility and a weir or flume for flow measurement are strongly recommended even for small installations.

REFERENCES

1. S. Arceivala, Simple Waste Treatment Methods, Middle East Tech. University, Ankara, Turkey, 1973.

2. W. W. Eckenfelder, Jr., Water Quality Engineering for Practicing Engineers, Barnes and Noble, Inc. N.Y., 1970.

3. R. McKinney and P. Benjes, Evaluation of two aerated lagoons, Jour. ASCE, San. Eng. Div., 1965.

4. E. Bartsch and C. Randall, Aerated lagoons: A report on the state of the art, Jour. WPCF, 43, 1971.

5. J. Mancini and E. Barnhart, Industrial waste treatment in aerated lagoons, in Advances in Water Quality Improvement, Vol. 1, Univ. of Texas Press, Austin, Texas, 1968.

6. C. Sawyer, New concepts in aerated lagoon design and operation, in Advances in Water Quality Improvement, Vol. 1, Univ. of Texas Press, Austin, Texas, 1968.

7. C. Adams and W. W. Eckenfelder, Jr., Eds., Process Design Techniques for Industrial Waste Treatment, Enviro Press, Nashville, Tennessee, 1974.

8. K. Murphy and A. Wilson, Characterization of mixing in aerated lagoons, Jour. ASCE, Env. Eng. Div., 1974.

9. S. Arceivala, Use of dispersed flow model in designing waste treatment units, in Theory and Practice of Biological Waste Treatment, NATO Advanced Study Institute, Istanbul, Noordhoff Int. Pub., Leiden, Netherlands, 1976.

10. W. W. Eckenfelder, McGee, and Adams, A rational design procedure for aerated lagoons, in <u>Advances in Water Pollution Research</u>, 6th International Conference, Jerusalem, Israel, 1976.

11. C. Adams and W. W. Eckenfelder, Personal communication, 1976.

12. D. Thirumurti, Design criteria for aerobic aerated lagoons, <u>Jour. ASCE, Env. Eng. Div.</u>, February, 1979.

13. M. Goronszy, Intermittent operation of extended aeration process for small systems, <u>Jour. Water Pollution Control Fed</u>., 51 (2): 274, 1979.

Chapter 16

WASTE STABILIZATION PONDS

16.1 INTRODUCTION

Waste stabilization ponds are the simplest of all waste treatment
techniques available for sewered wastewaters. They involve holding the
wastewaters in shallow ponds for a long enough detention period to enable
the natural stabilization of organic matter to occur through microbial
activity. As pond systems operate under entirely natural conditions without the benefit of any man-made accelerating devices (e.g., heating and
mechanical aeration), from this is immediately derived their advantages and
disadvantages.

Their advantages stem from their extreme simplicity and reliability of
operation. Nature cannot go wrong; there is no equipment to fail and no
tricks to successful operation. But nature is slow, requiring long detention periods which in turn imply large land requirements. Biological activity is also considerably affected by temperature, more so in the pond's
natural conditions. Thus, waste stabilization ponds are most appropriate
where land is inexpensive, climate favorable, and a simple method of treatment is desired that does not require equipment and operating skills.

The costs of waste stabilization ponds are generally very competitive
as long as land costs are not excessive. Construction is simple, involving
mostly earthwork, and operating costs are practically negligible compared
with other methods of treatment. Like all other treatment systems, ponds
also have their limitations, although in terms of performance they are almost like a conventional treatment system. Their low cost has sometimes
led some people to feel that ponds may be inferior in their performance.
But this is not generally the case, and often ponds have proved more dependable than other methods in the long run.

By 1975, engineered ponds were in use in over 50 countries, covering
a wide variety of climatic conditions ranging from equatorial to cold, and
including countries from Sweden and Canada in the north to New Zealand in

the south. A large number of these ponds can also be found in Asia, Africa, and Latin America [1].

Over the years, engineering design criteria based on field observations and basic research have come to be established. Current refinements in design enable ponds to be approached as reactors affected by reaction kinetics and flow patterns. Performance criteria have widened in many cases to include the removal of microorganisms and nutrients, besides biochemical oxygen demand.

16.2 TYPES OF PONDS

The term waste stabilization pond* in its simplest form is applied to a body of water, artificial or natural, used for retaining wastewaters until the wastes are rendered stable and inoffensive for discharge to receiving waters or land. Stabilization is assisted by various physical, chemical, and biological processes commonly referred to as "self purification" and, as a result of which, organics are broken down through bacterial activity to more stable end products.

Aerobic waste stabilization ponds are shallow ponds of about 0.3 m depth or less designed to maximize light penetration and the growth of algae through photosynthetic action. Aerobic conditions are maintained throughout the depth of the pond at all times, and the wastes are stabilized entirely through aerobic microorganisms. Such ponds are useful where the ultimate harvesting of algae is desired, but their use in waste treatment has not been widespread.

Anaerobic waste stabilization ponds require no dissolved oxygen for microbial activity as the anaerobic and facultative organisms use oxygen from compounds such as nitrates and sulfates as their hydrogen acceptors, and give end products such as methane and CO_2. Such ponds can therefore accept higher organic loadings and operate without algal photosynthesis. Light penetration is unimportant and they can be built deeper, 3.0 to 4.0 m depth being more common. Anaerobic ponds can be likened to unheated and

---

*This term is widely adopted as it is descriptive of the function, and includes aerobic and anaerobic modes of stabilization. The term "waste" includes sewage and industrial wastes. Neither of the terms "oxidation ponds" and "sewage lagoons" strictly includes all these aspects, although they are often used.

unstirred, open digesters. Their effluent is generally not fit for discharge without further treatment. Thus, they are often provided in tandem
with facultative ponds which follow them (see Sec. 16.18).

Facultative waste stabilization ponds are neither fully aerobic nor
fully anaerobic. They are often about 1 to 2 m in depth and favor algal
growth along with the growth of aerobic, anaerobic, and facultative microorganisms. Such ponds are predominantly aerobic during daylight as well as
for some hours of the night. In the few remaining hours, the pond bottom
waters may turn anaerobic. Benthic deposits are generally anaerobic beyond
the first few millimeters from the solids-water interface. Most of the existing waste stabilization ponds in the world are of the facultative type
with varying degrees of aerobicity and anaerobicity (Figure 16.1). In this
chapter therefore, reference to waste stabilization ponds implies facultative ponds unless stated otherwise.

Ponds receiving untreated wastewaters are referred to as raw or primary
waste stabilization ponds. Those receiving primary treated or biologically
treated wastewaters for further treatment are called secondary waste stabilization ponds.

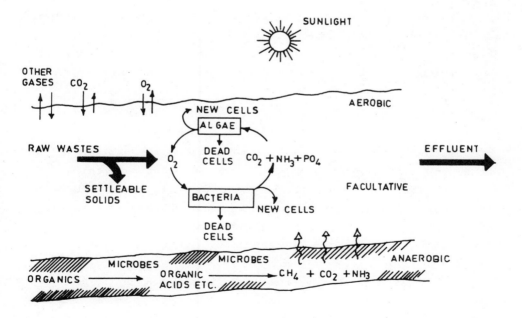

Figure 16.1 Algal and bacterial activity in a facultative pond.

Another type of pond is the <u>maturation pond</u> in which wastewaters, pre-treated either in ponds or conventional treatment plants, are exposed for a further period of time with the main purpose of achieving natural bacterial die off to desired levels. The ponds are lightly loaded in terms of organic matter. In warm climates, they may constitute an economically feasible alternative to disinfection by chlorine.

A <u>fish pond</u> is a secondary or a maturation pond, or an altogether separate pond in which fish are reared (Sec. 16.17).

16.3 TYPICAL FLOWSHEETS

Figure 16.2 gives a few of the typical alternative pathways that could be followed when using ponds. The preferred flowsheet depends mainly on the nature of the wastewater and the degree of treatment required, which in turn depends on the mode of final disposal (e.g., to a watercourse or land). A flowsheet which allows some flexibility in operation is desired, especially where wastewater characteristics fluctuate. (See also Figure 16.17.)

Industrial wastes may need suitable pre- or posttreatment when pond systems are used (e.g., nutrient addition).

The performance of the different types of ponds and flowsheets used can be judged for their efficiencies on three counts: (1) BOD removal, (2) microorganism reduction, and (3) nutrient removal. Performance must also be judged in the light of possible nuisance development (e.g., odors).

16.4 THE POND ECOSYSTEM

The waste stabilization pond can be regarded as an ecosystem or, better, as an ecological subsystem having its biotic and abiotic components. The principal members of the abiotic component are O_2, CO_2, water, light, and nutrients, whereas the biotic component includes algae, bacteria, protozoa, and a variety of other organisms (the pond community). In response to various interactions, a relatively stable community is evolved in accordance with the principles governing ecological succession.

Briefly, in a facultative waste stabilization pond in the presence of light and sufficient nutrients contained in the incoming waste, a healthy bloom of algae flourishes together with a large number of aerobic and facultative microorganisms. At the same time, benthic deposits are also present

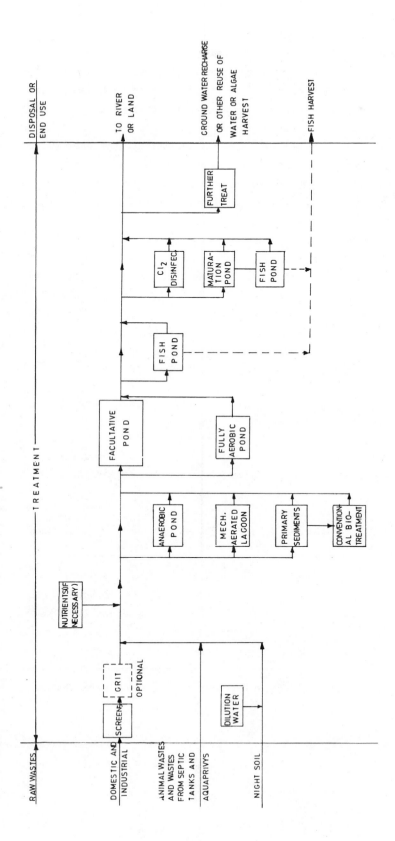

Figure 16.2 Some common alternatives for the treatment of wastes in waste stabilization ponds.

with anaerobic and facultative microorganisms. The growth patterns of algae
and bacteria depend upon the nature of the incoming wastes and on various
environmental conditions. Figure 16.1 shows the different activities and
their interrelationships in a typical facultative pond.

Algal concentrations in ponds are often less than 200 mg/liter (40 to
200 mg/liter) expressed as dry weight, although in terms of numbers they
could be in the range of 10^4 to 10^6 per ml and show seasonal and other fluctuations. Diurnal differences do not seem to follow regular patterns [22].

Algae can be expressed by the empirical formula $C_{106}H_{180}O_{45}N_{16}P_1$.
Their contents of carbon, nitrogen, and phosphorus, however, vary somewhat
with species and age; the range is as given in column 3 of Table 16.1. The
available concentrations of C, N, and P in domestic sewage, and therefore
the algal growth potential in ponds, are also shown in this table. The
table neglects the possible recycling of nutrients during the long detention
periods normally experienced in ponds and the possible gains and losses of
nutrients through various possible mechanisms. It is clear that, as far as
domestic sewage treatment is concerned, carbon is limiting or, in other
words, when all the carbon is used up for algal growth, N and P are still
left over in excess. Their removal would need another strategy as discussed
later.

Table 16.1 Algal Growth Potential in Ponds Treating Domestic Sewage

Element	Available concentration in typical urban sewage[a] (mg/liter)	Percent dry weight content in algae	Approximate range of algal growth potential in ponds (mg/liter, dry wt.)
C	~125	40-60 as C	200-250-300
N	40	5.5-12.5 as N	300-500-700
P	11	0.25-1.3 as P	700-1200-2200
K	>5	~0.5 as K	~1000
Mg	>5	~1.0 as Mg	~500

[a]Containing approximately BOD_5 = 250 mg/liter, N = 8.6 g/capita-day,
P = 2.4 g/capita-day.
   Source: Ref. 14.

16.5 CELL WASHOUT TIME

The algal growth rate in ponds may be stated as follows:

Observed rate of increase of algal biomass in pond	=	growth rate of algae within pond	−	engodenous decay rate of algae within pond	−	rate of removal of algae by flushing out of pond

Thus,

$$\left(\frac{dc}{dt}\right)V = \mu(CV) - K_d(CV) - CQ \tag{16.1}$$

where

    C = algal concentration in pond

    V = pond volume

    μ = specific growth rate constant, t^{-1}

    K_d = endogenous decay rate constant, t^{-1}

    Q = flow rate through pond

    t = time

Dividing throughout by V gives

$$\frac{dc}{dt} = (\mu - K_d)c - c\left(\frac{Q}{V}\right) \tag{16.2}$$

or

$$\frac{dc}{dt} = c\left(\mu_{obs} - \frac{1}{t}\right) \tag{16.3}$$

The term μ_{obs} is, in fact, the net observed growth rate of algae in the pond and the term $1/t$ is referred to as the "dilution rate."

It is evident that the algal concentration will increase only if the growth rate is greater than the dilution rate. At steady state, $dc/dt = 0$ and we get $\mu_{obs} = 1/t$. If $\mu_{obs} < 1/t$, all algae will be gradually flushed out from the pond. In polishing lagoons studied by Toms et al. [13] in the U.K., the values of μ_{obs} were found to range from about 0.05 to 0.7 per day, but on only six occasions in an 18 month period was μ_{obs} less than 0.48 per day. This is in general agreement with the observation in colder climates that very low algal concentrations are to be seen even in well-fertilized ponds if their detention times happen to be less than 2.0 to 2.5 days.

No significant correlation between incident radiation and algal growth rates has been found. In most cases, their light saturation level is lower than the radiation usually received.

16.6 FACTORS AFFECTING DEPTH OF PONDS

Light penetration is an important factor affecting photosynthesis and algal growth. Light is absorbed quickly in its passage through water so that penetration does not extend significantly beyond about 50 cm below the pond surface. The gradual attenuation of light as it penetrates through water follows the well-known Beer-Lambert law, which is given by the typical first-order plug-flow-type equation

$$I_d = I_i e^{-C_c \alpha d} \qquad (16.4)$$

where

I_i = incident light intensity at surface, ergs/cm^2-sec*

I_d = light intensity at any depth d, ergs/cm^2-sec*

C_c = algal concentration, mg/cm^3

α = specific absorption coefficient, cm^2/mg

d = depth of penetration, cm

The light compensation level is the depth at which photosynthetic production just balances algal respiration. The intensity of light I_d at the light compensation level is generally taken in approximate calculations of this type as 1000 ergs/cm^2-sec (27 footcandles). The value is utilized to estimate the depth up to which algae can grow.

Based on experimental work, Oswald [14] recommends that, to grow a specific concentration of algae, the permissible depth of a pond should not exceed by threefold the depth predicted by using Eq. (16.4) for the desired algal concentration.

Thus, Eq. (16.4) can be rewritten to apply to stabilization pond depth as under [14]:

$$d = 3 \left(\frac{\ln I_i - \ln I_d}{C_c \alpha} \right) \qquad (16.5)$$

---

*1 erg = 2.389 x 10^{-8} g-cal = 10^{-7} joules.

41.86 ergs/cm^2-sec = 1 footcandle = 10.82 lux units.

1 langley = 1 g-cal.

If d is made greater than the value predicted by Eq. (16.5), the value of C_c will be less than expected. This, of course, does not imply that the deeper pond will fail to perform, only that the algal concentration will reduce on the average. In fact, deep ponds may have other advantages as we shall see later.

Some interesting conclusions can be drawn from an inspection of Eq. (16.5) concerning pond depths.

1. The permissible pond depth "d" for a desired algal concentration varies with the logarithm of incident light as the value of ln I_i. Hence, wide variations in light cause relatively little effect on the depth of penetration. A tenfold decrease in the value of I_i gives only a 30% decrease in the value of depth d. Daylight intensities can vary from about 10,000 to over 400,000 ergs/cm²-sec* (see Table 16.3).

2. Depth is inversely proportional to the specific absorption coefficient α. The value of α depends on the presence and nature of algae, color, turbidity, or other factors which affect light absorption in water. Thus, treatment of particularly colored wastes (e.g., textiles and milk) must be limited to relatively shallow ponds. Generally, in domestic sewage treatment ponds, Oswald states that the value of α averages 1.5 x 10^{-3} cm²/mg. Some workers have reported values as high as 3.8 x 10^{-3} cm²/mg. These values are roughly three orders of magnitude greater than those for distilled water.

3. Depth is inversely proportional to the concentration of algal cells. This implies that algal growth can be self-limiting; the greater the growth, the less the light penetration.

Other factors that affect the depth of facultative stabilization ponds are as follows:

1. Pond depths greater than about 0.9 m are preferred in order to minimize the growth of emergent-type vegetation which would otherwise abound in very shallow ponds, especially at the periphery. Such vegetation may harbor mosquito larvae.

2. Very shallow ponds would be more readily affected by changes in ambient temperatures. They may reach high temperatures at noon in warm climates, which may adversely affect algal growth and retard methane fermentation in the lower part of the ponds. Ponds only 0.75 m deep do not provide the best stabilization and are likely to suffer from unpleasant odors during the warm seasons.

In hot weather, the surface of a very shallow pond may reach a temperature of 35° C or even more and a pH level of about 10.5 to 11.0. This

---

*Studies by Oswald and Gotaas [4] have shown that the mean light intensity in footcandles at any site for the entire 24 hr period of a day is about ten times the total insolation expressed in langleys per day multiplied by the fraction of time the sun is visible.

can be lethal to bacterial activity. It can also promote precipitation of
phosphates, calcium, etc., which act as flocculating agents assisting the
settling of algae and other solids to the pond bottom. The water having
been clarified, light penetration increases, thus enabling photosynthetically
produced oxygen to penetrate deeper and arrest methane fermentation which is
sensitive to the presence of oxygen and increased redox potential. With the
stoppage of methane fermentation, accumulation of volatile acids and other
compounds may lead to odor nuisances. The previously settled algae may
eventually float up and form scum layers. Oswald [28] refers to such a sequence of events as pond failure since no dependable biological activity
persists through such a period.

Very shallow ponds are desirable if the objective of treatment is to
maximize algal production in short-detention, controlled-velocity ponds for
nutrient stripping and/or algae harvesting.

Facultative waste stabilization ponds are preferably kept at least 1.2
to 1.5 m deep, and often even more. Sludge storage space, if provided, is
additional. Deeper ponds (2 to 3 m depth) have been noted to have several
advantages [28].

Generally, speaking, the deeper the pond, the more stable is its performance. It tends to give a more even quality of effluent throughout the
year despite diurnal and seasonal variations in temperature [2,28]. For a
given surface area (and areal loading rate, kg BOD/ha-day), the greater the
depth, the longer the detention time and thus, generally speaking, the
better the effluent quality with regard to BOD removal, microorganism die off,
and nitrogen removal. Deeper ponds have been reported to give fewer operational problems with regard to the breeding of <u>Chironomas</u> (bloodworm) larvae
[28]. On the other hand, anaerobicity is known to retard biochemical stabilization rates and also bacterial die off in ponds. Currently, it is not
possible to optimize for depth in pond design owing to a limited knowledge
of the interplay between the complex factors involved. However, depending
on climatic conditions and wastewater characteristics, there is a strong
case for considering deeper ponds than were generally used hitherto (also
see Figure 16.7).

Oswald [28] and Wachs and Berend [22] have reported on the operation
of deeper ponds, the latter from Israel with ponds up to 5 m depth.

<u>Example 16.1</u> Find the permissible depth to be provided in a pond so
that an algal concentration of 80 mg/liter is likely in the presence of

sufficient nutrients and incident light averaging 83,700 ergs/cm²-sec (=
2000 footcandles). Assume I_d = 1000 ergs/cm²-sec and α = 1.5 x 10⁻³ cm²/mg.
From Eq. (16.5), depth d is given by

$$d = 3\left(\frac{\ln I_i - \ln I_d}{C_c \alpha}\right)$$

$$= 3\left[\frac{\ln 83,700 - \ln 1000}{(80)(1.5 \times 10^{-3})}\right]$$

$$= 1.10 \text{ m}$$

16.7 ESTIMATION OF POND TEMPERATURES

The pond temperature which results from various factors such as ambient
temperatures, wind, humidity, area exposed, etc., invariably needs to be
estimated in advance for design purposes. All physical, chemical, and biological processes involved in pond treatment are considerably affected by
temperature.

A method of pond temperature estimation was given in Sec. 12.11 to
which the reader is referred. The use of this method is illustrated later
in Example 16.4.

16.8 SULFIDE FORMATION IN PONDS

A certain amount of sulfide formation often occurs in facultative ponds,
but when this is excessive it may lead to problems of malodors and even inhibition of algal activity. Generally, sulfide formation is limited to the
anaerobic bottom layers of a pond, and is chemically or biochemically oxidized as it diffuses upward into the aerobic layers. But, if sulfide production is excessive or if the upper layers are not sufficiently oxygenated
owing to inhibition of algal activity or any other reason, odor problems
may result.

The sulfates contained in the wastewater are reduced to sulfides by
those bacteria which utilize sulfate as a hydrogen acceptor in their cata-
bolic reactions. Typical of this group of bacteria are the Desulfovibrio
sp., for which favorable conditions imply a dissolved oxygen concentration
less than about 0.16 mg/liter, namely, practically anaerobic conditions
and temperatures above 15°C.

The sulfides so formed are readily oxidized by dissolved oxygen. They can also be oxidized biochemically by two different groups of microorganisms. The first group consists of sulfur bacteria, a group of strict aerobes which utilize sulfides but only under aerobic conditions (in which chemical oxidation by DO occurs more readily). Consequently, they are not generally found in stabilization ponds. The second group of microorganisms which oxidize sulfides are the photosynthetic sulfur bacteria which are often present in large numbers in stabilization ponds when sulfides are available to them. These bacteria require both light and sulfides, and utilize CO_2 as the hydrogen acceptor. Oxidation occurs as follows:

$$CO_2 + 2H_2S \xrightarrow{\text{photosynthesis}} (CH_2O) + 2S + H_2O \qquad (16.6)$$

$$3CO_2 + 2S + 5H_2O \xrightarrow{\text{photosynthesis}} 3(CH_2O) + 2H_2SO_4 \qquad (16.7)$$

Thus, photosynthetic sulfur bacteria help to reduce odor problems by oxidizing the sulfides. When they are present in large numbers, ponds show a typical brownish red color. Unlike the photosynthetic algae, these bacteria do not produce any oxygen and hence do not help in BOD removal.

Gloyna and Espino [12a] studied sulfide production in laboratory-scale stabilization ponds under different loading conditions. From their factorially designed experiments, they developed an empirical relation which could be stated as follows at 25°C:

$$\underset{\text{(mg/liter)}}{S^{2-}} = [0.0001058 \underset{\text{(kg BOD}_5\text{/ha-day)}}{(L)} - 0.001655 \underset{\text{(detention, days)}}{(t)} + 0.0553] \times \underset{\text{(mg/liter)}}{SO_4^{2-}}$$

$$(16.8)$$

Equation (16.8) enables an estimation of the average 24 hr sulfide concentration in a facultative stabilization pond at 25°C if the BOD loading, the detention time in the pond, and the incoming SO_4^{2-} concentration in the wastewater are known. The coefficients would evidently change at other temperatures for which data are not available at present. However, the relationship is likely to be useful in estimating sulfides during the critical summertime operation of ponds in many regions where pond temperatures would be around 25°C (see Example 16.4).

Inhibition of algal activity may occur if the average S^{2-} concentration exceeds about 4 mg/liter. Disappearance of algae has been observed by Gloyna and Espino and others [12a] at about 6.5 to 8.4 mg/liter.

ORP measurements have been used to detect the early formation of sulfides in ponds because relatively small increases in S^{2-} concentration cause
large decreases in ORP [1]. The S^{2-} concentration generally shows a diurnal
variation in ponds, reaching the maximum before daybreak and diminishing
thereafter as photosynthetic oxygenation picks up.

16.9 DESIGN OF FACULTATIVE PONDS

16.9.1 BOD Removal Efficiency

BOD removal efficiency of properly designed facultative waste stabilization ponds is often in the range of 80 to 90%. Even at overloaded conditions, their efficiency may be quite substantial. In this respect, pond
performance is practically equal to that of conventional waste treatment
methods using trickling filters or the activated sludge process. A study
of several ponds in India under warm climatic conditions showed that even
overloaded ponds gave efficiencies over 70% at all times (see Figure 16.3).
Some of the ponds were loaded as high as 700 to 800 kg/ha-day, but the ma-
jority were under 500 kg/ha-day [2].

McGarry [3] reviewed the performance data from 143 facultative pond
systems operating in tropical and temperate zones. He found similarly that
over 72.5% of the BOD was removed in the wide range of 30-500 lb/acre-day*
of raw BOD load on the ponds studied (see Figure 16.4).

---

*1 lb/acre-day = 1.12 kg/ha-day.

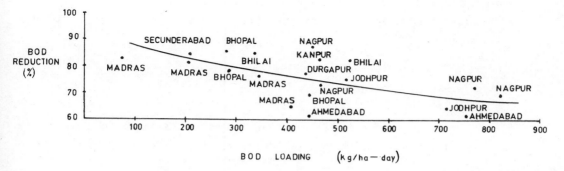

Figure 16.3 The performance of ponds at different loadings in India
(from Ref. 2).

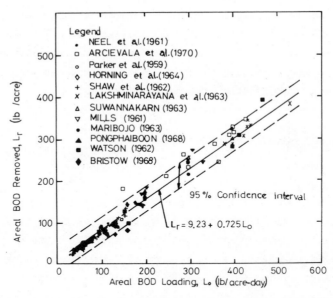

Figure 16.4 Areal BOD removal as a function of BOD loading (from Ref. 2).

Pond effluents generally contain some algae which add to the oxygen consumption in the BOD bottle. This can be compensated for by performing the BOD test in lighted conditions.* If a sample is filtered to remove the algae before the test, its BOD is somewhat reduced and a higher efficiency of BOD removal is consequently obtained. Unfortunately, filtering may remove some nonliving organic matter whose oxygen demand is a valid part of BOD. The BOD removal values given above are for unfiltered samples.

It is necessary to note here that the performance of a stabilization pond is much affected by temperature, and thus the effluent BOD varies from season to season. This is illustrated in Figures 16.9 and 16.15. Thus, efficiency data become more meaningful when their corresponding temperatures are also quoted.

16.9.2 Algae in Effluent

Normally, algae are <u>not</u> filtered out before the discharge of pond effluent to a watercourse. The effect on the receiving water depends upon

---

*Some prefer to incubate samples under alternating light and dark conditions to simulate natural conditions as far as possible.

whether the algae die, serve as food for zooplankton, or survive and grow in the watercourse. This depends on whether the conditions in the watercourse are favorable for the algal species involved. If algae die, aerobic bacterial breakdown of dead algal matter requires about 1.4 g of O_2 per gram of algal matter. More often than not, however, algae are consumed by zooplankton and enter the food chain. This can, in fact, be profitable if the food is harvestable (e.g., fish). The provision of tertiary ponds with adequate retention time can achieve this in a controlled manner. The algae not consumed may survive and multiply downstream.

If algae continue to multiply in the receiving waters, they may have a beneficial effect on the oxygen balance of the watercourse since they produce much more oxygen in photosynthesis during daylight hours than they consume by respiration during all 24 hr. For algae contained in the effluent of a polishing or tertiary lagoon, Toms et al. [13] found the respiratory demand to be about 0.007 mg O_2 per mg ss per hour, and oxygen production to be about 0.1 mg O_2 per mg ss per hour, namely, about 15 times more than the amount consumed.

Algal cells can also vegetate for long periods. Cultures of algae having concentrations of 200 to 300 mg/liter have been held in the absence of nutrients for 3 months without putrefaction as long as light and atmospheric CO_2 were available. In the dark, cells have remained alive for only a few days at temperatures above $25°$ C and a couple of weeks at near freezing conditions.

Often, especially in the developing countries, stabilization pond effluent is used for land irrigation. Algal content may then be beneficial. Blue-green algae assist in nitrogen fixation. Other algae when they die release nutrients for subsequent plant uptake. Excessive algal content may affect soil porosity, although no such cases have been reported as a result of the use of pond effluents for irrigation.

16.9.3 Oxygen Considerations

The role of algae in waste stabilization ponds is principally to supply the oxygen required for aerobic bacterial activity. As long as the algae can provide an excess of oxygen above that required by bacteria, an aerobic environment is maintained in the pond. A certain amount of oxygen transfer

from the atmosphere also occurs, but is relatively small compared with the
algal production of oxygen.

Algae can be expressed by the general empirical formula $C_{106}H_{180}O_{45}N_{16}P_1$,
and the production of every gram of algae is accompanied by the net production of 1.0 to 1.7 g of O_2.

The question that arises is, can we predict how much algal growth will
occur under a given set of conditions? If we can do that, we can also estimate how much oxygen will be produced simultaneously.

Algal growth is, in fact, a conversion of solar energy to chemical energy in the organic form. Empirical studies have shown that, generally,
about 4 to 6% of visible light energy can be converted to algal energy in
controlled laboratory experiments aimed at finding the optimum temperature
at which the light conversion efficiency is maximum [4].

The light conversion efficiency is maximum when the algal growth rate
is maximum. Apparent light conversion efficiencies estimated from field
studies have given higher values of 6 to 8% [2,6a]. Oswald [14] states that
fixation of light energy by algae is normally at less than 10% efficiency.

The chemical energy contained in an algal cell can be determined by use
of a bomb calorimeter to find unit heat of combustion and expressed as
cal/g of algae. Most algal communities of interest in waste stabilization
ponds exhibit a value around 6000 cal/g on an ash-free basis [2].

The light energy received on a horizontal surface varies with latitude
and cloud cover and is stated as g-cal per cm^2 per day (i.e., langleys/day).
It may be received from both direct and diffuse sources. Radiation of all
wavelengths in the solar spectrum is referred to as "total" radiation,
whereas the "visible" radiation which penetrates the water surface is limited to wavelengths of 4000 to 7000 Å. Table 16.2 gives some typical values
from data published by the U.S. Weather Bureau for a few latitudes. The
critical months are in the winter when the radiation received is least. In
the northern hemisphere, December has the least and June the highest radiation.

In Table 16.2, maximum and minimum values are shown for each month.
Maximum values occur when the sky is clear, and minimum values occur when
the sky is cloudy. Generally, in many countries, meteorological data are
readily available, giving radiation values and the percent of time the
weather is likely to be clear or cloudy in a month. The number of clear
hours could vary from, for example, 10% to 90% of the possible clear sunny
hours in a month. The average radiation received at a location can be

Table 16.2 Probable Average Values of Insolation (Direct and Diffuse) on a Horizontal Surface at Sea Level in cal/cm^2-day (U.S. Weather Bureau)

Latitude (degrees)	December			
	Visible		Total	
	Max	Min	Max	Min
60	5	1	75	7
50	26	7	180	24
40	66	24	270	65
30	126	70	320	210
20	182	120	480	320
10	225	162	570	420
0	253	195	685	510

Note: 1. Values given apply to latitudes north of the equator. Identical values apply south of the equator but with the months reversed.

2. 1 g-cal/cm^2-day = 1 langley/day.

3. Visible: 4000-7000 Å wavelength.
    Total: All wavelengths.

4. Maximum: Value which will not normally be exceeded.
    Minimum: Lowest values during 10-year record.

Source: Ref. 4.

estimated if the maximum and minimum values are known along with the sky clearance factor [4]. Thus,

Average radiation

= minimum + [(maximum - minimum) x sky clearance factor]   (16.9)

Visible radiation values are used when estimating algal production in ponds. Total radiation values help to estimate light intensity in terms of footcandles.

For towns located at high elevations, the sea level values given in Table 16.2 need to be corrected. The approximate corrections for elevations up to about 3000 m are:

Corrected radiation = (sea level value)(1 + ah)   (16.10)

where

h = elevation in thousands of meters

a = 0.0607 for total radiation and 0.0303 for visible radiation

Solar energy received at a pond can be somewhat diminished by the microclimate of the area, for example, by excessive amounts of dust particles from air pollution, smog, etc. Local measurements can be made if desired using a clockwork pyrheliometer.

With the help of data such as given above, the theoretical oxygen production per day can be estimated for a pond as illustrated in Example 16.2. The ultimate oxygen demand (BOD_u) of the wastewater proposed to be treated in the pond should be less than the oxygen produced if aerobic conditions are to be ensured at all times.

Example 16.2 Compute the theoretical oxygen production per day per hectare of pond surface for a location at $40°$N (Ankara, Turkey) where the sky clearance factor in December is 0.38. Make suitable assumptions.

From Eq. (16.9) and assuming that the radiation values given in Table 16.2 apply,

Average radiation received in December $= 24 + [(66 - 24) \times 0.38]$
$$= 40 \text{ cal/cm}^2\text{-day}$$
$$= 40 \times 10^4 \text{ cal/m}^2\text{-day}$$

Assume algal conversion efficiency $= 6\%$

Weight of algae produced at 6000 cal/g of algae

$$= \frac{40 \times 10^4 (0.06)}{6000}$$
$$= 4 \text{ g/m}^2\text{-day}$$

If the weight of oxygen produced is, for example, 1.7 times the weight of algae produced, we get

O_2 production $= 6.8 \text{ g/m}^2\text{-day}$
$$= 68 \text{ kg/ha-day}$$

From Example 16.2 we see how the oxygen production in a pond can be estimated. As stated earlier, to ensure aerobic conditions at all times, the ultimate first-stage oxygen demand (BOD_u) of the organic matter removed in the pond is kept equal to or less than the oxygen produced. Thus, for the pond in Example 16.2, the BOD_u removed in the pond should not exceed 68 kg/ha-day. If the pond efficiency is, for instance, 80%, the BOD_u load placed on the pond should not exceed 68/0.8, i.e., 85 kg/ha-day.

The BOD_u loading could also be stated in terms of radiation received at the pond surface. For the above case, with the average radiation in December equal to 40 langleys/day, we get 0.47 langley/day per kg BOD_u/ha-day (40/85 = 0.47).

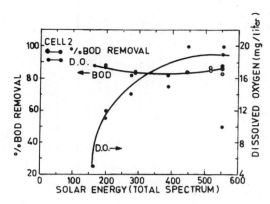

Figure 16.5 The DO concentration and BOD removal in a pond (cell 2) in Fayette, Missouri (from Ref. 6).

If a pond is loaded at a higher rate than the one indicated by the above type of calculations, it would give a lower value than 0.47. One would then like to know what happens. Is treatment efficiency reduced? Are any problems created?

Data from Neel et al. [5] for ponds at Fayette, Missouri, suggest that, as the solar radiation decreases in winter, the DO concentration in a pond effluent decreases; however, the BOD removal efficiency does not drop correspondingly. Figure 16.5 shows the monthly averages of BOD and DO versus radiation for these ponds.

The dissolved oxygen concentration is, of course, affected by several factors. At winter temperatures, the bacterial degradation rate, and thus the corresponding oxygen utilization rate, in fact slows down. Therefore, the reduction in DO observed must be attributed to a reduced production of oxygen either because of reduced radiation in winter, reduced algal conversion efficiency at the low temperature, or both.

Neel et al. concluded that, if ice cover need not be considered, an average solar radiation level of about 40 to 45 langleys/day (visible spectrum) would suffice to maintain oxygen in ponds loaded up to 112 kg of BOD_u/ha-day. This corresponds to about 0.38 langley per kg of BOD_u/ha.

In actual DO measurements made under different loadings in ponds in Nagpur, India, Arceivala et al. [2] have reported the results as shown in Figure 16.6, from which it is evident that as pond loading increases, the depth of the aerobic zone tends to diminish during both day and night. From operational experience (temperature range 24-33° C), reasonably nuisance-free conditions were obtained between 0.42 and 0.55 langley per day per kg BOD_u/ha-day.

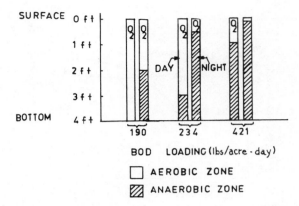

Figure 16.6 The depths of the aerobic zones at different loadings of ponds in Nagpur, India (from Ref. 2).

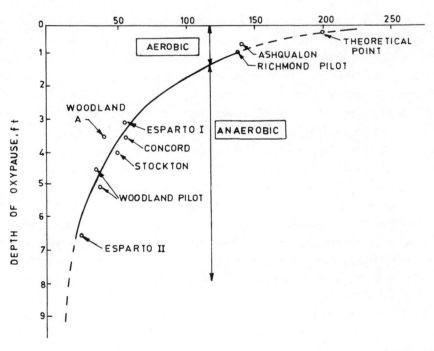

Figure 16.7 The influence of BOD loading on the depth of oxypause in waste stabilization ponds of average condition. (Adapted from Ref. 28.)

The available data from a number of sources on dissolved oxygen penetration and BOD loading have been assembled by Oswald [28] for several ponds in the California area. Figure 16.7 shows that during autumn conditions the average depth of the aerobic layer is larger for the more lightly loaded ponds than for the more heavily loaded ones. This is similar to the results from India just given.

Thirumurti in Canada [6] attempted to determine by laboratory experiments, using cool-white fluorescent lamps, the minimum solar radiation below which a reduction in BOD removal efficiency may occur as light would be limiting. He observed that the critical energy level was a function of the depth of the pond and, as the depth increased, the minimum solar radiation required per unit load also increased. Although the data are not conclusive, he suggests that for ponds deeper than about 1 m, a minimum visible radiation of about 0.18 langley or more per kg of BOD_u/ha is needed to ensure that light is not a limiting factor.

Generally, as a result of various other considerations, ponds are not loaded to such an extent that light may become limiting. From the point of view of dissolved oxygen, however, one may want to limit the pond loading so that a pond is aerobic at all times of day and night and the nuisance potential is minimized. This requires that the minimum visible radiation available is not less than about 0.4 langley per kg of BOD_u per hectare based on the computed and field data given above.

When the objective of waste treatment is mainly BOD removal, economy dictates that a pond be loaded up to the extent to which it can handle degradation of organic matter without impairing effluent quality or creating nuisance conditions. In warm climates especially, anaerobic activity can be counted upon to supplement aerobic activity in a pond. In fact, an ideal situation would be one in which a deep enough pond could be provided with anaerobic activity occurring continually in the lower part and aerobic activity in the upper part, the products of anaerobic decomposition becoming inoffensive as they rise up through the aerobic layer and oxidize.

As long as pond temperatures are above $15°C$, anaerobic activity occurs giving liquefaction of solids followed by gasification and production of CH_4 and CO_2. Some products of decomposition, of course, rise up into the aerobic layer and consume oxygen. To the extent that carbon is removed from the system in the form of methane and carbon dioxide, there is less left to exert a carbonaceous oxygen demand (BOD). Thus, gasification is an indirect measure of the BOD removal anaerobically.

In field experiments in Nagpur, India, Siddiqi et al. [7] used a floating device for gas collection from facultative ponds. The gas was then analyzed for CH_4 and CO_2 and an estimate was obtained of the COD thus removed
anaerobically. Under loading conditions usually adopted in India, about 25
to 50% of the applied BOD is removed anaerobically, the more heavily loaded
ponds often being anaerobic over a considerable portion of the 24-hr cycle
with some odor nuisance in their vicinity. Even in ponds designed to operate
aerobically, there is always some anaerobic activity in the bottom muds and
sediments, but there is generally no nuisance. Some of the residual products
of anaerobic degradation diffuse upward and generate an oxygen demand in the
upper layers. Some of the settleable organic material may be resuspended by
wind action. Thus, anaerobic activity is not taken into account when designing facultative ponds, and designs are made on the safer assumption that all
the BOD is to be removed aerobically. The provision of a separate anaerobic
pond prior to the facultative pond would give a more functional design.

In the case of colder temperatures where ponds may be below $15°C$ for
a considerable period of time, little anaerobic breakdown takes place in the
pond bottom, which becomes primarily a sludge storage area during that time.
The decrease in gas evolution is over fourfold for every $5°C$ decrease in
temperature over the range normally encountered in ponds (see Sec. 16.18).
Eventually, during the spring as the temperature of the pond rises, anaerobic activity picks up. But, as stated earlier, some of the products of anaerobic fermentation are capable of oxidation and thus place an oxygen demand
on the aerobic waters above. The effluent BOD thus tends to rise seasonally
in such ponds as shown in Figure 16.8. Therefore, it is customary where pond
temperatures fall below $15°C$ to design the pond neglecting anaerobic removal

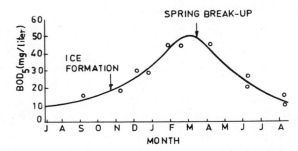

Figure 16.8 The seasonal variations in BOD of the effluent from
waste stabilization ponds.

Table 16.3 Generalized BOD Loading per Unit Area per Day
Under Various Climatic Conditions

Surface loading (kg BOD$_5$/ha-day)[a]	Population (per ha)[b]	Detention time (days)[c]	Environmental conditions
Less than 10	Less than 200	More than 200	Frigid zones; seasonal ice cover; uniformly low water temperatures; variable cloud cover
10-50	200-1000	200-100	Cold seasonal climate; seasonal ice cover and temperate summer temperatures for short season
50-150	1000-3000	100-33	Temperate to semitropical; occasional ice cover; no prolonged cloud cover
150-350	3000-7000	33-17	Tropical; uniformly distributed sunshine and temperature; no seasonal cloud cover

[a]These estimates are based on the assumption that the effluent volume is equal to the influent volume, i.e., the sum of the evaporative and seepage losses is not greater than the rainfall.

[b]Assuming a contribution of 50 g BOD$_5$ per person per day in developing areas.

[c]Based on an influent volume of 100 liters of waste per person per day.

Source: Ref. 1.

of organic matter and, in fact, to hold the BOD load within limits so as not to have the whole pond turn anaerobic during late spring (see Table 16.3).

16.9.4 Detention Time Requirement

Although the factors affecting oxygen production and distribution in a pond were seen in the previous discussion, the stabilization of organic matter is a time-rate process greatly affected by temperature and the treatability of the waste.

For domestic sewage, which is relatively readily treatable, the main factor affecting the stabilization rate is temperature, and therefore, for BOD removal, an adequate detention time must be provided in the pond depending on the temperature. Winter is again critical in this respect, just as it is for radiation.

Experience has shown that in most cases detention time in the pond is
more important than radiation as a design criterion for both BOD and bacterial removal. Generally, the detention times (volumes) provided are such
that, with the usual depths adopted, enough surface areas are available
from the point of view of radiation at most locations.

The microclimate of an area (temperature, sunlight, etc.) and the
nature of the wastewater to be treated (presence of toxic substances, color,
fluctuations, etc.) have a considerable effect on pond loadings and must be
taken into account when preparing specific designs. Gloyna [1] has summarized pond loadings and detention times for BOD removal under different
climatic conditions as given in Table 16.3. As a first approximation to
pond designs in India, the loadings shown in Figure 16.9 are used according
to the appropriate latitude of the proposed site. With the usual depths of
1.2 to 1.5 m provided in ponds, they give sufficient detention times for
ensuring BOD removal efficiencies upward of 80 to 85% [2].

Instead of arbitrarily selecting a pond loading from the range of
values given in tables or charts, it is advisable to prepare a more specific design for local conditions. This is especially more necessary in
the case of industrial wastes and large domestic sewage disposal projects.
The method of design is discussed further in the next section.

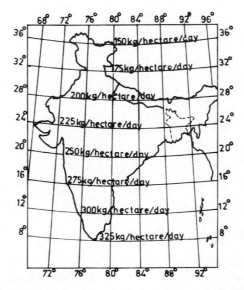

Figure 16.9 Recommended pond loadings in India (from Ref. 2).

16.10 BOD REMOVAL KINETICS

From the discussions presented earlier in Chapter 11 concerning reaction kinetics and reactor design, it will be recalled that the removal of organic matter generally follows first-order kinetics. Thus, the rate of substrate removal can be stated as

$$\frac{ds}{dt} = -K_p S \tag{16.11}$$

where

S = substrate mass

K_p = overall substrate removal rate constant per unit time in facultative-type waste stabilization ponds

Also, as seen earlier in Chapter 12, the following equations describe the actual substrate removal likely to occur using some of the more commonly used models describing mixing conditions in reactors:

For plug flow:

$$S = S_0 e^{-K_p t} \tag{16.12}$$

where

t = detention time

S_0 and S = initial and final concentration or mass, respectively

For complete mixing (single cell):

$$S = \frac{S_0}{1 + K_p t} \tag{16.13}$$

For complete mixing (equal cells in series):

$$S = \frac{S_0}{[1 + K_p t']^n} \tag{16.14}$$

where

n = number of equal-sized cells in series

t' = detention time per cell

For intermixing or dispersed flow:

$$S = S_0 \left[\frac{4ae^{1/2d}}{(1 + a)^2 e^{a/2d} - (1 - a)^2 e^{-a/2d}} \right] \tag{16.15}$$

where

$$a = \sqrt{1 + 4K_p td}$$
$$d = D/UL \text{ (dispersion number)}$$

Most designers of waste stabilization ponds have often followed either the ideal plug flow equation [2,6] or the ideal complete mixing equation [1,8,9,33,34], although some have also suggested that greater consideration be shown for mixing conditions by using the dispersed flow model when designing waste stabilization ponds [6a,8,34].

In reality, waste stabilization ponds exhibit neither ideal plug flow nor ideal complete mixing conditions; they fall somewhere in between. No doubt, ponds go through periods of gentle mixing owing to wind and temperature effects; the extent of mixing depends on their magnitude and on the pond geometry. But ponds are also known to remain stratified over a number of days at a time [9]. Furthermore, the design of inlets and outlets, the provision of baffles to lengthen the path of flow, and the provision of cells in series also have a considerable effect on the mixing conditions in a pond and the actual residence time for treatment. The mixing conditions also fluctuate with flow, wind, and other factors.

Consequently, some ponds exhibit more plug flow type conditions and some more complete mixing type, but, as stated earlier, they all fall somewhere between and are therefore better characterized by the dispersed flow model which takes into account the actual mixing conditions through the use of the term D/UL, the dispersion number for the pond system in question.

Table 11.1 gives some typical values of the dispersion number D/UL expected for different types of waste treatment units. It will be seen that in single ponds, generally with no baffles, values of D/UL range from 1.0 to 4.0 and over, signifying well-mixed conditions. With proper tank geometry, inlet-outlet design, and introduction of around-the-end-type baffles to reduce the width of flow (see Sec. 11.10), the value of D/UL in single ponds can be brought down to 1.0 or so, if desired.

Multiple ponds in series display overall values of D/UL much less than 1.0 depending on the number of cells in series. As seen earlier in Sec. 11.10, the provision of two cells in series can be expected to give overall values of D/UL ranging from 0.2 to 0.7, whereas the provision of four cells in series consistently gives overall values of D/UL less than 0.1. Thus, multiple ponds in series are preferred where more plug flow type conditions are desired.

Where for some reason it is desired to avoid plug-flow conditions and, rather, build into the system complete-mixing conditions, it would be advantageous to provide just a single basin of relatively low length/width ratio, preferably with an inlet protruding up to one third or more of the length of the pond. Such units, especially large ones, would show values of D/UL of about 4.0 or higher, indicating well-mixed conditions.

Single-celled waste stabilization ponds, with or without the provision of around-the-end-type baffles, can be designed using the dispersion model. Equations (11.21) to (11.23) enable one to estimate the likely value of the dispersion coefficient D in m^2/hr as a result of the assumed pond width W in meters. This in turn enables an estimation of D/UL for any desired pond geometry and the corresponding efficiency of treatment in a given detention time. By trial and error, one can estimate the pond dimensions required to attain any desired value of D/UL.

When the observed value of D/UL for a single pond is around 4.0 or larger, the use of the complete-mixing model gives predictions of substrate removal efficiency that are fairly equal to those obtained from use of the dispersed flow model. But, when the value of D/UL is low, the complete-mixing model underestimates the efficiency of removal. If the value of D/UL is around 0.2 or less, the results predicted by the plug flow model are close to those predicted by use of the dispersed flow model.

Multicelled ponds can also be designed using the dispersion model and a value of D/UL for each cell individually in a stepwise calculation since the effluent of the first cell becomes the influent to the second cell and so on. The equal cells-in-series model has also often been used with success, but the dispersed flow model with individual design of each cell has the advantage of flexibility since the cells can be made unequal in size and geometry, promoting good mixing, for example, in the first cell and more plug flow conditions in later cells. This flexibility may be essential with certain industrial wastes.

Laboratory model studies can be set up to establish the applicable values of D/UL for both single and multicelled ponds if the size of the installation or requirements of very high efficiency warrant such a study. The possible consequences of compartmentalization on pond operation are discussed in Sec. 16.12.

Once the value of D/UL is estimated for the prototype, the efficiency of treatment is readily computed using the Wehner-Wilhem equation if the

value of K_p is known. Example 16.4 illustrates how such a computation is
made.

16.10.1 BOD Removal Rate Constant K_p

      As organic substrate removal is generally force-fitted to first-order
kinetics, it is observed that the value of the removal rate constant K_p
increases as the organic load placed on the system increases and vice versa.
However, in terms of our earlier discussion, various limiting factors come
into play as the load increases and the value of K_p tends to level off.
This is also evident from Figure 16.10 which gives a plot of K_p values
observed at different BOD loadings on ponds treating domestic sewage under
predominantly aerobic conditions.

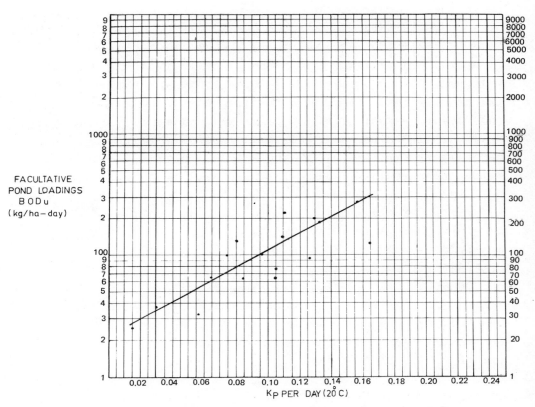

Figure 16.10 The values of the BOD removal rate constant K_p at different pond loadings that were observed using a dispersed flow model.

Ponds which are lightly loaded (more detention time) show a lower average value of the BOD removal rate constant K_p, whereas ponds with heavier
loading show larger values of K_p per day.

Very few values of K_p are available for domestic sewage after adjustment for actual dispersion conditions in the ponds studied. Some of these
are shown in Figure 16.10 which gives a semilog plot of the ultimate BOD
load on ponds versus K_p values computed after taking into account the actual
dispersion conditions in the experimental ponds and correcting for temperature at $20°$ C. A few values have been included based on best estimates of
dispersion conditions. The equation describing the line of best fit within
the range of loading of 25 to 300 kg BOD_u/ha-day is given by Arceivala [2]
as:

$$K_{p(20°C)} = 0.132 \log L - 0.169 \qquad\qquad (16.16)$$

where

   K_p = overall BOD removal rate constant per day at $20°$ C for facultative
        ponds treating predominantly domestic sewage

   L = ultimate BOD of organic load on the pond, kg/ha-day, $20°$ C

Values of K_p at other temperatures can be estimated using Eq. (12.32)
with an appropriate* value of θ. Figure 16.10 includes data from ponds in
India, Israel, Turkey, and the U.S. There is considerable scatter of the
data. This is possibly due to variations in the nature of the waste, algal
activity, flows, and possible diurnal and seasonal variations in values of
the dispersion number D/UL itself, although this is assumed to be constant
over the period of study. Effluent analysis methods have also varied in
that some have accounted for the presence of algae in the effluent, whereas
others have measured effluent BOD in the conventional manner.

Prior knowledge of BOD loading rate and temperature is necessary to
estimate K_p for design purposes, but loading rate itself depends on the
choice of K_p. Hence, the design can be adjusted by trial and error, or by
the choice of a K_p value based on an approximate idea of loading rates
generally valid for a given location. In Eq. (16.16), the logarithm of the
loading rate L is used; thus, a small variation in the value of L does not

---

*Recently, Thirumurti [6] has shown that $\theta = 1.035$ gives a better fit to
the data of Suvannakarn and Gloyna [1] from which the original value of
$\theta = 1.085$ was derived by Marais. This difference is due to the use of the
plug flow model by the former and the complete-mixing model type by the
latter.

have a significant effect on the K_p value calculated from it. Examples
16.3 and 16.4 illustrate applications of the method just described.

Values of K_p derived in the above manner are to be used only if the
dispersed flow model is used for design purposes.

If, on the other hand, the complete mixing model is used [Eq. (16.13)],
a different set of K_p values must be used. Data obtained from Suvannakarn
and Gloyna [1] from lab-scale ponds treating synthetic sewage have given the
following K_p values which are often used with the complete-mixing model, and
$\theta = 1.085$ for temperature correction:

Temperature, °C:	5	10	15	20	25	30	35
K_p per day: (complete-mixing model)	0.103	0.12	0.24	0.35	0.53	0.80	1.2

16.10.2 Determination of Waste Treatability

Laboratory tests are sometimes necessary to establish the treatability
of certain wastes, especially industrial ones containing toxic or inhibiting
substances, in order to develop design criteria including K_p values. Such
tests are often time consuming.

One method is to run several lab-scale units in parallel using the test
waste. After the units reach steady-state conditions, tests are made for
determining the algal count or percent light transmittance to see if there
is any significant reduction in algal growth beyond a certain concentration
of the wastewater. Similarly, measurements can be made for BOD, COD, DO,
redox, pH, etc., to reveal treatment characteristics.

A graphical plot of BOD removal per unit time versus the effluent BOD
can be developed from which the slope of the line gives a K_p value at the
test conditions of pond loading and temperature. At times, pretreatment
of the waste or control of a toxic constituent at its source may be indicated. Sometimes dilution may be desirable before treatment.

Example 16.3 A facultative pond is likely to be loaded with domestic
sewage at about 180 kg BOD_u/ha-day based on radiation and oxygen production
considerations at the location. Estimate the efficiency of BOD removal that
can be expected in the winter at a pond temperature of $10°$ C for different
values of the dispersion number D/UL equal to 0, 0.2, 0.5, 1.0, and ∞ in a
pond of 40 days' detention time. Also estimate the efficiency using the
complete-mixing model.

1. From Eq. (16.16) we have

$$K_{p(20)} = 0.132 \log(180) - 0.169 = 0.13 \text{ per day}$$

Assuming $\theta = 1.035$, $K_{p(10)} = 0.10(1.035)^{10-20} = 0.092$ per day. Thus,

$$K_{p(10)} \times t = 0.092 \times 40 = 3.686$$

From Figure 11.5, referring to the appropriate curve for the value of D/UL, we can estimate the efficiencies as tabulated below:

D/UL		BOD removal efficiency (%)
0	(plug flow)	97.5
0.2		93
0.5		88
1.0		84
∞	(completely mixed)	81

2. With the complete-mixing model, use $K_p = 0.12$ per day at $10°C$. Thus, Eq. (16.13) gives

$$\frac{S}{S_0} = \frac{1}{1 + K_p t} = \frac{1}{1 + (0.12)(40)} = 0.18$$

Hence, the efficiency of removal = 82%.

16.11   DESIGN OF PONDS FOR MICROORGANISM REMOVAL

The potential that ponds possess, especially in warm climates, for microorganism destruction must not be lost sight of when designing ponds. In warm climates, ponds may even be an economically feasible alternative to the chlorination of effluents.

The mechanism of microorganism die off in ponds has sometimes aroused discussion. Some have suggested that algae produce substances toxic to bacteria. But no evidence of such production by algae has been found [4]. To investigate this, Toms et al. [13] incubated, in daylight and in darkness, samples of wastewater mixed with water rich in algae, and others with borehole water devoid of algae. On exposure to daylight, bacteria in the borehole water were, in fact, killed more rapidly than those in the samples

rich in algae (probably because the algae shaded the bacteria from light).
Solar radiation is no doubt an important factor affecting bacterial die off
rates [16a].

Based on laboratory studies, Parhad and Rao [25] demonstrated that an
increase in pH, which was directly related to algal growth in ponds, was
responsible for rapid bacterial die off. When the sample was buffered to
hold the pH to 7.5, E. coli was not eliminated even in the presence of good
algal growth, but in unbuffered samples, the pH rose above 9.0, and sometimes as high as 10.4, with a consequent rapid die off of E. coli.

It would appear that bacterial die off in ponds occurs under natural
conditions mainly because of competition for food, predator grazing, and
various environmental factors. But from the above two studies it could be
said that the natural die off rate in ponds is somewhat retarded by the
shade or protection provided by algae and is accelerated at other times when
the pH in the pond is above 9.3 (e.g., at noon). The effect of temperature
on the natural die off rate is also important.

16.11.1 Removal Rate Constant K_b

Table 16.4 gives some observed values of the coliform and E. coli die
off rate constant K_b per day as observed in different pond systems under
known mixing conditions. Generally speaking, these values are similar to
those observed in natural waters including rivers.

The values of K_b obtained from batch-type studies made in the U.K.
show interesting differences between the values obtained in clear bottles
suspended at the top and bottom of a pond and those obtained using dark
bottles suspended at any depth. Assuming that the average daytime value
of K_b is the mean of the top and bottom clear bottle values, and the average
nighttime value corresponds to the dark bottle value, the overall value of
K_b can be estimated if the fraction of daylight hours is known. As a rough
approximation, the overall value of K_b for the batch studies (D/UL = 0)
made in the polishing lagoon in the U.K. can thus be computed as 1.0 to 1.5
per day at $20°$ C using $\theta = 1.19$. This value compares well with the value of
1.0 to 1.2 obtained at $20°$ C for the three-celled pond in Turkey after adjusting for dispersed flow conditions. Thus, a value of about 1.2 per day would
appear reasonable to adopt when the dispersed flow model is used for estimating the coliform/E. coli removal efficiency of facultative waste stabilization ponds. More data in this regard would evidently be welcome.

Table 16.4 Bacterial Die off Rate in Facultative Stabilization Ponds

Organism	Pond and its mixing conditions	K_b per day (average)	Pond temp. (°C)	Reference
Coliforms, fecal coliforms and fecal streptococci	Several ponds in S. Africa, single and multicelled (assumed as complete-mixing cells in series) (see Figure 16.11)	1.6-2.6	2-21	Marais [9]
E. coli	Sewage effluent polishing lagoon, U.K. (batch-type studies using light and dark bottles suspended in lagoon)			
	1. Light bottles at surface	0.8	12	Toms et al. [13]
	2. Light bottles at surface	0.96	17	
	3. Light bottles at bottom	0.50	17	
	4. Dark bottles at all depths	0.245-0.421	12-17	
Coliforms	Three-celled field pond, Ankara, Turkey	1.0-1.2	20	[11][a]

[a]Estimated by Arceivala from Ref. 11 after adjusting for pond mixing conditions.

Studies reported by Marais [9] from South Africa have shown that there is no significant difference between the die off rates of coliforms, fecal coliforms, and fecal streptococci in ponds and, therefore, any of the three organisms can be considered equally useful in determining the progress of fecal bacterial reduction in ponds. The South African studies have also shown that the value of the temperature correction coefficient θ in Eq. (12.32) is 1.19 within the range of 2° C to 21° C.

Above approximately a 21° C pond temperature, Marais has observed a tendency toward stratification in the pond and continued anaerobicity in the lower layers with a consequent reduction in bacterial removal and the value of K_b. The bacterial die off rate is slower under anaerobic conditions. In this manner, he explains the difference in the behavior of cold-climate and warm-climate ponds. In cold climates, with aerobic conditions over the whole pond depth in summer, better bacterial removals are observed in summer than in winter; in warm climates with anaerobicity in lower layers, poorer bacterial removals are actually observed in the summer and better removals in their (mild) winter.

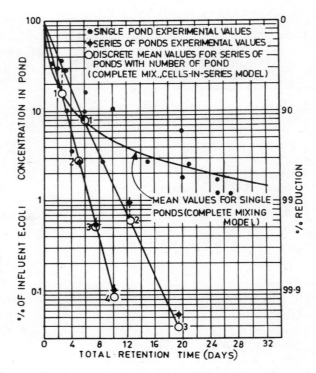

Figure 16.11 The experimental and theoretical reduction of E. coli in single ponds with various retention times, and in a series of ponds in southern Africa (from Ref. 9).

At Cape Town, South Africa, experimental use of a pontoon-mounted stirrer (not an aerator) has been made to keep a pond in a state of gentle mixing to prevent stratification and thus sustain aerobic conditions during the hot weather [9].

Raw sewage coliform concentrations generally average around 10^6 to 10^9 coliforms/100 ml, and if they are to be reduced, for instance, to 1000 coliforms/100 ml for restricted irrigation purposes or, for instance, to 100 coliforms/100 ml for unrestricted irrigational use (Chapter 9), the coliform removal efficiencies required are on the order of 99.9 to 99.999% and higher.

When such high removal efficiencies are contemplated at prevailing temperatures in a pond, flow mixing characteristics become extremely important. An examination of Figure 11.5 will immediately make it evident that high efficiencies demand that the dispersion number D/UL be held as low as possible since the curves tend to flare out. A single pond with complete-mixing flow, for example, can rarely give better than 90 to 95% removal.

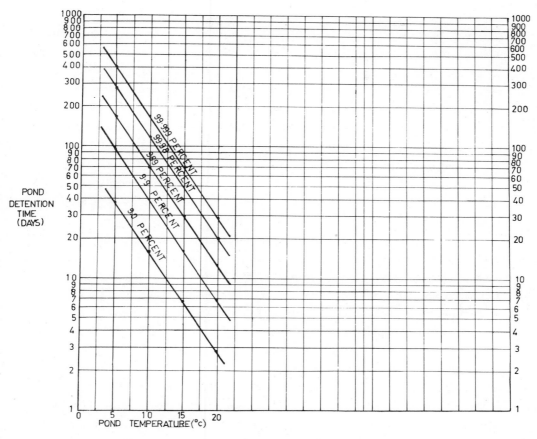

Figure 16.12 The effects of pond retention time and temperature on the efficiency of coliform removal. The relationship is based on $K_p = 1.2$ per day dispersion, $D/UL = 0.20$, and $\theta = 1.19$.

An efficiency requirement of 99.9% or more requires careful design for plug flow (e.g., cells in series) and a favorable temperature.

It is interesting to observe as a first approximation from Figure 16.12 that a pond of about 8 days' detention time, such as a maturation pond or a polishing lagoon, meeting the above assumed conditions can be expected to give about 90 to 99% coliform removal, provided temperatures are between 15° and 20° C. Higher removals from such ponds could be obtained only if dispersion conditions are even more prone to plug flow than implied in the assumption of $D/UL = 0.2$, or if temperatures were higher without inducing anaerobicity, or both. Random fluctuations could no doubt occur, giving both higher and lower values at times.

Similarly, a facultative waste stabilization pond of, for instance, 20 days' detention time and D/UL = 0.2 should be able to give, for example, 99% coliform removal at about 15° C to 99.99% at 20° C. An even wider range of values could be expected depending on summer and winter temperatures and prevailing mixing conditions. Series ponds would naturally give consistently better results than single ponds throughout the year.

The futility of expecting high coliform removal efficiencies in colder climates is clearly brought out from the values shown. Unrealistically high detention times would be needed to obtain good coliform removals, even at low values of the dispersion number D/UL obtained by providing cells in series. In such climates effluent chlorination may perhaps be a more feasible alternative.

Against this backdrop, let us see a few values of bacterial removal efficiencies reported from field studies, bearing in mind that in some cases pond temperatures were not recorded at the time of sampling and in almost all cases the pond mixing conditions were unknown. Generally speaking, however, series ponds were invariably found to give higher efficiencies than were single ponds, as would be expected from the previous discussion. Reported results are summarized in Table 16.5.

Experience with polishing lagoons of short detention time used in the U.K. to give tertiary treatment to effluents from conventional secondary sewage treatment plants has shown them to be very effective in reducing the numbers of fecal bacteria [13]. E. coli removals of 67 to 99.9% have been reported from such lagoons with 2.5 to 20 days' detention time and temperatures ranging from 5 to 21° C.

Neel et al. [5] have reported upward of 99.99% removal of coliforms from their studies on waste stabilization ponds at Fayette, Missouri, operated as two cells in series in the temperature range of 2 to 35° C (average about 17° C) and at a detention of 85 to 155 days. Good removals have also been reported from other countries [1] which would support the general procedure for estimation presented in this chapter.

16.11.2 Removal of Other Organisms

Besides the indicator organisms such as coliforms, interest in microorganism removal in ponds extends to members of the salmonella group, the helminths, the viruses, and others.

Table 16.5   Bacterial Efficiencies Observed in Some Field Studies

Stabilization pond Location	Cells	Detention time (days)	Temperature (°C)	Bacterial efficiency (%)	Organism	Reference
Israel (several)		5		90-99	Coliforms	15
Herzilia, Israel	Two, in series	20		99.99	Coliforms	1
India (several)	Single	12-20	22-32	80-99	Coliforms	2
Nagpur, India	Three, in series	7	18-32	99.99-99.999	Coliforms	2
Polishing ponds, U.K.		2.5-20	5-21	67-99.9	_E. coli_	13
Fayette, Missouri	Two, in series	85-155	2-33	99.99	Coliforms	5
South Africa	Single	2-26		60-98	_E. coli_	9
South Africa	Cells in series	5-20		97-99.95	_E. coli_	9

As stated earlier, the South African studies reported by Marais [9] have shown that there is no significant difference between the die off rates of coliforms, fecal coliforms, and fecal streptococci in ponds, and therefore any of the three organisms could be considered equally useful in determining the progress of fecal bacterial reduction in ponds.

There are, however, some studies which show that there are significant differences in the die off rates of E. coli/coliforms on the one hand and certain other organisms. For example, from the U.K., Toms et al. [13] report that fecal strepococci die off rates are higher than those of E. coli, both in the light and in the dark (about 50% higher on the average for samples exposed to sunlight). Coetzee et al. [16] have shown that the die off rate constant for S. typhi is only about 40% of the value for E. coli in ponds under similar conditions. Thus, it appears that while the general pattern of die off may be similar between different organisms, their rates may be different from those of E. coli.

Wherever studies have been made, complete removal of helminths, cysts or ova, have been reported irrespective of their concentration in the raw sewage [1,8].

Relatively little data are available as yet with regard to the virus removal efficiency of stabilization ponds measured in the field. In a field study of virus removal in a full-scale pond at Bhilai, India, treating 26,000 m^3/day flow, the raw sewage was monitored for nearly a year and was found to show between 10 and 1100 PFU (plaque-forming units) per liter, while the effluent showed a concentration of 8 to 80 PFU per liter, thus giving an average removal of 85% [17].

Hodgson [12] reports that snail vectors of bilharziasis transplanted into stabilization ponds in South Rhodesia died in every case within 10 weeks. The death was not due to lack of food, but presumably to adverse factors in the pond environment.

16.12 MEASURES FOR IMPROVING POND PERFORMANCE

It is possible to take several measures to improve the performance of stabilization ponds. Essentially, they depend on the nature of the waste, treatment kinetics, and the need to promote the desired type of mixing conditions in the pond. The measures generally involve (1) pond arrangement and geometry and (2) design of inlets and outlets.

Pond Arrangement and Geometry

Sewage treatment with its first-order kinetics is most efficient under plug flow conditions unless contraindicated by other factors discussed below. Generally, it is desirable that the value of D/UL be kept as low as possible, either by introducing baffles to reduce the width and increase the length and velocity of flow or by providing cells in series (see Sec. 11.10). This is particularly necessary to give consistently high efficiency of removal of BOD, coliforms, nitrogen, etc.

Plug flow conditions are of no special benefit for zero-order reactions or first-order reactions which have a relatively very low value of the removal rate constant K. Here, K would control, rather than the detention time t.

Series operation would not be desirable in cases where shock loads are expected from time to time. The receipt of toxic substances or just high loads of organic matter may upset the first one or two cells in a series, and it may be better to provide for a more well-mixed single pond of equivalent volume which would readily disperse the incoming slug of material and dilute it. This may be particularly necessary in the case of certain industrial wastes. Even a relatively readily treatable one such as milk waste may preferably be treated in a well-mixed pond owing to the likely receipt of shock loads from dumping of spoiled milk and milk products.

The provision of compartments in series may have some undesirable consequences also. For example, the earlier compartments in the train, especially the first one, may tend to turn anaerobic owing to the receipt of higher organic loads per unit area than the later compartments, although on an average, the whole system may receive organic loads within desired limits. Anaerobic conditions would tend to slow down both BOD and bacterial removal and, consequently, the very purpose of providing compartments may not be fully met. Nuisance conditions may also be created.

For this reason, some compromises may be called for. For example, even though maximum efficiency is theoretically attainable if all cells in a series are equal in size and have identical mixing conditions (Sec. 11.12), it may be advantageous to make the first cell larger than the others to prevent or minimize anaerobic conditions. This is often done, as long as the subsequent cells are not made too small so as to adversely affect the dispersion conditions in each cell (the smaller the cell size for a given flow, the larger may be the value of D/UL) or be less than the algal cell washout time at winter temperatures.

Other modes of getting around the difficulty of likely anaerobicity in the first cell are (1) incremental feeding or stepwise loading of the pond cells or (2) recycling of flow from a subsequent cell to an earlier one. Both methods (see Figure 11.8) help to reduce or eliminate anaerobic conditions in the first cell but could have an adverse effect on the effluent quality. Incremental feeding exposes increasing fractions of the influent to a decreasing number of ponds in the series.

Recycling, although helpful in other ways, increases the flow through each cell and thus can have a partly favorable and a partly unfavorable effect on the system and its overall efficiency. At very high recirculation rates, the set of cells tends to behave like one single large cell giving a higher value of D/UL than without recirculation. In this regard, Marais [9] suggests that recirculating from the second cell to the first cell may be preferable to recirculating the final effluent to the first cell since the former would limit the adverse effect to only the first two cells.

Design of Inlets and Outlets

Careful attention has to be paid to the design of inlets and outlets to promote the desired mixing conditions in the pond and to prevent short-circuiting. For example, tracer tests have shown that the practice of extending the inlet pipe to one quarter or one third of the pond length promotes complete-mixing conditions. If plug flow conditions are desired, the inlet should be restricted to the head end of the pond, preferably provided as a horizontal or vertical diffuser pipe. Inlet and outlet positioning also has a considerable effect on short circuiting. It is not advisable to locate them facing one another even though at long distance in a pond.

The provision of a baffle just upstream of the effluent weir is desirable as it prevents some loss of algae which are most abundant in the upper layers of the water (see Figure 16.13).

From tracer tests on models, based on a Reynolds number similarity between the prototype and the model, Mangelson and Watters [12] have demonstrated the likely effects of pond hydraulics on performance. Their studies showed that as the depth of the pond increased, the dead spaces in the pond increased. Thus, to realize the full benefit of increased depth, especially in deeper ponds, greater attention must be paid to inlet and outlet design.

Greater use of tracer tests on both lab-scale and field ponds seems desirable to enable a proper prediction of dispersion conditions under varying conditions.

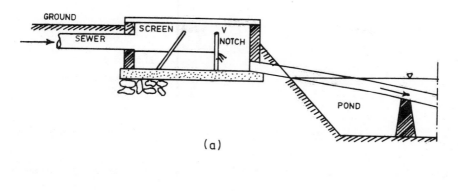

(a)

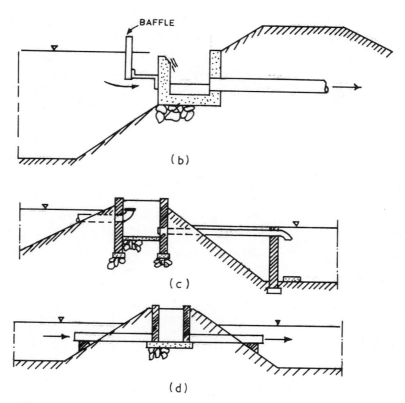

(b)

(c)

(d)

Figure 16.13 Typical pond inlet and outlet and interconnections
(from Ref. 2).

16.13. SLUDGE ACCUMULATION IN PONDS

A certain amount of sludge accumulation occurs in stabilization ponds over a period of time. The additional depth to be provided in a pond should be sufficient to take care of the likely accumulation by the time the pond becomes due for desludging.

The recommended frequency of cleaning is, on the order of once in six years in India where pond loadings are high and inflows tend to carry heavier grit loads [2]. The accumulation in the relatively lightly loaded ponds of U.S. municipalities has been reported to be only on the order of 12 cm/ha over a period of from 25 to 30 years [18].

Gloyna [1] gives sludge accumulation in ponds at the rate of 0.03 to 0.05 m^3 per person per year. These values are somewhat lower than the accumulation observed in India where, on the basis of studies on four ponds, they are reported to level off at 0.08 m^3 per person per year [2].

Other characteristics of accumulated sludge studied after extracting core samples from the Indian ponds showed:

Parameter	Percent
Moisture content	70-80
Volatile fraction	20-30
Total nitrogen	0.74-1.30
Phosphates	1.06-1.30

It is interesting to note that phosphate accumulations were higher than nitrogen. The possible precipitation of phosphates at high pH in the ponds has been discussed in Sec. 16.14.

Three factors normally contribute to the accumulation of sludge: (1) the grit content of incoming wastewater, (2) the settleable inorganic solids contained in the wastewater, and (3) the inert residues from the stabilization of organic matter in the pond.

Grit is generally made up of inorganic solids equal to or greater than 0.2 mm size. Their specific gravities range from 2.0 to 2.67. The average content of dry grit in sewage varies considerably between communities, for example:

Country	Grit in sewage $(m^3/1000\ m^3)$
U.S.	0.025
South Africa	0.05
India	0.06

Peak loads in all cases could amount to 0.1 $m^3/1000\ m^3$ sewage flow, or over. Assuming an average sewage flow of 140 liters/capita-day and grit content of 0.05 $m^3/1000\ m^3$ sewage flow, the total grit accumulation amounts to 0.00256 m^3/capita-year, which constitutes only about 5% of the total sludge accumulation in the pond.

Thus, prior removal of grit would not seem to be warranted before pond treatment but for the fact that most of the accumulation takes place around the inlet and in the first cell of a series of ponds. Large treatment installations should be provided with grit removal arrangements unless periodic cleaning of grit accumulations is envisaged or adequate storage space is provided around inlets in the first cell.

Example 16.4 A facultative stabilization pond is to be proposed for a town whose present sewered population is 70,000 persons and water supply is 184 liters/sec. River flow is negligible in the summer, but a good potential for agricultural use of the wastewater exists at that time. The rest of the year, river flow is sufficient to give dilution; hence, the treated effluent BOD_5 is proposed to be kept at 70 mg/liter. Groundwater pollution is not expected to present any problem. Suitable land is available at the proposed plant site.

The town is located at a latitude of 36° N. In the critical month of January, average daily radiation has been estimated at 117 cal/cm^2-day, the ambient temperature at 1.7° C, and the wastewater temperature at 20° C. In summer the pond temperatures may be assumed to be equal to or greater than 20° C.

The sulfate concentration in the wastewater is 75 mg/liter. The coliform concentration in the influent may be taken as 10^6 coliforms/100 ml. The effluent concentration has to be held below 100 coliforms/100 ml for "unrestricted" irrigational use and 1000 coliforms/100 ml for "restricted" irrigational use.

Wastewater flow and characteristics

Present water supply = 184 liters/sec = 15,500 m³/day

(= 221 liters/capita-day)

Estimated sewage flow = 80% of supply = 12,400 m³/day

Estimated BOD_5 = 50 g/capita-day = 3500 kg/day

$$\frac{3,500 \times 10^6}{12,400 \times 10^3} = 282 \text{ mg/liter} \quad \text{(or about 300 mg/liter)}$$

Actual sampling confirmed that the BOD_5 estimate was reasonable. The pH of the wastewater was between 7.8 and 8.0. There were no harmful or toxic substances likely to be present in this predominantly domestic type of wastewater which would affect bacterial or photosynthetic activity.

BOD removal efficiency

$$\text{Efficiency required} = \frac{300 - 70}{300} = 77\%$$

Radiation and production of oxygen

Average radiation received per hectare = $117(10^8)$ cal/day

Algal production at assumed conversion efficiency of 6% and unit heat of combustion of 6000 cal/g of algae produced,

$$\frac{(117 \times 10^8)(0.06)}{6000 \times 10^3} = 117 \text{ kg of algae/ha-day}$$

Weight of oxygen produced equals, for example, 1.7 times the weight of algae produced. Thus,

O_2 production = 1.7 x 117 = 200 kg O_2/ha-day

BOD_u removed is, for instance, 1.4 times the BOD_5 removed, or

BOD_u removed = (1.4)(0.77)(3500) = 3773 kg/day

Assuming O_2 production equals O_2 consumption, the minimum pond area required to ensure generally aerobic conditions,

$$\frac{3773}{200} = 18.86 \text{ ha}$$

This would give

$$BOD_u \text{ loading rate} = \frac{(1.4)(3500)}{18.86} = 260 \text{ kg/ha-day}$$

In terms of radiation, this implies $117/260 = 0.45$ langley per kg BOD_u applied per hectare, which appears sufficient. However, this area needs to be checked from the point of view of the time required for stabilization at a minimum pond temperature in January.

Estimation of K_p

In order to estimate the BOD removal rate constant K_p per day at pond conditions, let us assume, as a trial, a pond loading of only 200 kg BOD_u per ha-day. Then, by Eq. (16.16):

$$K_{p(20°C)} = 0.132 \log L - 0.169$$
$$= 0.132 \log 200 - 0.169 = 0.13 \text{ per day}$$

As a trial assumption, let the pond temperature equal 4°C in January (this will be checked later). Hence, from Eq. (12.30):

$$K_{p(4°C)} = K_{p(20°C)}(1.035)^{4-20}$$
$$= (0.13)(0.577) = 0.075 \text{ per day}$$

Dispersion number D/UL

It is proposed to construct the ponds in such a manner that the value of D/UL will be equal to or less than 0.2 at all times (see later). Hence, this value will be used in applying the Wehner-Wilhem equation (Figure 11.5).

Estimation of detention time

For the purposes of entering the chart shown in Figure 11.5,

$$\frac{S}{S_0} = \frac{70 \text{ mg/liter}}{300 \text{ mg/liter}} = 23.3\%$$

This gives a value of $K \times t \cong 2.0$ at D/UL = 0.2. Hence, at winter conditions (4°C):

$$t = \frac{2.0}{0.075} = 27 \text{ days}$$

(incidentally, t = 40 days if D/UL = 4, as in a well-mixed pond). At summer conditions (20°C), t = 2.0/0.13 = 15.4 days. However, provide for 27 days' detention. This gives a pond volume = 27 × 12,400 m³/day = 334,800 m³. If pond depth = 1.7 m (for instance),

$$\text{Pond area} = \frac{334,800}{1.7} = 197,000 \text{ m}^2$$

However, owing to side slopes, the area of water surface = 205,000 m^2
= 20.5 ha.

Check for pond loading

$$L = \frac{4900}{20.5} = 240 \text{ kg BOD}_u/\text{ha-day}$$

Retaining the value of K_p calculated earlier, we may adopt L = 240 kg
BOD$_u$/ha-day as it is lower than the limit of 260 kg/ha-day estimated earlier.
(If the pond depth had been assumed as 1.86 m, the pond loading would have
worked out to 260 kg/ha-day, at which the oxygen consumption would have just
equaled the oxygen production.)

Check for pond temperature

From Eq. (11.31) we get

$$\frac{t}{h} = \frac{T_i - T_L}{f(T_L - T_a)}$$

$$\frac{27}{1.7} = \frac{20^\circ C - T_L}{0.4(T_L - 1.7^\circ C)} \quad \text{and} \quad T_L = 4^\circ C \quad \text{(as assumed)}$$

Effluent BOD

The effluent BOD will vary from month to month owing to temperature
effects as shown in Figure 16.14, the maximum being 70 mg/liter in the month
of January. Note the low BOD in the irrigation season.

Coliform removal

For the purpose of estimating the coliform removal using the chart
given in Figure 11.5, which is based on the Wehner-Wilhem equation, let us
assume K_b = 1.2 per day at 20° C and compute K_b at other temperatures using

$$K_{b(T^\circ C)} = K_{b(20^\circ C)}(1.19)^{T-20}$$

The likely removal of coliforms in each month is found for the dispersion number D/UL = 0.2, and the value of $K_b t$ corresponding to the month
(t = 27 days from above). The calculated values of coliforms/100 ml in the
effluent are plotted for each month in Figure 16.14, along with the permissible thresholds for restricted and unrestricted irrigation from which
it is evident that the effluent is fit for restricted irrigation from mid-
April to mid-September but unrestricted irrigation only from mid-May to
mid-August.

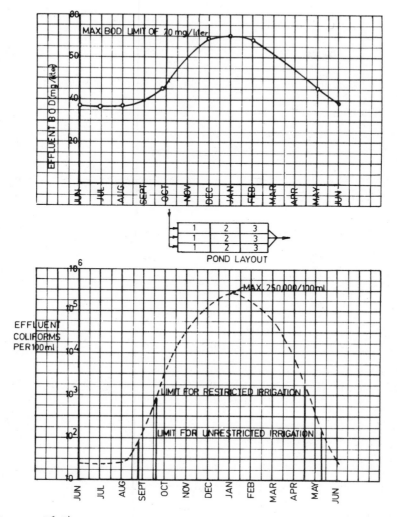

Figure 16.14 Estimated seasonal variations in pond effluent BOD and coliforms (see Example 16.4).

If the irrigation season is longer, or if unrestricted irrigation is desired over a longer period, two alternatives are available: either the pond detention time must be suitably increased or disinfection of effluent must be practiced.

Check for sulfide concentration in the pond

To check if excessive sulfide formation may inhibit algal activity in summer, assume that Eq. 16.5 applies at the pond's summer temperatures.

$$S^{2-} = [(0.0001058)(BOD_5 \text{ load}) - (0.001655)(\text{detention time})$$
$$+ (0.0553)] \times SO_4^{2-}$$
$$= [(0.0001058)(170 \text{ kg/ha-day}) - (0.001655)(27 \text{ days})$$
$$+ (0.0553)] \times 75 \text{ mg/liter}$$
$$= (0.0286)(75 \text{ mg/liter}) = 2.14 \text{ mg/liter} \quad (\text{acceptable})$$

Pond arrangement

To ensure that D/UL will be equal to or less than 0.2, we may estimate its value for some alternative configurations:

1. Let the pond dimensions be 900 m x 228 m x 1.7 m (L/W ≅ 4).

   Flow velocity $U = \dfrac{12,400 \text{ m}^3/\text{day}}{(228 \times 1.7)24} = 1.33 \text{ m/hr}$

From Eq. (11.21), the dispersion coefficient D = 16.7W, or

$$D = (16.7)(228) \cong 3944 \text{ m}^2/\text{hr}$$

Thus,

$$\frac{D}{UL} = \frac{3944}{1.33 \times 900} = 3.3 \quad (\text{well mixed})$$

2. Pond in three lengthwise compartments

Now, W = 76 m and L = 900 m (L/W ≅ 12). Flow will be divided equally between the three lengthwise compartments in parallel and U = 1.33 m/hr. Thus, D = (16.7)(76) = 1315 m²/hr and

$$\frac{D}{UL} = \frac{1315}{1.33 \times 900} = 1.1$$

(In this manner, if about 16 lengthwise compartments in parallel are formed, each of about 14 m width, D/UL will be 0.2).

3. Pond with two around-the-end baffles (lengthwise)

Another alternative will be to provide two lengthwise baffles of the around-the-end type. This gives

$$W = \frac{228}{3} = 76 \text{ m} \quad \text{and} \quad L \cong 900 \times 3 = 2700 \text{ m}$$

Hence,

$$U = \frac{12.400}{(76 \times 1.7)24} = 4.0 \text{ m/hr}$$

From Eq. (11.20) for baffled ponds

$$D = 33W = 2340 \text{ m}^2/\text{hr}$$

and

$$\frac{D}{UL} = \frac{2340}{4.0 \times 2700} = 0.21 \quad \text{(acceptable)}$$

4. **Pond with three compartments, each with three cells in series**

It is possible to compute sequentially the value of D/UL for this configuration in the above manner and find it capable of giving 0.2 or less. Also, since this arrangement suits the land topography at the proposed site, it is selected in preference to alternative 3 above (see Figure 16.14 for the layout).

Sludge accumulation

Assume an accumulation of 0.03 m³ sludge/capita-year and compute the space required for 10 years' accumulation.

The present population is 70,000.
The population after 10 years is 94,000 (estimated).
The average population thus equals 82,000.
Sludge volume = 82,000 × 0.03 × 10 years = 24,600 m³.

This will be accommodated by providing 2 m depth instead of 1.7 m in the first cell of each series.

Gross area requirement

Provide three cells per series and three such sets, giving 9 cells altogether as shown in Figure 16.14.

Providing roads, embankments, etc., as discussed earlier, the gross area required works out to 26 ha against 20.5 ha water surface area; i.e., it is about 25% more than the net area. The gross area requirement in terms of population is 3.7 m²/capita (= 260,000/70,000).

16.14 DESIGN OF PONDS FOR NUTRIENT REMOVAL

Besides BOD and microorganisms, a certain amount of nutrients (nitrogen and phosphorus) are also removed in waste stabilization ponds. Where removal is incidental to the BOD, removal efficiencies of about 50% or less are reported for nitrogen and phosphorus. If higher efficiencies are desired, modifications in treatment along with some additional treatment may be required.

Nitrogen Removal

Organic nitrogen contained in the wastewater undergoes ammonification, and the resulting ammonia nitrogen is then partly used in new cell growth and partly converted to nitrate-nitrogen if conditions are favorable. When cells die their nutrient content is released by bacterial breakdown and is recycled. The different forms of nitrogen are also partly flushed out with the effluent and the various cells contained in it. On the other hand, some nitrogen is gained by the pond through nitrogen fixation by blue-green algae, and some is lost through denitrification. Thus, a materials balance is difficult to prepare for a given pond system. In the absence of algal harvesting, the major mechanism for nitrogen removal from ponds is denitrification.

Removal of total nitrogen (comprising organic N, NH_3-N, NO_2N, and NO_3-N) may not correlate with the pond BOD loadings as several factors are involved. Field ponds often show wide variations in nitrogen removal efficiencies; the removals are generally expected to be greater at warmer temperatures and higher loading rates. Sometimes negative removals are reported, implying that effluents show higher values than influents over some periods of time [36]. A few results are summarized below.

Pond site	Load (kg BOD_5/ha-day)	Temp. (°C)	Total N removal (%)	(g/m²-day)	Reference
Missouri, USA	45	2-33	90±3	2.56	[5]
Missouri, USA	67	2-33	90±3	3.84	[5]
Missouri, USA	112	2-33	90±3	6.40	[5]
Paraiba, Brazil	80	25	85.7	1.32	[37]
Nagpur, India	450	29	50-60	5.62	[2]
Madras, India	206	30	50-60	2.60	[2]
California, USA	Deep ponds		~60	2.50	Estimated from [14]

Several workers including Neel et al. have commented on the general observation that nitrate is not produced in ponds. Nitrite is also absent, signifying a lack of nitrification. Organic nitrogen in the waste is converted to ammonia which is readily used by algae and generally not oxidized further to nitrite or nitrate.

In fact, some nitrate present in the influent may be removed in ponds. In their studies on polishing lagoons in the U.K., Toms et al. [13] observed a maximum nitrate-nitrogen loss of 1 g of N/m^2-day. The loss of nitrate was due partly to algal uptake and partly to denitrification. For most of the year, algal uptake accounted for less than 20% of the nitrate-nitrogen loss; much of the remainder was probably lost by denitrification.

In lagoons with short retention times of 2.5 to 3.5 days and little growth of phytoplankton, seasonal variations in nitrate losses were less marked, but an overall loss was still observed. There was a progressive loss of nitrate with time, the rate of loss being about 0.8 g of N_2/m^2-day. Toms et al. [13] refer to the work of others who have observed similar losses from short retention lagoons and river muds.

In the above studies, Pseudomonas sp. were the main denitrifying organisms found, and Clostridium perfringens were the main nitrate reducers. The former were between 2×10^4 and 2×10^5 per ml of lagoon water. Laboratory experiments in which aerated sewage effluent was allowed to stand over lagoon mud showed that even under well-oxygenated conditions denitrification occurred if the mud was present (see also Sec. 12.22).

Oswald [14] notes that in deep facultative ponds the influent nitrogen is converted to nitrogen gas by a mechanism not yet fully explored. Toms et al. note the importance of the water-mud interface and recommend exposure of the water to the maximum surface of mud to attain nitrate removal rates of 0.8 g of N/m^2-day.

Further studies are called for in this regard to understand more fully the effect of various factors. Temperature effects are one such instance where data are needed to enable application of design criteria under different climatic conditions. The effect of operation of cells in series on nitrogen removal is another. Gates and Borchardt [29] found 50% removal of nitrogen by harvesting the algal cultures grown in experimental ponds, and a similar operation with two-stage ponds produced 98% removal. Nitrogen removal is promoted by the use of cells in series.

In pond seepage, nitrates are readily passed through soil, but ammonia is not (Chapter 9). Fortunately, nitrates are not formed in stabilization ponds, and nitrates in the inflow are diminished as just discussed.

Ammonia Removal

The total inorganic ammonia concentration in a waterbody is distributed between ammonia (NH_3) and the ammonium ion (NH_4^+), depending on the pH of the

water. The concentration of NH_3 increases rapidly as the pH rises, and this
brings about a corresponding loss to the atmosphere by volatilization.

With algal growth the pH value of pond water may rise above 9.5 at
noon and a considerable loss of NH_3 can then occur by volatilization. Thus,
the NH_3 removal in ponds requires that the pH be raised, either artificially
(by addition of lime, for example) or naturally by algal activity. Photosynthetic activity also helps by the direct uptake of NH_3 for algal growth.

Ammonia removal from holding ponds was studied in Israel in connection
with the Dan Region Renovation project [23]. The effluent from a stabilization pond was treated with lime to raise the pH to 11.0 and exposed to air
in six cells in series operated with continuous flow and controlled wind
velocities (in a wind tunnel). The rate of ammonia release from the cells
was found to be influenced by the temperature of the liquid, the depth of
liquid in the cell, and the wind velocity. The initial concentration ranged
between 50 and 90 mg/liter as N. Thus it was concluded by the experimenters
that in the coastal areas of Israel (average wind velocity of 8 km/hr) and
under ordinary winter conditions, it should be possible to effect 90% removal of ammonia in 15 days' detention by flowing lime-treated effluent
(pH 11.0) through a series of 10 ponds, each with 1.5 days' residence time
(to promote plug flow conditions).

In Baroda, India, the use of an experimental single-celled pond showed
ammonia nitrogen removal ranging from about 40 to 60% in one day when the
pH had been raised earlier by lime addition to 11.0. The initial concentration of ammonia was in the range of 700 to 900 mg/liter as the effluent came
from a fertilizer factory. The results showed that phosphorus, fluorides,
etc., were also removed substantially owing to precipitation in the pond [19].

Ammonia release without the aid of lime has also been reported upon
from stabilization ponds, lakes, and streams. Toms et al. [13] quote instances of lagoons from which 19.7 and 22.4% losses of ammonia have been
noted after 2.5 and 3.25 days, respectively, and of a lake from which there
was no loss of ammonia in winter but 14% in summer (1.1 mg N/liter).
Stratton [24] studied ammonia losses from two impoundments, one a small
lake of about 2.27 ha with a depth ranging from 0.6m to 4.6 m, and the
other an effluent holding pond of 2.45 ha with a depth of only 1.37 m. The
average rate of ammonia loss for the lake was 51 mg/m^2-day at 29° C with pH
$\sim$9.8 and an ammonia concentration of about 0.47 mg/liter in the surface
water. The effluent holding lagoon at pH 9.1 and a temperature of 31.5
gave an ammonia loss rate of about 97 mg/m^2-day near its inlet and where

the ammonia concentration in the surface water was 1.75 mg/liter. Rates
of loss four to six times higher were anticipated under favorable conditions
of mixing and turbulence. Higher losses would also occur with a greater
concentration of ammonia in the water as the rate of loss is first order.

Phosphorus Removal

The total phosphorus present in wastes is made up mostly of organic
phosphorus and phosphates, the latter being 2 to 3 times more abundant.
Phosphorus removal in waste stabilization ponds generally occurs as a result of loss of the organic phosphorus contained in algal and bacterial
cells flushed out with the effluent, and as a result of precipitation of
phosphates at the high pH values often reached in ponds.

The algal growth potential based on the usual phosphorus content of
domestic sewage has been given in Table 16.1. It is evident that carbon
and nitrogen may become limiting before the phosphorus is fully ultized.
The phosphorus contained in algae generally varies from 0.25 to 1.3% as P
on a dry weight basis, although there have been some reports of a two to
three times higher percentage content [13]. Assuming a phosphorus content
of, for instance, 1.3% and an algal concentration of 80 mg/liter in the
pond effluent, algae would account for about 1.0 mg/liter loss of phosphorus.
A further loss would be due to the flushing out of bacteria, zooplankton,
and any inorganic suspension of phosphorus in the effluent; this is small
indeed.

A considerable proportion of the phosphorus loss can be accounted for
by the precipitation of hydroxyapatite [$Ca_5(PO_4)_3OH$] at high pH values
caused by the photosynthetic uptake of CO_2 from HCO_3^-. Toms et al. [13]
have shown an interesting correlation between the soluble phosphorus content in a lagoon and its pH. Figure 16.15 shows that **precipitation** begins
at pH 8.2, and the soluble phosphorus concentration decreases by a factor
of ten for each further unit increase in pH value. In this manner, precipitation was found to occur in their lagoon for about 6 months each year.
Other workers have also commented on the greater phosphate removals observed
in summer months.

Toms et al. also examined the total and soluble phosphorus content of
the lagoon muds and their interstitial waters and concluded that the phosphorus movement was from the water to the mud with no release back to the
lagoon water. Other workers have reported that soluble phosphorus was re-
generated overnight and in winter. The reader is referred to Sec. 6.4.1

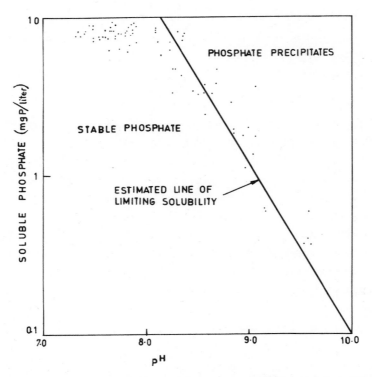

Figure 16.15 The effect of pH on the solubility of phosphate in rye meads effluent; total hardness is approximately 315 mg/liter as CaCO$_3$ (from Ref. 13).

concerning the mean residence time in lakes based on phosphorus. It would appear that stabilization ponds located in tropical climates should, with their higher pH values in all seasons, give better and more consistent removals of phosphorus and nitrogen than ponds in the colder climates.

More design criteria are needed to enable ponds to be so designed as to optimize for nutrient removal along with BOD and microorganism removal.

16.15 COMPLETE WASTE TREATMENT IN PONDS

From the foregoing discussions it should be evident that stabilization ponds have the potential to be complete treatment processes achieving BOD, bacterial, nitrogen, and phosphorus removal all at one stroke, provided the algae in the effluent can be harvested before discharge to a watercourse. (Algae harvesting is not necessary if the effluent is used for irrigation.)

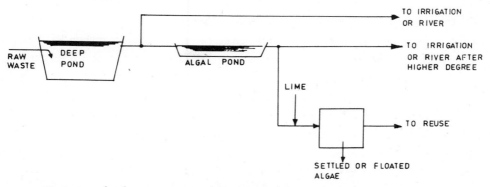

Figure 16.16 Complete waste treatment in ponds (from Ref. 14).

Oswald [14] quotes the example of Saint Helena, California, where the
above concept has been applied to the treatment of community wastewater
amounting to about 1600 m³/day. The system shown in Figure 16.16 includes
a nearly 3 m deep, 20-day detention primary pond followed by a 0.9 m deep,
10-day detention algal growth pond which is mixed at a velocity of 15 cm/sec.
The possibility of further treatment by chemical dosing and sedimentation
(or flotation) can also be envisaged as shown in the figure. The results
are given in the following informal table, from which it is seen that the
flowsheet is comparable in overall efficiency to a tertiary waste treatment
plant, but is estimated to cost only half as much.

| Item | Percent removal in pond treatment | |
	Without chemical separation of algae	After chemical separation of algae[a]
BOD	97	99.8
COD	93	
Carbon	78	
Nitrogen	92	93.3
Phosphate	64	92.0

[a]Using ~150 mg/liter CaO for coagulation and sedimentation in batch tests.

About 60% of the nitrogen is converted to nitrogen gas in the deep
pond and is removed from the system. Hence, deep impoundment first brings

nitrogen and carbon into balance with each other while considerable algal
growth occurs in the second pond. Phosphorus left over, together with the
surplus algae, can finally be removed by chemically aided sedimentation or
flotation. Of course sedimentation would be simpler to install and operate,
which may be a factor in less developed areas.

It may again be mentioned here that if sufficient land is available,
the effluent can be used for crop irrigation without algal separation or a
high degree of treatment. Waste stabilization ponds in conjunction with
land irrigation can give equivalent treatment to the method just described.
Such use would obviously recommend itself in warm and arid climates. It
would also be more compatible with the principles of conservation and ecology.

Algae harvesting has attracted much attention owing to their high
specific growth rates and protein food value. Several workers have tried
harvesting sewage-grown algae, using methods which would not harm its potential food value. Centrifuging has been most satisfactory in this regard.
To save on power costs and the size of the installation, centrifuges have
sometimes been preceded by chemically aided sedimentation or flotation [26,
27].

The objectives of commercial algal culturing do not quite match those
of wastewater treatment. The former demands that algae production be
maximized, which in turn requires relatively short retention periods and
shallow depths. The units also need to be operated at certain minimum
velocities of flow to prevent settlement and ensure aerobic conditions
throughout. As seen earlier, a detention time of about 2 to 3 days gives
maximum net production of algae per unit area and time. Sewage treatment
demands much longer detention times. Thus, where a high degree of sewage
treatment is aimed at, the flowsheet suggested in Figure 16.16 would be
more desirable.

16.16 CONSTRUCTION, OPERATION, AND MAINTENANCE

To summarize from previous discussions, data to be obtained for design
purposes include the nature of the waste (to determine its treatability) and
flow and its usual fluctuations, as well as climatic data and mode of final
disposal. Thereafter, the question of site selection arises, followed by
construction, operation, and maintenance considerations. Costs are discussed in Chapter 10.

Site Selection

Major factors to be considered in the choice of a site can be summarized as follows: topography, land availability, and soil conditions.

Topography of the area generally dictates the location of first choice. Every effort is made to avoid wastewater pumping and to be as close as possible to the source of the wastewater (e.g., town or industry) and the final disposal site (e.g., river or irrigable lands). Again, in disposal, pumping should not be necessary as far as possible.

Among the likely sites, unavailability of the land can often be a source of frustration. Either the land may be very expensive or may entail lengthy legal acquisition proceedings, besides becoming a source of poor public relations if the land has agricultural or other potential. This may force one to select less favorable sites or to adopt other methods of treatment requiring much less land.

Another consideration of importance is soil conditions. Rock would increase sharply all construction costs, whereas porous soil may need special treatment to prevent excessive seepage. A site entailing minimum earthwork costs would be desirable. Other considerations include the direction of the prevailing wind (which may carry occasional odors to neighboring developments), and groundwater pollution possibilities. Ponds are generally located not less than 300 m from residences and drinking water wells, although some exist much closer.

Construction of Ponds

Construction features are similar in many respects to those described in Chapter 15 with regard to mechanically aerated lagoons as, in both cases, the major construction work involved is earthwork.

Pond elevation is affected by one or more of the following: groundwater level, high flood level in case of river discharge, and farm level in case of irrigation. Where the groundwater table is high, the pond may need to be located above this level, thus implying more earthwork. Sewage pumping may also become necessary. A high groundwater table may also cause difficulties in excavation and may reduce pond efficiency because of a reduction in the detention time brought about by the entry of groundwater.

A pond need not necessarily have a distinct geometric shape. It could be formed to follow contours as long as the inlet and outlet designs are made to avoid short circuiting and the creation of idle pockets, and a path of travel is lengthened to obtain low values of dispersion number. For this reason, multicelled ponds are often formed in keeping with the contours.

Leveling of the pond area is desirable to keep the floor within 10 cm of the designed elevation. Compaction of soil may be needed in the case of relatively pervious soils to avoid percolation and groundwater pollution. In rare instances, clayey soil may need to be spread over porous limestone and sandy gravel formations, or plastic or soil-cement lining be used, all of which can steeply increase construction costs. In most cases, percolation tends to diminish naturally with time as deposited solids themselves help to seal the bottom. Under favorable conditions, seepage losses may be on the order of less than 10% of the hydraulic load on the pond, and often less than 1%.

The outer slopes of the embankment may range from 2.0 to 2.5 horizontal to 1.0 vertical. The inner slopes may also be kept the same, but, where the embankment is pitched with stones or brick or lined with asphalt or concrete or soil-cement, steeper slopes of 1.0 to 1.5 horizontal to 1.0 vertical may be used. Even with flatter slopes, limestone or brick pitching is always recommended about 30 cm below and above the water in order to protect the embankment from erosion by wave action. An attempt is always made to balance the cut and fill volumes to keep the earthwork to a minimum. Surface runoff must be prevented from entering the pond.

Recommended arrangements for inlets and outlets are shown in Figure 16.13. It is advisable to keep the inlet submerged to minimize short-circuiting and localized odors in the vicinity of the outfall. The outlet is easily provided over a weir protected by a baffle to reduce the overflow of algae which, as stated earlier, concentrates most in the top layer of water. Figure 16.13 also shows some possible arrangements for interconnecting two ponds when the water is at the same or at different levels in the two ponds. The installation of a simple flow measurement weir ahead of every inlet is highly recommended. Flow bypassing and recirculation arrangements can be readily provided if needed.

Land Requirement

The land requirement of ponds must take into account the required water surface areas and add about 15 to 30% more area to accommodate embankments, roads, etc. Pond volumes, and thus pond areas, are considerably affected by temperature and climatic conditions as seen earlier. The overall land requirements, including embankments, etc., generally vary from 1.0 to 3.0 m^2/capita in warmer areas and from 3.0 to 12.0 m^2/capita or even more in colder areas, depending on ice cover conditions.

The land requirements can be reduced considerably if ponds are operated in conjunction with (1) mechanically aerated lagoons (Chapter 15) or (2) anaerobic ponds (Sec. 16.18).

Operation and Maintenance

In order to bring a new pond into operation, it is generally desirable to fill it to a small depth with dilute sewage and allow a spontaneous growth of algae to establish itself in the pond within a short time. The pond is then operated with full flow.

Maintenance work generally consists of good "housekeeping" which mainly entails regular weed and grass trimming and other usual measures to prevent mosquito breeding. Floating scums and mats must be removed. Occasionally, embankments may need dressing and, once every few years, some desludging may be necessary.

Periodic visits by an operator are desirable to ensure that the pond is operating satisfactorily. When the characteristic green color of a pond changes to gray, brown, or pink, it is generally a sign of overloading and anaerobic conditions, often accompanied by malodors. When this occurs as a result of temporary overload, the wastewater may be bypassed for some time to allow the pond to restore itself. Repeated occurrences can be corrected by taking more permanent measures such as enlarging the existing ponds to cater to increased load. In this regard, periodic sampling and the installation of a simple measuring weir in the inlet chamber proves helpful to assess whether flow conditions and/or BOD or other characteristics of the wastewater have changed.

In the case of certain industrial wastewaters, nutrient supplementation may be necessary. Prior removal of grease may also be necessary.

16.17 FISH PONDS

Raw and treated domestic and municipal sewage have often been used to fertilize fish ponds with a view toward increasing their fish productivity. This is done on the basis that nutrients contained in sewage encourage primary productivity in the pond, thus providing more fish food. It should be evident from previous discussions that the success of such a venture would depend not only on the production of more algal food for fish, but also on the suitability of the fish for the food (e.g., type of algae) and for the environmental conditions present in the pond by way of temperature, dissolved oxygen, light penetration, pH, etc. Thus, fish ponds serve as a good example of the need for an ecosystem approach as described in Chapter 3.

In experiments performed under controlled conditions at Nagpur, India, Krishnamoorthi et al. [32] used six experimental fish ponds, each of 100 m^2 size (5.5 m x 10.5 m) and 1 m depth. One of these ponds was kept as a control pond with only plain tap water being used, while the others were filled with stabilization pond effluent diluted 50, 33, and 25% with tap water to be able to run all these ponds in parallel with different influents. The ponds were not operated on a continuous flow system, but only topped up each day to make up for water lost by seepage. Each pond was stocked with about 50 to 100 fingerlings of Cyprinus carpio and Labeo rohita. Various parameters such as DO, primary productivity, NH_3, and phosphorus were measured regularly along with phytoplankton, zooplankton, and bottom fauna.

The ponds were operated over different seasons, and pond temperatures ranged between 18 and 35°C. Fish were netted about once in 2 months and the average weights of at least 10 random specimens of each species were recorded. The results showed that where algal production was high, the fish production was also correspondingly high. However, algal production was highest in the pond receiving the effluent with 33% dilution and not in the one receiving it undiluted. The latter tended to give a dense algal growth with a correspondingly shallow euphotic zone. Thus, in designing fish ponds it is desirable to provide arrangements enabling the proper choice of a dilution ratio, depending on the treated sewage quality, to optimize for algal and fish productivity.

Cyprinus carpio, an omnivorous fish, particularly feeding on Microcystis (which happened to be the dominant algal form present in the above ponds), attained an average individual weight of 350 g in 6.5 months, which was about 4 times that in the control pond using plain water. Thus, the fish production rate in the fertilized pond was approximately 2 g net weight/m^2-day (about 730 kcal/m^2-day). The same kind of result could not be claimed for Labeo rohita, where the average weight leveled off at 75 g in 4 months, compared with 60 g in the control ponds. This, according to the authors, was due to the fact that this fish prefers diatoms and desmids for food rather than Microcystis.

In terms of energy flow through the pond ecosystem, assuming 3680 cal are fixed in algal form for each gram of oxygen liberated, it was estimated

for the ponds at Nagpur that 3.86% of the incident solar energy was fixed
through primary production and only 0.1% could be harvested in the form of
fish. Secondary productivity is always much less than primary productivity.

In the above studies it was also noted that sudden fish mortalities
occurred sometimes, especially in the warm months when pond temperatures
were high (up to 35° C); the heavy demand on oxygen during the night to meet
the respiration needs of the relatively profuse biological growths in the
fish ponds depleted them of dissolved oxygen below the minimum desirable
(about 2 mg/liter). The pH range of the ponds was 7.2 to 9.6, the high
values occurring in the afternoon and the low in the predawn hours. Ammonia nitrogen averaged well below 2 mg/liter, except where undiluted effluent
was used, in which case it ranged from 4 to 6 mg/liter. Ammonia toxicity
at high pH is well known. However, in this study it was observed that fish
mortality could not be accounted for by high pH nor ammonia toxicity since
high values were reached in the afternoons while fish mortalities always
occurred in predawn hours, thus implicating low DO as the major cause.

An interesting observation was that a PO_4^{3-} concentration greater than
about 4 mg/liter gave a very large phytoplankton crop in the ponds. This
gave supersaturation with oxygen in the daytime, but very low DO during
the predawn hours owing to the increased respiration needs of the algae.

Thus, it is possible to conclude that except for surface-breathing
fish, successful pisciculture demands that fish be cultured in separate
fish ponds rather than in regular waste stabilization ponds so that dilution and other conditions (e.g., PO_4 and NH_3) can be better controlled,
not only to protect the fish but also to optimize their growth. Separate
fish ponds could, of course, be of the static or continuous flow type.

Alternatively, fish could be introduced into the later part of a maturation pond where algal growths are not heavy and DO depletion at predawn
hours need not be feared. Evidently, fish ponds in warmer climates are
more precariously placed than those in colder ones.

Surface breathers (Ophiocephalus) apparently have a better chance of
survival under low DO conditions as they can come above the water surface
to breathe. In some areas they command a good market owing to their richer
iron content [2].

The possibility of building up objectionable concentrations of certain
organics and trace metals in fish always exists, especially where municipal
wastewaters with industrial wastes are used to fertilize the fish ponds.

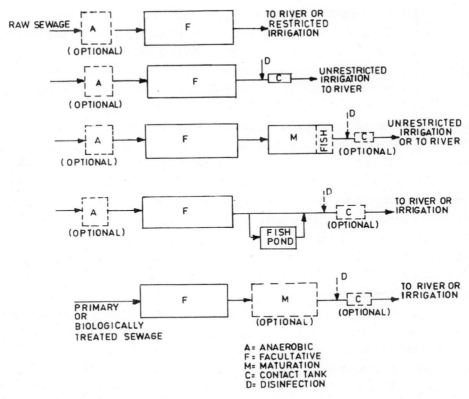

Figure 16.17 Some typical layouts for waste stabilization ponds (see also Figure 15.8).

However, the use of domestic sewage does not present the same sort of hazard.

Fish ponds have come to be associated with waste stabilization ponds owing to the ready availability of algae in the effluents of the latter. However, there is no reason why they should be limited to use after stabilization ponds alone. Fish ponds could as well utilize effluents from aerated lagoons and all other types of waste treatment plants producing aerobic effluents. Irrigation return flow can also be utilized. Figure 16.17 shows a few typical layouts.

16.18 ANAEROBIC PONDS AND LAGOONS

Anaerobic ponds and lagoons for wastewater treatment have a special role to play in the warmer climates where they have been used with success.

In the context of present-day high energy costs, units which require no
power to drive motors, pumps, and aerators are most welcome.

As anaerobic units depend neither on algal activity nor aeration, they
give considerable freedom to the designer to provide them as deeper units
(3 to 5 m deep) as compared with facultative ponds.

Anaerobic lagoons and ponds have been successfully used in the treatment of sewage and various industrial wastes, particularly from the food,
beverage, and fermentation industries (e.g., canning, sugar, alcohol distilleries, etc.). Anaerobic lagoons have also been reported as being useful
in treating pulp and paper mill wastes.

Effluents from anaerobic lagoons are generally not fit for direct discharge to a watercourse. They are often devoid of oxygen, are odorous, and
may contain various substances in a reduced state, such as NH_3, sulfides,
etc., which have much immediate oxygen demand. Therefore, these effluents
need to be given further treatment in aerated or facultative lagoons, ponds,
or other biological waste treatment units. Anaerobic lagoons help to reduce the BOD load on subsequent units by as much as 30 to 70%, thus helping to economize on overall land requirements for treatment. Anaerobic
activity also affects the nature of the solids so that when they settle in
the facultative pond, they exhibit a reduced tendency to ferment and float
up and break down more readily. Anaerobic activity also assists denitrification.

However, anaerobic ponds and lagoons have the disadvantage of odor
nuisance at times. The reader is referred to Sec. 16.8 which is concerned
with the formation of sulfides in ponds. Anaerobic units also need many
months for start-up even in mildly cold climates. They are recommended for
location at least 0.5 to 1.0 km from residences, preferably on the downwind
side. Anaerobic ponds have sometimes been disparagingly regarded as open
digesters, septic tanks, or aquaprivies since they are generally constructed
as open earthen basins. In fact, the mechanism of the breakdown of organic
matter is the same in all of these, and experience in the design of the
former can be extrapolated to such ponds to some extent.

Earlier it was explained that anaerobic activity progresses in two
steps, namely, first liquefaction and acid formation, and then methane fermentation. During digestion, the applied organic matter is converted primarily to methane, CO_2, N_2, and other gases. An average gas production
rate for ponds receiving domestic sewage is 0.6 to 0.75 m^3/kg BOD applied.
Thus, gas production implies a reduction of BOD to that extent.

It is most important to ensure methane fermentation if any BOD reduction is to be achieved. Two factors affect methane fermentation very
strongly: temperature and pH. Another factor is the presence of DO. For
methane fermentation to occur successfully, the pH should be more or less
around 7.0, free dissolved oxygen should be absent, and the temperature
should be above 15° C.

Figure 16.18 shows the effect of temperature on the gas production
rate in ft³ per acre per day from a pond loaded at 425 lb/acre-day, as observed by Oswald [28]. Very little gas production occurs at temperatures
less than 15° C.

Gloyna [1] has quoted from Marais that the gas production rate, in
kg/day, is affected by temperature as follows:

$$K_{g(T°C)} = 0.002(1.35)^{T-20} \qquad (16.17)$$

which shows the very substantial effect of temperature on the rate at which
gas evolves from the anaerobic breakdown of organic matter in ponds. Thus,
it is evident that anaerobic ponds have a place in waste treatment mainly
where climatic conditions are such that winter temperatures in the ponds
are above 15° C. This is possible in warmer climates or where the wastewater
itself is at a high temperature so that the resultant temperature in the
pond is favorable.

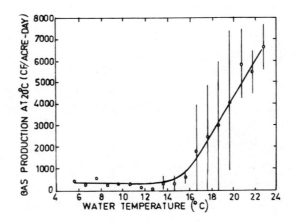

Figure 16.18 The relationship between temperature and gas production
in ponds. BOD load is constant and about 425 lb/acre-day.

In a properly operating anaerobic unit, the pH will be between 6.8 and 7.2, which is its optimum range. The unit will also have a considerable buffer capacity, thus permitting the treatment of raw wastes whose pH may be somewhat lower, even as low as 4.5 to 5.0 in the case of some fermentation wastes. Overloading would, of course, drop the lagoon pH and disturb its normal anaerobic operation.

The presence of dissolved oxygen is detrimental to the anaerobic digestion process, and thus the deeper the lagoon the better the performance it gives. A deeper lagoon also gives less temperature variations.

For a rational design of anaerobic lagoons, if the removal rate constant K per unit time for any wastewater in batch digestion is known, the required detention time in a full-scale lagoon can be estimated using an appropriate value of the dispersion number D/UL as discussed earlier for the case of facultative ponds. From pilot plant studies with domestic sewage Batachar [38] has observed K values of about 0.15 per day at 20° C.

Applying the principles of reactor design discussed earlier in Chapter 11, it is evident that better efficiency would be obtained in the treatment of wastes following first-order kinetics if plug-flow conditions are ensured. The provision of baffles or cells in series would be of advantage.

Start-up difficulties are sometimes experienced with anaerobic units owing to low pH values and a high accumulation of volatile acids. These can be overcome by lime addition. Once methane fermentation is established, lime addition is unnecessary and the unit functions effectively.

The performance characteristics of anaerobic ponds and lagoons in some typical applications are discussed below along with the design criteria followed. Applications have also included fully enclosed digesters with "bio-gas" collection and usage facilities to benefit from the fuel value of methane.

16.19 APPLICATIONS OF ANAEROBIC LAGOONS

Domestic Sewage

Gloyna has reported upon experiences from Zambia, South Africa, Australia, and Israel [1]. The seemingly wide variations in results are naturally due to the influence of temperature, pond mixing conditions, and other factors. Experience in India with domestic sewage ranges over detention times of about 2 to 3 days, giving 60 to 70% BOD reduction. Drawing

Table 16.6 Performance of Anaerobic Ponds Treating Domestic Sewage
at Less than 5 Days' Detention Time

Anaerobic pond temperature (°C)	Detention time (days)	Likely efficiency of BOD removal (%)
10-15	4-5	30-40
15-20	3-4	40-50
20-25	2.5-3	40-60
25-30	2-5	60-70

upon all these experiences, Table 16.6 has been prepared to indicate the
likely efficiency of BOD removal to be expected at different detention times
and temperatures. A detention time beyond 5 days is generally not recommended for domestic sewage because the pond may then tend to act as a facultative unit.

Under favorable environmental conditions of pH, temperature, etc.,
once methane fermentation is established, about 70% BOD reduction can be
expected in anaerobic units.

The above discussion may lead the reader to the conclusion that anaerobic pond or lagoon design is yet in the rule-of-thumb stage. This is
somewhat true, although one could regard such units as digesters and use
the criteria often followed in their design. For example, in sludge digesters operating under controlled conditions, conventional designs have
provided for average loadings of 0.5 to 0.65 kg VSS/m^3-day. High-rate digesters are designed to handle even three to five times more load than
conventional ones.

Anaerobic lagoons of 2 to 5 days' detention time treating domestic
sewage are loaded at about 0.10 to 0.30 kg VSS/m^3-day which places them in
the same range as septic tanks. These loadings are not necessarily limiting loadings, and higher rates may be justifiable with more concentrated
wastes. With first-order reactions, higher loading rates would be advantageous.

The regeneration time for methane bacteria has been reported to vary
from 2 to 11 days depending on temperature and species. Thus, anaerobic
units of less than this detention time would suffer from likely cell washout before the cell has had sufficient time to multiply at the prevailing

temperature. Consequently, a lagoon that is limited to 2 days' detention time in treating usual domestic sewage cannot possibly be loaded much higher than 0.30 kg VSS/m^3-day.

Industrial Wastes

For a strong industrial waste such as that from an alcohol distillery (using cane molasses) with a spent-wash BOD up to 45,000 mg/liter, Subba Rao [35] observed from the results of 10 full-scale plants in India that anaerobic lagoons (cells-in-series type) loaded as high as 0.6 kg BOD/m^3-day (~11,000 kg BOD/ha-day) at a detention time of 66 days can give a 90% BOD reduction. From separate lab-scale batch studies he found that the minimum hydraulic retention time required at the above loading (temperature range 18-27° C) was 13 days, below which performance deteriorated sharply. All these plants, of course, suffered from odor nuisance as sulfates ranged from 2000 to 5000 mg/liter in the raw waste. In one such lagoon, total nitrogen, which ranged between 400 and 800 mg/liter, was found to be 50 to 60% removed.

Other reports from India on the use of anaerobic lagoons in industrial waste treatment include tanneries using the vegetable tanning process where 80% efficiency is claimed for anaerobic lagoons, 66% for milk wastes, and still less for pulp and paper mill wastes.

Animal Wastes

The confinement of large numbers of animals on animal farms produces waste which is particularly treatable in anaerobic stabilization ponds. The wastes are rich in organic matter and quite concentrated as relatively little water is used in proportion to the waste matter. Pond loadings are thus on the order of 1200 to 7000 kg/ha-day depending on the nature of the waste and climatic conditions. Loadings are, however, sometimes given in terms of m^3 per animal in these cases.

Some typical characteristics of animal wastes and pond loadings are given in Table 16.7 (adapted from Hart and Turner [30] and Loehr [30a]).

The anaerobic pond volumes recommended per animal in Table 16.7 are based on minimum odor nuisance requirements from experiences in different parts of the U.S., particularly in the California area. If these volume requirements are converted to pond loadings in terms of kg of volatile solids per m^3 pond volume per day, it is interesting to observe that the loading range is practically the same for poultry, swine, or cattle. Under

Table 16.7 Some Typical Animal Wastes and Pond Loadings[a]

Item	Poultry	Swine	Cattle
Total solids, g/animal-day	30	227-363	4740
Volatile solids, g/animal-day	23	159-280	3810
5-Day BOD, g/animal-day	6.8	90-150	700
COD/BOD ratio	3.3-3.8	2.2-5.2	2.5-5.5
Nitrogen (N), % of total solids	5.4	4.0	3.7
Phosphorus (P_2O_5), % of total solids	4.6	3.1	1.1
Potassium (K), % of total solids	2.1	0-1.4	3.0
Coliforms per 100 ml	Highly variable[b]		
Anaerobic pond volume, m^3/animal	0.17-0.34	2.5-5.0	23-43
Anaerobic pond loading, kg VS/m^3-day	0.067-0.135	0.06-0.12	0.08-0.16

[a]Based on 4.5-lb (~ 2-kg) chickens, 100-lb (45.5-kg) swine, and 1000-lb (454-kg) cattle. Cattle include dairy and beef animals.

[b]Coliform output per animal may range from 3 to 9 times that of man.

more favorable conditions of temperature and site location, higher loadings can be tolerated. Anaerobic lagoon loadings of even 1.0 to 3.0 kg VS/m^3-day have been used with livestock wastes [30a]. Anaerobically treated effluents need further treatment if they are to be discharged to watercourses. This may be given in waste stabilization ponds or aerated lagoons or a combination of an aerated lagoon followed by a stabilization pond as a polishing unit. Effluent recycling to both anaerobic lagoons and aerated lagoons has been recommended at times.

It will be observed from Table 16.7 that the COD/BOD ratios for the three animal wastes are much higher than those observed for domestic sewage. This is due to the fact that animal wastes contain considerable quantities of hay steams, grain hulls, and other such relatively nondegradable but volatile matter. Thus, it is sometimes felt that it is not appropriate to use the loading parameter of volatile solids per m^3 for animal wastes. Some prefer to use BOD load or COD load per unit volume of pond*, whereas others prefer to give the pond volume per animal of stated weight.

It is of advantage, as stated earlier, to make anaerobic ponds as deep as possible. It saves land area and ensures better digestion. Even then,

---

*BOD measurements are sometimes difficult owing to antibiotics included in some animal feeds.

pond volumes can be considerable. Make-up water may need to be added to maintain pond levels. Pond sizes can be reduced by collecting much of the manure as dry matter and applying it on land (generally after composting). This can reduce up to half the BOD load.

The effluent from an anaerobic lagoon giving around 70 to 80% efficiency of BOD removal may still exhibit a high BOD value since animal wastes are quite concentrated (very little flushing water is used to drain the raw wastes to the lagoons). This may call for additional treatment before discharge to a watercourse. Another objectionable feature often found with even well-treated animal wastes is the persistence of a yellowish brown color in the final effluent. This is due to the presence of soluble humic substances. Consequently, land disposal of such wastes is preferred. Lagoons can then be so designed as to enable holding the wastes when not required for agricultural purposes.

Not much data are available for judging the extent of nutrient removal and sulfide concentrations in anaerobic lagoons treating domestic, industrial, or animal wastes.

Sludge accumulations in anaerobic ponds treating domestic sewage should occur at a lesser rate than the 0.03 to 0.05 m^3/capita-year seen in the case of stabilization ponds where the efficiency of BOD removal is higher. For organic industrial wastes, the sludge accumulations can be approximated on the basis of an equivalent population estimated at 54 g BOD_5/person-day. For animal wastes, they range from 0.06 m^3/animal-year for poultry to 0.3-0.6 m^3/animal-year for swine and 7.0 m^3/animal-year for cattle. Enough sludge storage space should be provided.

The construction of anaerobic lagoons can be carried out in a manner similar to the one described for mechanically aerated lagoons. The site selection must particularly take into account the likelihood of a considerable odor nuisance in the vicinity.

The start-up of an anaerobic lagoon may require the introduction of some digested sludge as seeding material to hasten the process. In warm countries cow dung has been used successfully to initiate the process for industrial waste.

REFERENCES

1. E. Gloyna, Waste stabilisation ponds, WHO Monograph Series No. 60,
    World Health Organization, Geneva, Switzerland, 1971.

2. S. Arceivala, J. Laxminarayana, S. Alagarsamy, and C. Sastry, Waste Stabilization Ponds — Design, Construction and Operation in India, Nat. Env. Eng. Res. Inst., Nagpur, India, 1970.

3. M. McGarry, Optimal Facultative Pond Design Criteria, Asian Inst. Tech., Bangkok, Thailand, 1971.

4. W. Oswald and H. Gotaas, Photosynthesis in sewage treatment, Trans. ASCE, 122, 1957.

5. J. Neel, J. McDermott, and C. Monday, Experimental lagooning of raw sewage in Fayette, Missouri, Jour. WPCF, 33, 1961.

6. D. Thirumurti, Design criteria for waste stabilisation ponds, Jour. WPCF, 46, 1974.

6a. D. Thirumurti, Design of stabilisation ponds, Jour. ASCE, San. Eng. Div., 1969.

7. Siddiqi et al., Measurement of Anaerobic Activity in a Waste Stabilisation Pond, NEERI, Nagpur, India, 1972.

8. S. Arceivala, Simple Waste Treatment Methods, Middle East Tech. Univ., Ankara, Turkey, 1973.

9. G. Marais, Fecal bacterial kinetics in stabilisation ponds, Jour. ASCE, Env. Eng. Div., 100, 1974.

10. K. Saffet, Waste stabilization ponds, Master's thesis, Middle East Tech. Univ., Ankara, Turkey, 1973.

11. C. Yemliha, Waste treatment by stabilization ponds, Master's thesis, Middle East Tech. Univ., Ankara, Turkey, 1974.

12. K. Mangelson and G. Watters, Treatment efficiency of waste stabilisation ponds, Jour. ASCE, San. Eng. Div., 1972.

12a. E. Gloyna and E. Espino, Sulphide production in stabilisation ponds, Jour. ASCE, San. Eng. Div., 1969.

13. I. Toms, M. Owens, J. Hall, and M. Mindenhall, Observations on the performance of polishing lagoons, Jour. Inst. Water Poll. Cont., U.K., 1975.

14. W. Oswald, Complete waste treatment ponds, in Advances in Water Pollution Research, 6th International Conference, Jerusalem, Israel, 1972.

15. H. Shuval, Public health aspects of wastewater utilisation in Israel, 17th Industrial Waste Conference, Purdue Univ., Lafayette, Indiana, 1962.

16. O. Coetzee and N. Fourie, Efficiency of ponds with respect to survival of organisms, Jour. Inst. Sewage Purif., 3, 1965.

16a. A. L. H. Gameson, Ed., Discharge to Sea from Sewage Outfalls, Water Res. Centre Int. Symp., Pergamon Press, N.Y., 1975.

17. N. U. Rao, N. Parhad, et al., Virus Removal in Waste Stabilisation Ponds, NEERI, Tech. Digest No. 19, Nagpur, India, 1971.

18. E. Middlebrooks et al., Sludge accumulation in municipal lagoons, Jour. Water and Sewage Works, 63, 1965.

19. Personal communication.

20. K. Murphy and A. Wilson, Characterisation of mixing of aerated lagoons, Jour. ASCE, Env. Eng. Div., 100, 1974.

21. H. Hodgson, Stabilisation ponds for a small African urban area, Jour. WPCF, 36, 1964.

22. A. Wachs and A. Berend, Extra-deep ponds, in Advances in Water Quality Improvement, Vol. I, Univ. of Texas Press, Austin, Texas, 1968.

23. Y. Folkman and A. Wachs, Nitrogen removal through ammonia release from holding ponds, 6th Int. Conf. Water Poll. Res., Jerusalem, Israel, 1972.

24. F. Stratton, Ammonia losses from streams, Jour. ASCE, 94, 1968.

25. N. Parhad and N. U. Rao, The effect of algal growth on the survival of E. coli in sewage, Ind. Jour. Env. Health, 14, 1972.

26. M. McGarry, C. Lin, and J. Merto, Photosynthetic yields and byproduct receiving from sewage oxidation ponds, in Advances in Water Pollution Research, 6th International Conference, Jerusalem, Israel, Pergamon Press, 1972.

27. Golueke and J. Oswald, Harvesting and processing of sewage grown algae, Jour. WPCF, 37, 1965.

28. W. Oswald, Advances in anaerobic pond systems design, in Advances in Water Quality Improvement, Vol. I, Univ. of Texas Press, Austin, Texas, 1968.

29. W. Gates and J. Borchardt, N and P extraction from domestic wastewater treatment plant effluents by controlled algal culture, Jour. WPCF, 36, 1964.

30. S. Hart and M. Turner, Waste stabilisation ponds for agricultural wastes, in Advances in Water Quality Improvement, Vol. I, Univ. of Texas Press, Austin, Texas, 1968.

30a. R. Loehr, Animal wastes, Jour. ASCE, San. Eng. Div., 1969.

31. J. Hemens and G. Stander, Nutrient removal from sewage effluents by algal activity, in Advances in Water Pollution Research, 4th International Conference, Prague, Czechoslovakia, 1969.

32. K. Krishnamoorthi, M. Abdulappa, R. Sarkar, and R. Siddiqi, Productivity of sewage fertilised fish ponds, in Water Research, Vol. 9, Pergamon Press, 1975.

33. W. W. Eckenfelder, Jr., Water Quality Engineering for Practising Engineers, Barnes and Noble, Inc., N.Y., 1970.

34. Metcalf and Eddy, Inc., Wastewater Engineering, McGraw-Hill, N.Y., 1972.

35. R. B. Subba, Personal communication.

36. R. Loehr and R. Stephenson, An oxidation pond as tertiary treatment device, Jour. ASCE, 91, 1965.

37. D. Mara, S. Silva, and B. de Ceballos, Technical notes, Jour. ASCE, Env. Eng. Div., 105: 151, 1979.

38. B. Batachar, Ph.D. thesis submitted to Engineering Faculty, Ege University, Izmir, Turkey (1980).

Chapter 17

SOLIDS SETTLING AND DISPOSAL

17.1 INTRODUCTION

The sludge generated in wastewater treatment by biological methods may amount to 1 to 2% of the raw wastewater flow. But this quantity may cause considerable materials handling problems and account for a good part of the total operating and maintenance costs of a plant. These costs are reported to average 21-50% in the United States [2a].

Solids settling occurs in conventional treatment plants at two or three positions: grit removal channels or chambers, and primary, intermediate, and/or final clarifiers.

While grit may be disposed of directly, or after washing to remove entrained organics, the settled sludge from the clarifiers generally needs further treatment in aerobic or anaerobic digesters (except in the case of the extended aeration process). After digestion the stabilized sludge is dewatered or dried for final disposal.

In aerated lagoons and waste stabilization ponds, settling occurs in the lagoon or pond itself where they accumulate unless removed.

Common methods of sludge dewatering include the following of which only the natural dewatering methods are discussed at length in this chapter:

1. Natural dewatering in (1) open or covered sand beds or (2) sludge lagoons.

2. Mechanical dewatering in (1) gravity-type sludge thickeners or (2) mechanical devices such as the solid bowl centrifuges, vacuum filters, and filter presses (after chemical preconditioning of sludge).

3. Physical methods such as heat drying, incineration, and recently, irradiation techniques.

Final disposal of sludge is, generally, either to land or to sea, in one of the following ways:

1. Agricultural use of dried or wet sludge

2. Use of dried sludge as landfill material in absence of agricultural demands

3. Spreading wet sludge on waste (or eroded) land, contouring the field so to speak, so as to gradually build up a topsoil of agricultural value

4. Disposing of wet sludge along with solid wastes for (a) composting or (b) sanitary landfill, and

5. Shipping it out to the sea for dumping into coastal waters. (This method of disposal is now becoming increasingly difficult.)

17.2 SEDIMENTATION

Solids settling is a commonly used unit operation in waste treatment. It involves the settling of inorganic solids such as sand, silt, and grit, and organic solids in raw and biologically flocculated wastewaters. There are four types of settling conditions recognized in waste treatment: plain, flocculent, zone, and compressive settling. They are not mutually exclusive, and often more than one type may occur in a given case.

17.2.1 Plain Settling

Grit removal constitutes a good example of plain settling. It generally involves individual (discrete) particles in a relatively dilute suspension. In continuously flowing tanks, the profile theoretically traced by a settling particle is a straight line of some slope (Figure 17.1) and if the tank is long enough compared with the depth, the particle eventually enters the sludge zone. The settling velocity v_o of the particle (and, therefore, the diameter), which is possible to remove in such a case, depends only on the "surface loading" or "overflow rate" Q/A since, theoretically, in terms of notations defined in Figure 17.1a:

$$v_o = \frac{h}{t} \quad \text{where} \quad t = \frac{V}{Q}$$

or

$$v_o = \frac{h}{V/Q} = \frac{h}{Ah/Q} = \frac{Q}{A} \text{ , m/day or cm/sec} \tag{17.1}$$

The actual surface loading rate should be less than the one corresponding to v_o determined from charts or equations for any desired diameter of particles to be removed at a given temperature [1]. This is because the actual performance of a tank is poorer owing to short-circuiting, currents, density differences, etc. Thus, a suitable value of "scale-up factor" is

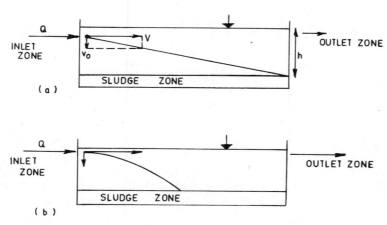

Figure 17.1 The profiles traced in a continuously flowing settling tank by (a) discrete particles and (b) flocculent particles; Q — inflow rate, v — tank volume, A — surface area, t — V/Q — detention time, h — tank depth up to sludge zone.

used and the area actually provided is more than the theoretically required one.

These principles of design are applied in sewage treatment in the conventional grit removal channel. Grit channels are generally designed to remove all inorganic grit particles equal to or larger than 0.2 mm dia., while often removal down to 0.15 mm dia. is aimed at. Grit specific gravity often ranges from 2.0 to 2.65. Thus, the theoretical settling velocities of grit particles range from about 10 to 20 cm/sec. Scale-up factors for grit channels may be even 5.0 or more to obtain 95% grit removal.

The cross-sectional area of a grit channel depends on the flow. A horizontal flow-through velocity of about 30 cm/sec in the channel is found to be desirable from experience as it permits grit settling without much organics. Hydraulic control devices (e.g., flumes and proportional flow weirs) are provided to keep the velocity relatively constant over expected flow fluctuations. Once the channel depth is found at peak flow, its length can be computed such that the grit will reach the channel bottom at its settling velocity before the flow reaches the outlet end. Channel lengths of 20 to 30 m are common.

Grit quantities expected in municipal sewage flows range from 0.025 to 0.1 m³/1000 m³ flow.

Example 17.1 Estimate the grit channel length necessary if the maximum depth of flow in the channel is likely to be 1.7 m. Assume grit settling velocity = 10 cm/sec and horizontal flow velocity = 30 cm/sec. Use a scale-up factor of 5.0.

$$\text{Channel length required} = \frac{1.7 \times 10^2 \text{ cm}}{10 \text{ cm/sec}} \times \frac{30 \text{ cm/sec}}{10^2} \times 5.0$$

$$= 25.5 \text{ m}$$

17.2.2 Flocculent Settling

When particles in relatively dilute suspension coagulate or agglomerate during settling, they exhibit different settling characteristics compared with discrete particles since the former increase in diameter with time and settle at an ever-increasing velocity. Consequently, they trace a curved profile (Figure 17.1a). The settling tank design in such cases depends on both surface loading and detention time. Typical examples from waste treatment are primary clarifiers for all types of plants and even final clarifiers in case of trickling filter and rotating disk plants.

Long tube settling tests can be performed in order to estimate specific values of surface loading and detention time for a desired efficiency of clarification for a given industrial wastewater using recommended methods of testing [1a]. Scale-up factors used in this case range from 1.25 to 1.75 for the overflow rate and 1.5 to 2.00 for the detention time when converting laboratory results to prototype design.

For primary settling tanks treating domestic sewage, laboratory tests are generally not necessary and the recommended design values range from 25 to 35 m^3/m^2-day overflow rate and 2 to 3 hr detention time. BOD removals of about 30 to 40% may be expected. Primary settling tanks for domestic sewage are also usually provided with a skimming device to remove the scum which floats up with grease and oil contained in the waste.

Various types of settling tanks are used, some circular and some rectangular, for which descriptions are available in the literature [1,2,15]. The possible use of the more efficient "tube and plate settlers" is also being investigated for use with wastewaters.

With circular and rectangular settling tanks, the average depth (side water depth) is generally kept around 2.5 to 3 m where mechanical sludge scrapers are installed. Using an appropriate value of surface loading rate from the range just stated, the required tank area is computed. Knowing the average depth, the detention time is then computed. An excessively high

Table 17.1 Secondary Clarifier Loadings for Activated Sludge

Secondary clarifier after	Overflow rate (m^3/m^2-day)		Solids loading rate (kg/m^2-day)	
	Average	Peak	Average	Peak
Activated sludge (air)	16-32	40-48	98-147	<245
Extended aeration	8-16	32	98-147	<245

Source: Ref. 15.

detention time must be avoided especially in warm climates where anaerobicity can be quickly induced. For the same reason, tanks must not be over-
designed to serve future expected flows.

17.2.3 Zone and Compression Settling

Zone settling occurs beyond a certain concentration when the particles
are close enough together to hinder free settling. In fact, interparticu-
late forces may hold the particles stationary relative to one another so
that the whole mass tends to settle as a single layer or "blanket" of sludge.
The top of the suspension layer is sharply defined since it entraps and
holds the finer sized particles which would otherwise have settled more
slowly. Thus, the whole suspension settles as though it were composed of
the fastest particles in it.

The rate at which a sludge blanket settles can be determined by timing
its position in a settling column test [2b] whose results can be plotted as
shown in Figure 17.2.

Compression settling may occur at the bottom of a tank if particles
are in such a concentration as to be in physical contact with one another.
The weight of particles is then partly supported by the lower layers of
particles leading to progressively greater compression with depth, and
thickening of the sludge. For this reason, the settling rate of the sludge
interface diminishes gradually with time as shown in the figure. In a
continuously flowing tank, the rate of sludge withdrawal from the compression
zone determines the solids concentration in it. From the settling column
tests, the limiting solids flux required to reach any desired underflow
concentration can be estimated as illustrated in Figure 17.2a, b, and c
from which the required tank area can be computed. The result is generally

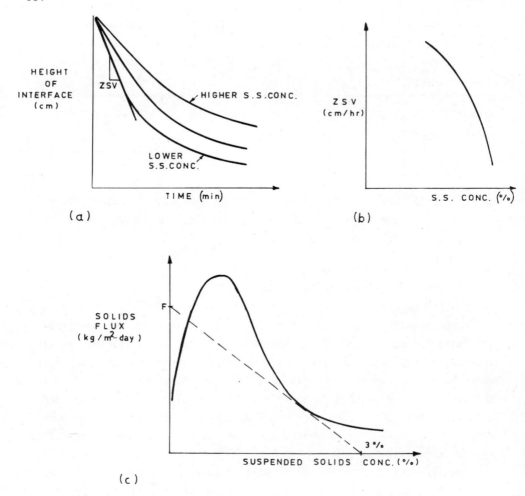

Figure 17.2 (a) Some typical interface settling curves at different
initial suspended solids (SS) concentrations in sludge. The zone settling
velocity (ZSV) is calculated from the slope of the straight portion of the
curve. (b) The observed ZSV at different SS concentrations; this is a log-
log plot. (c) A curve constructed by calculating the batch solids flux at
different SS concentrations. For example, if ZSV = 20 cm/hr at an SS concentration of 1.0%, then the batch flux rate = $(0.01)(1000 \text{ kg/m}^3)(20 \times 10^{-2})$
$\times 24 = 48$ kg/m^2-day. If the desired underflow concentration is, for example, 3%, then a tangent can be drawn to the curve as shown to cross the
ordinate at the limiting flux value of F, which can be used for design purposes.

expressed in terms of a "solids loading rate" (kg/m^2-day). It may be necessary to consider peak loadings on an hourly basis, especially with some industrial wastes.

The importance of this type of settling is now better recognized since
it has been demonstrated that secondary clarifiers in activated sludge
treatment plants often fall in this category. They must therefore be designed to fulfill not only the traditional criteria for effluent clarification in the form of adequate surface loading, but also to fulfill the solids
loading rate limitation. Whichever criteria give a larger clarifier surface
area must be adopted.

Secondary clarifiers may be designed on the basis of surface loading
criteria (Table 17.1) as long as MLSS concentrations are less than 2000 mg
per liter. These criteria are generally sufficient in the older activated
sludge plants, and also with trickling filter and rotating disk installations. At higher MLSS concentrations, which are often adopted in modern,
complete mixing, activated sludge aeration systems, the sizing of secondary
clarifiers is generally governed by sludge thickening requirements [3a,15].

Generally, the sludge is withdrawn as "underflow" from the bottom of
the final clarifier and recycled to the aeration tank. If the wastewater
average flow is Q, the recycled flow is R, and the solids concentration in
the aeration tank is X, the solids load on the clarifier is estimated in
terms of $(Q + R)X$, while the clarifier surface loading is estimated on the
basis of Q only (not Q + R) since the quantity R does not contribute to
the overflow from the tank.

Dick [3a] suggests the use of the batch flux technique just mentioned
to determine the limiting flux rate and clarifier area to transmit the applied solids mass loading to the clarifier floor. Loading rates adopted in
the United States are reported in Table 17.1 for activated sludge systems
[15].

Presently, the trend in the United States is also in favor of providing
deeper sidewater depths (up to 6 m) to allow for greater capacity to hold
solids displaced from the aeration tank at peak flows. This practice may
not be wise to recommend in warmer countries where anoxic conditions in
final clarifiers could quickly lead to denitrification and nitrogen gas
bubbles formation leading to rising sludge.

Another trend at present is to provide a surface skimming device in
final clarifiers similar to those provided in primary ones [15].

In the interest of economy, it is desirable to carry as high an MLSS
concentration in the activated sludge aeration tank as is possibly consistent with available aeration devices, mixing kinetics, and sludge recycling
rates. But high MLSS concentrations and high recycling rates bring greater

solids loads and require larger sized final clarifiers whose costs increase proportionately. Sludge recycling rates also affect pumping costs. Thus, more thickened underflow from final clarifiers is desirable. A cost optimization approach must be followed to arrive at an optimum design.

Example 17.2 Based on the data given in Example 14.1 for an activated sludge plant of 9000 m^3/day average flow, determine the diameter and sidewater depth of the primary and final clarifiers (MLSS in aeration tank = 4000 mg/liter. R/Q = 0.67 for 1% solids concentration in underflow).

1. For primary clarifier, assume surface loading at 28 m^3/m^2-day (Table 17.1). Hence, surface area required = 321 m^2 or diameter = 20 m. Take sidewater depth = 3 m. Hence,

$$\text{Detention time} = \frac{(321)(3)}{9000} = 2.64 \text{ hr}$$

2. For final clarifier,

$$\text{Incoming solids} = (Q + R)\text{MLSS}$$

$$= [9000 + 0.67(9000)] \frac{4000}{10^3} \text{ mg/liter}$$

$$= 60,120 \text{ kg/day}$$

If solids loading is not to exceed 110 kg/m^2-day at average conditions (Table 17.1), clarifier surface area required = 60,120 = 547 m^2, or diameter = 26.4 m. Corresponding overflow rate = 9000/547 = 16.4 m^3/m^2-day. For sidewater depth, see text.

17.3 SLUDGE CHARACTERISTICS

Moisture Content

The moisture content of sludge varies according to the source of the sludge and the nature of the treatment received by it. The moisture content of most sludges at their source is very high, ranging from 93 to 99% by weight.

In warmer climates, a more rapid withdrawal of sludge even with higher moisture content may be necessary in order to avoid septicity.

In most cases, the specific gravity of sludge is very close to that of water itself, that is, 1.01 to 1.005.

Source	Moisture content (% by weight)
Primary sedimentation	95-97
Final sedimentation underflow in activated sludge process	98.5-99
Mixed liquor from aeration tank	99.6-99.8
Underflow from mechanical thickeners	92-95

Thus, the weight of sludge is several times more than the weight of solids contained in it since

$$\text{Sludge weight} = \frac{\text{dry weight of solids in sludge}}{\text{fraction of solids by weight in sludge}} \qquad (17.2)$$

$$\text{Sludge volume} = \text{sludge weight} \times \frac{1.0 \text{ kg/liter}}{\text{sludge specific gravity}} \qquad (17.3)$$

Moisture content has an enormous effect on sludge weight and volume, and every effort should normally be made to withdraw the sludge from such a treatment step so that the moisture content is minimum. The following example illustrates this aspect.

Example 17.3 The surplus solids to be removed daily from an oxidation ditch (extended aeration process) amount to 30 g/person-day. Estimate the sludge volume if the point of withdrawal is (1) an aeration unit with an MLSS of 4000 mg/liter, and (2) the underflow of the final clarifier whose solids concentration is 10,000 mg/liter. Assume specific gravity of sludge = 1.000.

Case (1)

$$\text{Sludge volume} = \frac{30}{0.004} = 7500 \text{ g} = 7.5 \text{ liters/person-day}$$

Case (2)

$$\text{Sludge volume} = \frac{30}{0.01} = 3000 \text{ g} = 3 \text{ liters/person-day}$$

Thus, withdrawal of surplus sludge from the aeration tank instead of from the clarifier would entail handling 250% more volume. For this reason, the selection of the sludge withdrawal point is important, and also pre-thickening of sludge before disposal is often advantageous.

17.3.2 Stepwise Reduction in Dewatering

It will be interesting to observe how much water is removed at each step as sludge is taken from a treatment unit, passed through a thickener, and spread on an open sand drying bed. As soon as the wet sludge is spread on a sand bed, a lot of moisture drains off freely as underdrainage within the first few hours or one or two days. The balance of the water which is held in the sludge due to capillary and hygroscopic forces is only slowly reduced by evaporation from the surface. The moisture content of the sludge at each stage ranges around the typical values shown in Table 17.2 where for the sake of illustration an extended aeration sludge has been taken.

If the dry weight of the solids in the sludge is, for instance, 30 g per person per day, Table 17.2 shows the reduction in water weight and total weight at each step. The solids weight is naturally unaffected as the sludge dewaters.

It is evident that the bulk of the water is removed in the thickener. Thereafter, the bulk of the remaining moisture is removed in free drainage. Evaporation removes the least, but generally takes the longest time. The final "dried" sludge still has considerable moisture in it, but the sludge can now be "handled" or lifted with a spade. Further drying beyond that required to handle the sludge is resorted to if packaging in bags and transport are involved.

Table 17.2 Stepwise Reductions in Dewatering an Extended Aeration Sludge

Sludge	Moisture content (% by weight)	Weight (g/person-day)		
		Solids	Water	Total
1. Initial (from final settling tank)	99	30	2970	3000
2. After thickening	96	30	720	750
3. After initial percolation (free drainage)	90	30	270	300
4. At final lifting	60	30	45	75

17.3.3 Sludge Specific Resistance

The rate at which moisture from a mass of sludge drains out per unit area is directly proportional to the applied pressure and inversely proportional to the viscosity of the liquid, the weight of filterable solids present, and the resistance to dewatering. The specific resistance r_s is measured by means of the Buchner funnel test described in standard texts [14] and gives a good idea of the drainage characteristics of a sludge. (Also see Example 17.3.)

This test is no doubt useful in the design of vacuum filters, drum filters, and filter presses. In recent years it has also been used in the design of open sand beds [4]. The value of specific resistance is calculated from

$$r_s = \frac{2bPA^2}{\mu C} \qquad\qquad (17.4)$$

where

r_s = specific resistance, m/kg

b = slope of line obtained experimentally by measuring volume filtered at different filtration times, sec/m^2 (Figure 17.3)

P = applied pressure, N/m^2 (generally 49 kN/m^2)

A = filter area, m^2

μ = absolute viscosity, (N)(sec)/m^2

C = weight of solids deposited on filter per unit volume of filtrate, kg/m^3

Typical values of sludge specific resistance are given below, from which it will be seen that the more organic sludge gives higher resistance to dewatering.

Sludge	Typical values of r_s (m/kg) at 49 kN/m^3
Raw and primary settled	10^{13}-10^{15}
Activated sludge	10^{12}-10^{14}
Anaerobically digested	10^{12}-10^{13}
Aerobically stabilized	10^{11}-10^{12}
Chemical	10^{11}-10^{12}

Empirically, the specific resistance varies with pressure as follows:

$$\frac{r_s \text{ at pressure } P_2}{r_s \text{ at pressure } P_1} = \left(\frac{P_2}{P_1}\right)^S \tag{17.5}$$

where

P_1 and P_2 = applied pressure, N/m^2

S = coefficient of compressibility ($S = 0$ denotes incompressible sludge)

The value of S is calculated as the slope of the straight line generally obtained when a logarithmic plot is made of specific resistance against the pressure.

The coefficient S has been observed to vary from about 0.55 for primary sludge to about 0.88 for activated sludge and even about 1.0 for chemically conditioned biological sludges.

In studying the dewatering characteristics of sludge, the method used by the Water Pollution Research Laboratory, U.K. [4] may also be followed. A 5-cm diameter glass tube is used with sand supported on woven wire as the filtering medium. The depth of liquid sludge at the start may be, for example, 20 cm. The pressure across the filter corresponds to the average of the depths of sludge before and after the experiment. The specific resistance value obtained by this method agrees well with the value found by vacuum filtration through a Buchner funnel after adjusting for pressure.

Example 17.4   From the results obtained using the above method of testing on a sludge, find the value of the slope "b" by plotting log (t/V) against log V (see Figure 17.3).

Time t (sec)	Filtrate volume V (ml)	t/V
25	50	0.5
70	100	0.7
160	150	1.06
280	200	1.04
510	300	1.70

From the plot shown in Figure 17.3 we get $b = 0.42 \times 10^{10}$ sec/m^6.

Example 17.5   In a Buchner funnel test on sludge from an extended aeration unit, the slope b was found to be 0.725×10^8 sec/m^6. The test was conducted at $24°$ C with a pressure of 1.7 kN/m^2 using a filter area of

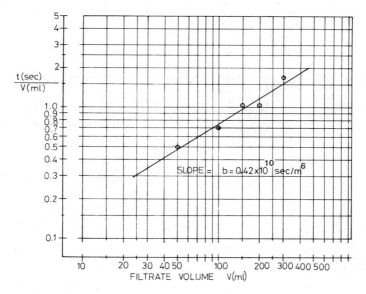

Figure 17.3 A specific resistance test.

0.0852 m^2. The weight of solids deposited on the filter was 4.61 kg/m^3 (absolute viscosity = 0.916 x 10^{-3} (N)(sec)/m^2 at 24° C). Find the specific resistance.

$$r_s = \frac{2bPA^2}{\mu C}$$

$$= \frac{(2)(0.725 \times 10^8)(1700)(0.0852)^2}{(0.916 \times 10^{-3})(4.61)}$$

$$= 4.2 \times 10^{11} \text{ m/kg at the test pressure}$$

Example 17.6 Compute the value of r_s at a pressure of 49 kN/m^2 from the above data, assuming the coefficient of compressibility S equals 0.80.

$$\frac{r_s \text{ at } 49}{r_s \text{ at } 1.7} = \left(\frac{49}{1.7}\right)^{0.80} = 14.8$$

Hence,

$$r_s \text{ at } 49 \text{ kN/m}^2 = (4.2 \times 10^{11})14.8$$
$$= 6.2 \times 10^{12} \text{ m/kg}$$

17.4 SAND BEDS FOR SLUDGE DRYING

Sand beds are generally constructed as shown in the typical cross-sectional view given in Figure 17.4. Instead of the brick wall shown in the figure, earth embankments or precast walls may be used, if desired.

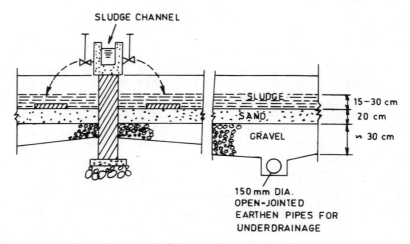

Figure 17.4 A sectional view through a typical open sand bed for
sludge drying.

Sludge is generally spread to a thickness of 15 to 30 cm over the
sand, which is kept about 20 cm deep. The sand is supported on a gravel
bed about 30 cm deep, through which are laid lines of open-jointed earthen
pipes of 15 cm diameter, spaced about 3 m apart and sloping at 1 in 150
toward the filtrate sump. Sand and gravel sizes are similar to those used
in coarse sand filters. The drying beds are often subdivided into smaller
units sized to receive the quantities expected at each withdrawal of sludge
from the treatment unit. Each bed may be 5 to 8 m wide and 15 to 50 m long.

The drying time averages about 1 to 2 weeks in warmer climates and 3
to 6 weeks or even more in unfavorable ones. In the northern U.S., an
average of 5 dryings per year are reported for open beds and about 7 to 10
dryings per year for covered beds [1]. Table 17.3 gives the typical range
of area requirements for open beds.

Table 17.3 Area Requirements for Open Sand Beds for Sludge Drying

Type of sludge and region	Area (m²/person)
Oxidation ditch sludge (Netherlands)	0.16-0.33
Extended aeration waste sludge (South Africa)	0.03-0.04
Oxidation ditch sludge and conventional digested sludge (India)	0.05-0.10
Conventional digested sludges (U.S.)	0.11-0.28

Source: Refs. 2 and 3.

From the area requirements given in Table 17.3 it is seen that they can vary by a factor of over six. This is due partly to climate and partly to the sludge characteristics, which substantially affect drying time (see below). A rational method for estimating the drying time is presented in the following section.

The performance of sand beds for sludge drying is often more meaningful and comparable if stated in terms of solids loading or "bed yield," expressed as kg solids/m^2-year, rather than just m^2/person. Example 17.9 illustrates how this is calculated. The average bed yield reported from the United States is about 98 kg/m^2-year (20 lb/ft^2-year) [2a].

Sludge drying on open sand beds may be difficult at certain times of the year (e.g., during the wet season or winter freezing). A covering or roofing over the beds may then help, provided the sides are kept open for natural wind movement which is essential as convection is important for sludge drying. Roofing does obstruct the drying process somewhat, and transparent covering (e.g., glass) is preferred over an opaque one as it significantly improves the bed yield [4]. Roofing increases the construction cost, but reduces land requirement, and an optimum design can be prepared using the general procedure outlined in the following section to find, monthwise, the drying time required with and without roofing.

When sludge drying is not feasible over a certain time period, one of the following alternatives may be possible depending on local conditions:

1. Provide a sludge "holding lagoon" to hold the sludge during the non-drying period. Later, the stored sludge can be gradually withdrawn for open drying along with the sludge regularly received.

2. Provide a sludge holding lagoon in combination with covered sand beds. In this manner, the lagoon required is of a smaller capacity since the covered drying bed may be able to operate over a longer part of the year than an open bed would.

3. Omit drying beds and provide only a sludge lagoon of adequate size (see Sec. 17.6).

4. Explore alternative methods of sludge discharge during nondrying periods. For example, discharge sludge to a river which may be at high flood during the wet season.

17.5 RATIONAL DESIGN OF DRYING BEDS

Traditionally, drying beds have been designed on a rule-of-thumb basis depending upon the local climate and experience. However, the present trends

are to develop rational design criteria capable of universal application
and permitting optimization in design.

The various factors affecting the overall drying time of sludge include
the following:

1. Extent of free drainage. The greater the fraction of total drainable
   water, the less the moisture left to be evaporated.
2. Climatic factors insofar as they affect:
   a. Evaporation rate from the sludge surface. This rate is slow
      and generally controls the overall drying time.
   b. Moisture added to the sludge by rainfall.
3. Permissible moisture content in the outgoing sludge.

The overall drying time can be computed by using a simple model given
by Arceivala [3] in which a materials balance of all incoming and outgoing
moisture masses is made. The model is applicable to cases where the evaporation time controls the overall drying time. Briefly, for a given mass
of sludge spread over a unit area which has been exposed to rainfall (see
Figure 17.5) the moisture to be evaporated is given by:

$$q_e = (1 - f_i)q_i + (1 - f_r)q_r - q_d \qquad (17.6)$$

where

q_e = moisture to be evaporated, mass/area

q_i = moisture initially present in sludge, mass/area

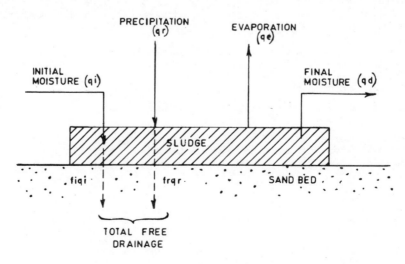

Figure 17.5 A model for estimating overall drying time (from Ref. 3).

q_r = moisture received through rainfall, mass/area

q_d = moisture remaining in dried sludge, mass/area

f_i and f_r = fraction of q_i and q_r, respectively, that is <u>drained</u> from the bed. Thus, $(1 - f_i)$ and $(1 - f_r)$ are the fractions remaining in the sludge.

Generally, in order to suit the usual mode in which rainfall and evaporation data are given, the above terms can be converted from mass/area to volume/area, namely, m^3/m^2, or stated simply as meters or mm.*

The evaporation rate from the sludge surface is less than the free water surface evaporation rate which is determined using standard "pans." Thus, a suitable reduction factor is applied to the standard evaporation rate data generally available for a locality. The required sludge drying time is then obtained as

$$t = \frac{q_e}{f_e E_w} \qquad (17.7)$$

where

t = required drying time, days

E_w = evaporation rate from free water surface, mass/area/time or mm/month

f_e = reduction factor to account for reduced evaporation rate from sludge surface

Combining Eqs. (17.6) and (17.7), the required drying time can be stated as

$$t = \frac{(1 - f_i)q_i + (1 - f_r)q_r - q_d}{f_e E_w} \qquad (17.8)$$

From Eq. (17.8) it is evident that the overall drying time is affected by (1) the effective evaporation rate, (2) the rainfall received at site (for a covered drying bed, $q_r = 0$), (3) the permissible moisture content in dried sludge, and (4) the sludge characteristics insofar as they affect the coefficients f_i, f_r, and f_e.

The typical range of experimentally observed values of the coefficients for aerobic and anaerobic digested sewage sludges which can be applied to sand drying beds at depths of 15 to 30 cm are given in Table 17.4. Aerobic

---

*1 mm = 1 kg/m^2.

Table 17.4 Typical Values of Coefficients for Aerobically
and Anaerobically Treated Sludges[a]

Coefficients	Anaerobically digested sludges	Aerobically treated sludges
f_i	0.45-0.65	0.8-0.9
f_r	0.43	0.75
f_e	0.78	0.78

[a]In the general range of 97-99% moisture concentration, and for
sludges applied up to 15-30 cm depths on sand beds.
Source: Ref. 3.

sludges are those from extended aeration plants (Pasveer and carrousel-type
ditches) and aerobic digesters.

The range of values given in Table 17.4 are typical only as they are
affected by various factors. Observe that aerobic sludges drain better than
anaerobically digested ones. This has been observed by several workers [4-
9]. The value of f_r is very nearly equal to that of f_i, but is sometimes
less than f_i. The timing of rainfall seems to affect the fraction draining
through.

Working with different types of chemical sludges, Novak et al. [9]
have shown that the evaporation rate is slow at the start, but becomes
quite rapid later on. The effective rate is, therefore, a weighted average
of the two rates. For anaerobically digested sludge, f_e has been shown to
be 78% of the free water surface rate [10]. For aerobically treated sludges,
the same value of f_e has been assumed in Table 17.4.

Both coefficients f_i and f_r are affected by the nature of the sludge.
The more the organic content in the sludge, the lower the value of these
coefficients. Similarly, the presence of hygroscopic fibrous materials
helps the retention of moisture and, therefore, reduces the values of the
two coefficients. Arceivala [3] has observed an inverse correlation between the coefficient f_i and the specific resistance of the sludge; the
greater the specific resistance, the lower the value of f_i.

The fraction f_i can actually also be determined using a very simple
laboratory column of sand to simulate a typical sand bed on which the desired sludge may be applied at the same depth and moisture content likely
in the full-scale unit. Both the time rate of drainage and the amount of
moisture drained may be observed using the method suggested by Swanwick [4].

Of course, drainage time is not as important in the design of sand beds as it is in the design of vacuum filters since drainage can continue even while evaporation is proceeding. It is the mass of water that will <u>not</u> drain off $(1 - f_i)$ that is important in the case of sand beds. Hence, the purpose of the experiment just described should be to find the moisture content of the sludge sample after free drainage has occurred. In the absence of specific experimental data, the typical values of the coefficients given in Table 17.4 may prove useful especially when dealing with predominantly domestic sludges (see Example 17.7).

Another purpose for which the experiment just described can be used is to find the effect of the initial moisture content on the value of the fraction f_i. It is evident that benefit is to be gained from prethickening the sludge as long as it does not unfavorably affect the drainage moisture. Similarly, the depth of the sludge application can be optimized by observing its effect on the fraction f_i for a given type of sludge.

Under given climatic conditions, it will generally be found that aerobically stabilized sludges will require less drying time, and therefore less drying bed area than would anaerobically digested sludges. This may be kept in view when comparing the two treatment methods. Equation (17.8) can also be used to estimate the reduced size of drying beds required if covered over to exclude rainfall.

<u>Example 17.7</u> For an aerobically stabilized sludge (Pasveer-type oxidation ditch), estimate the drying time required in the month of April under the following conditions:

The depth of sludge application = 20 cm
Initial moisture in sludge after prethickening = 96% by weight
Final moisture acceptable = 60% by weight
Rainfall in month = 11 mm (kg/m^2).
Pan evaporation rate in month = 160 mm/month $(kg/m^2$-month)
Sludge density = 1 kg/liter

<u>Solution</u>

Liquid sludge applied over 1 m^2 area = 0.2 x 1.0 = 0.2 m^3

Weight applied = 200 kg

At 96% moisture content, the water contained in the applied sludge will be 192 kg and solids will be 8 kg. The weight of water in the dried sludge = 0.6(8 kg/0.4) = 12 kg. Assuming suitable values of the coefficients

f_i, f_r, and f_e from Table 17.4, the drying time can be computed using Eq. (17.8) to give over a unit area of 1 m^2:

$$t = \frac{(1 - f_i)q_i + (1 - f_r)q_r - q_d}{f_e E_w}$$

$$= \frac{(1 - 0.8)192 \text{ kg} + (1 - 0.75)11 \text{ kg} - 12 \text{ kg}}{(0.78)(160 \text{ kg/month})} = 7 \text{ days}$$

Example 17.8 Estimate the drying time for an anaerobically digested sludge assuming that all the data given in Example 17.7 are equally applicable here.

Using appropriate values of the coefficients (Table 17.4) for anaerobically digested sludges, for example, $f_i = 0.6$, $f_r = 0.43$, and $f_e = 0.78$, the drying time can be computed using Eq. (17.8) to give t = 15 days. Thus, it is possible to differentiate rationally between the drying time required for aerobic and anaerobic sludges.

Example 17.9 Estimate the "bed yield" for the two cases given in Examples 17.7 and 17.8. Assume 2 days are required for lifting the dried sludge and refilling the bed.

1. For the aerobic sludge (Example 17.7):

$$\text{Bed yield} = \frac{8 \text{ kg solids/m}^2}{(7 + 2) \text{ days}} = 0.88 \text{ kg/m}^2\text{-day in April}$$

2. For anaerobic sludge (Example 17.8):

$$\text{Bed yield} = \frac{8 \text{ kg solids/m}^2}{(15 + 2) \text{ days}} = 0.47 \text{ kg/m}^2\text{-day in April}$$

Note: Different rates will be observed for each month from which the annual average can be worked out.

17.6 SLUDGE LAGOONS

Sludge lagooning may be considered as an alternative to sludge drying, and may be the method of choice in certain circumstances. Sludge lagoons of the type discussed below are meant for receiving aerobically stabilized or anaerobically digested, but not raw, sludges. They are also not sludge holding lagoons which hold the sludge only until it can be dried when the climate becomes favorable, or discharged to a river at high flood. They are lagoons in which the solids accumulate and compact in the bottom over a long period of time, during which some additional degradation of organic matter may occur, while some supernatant is displaced continuously by the incoming flow.

Lagoons can be formed in natural depressions where topography permits. Odor problems do not occur as in the case of raw sludge lagoons, and are minimal where aerobically stabilized sludges are lagooned. In site selection, some consideration should be given to the possibility of groundwater pollution as in the case of other waste treatment ponds and lagoons. Surface runoff should be prevented from entering the lagoon. The supernatant displaced from the lagoon can be routed back to the plant or disposed of directly on land if feasible. In warm climates, evaporation losses may account for a considerable portion of the liquid volume reduction.

At the desired cleaning interval of a few years, the lagoon or one of its compartments can be bypassed, and the accumulated sludge removed and profitably used on agricultural fields. Thus, dried sludge is available in a large bulk once in a few years instead of small quantities available regularly from sludge drying beds.

Figure 17.6 shows that the moisture content of sludge in lagoons progressively reduces with time, although not much reduction occurs after 3 to 5 years when the accumulated sludge may compact to about 83 to 87% moisture content. Thus, a cleaning interval longer than about 3 years may not be economical.

During accumulation, the organic solids may undergo further degradation at a slow rate. Not much data are available in this regard to permit establishing a value of the rate constant K_v per unit time at the lagoon temperature and other conditions. Some estimates range from $K_v = 0.4$ to 0.6 per year [11]. Both solids degradation and accumulation progress with time. Assuming first-order kinetics, the net accumulation can be given by

$$W_t = \frac{W_O}{K_{vL}} \left(1 - e^{-K_{vL}t} \right)$$ (17.9)

where

W_t = weight of volatile solids at time t

W_O = incoming weight per unit time

K_{vL} = degradation rate constant, time^{-1}, under anaerobic conditions in lagoon

Okun and Ponghis [2] give sludge lagoon loadings within the range of 25 to 75 kg solids per m^2 per year. However, it is desirable to compute

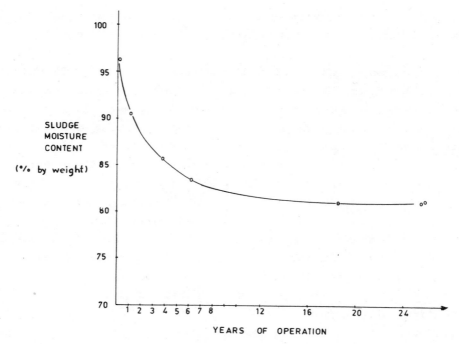

Figure 17.6 The gradual reduction in sludge moisture content in a sludge lagoon.

lagoon requirements based on local data [3] as illustrated in Example 17.10. In terms of land requirement, lagoons which are not relatively deep may not be very competitive with sand beds where the climate is favorable for drying. Not much saving in land may occur in warm climates. But construction costs may be lower where a suitable topography exists [3].

Example 17.10 A sludge lagoon is to be designed to receive the excess sludge from an extended aeration plant serving 25,000 people. The excess solids are expected to be 30 g/person-day with 55% of the solids volatile and the remaining fixed. The incoming sludge will have 99% moisture content (namely, 1% solids).

The lagoon effluent is likely to carry away a suspended solids concentration of 40 mg/liter. The cleaning interval is to be not more than 5 years. Neglect evaporation and seepage loss, assuming it is balanced by precipitation. Assume the degradation rate K_v for volatile solids equals 0.5/year at the average expected lagoon temperature. Estimate the lagoon size and its solids loading if the moisture content at the end of 5 years is likely to be 90%.

Incoming sludge solids = $= \dfrac{30 \times 25{,}000 \times 365}{10^3} = 273{,}750$ kg/year

Incoming volatile solids $= 0.55 \times 273{,}750 = 150{,}563$ kg/year

Incoming fixed solids $= (1 - 0.55)273{,}750 = 123{,}187$ kg/year

Incoming sludge volume $\cong \dfrac{273{,}750}{0.01} = 27{,}375$ m^3/year

After the lagoon is filled, the effluent volume also equals 27,375 m^3 per year.

Outgoing solids $= 27{,}375 \times 40 \times \dfrac{10^3}{10^6} = 1095$ kg/year

Assuming the relative proportion of volatile and fixed solids remains the same in the effluent and influent,

Outgoing volatile solids = 600 kg/year

Outgoing fixed solids   = 495 kg/year

Hence,

Net volatile solids = 150,563 - 600 = 149,963 kg/year

and

Net fixed solids = 123,187 - 495 = 122,692 kg/year

(The outgoing solids could as well have been neglected without affecting design in this case.)

The likely quantity of volatile solids accumulating after degradation over a 5 year period can be estimated from Eq. (17.9) as

$$W_t = \dfrac{W_0}{K_{vL}}\left(1 - e^{-K_{vL}t}\right)$$

$$= \dfrac{149{,}963 \text{ kg/year}}{0.5/\text{year}}\left(1 - e^{-(0.5)(5 \text{ years})}\right)$$

$$= 275{,}300 \text{ kg at the end of 5 years}$$

Total accumulation at the end of 5 years (volatile + fixed) = 275,300 + (5 × 122,692) = 888,760 kg.

If the moisture content of the accumulated sludge is 90% by weight, and if sludge specific gravity is nearly equal to 1.0,

Accumulated sludge volume $= \dfrac{888{,}760}{(1 - 0.90)10^3} = 8888$ m^3

Rate of accumulation $= \dfrac{8888 \text{ m}^3}{25{,}000 \text{ persons} \times 5 \text{ years}}$

$= 0.071$ m^3/person-year

If, for instance, the lagoon is filled up to 4 m with this volume, the area covered = 2222 m^2 = 0.09 m^2 (net)/person.

$$\text{Solids loading} = \frac{\text{incoming solids per year}}{\text{area covered}}$$

$$= \frac{273,750 + 122,692}{2222} = 123 \text{ kg/m}^2\text{-year}$$

17.7 LAND DISPOSAL OF SLUDGE

Some of the alternative ways of disposing wet or dried sludge on land were mentioned earlier and could be usefully explored when designing a new project, keeping in mind their health and ecological significance besides relative costs.

Agricultural Use of Dried Sludge

Dried sludge is a good "soil conditioner" as it permits more moisture to be retained in the soil and promotes higher bacterial activity. It also provides nutrients and "growth factors." Pathogens are not fully destroyed during digestion and air drying of sludge. Heat-dried sludge is considered pathogenically safe.

The volume of air-dried sludge (with 50 to 60% moisture) available from municipal plants is in the range of 50 to 100 m^3/1000 persons/year.

Use of Dried Sludge as Landfill

The use of dried sludge as landfill is to be considered only if agricultural use is not possible owing to lack of demand, long distance for transport, or objectionable nature of the sludge itself. A suitable site for landfill must be available.

Agricultural Use of Wet Sludge

Direct use of wet sludge, without drying, can be made as discussed more fully in Chapter 9. Both aerobically and anaerobically stabilized sludges have been used. Wet sludges have been piped or conveyed in channels or even transported in special tank cars for sprinkling over agricultural fields.

Irrigation rates depend on climate, crops, and soil and vary widely from 25 to 250 m^3/ha-day during the watering season. During nonwatering periods, the sludge may need to be held in a lagoon or disposed of in some other manner. This may create some problems and add to costs.

The nitrogen, phosphorus, and potassium contents of the liquid sludge, including solids, may be on the order of 5%, 3%, and 0.5%, respectively. To avoid sludge cake formation on topsoil, deep plowing of the soil is considered desirable between crop plantings. The presence of pathogens makes it necessary to adopt careful handling procedures and to restrict crops to nonedible items. In the case of industrial sludges, the presence of heavy metals may be of concern.

Use of Wet Sludge to Build up a Topsoil

Wet sludge can be spread on poor or eroded wastelands in thin layers of about 3 to 5 cm at a time (300-500 m^3/ha-day) and allowed to dry. Layer upon layer could be applied in this manner to gradually build up a good topsoil of agricultural value. Care should be taken to develop contour steppings if required in order to prevent erosion of the newly formed soil layers.

In this manner, sludge can be put to use in developing land resources and raising land values. Odor problems would be minimal if well-stabilized sludges are used. Some fly and mosquito nuisance is possible, especially in warm climates. Control may be feasible in the form of intermittent operation timed so as to drown out the larvae as soon as they are formed. Other management problems may include holding the sludge in a lagoon during the wet season, thereby guarding against surface and groundwater pollution. This method of using wet sludge can, however, be very rewarding where conditions are suitable.

Disposal of Wet Sludge with Solid Wastes

Wet sludge can be disposed of with solid waste either for composting or for sanitary landfilling.

Much experience is available in the use of both raw and digested sludges for composting along with presorted solid wastes. A proper admixture of the two is essential to obtain a desirable carbon-to-nitrogen ratio of between 17/1 and 13/1 and, for the purpose of aerobic composting, to maintain a moisture content of 35 to 45% [12]. Composting may be desirable only if a market for the compost exists in the area.

Sanitary landfill of sludge is feasible where a landfill site exists nearby. The wet sludge is absorbed by the solid waste, and hence no additional land is required for the sludge. As long as the absorptive capacity is not exceeded, there is no additional leachate from the site.

The absorptive capacity of a solid waste depends on its composition. In a study in the United States, the absorptive capacity of the solid waste of a town (typical of the U.S.) was found to be ranging from 0.6 to 1.8 kg of water per kg of dry solid waste [13]. In many developing countries, the capacity would be lower owing to a smaller content of paper, textiles, and food in the solid wastes. Furthermore, the extent of solid wastes available would also be less on a per capita basis in the developing countries (0.2 to 0.8 kg/person/day compared with even over 2 kg/person/day in the U.S.).

If the solid waste production is taken as 0.8 kg/person/day and if the absorptive capacity of solid waste is assumed as 100% (1 kg sludge/kg solid waste), which may be somewhat on the high side for a community in a developing country, it can count on only about 0.8 liter of sludge to be absorbed with its per capita solid waste.

Aerobically stabilized sludges containing solids produced at, for example, 30 g/person/day dry weight and having 99% moisture content would have a volume of 3 liters/person/day, which would then need to be thickened to about 96% moisture content to give a sludge volume of about 0.75 liter per person per day which is required to match the absorptive capacity of the available solid waste. Without thickening, only a part of the sludge can be disposed of with the solid waste, the remainder requiring drying or some other form of disposal.

REFERENCES

1. G. Fair, J. Geyer, and D. Okun, Water and Wastewater Engineering, John Wiley and Sons, N.Y., 1968.

1a. A. Zenoni and M. Blomquist, Column settling tests for flocculent suspensions, Jour. ASCE, Env. Eng. Div., 101, 1975.

2a. D. Okun and G. Ponghis, Community wastewater disposal, WHO Monograph Series, 1975.

2b. A. Vasilind, Treatment and Disposal of Wastewater Sludges, Ann Arbor Science, Ann Arbor, Michigan, 1974.

3. S. Arceivala, A Model for Estimation of Sludge Drying Time, in press.

3a. R. Dick, Role of activated sludge settling tanks, Jour. ASCE, 96, 1967.

4. Swanwick and Baskerville, Sludge Dewatering on Drying Beds, Institute of Sewage Purification, 1965.

5. Jennett and Santry, Characteristics of sludge drying, Jour. ASCE, San. Eng. Div., 1969.

6. Parker, Randall, and King, Biological conditioning for improved sludge filterability, Jour. WPCF, 44, 1972.

7. C. Randall and T. Koch, Dewatering characteristics of aerobically digested sludges, Jour. WPCF, 41, 1969.

8. R. Handa, Ph.D. thesis, Nagpur University, Nagpur, India.

9. J. Novak and G. Montgomery, Chemical sludge dewatering on sand beds, Jour. ASCE, Env. Eng. Div., 101, 1975.

10. J. Quon and T. Tamblyn, Intensity of radiation and rate of sludge drying, Jour. ASCE, San. Eng. Div., 91, 1965.

11. Adams and W. W. Eckenfelder, Process Design Techniques for Industrial Waste Treatment, Enviro Press, Nashville, Tennessee, 1974.

12. NEERI (CPHERI), Annual reports, Nagpur, India, 1972-1975.

13. Landfill disposal of liquid sewage sludge, Jour. ASCE, Env. Eng. Div., 101, 1975.

14. Standard Methods, APHA, U.S. Public Health Service, 1976.

15. Final clarifiers for activated sludge plants, Report of Task Committee, Environmental Engineering Division, Jour. ASCE, Env. Eng. Div., 105, 1979.